GUN TRADER'S GUIDE

23rd Edition — Completely Revised and Updated

STOEGER PUBLISHING COMPANY
Wayne, New Jersey

EDITORIAL STAFF:

Editor
Christopher R. Long

Production & Design
Publisher's First Choice

Research and Analysis
John Karns

Art
Keeler Chapman
John Karns

Cover Photographer
Ray and Matt Wells

Publisher
David C. Perkins

President
Brian T. Herrick

Published by:
Stoeger Publishing Company
5 Mansard Court
Wayne, New Jersey 07470

Manufactured in the United States of America

International Standard Book Number: 0-88317-219-4
Library of Congress Catalog Card No. 85-641040

In the United States, distributed to the book trade and to the sporting goods trade by:

Stoeger Industries
5 Mansard Court
Wayne, New Jersey 07470
973-872-9500 Fax: 973-872-2230

In Canada, distributed to the book trade and to the sporting goods trade by:
Stoeger Canada Ltd.
1801 Wentworth Street, Unit 16
Whitby, Ontario, L1N 8R6, Canada1

FRONT COVER

This year's *Gun Trader's Guide* front cover features the Winchester Model 1876. Both rifles are deluxe models. The upper rifle was produced in 1877 and uses the 45/75 caliber cartridge where the lower rifle used the 45-60 caliber cartridge which was first introduced in 1879 along with the 50/95 caliber cartridge. Both rifles feature the standard 28-inch octagon barrel, burl walnut stocks, checkered pistol grips, fitted crescent butt plates and sling swivels. The Model 1876 was considered the greatest repeater rifle of its time — producing around 63,900 rifles. There were basically three variations of the model 1876 — early intermediate and late with a varity of modifications taking place from 1877 to 1888, which included the development of the 1876 musket muzzle loader in 1888. The estimated cost of this firearm from 1877 to 1883 was about $83.00. Production was discontinued in 1898.

NOTE

All prices shown in this book are for guns in excellent condition (almost new) and the prices are retail (what a dealer would normally charge for them). A dealer will seldom pay the full value shown in this book. If a gun is in any condition other than excellent, the price in this book must be multiplied by the appropriate factor to obtain a true value of the gun in question.

INTRODUCTION

The Twenty-Third edition of *Gun Trader's Guide* provides the gun enthusiast with more specifications, photographs and line drawings than ever before. In the past 50 years, *Gun Trader's Guide* has grown tremendously — listing more than 6,000 firearms and variations.

The first edition of *Gun Trader's Guide* contained some 1,360 listings — accompanied by 100 illustrations. Now, after twenty-three revisions, the book has evolved into a complete catalog of rifles, shotguns and handguns of the twentieth century. The current edition of *Gun Trader's Guide* has been expanded to include more listings than ever before, with nearly 2700 illustrations. Hundreds of thousands of gun buffs have made *Gun Trader's Guide* their primary reference for firearm identification and comparison of sporting, military and law enforcement models including rare and unusual collectibles and commemoratives.

Simplified Structure

- Production data
- Specifications
- Variations of different models
- Dates of manufacture
- Current values
- Tabbed sections for user-friendly reference
- Complete INDEX of all firearms

The format of *Gun Trader's Guide* is simple and straightforward listing thousands of firearms manufactured since 1900, both in the United States and abroad. Most entries include complete specifications: model number and name, caliber/gauge, barrel length, overall length, weight, distinguishing features, variations, and the dates of manufacture (when they can be accurately determined). Many illustrative photos and drawings accompany the text to help the reader with identifications and comparisons.

Gun Trader's Guide is revised annually to ensure that its wealth of information is both current and comprehensive. The principal features that contribute to the uniqueness of this firearms reference guide include the extensive pictorial format and the accompanying comprehensive specifications. These combined features provide a convenient method and user friendly procedure for identifying vintage firearms while simultaneously determining and verifying the current value.

Values shown are based on national averages, obtained by conferring with hundreds of different gun dealers and auctioneers, not by applying an arbitrary mathematical formula that could produce unrealistic and non-applicable figures. The values listed accurately reflect the prices being charged nationwide at the time of publication. In other words, the published values accurately reflect the prices for which various models are sold — not arbitrary figures derived from a mathematical calculation intended to determine a depreciated value.

In some rare cases, however — like the Winchester Model 1873 "One of One Thousand" rifle or the Parker AA1 Special shotgun in 28 gauge — where very little (if any) trading took place, gun collectors were consulted to obtain current market values.

Organization of Listings

In the early editions of *Gun Trader's Guide*, firearms were frequently organized chronologically by date of production within manufacturers' listings. Firearms aficionados know that many gunmaking companies used the date that a particular model was introduced as the model number. For example, the Colt U.S. Model 1911 semiautomatic pistol was introduced in 1911; the French Model 1936 military rifle was introduced in 1936; and the Remington Model 32 shotgun debuted in 1932. However, during the first quarter of this century, gunmakers began assigning names and numbers that did not relate to the year that gun was introduced. As these more recent models and their variations multiplied through the years, it became increasingly difficult to track them by date — especially for the less experienced shooting enthusiasts.

Some Winchester and Remington firearms are grouped differently in this edition. For example, The Winchester Model 1894, in its many variations, has been produced since 1894 and is still being manufactured under "Model 94" by U.S. Repeating Arms Co. In general, Winchester used the year of introduction to name its firearms; that is, Model 1890, 1892, 1894, 1895, etc. Then shortly after World War I, Winchester dropped the first two digits and listed the models as 90, 92, 94, 95 etc. Furthermore, new models were given model numbers that had no relation to the date of manufacture. Marlin and several other manufacturers used a similar approach in handling model designations.

Consequently, Winchester rifles are grouped alphanumerically in two different groups: Early Winchesters that were manufactured before 1920 under the 4-digit model/date designations and those manufactured after 1920 with the revised model format designations. If any difficulty is encountered in locating a particular model, refer to the INDEX; the different models and their variations are cross-referenced.

Readers in the past have had difficulty in finding certain Remington rifles. In this edition, Remington rifles have been grouped according to action type; that is, single-shot rifles, slide-actions, autoloaders, etc. A survey revealed this to be the easiest way to locate a particular firearm. Again, use the INDEX if any difficulty is encountered.

In researching data for firearms, we have found through the years that not all records are available and therefore some information is unobtainable. For example, many early firearm records were destroyed in the fire that ravaged the Winchester plant. Some manufacturers' records have simply been lost, or just not maintained accurately. These circumstances result in some minor deviations in the presentation format of certain model listings. For example production dates may not be listed, when manufacturing records are unavailable. As an alternative, approximate dates may be listed to reflect "availability" from the manufacture or distributor. In such cases, these figures represent disposition dates indicating when that particular model was shipped to a distributor or importer. Frequently, with foreign manufacturers, production records are unavailable, therefore, availability information is often based on importation records that reflect domestic distribution only.

We wish to advise the reader of both the procedure and policy used regarding these published dates and further established the distinction between "production dates," which are based on manufacturers' records and the alternative, "availability dates," which are based on distribution records in the absence recorded production data. To further ensure that we have the most accurate information available, we encourage and solicit the users of *Gun Trader's Guide* to communicate with our research staff at the Stoeger offices and send in any verifiable information that you may have access to, especially in relation to older, out-of-production models.

Caution to Readers

To comply with new Federal regulations, all manufacturers who produce firearms that are intended for disposition to the general public and are designed to accept large-capacity ammunition feeding devices are required to redesign those models to limit their capacities to 10 rounds or less, or discontinue production or importation. This amendment to the Gun Control Act prohibits the manufacture, transfer or possession of all such devices manufactured after October 13, 1994. The "grandfather" clause of this amendment exempts all such devices lawfully possessed at the time the legislation became law. These "pre-ban" arms (those manufactured before October 13, 1994) may therefore be bought, sold or traded without any additional restrictions imposed by this law.

All "post-ban" feeding devices must meet the new requirements. For purposes of this book: Models previously designed to accept high-capacity feeding devices will be listed at their original specifications and capacities if only the feeding device was modified to reduce that capacity.

Regarding shotguns, the general public should be aware that shotgun barrels must be 18 inches or longer; anything shorter is illegal, except when used by military and law-enforcement personnel. A special permit from the BATF is required for all others.

Current Trends

We are witnessing an all-out assault on the firearms industry orchestrated by a well-organized coalition of politicians and legislators previously assembled to coordinate the action initiated against the tobacco industry. The political, judicial and legislative strategies employed in these litigation proceedings are uniquely similar to those that resulted in decisions rendered against the tobacco faction. Regardless of your position concerning judicial decisions addressing the tobacco industry issues, those legal proceedings have generated an atmosphere where political, judicial and legislative actions are viewed as corrective tools available to legislators and politicians to protect the American consumer and the general public from themselves.

Stimulated by the success of those earlier proceedings, the adversarial agenda has now been expanded to include the firearms industry. With the additional support of the precedence established by the court's decision concerning the tobacco industry, the perpetrators of these actions seek to benefit from both that momentum generated and the inferred parallel of *"Protecting Our Children"* as suggested by the same line of reasoning applied in the tobacco industry judgments.

Crusading municipalities continue to initiate lawsuits against firearms manufacturers, distributors and dealers under the guises of recovering millions of dollars in medical costs caused by the *illegal use* of firearms. As these accusations are being considered by the courts, the basic fallacy in this reasoning is becoming more apparent with each decision rendering such charges to be groundless. Consequently, more State Governments including; California, Maryland and Massachusetts are imposing legislation directed specifically at controlling and/or restricting the firearms industry.

The enactment of such restrictive legislation has caused a mass exodus of firearms manufacturers from the state of California into surrounding states. Due to both financial and legislative pressures, many domestic firearms manufacturers have sought additional protective measures including consolidations and mergers with larger business conglomerates frequently controlled by financial entities outside the United States. Recent legislation passed by the state of Maryland enacted additional gun control regulations in conjunction with the expansion of existing firearms restrictions, which are to be implemented in phases. Finally, Massachusetts became the first state in the nation to use consumer protection laws to enforce sweeping handgun safety regulations.

As these actions were being implemented by the various states (with additional legislation already proposed and more states soon to follow), the Federal government enacted *Project Exile* nationwide. This program, originally enacted by Virginia three years ago and endorsed by the National Rifle Association, has proven to be an effective law enforcement program (*designed to reward states from a $100 million grant fund when they impose five-year mandatory minimum sentences for using a gun to commit a crime and for felons caught carrying guns*). While President Clinton was engaged in "*the theatre of press conference politics*," the U.S. House of Representatives voted 358-60 to adopt *Project Exile* to supplement some of the "*unenforceable or unenforced laws*" that now exist as statutes.

This is a clear and exacting response from Congress to the Administration's calls for more and more gun control laws as the evidence continues to mount demonstrating that both the policy of this Administration's and the practice to enforce current law has been less and less productive. Therefore, Congress has apparently looked past the Administration's rhetoric to examine their record and found that referrals for prosecution of Federal law violations have declined from 9,885 to 5,510 in the last six years and Federal firearms law violations cases sent to Federal, State and local prosecutors declined from 12,084 to 6,470 in the same period, while *BATF-referred prosecutions in the state of Virginia led to 242 gun criminals being exiled to prison in 1998 alone*. Unbelievably, more armed criminals were sent to prison in Virginia in 1998 than in the far more populous states of California (70) and New York (140) and New Jersey (14) combined. The District of Columbia, which has effectively disarmed its law-abiding residents and has a violent crime rate about six times higher than Virginia, sent only two gun criminals to prison for breaking federal gun laws in 1998.

Although it appears that some logic has prevailed to support the nationwide enactment of this effective law enforcement legislation, it may well be too late to stabilize the firearms industry. While under the combined onslaught of stifling legislation and the threat of lawsuits from more than 30 cities, Smith & Wesson, America's largest gun manufacturer, capitulated to a government sponsored deal. The agreement mirrors the packaged arrangement accepted by the tobacco companies to settle lawsuits seeking reparations for government costs related to treating sick smokers.

Political analysts called "*The Deal*" a watershed in the ongoing gun control debate, which is increasingly a political topic as shooting deaths dominate the headlines. Additionally, Smith & Wesson's decision was interpreted as a break with the leadership of gun rights lobbies and further promoted the conclusion that other gunmakers would comply and offer safety concessions to avoid lawsuits.

Obviously, Smith & Wesson made their determination after much consideration by upper management, which certainly anticipated or expected some negative reaction. But perhaps they greatly underestimated their need for "*Damage Control*" considering that following the agreement, RSR Group Inc., a distributor of Smith & Wesson products for more than 20 years said, "*We have come to the difficult conclusion that we cannot continue to do business with Smith & Wesson under the problematic terms of their current agreement*."

In response to this reaction, Ken Jorgensen, a Smith & Wesson spokesman stated that some gun dealers (and obviously some distributors) opposed to the settlement will no longer carry the company's models. He also said the Springfield, Massachusetts-based company has received reports of attempts to pressure gun publications into refusing Smith & Wesson advertisements. This and other allegations have resulted in an antitrust investigation being initiated by Connecticut's attorney general to determine if gun manufacturers were illegally targeting Smith & Wesson in retaliation for its agreement.

It is now painfully apparent that the firearms industry no longer stands undivided and it is equally apparent that a stereotypical cliché may well be an appropriate epitaph for a disintegrating industry of trade — *Together We Stood, But divided We.........* perhaps the a more proverbial supplication would be more fitting prior to the pronouncement of extinction — *We Have Seen The Enemy and They Are Us!* But, in reality, an epitaph will be necessarily written by a dwindling few, while an invocation would be announced an a loud and meaningful voice by a determined and resolute assembly committed to eternal vigilance.

Acknowledgements

The Publisher wishes to express special thanks to the many collectors, dealers, manufacturers, shooting editors and other industry professionals who provide product information and willingly share their knowledge, which contributes to making this a better book.

A big thank you is extended, in particular, to the firearms firms and distributors — their public relations and production personnel, and all the research people that we work with throughout the year — we are especially grateful to all of you for your assistance and cooperation in compiling information for *Gun Trader's Guide* and for allowing us to reproduce photographs and illustrations of your firearms. Finally, thank you to all the dedicated readers who take the time to write in with comments, suggestions and queries about various firearms. We appreciate them all.

HOW TO USE THIS GUIDE

Buying or selling a used gun? Perhaps you want to establish the value of your own gun! Gun enthusiasts inevitably turn to *Gun Trader's Guide* to determine specifications, date of manufacture, and the value of specific firearms. Opening the book, he silently asks two questions: "How much will I be able to get (or expect to pay) for a particular gun?" and, "How was that price determined?"

Prices contained in this book are "retail"; that is, the price the consumer may expect to pay for a similar item. However, many variables must be considered when buying or selling any gun. In general, scarcity, demand, geographical location, the buyer's position and the gun's condition will govern the selling price of a particular gun. Sentiment may also affect the value of an individual's gun, but can be neither logically cataloged nor effectively marketed.

To illustrate how the price of a particular gun may fluctuate, let us consider the popular Winchester Model 94 and see what its value might be.

In general, the Model 1894 (94) is a lever-action, solid-frame repeater. Round or octagon barrels of 26 inches were standard when the rifle was first introduced in 1894. However, half-octagon barrels were available for a slight increase in price. Various magazine lengths were also available.

Fancy grade versions in all Model 94 calibers were available with 26-inch round, octagon or half-octagon nickel-steel barrels. This grade used a checkered fancy walnut pistol-grip stock and forearm, and was available with either shotgun or rifle-type buttplates.

In addition, Winchester produced this model in carbine-style with a saddle ring on the left side of the receiver. The carbine had a 20-inch round barrel and full- or half-magazine. Some carbines were supplied with standard grade barrels, while others were made of nickel steel. Trapper models were also available with shortened 14-, 16-, and 18-inch barrels.

In later years, the rifle and trapper versions were discontinued and only the carbine remained. Eventually, the saddle ring was eliminated from this model and the carbine buttstock was replaced with a shotgun-type buttstock and shortened forend. After World War II, the finish on Winchester Model 94 carbines changed to strictly hot-caustic bluing; thus, prewar models usually demand a premium over postwar models. Then in 1964, beginning with serial number 2,700,000, the action on Winchester Model 94s was redesigned for easier manufacturing. Many collectors and firearm enthusiasts consider this design change to be inferior to former models. Whether this evaluation is correct or not is not the issue; the main reason for a price increase of pre-1964 models was that they were no longer available. This diminished availability placed them immediately in the "scarce" class and made them increasingly more desirable to collectors.

Shortly after the 1964 transition, Winchester started producing Model 94 commemoratives in virtually endless numbers — further adding to the confusion of the concept of "limited production" as increased availability adversely affected the annual appreciation and price stability of these commemorative models. The negative response generated by this marketing practice was further stimulated when the Winchester Company was sold in the 1980s and the long established name of this firearms manufacturer was changed to U.S. Repeating Arms Co. This firm still manufactured the Model 94 in both standard carbine and Big Bore, and later introduced its "Angle-Eject" model to allow for the mounting of telescopic sights directly over the action.

With the above facts in mind, let's explore the *Gun Trader's Guide* to establish the approximate value of a particular Winchester Model 94. To illustrate, we will assume that you recently inherited a rifle that has the inscription "Winchester Model 94" on the barrel. In order to address this inquiry, you turn to the rifle section of this book and look under the W's until you find "Winchester." The Contents section will also indicate where Winchester Rifle Section begins. The Index in the back of the book is another possible means of locating your rifle. Since the listings in *Gun Trader's Guide* are arranged within each manufacturer's entry, first by Model Numbers in consecutive numerical order (followed by Model Names in alphabetical order), you may leaf through the pages until you come to "Model 94." At first glance, you realize there are two model designations that may apply — the original designation, Model 1894 or the revised shorter designation, Model 94. Which of these designations apply to your recently acquired Winchester?

The first step is to try to match the appearance of your model with an illustration in the book. They may all look similar at first, but close evaluation and careful attention to detail will soon enable you to eliminate the models that are not applicable. Further examination of your gun might reveal a curved, or crescent-shaped buttplate. By careful observation of your guns characteristics and close visual comparison of the photographic examples, you may logically conclude that your gun is the "Winchester Model 94 Lever Action Rifle."

You have now tentatively determined your model, but, to be sure, you should read through the specifications and further establish that the barrel on the pictured rifle is 26 inches long — noted to be either round, octagonal or half-octagonal. Upon measuring, you find that your rifle barrel is approximately 26 inches long, maybe a trifle under, and it is obviously round. (Please note that the guns are not always shown in proportion to one another; that is, a carbine might not appear shorter than a rifle.)

Additionally, your rifle is marked ".38-55"— the caliber designation. Since the caliber offerings in the listed specifications include 38-55, you are further convinced. You may read on to determine that this rifle was manufactured from 1894 to 1937. After that date, only the shorter-barreled carbine was offered by Winchester and then only in calibers 25-35, 30-30, and 32 Special.

At this point you know you have a Winchester Model 94 rifle manufactured before World War II. You read the value as $1550 and decide to take the rifle to your local dealer to initiate the sale. Here are some of the scenarios you may encounter:

Scenario I

If the rifle is in excellent condition — that is, if it retains at least 95 percent of its original finish on both the metal and wood and has a perfect bore — then the gun does have a retail value of $1550. However, the dealer is in business to make some profit. If he pays you $1550 for the gun, he will have to charge more than this when he sells it in order to realize a reasonable profit . If more than the fair market value is charged , then either the gun will not sell, or someone will pay more than the gun is worth. Therefore, a reputable dealer would offer you less than the published retail price for the gun to allow for a fair profit. However, the exact amount will vary. For example, if this dealer already has a dozen or so of the same model on his shelf and they have been slow moving, his offer will probably be considerable lower. On the other hand, if the dealer does not have any of this model in stock, and knows several collectors who want it, chances are the his offer will be considerable higher.

Scenario II

Perhaps you overlooked the rifle's condition. Although the gun apparently functions flawlessly, not much of the original bluing is left. Rather, there are several shiny bright spots mixed with a brown patina finish over the remaining metal. Also much of the original varnish on the wood has been worn away from extended use. Consequently, your rifle is not in "excellent" condition, and you will have to settle for lesser adjusted value than shown in this book.

Scenario III

So your Winchester Model 94 rifle looks nearly new, as if it were just out of the box, and the rifle works perfectly. Therefore, you are convinced that the dealer is going to pay you full value (less a reasonable profit of between 25% and 35%). But, when the dealer offers you about half of what you expected, you are shocked! Although the rifle looks new to you, the experienced dealer has detected that the gun has been refinished, Perhaps you did not notice the rounding of the formerly sharp edges on the receiver, or the slight funneling of some screw holes — all of which are a dead giveaway that the rifle has been refinished. Therefore, your rifle is not in excellent condition as you originally assumed and is ,therefore, worth less than "book" value.

While any gun's exterior condition is a big factor in determining its value, internal parts also play an important role in pricing.

A knowledgeable gun dealer will check the rifle to determine that it functions properly; the condition of interior parts will also be a factor in determining the value of any firearm.

Now if you are somewhat of an expert and know for certain that your rifle has never been refinished or otherwise tampered with, and there is still at least 95% of its original finish left, and you believe you have a firearm that is worth full book value, a dealer will still offer you from 25% to 35% less; perhaps even less if he is overstocked with this model.

Another alternative is to advertise in a local newspaper, selling the firearm directly to a private individual. However, this approach may prove both frustrating and expensive. In addition, there may be federal and local restrictions on the sale of firearms in your area. If you experience complications, chances are, the next time you have a firearm to sell, you will be more than happy to sell to a dealer and let him make his fair share of profit.

National Rifle Association Standards of Condtion

- **New:** Not previously sold at retail, in same condition as current factory production.
- **New, Discontinued:** Same as New, but discontinued model.
- **Perfect:** In new condition in every respect; sometimes referred to as mint.

- **Excellent:** New condition, used very little, no noticeable marring of wood or metal, bluing perfect (except at muzzle or sharp edges).
- **Very Good:** In perfect working condition, no appreciable wear on working surfaces, no corrosion or pitting, only minor surface dents or scratches.
- **Good:** In safe working condition, minor wear on working surfaces, no broken parts, no corrosion or pitting that will interfere with proper functioning.
- **Fair:** In safe working condition, but well worn, perhaps requiring replacement of minor parts or adjustments that should be indicated in advertisement; no rust, but may have corrosion pits that do not render the gun unsafe or inoperable.
- **Poor:** Badly worn, rusty and battered, perhaps requiring major adjustment or repairs to place in operating condition.

Condition

The condition of a firearm is a big factor in determining its value. In some rare and unusual models, a variation in condition from excellent to very good can mean a value difference of several thousand dollars. Therefore, you must be able to determine condition before you can accurately evaluate the value of the firearm. Several sets of standards are available, with the National Rifle Association Standards of Condition of Modern Firearms probably being the most popular. However, in recent years, condition has been specified by percentage of original finish remaining on the firearm — both on the wood and metal. Let's see how these standards stack up against each other.

When a collectable firearm has been expertly refinished to "excellent" condition, a rule of thumb is to deduct 50% from the value indicated in this book; if poorly done, deduct 80%.

For the purpose of assigning comparative values as a basis for trading, firearms listed in this book are assumed to be in "excellent" condition, with 95% or better remaining overall finish, no noticeable marring of wood or metal, and bores excellent with no pits. To the novice, this means a practically new gun, almost as though you had removed it from the factory carton. The trained eye, however, will know the difference between "new or mint" condition, and "excellent."

Any other defects, regardless of how minor, will lower the value of the subject firearm below those listed in this book. For example, if more than 5% of the original finish is gone and there are minor surface dents or scratches — regardless of how small — the gun is no longer in "excellent" condition, rather it takes on the condition of "very good," provided the gun is in perfect working order. Even in this state, other than for the minor defects, the gun will still look relatively new to the uninitiated gun buyer.

If the gun is in perfect working condition — functions properly, does not jam and is accurate — but has minor wear on working surfaces, perhaps some bad scratches on the wood or metal, etc., then the gun takes on the condition of "good," one grade below "very good," according to the NRA Standards. Again, the price in this book for that particular firearm must be lowered more to obtain its true value.

The two remaining NRA conditions fall under the headings of "fair" and "poor," respectively.

Previous editions of *Gun Trader's Guide* gave multiplication factors to use for firearms in other than "excellent" condition. While these factors are still listed below, please be aware that they are not infallible. They are only one "rough" means of establishing a value.

For guns in other than "excellent" condition, multiply the price in this book for the model in question by the following appropriate factors:

Multiplication Factors for Guns in Other Than Excellent Condition

Condition	Multiplication Factor
Mint or New	1.25
Excellent	1.00
Very Good	0.85
Good	0.68
Fair	0.35
Poor	0.15

Remember, the word "Guide" in *Gun Trader's Guide* should be taken literally. It is a guide only, not gospel. We sincerely hope, however, that this publication is helpful when you do decide to buy or sell a used firearm.

CONTENTS

Section II
RIFLES

Section III
SHOTGUNS

Section I
HANDGUNS

AA ARMS
Monroe, North Carolina

AA Arms AP-9 Series
Semiautomatic recoil-operated pistol w/polymer integral grip/frame design. Fires from a closed bolt. Caliber: 9mm Parabellum. 10- or 20*-shot magazine. 3-, 5- or 11-inch bbl. 11.8 inches overall w/5-inch bbl. Weight: 3.5 lbs. Fixed blade, protected post front sight adjustable for elevation, winged square notched rear. Matte phosphate/blue or nickel finish. Checkered polymer grip/frame. Made 1988 to date.

AP9 Model* (pre-94 w/ventilated bbl. shroud)	$325
AP9 Mini Model (post-94 w/o bbl. shroud)	195
AP9 Target Model* (pre-94 w/11-inch bbl.)	350
Electroless Nickel Finish, **add**	25

ACCU-TEK
Chino, California

**Accu-Tek
Model AT-380**

Accu-Tek Model AT-9 Auto Pistol
Caliber: 9mm Para. 8-shot magazine. Double action only. 3.2-inch bbl. 6.25 inches overall. Weight: 28 oz. Fixed blade front sight, adj. rear w/3-dot system. Firing-pin block with no external safety. Stainless or black over stainless finish. Checkered black nylon grips. Announced 1992, but made 1995 to date.

Satin Stainless Model	$175
Matte Black Stainless	185

Accu-Tek Model AT-25 Auto Pistol
Similar to Model AT-380, except chambered 25 ACP w/7-shot magazine. Made 1992-96.

Lightweight w/aluminum frame	$115
Bright Stainless (Disc. 1991)	120
Satin Stainless Model	115
Matte Black Stainless	120

Accu-Tek Model AT-32 Auto Pistol
Similar to Model AT-380, except chambered 32 ACP. Made 1990 to date.

Lightweight w/aluminum Frame (Disc. 1991)	$100
Satin Stainless Model	125
Matte Black Stainless	130

Accu-Tek Model AT-40 DA Auto Pistol
Caliber: 40 S&W. 7-shot magazine. 3.2-inch bbl. 6.25 inches overall. Weight: 28 oz. Fixed blade front sight, adj. rear w/3-dot system. Firing-pin block with no external safety. Stainless or black over stainless finish. Checkered black nylon grips. Announced 1992, but made 1995-96.

Satin Stainless Model	$185
Matte Black Stainless	195

Accu-Tek Model AT-380 Auto Pistol
Caliber: 380 ACP. 5-shot magazine. 2.75-inch bbl. 5.6 inches overall. Weight: 20 oz. External hammer w/slide safety. Grooved black composition grips. Alloy or stainless frame w/steel slide. Black, satin aluminum or stainless finish. Made 1992 to date.

Standard Alloy Frame (Disc. 1992)	$125
Satin Stainless Model	130
Matte Black Stainless	140

**Accu-Tek
Model BL-9**

Accu-Tek Models BL-9, BL 380 .... $135
Ultra compact DAO semiautomatic pistols. Calibers: 380 ACP, 9mm Para. 5-shot magazine. 3-inch bbl. 5.6 inches overall. Weight: 24 oz. Fixed sights. Carbon steel frame and slide w/black finish. Polymer grips. Made 1997 to date.

**Accu-Tek
HC-380SS**

Accu-Tek Models CP-9, CP-40, CP-45

Compact, double action only, semiautomatic pistols. Calibers: 9mm Parabellum, 40 S&W, 45 ACP. 8-, 7- or 6-shot magazine. 3.2-inch bbl. 6.25 inches overall. Weight: 28 oz. Fixed blade front sight, adj. rear w/3-dot system. Firing-pin block with no external safety. Stainless or black over stainless finish. Checkered black nylon grips. Made 1997 to date.

Black Stainless Model $135
Satin Stainless Model 145

Accu-Tek Model HC-380SS Auto Pistol $125

Caliber: 380 ACP. 13-shot magazine. 2.75-inch bbl. 6 inches overall. Weight: 28 oz. External hammer w/slide safety. Checkered black composition grips. Stainless finish. Made 1993 to date.

**Advantage Arms
Model 422**

ACTION ARMS
Philadelphia, Pennsylvania

See also listings under CZ pistols. Action Arms stopped importing firearms in 1994.

**Action Arms AT-84
with Prototype of Model AT-84P in
background**

Action Arms AT-84 DA Automatic Pistol $395

Caliber: 9mm Para. 15-shot magazine. 4.75-inch bbl. 8 inches overall. Weight: 35 oz. Fixed front sight, drift-adj. rear. Checkered walnut grips. Blued finish. Made in Switzerland 1987-89.

Action Arms AT-84P DA Auto Pistol . . . . (SRP) $498

Compact version of the Model AT-84. Only a few prototypes were manufactured in 1985. No resale value established.

Action Arms AT-88S DA Automatic Pistol $425

Calibers: 9mm Para. or .41 Action Express, 10-shot magazine. 4.6-inch bbl., 8.1 inches overall. Weight: 35.3 oz. Fixed blade front sight, adj. rear. Checkered walnut grips. Imported 1989-91.

Action Arms AT-88P DA Auto Pistol . . . . (SRP) $498

Compact version of the AT-88S w/3.7-inch bbl. Only a few prototypes of this model were manufactured in 1985. No resale value established.

ADVANTAGE ARMS
St. Paul, Minnesota

Advantage Arms Model 422 . . . . $185

Hammerless, top-break, 4-bbl. derringer w/rotating firing pin. Calibers: 22 LR and 22 Mag. 4-cartridge capacity. 2.5 inch bbl. 4.5 inches overall. Weight: 15 oz. Fixed sights. Walnut grips. Blued, nickel or PDQ matte black finish. Made 1985-87.

S. A. ALKARTASUNA FABRICA DE ARMAS
Guernica, Spain

Alkartasuna "Ruby" Automatic Pistol $250

Caliber: 32 Automatic (7.65mm). 9-shot magazine. 3.63-inch bbl. 6.38 inches overall. Weight: about 34 oz. Fixed sights. Blued finish. Checkered wood or hard rubber grips. Made 1917-22. *Note:* Mfd. by a number of Spanish firms, the Ruby was a secondary standard service pistol of the French Army in World Wars I and II. Specimens made by Alkartasuna bear the "Alkar" trademark.

AMERICAN ARMS
Kansas City, Missouri

American Arms Bisley SA Revolver $295

Umberti reproduction of Colt's Bisley. Caliber: 45 LC. 6-shot cylinder. 4.75-, 5.5- or 7.7-inch bbl. Case-hardened steel frame. Fixed blade front sight, grooved top strap rear. Hammer block safety. Imported 1997 to date.

American Arms Buckhorn SA Revolver

Similar to Regulator Model, except chambered 44 Mag. w/4.75-, 6- or 7.7-inch bbl. Fixed or adjustable sights. Hammer block safety. Imported 1993-96.

Buckhorn Model (standard sights) $250
W/Adjustable Sights, **add** 15

American Arms CX-22

American Arms CX-22 DA Automatic Pistol

Similar to Model PX-22, except w/8-shot magazine. 3.33-inch bbl. 6.5 inches overall. Weight: 22 oz. Made 1990 to 1995.

Standard w/Chrome Slide (Disc. 1990) $135
Classic Model 140

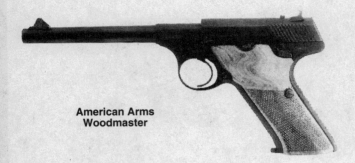

**American Arms
Woodmaster**

American Arms EP-380 DA Automatic Pistol $275
Caliber: 380 Automatic. 7-shot magazine. 3.5-inch bbl. 6.5 inches overall. Weight: 25 oz. Fixed front sight, square-notch adj. rear. Stainless finish. Checkered wood grips. Made 1989-91.

American Arms Escort DA Auto Pistol $225
Caliber: 380 ACP. 7-shot magazine. 3.38-inch bbl. 6.13 inches overall. Weight: 19 ounces. Fixed, low-profile sights. Stainless steel frame, slide, and trigger. Nickel-steel bbl. Soft polymer grips. Loaded chamber indicator. Made 1995-97 .

American Arms Mateba Auto Revolver
Unique combination action design allows both slide and cylinder to recoil together causing cylinder to rotate. Single or double action mode. Caliber: 357 Mag. 6-shot cylinder. 4- or 6-inch bbl. 8.77 inches overall w/4-inch bbl. Weight: 2.75 lbs. Steel/alloy frame. Ramped blade front sight, adjustable rear. Blue finish. Smooth walnut grips. Imported 1997 to date.
Mateba Model (w/4-inch bll.) $795
Mateba Model (w/6-inch bll.) 850

American Arms P-98 DA Automatic Pistol $135
Caliber: 22 LR. 8-shot magazine. 5-inch bbl. 8.25 inches overall. Weight: 25 oz. Fixed front sight, square-notch adj. rear. Blued finish. Serrated black polymer grips. Made 1989-96.

American Arms PK-22 DA Automatic Pistol $130
Caliber: 22 LR. 8-shot magazine. 3.33-inch bbl. 6.33 inches overall. Weight: 22 oz. Fixed front sight, V-notch rear. Blued finish. Checkered black polymer grips. Made 1989-96.

American Arms PX-22 DA Automatic Pistol $125
Caliber: 22 LR. 7-shot magazine. 2.75-inch bbl. 5.33 inches overall. Weight: 15 oz. Fixed front sight, V-notch rear. Blued finish. Checkered black polymer grips. Made 1989-96.

American Arms PX-25 DA Automatic Pistol $135
Same general specifications as the Model PX-22, except chambered for 25 ACP. Made 1991-92.

American Arms Regulator SA Revolver
Similar in appearance to the Colt Single-Action Army. Calibers: 357 Mag., 44-40, 45 Long Colt. 6-shot cylinder. 4.75- or 7.5-inch bbl. Blade front sight, fixed rear. Brass trigger guard/backstrap on Standard model. Casehardened steel on Deluxe model. Made 1992 to date.
Standard Model $220
Standard Combo Set
  (45 LC/45 ACP & 44-40/44 Spec.) 265
Deluxe Model 245
Deluxe Combo Set
  (45 LC/45 ACP & 44-40/44 Spec.) 285
Stainless Steel 275

American Arms Sabre DA Automatic Pistol
Calibers: 9mm Para., 40 S&W. 8-shot magazine (9mm), 9-shot (40 S&W). 3.75-inch bbl. 6.9 inches overall. Weight: 26 oz. Fixed blade front sight, square-notch adj. rear. Black polymer grips. Blued or stainless finish. Advertised 1991, but not imported.
Blued Finish . . . . (SRP) $309
Stainless Steel . . . . (SRP) 339

American Arms Spectre DA Auto Pistol
Blowback action, fires closed bolt. Calibers: 9mm Para., 40 S&W, 45 ACP. 30-shot magazine. 6-inch bbl. 13.75 inches overall. Weight: 4 lbs. 8 oz. Adj. post front sight, fixed U-notch rear. Black nylon grips. Matte black finish. Imported 1990-94.
9mm Para. $295
40 S&W (Disc. 1991) 320
45 ACP 350

American Arms Woodmaster SA Auto Pistol $175
Caliber: 22 LR. 10-shot magazine. 5.88-inch bbl. 10.5 inches overall. Weight: 31 oz. Fixed front sight, square-notch adj. rear. Blued finish. Checkered wood grips. Discontinued 1989.

American Arms 454 SSA Revolver $575
Umberti SSA chambered 454. 6-shot cylinder. 6-inch solid raised rib or 7.7-inch top-ported bbl. Satin nickel finish. Hammer block safety. Imported 1996-97.

AMERICAN DERRINGER CORPORATION
Waco, Texas

**American Derringer
Model 1**

American Derringer Model 1, Stainless
Single-action pocket pistol similar to the Remington O/U Derringer. 2-shot capacity. More than 60 calibers from 22 LR to 45-70. 3- inch bbl. 4.82 inches overall. Weight: 15 oz. Automatic bbl. selection. Satin or high-polished stainless steel. Rosewood grips. Made 1980 to date.
45 Colt, 44-40 Win., 44 Special, .410 × 2.5 Inches $275
45-70, 44 Mag., 41 Mag., 30-30 Win., 223 Rem. 225
357 Max., 357 Mag., 45 Win Mag., 9mm Para. 195
38 Special, 38 Super, 32 Mag., 22 LR, 22 Rim Mag. 165

American Derringer Model 2 Steel "Pen" Pistol
Calibers: 22 LR, 25 Auto, 32 Auto (7.65mm). single-shot. 2-inch bbl. 5.6 inches overall (4.2 inches in pistol format). Weight: 5 oz. Stainless finish. Made 1993-94.
22 Long Rifle $185
25 Auto 195
32 Auto 225

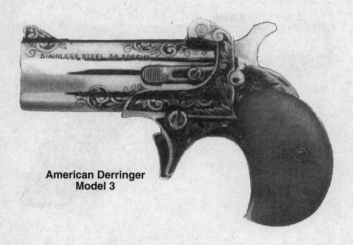

American Derringer Model 3

American Derringer Model 38 DA

American Derringer Model 3 Stainless Steel $65
Single-shot. Calibers: 32 Mag. or 38 Special. 2.5-inch bbl.. 4.9 inches overall. Weight: 8.5 oz. Rosewood grips. Made 1984-95.

American Derringer Model 4 Double Derringer
Calibers: 357 Mag., 357 Max., 44 Mag., 45 LC, 45 ACP (upper bbl. and 3-inch .410 shotshell (lower bbl.). 4.1-inch bbl. 6 inches overall. Weight: 16.5 oz. Stainless steel. Stag horn grips. Made 1984 to date.

357 Mag., 357 Max. $245
44 Mag., 45 LC, 45 ACP 335
Engraved, **add**. 1100

American Derringer Model 6
Caliber: 22 Mag., 357 Mag., 45 LC, 45 ACP or 45 LC/.410 or 45 Colt. Bbl.: 6 inches. 8.2 inches overall. Weight: 22 oz. Satin or high-polished stainless steel w/rosewood grips. Made 1986 to date.

22 Magnum. $195
357, 45 ACP, 45 LC 225
45LC/410 O/U 275
Engraved add 1295

American Derringer Model 7
Same general specifications as the Model 1, except high-strength aircraft aluminum is used to replace some of the stainless steel parts, which reduces its weight to 7.5 oz. Made 1986 to date.

22 Long Rifle, 22 WMR. $135
Calibers 32, 38 and 44 150

American Derringer Model 10
Same general specifications as the Model 7, except chambered for 45 ACP or 45 Long Colt.

45 ACP $165
45 Long Colt 175

American Derringer Model 11 .... $145
Same general specifications as Model 7, except chambered for 38 and 45 calibers only, and weighs 11 oz. Made 1980 to date.

American Derringer 25 Automatic Pistol
Calibers: 25 ACP or 250 Mag. Bbl.: 2.1 inches. 4.4 inches overall. Weight: 15.5 oz. Smooth rosewood grips. Limited production.

25 ACP Blued (est. production 50). $400
25 ACP Stainless (est. production 400) 350
250 Mag. Stainless (est. production 100) 425

American Derringer Model 38 DA Derringer
Hammerless, double action, double bbl (over/under). Calibers: 38 Special, 9mm Para., 357 Mag., 40 S&W. 3-inch bbl. Weight: 14.5 oz. Made 1990 to date.

38 Special $165
9mm Para. 175
357 Mag. 215
40 S&W 225

American Derringer Alaskan Survival Model $285
Same general specifications as the Model 4, except upper bbl. chambered 45-70 or 3-inch .410 and 45 Colt lower bbl. Also available in 45 Auto, 45 Colt, 44 Special, 357 Mag. and .357 Max. Made 1985 to date.

American Derringer Cop DA Derringer ...... (SRP) $425
Hammerless, double-action, four-bbl. derringer. Caliber: 357 Mag. 3.15-inch bbl. 5.5 inches overall. Weight: 16 oz. Blade front sight, open notched rear. Rosewood grips. Intro. 1990 but only limited production occurred. No resale value established.

American Derringer Lady Derringer

American Derringer Lady Derringer
Same general specifications as Model 1, except w/custom-tuned action fitted w/scrimshawed synthetic ivory grips. Calibers: 32 H&R Mag., 32 Special, 38 Special (additional calibers on request). Deluxe Grade engraved and highly polished w/French fitted jewelry box. Made 1991 to date.

Lady Derringer $175
Deluxe. 185
Engraved 550

American Derringer Mini-Cop DA Derringer $200
Same general specifications as the American Derringer Cop, except chambered for 22 Magnum. Made 1990-95.

**American Derringer
Semmerling LM-4**

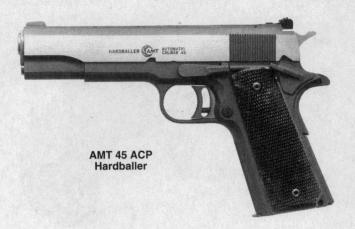

**AMT 45 ACP
Hardballer**

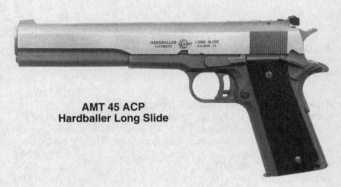

**AMT 45 ACP
Hardballer Long Slide**

American Derringer Semmerling LM-4
Manually operated repeater. Calibers: 45 ACP or 9mm. 5-shot (45 ACP) or 7-shot magazine (9mm). 3.6-inch bbl. 5.2 inches overall. Weight: 24 oz. Made 1997 to date. Limited availability.
Blued Finish **$1995**
Stainless Steel . . . . **(SRP) 2650**

American Derringer Texas Commemorative
Same general specifications as Model 1, except w/solid brass frame, stainless bbls. and rosewood grips. Calibers: 38 Special, 44-40 Win. or 45 Colt. Made 1991 to date.
38 Special **$175**
44-40 or 45 Colt **265**

AMERICAN FIREARMS MFG. CO., INC.
San Antonio, Texas

American 25 Auto Pistol
Caliber: 25 Auto. 8-shot magazine. 2.1-inch bbl. 4.4 inches overall. Weight: 14.5 oz. Fixed sights. Stainless or blued ordnance steel. Smooth walnut grips. Made 1966-74.
Stainless Steel Model **$175**
Blued Steel Model **150**

American 380 Auto Pistol **$345**
Caliber: 380 Auto. 8-shot magazine. 3.5-inch bbl. 5.5 inches overall. Weight: 20 oz. Stainless steel. Smooth walnut grips. Made 1972-74.

AMT (ARCADIA MACHINE & TOOL)
Irwindale, California
(Previously Irwindale Arms, Inc.)

> **NOTE**
>
> The AMT Backup II Automatic Pistol was introduced in 1993 as a continuation of the original 380 backup with the traditional double action function and a redisigned double safety.

AMT 45 ACP Hardballer
Caliber: 45 ACP. 7-shot magazine. 5-inch bbl. 8.5 inches overall. Weight: 39 oz. Adj. or fixed sights. Serrated matte slide rib w/loaded chamber indicator. Extended combat safety, adj. trigger and long grip safety. Wraparound Neoprene grips. Stainless steel. Made 1978 to date.
45 ACP Hardballer **$275**
Long Slide Conversion Kit (Disc. 1997), **add** **295**

AMT 45 ACP Hardballer Long Slide
Similar to the standard AMT Hardballer except w/2-inch longer bbl. and slide. Also chambered 400 Cor-Bon. Made 1980 to date.
45 ACP Long Slide **$325**
400 Cor-Bon Long Slide (Intro.1998) **375**
5-inch Conversion Kit (Disc. 1997), **add** **295**

AMT 1911 Government Model Auto Pistol **$250**
Caliber: 45 ACP. 7-shot magazine. 5-inch bbl. 8.5 inches overall. Weight: 38 ounces. Fixed sights. Wraparound Neoprene grip. Made 1979 to date.

AMT Automag II Automatic Pistol **$225**
Caliber: 22 Mag. 7- or 9-shot magazine. Bbl. lengths: 3.38-,4.5-, 6-inch. Weight: 32 oz. Fully adj. Millett sights. Stainless finish. Smooth black composition grips. Made 1986 to date.

AMT Automag III Automatic Pistol **$275**
Calibers: 30 M1 and 9mm Win. Mag., 8-shot magazine. 6.38-inch bbl. 10.5 inches overall. Weight: 43 ounces. Millet adj. sights. Stainless finish. Carbon fiber grips. Made 1989 to date.

AMT Automag IV Automatic Pistol **$350**
Calibers: 10mm Mag., 45 Win. Mag. 8- or 7-shot magazine. 6.5- or 8.63-inch bbl. 10.5 inches overall. Weight 46 oz. Millet adj. sights. Stainless finish. Carbon fiber grips. Made 1990 to date.

AMT Automag V Automatic Pistol **$675**
Caliber: 50 A.E. 5-shot magazine. 7-inch bbl. 10.5 inches overall. Weight: 46 oz. Custom adj. sights. Stainless finish. Carbon fiber grips. Made 1994-95.

AMT Backup

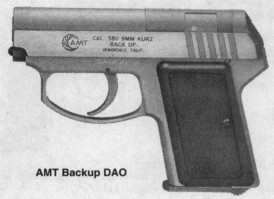

AMT Backup DAO

AMT Backup Automatic Pistol
Caliber: 22LR, 380 ACP. 8-shot (22LR) or 5-shot (380 ACP) magazine. 2.5-inch bbl. 5 inches overall. Weight: 18 oz. Open sights. Carbon fiber or walnut grips. Stainless steel finish. Made 1990 to date.
22 LR (Discontinued 1987) **$185**
380 ACP (Discontinued 1993) **175**

AMT Backup II Automatic Pistol **$180**
Caliber: 380 ACP, 5-shot magazine, 2.5-inch bbl., 5 inches overall. Weight: 18 oz. Open sights. Stainless steel finish. Carbon-fiber grips. Made 1993 to date.

AMT Backup DAO Auto Pistol
Calibers: 380 ACP, 38 Super, 9mm Para., 40 Cor-Bon, 40 S&W, 45 ACP. 6-shot (380, 38 Super 9mm) or 5-shot (40 Cor-Bon, 40 S&W, 45 ACP) magazine. 2.5-inch bbl. 5.75-inches overall. Weight: 18 oz. (380 ACP) or 23 oz. Open fixed sights. Stainless steel finish. Carbon fiber grips. Made 1992 to date.
380 ACP **$175**
38 Super, 9mm **185**
40 Cor-Bon, 40 S&W, 45 ACP. **195**

AMT Bull's Eye Target Model **$325**
Caliber: 40 S&W. 8-shot magazine. 5-inch bbl. 8.5 inches overall. Weight: 38 oz. Millet adjustable sights. Wide adj. trigger. Wraparound Neoprene grips. Made 1990-92.

AMT Javelina **$425**
Caliber: 10mm. 8-shot magazine. 7-inch bbl. 10.5 inches overall. Weight: 48 oz. Long grip safety, beveled magazine well, wide adj. trigger. Millet adj. sights. Wraparound Neoprene grips. Stainless finish. Made 1991-93.

AMT Bull's Eye Target

AMT Lightning Auto Pistol
Caliber: 22 LR. 10-shot magazine. 5-, 6.5-, 8.5-, 10-inch bbl. 10.75 inches overall (6.5-inch bbl.). Weight: 45 oz. (6.5-inch bbl.). Millett adj. sights. Checkered rubber grips. Stainless finish. Made 1984-87.
Standard Model. **$175**
Bull's-Eye Model **315**

AMT On Duty DA Pistol
Calibers: 40 S&W, 9mm Para., 45 ACP. 15-shot (9mm), 13-shot (40 S&W) or 9-shot (45 ACP) magazine. 4.5-inch bbl. 7.75 inches overall. Weight: 32 oz. Hard anodized aluminum frame. Stainless steel slide and bbl. Carbon fiber grips. Made 1991-94.
9mm or 40 S&W **$275**
45 ACP **350**

AMT Skipper

AMT Skipper Auto Pistol **$295**
Calibers: 40 S&W and 45 ACP. 7-shot magazine. 4.25-inch bbl. 7.5 inches overall. Weight: 33 oz. Millett adj. sights. Walnut grips. Matte finish stainless steel. Made 1990-92.

ANSCHUTZ PISTOLS
Ulm, Germany
Mfd. by J.G. Anschutz GmbH Jagd und Sportwaffenfabrik

Anschutz Model 64P
Calibers: 22 LR or 22 Magnum. 5- or 4-shot magazine. 10-inch bbl. 64MS action w/2-stage trigger. Target sights optional. Rynite black synthetic stock. Imported 1998 to date.
22 LR **$325**
22 Mag. **350**
W/Tangent Sights, **add** **80**

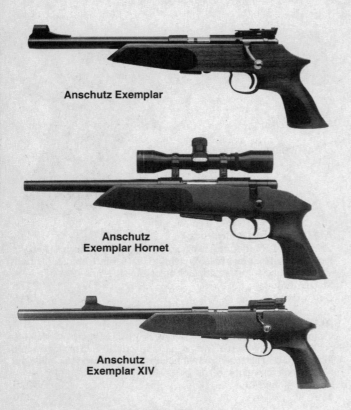

Anschutz Exemplar

Anschutz Exemplar Hornet

Anschutz Exemplar XIV

Anschutz Exemplar Bolt-Action Pistol
Caliber: 22 LR, Single-shot or 5-shot clip. 7- or 10-inch bbl. 19 inches overall (10-inch bbl.). Weight: 3.33 lbs. Match 64 action. Slide safety. Hooded ramp post front sight, adjustable open notched rear. European walnut contoured grip. Exemplar made 1987-95 and 1400 series made 1997 to date. *Note:* The 22 WMR chambering was also advertised, but never manufactured.

Exemplar w/7- or 10-inch bbl $325
Left-Hand Model (Disc. 1997) 395
Model 1451P (single-shot) 325
Model 1416P (5-shot repeater) 315

Anschutz Exemplar Hornet $595
Based on the Anschutz Match 54 action, tapped and grooved for scope mounting with no open sights. Caliber: 22 Hornet, 5-shot magazine. 10-inch bbl. 20 inches overall. Weight: 4.35 lbs. Checkered European walnut grip. Winged safety. Made 1990 to 1995.

Anschutz Exemplar XIV $345
Same general specifications as the standard Exemplar Bolt Action Pistol, except w/14-inch bbl., weight 4.15 lbs. Made 1989-95.

ARMSCOR (Arms Corp.)
Manila, Philippines
Currently imported by K.B.I., Harrisburg, PA.
(Imported 1991-95 by Ruko Products, Inc., Buffalo NY.
Previously by Armscor Precision, San Mateo, CA.)

Armscor Model M1911-A1 Automatic Pistol $295
Caliber: 45 ACP. 8-shot magazine. 5-inch bbl. 8.75 inches overall. Weight: 38 oz. Blade front sight, drift adjustable rear w/3-Dot system. Skeletonized tactical hammer and trigger. Extended slide release and beavertail grip safety. Parkerized finish. Checkered composition or wood stocks. Imported 1996-97.

Armscor Model 200DC/TC DA Revolver $125
Caliber: 38 Special. 6-shot cylinder, 2.5-, 4-, or 6-inch bbl., 7.3, 8.8, or 11.3 inches overall. Weight: 22, 28, or 34 oz. Ramp front and fixed rear sights. Checkered mahogany or rubber grips. Imported 1996 to date.

ASTRA PISTOLS
Guernica, Spain
Manufactured by Unceta y Compania

Astra Model 41 DA Revolver $210
Same general specifications as Model 44 (below), except in 41 Mag. Imported 1980-85.

Astra Model 44 DA

Astra Model 44 DA Revolver
Similar to Astra 357, except chambered for 44 Magnum. 6- or 8.5-inch bbl. 11.5 inches overall (6-inch bbl.). Weight: 44 oz. (6-inch bbl.). Imported 1980-93.
Blued Finish (Disc. 1987). $225
Stainless Finish (Disc. 1993) 285

Astra Model 45 DA Revolver $225
Similar to Astra 357, except chambered for 45 Colt or 45 ACP. 6- or 8.5-inch bbl. 11.5 inches overall (6-inch bbl.). Weight: 44 oz. (6-inch bbl.). Imported 1980-87.

Astra Model 200 Firecat Vest Pocket
Auto Pistol $185
Caliber: 25 Automatic (6.35mm). 6-shot magazine. 2.25-inch bbl. 4.38 inches overall. Weight: 11.75 oz. Fixed sights. Blued finish. Plastic grips. Made 1920 to date. U.S. importation discontinued in 1968.

**Astra Model 202
Firecat Vest Pocket**

Astra Model 202 Firecat Vest Pocket Auto Pistol. $395
Same general specifications as the Model 200 except chromed and engraved w/pearl grips. U.S. importation disc. 1968.

Astra Model 357 DA Revolver **$195**
Caliber: 357 Magnum. 6-shot cylinder. 3-, 4-, 6-, 8.5-inch bbl.
11.25 inches overall (6-inch bbl.). Weight: 42 oz. (6-inch bbl.).
Ramp front sight, adj. rear sight. Blued finish. Checkered wood
grips. Imported 1972-88.

Astra Model 400 Auto Pistol **$275**
Caliber: 9mm Bayard Long (38 ACP, 9mm Browning Long, 9mm
Glisenti, 9mm Para. and 9mm Steyr cartridges may be used inter-
changeably in this pistol because of its chamber design).9-shot
magazine. 6-inch bbl.10 inches overall. Weight: 35 oz. Fixed sights.
Blued finish. Plastic grips. Made 1922-45. *Note:* This pistol, as well
as Astra Models 600 and 3000, is a modification of the Browning
Model 1912.

Astra Model 600 Mil./Police Type Auto Pistol **$300**
Calibers: 32 Automatic (7.65mm), 9mm Para. Magazine: 10- (32
cal.) or 8-shot (9mm). 5.25-inch bbl. 8 inches overall. Weight:
about 33 oz. Fixed sights. Blued finish. Checkered wood or plastic grips. Made 1944-45.

Astra Model 800 Condor Military Auto Pistol **$1050**
Similar to Models 400 and 600, except has an external hammer.
Caliber: 9mm Para. 8-shot magazine. 5.25-inch bbl. 8.25 inches
overall. Weight: 32.5 oz. Fixed sights. Blued finish. Plastic grips.
Imported 1958-65.

Astra Model 2000 Camper Automatic Pistol **$245**
Same as Model 2000 Cub, except chambered for 22 Short only, has
4-inch bbl, overall length, 6.25 inches, weight, 11.5 oz. Imported
1955-60.

**Astra
Model 3003 Pocket**

**Astra
Model 4000 Falcon**

Astra Model 2000 Cub

Astra Model 2000 Cub Pocket Auto Pistol **$195**
Calibers: 22 Short, 25 Auto. 6-shot magazine. 2.25-inch bbl. 4.5
inches overall. Weight: about 11 oz. Fixed sights. Blued or
chromed finish. Plastic grips. Made 1954 to date. U.S. importation disc. 1968.

Astra Model 3000 Pocket Auto Pistol **$325**
Calibers: 22 LR, 32 Automatic (7.65mm), 380 Auto (9mm Short).
10-shot magazine (22 cal.), 7-shot (32 cal.), 6-shot (380 cal.). 4-inch
bbl. 6.38 inches overall. Weight: about 22 oz. Fixed sights. Blued
finish. Plastic grips. Made 1947-56.

Astra Model 3003 Pocket Auto Pistol **$625**
Same general specifications as the Model 3000 except chromed and
engraved w/pearl grips. Disc. 1956.

Astra Model 4000 Falcon Auto Pistol **$365**
Similar to Model 3000, except has an external hammer. Calibers:
22 LR, 32 Automatic (7.65mm), 380 Auto (9mm Short). 10-shot
magazine (22 LR), 8-shot (32 Auto), 7-shot (380 Auto), 3.66-inch
bbl. 6.5-inches overall. Weight: 20 oz. (22 cal.) or 24.75 oz. (32
and 380). Fixed sights. Blued finish. Plastic grips. Made 1956-71.

Astra Model A-60 DA Automatic Pistol **$275**
Similar to the Constable, except in 380 only, w/13-shot magazine and
slide-mounted ambidextrous safety. Blued finish only. Imported 1980-
91.

Astra Model A-70 Compact Auto Pistol
Calibers: 9mm Para., 40 S&W. 8-shot (9mm) or 7-shot (40 S&W)
magazine. 3.5-inch bbl. 6.5 inches overall. Blued, nickel or stainless
finish. Weight: 29.3 oz. Imported 1992-96.
Blued Finish **$235**
Nickel Finish. **265**
Stainless Finish. **295**

Astra Model A-75 Decocker Auto Pistol
Similar to the Model 70, except in 9mm, 40 S&W and 45 ACP w/de-
cocking system and contoured pebble-textured grips. Imported
1993-97.
Blued Finish, 9mm or 40 S&W **$225**
Nickel Finish, 9mm or 40 S&W. **235**
Stainless, 9mm or 40 S&W. **295**
Blued Finish, 45 ACP. **235**
Nickel Finish, 45 ACP **250**
Stainless, 45 ACP. **335**

Astra Model A-75 Ultralight **$255**
Similar to the standard Model 75, except 9mm only w/24-oz. alloy
frame. Imported 1994-97.

**Astra
Model A-80**

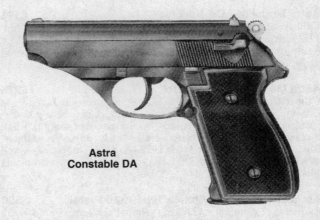

**Astra
Constable DA**

Astra Model A-80 Auto Pistol **$295**
Calibers: 9mm Para., 38 Super, 45 ACP. 15-shot magazine or 9-shot (45 ACP). Bbl.: 3.75 inches. 7 inches overall. Weight: 36 oz. Imported 1982-89.

Astra Model A-90 DA Automatic Pistol **$325**
Calibers: 9mm Para., 45 ACP. 15-shot (9mm) or 9-shot (45 ACP) magazine. 3.75-inch bbl. 7 inches overall. Weight: about 40 oz. Fixed sights. Blued finish. Checkered plastic grips. Imported 1985-90.

Astra Model A-100 DA Auto Pistol
Same general specifications as the Model A-90, except selective double action chambered for 9mm Para., 40 S&W or 45 ACP. Imported in 1991-97.
Blued Finish **$295**
Nickel Finish. 325
For Night Sights, **add** 85

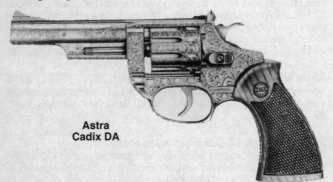

**Astra
Cadix DA**

Astra Cadix DA Revolver
Calibers: 22 LR, 38 Special. 9-shot (22 LR) or 5-shot (38 cal.) cylinder. 4- or 6-inch bbl. Weight: about 27 oz. (6-inch bbl.). Ramp front sight, adj. rear sight. Blued finish. Plastic grips. Imported 1960-68.
Standard Model . . . . **$ 155**
Lightly Engraved Model. 250
Heavily Engraved Model (shown) 495

Astra Constable DA Auto Pistol
Calibers: 22 LR, 32 Automatic (7.65mm), 380 Auto (9mm Short). Magazine capacity: 10-shot (22 LR), 8-shot (32), 7-shot (380). 3.5-inch bbl. 6.5 inches overall. Weight: about 24 oz. Blade front sight, windage adj. rear. Blued or chromed finish. Imported 1965-92.
Stainless Finish. **$275**
Blued Finish 235
Chrome Finish (Disc. 1990) 250

AUTO-ORDNANCE CORPORATION
West Hurley, New York

Auto-Ordnance 1911A1 Government Auto Pistol
Copy of Colt 1911A1 semiautomatic pistol. Calibers: 9mm Para., 38 Super, 10mm, 45 ACP. 9-shot (9mm, 38 Super) or 7-shot (10mm, 45 ACP) magazine. 5-inch bbl. 8.5 inches overall. Weight: 39 oz. Fixed blade front sight, rear adj. Blued, satin nickel or Duo-Tone finish. Checkered plastic grips. Made 1983-99.
45 ACP Caliber. **$275**
9mm, 10mm, 38 Super 305

Auto-Ordnance 1911A1 40 S&W Pistol **$315**
Similar to the Model l911A1, except has 4.5-inch bbl. w/7.75-inch overall length. 8-shot magazine. Weight: 37 oz. Blade front and adj. rear sights w/3-dot system. Checkered black rubber wraparound grips. Made 1991-99.

Auto-Ordnance 1911 "The General" **$275**
Caliber: 45 ACP. 7-shot magazine. 4.5-inch bbl. 7.75 inches overall. Weight: 37 oz. Blued nonglare finish. Made 1992-99.

**Auto-Ordnance
1927 A-5 w/drum
magazine**

Auto-Ordnance 1927 A-5 Semiautomatic Pistol
Similar to Thompson Model 1928A submachine gun, except has no provision for automatic firing and does not have detachable buttstock. Caliber: 45 ACP, 5-, 15-, 20- and 30-shot detachable box magazines. 30-shot drum also available. 13-inch finned bbl. 26 inches overall. Weight: about 6.75 lbs. Adj. rear sight, blade front. Blued finish. Walnut grips. Made 1977-94.
W/box magazine **$695**
W/drum magazine (illustrated) 950

**Auto-Ordnance
ZG-51 Pit Bull**

Auto-Ordnance ZG-51 Pit Bull Automatic Pistol . . . $285
Caliber: 45 ACP. 7-shot magazine. 3.5-inch bbl. 7 inches overall. Weight: 32 oz. Fixed front sight, square-notch rear. Blued finish. Checkered plastic grips. Made 1991-99.

AUTAUGA ARMS
Prattville, Alabama

Autauga Model 32 DAO Automatic Pistol $250
Caliber: 32 ACP. 6-shot magazine. 2-inch bbl. Weight: 11.36 oz.. Double action only. Stainless steel. Black polymer grips. Made 1997 to date.

LES BAER
Hillsdale, Illinois

Baer 1911 Concept Series Automatic Pistol
Similar to Government 1911 built on steel or alloy full-size or compact frame. Caliber: 45 ACP. 7-shot magazine. 4.25- or 5-inch bbl. Weight: 34 to 37 oz. Adjustable low mount combat or BoMar target sights. Blued, matte black, Two-Tone or stainless finish. Checkered wood grips. Made 1996 to date.
Concept Models I & II $895
Concept Models III, IV & VII 985
Concept Models V, VI & VIII 1025
Concept Models IX & X 1035

Baer 1911 Premier Series Automatic Pistol
Similar to the Concept Series, except also chambered 38 Super, 9x23 Win., 400 Cor-Bon and 45 ACP. 5- or 6-inch bbl. Weight: 37 to 40 oz. Made 1996 to date.
Premier II (9x23 w/5-inch bbl.) $1095
Premier II (400 Cor-Bon w/5-inch bbl.) 995
Premier II (45 ACP w/5-inch bbl.) 925
Premier II (45 ACP S/S w/5-inch bbl.) 1025
Premier II (45/400 Combo w/5-inch bbl.) 1150
Premier II (38 Super w/6-inch bbl.) 1295
Premier II (400 Cor-Bon w/6-inch bbl.) 1165
Premier II (45 ACP w/6-inch bbl.) 1095

Baer S.R.P. Automatic Pistol
Similar to F.B.I. Contract "Swift Response Pistol" built on a (customer supplied) Para-Ordance over-sized frame or a 1911 full-size or compact frame. Caliber: 45 ACP. 7-shot magazine. 5-inch bbl. Weight: 37 oz. Ramp front and fixed rear sights, w/Trijicon night insert.
SRP 1911Government or Commanche Model $1595
SRP P-12 Model 1825
SRP P-13 Model 1650
SRP P-14 Model 1550

Baer 1911 Ultimate Master Combat Series
Model 1911 in Combat Competition configuration. Calibers: 38 Super, 9x23 Win., 400 Cor-Bon and 45 ACP. 5- or 6-inch NM bbl. Weight: 37 to 40 oz. Made 1996 to date.
Ultimate MC (38 or 9x23 w/5-inch bbl.) $1650
Ultimate MC (400 Cor-Bon w/5-inch bbl.) 1525
Ultimate MC (45 ACP w/5-inch bbl.) 1450
Ultimate MC (38 or 9x23 w/6-inch bbl.) 1695
Ultimate MC (400 Cor-Bon w/6-inch bbl.) 1575
Ultimate MC (45 ACP w/6-inch bbl.) 1495
Ultimate "Steel Special" (38 Super Bianchi SPS) 1895
Ultimate "PARA" (38, 9x23 or 45 IPSC Comp) 1925
W/Triple-Port Compemsator, **add** 95

Baer 1911 Custom Carry Series Automatic Pistol
Model 1911 in Combat Carry configuration built on steel or alloy full-size or compact frame. 4.5- or 5-inch NM bbl., chambered 45 ACP. Weight: 34 to 37 oz.
Custom Carry (Steel Frame w/4.24- or 5-inch bbl.) $1050
Custom Carry (Alloy Frame w/4.24-inch bbl.) 1225

BAUER FIREARMS CORPORATION
Fraser, Michigan

Bauer 25 Automatic Pistol $140
Stainless steel. Caliber: 25 Automatic. 6-shot magazine. 2.13-inch bbl. 4 inches overall. Weight: 10 oz. Fixed sights. Checkered walnut or simulated pearl grips. Made 1972-84.

BAYARD PISTOLS
Herstal, Belgium
Mfd. by Anciens Etablissements Pieper

Bayard Model 1908 Pocket Automatic Pistol $225
Calibers: 25 Automatic (6.35mm), 32 Automatic (7.65mm), 380 Automatic (9mm Short). 6-shot magazine. 2.25-inch bbl. 4.88 inches overall. Weight: about 16 oz. Fixed sights. Blued finish. Hard rubber grips. Intro. 1908. Discontinued 1923.

Bayard Model 1923 Pocket 25 Automatic Pistol $195
Caliber: 25 Automatic (6.35mm). 2.13-inch bbl. 4.31 inches overall. Weight: 12 oz. Fixed sights. Blued finish. Checkered hard-rubber grips. Intro. 1923. Discontinued 1930.

Bayard Model 1923 Pocket Automatic Pistol $235
Calibers: 32 Automatic (7.65mm), 380 Automatic (9mm Short). 6-shot magazine. 3.31-inch bbl. 5.5 inches overall. Weight: about 19 oz. Fixed sights. Blued finish. Checkered hard-rubber grips. Intro. 1923. Discontinued 1940.

Bayard Model 1930 Pocket 25 Automatic Pistol $225
This is a modification of the Model 1923, which it closely resembles.

BEEMAN PRECISION ARMS, INC.
Santa Rosa, California

Beeman P08 Automatic Pistol $265
Caliber: 22 LR. 10-shot magazine. 3.8-inch bbl. 7.8 inches overall. Weight: 25 oz. Fixed sights. Blued finish. Checkered hardwood grips. Imported 1969-91.

Beeman Mini P08 Automatic Pistol $285
Caliber: Same general specifications as P08, except shorter 3.5-inch bbl., 7.4 inches overall and weight of 20 oz. Imported 1986-91.

**Beeman P08
Automatic Pistol**

Beeman SP Metallic Silhouette Pistols

Caliber: 22 LR. Single-shot. Bbl. 6-, 8-, 10- or 15-inches. Adj. rear sight. Receiver contoured for scope mount. Walnut target grips w/adj. palm rest. Models SP made 1985-86 and SPX 1993-94.

SP Standard W/8-or 10-inch bbl.	**$195**
SP Standard W/12-inch bbl.	**215**
SP Standard W/15-inch bbl.	**235**
SP Deluxe W/8-or 10-inch bbl.	**225**
SP Deluxe W/12-inch bbl.	**245**
SP Deluxe W/15-inch bbl.	**255**
SPX Standard W/10-inch bbl.	**495**
SPX Deluxe W/10-inch bbl.	**675**

BEHOLLA PISTOL
Suhl, Germany
Mfd. by both Becker and Holländer and Stenda-Werke GmbH

Beholla Pocket Automatic Pistol **$175**
Caliber: 32 Automatic (7.65mm). 7-shot magazine. 2.9-inch bbl. 5.5 inches overall. Weight: 22 oz. Fixed sights. Blued finish. Serrated wood or hard rubber grips. Made by Becker and Hollander 1915-1920, by Stenda-Werke circa 1920-25. *Note:* Essentially the same pistol was manufactured concurrently w/the Stenda version as the "Leonhardt" by H. M. Gering and as the "Menta" by August Menz.

BENELLI PISTOLS
Urbino, Italy
Currently imported by Benelli USA

Benelli MP90S World Cup Target Pistol

Semiautomatic blowback action. Calibers: 22Short, 22LR, 32 W.C. 5-shot magazine. 4.33-inch fixed bbl. 6.75 inches overall. Weight: 36 oz. Post front sight, adjustable rear. Blue finish. Anatomic shelf-style grip. Imported 1992 to date.

MP90S (22LR)	**$825**
MP90S (22Short, Disc. 1995)	**895**
MP90S (32WC)	**1050**
W/Conversion Kit, **add**	**550**

Benelli MP95E Sport Target Pistol

Similar to the MP90S, except with 5- or 9-shot magazine. 4.25-inch bbl. Blue or chrome finish. Checkered target grip. Imported 1994 to date.

Blue MP95 (22LR)	**$475**
Blue MP95 (32WC)	**550**
Chrome, **add**	**85**

BERETTA USA CORP.
Accokeek, Maryland

Beretta pistols are manufactured by Fabbrica D'Armi Pietro Beretta S. p. A. in the Gardone Val Trompia (Brescia), Italy. This prestigious firm has been in business for over 474 years (since 1526). Since the late 1970s, many of the models sold in the U.S. have been made at the Accokeek (MD) plant.

Beretta Model 20 Double-Action Auto Pistol **$135**
Caliber: 25 ACP. 8-shot magazine. 2.5-inch bbl. 4.9 inches overall. Weight: 10.9 oz. Plastic or walnut grips. Fixed sights. Made 1984-85.

Beretta Model 21

Beretta Model 21 Double-Action Auto Pistol

Calibers: 22 LR and 25 ACP. 7-shot (22 LR) or 8-shot (25 ACP) magazine. 2.5-inch bbl. 4.9 inches overall. Weight: about 12 oz. Blade front sight, V-notch rear. Walnut grips. Made 1985 to date.

Blued Finish	**$150**
Nickel Finish (22 LR only)	**175**
Model 21EL Engraved Model	**225**

Beretta Model 70

Beretta Model 70 Automatic Pistol **$175**
Improved version of Model 1935. Steel or lightweight alloy. Calibers: 32 Auto (7.65mm), 380 Auto (9mm Short). 8-shot (32) or 7-shot (380) magazine. 3.5-inch bbl. 6.5 inches overall. Weight: steel, 22.25 oz., alloy, 16 oz. Fixed sights. Blued finish. Checkered plastic grips. Made 1959-85. *Note:* Formerly marketed in U.S. as "Puma" (alloy model in 32) and "Cougar" (steel model in 380). Discontinued.

Beretta Model 70

Beretta Model 71

Beretta Model 72

Beretta Model 76

Beretta Model 70S $195
Similar to Model 70T, except chambered for 22 Auto and 380 Auto. Longer bbl. guide and safety lever blocking hammer. Front blade and rear sight fixed on breechblock. Weight: 1 lb. 7 oz. Made 1977-85.

Beretta Model 70T Automatic Pistol $245
Similar to Model 70. Caliber: 32 Automatic (7.65mm). 9-shot magazine. 6-inch bbl. 9.5 inches overall. Weight: 19 oz. adj. rear sight, blade front sight. Blued finish. Checkered plastic grips. Intro. in 1959. Discontinued.

Beretta Model 71 Automatic Pistol $185
Same general specifications as alloy Model 70. Caliber: 22 LR. 6-inch bbl. 8-round magazine. Adj. rear sight frame. Single action. Made 1959-89. *Note:* Formerly marketed in U.S. as the "Jaguar Plinker."

Beretta Model 72 .... $190
Same as Model 71, except has 6-inch bbl., weighs 18 oz. Intro. in 1959. Discontinued. *Note:* Formerly marketed in U.S as "Jaguar Plinker."

Beretta Model 76 Auto Target Pistol
Caliber: 22 LR. 10-shot magazine. 6-inch bbl. 8.8 inches overall. Weight: 33 oz. Adj. rear sight, front sight w/interchangeable blades. Blued finish. Checkered plastic or wood grips (Model 76W). Made 1966-85. *Note:* Formerly marketed in the U.S. as the "Sable."
Model 76 w/Plastic Grips $295
Model 76W w/Wood Grips 345

Beretta Model 81

Beretta Model 81 Double-Action Auto Pistol $245
Caliber: 32 Automatic (7.65mm). 12-shot magazine. 3.8-inch bbl. 6.8 inches overall. Weight: 23.5 oz. Fixed sights. Blued finish. Plastic grips. Made principally for the European market 1975-84, w/similar variations to the Model 84.

Beretta Model 84 Double-Action Auto Pistol $235
Same as Model 81, except made in caliber 380 Automatic w/13-shot magazine. 3.82-inch bbl. 6.8 inches overall. Weight: 23 oz. Fixed front and rear sights. Made 1975-82.

Beretta Model 84

Beretta Model 84B DA Auto Pistol $245
Improved version of Model 84 w/strengthened frame and slide, and firing-pin block safety added. Ambidextrous reversible magazine release. Blued or nickel finish. Checkered black plastic or wood grips. Other specifications same. Made c. 1982-84.

Beretta Model 85

Beretta Model 85BB

Beretta Model 86

Beretta Model 84(BB) Double-Action Auto Pistol
Improved version of Model 84B, w/further strengthened slide, frame and recoil spring. Caliber: 380 ACP. 13-shot magazine. Bbl.: 3.82 inches. 6.8 inches overall. Weight: 23 oz. Checkered black plastic or wood grips. Blued or nickel finish. Fixed sights. Made c. 1984-94.
Blued w/Plastic Grips $235
Blued w/Wood Grips 255
Nickel Finish **add** 35

Beretta Model 85 Double-Action Auto Pistol $295
This is basically the same gun as the Model 84 and has seen a similar evolution. However, this 8-shot version has no ambidextrous magazine release, has a slightly narrower grip and weighs 21.8 oz. It was intro. a little later than the Model 84.

Beretta Model 85B DA Auto Pistol $315
Improved version of the Model 85. Made 1982-85.

Beretta Model 85BB Double-Action Pistol
Improved version of the Model 85B, w/strengthened frame and slide. Caliber: 380 ACP. 8-shot magazine. 3.82 inch bbl. 6.8 inches overall. Weight: 21.8 oz. Blued or nickel finish. Checkered black plastic or wood grips. Made 1985 to date.
Blued Finish w/Plastic Grips $295
Blued Finish w/Wood Grips 335
Nickel Finish **add** 40

Beretta Model 85F Double-Action Pistol
Similar to the Model 85BB, except has re-contoured trigger guard and manual ambidextrous safety w/decocking device. Made in 1990.
Blued Finish w/Plastic Grips $300
Blued Finish w/Wood Grips 325
Nickel Finish w/Wood Grips 365

Beretta Model 86 Double-Action Auto Pistol $325
Caliber: 380 auto. 8-shot magazine. Bbl.: 4.33 inches, tip-up. 7.33 inches overall. Weight: 23 oz. Made 1986-89. (Reintroduded 1990 in the Cheetah Series.)

Beretta Model 87

Beretta Model 87 Auto Pistol
Similar to the Model 85, except in 22 LR w/8-shot magazine and optional extended 6-inch bbl. (Target in single action). Overall length: 6.8 inches. 8.9 (6-inch bbl.). Weight: 20.8 oz., 23 oz. (6-inch bbl.). Checkered wood grips. Made 1987 to date.
Blued Finish (Double-Action) $325
Target Model (Single Action) 345

Beretta Model 89 Target Automatic Pistol $435
Caliber: 22 LR. 8-shot magazine. 6-inch bbl. 9.5 inches overall. Weight: 41 oz. Adj. target sights. Blued finish. Target-style walnut grips. Made 1988 to date.

Beretta Model 90 DA Auto Pistol $195
Caliber: 32 Auto (7.65mm). 8-shot magazine. 3.63-inch bbl. 6.63 inches overall. Weight: 19.5 oz. Fixed sights. Blued finish. Checkered plastic grips. Made 1969-83.

Beretta Model 90

Beretta Model 92

Beretta Model 92F

Beretta Model 92 SB-F

Beretta Model 96

Beretta Model 92 DA Auto Pistol (1st series) $425
Caliber: 9mm Para. 15-shot magazine. 4.9-inch bbl. 8.5 inches over-all. Weight: 33.5 oz. Fixed sights. Blued finish. Plastic grips. Initial production of 5,000 made in 1976.

Beretta Model 92D DA Auto Pistol
Same general specifications as Model 92F, except DA only w/bobbed hammer and 3-Dot Sight. Made 1992 to date.
Model 92D **$325**
Model 92D w/Triticon Sight System **395**

Beretta Model 92F Compact DA Auto $425
Caliber: 9mm Para. 12-shot magazine. 4.3-inch bbl. 7.8 inches overall. Weight: 31.5 oz. Wood grips. Square-notched rear sight, blade front integral w/slide. Made 1986-93.

Beretta Model 92F DA Auto Pistol
Same general specifications as Model 92, except w/slidemounted safety and repositioned magazine release. Replaced Model 92SB. Blued or stainless finish. Made 1985 to date.
Blued Finish **$415**
Stainless Finish **395**
Model 92F-EL Gold **495**

Beretta Model 92S DA Auto Pistol (2nd series) $375
Revised version of Model 92 w/ambidextrous slide-mounted safety modification intended for both commercial and military production. Evolved to Model 92S-1 for U.S. Military trials. Made 1980-85.

Beretta Model 92SB DA Auto Pistol (3rd series) . . . $395
Same general specifications as standard Model 92, except has slide-mounted safety and repositioned magazine release. Made 1981-85.

Beretta Model 92 SB-F DA Auto Pistol $415
Caliber: 9mm Para. 15-shot magazine. Bbl.: 4.9 inches. 8.5 inches overall. Weight: 34 oz. Plastic or Beretta Model 92 SB-F DA Auto Pistol wood grips. Square-notched rear sight, blade front sight integral w/slide. This model, also called **Model 92S-1**, is the standard-issue sidearm for the U.S. Armed Forces. Made 1985 to date.

Beretta Model 96 Double-Action Auto Pistol
Same general specifications as Model 92F, except in 40 S&W. 10-shot magazine (9-shot in Compact Model). Made 1992 to date.
Model 96 D (DA only) **$325**
Model 96 Centurion (Compact) **335**
W/Triticon Sights, **add** **50**
W/Triticon Sights System, **add** **95**

Beretta Model 950B

Beretta Model 950CC

Beretta Model 950-BS-EL

**Beretta Model 950CC
Special**

Beretta Model 101 . . . . **$215**
Same as Model 70T, except caliber 22 LR, has 10-shot magazine. Intro. in 1959. Discontinued.

Beretta Model 318 (1934) Auto Pistol **$235**
Caliber: 25 Automatic (6.35mm). 8-shot magazine. 2.5-inch bbl. 4.5 inches overall. Weight: 14 oz. Fixed sights. Blued finish. Plastic grips. Made 1934 to c. 1939.

Beretta Model 949 Olimpionico Auto Pistol **$525**
Calibers: 22 Short, 22 LR. 5-shot magazine. 8.75-inch bbl. 12.5 inches overall. Weight: 38 oz. Target sights. Adj. bbl. weight. Muzzle brake. Checkered walnut grips w/thumbrest. Made 1959-64.

Beretta Model 950B Auto Pistol **$125**
Same general specifications as Model 950CC, except caliber 25 Auto, has 7-shot magazine. Made 1959 to date. *Note:* Formerly marketed in the U.S. as "Jetfire."

Beretta Model 950 BS Single Action Semiautomatic
Calibers: 25 ACP or 22 Short. Magazine capacity: 7 rounds (22 short), 8 rounds (25 ACP). 2.5- or 4-inch bbl. 4.5 inches overall (2.5-inch bbl.) Weight: about 10 oz. Blade front sight, V-notch in rear. Checkered black plastic grips. Made 1987 to date.
Blued Finish **$125**
Nickel Finish 165
W/4-inch bbl. (22 short) 145
Model 950 EL gold-etched version 225

Beretta Model 951 (1951)

Beretta Model 950CC Auto Pistol **$125**
Caliber: 22 Short. 6-shot magazine. Hinged, 2.38-inch bbl. 4.75 inches overall. Weight: 11 oz. Fixed sights. Blued finish. Plastic grips. Made 1959 to date. *Note:* Formerly marketed in the U.S. as "Minx M2."

Beretta Model 950CC Special Auto Pistol **$130**
Same general specifications as Model 950CC Auto, except has 4-inch bbl. Made 1959 to date. *Note:* Formerly marketed in the U.S. as "Minx M4."

Beretta Model 951 (1951) Military Auto Pistol **$245**
Caliber: 9mm Para. 8-shot magazine. 4.5-inch bbl. 8 inches overall. Weight: 31 oz. Fixed sights. Blued finish. Plastic grips. Made 1952 to date. *Note:* This is the standard pistol of the Italian Armed Forces, also used by Egyptian and Israeli armies and by the police in Nigeria. Egyptian and Israeli models usually command a premium. Formerly marketed in the U.S. as the "Brigadier."

Beretta Model 1915

Beretta Model 1923

Beretta Model 1935

Beretta Model 1915 Auto Pistol **$525**
Calibers: 9mm Glisenti and 32 ACP (7.65mm). 8-shot magazine. 4-inch bbl. 6.7 inches overall (9mm), 5.7 inches (32 ACP). Weight: 30 oz. (9mm), 20 oz. (32 ACP). Fixed sights. Blued finish. Wood grips. Made 1915-1922. An improved postwar 1915/1919 version in caliber 32 ACP was later offered for sale in 1922 as the Model 1922.

Beretta Model 1923 Auto Pistol **$575**
Caliber: 9mm Glisenti (Luger). 8-shot magazine. 4-inch bbl. 6.5 inches overall. Weight: 30 oz. Fixed sights. Blued finish. Plastic grips. Made c. 1923-36.

Beretta Model 1934 Auto Pistol
Caliber: 380 Automatic (9mm Short). 7-shot magazine. 3.38-inch bbl. 5.88 inches overall. Weight: 24 oz. Fixed sights. Blued finish. Plastic grips. Official pistol of the Italian Armed Forces. Wartime pieces not as well made and finished as commercial models. Made 1934-59.
Commercial Model **$295**
War Model 275

Beretta Model 1935 Auto Pistol
Caliber: 32 ACP (7.65mm). 8-shot magazine. 3.5-inch bbl. 5.75 inches overall. Weight: 24 oz. Fixed sights. Blued finish. Plastic grips. A roughly finished version of this pistol was produced during WW II. Made 1935-1959.
Commercial Model **$275**
War Model 235

Beretta Model 3032 DA Semiautomatic Tomcat
Caliber: 32 ACP. 7-shot magazine. 2.45-inch bbl. 5 inches overall. Weight: 14.5 oz. Fixed sights. Made 1996 to date.
Matte Blue **$175**
Polished Blue 225

Beretta Model 8000/8040 Cougar DA Pistol
Short recoil action w/rotating barrel. Caliber: 9mm, 40 S&W. 10- or 15-shot magazine. 3.6-inch bbl. 7.0 inches overall. Weight: 33.5 oz. Fixed sights w/3-dot Tritium system. Textured black composition grips. Matte black Bruniton finish w/alloy frame. Made 1994 to date.
DA Model **$435**
DAO Model 415

Beretta Model 8000/8040 Mini-Cougar DA Pistol
Similar to standard Cougar, except w/one-inch shorter grip. 8- or 10-shot w/extended magazine. Weight 27 oz. Checkered black plastic grips. Made 1998 to date.
DA Model **$425**
DAO Model 395

VINCENZO BERNARDELLI, S.P.A.
Gardone V. T. (Brescia), Italy

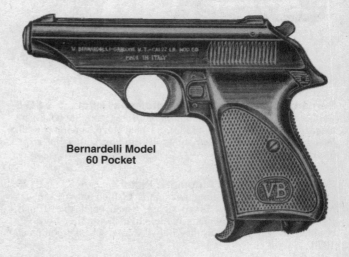

**Bernardelli Model
60 Pocket**

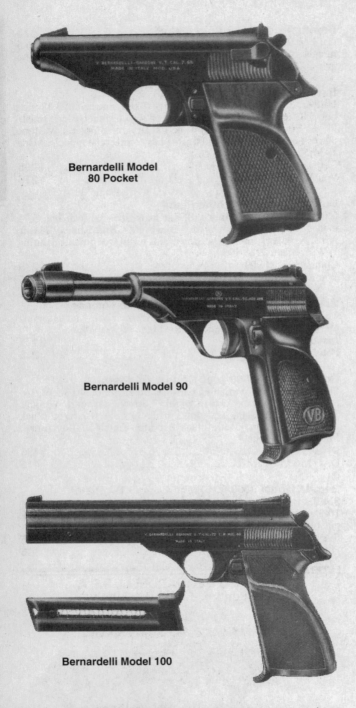

Bernardelli Model 80 Pocket

Bernardelli Model 90

Bernardelli Model 100

Bernardelli Model 60 Pocket Automatic Pistol $195
Calibers: 22 LR, 32 Auto (7.65mm), 380 Auto (9mm Short). 8-shot magazine (22 and 32), 7-shot (380). 3.5-inch bbl. 6.5 inches overall. Weight: about 25 oz. Fixed sights. Blued finish. Bakelite grips. Made 1959-90.

Bernardelli Model 68 Automatic Pistol $125
Caliber: 6.35. 5- and 8-shot magazine. 2.13-inch bbl. 4.13 inches overall. Weight: 10 oz. Fixed sights. Blued or chrome finish. Bakelite or pearl grips. This model, like its smaller bore 22-counterpart, was known as the "Baby" Bernardelli. Discontinued 1970.

Bernardelli Model 69 Automatic Target Pistol $425
Caliber: 22 LR. 10-shot magazine. 5.9-inch bbl. 9 inches overall. Weight: 2.2 lbs. Fully adj. target sights. Blued finish. Stippled right- or left-hand wraparound walnut grips. Made 1987 to date. *Note:* This was previously Model 100.

Bernardelli Model 80 Automatic Pistol $165
Calibers: 22 LR, 32 ACP (7.65mm), 380 Auto (9mm Short). Magazine capacity: 10-shot (22), 8-shot (32), 7-shot (380). 3.5-inch bbl. 6.5 inches overall. Weight: 25.6 oz. adj. rear sight, white dot front sight. Blued finish. Plastic thumbrest grips. *Note:* Model 80 is a modification of Model 60 designed to conform w/U.S. import regulations. Made 1968-88.

Bernardelli Model 90 Sport Target $175
Same as Model 80, except has 6-inch bbl., is 9 inches overall, weighs 26.8 oz. Made 1968-90.

Bernardelli Model 100 Target Automatic Pistol $315
Caliber: 22 LR. 10-shot magazine. 5.9-inch bbl. 9 inches overall. Weight: 37.75 oz. Adj. rear sight, interchangeable front sights. Blued finish. Checkered walnut thumbrest grips. Made 1969-86. *Note:* Formerly Model 69.

Bernardelli Model AMR Auto Pistol $295
Simlar to Model USA, except with 6-inch bbl. And target sights. Imported 1992-94.

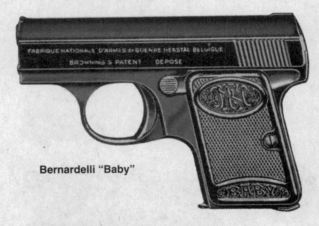

Bernardelli "Baby"

Bernardelli "Baby" Automatic Pistol $165
Calibers: 22 Short, 22 Long. 5-shot magazine. 2.13-inch bbl. 4.13 inches overall. Weight: 9 oz. Fixed sights. Blued finish. Bakelite grips. Made 1949-68.

Bernardelli Model P010 Automatic Pistol $525
Caliber: 22 LR. 5- and 10-shot magazine. 5.9-inch bbl. w/7.5-inch sight radius. Weight: 40.0 oz. Interchangeable front sight, adj. rear. Blued finish. Textured walnut grips. Made 1988-92 and 1995-97.

Bernardelli P018 Compact Model $375
Slightly smaller version of the Model P018 standard DA automatic, except has 14-shot magazine and 4-inch bbl. 7.68 inches overall. Weight: 33 oz. Walnut grips only. Imported 1987-96.

Bernardelli P018 Double-Action Automatic Pistol
Caliber: 9mm Para. 16-shot magazine. 4.75-inch bbl. 8.5 inches overall. Weight: 36 oz. Fixed combat sights. Blued finish. Checkered plastic or walnut grips. Imported 1987-96.
W/Plastic Grips $350
W/Walnut Grips 385

Bernardelli Model P010

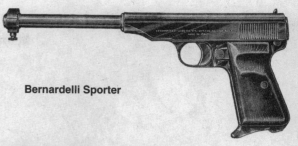

Bernardelli Sporter

Bernardelli Model USA Auto Pistol
Single-action, blowback. Calibers:22LR, 32ACP, 380ACP. 7-shot magazine or 10-shot magazine (22LR). 3.5-inch bbl. 6.5 inches overall. Weight: 26.5 oz. Ramped front sight, adjustable rear. Blue or chrome finish. Checkered black bakelite grips w/thumbrest. Imported 1991-97.

Model USA Blue Finish **$265**
Model USA Chrome Finish **325**

Bernardelli Vest Pocket

Bernardelli Model P018

Bernardelli Vest Pocket Automatic Pistol **$185**
Caliber: 25 Auto (6.35mm). 5- or 8-shot magazine. 2.13-inch bbl. 4.13 inches overall. Weight: 9 oz. Fixed sights. Blued finish. Bakelite grips. Made 1945-68.

Bernardelli P. One DA Auto Pistol
Caliber: 9mm Parabellun or 40 S&W. 10- or 16-shot magazine. 4.8-inch bbl. 8.35 inches overall. Weight: 34 oz. Blade front sight, adjustable rear w/3-dot system. Matte black or chrome finish. Checkered walnut or black plastic grips. Imported 1993-97.

Model P One Blue Finish **$425**
Model P One Chrome Finish **465**
W/Walnut Grips, **add** **35**

Bernardelli P. One Practical VB Auto Pistol
Similar to Model P One, except chambered 9x21mm w/2-, 4- or 6-port compensating system for IPSC compition. Imported 1993-97.

Model P One Practical (2 port) **$795**
Model P One Practical (4 port) **850**
Model P One Practical (6 port) **1150**
W/Chrome Finish, **add** **50**

Bernardelli Sporter Automatic Pistol **$250**
Caliber 22 LR. 8-shot magazine. Bbl. lengths: 6-, 8- and 10-inch. 13 inches overall (10-inch bbl.). Weight about 30 oz. (10-inch bbl. Target sights. Blued finish. Walnut grips. Made 1949-68.

BERSA PISTOLS
Argentina
Currently imported by Eagle Imports, Wanamassa, NJ (Previously by Interarms & Outdoor Sports)

Bersa Model 85

**Bersa
Thunder 380**

**Bersa
Thunder 380 Deluxe**

Bersa Model 83 Double-Action Auto Pistol
Similar to the Model 23 except for the following specifications: Caliber: 380 ACP. 7-shot magazine. 3.5-inch bbl. Front blade sight integral on slide, square-notch rear adj. for windage. Blued or satin nickel finish. Custom wood grips. Imported 1988-94.
Blued Finish $165
Satin Nickel 195

Bersa Model 85 Double-Action Auto Pistol
Same general specifications as Model 83, except 13-shot magazine. Imported 1988-94.
Blued Finish $225
Satin Nickel 265

Bersa Model 86 Double-Action Auto Pistol
Same general specifications as Model 85, except available in matte blued finish and w/Neoprene grips. Imported 1992-94.
Matte Blued Finish $235
Nickel Finish. 275

Bersa Model 95 DA Automatic Pistol
Caliber: 380 ACP. 7-shot magazine. 3.5-inch bbl. Weight: 23 oz. Wraparound rubber grips. Blade front and rear notch sights. Imported 1995 to date.
Blued Finish $155
Nickel Finish. 175

Bersa Model 97 Auto Pistol $275
Caliber: 380 ACP. 7-shot magazine. 3.3-inch bbl. 6.5 inches overall. Weight: 28 oz. Intro. 1982. Discontinued.

Bersa Model 223
Same general specifications as Model 383, except in 22 LR w/10-round magazine capacity. Discontinued 1987.
Double Action $165
Single Action 150

Bersa Model 224
Caliber: 22 LR. 10-shot magazine. 4-inch bbl. Weight: 26 oz. Front blade sight, square-notched rear adj. for windage. Blued finish. Checkered nylon or custom wood grips. Made 1984. SA. Discontinued 1986.
Double-Action $170
Single Action 165

Bersa Model 226
Same general specifications as Model 224, but w/6-inch bbl. Discontinued 1987.
Double Action $165
Single Action 150

**Bersa
383**

Bersa Model 383 Auto Pistol
Caliber: 380 Auto. 7-shot magazine. 3.5-inch bbl. Front blade sight integral on slide, square-notched rear sight adj. for windage. Custom wood grips on double-action, nylon grips on single action. Blued or satin nickel finish. Made 1984. SA. Discontinued 1989.
Double-Action $135
Single Action 115
Satin Nickel 145

Bersa Model 622 Auto Pistol $125
Caliber: 22 LR. 7-shot magazine. 4- or 6-inch bbl. 7 or 9 inches overall. Weighs 2.25 lbs. Blade front sight, square-notch rear adj. for windage. Blued finish. Nylon grips. Made 1982-87.

Bersa Model 644 Auto Pistol $135
Caliber: 22 LR. 10-shot magazine. 3.5-inch bbl. Weight: 26.5 oz. 6.5 inches overall. Adj. rear sight, blade front. Contoured black nylon grips. Made 1980-88.

Bersa Thunder 9 Auto Pistol $295
Caliber: 9mm Para. 15-shot magazine. 4-inch bbl. 7.38 inches overall. Weight: 30 oz. Blade front sight, adj. rear w/3-dot system. Ambidextrous safety and decocking devise. Matte blued finish. Checkered black polymer grips. Made 1993-96.

Bersa Thunder 22 Auto Pistol (Model 23)
Caliber: 22 LR, 10-shot magazine. 3.5-inch bbl., 6.63 inches overall. Weight: 24.5 oz. Notched-bar dovetailed rear, blade integral w/slide front. Black polymer grips. Made 1989 to date.
Blued Finish **$165**
Nickel Finish. **185**

Bersa Thunder 380 Auto Pistol
Caliber: 380 ACP, 7-shot magazine. 3.5-inch bbl., 6.63 inches overall. Weight: 25.75 oz. Notched-bar dovetailed rear, blade integral w/slide front. Blued, satin nickel, or Duo-Tone finish. Made 1995 to date.
Blued Finish **$170**
Satin Nickel Finish **190**
Duo-Tone Finish **180**

Bersa Thunder 380 Plus Auto Pistol
Same general specifications as standard Thunder 380 except has 10-shot magazine and weighs 26 oz. Made 1995-97.
Matte Finish **$175**
Satin Nickel Finish **215**
Duo-Tone Finish. **200**

BRNO PISTOLS
Manufactured in Czechoslovakia

Brno ZBP 99 DA Automatic Pistol . . . . (SRP) $TBA
Calibers: 9mm Parabellum, 40 S&W. 10-shot magazine. 4-inch bbl. 6.75 inches overall. Weight: 27.5 oz. Fixed sights. Blued finish. Checkered black plastic stocks. Announced 1998.

BROLIN ARMS
La Verne, California

Brolin Arms "Legend Series" SA Automatic Pistol
Caliber: 45 ACP. 7-shot magazine. 4- or 5-inch bbl. Weight 32-36 oz. Walnut grips. Single action, full size, compact, or full size frame compact slide. Matte blued finish. Lowered and flared ejection port. Made 1995 to date.
Model L45 . . . . **$325**
Model L45C **345**
Model L45T **345**

Brolin Arms "Patriot Series" SA Automatic Pistol
Caliber: 45 ACP. 7-shot magazine. 3.25- and 4-inch bbl. Weight: 33-37 oz. Wood Grips. Fixed rear sights. Made 1996 to date.
Model P45. . . . . **$425**
Model P45C (Disc. 1997) **445**
Model P45T (Disc. 1997) **445**

Brolin Arms "Pro-Stock and Pro-Comp" SA Automatic Pistol
Caliber: 45 ACP. 8-shot magazine. 4- or 5-inch bbl. Weight 37 oz. Single action, blued or two-tone finish. Wood grips. Bomar adjustable sights. Made 1996-97.
Model Pro Comp. **$545**
Model Pro Stock **495**

Brolin Arms TAC Series
Caliber: 45 ACP. 8-shot magazine. 5-inch bbl. 8.5 inches overall. Weight: 37 oz. Low profile combat or Tritium sights. Beavertail grip safety. Matte blue, chrome or Two-Tone finish. Checkered wood or contoured black rubber grips. Made 1997 to date.
Model TAC 11 Service. **$435**
Model TAC 11 Compact **445**
W/Tritium Sights, add **95**

BRONCO PISTOL
Eibar, Spain
Manufactured by Echave y Arizmendi

Bronco Model 1918 Pocket Automatic Pistol **$115**
Caliber: 32 ACP (7.65mm). 6-shot magazine. 2.5-inch bbl. 5 inches overall. Weight: 20 oz. Fixed sights. Blued finish. Hard rubber grips. Made c.1918-25.

Bronco Semiautomatic Pistol **$130**
Caliber: 25ACP, 6-shot magazine. 2.13-inch bbl., 4.13 inches overall. Weight: 11 oz. Fixed sights. Blued finish. Hard rubber grips. Made 1919-35.

BROWNING PISTOLS
Morgan, Utah

The following Browning pistols have been manufactured by Fabrique Nationale d'Armes de Guerre (now Fabrique Nationale Herstal) of Herstal, Belgium, by Arms Technology Inc. of Salt Lake City and by J. P. Sauer & Sohn of Eckernforde, W. Germany. (See also FN Browning and J.P. Sauer & Sohn listings.)

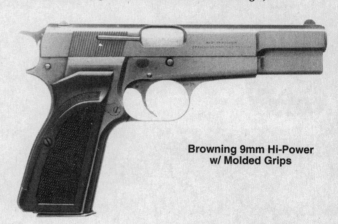

**Browning 9mm Hi-Power
w/ Molded Grips**

Browning Hi-Power Automatic Pistol
Same general specifications as FN Browning Model 1935, except chambered 9mm Para., 30 Luger and 40 S&W. 10- or 13-shot magazine. 4.63-inch bbl. 7.75 inches overall. Weight: 32 oz. (9mm) or 35 oz. (40 S&W). Fixed sights, also available w/rear sight adj. for windage and elevation and ramp front sight. Ambidextrous safety added after 1989. Standard Model blued, chrome-plated or Bi-Tone finish. Checkered walnut, contour-molded Polyamide or wraparound rubber grips. Renaissance Engraved Model, chrome-plated, Nacrolac pearl grips. Made by FN from 1955 to date.
Standard Model, fixed sights, 9mm **$345**
Standard Model, fixed sights, 40 S&W (intro. 1986). **350**
Standard Model, 30 Luger (1986-89) **565**
Renaissance Model, fixed sights. **750**
W/Adjustable Rear Sight, add **50**
W/Ambidextrous Safety, add **95**
W/Moulded Grips, deduct **40**
W/Tangent Rear Sight (1965-78), add* **295**
W/T-Slot Grip & Tangent Sight (1965-78), add* **595**
*Check SN range to certify value

Browning Hi-Power Capitan Automatic **$395**
Similar to the standard Hi-Power, except fitted w/adj. 500-meter tangent rear sight and rounded serrated hammer. Made 1993 to date.

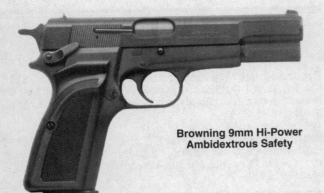

**Browning 9mm Hi-Power
Ambidextrous Safety**

Browning Hi-Power Practical Automatic Pistol

Similar to the standard Hi-Power, except has silver-chromed frame and blued slide w/Commander-style hammer. Made 1991 to date.

W/Fixed Sights **$375**
W/adj. Sights **425**

Browning Hi-Power 9mm Classic

Limited Edition 9mm Hi-Power, w/silver gray finish, high-grade engraving and finely checkered walnut grips w/double border. Proposed production of the Classic was 5000 w/less than half that number produced. Gold Classic limited to 500 w/two-thirds proposed production in circulation. Made 1985-86.

Gold Classic **$1795**
Standard Classic **895**

**Browning
Hi-Power Capitan**

**Browning 25 Automatic Standard
Model**

**Browning
9mm Classic**

Browning 25 Automatic Pistol

Same general specifications as FN Browning Baby (*see* separate listing). Standard Model, blued finish, hard rubber grips. LightModel, nickel-plated, Nacrolac pearl grips. Renaissance Engraved Model, nickel-plated, Nacrolac pearl grips. Made by FN 1955-69.

Standard Model **$265**
Lightweight Model **355**
Renaissance Model **750**

**Browning 380 Automatic Pistol,
1955 Type**

**Browning
Hi-Power Practical**

Browning 380 Automatic Pistol, 1955 Type

Same general specifications as FN Browning 380 Pocket Auto. Standard Model, Renaissance Engraved Model, as furnished in 25 Automatic. Made by FN 1955-69.

Standard Model **$325**
Renaissance Model **895**

Browning BDA 380

**Browning BDA 380
Nickel**

**Browning BDM 9mm DA
Automatic**

Browning 380 Automatic Pistol, 1971 Type

Same as 380 Automatic, 1955 Type, except has longer slide, 4.44-inch bbl., is 7.06 inches overall, weighs 23 oz. Rear sight adj. for windage and elevation, plastic thumbrest grips. Made 1971-75.

Standard Model. $295
Renaissance Model 795

Browning BDA DA Automatic Pistol

Similar to SIG-Sauer P220. Calibers: 9mm Para., 38 Super Auto, 45 Auto. 9-shot magazine (9mm and 38),7-shot (45 cal). 4.4-inch bbl. 7.8 inches overall. Weight: 29.3 oz. Fixed sights. Blued finish. Plastic grips. Made 1977-79 by J. P. Sauer.

BDA Model, 9mm, 45 ACP $395
BDA Model, 38 Super $525

Browning BDA-380 DA Automatic Pistol

Caliber: 380 Auto. 10- or 13-shot magazine. Bbl. length: 3.81 inches. 6.75 inches overall. Weight: 23 oz. Fixed blade front sight, square-notch drift-adj. rear sight. Blued or nickel finish. Smooth walnut grips. Made 1982-97 by Beretta.

Blued Finish $295
Nickel Finish. 325

Browning BDM 9mm DA Automatic Pistol

Caliber: 9mm Para. 10- or 15-shot magazine. 4.73-inch bbl. 7.85 inches overall. Weight: 31 oz. Low-profile removable blade front sight and windage-adj. rear sight w/3-dot system. Matte blued, Bi-Tone or silver chrome finish. Selectable shooting mode and decocking safety lever. Made 1991 to date.

Blued Finish $395
Bi-Tone Finish 400
Nickel Finish. 425

**Browning
Buck Mark 22 Bulls Eye**

**Browning
Buck Mark 22 Micro**

Browning Buck Mark 22 Automatic Pistol

Caliber: 22 LR. 10-shot magazine. Bbl.: 5.5 inches. 9.5 inches overall. Weight: 32 oz. Black molded grips. Adj. rear sight. Blued or nickel finish. Made 1985 to date.

Blued Finish $145
Nickel Finish. 195

Browning Buck Mark 22 Bulls Eye $215

Same general specifications as the standard Buck Mark 22, except w/7.25-inch fluted barrel. 11.3 inches overall. Weight: 36 oz. Adjustable trigger. Undercut post front sight, click-adjustable Pro-Target rear. Laminated, Rosewood or black rubber grips. Made 1997 to date.

**Browning Buck Mark 22
Silhouette**

**Browning Challenger
Standard Model**

Browning Buck Mark 22 Micro
Same general specifications as standard Buck Mark 22 except w/4-inch bbl. 8 inches overall. Weight: 32 oz. Molded composite grips. Ramp front sight, Pro Target rear sight. Made 1992 to date.
Blued Finish **$145**
Nickel finish **195**

Browning Buck Mark 22 Plus
Same general specifications as standard Buck Mark 22 except for black molded, impregnated hardwood grips. Made 1987 to date.
Blued Finish **$165**
Nickel finish **205**

Browning Buck Mark 22 Silhouette **$250**
Same general specifications as standard Buck Mark 22, except for 9.88-inch bbl., 53-oz. weight, target sights mounted on full-length scope base, and laminated hardwood grips and forend. Made 1987 to date.

**Browning Challenger
Renaissance Model**

Browning Buck Mark 22 Target 5.5
Same general specifications as Buck Mark 22, except 5.5-inch bbl., 35.5 oz. weight and target sights mounted on full-length scope base. Made 1989 to date.
Standard **$235**
Gold Model **255**

Browning Buck Mark 22 Unlimited Silhouette **$285**
Same general specifications as standard Buck Mark 22 Silhouette, except w/14-inch bbl. 18.69 inches overall. Weight: 64 oz. Interchangeable post front sight and Pro Target rear. Nickel finish. Made 1992 to date.

**Browning Challenger III
Sporter 22**

Browning Buck Mark 22 Varmint Auto Pistol **$225**
Same general specifications as standard Buck Mark 22, except for 9.88-inch bbl., 48-oz. weight, no sights, full-length scope base, and laminated hardwood grips. Made 1987 to date.

Browning Challenger Automatic Pistol
Caliber: 22 LR. 10-shot magazine. Bbl. lengths: 4.5 and 6.75-inch. 11.44 inches overall (6.75-inch bbl.). Weight: 38 oz. (6.75-inch bbl.). Removable blade front sight, screw adj. rear. Standard finish, blued, also furnished gold inlaid (Gold Model) and engraved and chrome-plated (Renaissance Model). Checkered walnut grips. Finely figured and carved grips on Gold and Renaissance Models. Standard made by FN 1962-75, higher grades. Intro. 1971. Discontinued.
Standard Model . . . . **$ 275**
Gold Model **995**
Renaissance Model **1095**

Browning Challenger II Automatic Pistol **$180**
Same general specifications as Challenger Standard Model w/6.75-inch bbl., changed grip angle and impregnated hardwood grips. Original Challenger design modified for lower production costs. Made by ATI 1976-83.

**Browning International
Medalist**

Browning Challenger III Automatic Pistol **$175**
Same general description as Challenger II, except has 5.5 inch bull bbl., alloy frame and new sight system. Weight: 35 oz. Made 1982-84. **Sporter Model** w/6.75-inch bbl.. Made 1984-86.

Browning International Medalist Automatic Target Pistol **$575**
Modification of Medalist to conform w/International Shooting Union rules. 5.9-inch bbl. Smaller grip with no forearm. Weight: 42 oz. Made 1970-73.

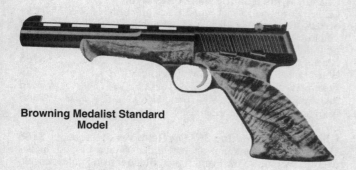

Browning Medalist Standard Model

Browning Medalist Renaissance Model

Browning Nomad

Browning Medalist Automatic Target Pistol

Caliber: 22 LR. 10-shot magazine. 6.75-inch bbl. w/vent rib. 11.94 inches overall. Weight: 46 oz. Removable blade front sight, click-adj. micrometer rear. Standard finish, blued also furnished gold-inlaid (Gold Model) and engraved and chrome-plated (Renaissance Model). Checkered walnut grips w/thumbrest (for right- or left-handed shooter). Finely figured and carved grips on Gold and Renaissance Models. Made by FN 1962-75. Higher grades. Intro. 1971.

Standard Model . . . . $ 595
Gold Model 1200
Renaissance Model 1795

Browning Nomad Automatic Pistol $225

Caliber: 22 LR. 10-shot magazine. Bbl. lengths: 4.5 and 6.75-inch. 8.94 inches overall (4.5-inch bbl.). Weight: 34 oz. (4.5-inch bbl.). Removable blade front sight, screw adj. rear. Blued finish. Plastic grips. Made by FN 1962-74.

Browning Renaissance 9mm, 25 Auto and 380 Auto (1955) Engraved Models, Cased Set $3595

One pistol of each of the three models in a special walnut carrying case, all chrome-plated w/Nacrolac pearl grips. Made by FN 1955-69.

BRYCO ARMS INC.
Irvine, California
Distributed by Jennings Firearms. Inc.
Carson City, NV

Bryco/Jennings Models J22, J25 Auto Pistol

Calibers: 22 LR, 25 ACP. 6-shot magazine. 2.5-inch bbl. About 5 inches overall. Weight: 13 oz. Fixed sights. Chrome, satin nickel or black Teflon finish. Walnut, grooved black Cycolac or resin-impregnated wood grips. Made from 1981 to date.

Model J-22 . . . . $65
Model J-25 (Disc. 1995) 60

Bryco/Jennings Models M25, M32, M38 Auto Pistol

Calibers: 25 ACP, 32 ACP, 380 ACP. 6-shot magazine. 2.81-inch bbl. 5.31 inches overall. Weight: 11oz. to 15 oz. Fixed sights. Chrome, satin nickel or black Teflon finish. Walnut, grooved, black Cycolac or resin-impregnated wood grips. Made from 1988 to date.

Model M25 (Disc.) . . . . $ 65
Model M32 75
Model M38 85

Bryco/Jennings Model M48 Auto Pistol $95

Calibers: 22 LR, 32 ACP, 380 ACP. 7-shot magazine. 4-inch bbl. 6.69 inches overall. Weight: 20 oz. Fixed sights. Chrome, satin nickel or black Teflon finish. Smooth wood or black Teflon grips. Made from 1989-95.

Bryco/Jennings Model M58 Auto Pistol $105

Caliber: 380 ACP. 10-shot magazine. 3.75-inch bbl. 5.5 inches overall. Weight: 30 oz. Fixed sights. Chrome, satin nickel, blued or black Teflon finish. Smooth wood or black Teflon grips. Made 1993-95.

Bryco/Jennings Model M59 Auto Pistol $115

Caliber: 9mm Para. 10-shot magazine. 4-inch bbl. 6.5 inches overall. Weight: 33 oz. Fixed sights. Chrome, satin nickel, blued or black Teflon finish. Smooth wood or black Teflon grips. Made 1994-96.

Bryco/Jennings Model Nine SA Auto Pistol $95

Similar to Bryco/Jennings Model M59, except w/redesigned slide w/loaded chamber indicator and frame mounted ejector. Weight: 30 oz. Made 1997 to date.

Bryco/Jennings Model T-22 Target Pistol $90

Similar to Bryco/Jennings Model M48, except chambered 22 LR only, w/target configuration sights and redesigned slide w/loaded chamber indicator and hold open. Made 1997 to date.

BUDISCHOWSKY PISTOL
Mt. Clemens, Michigan
Mfd. by Norton Armament Corporation

Budischowsky TP-70 DA Automatic Pistol

Calibers: 22 LR, 25 Auto. 6-shot magazine. 2.6-inch bbl. 4.65 inches overall. Weight: 12.3 oz. Fixed sights. Stainless steel. Plastic grips. Made 1973-77.

22 Long Rifle $365
25 Automatic 250

CALICO LIGHT WEAPONS SYSTEMS
Bakersfield, California

Calico Model 110 Auto Pistol **$325**
Caliber: 22 LR. 100-shot magazine. 6-inch bbl. 17.9 inches overall. Weight: 3.75 lbs. Adj. post front sight, fixed U-notch rear. Black finish aluminum frame. Molded composition grip. Made 1986-94.

Calico Model M-950 Auto Pistol **$355**
Caliber: 9mm Para. 50- or 100-shot magazine. 7.5-inch bbl. 14 inches overall. Weight: 2.25 lbs. Adj. post front sight, fixed U-notch rear. Glass-filled polymer grip. Made 1989-94.

CHARTER ARMS CORPORATION
Stratford, Connecticut

Charter Arms Model 40 Automatic Pistol **$235**
Caliber: 22 RF. 8-shot magazine. 3.3-inch bbl. 6.3 inches overall. Weight: 21.5 oz. Fixed sights. Checkered walnut grips. Stainless steel finish. Made 1985-86.

Charter Arms Model 79K DA Automatic Pistol **$295**
Calibers: 380 or 32 Auto. 7-shot magazine. 3.6-inch bbl. 6.5 inches overall. Weight: 24.5 oz. Fixed sights. Checkered walnut grips. Stainless steel finish. Made 1985-86.

Charter Arms Clyde

Charter Arms Bonnie

Charter Arms Bonnie and Clyde Set **$375**
Matching pair of shrouded 2.5-inch bbl. revolvers, chambered for 32 Magnum and 38 Special. Blued finish w/scrolled name on bbls.. Made 1989-90.

Charter Arms Bulldog 44 DA Revolver
Caliber: 44 Special. 5-shot cylinder. 2.5- or 3-inch bbl. 7 or 7.5 inches overall. Weight: 19 or 19.5 oz. Fixed sights. Blued, nickel-plated or stainless finish. Checkered walnut Bulldog or square buttgrips. Made from 1973-96.
Blued Finish/Pocket Hammer (2.5") **$165**
Blued Finish/Bulldog Grips (3") (Disc. 1988) 155
Electroless Nickel 185
Stainless Steel/Bulldog Grips (Disc. 1992) 145
Neoprene Grips/Pocket Hammer 160

Charter Arms Bulldog 357 DA Revolver **$165**
Caliber: 357 Magnum. 5-shot cylinder. 6-inch bbl. 11 inches overall. Weight: 25 oz. Fixed sights. Blued finish. Square, checkered walnut grips. Intro. 1977-96.

Charter Arms Bulldog New Police DA Revolver
Same general specifications as Bulldog Police, except chambered for 44 Special. 5-shot cylinder. 2.5- or 3.5-inch bbl. Made 1990-92.
Blued Finish **$165**
Stainless Finish (2.5-inch bbl. only) 175

Charter Arms Bulldog Police DA Revolver
Caliber: 38 Special or 32 H&R Magnum. 6-shot cylinder. 4-inch bbl. 8.5 inches overall. Weight: 20.5 oz. Adj. rear sight, ramp front. Blued or stainless finish. Square checkered walnut grips. Made 1976-93. Shroud dropped on new model.
Blued Finish **$155**
Stainless Finish 145
32 H&R Magnum (Disc.1992) 160

**Charter Arms
Bulldog Pug**

Charter Arms Bulldog Pug DA Revolver
Caliber: 44 Special. 5-shot cylinder. 2.5 inch bbl. 7.25 inches overall. Weight: 20 oz. Blued or stainless finish. Fixed ramp front sight, fixed square-notch rear. Checkered Neoprene or walnut grips. Made 1988-93.
Blued Finish **$165**
Stainless Finish 185

Charter Arms Bulldog Target DA Revolver
Calibers: 357 Magnum, 44 Special (latter intro. in 1977). 4-inch bbl. 8.5 inches overall. Weight: in 357, 20.5 oz. Adj. rear sight, ramp front. Blued finish. Square checkered walnut grips. Made 1976-92.
Blued Finish (Disc.1989) **$145**
Stainless Steel 155

**Charter Arms Police
Bulldog 38 Special**

**Charter Arms Bulldog
Target**

**Charter Arms Bulldog
Tracker**

**Charter Arms Explorer II
Standard Model**

**Charter Arms Explorer II Silvertone
w/optional Barrels**

Charter Arms Pathfinder

Charter Arms Bulldog Tracker DA Revolver **$145**
Caliber: 357 Mag. 5-shot cylinder. 2.5-, 4- or 6-inch bbl. 11 inches
overall (6-inch bbl.). Weight: 21 oz. (2.5-inch bbl.). Adj. rear sight,
ramp front. Checkered walnut grips. Blued finish. 4- or 6-inch bbl.
Disc.1986. Reintroduced 1989-92.

Charter Arms Explorer II Semiauto Survival Pistol
Caliber: 22 RF. 8-shot magazine. 6-, 8- or 10-inch bbl. 13.5 inches
overall (6-inch bbl.). Weight: 28 oz. Finishes: black, heat cured,
semigloss textured enamel or silvertone anticorrosion. Disc.1987.
Standard Model. **$75**
Silvertone (w/optional 6- or 10-inch bbl.) **85**

Charter Arms Off-Duty DA Revolver
Calibers: 22 LR or 38 Special. 6-shot (22 LR) or 5-shot (38 Spec.) cyl-
inder. 2-inch bbl. 6.25 inches overall. Weight: 16 oz. Fixed rear sight,
Patridge-type front sight. Plain walnut grips. Matte black, electroless
nickel or stainless steel finish. Made 1992-96.
Matte Black Finish **$130**
Electroless Nickel **175**
Stainless Steel **165**

Charter Arms Pathfinder DA Revolver
Calibers: 22 LR, 22 WMR. 6-shot cylinder. Bbl. lengths: 2-, 3-, 6-
inch. 7.13 inches overall (3-inch bbl.). and regular grips. Weight:
18.5 oz. (3-inch bbl.). Adj. rear sight, ramp front. Blued or stainless
finish. Plain walnut regular, checkered Bulldog or square buttgrips.
Made 1970 to date. *Note:* Originally designated "Pocket Target,"
name was changed in 1971 to "Pathfinder." Grips changed in 1984.
Disc.1993.
Blued Finish **$135**
Stainless Finish **150**

Charter Arms Pit Bull DA Revolver
Calibers: 9mm, 357 Magnum, 38 Special. 5-shot cylinder. 2.5-, 3.5-
or 4-inch bbl. 7 inches overall (2.5-inch bbl.). Weight: 21.5 to 25 oz.
All stainless steel frame. Fixed ramp front sight, fixed square-notch
rear. Checkered Neoprene grips. Blued or stainless finish. Made
1989-93.
Blued Finish **$165**
Stainless Finish **175**

Charter Arms Undercover DA Revolver
Caliber: 38 Special. 5-shot cylinder. bbl. lengths: 2-, 3-, 4-inch. 6.25
inches overall (2-inch bbl.). and regular grips. Weight: 16 oz. (2-inch
bbl.). Fixed sights. Plain walnut, checkered Bulldog or square buttgrips. Made 1965-96.
Blued or Nickel-plated Finish. **$155**
Stainless Finish **165**

**Charter Arms
Undercover Stainless**

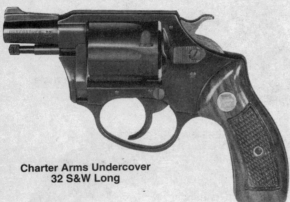

**Charter Arms Undercover
32 S&W Long**

**Charter Arms Police
Undercover 32 H&R Magnum**

Charter Arms Undercover 32 H&R Magnum or S&W Long

Same general specifications as standard Undercover, except chambered for 32 H&R Magnum or 32 S&W Long, has 6-shot cylinder and 2.5-inch bbl.

32 H&R Magnum (Blued)	**$130**
32 H&R Magnum (Electroless Nickel)	155
32 H&R Magnum (Stainless)	145
32 S&W Long (Blued) Disc. 1989	125

Charter Arms Undercover Pocket Police DA Revolver

Same general specifications as standard Undercover, except has 6-shot cylinder and pocket-type hammer. Blued or stainless steel finish. Made 1969-81.

Blued Finish	**$135**
Stainless Steel	150

Charter Arms Undercover Police DA Revolver

Same general specifications as standard Undercover, except has 6-shot cylinder. Made 1984-89. Reintroduced 1993.

Blued, 38 Special	**$145**
Stainless, 38 Special	155
32 H&R Magnum	140

Charter Arms Undercoverette DA Revolver $125

Same as Undercover model w/2-inch bbl., except caliber 32 S&W Long, 6-shot cylinder, blued finish only. Weighs 16.5 oz. Made 1972-83.

CIMMARRON F.A. CO.
Fredricksburg, Texas

Cimarron El Pistolero

Cimarron El Pistolero Single-Action Revolver $265

Calibers: .357 Mag., 45 Colt. 6-shot cylinder. 4.75- 5.5- or 7.5-inch bbl. Polished brass backstrap and triggerguard. Otherwise, same as Colt Single-Action Army revolver w/parts being interchangeable. Made 1997-98.

COLT MANUFACTURING CO., INC.
Hartford, Connecticut

Previously Colt Industries, Firearms Division. Production of some Colt handguns spans the period before World War II to the postwar years. Values shown for these models are for earlier production. Those manufactured c. 1946 and later generally are less desirable to collectors, and values are approximately 30 percent lower.

AUTOMATIC PISTOLS

> **NOTE**
>
> For ease in finding a particular firearm, Colt handguns are grouped into three sections: Automatic Pistols, Single Shot Pistols, Deringers and Revolvers. For a complete listing, please refer to the Index.

Colt Model 1900 38 Automatic Pistol $4850

Caliber: 38 ACP (modern high-velocity cartridges should not be used in this pistol). 7-shot magazine. 6-inch bbl. 9 inches overall. Weight: 35 oz. Fixed sights. Blued finish. Plain walnut grips. Sharp-spur hammer. Combination rear sight and safety. Made 1900-03.

Colt Model 1902 Military 38 Automatic Pistol $1595

Caliber: 38 ACP (modern high-velocity cartridges should not be used in this pistol). 8-shot magazine. 6-inch bbl. 9 inches overall. Weight: 37 oz. Fixed sights, knife-blade and V-notch. Blued finish. Checkered hard rubber grips. Round back hammer, changed to spur type in 1908. No safety. Made 1902-29.

Colt Model 1902 Sporting

Colt Model 1903 Pocket Hammer

Colt Model 1903 Pocket Hammerless

Colt Model 1905 Military

Colt Model 1902 Sporting 38 Automatic Pistol $1950

Caliber: 38 ACP (modern high-velocity cartridges should not be used in this pistol). 7-shot magazine. 6-inch bbl. 9 inches overall. Weight: 35 oz. Fixed sights, knife-blade and V-notch. Blued finish. Checkered hard rubber grips. Round back hammer. No safety. Made 1902-08.

Colt Model 1903 Pocket 38 Hammer Automatic Pistol $895

Caliber: 38 ACP (modern high-velocity cartridges should not be used in this pistol). Similar to Model 1902 Sporting 38, but w/4.5-inch bbl. 7.5 inches overall. Weight: 31 oz. Fixed sights, knife-blade and V-notch. Blued finish. Checkered hard rubber grips. Round back hammer, changed to spur type in 1908. No safety. Made 1903-1929.

Colt Model 1903 Pocket 32 Hammerless Automatic Pistol $450

Caliber: 32 ACP. 8-shot magazine. Similar to Model 1903 Pocket 38, except hammerless and equipped w/slide-lock and grip safety. 3.75-inch bbl. Weight: 24 oz. Fixed sights, knife-blade and V-notch. Blued finish. Checkered hard rubber grips. Made 1903-45.

Colt Model 1905 45 Automatic Pistol $2795

Caliber: 45 Automatic. 7-shot magazine. 5-inch bbl. 8 inches overall. Weight: 32.5 oz. Fixed sights, knife-blade and V-notch. Blued finish. Checkered walnut grips. Similar to Model 1902 38 Auto Pistol. Made 1905-11.

Colt Model 1911 Automatic Pistol

Caliber: 45 Auto. 7-shot magazine. 5-inch bbl. 8.5 inches overall. Weight: 39 oz. Fixed sights. Blued finish on Commercial Model Parkerized or similar finish on most military pistols. Checkered walnut grips (early production), plastic grips (later production). Checkered, arched mainspring housing and longer grip safety spur adopted in 1923 (on M1911A1).

Model 1911 Commercial (C-series)	**$1350**
Model 1911A1 Commercial (Pre-WWII)	**1195**

Colt Model 1911

U.S. Government Model 1911

Colt manufacture	$ 995
North American manufacture	16,950
Remington-UMC manufacture	1,725
Springfield manufacture	1,625
Navy Model M1911 Type	2,195

U.S. Government Model 1911A1

Singer manufacture	$17,950
Colt, Ithaca, Remington-Rand manufacture	695
Union Switch & Singal manufacture	950

Colt 1991 A1

Colt All American Model 2000

NOTE

During both World Wars, Colt licensed other firms to make these pistols under government contract: Ithaca Gun Co., North American Arms Co. Ltd. (Canada) Remington-Rand Co., Remington-UMC, Singer Sewing Machine Co., and Union Switch & Signal Co.; M1911 also produced at Springfield Armory.

Colt Model M1991 A1 Semiauto Pistol
Reissue of Model 1911A1 (*see* above) w/a continuation of the original serial number range 1945. Caliber: 45 ACP. 7-shot magazine. 5-inch bbl. 8.5 inches overall. Weight: 39 oz. Fixed blade front sight, square notch rear. Parkerized finish. Black composition grips. Made 1991 to date. (Commander and Compact variations intro. 1993).

Standard Model **$350**
Commander w/4.5-inch bbl. **355**
Compact w/3.5-inch bbl. (6-shot) **360**

Colt Cadet 22

Colt 22 Cadet Automatic Pistol **$185**
Caliber: 22LR. 10-shot magazine. 4.5-inch vent rib bbl. 8.63 inches overall. Weight: 33.5 oz. Blade front sight, dovetailed rear. Stainless finish. Textured black polymer grips w/Colt medallion. Made 1993-95. Note: The Cadet Model name was discontinued under litigation but the manufacturer continued to produce this pistol configuration as the Model "Colt 22". For this reason the "Cadet" model will command slight premiums.

Colt 22 Sport Automatic Pistol **$180**
Same specifications as Cadet Model, except renamed Colt 22 w/composition monogrip or wraparound black rubber grip. Made 1995-98.

Colt 22 Target Pistol **$195**
Similar to Colt 22 Sport Model, except w/6-inch vent rib bbl. 10.12 inches overall. Weight: 40.5 oz. Partridge style front sight, adjustable white outline rear on full length grooved rib. Made 1995 to date.

Colt Ace Automatic Pistol
Caliber: 22 LR (regular or high speed). 10-shot magazine. Built on the same frame as the Government Model 45 Auto, w/same safety features, etc. Hand-honed action, target bbl., adj. rear sight. 4.75-inch bbl. 8.25 inches overall. Weight: 38 oz. Made 1930-40.

Commercial Model **$1395**
Service Model (1938-42) **1795**

Colt All American Model 2000 DA Pistol **$550**
Hammerless semiautomatic w/blued slide and polymer or alloy receiver fitted w/roller bearing trigger. Caliber: 9mm Para. 15-shot magazine. 4.5-inch bbl. 7.5 inches overall. Weight: 29 oz. (Polymer) or 33 oz (Alloy). Fixed blade front sight, square-notch rear w/3-dot system. Matte blued slide w/black polymer or anodized aluminum receiver. Made 1992-94.

Colt Challenger Automatic Pistol **$275**
Same basic design as Woodsman Target, Third Issue, but lacks some of the refinements. Fixed sights. Magazine catch on butt as in old Woodsman. Does not stay open on last shot. Lacks magazine safety. 4.5- or 6-inch bbl. 9 to 10.5 inches overall. Weight: 30 oz. (4.5-inch bbl.) or 31.5 oz. (6-inch bbl.) Blued finish. Checkered plastic grips. Made 1950-55.

Colt Combat Commander Automatic Pistol
Same as Lightweight Commander, except has steel frame w/blued or nickel-plated finish. Weighs 36 oz. Made 1950-76.

9mm Para. **$495**
38 Super, 45 ACP **575**

Colt Commander Lightweight Automatic Pistol
Same basic design as Government Model, except w/shorter 4.25-inch bbl. and a special lightweight "Coltalloy" receiver and mainspring housing. Calibers: 45 Auto, 38 Super Auto, 9mm Para. 7-shot magazine (45 cal.), 9-shot (38 Auto and 9mm). 8 inches overall. Weight:26.5 oz. Fixed sights. Round spur hammer. Improved safety lock. Blued finish. Checkered plastic or walnut grips. Made 1950-76.

9mm Para. **$485**
38 Super, 45 ACP **525**

Colt Conversion Unit—22-45 . . . . **$2100**
Converts Service Ace 22 to National Match 45 Auto. Unit consists of match-grade slide assembly and bbl., bushing, recoil spring, recoil spring guide and plug, magazine and slide stop. Made 1938-42.

Colt Delta Gold Cup

**Colt Gold Cup
National Match**

Colt Conversion Unit — 45-22 . . . . **$350**
Converts Government Model 45 Auto to a 22 LR target pistol. Unit consists of slide assembly, bbl., floating chamber (as in Service Ace), bushing, ejector, recoil spring recoil spring guide and plug, magazine and slide stop. The componet parts differ and are not interchangable between post war, series 70, series 80, ACE I and ACE II units. Made 1938 to date.

Colt Delta Elite Semiauto Pistol
Caliber: 10 mm. 5-inch bbl. 8.5 inches overall. 8-round magazine. Weight: 38 oz. Checkered neoprene combat grips w/Delta medallion. 3-dot, high-profile front and rear combat sights. Blued or stainless finish. Made 1987-96.
First Edition (500 Ltd. Edition) **$895**
Blued Finish 475
Matte Stainless Finish. 495
Ultra Stainless Finish 575

Colt Delta Gold Cup Semiauto Pistol
Same general specifications as Delta Elite, except in match configuration w/Accro adjustable rear sight. Made 1989-93 and 1995-96.
Blued Finish (disc 1991). **$545**
Stainless Steel Finish 635

Colt Gold Cup Mark III National Match **$795**
Similar to Gold Cup National Match 45 Auto, except chambered for 38 Special mid-range. 5-shot magazine. Made 1961-74.

Colt Gold Cup National Match 45 Auto **$675**
Match version of Government Model 45 Auto w/same general specifications, except: match grade bbl. w/new design bushing, flat mainspring housing, long wide trigger w/adj. stop, handfitted slide w/improved ejection port, adj. rear sight, target front sight checkered walnut grips w/gold medallions. Weight: 37 oz. Made 1957-70.

Colt Government Model 1911/1911A1
See Colt Model 1911.

Colt Huntsman **$275**
Same specifications as the Challenger. Made 1955-76.

Colt MK I & II/Series '90 Double Eagle
Combat Commander **$455**
Calibers: 40 S&W, 45 ACP. 7-shot magazine. 4.25-inch bbl. 7.75 inches overall. Weight: 36 oz. Fixed blade front sight, square-notch rear. Checkered Xenoy grips. Stainless finish. Made 1992-96.

Colt MK II/Series '90 Double Eagle DA
Semiauto Pistol
Calibers: 38 Super, 9mm, 40 S&W, 10mm, 45 ACP. 7-shot magazine. 5-inch bbl. 8.5 inches overall. Weight: 39 oz. Fixed or Accro adj. sights. Matte stainless finish. Checkered Xenoy grips. Made 1991-96.
38 Super, 9mm, 40 S&W (Fixed Sights) **$450**
45 ACP (Adjustable Sights) 475
45 ACP (Fixed Sights) 455
10mm (Adjustable Sights) 470
10mm (Fixed Sights) 450

**Colt MK II/Series '90 Double
Eagle Officer's ACP**

Colt MK II/Series '90 Double Eagle Officer's ACP **$460**
Same general specifications as Double Eagle Combat Commander, except chambered for 45 ACP only. 3.5-inch bbl. 7.25 inches overall. Weight: 35 oz. Also available in lightweight (25 oz.) w/blued finish (same price). Made 1990-93.

Colt MK IV/Series '70 Combat Commander
Same general specifications as the Lightweight Commander, except made 1970-83.
Blued Finish **$425**
Nickel Finish. 495

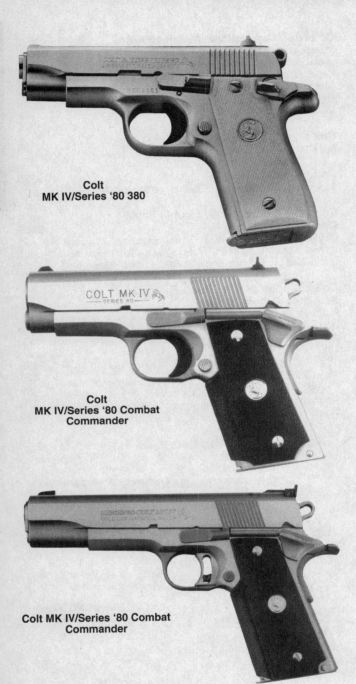

**Colt
MK IV/Series '80 380**

**Colt
MK IV/Series '80 Combat
Commander**

**Colt MK IV/Series '80 Combat
Commander**

Colt MK IV/Series '70 Gold Cup National Match 45 Auto $625
Match version of MK IV/Series '70 Government Model. Caliber: 45 Auto only. Flat mainspring housing. Accurizor bbl. and bushing. Solid rib, Colt-Elliason adj. rear sight undercut front sight. Adj. trigger, target hammer. 8.75 inches overall. Weight: 38.5 oz. Blued finish. Checkered walnut grips. Made 1970-84.

Colt MK IV/Series '70 Gov't. Auto Pistol $485
Calibers: 45 Auto, 38 Super Auto, 9mm Para. 7-shot magazine in 45, 9-shot in 38 and 9mm. 5-inch bbl. 8.38 inches overall. Weight: 38 oz., 45, 39 oz., 38 and 9mm. Fixed rear sight and ramp front sight. Blued or nickel-plated finish. Checkered walnut grips. Made 1970-84.

Colt MK IV/Series '80 380 Automatic Pistol
Caliber: 380 ACP. 3.29-inch bbl. 6.15 inches overall. Weight: 21.8 oz. Composition grips. Fixed sights. Made 1984 to date.

Blued Finish (Disc.1997)	**$245**
Bright Nickel (Disc.1995)	**285**
Satin Nickel Electroless/Coltguard (Disc.1989)...........	**265**
Stainless Finish......................................	**290**

Colt MK IV/Series '80 Combat Commander
Updated version of the MK IV/Series '70 w/same general specifications. Blued, two-tone or stainless steel w/"pebbled" black Neoprene wraparound grips. Made 1979 to date.

Blued Finish (Disc.1996)	**$405**
Satin Nickel (Disc.1987)............................	**425**
Stainless Finish....................................	**435**
Two-Tone Finish	**455**

Colt MK IV/Series '80 Combat Elite
Same general specifications as MK IV/Series '80 Combat Commander, except w/Elite enhancements. Calibers: 38 Super, 40 S&W, 45 ACP. Stainless frame w/blued steel slide. Accro adj. sights and beavertail grip safety. Made 1992-96.

38 Super, 45 ACP....................................	**$495**
40 S&W	**475**

Colt MK IV/Series '80 Gold Cup National Match
Same general specifications as Match '70 version, except w/additional finishes and "pebbled" wraparound Neoprene grips. Made 1983-96.

Blued Finish	**$505**
Bright Blued Finish.................................	**575**
Stainless Finish....................................	**525**

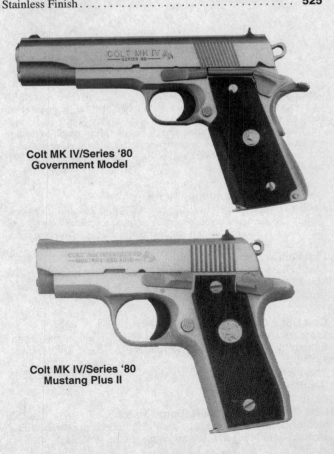

**Colt MK IV/Series '80
Government Model**

**Colt MK IV/Series '80
Mustang Plus II**

Colt MK IV/Series '80 Mustang Pocketlite

Colt MK IV/Series '80 Officer's ACP

Colt MK IV/Series '80 Government Model
Same general specifications as Government Model Series '70, except also chambered in 40 S&W, w/"pebbled" wraparound Neoprene grips, blued or stainless finish. Made 1983 to date.

Blued Finish **$405**
Bright Blued Finish. **445**
Bright Stainless Finish **475**
Matte Stainless Finish. **430**

Colt MK IV/Series '80 Lightweight Commander **$425**
Updated version of the MK IV/Series '70 w/same general specifications.

Colt MK IV/Series '80 Mustang 380 Automatic
Caliber: 380 ACP. 5- or 6-shot magazine. 2.75-inch bbl. 5.5 inches overall. Weight: 18.5 oz. Blued, nickel or stainless finish. Black composition grips. Made 1986-97.

Blued Finish **$265**
Nickel Finish (Disc.1994). **285**
Satin Nickel Electroless/Coltguard (Disc.1988). **275**
Stainless Finish. **285**

Colt MK IV/Series '80 Mustang Plus II
Caliber: 380 ACP, 7-round magazine. 2.75-inch bbl. 5.5 inches overall. Weight: 20 oz. Blued or stainless finish w/checkered black composition grips. Made 1988-96.

Blued Finish **$265**
Stainless Finish **285**

Colt MK IV/Series '80 Mustang Pocketlite
Same general specifications as the Mustang 30, except weighs only 12.5 oz. w/aluminum alloy receiver. Blued, chrome or stainless finish. Optional wood grain grips. Made 1988 to date.

Blued Finish. **$265**
Lady Elite (Two-Tone) Finish **345**
Stainless Finish. **285**
Teflon/Stainless Finish **295**

Colt MK IV/Series '80 Officer's ACP Automatic Pistol
Calibers: 40 S&W and 45 ACP. 3.63-inch bbl. 7.25 inches overall. Weight: 34 oz. Made 1984-97. 40 S&W. Disc.1992.

Blued Finish (Disc.1996) **$395**
Matte Finish **385**
Satin Nickel Finish **425**
Stainless Steel **435**

Colt MK IV/Series '80 SA Lightweight
Concealed Carry Officer **$465**
Caliber: 45 ACP. 7-shot magazine. 4,25-inch bbl. 7.75 inches overall. Weight: 35 oz. Aluminum alloy receiver w/stainless slide. Dovetailed low-profile sights w/3-Dot system. Matte stainless finish w/blued receiver. Wraparound black rubber grip w/finger grooves. Made 1998 to date.

Colt MK IV/Series '90 Defender SA Lightweight **$475**
Caliber: 45 ACP. 7-shot magazine. 3-inch bbl. 6.75 inches overall. Weight: 22.5 oz. Aluminum alloy receiver w/stainless slide. Dovetailed low-profile sights w/3-Dot system. Matte stainless finish w/Nickle-Teflon receiver. Wraparound black rubber grip w/finger grooves. Made 1998 to date.

Colt MK IV/Series 90 Pony DAO Pistol **$325**
Caliber: 380ACP. 6-shot magazine. 2.75-inch bbl. 5.5 inches overall. Weight: 19 oz. Ramp front sight, dovetailed rear. Stainless finish. Checkered black composition grips. Made 1997 to date.

Colt MK IV/Series 90 Pony Pocketlite **$320**
Similar to standard weight Pony Model, except w/aluminum frame. Brushed stainless and Teflon finish. Made 1997 to date.

Colt National Match Automatic Pistol
Identical w/the Government Model 45 Auto, but w/hand-honed action, match-grade bbl., adj. rear and ramp front sights or fixed sights. Made 1932-40.

W/Adjustable Sights **$1995**
W/Fixed Sights **1495**

Colt NRA Centennial 45 Gold Cup
National Match **$895**
2500 produced in 1971.

Colt Pocket Model 25 (1908) Auto Pistol **$395**
Caliber: 25 Auto. 6-shot magazine. 2-inch bbl. 4.5 inches overall. Weight: 13 oz. Flat-top front, square-notch rear sight in groove. Blued or nickel finish. Checkered hard rubber grips on early models, checkered walnut on later type, special pearl grips illustrated. Disconnector added in 1916 at pistol No. 141000. Made 1908-41.

Colt Pocket Model 32 (1903) Automatic
Pistol, First Issue **$425**
Caliber: 32 Auto. 8-shot magazine. 4-inch bbl. 7 inches overall. Weight: 23 oz. Fixed sights. Blued finish. Checkered hard rubber grips. Hammerless. Slide lock and grip safeties. Barrel-lockbushing similar to that on Government Model 45 Auto. Made 1903-11. *See* Colt Model 1903.

Colt Pocket Model 25

Colt Pocket Junior

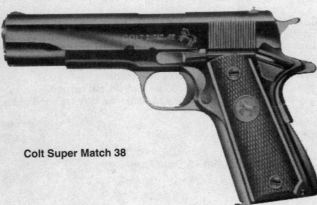

Colt Super Match 38

Colt Pocket Model 380 (1908) Automatic Pistol
Same as Pocket 32 Auto, First, Second and Third Issues, respectively, except chambered for caliber 380 Auto w/7-shot magazine.
First Issue (Made 1908-11). **$425**
Second Issue (Made 1911-26) **475**
Third Issue (Safety disconnector on all pistols
  above No. 92,894 Made 1926-45) **400**

Colt Pocket Junior Model Automatic Pistol **$255**
Made in Spain by Unceta y Cia (Astra). Calibers: 22 Short, 25 Auto. 6-shot magazine. 2.25 inch bbl. 4.75 inches overall. Weight: 12 oz. Fixed sights. Checkered walnut grips. Made 1958-68.

Colt Super 38 Automatic Pistol
Identical w/Government Model 45 Auto, except for caliber and magazine capacity. Caliber: 38 Automatic. 9-shot magazine. Made 1928-70.
Pre-war **$1795**
Post-war **695**

Colt Super Match 38 Automatic Pistol
Identical to Super 38 Auto, but w/hand-honed action, match grade bbl., adjustable rear sight and ramp front sight or fixed sights. Made 1933-46.
W/Adjustable Sights. **$4595**
W/Fixed Sights **3295**

Colt Targetsman

Colt Targetsman **$325**
Similar to Woodsman Target but has "economy" adj. rear sight, lacks automatic slide stop. Made 1959-76.

**Colt Woodsman
Match Target First Issue**

Colt Pocket Model 32 Automatic
Pistol, Second Issue **$445**
Same as First Issue but without barrel-lock bushing. Made 1911-26.

Colt Pocket Model 32 Automatic
Pistol, Third Issue **$375**
Caliber: 32 Auto. Similar to First and Second Issues, but has safety disconnector on all pistols above No. 468097 which prevents firing of cartridge in chamber if magazine is removed. 3.75-inch bbl. 6.75 inches overall. Weight: 24 oz. Fixed sights. Blued or nickel finish. Checkered walnut grips. Made 1926-45.

Colt Woodsman Match Target Automatic
Pistol, First Issue **$1095**
Same basic design as other Woodsman models. Caliber: 22 LR. 10-shot magazine. 6.5-inch bbl., slightly tapered w/flat sides. 11 inches overall. Weight: 36 oz. Adjustable rear sight. Blued finish. Checkered walnut one-piece grip w/extended sides. Made 1938-42.

Colt Woodsman Match Target Auto Pistol, Second Issue **$650**

Same basic design as Woodsman Target, Third Issue. Caliber: 22 LR (reg. or high speed). 10-shot magazine. 6-inch flat-sided heavy bbl. 10.5 inches overall. Weight: 40 oz. Click adj. rear sight, ramp front. Blued finish. Checkered plastic or walnut grips. Made 1948-76.

Colt Woodsman Match Target "4½" Automatic Pistol **$495**

Same as Match Target, 2nd Issue, except w/4.5-inch bbl. 9 inches overall. Weight: 36 oz. Made 1950-76.

Colt Woodsman Sport Model Automatic Pistol, First Issue **$695**

Caliber: 22 LR (reg. or high speed). Same as Woodsman Target, Second Issue, except has 4.5-inch bbl. Adjustable rear sight w/fixed or adjustable front sight. Weight: 27 oz. 8.5 inches overall. Made 1933-48.

Colt Woodsman Sport Model Automatic Pistol, Second Issue **$525**

Same as Woodsman Target, 3rd Issue, but w/4.5-inch bbl. 9 inches overall. Weight: 30 oz. Made 1948-76.

Colt Woodsman Target Model Automatic, First Issue **$495**

Caliber: 22 LR (reg. velocity). 10-shot magazine. 6.5-inch bbl. 10.5 inches overall. Weight: 28 oz. Adjustable sights. Blued finish. Checkered walnut grips. Made 1915-32. *Note:* The mainspring housing of this model is not strong enough to permit safe use of high-speed cartridges.

Change to a new heat-treated mainspring housing was made at pistol No. 83,790. Many of the old models were converted by installation of new housings. The new housing may be distinguished from the earlier type by the checkering in the curve under the breech. The new housing is grooved straight across, while the old type bears a diagonally checkered oval.

Colt Woodsman Target Model Automatic, Second Issue **$475**

Caliber: 22 LR (reg. or high speed). Same as original model except has heavier bbl. and high-speed mainspring housing. *See* note under Woodsman, First Issue. Weight: 29 oz. Made 1932-48.

Colt Woodsman Target Model Automatic, Third Issue **$395**

Same basic design as previous Woodsman pistols, but w/longer grip, magazine catch on left side, larger thumb safety, slide stop, slide stays open on last shot, magazine disconnector thumbrest grips. Caliber: 22 LR (reg. or high speed). 10-shot magazine. 6-inch bbl. 10.5 inches overall. Weight: 32 oz. Click adjustable rear sight, ramp front sight. Blued finish. Checkered plastic or walnut grips. Made 1948-76.

Colt Woodsman Sport Model First Issue

Colt Woodsman Sport Model Second Issue

Colt Woodsman Target Model First Issue

Colt Woodsman Target Model Second Issue

Colt Woodsman Match Target Second Issue

Colt Woodsman Target Model Third Issue

**Colt World War II D-Day Invasion
Commemorative**

**Colt World War II 50th Anniversary
Commemorative**

**Colt World War II Golden
Anniversary V-J Day Tribute**

Colt World War I 50th Anniversary Commemorative Series 45 Auto

Limited production replica of Model 1911 45 Auto engraved w/battle scenes, commemorating Battles at Chateau Thierry, Belleau Wood Second Battle of the Marne, Mouse Argonne. In special presentation display cases. Production: 7,400 standard model, 75 deluxe, 25 special deluxe grade. Match numbered sets offered. Made in 1967, 68, 69. Values indicated are for commemoratives in new condition.

Standard Grade . . . . $ 650
Deluxe Grade 1395
Special Deluxe Grade 2895

Colt World War II Commemorative 45 Auto $725

Limited production replica of Model 1911A1 45 Auto engraved w/respective names of locations where historic engagements occurred during WW II, as well as specific issue and theater identification. European model has oak leaf motif on slide, palm leaf design frames the Pacific issue. Cased. 11,500 of each model were produced. Made in 1970. Value listed is for gun in new condition.

Colt World War II 50th Anniversary Commemorative 45 Auto $1395

Same general specifications as the Colt World War II Commemorative 45 Auto, except slightly difference scroll engraving, 24-karat gold-plate trigger, hammer, slide stop, magazine catch, magazine catch lock, safety lock, and four grip screws. Made in 1995 only.

Colt World War II D-Day Invasion Commemorative $800

High-luster and highly decorated version of the Colt Model 1911A1. Caliber: 45 ACP. Same general specifications as the Colt Model 1911 except for 24-karat gold-plated hammer, trigger, slide stop, magazine catch, magazine catch screw, safety lock and four grip screws. Also has scrolls and inscription on slide. Made in 1991 only.

Colt World War II Golden Anniversary V-J Day Tribute 45 Auto $1495

Basic Colt Model 1911A1 design w/highly-polished bluing and decorated w/specialized tributes to honor V-J Day. Two 24-karat gold scenes highlight the slide. 24-karat gold-plated hammer. Checkered wood grips w/gold medallion on each side. Made 1995.

NOTE

For ease in finding a particular firearm, Colt handguns are grouped into three sections: Automatic Pistols (which precedes this one), this section, and Revolvers, which follows. For a complete listing, please refer to the Index.

COLT SINGLE-SHOT PISTOLS & DERINGERS

Colt Camp Perry First Issue

Colt Camp Perry Second Issue

Colt Camp Perry Model Single-Shot Pistol, First Issue $1250
Built on Officers' Model frame. Caliber: 22 LR (embedded head chamber for high-speed cartridges after 1930). 10 inch bbl. 13.75 inches overall. Weight: 34.5 oz. Adj. target sights. Hand-finished action. Blued finish. Checkered walnut grips. Made 1926-34.

Colt Camp Perry Model Second Issue $1050
Same general specifications as First Issue, except has shorter hammer fall and 8-inch bbl. 12 inches overall. Weight: 34 oz. Made 1934-41 (about 440 produced).

Colt Civil War Centennial Model Pistol
Single-shot replica of Colt Model 1860 Army Revolver. Caliber: 22 Short. 6-inch bbl. Weight: 22 oz. Blued finish w/gold-plated frame, grip frame, and trigger guard, walnut grips. Cased. 24,114 were produced. Made in 1961.
Single Pistol $195
Pair w/consecutive serial numbers 450

Colt Deringer No. 4
Replica of Deringer No. 3 (1872). Single-shot w/sideswing bbl. Caliber: 22 Short. 2.5-inch bbl. 4.9 inches overall. Weight: 7.75 oz. Fixed sights. Gold-finished frame, blued bbl., walnut grips, also nickel-plated w/simulated ivory grips. Cased. Made 1959-63.
Single Pistol $125
Pair w/consecutive serial numbers 250

Colt Deringer No. 4 Commemorative Models
Limited production version of 22 Deringer issued, w/appropriate inscription, to commemorate historical events.

1961 Issue

Geneseo, Illinois, 125th Anniversary (104 produced)...... **$695**

1962 Issue

Fort McPherson, Nebraska, Centennial (300 produced) **$425**

Colt Lord and Lady Deringers
Same as Deringer No. 4. Lord model is blued w/gold-plated frame and walnut grips. Lady model is gold-plated w/simulated pearl grips. Furnished in cased pairs. Made 1970-72.
Lord Deringer, pair in case **$200**
Lady Deringer, pair in case 200
Lord and Lady Deringers, one each, in case............. 225

Colt Rock Island Arsenal Centennial Pistol $250
Limited production (550 pieces) version of Civil War Centennial Model single-shot 22 pistol, made exclusively for Cherry's Sporting Goods, Geneseo, Illinois, to commemorate the centennial of the Rock Island Arsenal in Illinois. Cased. Made in 1962.

NOTE

This section of Colt handguns contains only revolvers. For automatic pistols or single-shot pistols and derringers, please see the two sections that precede this. For a complete listing, please refer to the Index.

REVOLVERS

Colt Agent First Issue

Colt 38 DS II Revolver $265
Caliber: 38 Special. 6-shot cylinder. 2-inch bbl. 7 inches overall. Weight: 21 oz. Ramp front sight, fixed notch rear. Satin stainless finish. Black rubber combat grip w/finger grooves. Made 1997 to date.

Colt Agent DA Revolver, First Issue $335
Same as Cobra, First Issue, except has short-grip frame 38 Special only, weighs 14 oz. Made 1955-72.

Colt Agent (LW) DA Revolver, Second Issue $295
Same as Colt Agent, first issue, except has shrouded ejector rod and alloy frame. Made 1973-86.

Colt Anaconda

Colt Bankers' Special

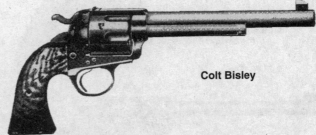

Colt Bisley

Colt Buntline Scout

Colt Anaconda DA Revolver

Calibers: 44 Mag., 45 Colt. Bbl. lengths: 4, 6 or 8 inches. 11.63 inches overall (6-inch bbl.). Weight: 53 oz. (6-inch bbl.). Adj. white outline rear sight, red insert rampstyle front. Matte stainless or Realtree Grey camo finish. Black Neoprene combat grips w/finger grooves. Made 1990 to date.

Matte Stainless	**$375**
Realtree Grey Camo Finish (Disc. 1996)	**495**
Custom Model (44 Mag w/ported bbl.)	**475**
First Edition Model (Ltd Edition 1000)	**650**
Hunter Model (44 Mag w/2X Scope)	**750**

Colt Anaconda Titanium DA Revolver

Same general specifications as the standard Anaconda except, chambered 44 Mag. only w/titanium-plated finish, gold-plated trigger, hammer and cylinder release. Limited edition of 1,000 distributed by American Historical Foundation w/personalized inscription. Made 1996.

One of 1000	**$1800**
Presentation Case **add**	**200**

Colt Army Special DA Revolver **$465**

41-caliber frame. Calibers: 32-20, 38 Special (41 Colt). 6-shot cylinder, right revolution. Bbl. lengths: 4-, 4.5-, 5-, and 6-inch. 9.25 inches overall (4-inch bbl.). Weight: 32 oz. (4-inch bbl.). Fixed sights. Blued or nickel-plated finish. Made 1908-27. *Note:* This model has a somewhat heavier frame than the New Navy, which it replaced. Serial numbers begin w/300,000. The heavy 38 Special High Velocity loads should not be used in 38 Special arms of this model.

Colt Bankers' Special DA Revolver

This is the Police Positive w/a 2-inch bbl., otherwise specifications same as that model, rounded butt intro. in 1933. Calibers: 22 LR (embedded head-cylinder for high speed cartridges intro. 1933), 38 New Police. 6.5 inches overall. Weight: 23 oz. (22 LR), 19 oz. (38). Made 1926-40.

38 Caliber	**$ 775**
22 Caliber	**1495**

Colt Bisley Model SA Revolver

Variation of the Single-Action Army, developed for target shooting w/modified grips, trigger and hammer. Calibers: general specifications same as SA Army. Target Model made w/flat-topped frame and target sights. Made 1894-1915.

Standard Model	**$6250**
Target Model (Flat-Top)	**9200**

Colt Buntline Scout **$350**

Same as Frontier Scout, except has 9.5-inch bbl. Made 1959-71.

Colt Buntline Special 45 . . . . **$925**

Same as standard SA Army, except has 12-inch bbl., caliber 45 Long Colt. Made 1957-75.

Colt Cobra DA Revolver, Round Butt, First Issue **$355**

Lightweight Detective Special w/same general specifications as that model, except w/Colt-alloy frame. 2-inch bbl. Calibers: 38 Special, 38 New Police, 32 New Police. Weight: 15 oz., 38 cal. Blued finish. Checkered plastic or walnut grips. Made 1951-73.

Colt Cobra DA Revolver, Second Issue **$295**

Lightweight version of Detective Special, Second Issue has aluminum alloy frame. 16.5 oz. Made 1973-81.

Colt Cobra DA Revolver Square Butt **$315**

Lightweight Police Positive Special w/same general specifications, except has Colt-alloy frame. 4-inch bbl. Calibers: 38 Special, 38 New Police, 32 New Police. Weight: 17 oz. in 38 caliber. Blued finish. Checkered plastic or walnut grips. Made 1951-73.

Colt Commando Special DA Revolver **$285**

Caliber: 38 Special. 6-shot cylinder. 2-inch bbl., 6.88 inches overall. Weight: 21.5 oz. Fixed sights. Low-luster blued finish. Made 1982-86.

**Colt Cobra, Round Butt
First Issue**

**Colt Detective Special
First Issue**

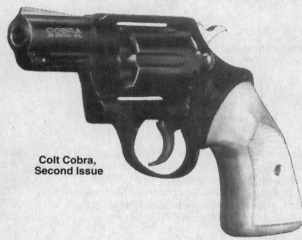

**Colt Cobra,
Second Issue**

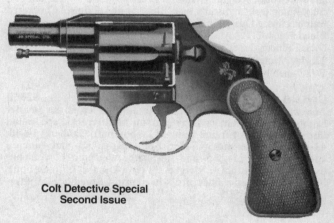

**Colt Detective Special
Second Issue**

Colt Commando Special

Colt Detective Special DA Revolver, First Issue
Similar to Police Positive Special w/2-inch bbl., otherwise specifications same as that model, rounded butt intro. 1933. 38 Special only in pre-war issue. Blued or nickel-plated finish. Weight: 17 oz. 6.75 inches overall. Made 1926-46.
Blued Finish **$550**
Nickel Finish. **625**

Colt Detective Special DA Revolver, 2nd Issue
Similar to Detective Special, First Issue, except w/2- or 3-inch bbl. and also chambered 32 New Police, 38 New Police. Wood, plastic or over-sized grips. Made 1947-72.
Blued Finish **$350**
Nickel Finish. **395**
W/3-inch bbl., **add** **95**

Colt Detective Special DA Revolver, 3rd Issue
"D" frame, shrouded ejector rod. Caliber: 38 Special. 6-shot cylinder. 2-inch bbl. 6.88 inches overall. Weight: 21.5 oz. Fixed rear sight, ramp front. Blued or nickel-plated finish. Checkered walnut wraparound grips. Made 1973-84.
Blued Finish **$295**
Nickel Finish. **350**
W/3-inch bbl., **add** **75**

Colt Detective Special DA Revolver, 4th Issue
Similar to Detective Special, Third Issue, except w/alloy frame. Blued or chrome finish. Wraparound black neoprene grips w/Colt medallion. Made 1993-95.
Blued Finish **$265**
Chrome Finish **325**
DAO Model (bobbed hammer). **365**

Colt Diamondback

Colt Diamondback DA Revolver
"D" frame, shrouded ejector rod. Calibers: 22 LR, 22WRF, 38 Special. 6-shot cylinder. 2.5-, 4- or 6-inch bbl. w/vent rib. 9 inches overall (4-inch bbl). Weight: 31.75 oz. (22 cal., 4-inch bbl.), 28.5 oz. (38 cal.). Ramp front sight, adj. rear. Blued or nickel finish. Checkered walnut grips. Made 1966-84.

Blued Finish **$365**
Nickel Finish. 415
22 Mag. Model 435
W/2.5-inch bbl., **add** 75

Colt DA Army Revolver **$2650**
Also called DA Frontier. Similar in appearance to the smaller Lightning Model, but has heavier frame of different shape, round disc on left side of frame, lanyard loop in butt. Calibers: 38-40, 44-40, 45 Colt. 6-shot cylinder. Bbl. lengths: 3.5- and 4-inches (w/o ejector), 4.75-, 5.5- and 7.5-inches w/ejector. 12 .5 inches overall (7.5-inch bbl.). Weight: 39 oz. (45 cal., 7.5-inch bbl.). Fixed sights. Hard rubber bird's-head grips. Blued or nickel finish. Made 1878-1905.

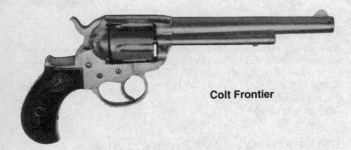

Colt Frontier

Colt Frontier Scout Revolver
SA Army replica, ⅞ scale. Calibers: 22 Short, Long, LR or 22 WMR (interchangeable cylinder available). 6-shot cylinder. 4.75-inch bbl. 9.9 inches overall. Weight: 24 oz. Fixed sights. Plastic grips. Originally made w/bright alloy frame. Since 1959 w/steel frame and blued finish or all-nickel finish w/wood grips. Made 1958-71.

Blued Finish, plastic grips **$295**
Nickel Finish, wood grips 315
Extra interchangeable cylinder 65

Colt Frontier Scout Revolver Commemorative Models
Limited production versions of Frontier Scout issued, w/appropriate inscription, to commemorate historical events. Cased. *Note:* Values indicated are for commemoratives in new condition.

1961 Issues

Kansas Statehood Centennial (6201 produced) **$425**
Pony Express Centennial (1007 produced). 475

1962 Issues

Columbus, Ohio, Sesquicentennial (200 produced) . . . . . . **$ 595**
Fort Findlay, Ohio, Sesquicentennial (130 produced) 730
Fort Findlay Cased Pair, 22 Long Rifle and 22
   Magnum (20 produced) 2950
New Mexico Golden Anniversary 425
West Virginia Statehood Centennial (3452 produced). 425

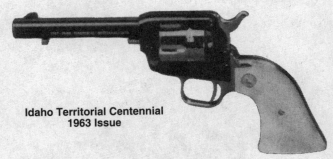

**Idaho Territorial Centennial
1963 Issue**

1963 Issues

Arizona Territorial Centennial (5355 produced) **$395**
Battle of Gettysburg Centennial (1019 produced) 425
Carolina Charter Tercentenary (300 produced) 395
Fort Stephenson, Ohio, Sesquicentennial (200 produced) 595
General John Hunt Morgan Indiana Raid. 675
Idaho Territorial Centennial (902 produced) 395

**General Hood Centennial
1964 Issue**

1964 Issues

California Gold Rush (500 produced) **$395**
Chamizal Treaty (450 produced) 415
General Hood Centennial (1503 produced) 395
Montana Territorial Centennial (2300 produced). 385
Nevada "Battle Born" (981 produced) 425
Nevada Statehood Centennial (3984 produced) 375
New Jersey Tercentenary (1001 produced) 395
St. Louis Bicentennial (802 produced). 425
Wyoming Diamond Jubilee (2357 produced). 395

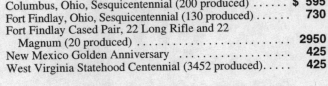

New Jersey Tercentenary — 1964 Issue

1965 Issues

Appomattox Centennial (1001 produced) **$395**
Forty-Niner Miner (500 produced) 425
General Meade Campaign (1197 produced) 395
Kansas Cowtown Series—Wichita (500 produced) 415
Old Fort Des Moines Reconstruction (700 produced) 395
Oregon Trail (1995 produced) 385
St. Augustine Quadricentennial (500 produced) 425

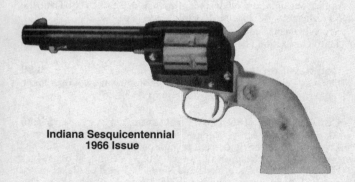

**Indiana Sesquicentennial
1966 Issue**

1966 Issues

Colorado Gold Rush (1350 produced) **$395**
Dakota Territory (1000 produced) 425
Indiana Sesquicentennial (1500 produced) 385
Kansas Cowtown Series—Abilene (500 produced) 425
Kansas Cowtown Series—Dodge City (500 produced) 395
Oklahoma Territory (1343 produced) 365

1967 Issues

Alamo (4500 produced) **$415**
Kansas Cowtown Series—Coffeyville (500 produced) 385
Kansas Trail Series—Chisholm Trail (500 produced) 415
Lawman Series—Bat Masterson (3000 produced) 395

1968 Issues

Kansas Trail Series—Santa Fe Trail (501 produced) **$395**
Kansas Trail Series—Pawnee Trail (501 produced) 385
Lawman Series—Pat Garrett (3000 produced) 415
Nebraska Centennial (7001 produced) 385

**Golden Spike Centennieal
1969 Issue**

1969 Issues

Alabama Sesquicentennial (3001 produced) **$375**
Arkansas Territory Sesquicentennial (3500 produced) 385
California Bicentennial (5000 produced) 365
General Nathan Bedford Forrest (3000 produced) 375
Golden Spike (11,000 produced) 385
Kansas Trail Series—Shawnee Trail (501 produced) 395
Lawman Series—Wild Bill Hickock (3000 produced) 425

1970 Issues

Kansas Fort Series—Fort Larned (500 produced) **$385**
Kansas Fort Series—Fort Hays (500 produced) 375
Kansas Fort Series—Fort Riley (500 produced) 385
Lawman Series—Wyatt Earp (3000 produced) 495
Maine Sesquicentennial (3000 produced) 375
Missouri Sesquicentennial (3000 produced) 385

1971 Issues

Kansas Fort Series—Fort Scott (500 produced) **$385**

1972 Issues

Florida Territory Sesquicentennial (2001 produced) **$395**

1973 Issues

Arizona Ranger (3001 produced) **$375**

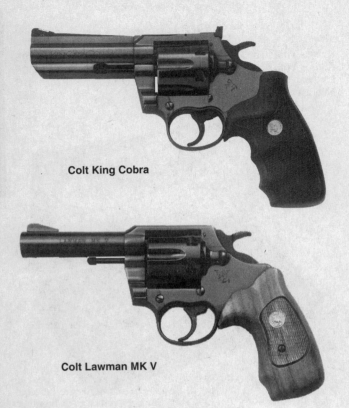

Colt King Cobra

Colt Lawman MK V

Colt King Cobra Revolver
Caliber: 357 Mag. Bbl. lengths: 2.5-, 4-, 6- or 8-inch. 9 inches overall (4-inch bbl.). Weight: 42 oz., average. Matte stainless steel finish. Black Neoprene combat grips. Made 1986 to date. 2.5-inch bbl., and "Ultimate" bright or blued finish. Made 1988-92.
Matte Stainless **$295**
Ultimate Bright Stainless **325**
Blued. **275**

Colt Lawman MK III DA Revolver
"J" frame, shrouded ejector rod on 2-inch bbl. only. Caliber: 357 Magnum. 6-shot cylinder. Bbl. lengths: 2-, 4-inch. 9.38 inches overall w/4-inch bbl. Weight: w/4-inch bbl., 35 oz. Fixed rear sight, ramp front. Service trigger and hammer or target trigger and wide-spur hammer. Blued or nickel-plated finish. Checkered walnut service or target grips. Made 1969-1982.
Blued Finish **$195**
Nickel Finish. **215**

Colt Lawman MK V DA Revolver
Similar to Trooper MK V. Caliber: 357 Mag. 6-shot cylinder. 2- or 4-inch bbl., 9.38 inches overall (4-inch bbl.). Weight: 35 oz. (4-inch bbl.). Fixed sights. Checkered walnut grips. Made 1983-85.
Blued Finish **$205**
Nickel Finish. **225**

Colt Magnum Carry DA Revolver **$265**
Similar to Model DS II, except chambered 357 Magnum. Made 1998 to date.

Colt Marine Corps Model (1905) DA Revolver **$2150**
General specifications same as New Navy, Second Issue, except this has round butt, was supplied only in 38 caliber (38 Short & Long Colt, 38 Special) w/6-inch bbl. Made 1905-09.

Colt Metropolitan MK III DA Revolver **$325**
Same as Official Police MK III, except has 4-inch bbl. w/service or target grips. Weighs 36 oz. Made 1969-72.

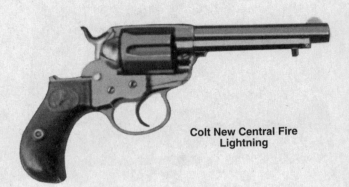

Colt New Central Fire Lightning

Colt New DA Central Fire Revolver **$1395**
Also called Lightning Model. Calibers: 38 and 41 Centerfire. 6-shot cylinder. Bbl. lengths: 2.5-, 3.5-, 4.5- and 6-inch without ejector, 4.5- and 6-inch w/ejector. 8.5 inches overall (3.5-inch bbl.). Weight: 23 oz. (38 cal., 3.5-inch bbl.), Fixed sights. Blued or nickel finish. Hard rubber bird's-head grips. Made 1877-09.

Colt New Frontier Buntline Special
Same as New Frontier SA Army, except has 12-inch bbl.
Second Generation (1962-75). **$995**
Third Generation (1976-92) **825**

Colt New Frontier SA Army Revolver
Same as SA Army, except has flat-top frame, adj. target rear sight, ramp front sight, smooth walnut grips. 5.5- or 7.5-inch bbl. Calibers: 357 Magnum, 44 Special, 45 Colt. Made 1961-92.
Second Generation (1961-75). **$925**
Third Generation (1976-92) **795**

Colt New Frontier SA 22 Revolver **$260**
Same as Peacemaker 22, except has flat-top frame, adj. rear sight, ramp front sight. Made 1971-76; reintro. 1982-86.

Colt New Navy (1889) DA, First Issue **$1295**
Also called New Army. Calibers: 38 Short & Long Colt, 41 Short & Long Colt. 6-shot cylinder, left revolution. Bbl. lengths: 3-, 4.5- and 6-inch. 11.25 inches overall (6-inch bbl.). Weight: 32 oz., 6-inch bbl. Fixed sights, knife-blade and V-notch. Blued or nickel-plated finish. Walnut or hard rubber grips. Made 1889-94. *Note:* This model, which was adopted by both the Army and Navy, was Colt's first revolver of the solid frame, swing-out cylinder type. It lacks the cylinder-locking notches found on later models made on this 41 frame, ratchet on the back of the cylinder is held in place by a double projection on the hand.

Colt New Navy (1892) DA, Second Issue **$1150**
Also called New Army. General specifications same as First Issue except has double cylinder notches and double locking bolt. Calibers: 38 Special added in 1904 and 32-20 in 1905. Made 1892-07. *Note:* The heavy 38 Special High Velocity loads should not be used in 38 Special arms of this model.

Colt New Pocket DA Revolver **$475**
Caliber: 32 Short & Long Colt. 6-shot cylinder. Bbl. lengths: 2.5, 3.5- and 6-inch. 7.5 inches overall w/3.5-inch bbl. Weight: 16 oz., 3.5-inch bbl. Fixed sights knife-blade and V-notch. Blued or nickel finish. Rubber grips. Made 1893-05.

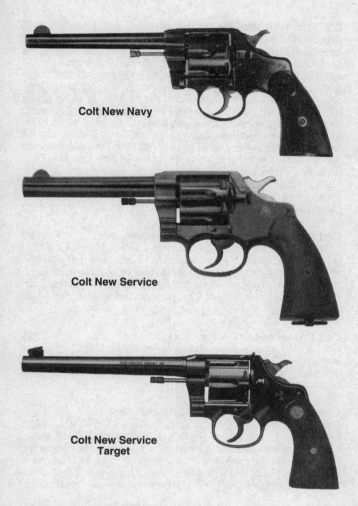

Colt New Navy

Colt New Service

Colt New Service Target

Colt New Police DA Revolver $395
Built on New Pocket frame, but w/larger grip. Calibers: 32 Colt New Police, 32 Short & Long Colt. Bbl. lengths: 2.5-, 4- and 6-inch. 8.5 inches overall (4-inch bbl.). Weight: 17 oz., 4-inch bbl. Fixed knife-blade front sight, V-notch rear. Blued or nickel finish. Rubber grips. Made 1896-05.

Colt New Police Target DA Revolver $695
Target version of the New Police w/same general specifications. Target sights. 6-inch bbl. Blued finish only. Made 1896-05.

Colt New Service DA Revolver
Calibers: 38 Special, 357 Magnum (intro. 1936), 38-40, 44-40, 44 Russian, 44 Special, 45 Auto, 45 Colt, 450 Eley, 455 Eley, 476 Eley. 6-shot cylinder. Bbl. lengths: 4-, 5- and 6-inch in 38 Special and 357 Magnum, 4.5-, 5.5- and 7.5 inch in other calibers. 9.75 inches overall (4.5-inch bbl.). Weight: 39 oz., 45 cal. (4.5-inch bbl.). Fixed sights. Blued or nickel finish. Checkered walnut grips. Made 1898-42. *Note:* More than 500,000 of this model in caliber 45 Auto (designated "Model 1917 Revolver") were purchased by the U.S. Govt. during WW I. These arms were later sold as surplus to National Rifle Association members through the Director of Civilian Marksmanship. Price was $16.15 plus packing charge. Supply exhausted during the early 1930s.

Commercial Model	$1295
Magnum	825
1917 Army	795

Colt New Service Target $1850
Target version of the New Service. Calibers: originally chambered for 44 Russian, 450 Eley, 455 Eley and 476 Eley, later models in 44 Special, 45 Colt and 45 Auto. 6- or 7.5-inch bbl. 12.75 inches overall (7.5-inch bbl.). Adj. target sights. Hand-finished action. Blued finish. Checkered walnut grips. Made 1900-40.

Colt Officers' Model Match $395
Same general design as Officers' Model revolvers. Has tapered heavy bbl., wide hammer spur, Accrued rear sight ramp front sight, large target grips of checkered walnut. Calibers: 22 LR, 38 Special. 6-inch bbl. 11.25 inches overall. Weight: 43 oz. (22 cal.), 39 oz. (38 cal.). Blued finish. Made 1953-70.

Colt Officers' Match

Colt Officers' Model Special $495
Target version of Officers' Model, Second Issue w/similar characters, except w/heavier, nontapered bbl., redesigned hammer. Ramp front sight, Colt Officers' Model Special "Coltmaster" rear sight adj. for windage and elevation. Calibers: 22 LR, 38 Special. 6-inch bbl. 11.25 inches overall. Weight: 39 oz. (38 cal.), 43 oz., (22 cal.). Blued finish. Checkered plastic grips. Made 1949-53.

Colt Officers' Model Target DA Revolver, First Issue $895
Caliber: 38 Special. 6-inch bbl. Hand-finished action. Adj. target sights. Checkered walnut grips. General specifications same as New Navy, Second Issue. Made 1904-08.

Colt Officers' Target Second Issue

Colt Officers' Model Target, Second Issue $750
Calibers: 22 LR (intro. 1930, embedded head-cylinder for high-speed cartridges after 1932), 32 Police Positive (made 1932-1942), 38 Special. 6-shot cylinder. Bbl. lengths: 4-, 4.5-, 5-, 6- and 7.5-inch (38 Special) or 6-inch only (22 LR and 32 PP). 11.25 inches overall (6-inch bbl., 38 Special). Adj. target sights. Blued finish. Checkered walnut grips. Hand-finished action. General features same as Army Special and Official Police of same date. Made 1908-49 (w/exceptions noted).

Colt Official Police

Colt Peacekeeper

Colt Pocket Positive

**Colt Police Positive
First Issue**

Colt Official Police DA Revolver

Calibers: 22 LR (intro. 1930, embedded head-cylinder for high-speed cartridges after 1932), 32-20 (discontinued 1942), 38 Special, 41 Long Colt (discontinued 1930). 6-shot cylinder. Bbl. lengths: 4-, 5-, and 6-inch or 2-inch and 6-inch heavy bbl. in 38 Special only, 22 LR w/4 and 6-inch bbl.s only. 11.25 inches overall. Weight: 36 oz. (standard 6-inch bbl.). in 38 Special. Fixed sights. Blued or nickel-plated finish. Checkered walnut grips on all revolvers of this model, except some of postwar production had checkered plastic grips. Made 1927-69. *Note:* This model is a refined version of the Army Special, which it replaced in 1928 at about serial number 520,000. The Commando 38 Special was a wartime adaptation of the Official Police made to Government specifications. Commando can be identified by its sandblasted blued finish. Serial numbers start w/number 1 (1942).
Commercial Model **$395**
Commando Model **350**

Colt Official Police MK III DA Revolver **$190**

"J" frame, without shrouded ejector rod. Caliber: 38 Special. 6-shot cylinder. Bbl. lengths: 4-, 5-, 6-inch. 9.25 inches overall w/4-inch bbl. Weight: 34 oz. (4-inch bbl.). Fixed rear sight, ramp front. Service trigger and hammer or target trigger and wide-spur hammer. Blued or nickel-plated finish. Checkered walnut service grips. Made 1969-75.

Colt Peacekeeper DA Revolver **$255**

Caliber: 357 Mag. 6-shot cylinder. 4- or 6-inch bbl. 11.25 inches overall (6-inch bbl.). Weight: 46 oz. (6-inch bbl.). Adj. white outline rear sight, red insert ramp-style front. Non-reflective matte blued finish. Made 1985-89.

Colt Peacemaker 22 Second Amendment
Commemorative **$425**

Peacemaker 22 SA Revolver w/7.5-inch bbl. Nickel-plated frame, bbl., ejector rod assembly, hammer and trigger, blued cylinder, backstrap and trigger guard. Black pearlite grips. Bbl. inscribed "The Right to Keep and Bear Arms." Presentation case. Limited edition of 3000 issued in 1977. Value is for revolver in new condition.

Colt Peacemaker 22 SA Revolver **$295**

Calibers: 22 LR and 22 WMR. Furnished w/cylinder for each caliber. 6-shot. Bbl.: 4.38-, 6- or 7.5-inch. 11.25 inches overall (6-inch bbl.). Weight: 30 .5 oz. (6-inch bbl.). Fixed sights. Black composite grips. Made 1971-76.

Colt Pocket Positive DA Revolver **$395**

General specifications same as New Pocket, except this model has positive lock feature (*see* Police Positive). Calibers: 32 Short & Long Colt (discontinued 1914), 32 Colt New Police (32 S&W Short & Long). Fixed sights, flat top and square notch. Made 1905-40.

Colt Police Positive DA, First Issue **$350**

Improved version of the New Police w/the "Positive Lock," which prevents the firing pin coming in contact w/the cartridge except when the trigger is pulled. Calibers: 32 Short & Long Colt (discontinued 1915), 32 Colt New Police (32 S&W Short & Long), 38 New Police (38 S&W). 6-shot cylinder. Bbl. lengths: 2.5- (32 cal. only), 4-, 5- and 6-inch. 8 .5 inches overall (4-inch bbl.). Weight 20 oz. (4-inch bbl.). Fixed sights. Blued or nickel finish. Rubber or checkered walnut grips. Made 1905-47.

Colt Police Positive DA, Second Issue **$335**

Same as Detective Special, Second Issue, except has 4-inch bbl., is 9 inches overall, weighs 26 .5 oz. Intro. in 1977. *Note:* Original Police Positive (First Issue) has a shorter frame, is not chambered for 38 Special.

Colt Police Positive Special DA Revolver $375
Based on the Police Positive w/frame lengthened to permit longer cylinder. Calibers: 32-20 (discontinued 1942), 38 Special, 32 New Police and 38 New Police (intro. 1946). 6-shot cylinder. Bbl. lengths: 4- (only length in current production), 5- and 6-inch. 8.75 inches overall (4-inch bbl.). Weight: 23 oz. (4-inch bbl., 38 Special) Fixed sights. Checkered grips of hard rubber, plastic or walnut. Made 1907-73.

Colt Police Positive Target DA Revolver $595
Target version of the Police Positive. Calibers: 22 LR (intro. 1910, embedded head-cylinder for high-speed cartridges after 1932), 22 WRF (1910-35), 32 Short & Long Colt. 1915), 32 New Police (32 S&W Short & Long). 6-inch bbl. Blued finish only. 10.5 inches overall. Weight: 26 oz. in 22 cal. Adj. target sights. Checkered walnut grips. Made 1905-40.

Colt Python DA Revolver
"I" frame, shrouded ejector rod. Calibers: 357 Magnum, 38 Special. 6-shot cylinder. 2.5-, 4-, 6- or 8-inch vent rib bbl. 11.25 inches overall (6-inch bbl.). Weight: 44 oz. (6-inch bbl.). Adj. rear sight, ramp front. Blued, nickel or stainless finish. Checkered walnut target grips. Made 1955 to date. Ultimate stainless finish made 1985 to date.

Blued Finish	$395
Royal Blued Finish	445
Nickel Finish	425
Stainless Finish	475
UltimateStainless Finish	525
Hunter Model (w/2x Scope)	765
Silhouette Model (w/2x Scope)	775

Colt Shooting Master DA Revolver $995
Deluxe target arm based on the New Service model. Calibers: originally made only in 38 Special, 44 Special, 45 Auto and 45 Colt added in 1933, 357 Magnum in 1936. 6-inch bbl. 11.25 inches overall. Weight: 44 oz., 38 cal. Adj. target sights. Hand-finished action. Blued finish. Checkered walnut grips. Rounded butt. Made 1932-41.

Colt SA Army Revolver
Also called Frontier Six-Shooter and Peacemaker. Calibers: 22 Rimfire (Short, Long, LR), 22 WRF, 32 Rimfire, 32 Colt, 32 S&W, 32-20, 38 Colt, 38 S&W, 38 Special, 357 Magnum, 38-40, 41 Colt, 44 Rimfire, 44 Russian, 44 Special, 44-40, 45 Colt, 45 Auto, 450 Boxer, 450 Eley, 455 Eley, 476 Eley. 6-shot cylinder. Bbl. lengths: 4.75, 5 .5 and 7.5 inches w/ejector or 3 and 4 inches w/o ejector. 10.25 inches overall (4.75-inch bbl.). Weight: 36 oz. (45 cal. w/4.75-inch bbl.). Fixed sights. Also made in Target Model w/flat top-strap and target sights. Blued finish w/casehardened frame or nickel-plated. One-piece smooth walnut or checkered black rubber grips.

S.A. Army Revolvers w/serial numbers above 165,000 (circa 1896) are adapted to smokeless powder and cylinder pin screw was changed to spring catch at about the same time. Made 1873-1942, but production resumed in 1955 w/serial number 1001SA. Current calibers: 357 Magnum, 44 Special, 45 Long Colt.

Frontier Six-Shooter, 44-40	$8,550
Storekeeper's Model, 3-inch/4-inch bbl., no ejector	10,500
Target Model, flat top strap, target sights	11,500
U.S. Artillery Model, 45 Colt, 5.5-inch bbl.	7,750
U.S. Cavalry Model, 45, 7.5-inch bbl.	14,500

(Above values apply only to original models, not to similar S.A.A. revolvers of recent manufacture.)

Standard Model, pre-1942	4750
Standard Model (1955-82)	1695
Standard Model (Reissued 1992)	825

Colt Police Positive Target

Colt Python

Colt SA Army

Colt SA Army Flat Top

Colt SA Army — 125th Anniversary $1395
Limited production deluxe version of SA Army issued in commemoration of Colt's 125th Anniversary. Caliber: 45 Long Colt. 7.5-inch bbl. Gold-plated frame trigger, hammer, cylinder pin, ejector rod tip, and grip medallion. Presentation case w/anniversary medallion. Serial numbers "50AM." 7368 were made in 1961.

Colt SA Army Commemorative Models
Limited production versions of SA Army 45 issued, w/appropriate inscription to commemorate historical events. Cased. *Note:* Values indicated are for commemorative revolvers in new condition.

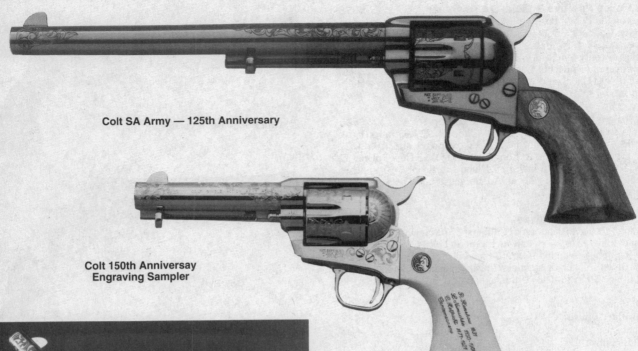

Colt SA Army — 125th Anniversary

Colt 150th Anniversay
Engraving Sampler

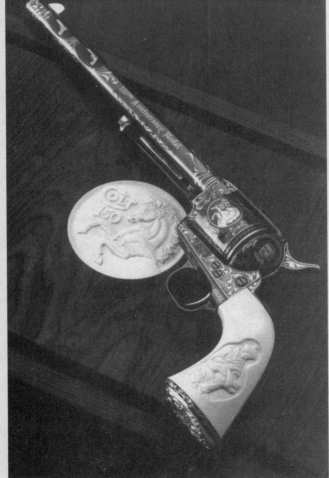

Colt 150th Anniversary Deluxe Model

1963 Issues

Arizona Territorial Centennial (1280 produced) $1295
West Virginia Statehood Centennial (600 produced).   1350

1964 Issues

Chamizal Treaty (50 produced) $1850
Colonel Sam Colt Sesquicentennial
Presentation (4750 produced). 1295
Deluxe Presentation (200 produced) 2595
Special Deluxe Presentation (50 produced) 4250
Montana Territorial Centennial (851 produced). 1295
Nevada "Battle Born" (100 produced) 1495
Nevada Statehood Centennial (1877 produced). 1250
New Jersey Tercentenary (250 produced) 1295
Pony Express Presentation (1004 produced) 1495
St. Louis Bicentennial (450 produced). 1250
Wyatt Earp Buntline (150 produced) 2395

1965 Issues

Appomattox Centennial (500 produced) **$1295**
Old Fort Des Moines Reconstruction (200 produced) **1350**

1966 Issues

Abercrombie & Fitch Trailblazer—Chicago
   (100 produced) **$1250**
Abercrombie & Fitch Trailblazer—New York
   (200 produced) **1225**
Abercrombie & Fitch Trailblazer—San Francisco
   (100 produced) **1250**
California Gold Rush (130 produced) **1425**
General Meade (200 produced) **1250**
Pony Express Four Square (4 guns) **6250**

1967 Issues

Alamo (1000 produced) **$1295**
Lawman Series—Bat Masterson (500 produced). **1395**

1968 Issues

Lawman Series—Pat Garrett (500 produced) **$1225**

1969 Issues

Lawman Series—Wild Bill Hickok (500 produced). **$1250**

1970 Issues

Lawman Series—Wyatt Earp (501 produced) **$2495**
Missouri Sesquicentennial (501 produced) **1195**
Texas Ranger (1000 produced) **2250**

1971 Issues

NRA Centennial, 357 or 45 (5001 produced) **$1250**

1975 Issues

Peacemaker Centennial 45 (1501 produced) **$1450**
Peacemaker Centennial 44-40 (1501 produced) **1550**
Peacemaker Centennial Cased Pair (501 produced) **2995**

1979 Issues

Ned Buntline 45 (3000 produced) **$995**

Colt Trooper

1986 Issues

Colt 150th Anniversary (standard) **$1695**
Colt 150th Anniversay (Engraved). **2695**

Colt SA Cowboy Revolver **$445**
SSA variant designed for "Cowboy Action Shooting". Caliber: 45
Colt. 6-shot cylinder. 5.5-inch bbl. 11 inches overall. Weight: 42 oz.
Blade front sight, fixed V-notch rear. Blued finish w/color casehard-
ened frame. Smooth walnut grips. Made 1999 to date.

Colt SA Sheriff's Model 45
Limited edition of replica of Storekeeper's Model in caliber 45 Colt,
made exclusively for Centennial Arms Corp. Chicago, Illinois.
Numbered "1SM." Blued finish w/casehardened frame or nickel-
plated. Walnut grips. Made in 1961.
Blued Finish (478 produced) **$2095**
Nickel Finish (25 produced). **4550**

Colt Three-Fifty-Seven DA Revolver
Heavy frame. Caliber: 357 Magnum. 6-shot cylinder. 4 or 6-inch bbl.
Quick-draw ramp front sight, Accro™ rear sight. Blued finish.
Checkered walnut grips. 9.25 or 11.25 inches overall. Weight: 36 oz.
(4-inch bbl.), 39 oz. (6 inch bbl.). Made 1953-61.
W/standard hammer and service grips **$425**
W/wide-spur hammer and target grips **445**

Colt Trooper DA Revolver
Same specifications as Officers' Model Match, except has 4-inch
bbl. w/quick-draw ramp front sight, weighs 34 oz. in 38 caliber.
Made 1953-69.
W/standard hammer and service grips **$315**
W/wide-spur hammer and target grips **365**

Colt Trooper MK III DA Revolver
"J"frame, shrouded ejector rod. Calibers: 22 LR, 22 Magnum, 38 Special, 357 Magnum. 6-shot cylinder. Bbl. lengths: 4-, 6-inches. 9.5 inches
overall (4-inch bbl.). Weight: 39 oz. (4-inch bbl.). Adj. rear sight, ramp
front. Target trigger and hammer. Blued or nickel-plated finish. Checkered walnut target grips. Made 1969-78.
Blued Finish **$195**
Nickel Finish. **215**

Colt Trooper MK V

Colt Trooper MK V Revolver
Re-engineered Mark III for smoother, faster action. Caliber: 357
Magnum. 6-shot cylinder. Bbl. lengths: 4-, 6-, 8-inch w/vent rib.
Adj. rear sight, ramp front, red insert. Checkered walnut grips. Made
1982-86.
Blued Finish **$280**
Nickel Finish. **305**

Colt Viper DA Revolver **$335**
Same as Cobra, Second Issue, except has 4-inch bbl., is 9 inches
overall, weighs 20 oz. Made 1977-84.

Colt U.S. Bicentennial Commemorative Set

Colt U.S. Bicentennial Commemorative Set **$2250**
Replica Colt 3rd Model Dragoon Revolver w/accessories, Colt SA
Army Revolver, and Colt Python Revolver. Matching roll-engraved
unfluted cylinders, blued finish, and rosewood grips w/Great Seal of
the United States silver medallion. Dragoon revolver has silver grip
frame. Serial numbers 0001 to 1776, all revolvers in set have same
number. Deluxe drawer-style presentation case of walnut, w/book
compartment containing a reproduction of "Armsmear." Issued in
1976. Value is for revolvers in new condition.

COONAN ARMS, INC.
St. Paul, Minnesota

Coonan Arms Model 357 Magnum Auto Pistol
Caliber: 357 Mag. 7-shot magazine. 5- or 6-inch bbl. 8.3 inches over-
all (5-inch bbl.). Weight: 42 oz. Front ramp interchangeable sight,
fixed rear sight, adj. for windage. Black walnut grips.
Model A Std. Grade w/o Grip Safety (Disc. 1991) **$775**
Model B Std. Grade w/5-inch Bbl. 450
Model B Std. Grade w/6-inch Bbl. 475
Model B w/5-inch Compensated Bbl. (Classic). 850
Model B w/6-inch Compensated Bbl 675

Coonan Arms 357 Magnum Cadet Compact
Similar to the standard 357 Magnum Model, except w/3.9-inch bbl.
on compact frame. 6-shot (Cadet), 7- or 8-shot magazine (Cadet II).
Weight: 39 oz., 7.8 inches overall. Made 1993 to date.
Cadet Model **$535**
Cadet II Model 565

CZ PISTOLS
Uhersky Brod (formerly Strakonice), Czechoslovakia
Mfd. by Ceska Zbrojovka-Nardoni Podnik (formerly Bohmische Waffenfabrik A. G.)

*Currently imported by CZ USA, Oakhurst, CA. Previously by Mag-
num Research and Action Arms. Vintage importation is by Century
International Arms.*

CZ Model 27 Auto Pistol **$365**
Caliber: 32 Automatic (7.65mm). 8-shot magazine. 4-inch bbl. 6
inches overall. Weight: 23.5 oz. Fixed sights. Blued finish. Plastic grips. Made 1927-51. *Note:* After the German occupation,
March 1939, Models 27 and 38 were marked w/manufacturer
code "fnh." Designation of Model 38 was changed to "Pistole
39(t)."

CZ Model 38 Auto Pistol (VZ Series)
Caliber: 380 Automatic (9mm). 9-shot magazine. 3.75-inch bbl. 7
inches overall. Weight: 26 oz. Fixed sights. Blued finish. Plastic
grips. After 1939 designated as T39. Made 1938-45.
CZ DAO Model **$295**
CZ SA/DA Model 895

CZ Model 50 DA Auto Pistol **$135**
Similar to Walther Model PP, except w/frame-mounted safety and
trigger guard not hinged. Caliber: 32 ACP (7.65mm), 8-round magazine. 3.13-inch bbl., 6.5 inches overall. Weight: 24.5 oz. Fixed
sights. Blued finished. Intro. in 1950. Discontinued. *Note:* "VZ50" is
the official designation of this pistol used by the Czech National Police ("New Model .006" was the export designation but very few
were released).

CZ Model 52 SA Auto Pistol
Roller-locking breech system. Calibers: 7.62mm or 9mm Para.
8-shot magazine. 4.7-inch bbl. 8.1 inches overall. Weight: 31 oz.
Fixed sights. Blued finish. Grooved composition grips. Made
1952-56.
7.62mm Model. **$145**
9mm Model. 185

CZ Model 70 DA Auto Pistol **$325**
Similar to Model 50, but redesigned to improve function and de-
pendability. Made 1962-83.

**CZ Model 75
Compact**

**CZ Model 75
Kadet**

**CZ Model 85
Combat**

CZ Model 75 DA/DAO Automatic Pistol
Calibers: 9mm Para. or 40 S&W w/selective action mode. 10-, 13- or 15-shot magazine. 3.9-inch bbl. (Compact) or 4.75-inch bbl. (Standard). 8 inches overall (Standard). Weight: 35 oz. Fixed sights. Blued, nickel, Two-Tone or black polymer finish. Checkered wood or high-impact plastic grips. Made 1994 to date.

Black Polymer Finish	**$295**
High-Polish Blued Finish	345
Matte Blued Finish	315
Nickel Finish	335
Two-Tone Finish	325
W/22 Kadet Conversion, **add**	250
Compact Model, **add**	35

CZ 82 DA Auto Pistol $225
Similar to the standard CZ 83 Model, except chambered in 9 ×18 Makarov. This model currently is the Czech military sidearm.

CZ 83 DA Automatic Pistol
Calibers: 32 ACP, 380 ACP. 15-shot (32 ACP) or 13-shot (380 ACP) magazine. 3.75-inch bbl. 6.75 inches overall. Weight: 26.5 oz. Fixed sights. Blued (standard); chrome and nickel (optionalspecial edition) w/brushed, matte or polished finish. Checkered black plastic grips. Made 1985 to date.

Standard finish	**$235**
Special Edition	355
Engraved (shown)	795

CZ 85 Automatic DA Pistol
Same as CZ 75, except w/ambidextrous slide release and safety. Calibers: 9mm Para., 7.65mm. Made 1986 to date.

Black Polymer Finish	**$315**
High-Polish Blued Finish	395
Matte Blued Finish	375

CZ 85 Combat DA Automatic Pistol
Similar to the standard CZ 85 Model, except w/13-shot magazine, combat-style hammer, fully adj. rear sight and walnut grips. Made 1986 to date.

Black Polymer Finish	**$325**
High-Polished Blued Finish	395
Matte Blued Finish	355

CZ Model 1945 DA Pocket Auto Pistol $185
Caliber: 25 Auto (6.35mm). 8-shot magazine. 2.5-inch bbl. 5 inches overall. Weight: 15 oz. Fixed sights. Blued finish. Plastic grips. Intro. 1945. Discontinued.

CZ Duo Pocket Auto Pistol $175
Caliber: 25 Automatic (6.35mm). 6-shot magazine. 2.13 inch bbl. 4.5 inches overall. Weight: 14 .5 oz. Fixed sights. Blued or nickel finish. Plastic grips. Made c.1926-60.

CZ Model 97B

CZ Model 97B DA Autoloading Pistol $395
Similar to the CZ Model 75 except chambered for the 45 ACP cartridge. 10-shot magazine. Frame-mounted thumb safety that allows single-action, cocked-and-locked carry. Made 1997 to date.

CZ Model 100

CZ Model 100 DA Automatic Pistol $2985
Caliber: 9mm, 40 S&W. 10-shot magazine. 3.8-inch bbl. Weight: 25 oz. Polymer grips w/fixed low profile sights. Made 1996 to date.

DAEWOO PISTOLS
Seoul, Korea
Mfd. by Daewoo Precision Industries Ltd.

Imported by Daewoo Precision Industries, Southhampton, PA, Previously by Nationwide Sports Distributors and KBI, Inc.

Daewoo DH40

Daewoo DH40 Auto Pistol **$295**
Caliber: 40 S&W. 12-shot magazine. 4.25-inch bbl. 7 inches overall. Weight: 28 oz. Blade front sight, dovetailed rear w/3-dot system. Blued finish. Checkered composition grips. DH/DP series feature a patented "fastfire" action w/5-6 lb. trigger pull. Made 1994-96.

Daewoo DH45 Auto Pistol . . . . (SRP)**$495**
Caliber: 45 ACP. 13-shot magazine. 5-inch bbl. 8.1 inches overall. Weight: 35 oz. Blade front sight, dovetailed rear w/3-dot system. Blued finish. Checkered composition grips. Announced 1994, but not imported.

Daewoo DP51 Auto Pistol **$275**
Caliber: 9mm Para. 13-shot magazine. 4.1-inch bbl. 7.5 inches overall. Weight: 28 oz. Blade front and square-notch rear sights. Matte black finish. Checkered composition grips. Made 1991-96.

Daewoo DP52 Auto Pistol **$245**
Caliber: 22 LR. 10-shot magazine. 3.8-inch bbl. 6.7 inches overall. Weight: 23 oz. Blade front sight, dovetailed rear w/3-dot system. Blued finish. Checkered wood grips. Made 1994-96.

DAKOTA/E.M.F. CO.
Santa Ana, California

Dakota Model 1873 SA Revolver
Calibers: 22 LR, 22 Mag., 357 Mag., 45 Long Colt, 30 M1 carbine, 38-40, 32-20, 44-40. Bbl. lengths: 3.5, 4.75, 5.5, 7.5 inches. Blued or nickel finish. Engraved models avail.
Standard Model. **$295**
W/Extra Cylinder **445**

Dakota Model 1875 Outlaw SA Revolver **$335**
Calibers: 45 Long Colt, 357 Mag., 44-40. 7.5-inch bbl. Casehardened frame, blued finish. Walnut grips. This is an exact replica of the Remington #3 revolver produced 1875-89.

Dakota Model 1890 Remington Police
Calibers: 357 Mag., 44-40, 45 Long Colt. 5.75-inch bbl. Blued or nickel finish. Similar to Outlaw w/lanyard ring and no bbl. web .
Standard Model. **$355**
Nickel Model **425**
Engraved Model **495**

Dakota Bisley SA Revolver
Calibers: 44-40, 45 Long Colt, 357 Mag. 5.5- or 7.5-inch bbl. Disc. 1992. Reintroduced 1994.
Standard Model. **$295**
Target Model **325**

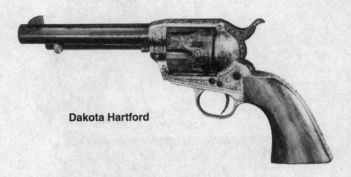

Dakota Hartford

Dakota Hartford SA Revolver
Calibers: 22 LR, 32-20, 357 Mag., 38-40, 44-40, 44 Special, 45 Long Colt. These are exact replicas of the original Colts, w/steel backstraps, trigger guards and forged frames. Blued or nickel finish. Imported 1990 to date.
Standard Model. **$295**
Engraved Model **495**
Hartford Artillery, U.S. Cavalry Models **320**

Dakota Sheriff's Model SA Revolver **$325**
Calibers: 32-20, 357 Mag., 38-40, 44 Special, 44-40, 45 LC. 3.5-inch bbl. Reintroduced 1994.

Dakota Target SA Revolver **$295**
Calibers: 45 Long Colt, 357 Mag., 22 LR. 5.5- or 7.5-inch bbl. Polished, blued finish, casehardened frame. Walnut grips. Ramp front, blade target sight, adj. rear sight.

CHARLES DALY HANDGUNS
Manila, Philippines
Currently imported by K.B.I., Harrisburg, PA.

Charles Daly Model M1911 Automatic Pistol
Caliber: 45 ACP. 8- or 10-shot magazine (Hi-Cap). 5-inch bbl. 8.75 inches overall. Weight: 38 oz. Blade front sight, drift adjustable rear w/3-Dot system. Skeletonized tactical hammer and trigger. Extended slide release and beavertail grip safety. Matte blue finish. Checkered composition or wood stocks. Imported 1996-97.
Model M1911-A1 . . . . **$325**
Model M1911-A2 (Hi-Cap) **450**

DAVIS INDUSTRIES, INC.
Chino, California

Davis Model D Derringer
Single-action double derringer. Calibers: 22 LR, 22 Mag., 25 ACP, 32 Auto, 32 H&R Mag., 9mm, 38 Special. 2-shot capacity. 2.4-inch or 2.75-inch bbl. 4 inches overall (2.4-inch bbl.). Weight: 9 to 11.5 oz. Laminated wood grips. Black Teflon or chrome finish. Made 1987 to date.
22 LR or 25 ACP **$50**
22 Mag., 32 H&R Mag., 38 Spec. **65**
32 Auto **70**
9mm Para. **85**

Davis Long Bore Derringer. **$80**
Similar to Model D, except in calibers 22 Mag., 32 H&R Mag., 38
Special, 9mm Para. 3.75-inch bbl. Weight: 16 oz. Made 1995 to date.

Davis Model P-32. . . . . **$65**
Caliber: 32 Auto. 6-round magazine. 2.8-inch bbl. 5.4 inches overall. Weight: 22 oz. Black teflon or chrome finish. Laminated wood
grips. Made 1987 to date.

Davis Model P-380. . . . . **$60**
Caliber: 380 Auto. 5-shot magazine. 2.8-inch bbl. 5.4 inches overall.
Weight: 22 oz. Black teflon or chrome finish. Made 1990 to date.

DESERT INDUSTRIES, INC.
Las Vegas, Nevada
(Previously Steel City Arms, Inc.)

Desert Industries Double Deuce DA Pistol. **$295**
Caliber: 22 LR. 6-shot magazine. 2.5-inch bbl. 5.5 inches overall.
Weight: 15 oz. Matte-finish stainless steel. Rosewood grips.

Desert Industries Two-Bit Special Pistol **$300**
Similar to the Double Deuce Model, except chambered in 25 ACP
w/5-shot magazine.

(NEW) DETONICS MFG. CORP
Phoenix, Arizona
(Previously Detonics Firearms Industries,
Bellevue, WA)

Detonics Combat Master

Detonics Combat Master
Calibers: 45 ACP, 451 Detonics Mag. 6-round magazine. 3.5-inch
bbl. 6.75 inches overall. Combat-type w/fixed or adjustable sights.
Checkered walnut grip. Stainless steel construction. Discontinued
1992.
MK I Matte Stainless, Fixed Sights, (Disc. 1981) **$425**
MK II Stainless Steel Finish, (Disc. 1979). **400**
MK III Chrome, (Disc. 1980) **375**
MK IV Polished Blued, Adj. Sights, (Disc. 1981). **415**
MK V Matte Stainless, Fixed Sights, (Disc. 1985). **530**
MK VI Polished Stainless, Adj. Sights, (Disc. 1989). **575**
MK VI in 451 Magnum, (Disc. 1986) **850**
MK VII Matte Stainless Steel, No Sights, (Disc. 1985). **750**
MK VII in 451 Magnum, (Disc. 1980). **995**

Detonics Pocket 9 . . . . **$350**
Calibers: 9mm Para., 380. 6-round magazine. 3-inch bbl. 5.88 inches
overall. Fixed sights. Double- and single-action trigger mechanism.
Discontinued 1986.

Detonics Scoremaster

Detonics Scoremaster **$825**
Calibers: 45 ACP, 451 Detonics Mag. 7-round magazine. 5- or 6-inch
heavyweight match bbl. 8.75 inches overall. Weight: 47 oz. Stainless steel construction, self-centering bbl. system. Discontinued
1992.

Detonics Service Master. **$595**
Caliber: 45 ACP. 7-round magazine. 4.25-inch bbl. Weight: 39 oz. Interchangeable front sight, millett rear sight. Discontinued 1986.

Detonics Service Master II **$675**
Same general specifications as standard Service Master, except
comes in polished stainless steel w/self-centering bbl. system.
Discontinued 1992.

DOWNSIZER CORPORATION
Santee, California

Downsizer Model WSP DAO Pistol. **$200**
Single-shot, tip-up pistol. Calibers: 22Mag., 32 Mag., 380 ACP.
9mm Parabullum, 357 Mag., 40 S&W, 45 ACP. 6-shot cylinder,
2.10-inch bbl. w/o extractor. 3.25 inches overall. Weight:11 oz. No
sights. Stainless finish. Synthetic grips. Made 1994 to date.

DREYSE PISTOLS
Sommerda, Germany
Mfd. by Rheinische Metallwaren und Maschinen-
fabrik ("Rheinmetall")

Dreyse Model 1907

Dreyse Model 1907 Automatic Pistol. **$200**
Caliber: 32 Auto (7.65mm). 8-shot magazine. 3.5-inch bbl. 6.25
inches overall. Weight: about 24 oz. Fixed sights. Blued finish. Hard
rubber grips. Made c. 1907-14.

HANDGUNS

Dreyse Vest Pocket Automatic Pistol $195
Conventional Browning type. Caliber: 25 Auto (6.35mm). 6-shot magazine. 2-inch bbl. 4.5 inches overall. Weight: about 14 oz. Fixed sights. Blued finish. Hard rubber grips. Made c. 1909-14.

DWM PISTOL
Berlin, Germany
Mfd. by Deutsche Waffen-und-Munitionsfabriken

DWM Pocket

DWM Pocket Automatic Pistol $595
Similar to the FN Browning Model 1910. Caliber: 32 Automatic (7.65mm). 3.5-inch bbl. 6 inches overall. Weight: about 21 oz. Blued finish. Hard rubber grips. Made c. 1921-31.

ED BROWN
Perry, Montana

Ed Brown "Class A Ltd" SA Automatic Pistol $1225
Caliber: 38 Super, 9mm, 9 X 23, 45 ACP. 7-shot magazine. 4.25- or 5-inch bbl. Weight: 34-39 oz. Rubber checkered or optional Hogue exotic wood grip. M1911 style single action pistol. Fixed front and rear Novak lo-mount or fully adjustable sights.

**Ed Brown "Classic Custom" SA
Automatic Pistol** $1495
Caliber: 45 ACP. 7-shot magazine. 4.25- or 5-inch bbl. Weight: 39 oz. Exotic Hogue wood grip w/modified ramp or post front and rear adjustable sights.

**Ed Brown "Special Forces" SA
Automatic Pistol** $935
Caliber: 45 ACP. 7-shot magazine. 4.25- or 5-inch bbl. Weight: 34-39 oz. Rubber checkered, optional exotic wood grips. Single action M1911 style pistol.

ENFIELD REVOLVER
Enfield Lock, Middlesex, England
Manufactured by Royal Small Arms Factory

Enfield (British Service) No. 2 MK 1 Revolver $195
Webley pattern. Hinged frame. Double action. Caliber: 380 British Service (38 S&W w/200-grain bullet). 6-shot cylinder. 5-inch bbl. 10.5 inches overall. Weight: about 27.5 oz. Fixed sights. Blued finish. Vulcanite grips. First issued in 1932, this was the standard revolver of the British Army in WW II. Now obsolete. *Note:* This model also produced w/spurless hammer as No. 2 Mk 1* and Mk 1**.

**Enfield (British Service) No. 2 MK 1
Revolver**

ENTERPRISE ARMS
Irwindale, California

Enterprise Arms Boxer SA Automatic Pistol $795
Similar to Medalist Model, except w/profiled slide configuration and fully adjustable target sights. Weight: 42 oz. Made 1997 to date.

Enterprise Arms "Elite" SA Automatic Pistol $495
Single action M1911 style pistol. Caliber: 45 ACP. 10-shot magazine. 3.25-, 4.25-, 5-inch bbl. (models P325, P425, P500). Weight: 36-40 oz. Ultraslim checkered grips. Tactical 2 high profile sights w/3-Dot system. Lightweight adjustable trigger. Blued or matte black oxide finish. Made 1997 to date.

Enterprise Arms "Medalist" SA Automatic Pistol
Similar to Elite Model, except machined to match tolerances and target configuration. Caliber: 45 ACP, 40 S&W. 10-shot magazine. 5-inch compensated bbl. w/dovetail front and fully adjustable rear Bomar sights. Weight: 40 oz. Made 1997 to date.
40 S&W Model $775
45 ACP Model 695

Enterprise Arms "Tactical" SA Automatic Pistol
Similar to Elite Model, except in combat carry configuration. De-horned frame and slide w/ambidextrous safety. Caliber: 45 ACP. 10-shot magazine. 3.25-, 4.25-, 5-inch bbl. Weight: 36-40 oz. Tactical 2 Ghost Ring or Noval Lo-mount sights.
Tactical 2 Ghost Ring Sights $645
Novak Lo-mount. 635
Tactical Plus Model 665

Enterprise Arms "TSM" SA Automatic Pistol
Similar to Elite Model, except in IPSC configuration. Caliber: 45 ACP, 40 S&W. 10-shot magazine. 5-inch compensated bbl. w/dovetail front and fully adjustable rear Bomar sights. Weight: 40 oz. Made 1997 to date.
TSM I Model $1550
TSM II Model 1375
TSM III Model 1845

ERMA-WERKE
Dachau, Germany

Erma-Werke Model ER-772 Match Revolver $750
Caliber: 22 LR. 6-shot cylinder. 6-inch bbl. 12 inches overall. Weight: 47.25 oz. Adjustable micrometer rear sight and front sight blade. Adjustable trigger. Interchangeable walnut sporter or match grips. Polished blued finish. Made 1991-94.

**Erma Model ER-772
Match Revolver**

**Erma-Werke
Model KGP69**

Erma-Werke Model ER-773 Match Revolver **$725**
Same general specifications as Model 772, except chambered for 32 S&W. Made 1991-95.

Erma-Werke Model ER-777 Match Revolver **$675**
Caliber: 357 Magnum. 6-shot cylinder. 4- or 5.5-inch bbl. 9.7 to 11.3 inches overall. Weight: 43.7 oz. (5.5-inch bbl.). Micrometer adj. rear sight. Checkered walnut sporter or match-style grip (interchangeable). Made 1991-95.

**Erma Model ESP-85A
Competition Pistol**

Erma-Werke Model ESP-85A Competition Pistol
Calibers: 22 LR and 32 S&W Wadcutter. 8- or 5-shot magazine. 6-inch bbl. 10 inches overall. Weight: 40 oz. Adj. rear sight, blade front sight. Checkered walnut grip w/thumbrest. Made 1991-97.
Match Model **$825**
Chrome Match 875
Sporting Model 825
Conversion Unit 22 LR. 765
Conversion Unit 32 S&W. 800

Erma-Werke Model KGP68 Automatic Pistol **$350**
Luger type. Calibers: 32 Auto (7.65mm), 380 Auto (9mm Short).6-shot magazine (32 Auto), 5-shot (380 Auto). 4-inch bbl. 7.38 inches overall. Weight: 22.5 oz. Fixed sights. Blued finish. Checkered walnut grips. Made 1968-93.

Erma-Werke Model KGP69 Automatic Pistol **$250**
Luger type. Caliber: 22 LR. 8-shot magazine. 4-inch bbl. 7.75 inches overall. Weight: 29 oz. fixed sights. Blued finish. Checkered walnut grips. Imported 1969-93.

EUROPEAN AMERICAN ARMORY
Hialeah, Florida

See **also listings under Astra Pistols.**

European American Armory European Model Auto Pistol
Calibers: 32 ACP (SA only), 380 ACP (SA or DA). 3.85-inch bbl. 7.38 overall. 7-shot magazine. Weight: 26 oz. Blade front sight, drift-adj. rear. Blued, chrome, blue/chrome, blue/gold, Duo-Tone or Wonder finish. Imported 1991 to date.
Blued 32 Caliber (Disc. 1995) **$125**
Blue/Chrome 32 Caliber (Disc. 1995) 145
Chrome 32 Caliber (Disc. 1995) 150
Blued 380 Caliber 130
Blue/chrome 380 Caliber (Disc. 1993) 145
DA 380 Caliber (Disc. 1994) 160
Lady 380 Caliber (Disc. 1995) 195
Wonder finish 380 Caliber 130

**European American Armory
Big Bore Bounty Hunter**

European American Armory Big Bore Bounty Hunter SA Revolver
Calibers: 357 Mag., 41 Mag., 44-40, 44 Mag.,45 Colt. Bbl. lengths: 4.63, 5.5, 7.5 inches. Blade front and grooved topstrap rear sights. Blued or chrome finish w/color casehardened or gold-plated frame. Smooth walnut grips. Imported 1992 to date.
Blued Finish **$195**
Blued w/Color-Casehardened Frame 215
Blued w/Gold-Plated Frame 235
Chrome Finish 240
Gold-Plated Frame, **add** 120

European American Armory Bounty Hunter SA Revolver

Calibers: 22 LR, 22 Mag. Bbl. lengths: 4.75, 6 or 9 inches. Blade front and dovetailed rear sights. Blued finish or blued w/gold-plated frame. European hardwood grips. Imported 1991 to date.

Blued Finish (4.75-inch bbl.)	$ 65
Blued 22 LR/22 WRF Combo (4.75-inch bbl.)	80
Blued 22 LR/22 WRF Combo (6-inch bbl.)	85
Blued 22 LR/22 WRF Combo (9-inch bbl.)	90

European American Armory EA22 Target $285

Caliber: 22 LR. 12-shot magazine. 6-inch bbl. 9.10 inches overall. Weight: 40 oz. Ramp front sight, fully adj. rear. Blued finish. Checkered walnut grips w/thumbrest. Made 1991-94.

European American Armory FAB 92 Auto Pistol

Similar to the Witness Model, except chambered in 9mm only w/slide-mounted safety and no cock-and-lock provision. Imported 1992-95.

FAB 92 Standard	$255
FAB 92 Compact	245

European American Armory Standard Grade Revolver

Calibers: 22 LR, 22 WRF, 32 H&R Mag., 38 Special. 2-, 4- or 6-inch bbl. Blade front sight, fixed or adj. rear. Blued finish. European hardwood grips w/finger grooves. Imported 1991 to date.

22 LR (4-inch bbl.)	$135
22 LR (6-inch bbl.)	145
22 LR Combo (4-inch bbl.)	195
22 LR Combo (6-inch bbl.)	225
32 H&R, 38 Special (2-inch bbl.)	140
38 Special (4-inch)	150
357 Mag	160

European American Armory Tactical Grade Revolver

Similar to the Standard Model, except chambered in 38 Special only. 2- or 4-inch bbl. Fixed sights. Available w/compensator. Imported 1991-93.

Tactical Revolver	$185
Tactical Revolver w/Compensator	255

**European American Armory
Windicator Target**

European American Armory Windicator Target Revolver $295

Calibers: 22 LR, 38 Special, 357 Magnum. 8-shot cylinder in 22 LR, 6-shot in 38 Special and 357 Magnum. 6-inch bbl. w/bbl. weights. 11.8 inches overall. Weight: 50.2 oz. Interchangeable blade front sight, fully adj. rear. Walnut competition-style grips. Imported 1991-93.

**European American
Armory Witness**

European American Armory Witness DA Auto Pistol

Similar to the Brno CZ-75 w/a cocked-and-locked system. Double or single action. Calibers: 9mm Para. 38 Super, 40 S&W, 10mm, 41 AE and 45 ACP. 16-shot magazine (9mm), 12 shot (38 Super/40 S&W), 10-shot (10mm/45 ACP). 4.75-inch bbl. 8.10 inches overall. Weight: 35.33 oz. Blade front sight, rear sight adj. for windage w/3-dot sighting system. Blued, satin chrome, blue/chrome or stainless finish. Checkered rubber grips. Imported 1991 to date.

9mm Blue	$280
9mm Chrome or Blue/Chrome	290
9mm Stainless	335
38 Super and 40 S&W Blued	295
38 Super and 40 S&W Chrome or Blue/Chrome	325
38 Super and 40 S&W Stainless	345
10mm, 41 AE and 45 ACP Blued	365
10mm, 41 AE and 45 ACP Chrome or Blue/Chrome	380
10mm, 41 AE and 45 ACP Stainless	425

European American Armory Witness Carry Comp

Double/Single action. Calibers: 38 Super, 9mm Parabellum, 40 S&W, 10mm, 45ACP. 10-, 12- or 16-shot magazine. 4.25-inch bbl. w/1-inch compensator. Weight: 33 oz. 8.10 inches overall. Black rubber grips. Post front sight, drift adjustable rear w/3-Dot system. Matte blue, Duo-Tone or Wonder finish. Imported 1992 to date.

9mm, 40 S&W	$285
38 Super, 10mm, 45 ACP	335
W/Duo-Tone Finish (disc.), **add**	25
W/Wonder Finish, **add**	50

European American Armory Witness Limited Class Auto Pistol $650

Single action. Calibers: 38 Super, 9mm Parabellum, 40 S&W, 45ACP. 10-shot magazine. 4.75-inch bbl. Weight: 37 oz. Checkered competition-style walnut grips. Long slide w/post front sight, fully adj. rear. Matte blue finish. Imported 1994 to date.

European American Armory Witness Subcompact DA Auto Pistol

Calibers: 9mm Para., 40 S&W, 41 AK, 45 ACP. 13-shot magazine in 9mm, 9-shot in 40 S&W. 3.66-inch bbl. 7.25 inches overall. Weight: 30 oz. Blade front sight, rear sight adj. for windage. Blued, satin chrome or blue/chrome finish. Imported 1995-97.

9mm Blue	$270
9mm Chrome or Blue/Chrome	295
40 S&W Blue	295
40 S&W Chrome or Blue/Chrome	325
41 AE Blue	350
41 AE Chrome or Blue/Chrome	375
45 ACP Blued	360
45 ACP Chrome or Blue/Chrome	385

European American Armory Witness Target Pistols

Similar to standard Witness Model, except fitted w/2- or 3-port compensator, competition frame and S/A target trigger. Calibers: 9mm Para., 9×21, 40 S&W, 10mm and 45 ACP. 5.25-inch match bbl. 10.5 inches overall. Weight: 38 oz. Square post front sight, fully adj. rear or drilled and tapped for scope. Blued or hard chrome finish. Low-profile competition grips. Imported 1992 to date.

Silver Team (Blued w/2-Port Compensator)	**$ 625**
Gold Team (Chrome w/3-Port Compensator)	**1295**

FAS PISTOLS
Malino, Italy
Currently imported by Nygord Precision Products (Previously by Beeman Precision Arms and Osborne's, Cheboygan, MI)

FAS 601 Semiautomatic Match Target Pistol

Caliber: 22 Short. 5-shot top-loading magazine. 5.6-inch ported and ventilated bbl. 11 inches overall. Weight: 41.5 oz. Removable, adj. trigger group. Blade front sight, open-notch fully adj. rear. Stippled walnut wrapaound or adj. target grips.

Right-Hand Model	**$695**
Left-Hand Model	**725**

FAS 602 Semiautomatic Match Target Pistol

Similar to Model FAS 601, except chambered 22 LR. Weight: 37 oz.

Right-Hand Model	**$650**
Left-Hand Model	**675**

FAS 603 Semiautomatic Match Target Pistol **$725**

Similar to Model FAS 601, except chambered 32 S&W (wadcutter).

FAS 607 Semiautomatic Match Target Pistol **$745**

Similar to Model FAS 601, except chambered 22 LR, w/removable bbl. weights. Imported 1995 to date.

FEATHER INDUSTRIES, INC.
Boulder, Colorado

**Feather Guardian
Angel Derringer**

Feather Guardian Angel Derringer

Double-action over/under derringer w/interchangeable drop-in loading blocks. Calibers: 22 LR, 22 WMR, 9mm, 38 Spec., 2-shot capacity. 2-inch bbl. 5 inches overall. Weight: 12 oz. Stainless steel. Checkered black grip. Made 1988-95.

22 LR, 22 WMR	**$75**
9mm, 38 Special (Disc. 1989)	**95**

FEG (FEGYVERGYAN) PISTOLS
Budapest, Soroksariut, Hungary
Currently imported by KBI, Inc. and Century International Arms (Previously by Interarms)

FEG Mark II AP-22

FEG Mark II AP-22 DA Automatic Pistol **$185**

Caliber: 22 LR. 8-shot magazine. 3.4-inch bbl. Weight: 23 oz. Drift-adj. sights. Double action, all-steel pistol. Imported 1997 to date.

FEG Mark II AP-380 DA Automatic Pistol **$190**

Caliber: 380. 7-shot magazine. 3.9-inch bbl. Weight 27 oz. Drift-adj. sights. Double action, all-steel pistol. Imported 1997 to date.

FEG Mark II APK-380 DA Automatic Pistol **$195**

Caliber: 380. 7-shot magazine. 3.4-inch bbl. Weight: 25 oz. Drift-adj. sights. Double action, all-steel pistol. Imported 1997 to date.

FEG Model GKK-9 (92C) Auto Pistol **$255**

Improved version of the double-action FEG Model MBK. Caliber: 9mm Para. 14-shot magazine. 4-inch bbl. 7.4 inches overall. Weight: 34 oz. Blade front sight, rear sight adj. for windage. Checkered wood grips. Blued finish. Imported 1992-93.

FEG Model GKK-45 Auto Pistol

Improved version of the double-action FEG Model MBK. Caliber: 45 ACP. 8-shot magazine. 4.1-inch bbl. 7.75 inches overall. Weight: 36 oz. Blade front sight, rear sight adj. for windage w/3-Dot system. Checkered walnut grips. Blued or chrome finish. Imported 1993-96.

Blued Model (Discontinued 1994)	**$245**
Chrome Model	**255**

FEG Model MBK-9HP Auto Pistol **$265**

Similar to the double-action Browning Hi-Power. Caliber: 9mm Para. 14-shot magazine. 4.6-inch bbl. 8 inches overall. Weight: 36 oz. Blade front sight, rear sight adj. for windage. Checkered wood grips. Blued finish. Imported 1992-93.

FEG Model PJK-9HP Auto Pistol

Similar to the single-action Browning Hi-Power. Caliber: 9mm Para. 13-shot magazine. 4.75-inch bbl. 8 inches overall. Weight: 21 oz. Blade front sight, rear sight adj. for windage w/3-Dot system. Checkered walnut or rubber grips. Blued or chrome finish. Imported 1992 to date.

Blued Model	**$195**
Chrome Model	**275**

FEG Model PJK-9HP

F.I.E. Model A27BW

F.I.E. Arminius

FEG Model PSP-25 Auto Pistol
Similar to the Browning 25. Caliber: 25 ACP. 6-shot magazine. 2.1-inch bbl. 4.1 inches overall. Weight: 9.5 oz. Fixed sights. Checkered composition grips. Blued or chrome finish.
Blued Model **$185**
Chrome Model **215**

FEG Model SMC-22 Auto Pistol **$190**
Same general specifications as FEG Model SMC-380, except in 22 LR. 8-shot magazine. 3.5-inch bbl. 6.1 inches overall. Weight: 18.5 oz. Blade front sight, rear sight adj. for windage. Checkered composition grips w/thumbrest. Blued finish.

FEG Model SMC-380 Auto Pistol **$165**
Similar to the Walther DA PPK w/alloy frame. Caliber: 380 ACP. 6-shot magazine. 3.5-inch bbl. 6.1 inches overall. Weight: 18.5 oz. Blade front sight, rear sight adj. for windage. Checkered composition grips w/thumbrest. Blued finish. Imported 1993 to date.

FEG Model SMC-918 Auto Pistol **$185**
Same general specifications as FEG Model SMC-380, except chambered in 9×18mm Makarov. Imported 1994-97.

FIALA OUTFITTERS, INC.
New York, New York

Fiala Repeating Pistol **$395**
Despite its appearance, which closely resembles that of the early Colt Woodsman and High-Standard, this arm is not an automatic pistol. It is hand-operated by moving the slide to eject, cock and load. Caliber: 22 LR. 10-shot magazine. Bbl. lengths: 3-, 7.5- and 20-inch. 11.25 inches overall (7.5-inch bbl.). Weight: 31 oz. (7.5-inch bbl.). Target sights. Blued finish. Plain wood grips. Shoulder stock was originally supplied for use w/20-inch bbl. Made 1920-23. Value shown is for pistol w/one bbl. and no shoulder stock.

F.I.E. CORPORATION
Hialeah, Florida

The F.I.E Corporation became QFI (Quality Firearms Corp.) of Opa Locka, Fl., about 1990, when most of F.I.E's models were discontinued.

F.I.E. Model A27BW "The Best" Semiauto **$95**
Caliber: 25 ACP. 6-round magazine. 2.5-inch bbl. 6.75 inches overall. Weight: 13 oz. Fixed sights. Checkered walnut grip. Discontinued 1990.

F.I.E. Arminius DA Standard Revolver
Calibers: 22 LR, 22 combo w/interchangeable cylinder, 32 S&W, 38 Special, 357 Magnum. 6, 7 or 8 rounds depending on caliber. Swingout cylinder. Bbl. lengths: 2-, 3-, 4, 6-inch. Vent rib on calibers other than 22. 11 inches overall (6-inch bbl.). Weight: 26 to 30 oz. Fixed or micro-adj. sights. Checkered plastic or walnut grips. Blued finish. Made in Germany. Discontinued.
22 LR . . . . **$ 75**
22 Combo **100**
32 S&W **105**
38 Special **110**
357 Magnum **135**

F.I.E. Buffalo Scout SA Revolver
Calibers: 22 LR, 22 WRF, 22 combo w/interchangeable cylinder. 4.75-inch bbl. 10 inches overall. Weight: 32 oz. Adjustable sights. Blued or chrome finish. Smooth walnut or black checkered nylon grips. Made in Italy. Discontinued.
Blued Standard . . . . **$ 50**
Blued Convertible **75**
Chrome Standard **65**
Chrome Convertible **85**

F.I.E. Hombre SA Revolver **$175**
Calibers: 357 Magnum, 44 Magnum, 45 Colt. 6-shot cylinder. Bbl. lengths: 6 or 7.5 inches. 11 inches overall (6-inch bbl.). Weight: 45 oz. (6-inch bbl.). Fixed sights. Blued bbl. w/color-casehardened receiver. Smooth walnut grips. Made 1979-90.

F.I.E. Little Ranger SA Revolver
Same as the Texas Ranger, except w/3.25-inch bbl. and bird's-head grips. Made 1986-90.
Standard **$80**
Convertible **95**

F.I.E. Super Titan II
Caliber: 32 ACP or 380 ACP. 3.25-inch bbl. Weight: 28 oz. Blued or chrome finish. Discontinued 1990.
32 ACP in Blue $130
32 ACP in Chrome 145
380 ACP in Blue 160
380 ACP in Chrome 175

F.I.E. Texas Ranger Single-Action Revolver
Calibers: 22 LR, 22 WRF, 22 combo w/interchangeable cylinder. Bbl. lengths: 4.75-, 6.5-, 9-inch. 10 inches overall (4.75-inch bbl.). Weight: 32 oz. (4.75-inch bbl.). Fixed sights. Blued finish. Smooth walnut grips. Made 1983-90.
Standard $60
Convertible 90

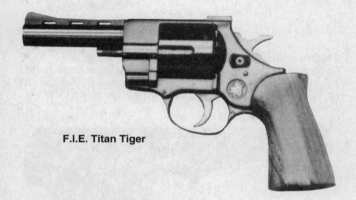

F.I.E. Titan Tiger

F.I.E. Titan Tiger Double-Action Revolver $100
Caliber: 38 Special. 6-shot cylinder. 2- or 4-inch bbl. 8.25 inches overall (4-inch bbl.). Weight: 30 oz. (4-inch bbl.). Fixed sights. Blued finish. Checkered plastic or walnut grips. Made in the U.S. Discontinued 1990.

F.I.E. Titan II

F.I.E. Titan II Semiautomatic
Caiibers: 22 LR, 32 ACP, 380 ACP. 10-round magazine. Integral tapered post front sight, windage-adjustable rear sight. European walnut grips. Blued or chrome finish. Discontinued 1990.
22 LR in Blue . . . . $ 95
32 ACP in Blued 135
32 ACP in Chrome 155
380 ACP in Blue 155
380 ACP in Chrome 175

F.I.E. Model TZ75

F.I.E. Model TZ75 DA Semiautomatic
Double action. Caliber: 9mm. 15-round magazine. 4.5-inch bbl. 8.25 inches overall. Weight: 35 oz. Ramp front sight, windage-adjustable rear sight. European walnut or black rubber grips. Imported 1988-90.
Blued Finish $315
Satin Chrome 335

F.I.E. Yellow Rose SA Revolver
Same general specifications as the Buffalo Scout, except in 22 combo w/interchangeable cylinder and plated in 24-karat gold. Limited Edition w/scrimshawed ivory polymer grips and American walnut presentation case. Made 1987-90.
Yellow Rose 22 Combo $100
Yellow Rose Ltd. Edition 250

FIREARMS INTERNATIONAL CORP.
Washington, D.C.

Firearms International Model D

Firearms Int'l. Model D Automatic Pistol $185
Caliber: 380 Automatic. 6-shot magazine. 3.3-inch bbl. 6.13 inches overall. Weight: 19.5 oz. Blade front sight, windage-adjustable rear sight. Blued, chromed, or military finish. Checkered walnut grips. Made 1974-77.

Firearms Int'l. Regent DA Revolver $90
Calibers: 22 LR, 32 S&W Long. 8-shot cylinder (22 LR). 7-shot (32 S&W). Bbl. Lengths: 3-, 4-, 6-inches (22 LR) or 2.5-, 4-inches (32 S&W). Weight: 28 oz. (4-inch bbl.). Fixed sights. Blued finish. Plastic grips. Made 1966-72.

**Firearms International
Regent**

FN BROWNING PISTOLS
Liege, Belgium
Mfd. by Fabrique Nationale Herstal

See also Browning Pistols.

**FN Browning 6.35mm
Pocket**

FN Browning 6.35mm Pocket Auto Pistol **$335**
Same specifications as Colt Pocket Model 25 Automatic.

**FN Browning Model
1900 Pocket**

FN Browning Model 1900 Pocket Auto Pistol **$325**
Caliber: 32 Automatic (7.65mm). 7-shot magazine. 4-inch bbl. 6.75 inches overall. Weight: 22 oz. Fixed sights. Blued finish. Hard rubber grips. Made 1899-10.

FN Browning Model 1903 Military Auto Pistol **$395**
Caliber: 9mm Browning Long. 7-shot magazine. 5-inch bbl. 8 inches overall. Weight: 32 oz. Fixed sights. Blued finish. Hard rubber grips. *Note:* Aside from size, this pistol is of the same basic design as the Colt Pocket 32 and 380 Automatic pistols. Made 1903-39.

**FN Browning Model
1910 Pocket**

FN Browning Model 1910 Pocket Auto Pistol **$325**
Calibers: 32 Auto (7.65mm), 380 Auto (9mm). 7-shot magazine (32 cal.), 6-shot (380 cal.). 3.5-inch bbl. 6 inches overall. Weight: 20.5 oz. Fixed sights. Blued finish. Hard rubber grips. Made 1910-54.

**FN Browning
Model 1922 Police/Military**

FN Browning Model 1922 Police/Military Auto **$250**
Calibers: 32 Auto (7.65mm), 380 Auto (9mm). 9-shot magazine (32 cal.), 8-shot (380 cal.). 4.5-inch bbl. 7 inches overall. Weight: 25 oz. Fixed sights. Blued finish. Hard rubber grips. Made 1922-59.

FN Browning Model 1935 Military Hi-Power Pistol
Variation of the Browning-Colt 45 Auto design. Caliber: 9mm Para.13-shot magazine. 4.63-inch bbl. 7.75 inches overall. Weight: about 35 oz. Adjustable rear sight and fixed front, or both fixed. Blued finish (Canadian manufacture Parkerized). Checkered walnut or plastic grips. *Note:* Above specifications in general apply to both the original FN production and the pistols made by John Inglis Company of Canada for the Chinese Government. A smaller version, w/shorter bbl. and slide and 10-shot magazine, was made by FN for the Belgian and Rumanian Governments about 1937-1940. Both types were made at the FN plant during the German Occupation of Belgium.
W/adjustable rear sight **$695**
FN manufacture, w/fixed rear sight 575
Inglis manufacture, w/fixed rear sight 650

FRAZER

**FN Browning Model 1935
Military Hi-Power**

FN Browning Baby

FN Browning Baby Auto Pistol **$335**
Caliber: 25 Automatic (6.35mm). 6-shot magazine. 2.13-inch bbl. 4 inches overall. Weight: 10 oz. Fixed sights. Blued finish. Hard rubber grips. Made 1931-83.

FOREHAND & WADSWORTH
Worcester, Massachusetts

Forehand & Wadsworth Revolvers
See listings of comparable Harrington & Richardson and Iver Johnson revolvers for values.

LE FRANCAIS PISTOLS
St. Etienne, France
Produced by Manufacture Francaise
d'Armes et Cycles

Le Francais Army Model Automatic Pistol **$395**
Similar in operation to the Le Francais 25 Automatics. Caliber: 9mm Browning Long. 8-shot magazine. 5-inch bbl. 7.75 inches overall. Weight: about 34 oz. Fixed sights. Blued finish. Checkered walnut grips. Made from 1928-38.

Le Francais Policeman Model Automatic Pistol **$295**
DA. Hinged bbl. Caliber: 25 Automatic (6.35mm). 7-shot magazine. 3.5-inch bbl. 6 inches overall. Weight: about 12 oz. Fixed sights. Blued finish. Hard rubber grips. Intro. 1914. Discontinued.

**Le Francais Staff Officer Model
Automatic Pistol** **$125**
Caliber: 25 Automatic. Similar to the "Policeman" Model except does not have cocking-piece head, bbl. is about an inch shorter and weight is an oz. less. Intro. 1914. Discontinued.

FREEDOM ARMS
Freedom, Wyoming

Freedom Arms Model FA-44 (83-44) SA Revolver
Similar to Model 454 Casull, except chambered in 44 Mag. Made from 1988 to date.
Field Grade **$725**
Premier Grade 795
Silhouette Class (w/10-inch bbl.) 695
Silhouette Pac (10-inch bbl., access.) 765
For Fixed Sights, deduct 95

Freedom Arms Model FA-45 (83-45) SA Revolver
Similar to Model 454 Casull, except chambered in 45 Long Colt. Made 1988-90.
Field Grade **$725**
Premier Grade 825
For Fixed Sights, deduct 95

**Freedom Arms Model FA-252
Silhouette Class**

Freedom Arms Model FA-252 (83-22) SA Revolver
Calibers: 22 LR w/optional 22 Mag. cylinder. Bbl. lengths: 5.13 and 7.5 (Varmint Class), 10 inches (Silhouette Class). Adjustable express or competition silhouette sights. Brushed or matte stainless finish. Black micarta (Silhouette) or black and green laminated hardwood grips (Varmint). Made 1991 to date.
Silhouette Class . . . . **$ 895**
Silhouette Class w/extra 22 Mag. cyl 1125
Varmint Class 750
Varmint Class w/extra 22 Mag. Cyl 995

**Freedom Arms Model FA-353
Field Grade**

Freedom Arms Model FA-353 SA Revolver
Caliber: 357 Mag. Bbl. lengths: 4.75, 6, 7.5 or 9 inches. Removable blade front sight, adjustable rear. Brushed or matte stainless finish. Pachmayr Presentation or impregnated hardwood grips.
Field Grade **$725**
Premier Grade 795
Silhouette Class (w/9-inch bbl.) 745

Freedom Arms Model FA-454 Casull

Freedom Arms Model FA-555

Freedom Arms Model FA-454AS Revolver
Caliber: 454 Casull (w/optional 45 ACP, 45 LC, 45 Win. Mag. cylinders). 5-shot cylinder. Bbl. lengths: 4.75, 6, 7.5 or 10 inches. Adjustable express or competition silhouette sights. Pachmayr presentation or impregnated hardwood grips. Brushed or matte stainless steel finish.

Field Grade	$735
Premier Grade	835
Silhouette Class (w/10-inch bbl.)	750
For Extra Cylinder, add	250

Freedom Arms Model FA-454 Field Grade

Freedom Arms Model FA-454FS Revolver
Same general specifications as Model FA-454AS, except w/fixed sight.

Field Grade	$625
Premier Grade	725

Freedom Arms Model FA-454GAS Revolver $255
Field Grade version of Model FA-454AS, except not made w/12-inch bbl. Matte stainless finish, Pachmayr presentation grips. Adj. Sights or fixed sight on 4.75-inch bbl.

Freedom Arms Model FA-555 Revolver
Similar to Model 454 Casull, except chambered in 50 AK. Made 1994 to date.

Field Grade	$750
Premier Grade	795

Freedom Arms Model FA-BG-22LR
Mini-Revolver $150
Caliber: 22 LR. 3-inch tapered bbl. Partial high-gloss stainless steel finish. Discontinued 1987.

Freedom Arms Model FA-BG-22M Mini-Revolver $175
Same general specifications as model FA-BG-22LR, except in caliber 22 WMR. Discontinued 1987.

Freedom Arms Model FA-BG-22P Mini-Revolver..... $165
Same general specifications as Model FA-BG-22LR, except in 22 percussion. Discontinued 1987.

Freedom Arms Model FA-L-22LR Mini-Revolver $95
Caliber: 22 LR. 1.75-inch contoured bbl. Partial high-gloss stainless steel finish. Bird's-head-type grips. Discontinued 1987.

Freedom Arms Model FA-L-22M Mini-Revolver $135
Same general specifications as Model FA-L-22LR, except in caliber 22 WMR. Discontinued 1987.

Freedom Arms Model FA-L-22P Mini-Revolver $115
Same general specifications as Model FA-L-22LR, except in 22 percussion. Discontinued 1987.

Freedom Arms Model FA-S-22LR Mini-Revolver $125
Caliber: 22 LR. 1-inch contoured bbl. Partial high-gloss stainless steel finish. Discontinued 1988.

Freedom Arms Model FA-S-22M Mini-Revolver $155
Same general specifications as Model FA-S-22LR, except in caliber 22 WMR. Discontinued 1988.

Freedom Arms Model FA-S-22P Mini-Revolver $125
Same general specifications as Model FA-S-22LR, except in percussion. Discontinued 1988.

FRENCH MILITARY PISTOLS
Cholet, France
Mfd. originally by Société Alsacienne de Constructions Mécaniques (S.A.C.M.). Currently made by Manufacture d'Armes Automatiques, Lotissement Industriel des Pontots, Bayonne

French Model 1935A Automatic Pistol $195
Caliber: 7.65mm Long. 8-shot magazine. 4.3-inch bbl. 7.6 inches overall. Weight: 26 oz. Two-lug locking system similar to the Colt U.S. M1911A1. Fixed sights. Blued finish. Checkered grips. Made 1935-45. *Note:* This pistol was used by French troops during WW II and in Indo-China 1945-54.

French Model 1935S Automatic Pistol $275
Similar to Model 1935A, except shorter (4.1-inch bbl. and 7.4 inches overall) and heavier (28 oz.). Single-step lug locking system.

French Model 1950 Automatic Pistol $325
Caliber: 9mm Para. 9-shot magazine. 4.4-inch bbl. 7.6 inches overall. Weight: 30 oz. Fixed sights, tapered post front and U-notched rear. Similar in design and function to the U.S. 45 service automatic, except no bbl. bushing.

French Model MAB F1 Automatic Pistol **$550**
Similar to Model MAB P-15, except w/6-inch bbl. and 9.6 inches
overall. Adjustable target-style sights. Parkerized finish.

French Model MAB P-8 Automatic Pistol **$385**
Similar to Model MAB P-15, except w/8-shot magazine.

French Model MAB P-15 Automatic Pistol **$475**
Caliber: 9mm Para. 15-shot magazine. 4.5-inch bbl. 7.9 inches overall.
Weight: 38 oz. Fixed sights, tapered post front and U-notched rear.

FROMMER PISTOLS
Budapest, Hungary
Mfd. by Fémáru-Fegyver-és Gépgyár R.T.

Frommer Baby Pocket Automatic Pistol **$195**
Similar to Stop model, except has 2-inch bbl. 4.75 inches overall.
Weight 17.5 oz. Magazine capacity is one round less than Stop
Model. Intro. shortly after WW I.

Frommer Liliput Pocket Automatic Pistol **$185**
Caliber: 25 Automatic (6.35mm). 6-shot magazine. 2.14-inch bbl.
4.33 inches overall. Weight: 10.13 oz. Fixed sights. Blued finish.
Hard rubber grips. Made during early 1920s. *Note:* Although similar
in appearance to the Stop and Baby, this pistol is blowback operated.

Frommer Stop Pocket Automatic Pistol **$180**
Locked-breech action, outside hammer. Calibers: 32 Automatic
(7.65mm), 380 Auto (9mm short). 7-shot (32 cal.) or 6-shot (380
cal.) magazine. 3.88-inch bbl. 6.5 inches overall. Weight: about 21
oz. Fixed sights. Blued finish. Hard rubber grips. Made 1912-20.

GALESI PISTOLS
Collebeato (Brescia), Italy
Mfd. by Industria Armi Galesi

**Galesi Model 6
Pocket**

Galesi Model 6 Pocket Automatic Pistol **$125**
Calibers: 22 Long, 25 Automatic (6.35mm). 6-shot magazine. 2.25-
inch bbl. 4.38 inches overall. Weight: about 11 oz. Fixed sights.
Blued finish. Plastic grips. Made from 1930 to date.

Galesi Model 9 Pocket Automatic Pistol
Calibers: 22 LR, 32 Auto (7.65mm), 380 Auto (9mm Short). 8-shot
magazine. 3.25-inch bbl. 5.88 inches overall. Weight: about 21 oz.
Fixed sights. Blued finish. Plastic grips. Made from 1930 to date.
Note: Specifications vary, but those shown for 32 Automatic are
commom.
22 Long Rifle or 380 Automatic **$135**
32 Automatic **150**

GLISENTI PISTOL
Carcina (Brescia), Italy
Mfd. by Societa Siderurgica Glisenti

**Glisenti Model 1910
Italian Service**

Glisenti Model 1910 Italian Service Automatic **$575**
Caliber: 9mm, Glisenti. 7-shot magazine. 4-inch bbl. 8.5 inches
overall. Weight: about 32 oz. Fixed sights. Blued finish. Hard rubber
or plastic grips. Adopted 1910 and used through WWII.

GLOCK, INC.
Smyrna, Georgia

> **NOTE**
>
> Models: 17, 19, 20, 21, 22, 23, 24, 31, 32, 33, 34 and 35 were
> fitted with a redesigned grip-frame in 1998. Models: 26, 27, 29,
> 30 and all "C" guns (compensated models) retained the origi-
> nal frame design.

Glock Model 17

Glock Model 17 DA Automatic Pistol **$395**
Caliber: 9mm Parabellum. 17-shot magazine. 4.5-inch bbl. 7.2
inches overall. Weight: about 22 oz., empty. Hi-tech polymer frame
and receiver; steel bbl., slide and springs. Fixed or adj. rear sights.
Matte, nonglare finish. Made of only 33 components, including 3 in-
ternal safety devices. This gun received the "Best Pistol Award of
Merit" in 1987 by the American Firearms Industry. Made in Austria.
Imported 1983 to date.

Glock Model 17L Competition **$525**
Same general specifications as Model 17, except weighs 23.35 oz.
with 6-inch bbl.; 8.85 inches overall. Imported 1988 to date.

Glock Model 17L

**Glock Model 19
Compact**

Glock Model 20

Glock Model 19 Compact $395
Same general specifications as Model 17, except smaller version with 4-inch bbl., 6.85 inches overall and 21-oz. weight. Imported 1988 to date.

Glock Model 20 DA Auto Pistol $450
Caliber: 10mm.15-shot. Hammerless. 4.6-inch bbl. 7.59 inches overall. Weight: 26.3 oz. Fixed or adj. sights. Matte, nonglare finish. Made from 1991 to date.

Glock Model 21 Automatic Pistol $445
Same general specifications as Model 17, except chambered in 45 ACP. 13-shot magazine. 7.59 inches overall. Weight: 25.2 oz. Imported 1991 to date.

Glock Model 22 Automatic Pistol $395
Same general specifications as Model 17, except chambered for 40 S&W. 15-shot magazine. 7.4 inches overall. Imported 1992 to date.

Glock Model 23 Automatic Pistol $390
Same general specifications as Model 19 except chambered for 40 S&W. 13-shot magazine. 6.97 inches overall. Imported 1992 to date.

Glock Model 24 Automatic Pistol $445
Caliber: 40 S&W, 10- and 15-shot magazines; the latter for law enforcement and military use only. 8.85 inches overall. Weight: 26.5 oz. Manual trigger safety; passive firing block and drop safety. Made 1995 to date.

Glock Model 26 DA Automatic Pistol $415
Caliber: 9mm, 10-round magazine. 3.47-inch bbl. 6.3 inches overall. Weight: 19.77 oz. Imported 1995 to date.

Glock Model 27 DA Auto Pistol
Similar to the Glock Model 22, except subcompact. Caliber: 40 S&W, 10-shot magazine. 3.5-inch bbl. Weight: 21.7 oz. Polymer stocks, fixed or fully adjustable sights. Imported 1995 to date.
Fixed Trijicon Sights $495
Fixed Meprolight 445
Adjustable Sights 415
Fixed Sights 375

Glock Model 29 DA Auto Pistol
Similar to the Glock Model 20, except subcompact. Caliber: 10mm. 10-shot magazine. 3.8-inch bbl. Weight: 27.1 oz. Polymer stocks, fixed or fully adjustable sights. Imported 1997 to date.
Fixed Trijicon Sights $545
Fixed Meprolight 505
Adjustable Sights 465
Fixed Sights 435

Glock Model 30

Glock Model 30 DA Auto Pistol
Similar to the Glock Model 21, except subcompact. Caliber: 45 ACP. 10-shot magazine. 3.8-inch bbl. Weight: 26.5 oz. Polymer stocks, fixed or fully adjustable sights. Imported 1997 to date.
Fixed Trijicon Sights $545
Fixed Meprolight 525
Adjustable Sights 470
Fixed Sights 445

Glock Model 34 Auto Pistol
Similar to Model 17, except w/redesigned grip-frame and extended slide-stop lever and magazine release. 5.32- inch bbl. 10-, 17- or 19-shot magazine. Weight: 22.9 oz. Fixed or adjustable sights. Imported 1998 to date.
Model 34 W/Fixed Sights	**$425**
W/Adjustable Sights, **add**	30
W/Trijicon Sights, **add**	90

Glock Model 35 Auto Pistol
Similar to Model 34, except 40 S&W. Imported 1998 to date.
Model 35 W/Fixed Sights	**$435**
W/Adjustable Sights, **add**	30
W/Trijicon Sights, **add**	90

Glock Desert Storm Commemorative $795
Same specifications as Model 17, except "Operation Desert Storm, January 16-February 27, 1991" engraved on side of slide w/list of coalition forces. Limited issue of 1,000 guns. Made in 1991.

GREAT WESTERN ARMS CO.
North Hollywood, California

Great Western Double Barrel Derringer $265
Replica of Remington Double Derringer. Caliber: 38 S&W. Double bbls. (superposed), 3-inch. Overall length: 5 inches. Fixed sights. Blued finish. Checkered black plastic grips. Made 1953-62.

Great Western SA Frontier Revolver............. $435
Replica of the Colt Single Action Army Revolver. Calibers: 22 LR, 357 Magnum, 38 Special 44 Special, 44 Magnum 45 Colt. 6-shot cylinder. Bbl. lengths: 4.75-, 5.5 and 7.5-inch. Weight: 40 oz., 22 cal. w/5.5-inch bbl. Overall length: 11.13 inches w/5.5-inch bbl. Fixed sights. Blued finish. Imitation stag grips. Made 1951-1962. *Note:* Value shown is for improved late model revolvers; early Great Westerns are variable in quality and should be evaluated accordingly. It should also be noted that, beginning about July 1956, these revolvers were offered in kit form, values of guns assembled from these kits will, in general, be of less value than for factory-completed weapons.

GRENDEL, INC.
Rockledge, Florida

Grendel Model P-12

Grendel Model P-10 Automatic Pistol
Hammerless, blow-back action. DAO with no external safety. Caliber: 380 ACP. 10-shot box magazine integrated in grip. 3-inch bbl. 5.3 inches overall. Weight: 15 oz. Matte blue, electroless nickel or green Teflon finish. Made 1988-91.
Blued Finish	**$120**
Nickel Finish	135
Teflon Finish	140
W/compensated bbl., **add**	45

Grendel Model P-12 DA Automatic Pistol
Caliber: 380 ACP. 11-shot Zytel magazine. 3-inch bbl. 5.3 inches overall. Weight: 13 oz. Fixed sights. Polymer DuPont ST-800 grip. Made 1991-95.
Standard Model	**$130**
Electroless Nickel	145

Grendel Model P-30 Automatic Pistol
Caliber: 22 WMR. 30-shot magazine. 5-or 8-inch bbl. 8.5 inches overall w/5-inch bbl. Weight: 21 oz. Blade front sight, fixed rear sight. Made 1991-95.
W/5-inch Bbl.	**$165**
W/8-inch Bbl.	195

Grendel Model P-31 Automatic Pistol $285
Caliber: 22 WMR. 30-shot Zytel magazine. 11-inch bbl. 17.3 inches overall. Weight: 48 oz. Adj. blade front sight, fixed rear. Checkered black polymer DuPont ST-800 grip and forend. Made 1991-95.

GUNSITE
Paulden, Arizona

Gunsite "Advanced Tactical" SA Auto Pistol
Manuafactured w/Colt 1991 or Springfield 1991 parts. Caliber: 45 ACP. 8-shot magazine. 3.5-, 4.25-, 5-inch bbl. Weight: 32-38 oz. Checkered or lazer engraved walnut grips. Fixed or Novak Lo-mount sights.
Stainless Finish	**$795**
Blued Finish	695

Gunsite "Custom Carry" SA Auto Pistol
Caliber: 45 ACP. 8-shot magazine. 3.5-, 4.25-, 5-inch bbl. Weight: 32-38 oz. Checkered or lazer engraved walnut grips. Fixed Novak Lo-mount sights. Single action, manufactured based on enhanced colt models.
Stainless Finish	**$825**
Blued Finish	795

H&R 1871, INC.
Gardner, Massachusetts

See listings under Harrington & Richardson, Inc.

HÄMMERLI AG JAGD-UND SPORTWAFFENFABRIK
Lenzburg, Switzerland

Currently imported by Sigarms Inc, Exeter, NH. Previously by Hammerli, USA; Beeman Precision Arms & Mandall Shooting Supplies.

Hämmerli Model 33MP Free Pistol............... $725
System Martini single-shot action, set trigger. Caliber: 22 LR. 11.5-inch octagon bbl. 16.5 inches overall. Weight: 46 oz. Micrometer rear sight, interchangeable front sights. Blued finish. Walnut grips, forearm. Imported 1933-49.

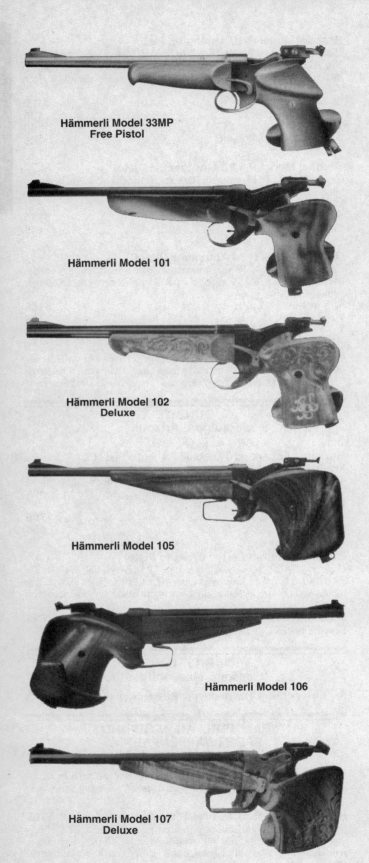

**Hämmerli Model 33MP
Free Pistol**

Hämmerli Model 101

**Hämmerli Model 102
Deluxe**

Hämmerli Model 105

Hämmerli Model 106

**Hämmerli Model 107
Deluxe**

Hämmerli Model 100 Free Pistol
Same general specifications as Model 33MP. Improved action and sights, redesigned stock. Standard model has plain grips and forearm, deluxe model has carved grips and forearm. Imported 1950-56.
Standard Model. **$645**
Deluxe Model **755**

Hämmerli Model 101 . . . . **$625**
Similar to Model 100, except has heavy round bbl. w/matte finish, improved action and sights, adj. grips. Weight: about 49 oz. Imported 1956-60.

Hämmerli Model 102
Same as Model 101, except bbl. has highly polished blued finish. Deluxe model (illustrated) has carved grips and forearm. Made 1956-60.
Standard Model. **$635**
Deluxe Model **735**

Hämmerli Model 103 . . . . **$650**
Same as Model 101, except has lighter octagon bbl. (as in Model 100) w/highly polished blued finish, grips and forearm of select French walnut. Weight: about 46 oz. Imported 1956-60.

Hämmerli Model 104 . . . . **$585**
Similar to Model 102, except has lighter round bbl., improved action redesigned grips and forearm. Weight: 46 oz. Imported 1961-65.

Hämmerli Model 105 . . . . **$675**
Similar to Model 103, except has improved action, redesigned grips and forearm. Imported 1961-65.

Hämmerli Model 106 . . . . **$595**
Similar to Model 104, except has improved trigger and grips. Made 1966-71.

Hämmerli Model 107
Similar to Model 105, except has improved trigger and stock. Deluxe model (illustrated) has engraved receiver and bbl., carved grips and forearm. Imported 1966-71.
Standard Model. **$675**
Deluxe Model **895**

**Hämmerli Model 120
Heavy Barrel**

Hämmerli Model 120 Heavy Barrel
Same as Models 120-1 and 120-2, except has 5.7-inch heavy bbl. Weight: 41 oz. Available w/standard or adj. grips. Imported 1972 to date.
W/standard grips. **$425**
W/adj. grips. **440**

Hämmerli Model 120-1 Single-Shot Free Pislol $485

Side lever-operated bolt action. Adj. single-stage or two-stage trigger. Caliber: 22 LR. 9.9-inch bbl. 14.75 inches overall. Weight: 44 oz. Micrometer rear sight, front sight on high ramp. Blued finish bbl. and receiver, lever and grip frame anodized aluminum. Checkered walnut thumbrest grips. Imported 1972 to date.

Hämmerli Model 120-2 .... $495

Same as Model 120-1, except has hand-contoured grips w/adj. palm rest (available for right or left hand). Imported 1972 to date.

Hämmerli Model 120-1

Hämmerli Models 150/151 Free Pistols

Improved Martini-type action w/lateral-action cocking lever. Set trigger adj. for weight, length and angle of pull. Caliber: 22 LR. 11.3-inch round bbl., free-floating. 15.4 inches overall. Weight: 43 oz. (w/extra weights, 49.5 oz.). Micrometer rear sight, front sight on high ramp. Blued finish. Select walnut forearm and grips w/adj. palm shelf. Imported 1972-93.

Model 150 (Disc. 1989) $1395
Model 151 (Disc. 1993) 1425

Hämmerli Models 150

Hämmerli Model 152 Electronic Pistol

Same general specifications as Model 150, except w/electronic trigger. Made 1990-92.

Right Hand $1400
Left Hand 1450

Hämmerli Models 160/162 Free Pistols

Caliber: 22 LR. Single-shot. 11.31-inch bbl. 17.5 inches overall. Weight: 46.9 oz. Interchangeable front sight blades, fully adj. match rear. Match-style stippled walnut grips w/adj. palm shelf and polycarbon fiber forend. Imported 1993 to date.

Model 160 w/Mechanical Set Trigger................. $1195
Model 162 w/Electronic Trigger 1425

Hämmerli Models 160

Hämmerli Model 208 Standard Auto Pistol...... $1195

Caliber: 22 LR. 8-shot magazine. 5.9-inch bbl.10 inches overall. Weight: 35 oz. (bbl. weight adds 3 oz.). Micrometer rear sight, ramp front. Blued finish. Checkered walnut grips w/adj. heel plate. Imported 1966-88.

Hämmerli Model 208S Target Pistol $1095

Caliber: 22 LR. 8-shot magazine. 6-inch bbl. 10.2 inches overall. Weight: 37.3 oz. Micrometer rear sight, ramp front sight. Blued finish. Stippled walnut grips w/adj. heel plate. Imported 1990 to date.

Hämmerli Model 208

Hämmerli Model 211 .... $1125

Same as Model 208, except has standard thumbrest grips. Imported 1966-91.

Hämmerli Model 212 Hunter's Pistol $895

Caliber: 22 LR. 4.88-inch bbl. 8.5 inches overall. Weight: 31 oz. Blade front sight, square-notched fully adj. rear. Blued finish. Checkered walnut grips. Imported 1984-93.

Hämmerli Model 215 .... $950

Similar to the Model 208 except w/heavier bbl. and fewer deluxe features. Imported 1990-93.

Hämmerli Model 230-1 Rapid Fire Auto Pistol $595

Caliber: 22 Short. 5-shot magazine. 6.3-inch bbl. 11.6 inches overall. Weight: 44 oz. Micrometer rear sight, post front. Blued finish. Smooth walnut thumbrest grips. Imported 1970-83.

Hämmerli Model 230-2 .... $635

Same as Model 230-1, except has checkered walnut grips w/adj. heel plate. Imported 1970-83.

Hämmerli Model 215

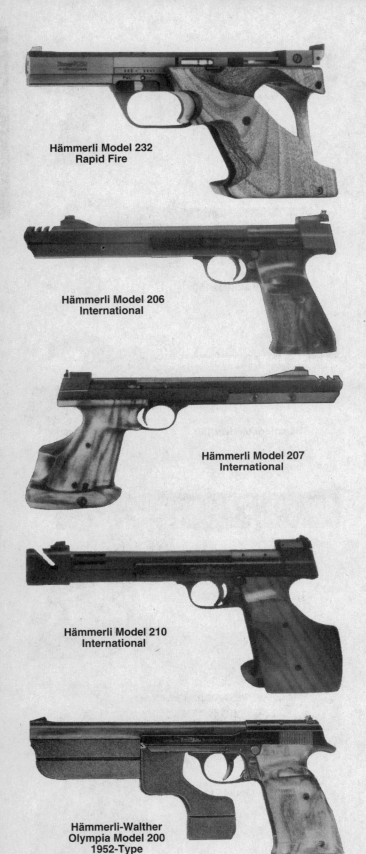

**Hämmerli Model 232
Rapid Fire**

**Hämmerli Model 206
International**

**Hämmerli Model 207
International**

**Hämmerli Model 210
International**

**Hämmerli-Walther
Olympia Model 200
1952-Type**

Hämmerli Model 232 Rapid Fire Auto Pistol $1025
Caliber: 22 Short. 6-shot magazine. 5.1-inch ported bbl. 10.5 inches overall. Weight: 44 oz. Fully adj. target sights. Blued finish. Stippled walnut wraparound target grips. Imported 1984-93.

Hämmerli Model 280 Target Pistol
Carbon-reinforced synthetic frame and bbl. housing. Calibers: 22 LR, 32 S&W Long WC. 6-shot (22 LR) or 5-shot (32 S&W) magazine. 4.5-inch bbl. w/interchangeable metal or carbon fiber counterweights. 11.88 inches overall. Weight: 39 oz. Micro-adj. match sights w/interchangeable elements. Imported 1988 to date.
22 Long Rifle. . . . . $ 925
32 S&W Long WC 1095

Hämmerli International Model 206 Auto Pistol $535
Calibers: 22 Short, 22 LR. 6-shot (22 Short) or 8-shot (22 LR) magazine. 7.1-inch bbl. w/muzzle brake. 12.5 inches overall. Weight: 33 oz. (22 Short), 39 oz. (22 LR) (supplementary weights add 5 and 8 oz.). Micrometer rear sight, ramp front. Blued finish. Standard thumbrest grips. Imported 1962-69.

Hämmerli International Model 207 . . . . $565
Same as Model 206, except has grips w/adj. heel plate, weighs 2 oz. more. Made 1962-69.

Hämmerli International Model 209 Auto Pistol $675
Caliber: 22 Short. 5-shot mag. 4.75-inch bbl. w/muzzle brake and gas-escape holes. 11 inches overall. Weight: 39 oz. (interchangeable front weight adds 4 oz.). Micrometer rear sight, post front. Blued finish. Standard thumbrest grips of checkered walnut. Imported 1966-70.

Hämmerli International Model 210 . . . . $695
Same as Model 209, except has grips w/adj. heel plate, is 0.8-inch longer and weighs 1 ounce more. Made 1966-70.

Hämmerli Virginian SA Revolver $395
Similar to Colt Single-Action Army, except has base pin safety system (SWISSAFE). Calibers: 357 Magnum, 45 Colt. 6-shot cylinder. 4.63-, 5.5- or 7.5-inch bbl. 11 inches overall (5.5-inch bbl.). Weight: 40 oz. (5.5-inch bbl.). Fixed sights. Blued bbl. and cylinder, case-hardened frame, chrome-plated grip frame and trigger guard. One-piece smooth walnut stock. Imported 1973-76 by Interarms, Alexandria, Va.

Hämmerli-Walther Olympia Model 200 Automatic Pistol, 1952-Type $595
Similar to 1936 Walther Olympia Funfkampf Model. Calibers: 22 Short, 22 LR. 6-shot (22 Short) or 10-shot (22 LR) magazine. 7.5-inch bbl. 10.7 inches overall. Weight: 27.7 oz. (22 Short, light alloy breechblock), 30.3 oz. (22 LR). Supplementary weights provided. Adj. target sights. Blued finish. Checkered walnut thumbrest grips. Imported 1952-58.

Hämmerli-Walther Olympia Model 200, 1958-Type $585
Same as Model 200, 1952 Type, except has muzzle brake, 8-shot magazine (22 LR). 11.6 inches overall. Weight: 30 oz. (22 Short), 33 oz. (22 LR). Imported 1958-63.

Hämmerli-Walther Olympia Model 201 . . . . $595
Same as Model 200, 1952 Type, except has 9.5-inch bbl. Imported 1955-57.

Hämmerli-Walther Olympia Model 202 . . . . $600
Same as Model 201, except has grips w/adj. heel plate. Imported 1955-57.

Hämmerli-Walther Olympia Model 203
Same as corresponding Model 200 (1955-Type lacks muzzle brake), except has grips w/adj. heel plate. Imported1955-63.
1955-Type **$595**
1958-Type **625**

Hämmerli-Walther Olympia Model 204
American Model. Same as corresponding Model 200 (1956-Type lacks muzzle brake), except in 22 LR only, has slide stop and micrometer rear sight. Imported 1956-63.
1956-Type **$650**
1958-Type **615**

Hämmerli-Walther Olympia Model 205
American Model. Same as Model 204, except has grips w/adj. heel plate. Imported 1956-63.
1956-Type **$595**
1958-Type **625**

SIG-Hämmerli Model P240 Target Auto Pistol
Calibers: 32 S&W Long (Wadcutter), 38 Special (Wadcutter). 5-shot magazine. 5.9-inch bbl. 10 inches overall. Weight: 41 oz. Micrometer rear sight, post front. Blued finish/ Smooth walnut thumbrest grips. Accessory 22 LR conversion unit available. Imported 1975-86.
32 S&W Long **$1095**
38 Special **1695**
22 LR conversion unit, add **495**

HARRINGTON & RICHARDSON, INC.
Gardner, Massachusetts
Now H&R 1871, Inc., Gardner, Mass.

Formerly Harrington & Richardson Arms Co. of Worcester, Mass. One of the oldest and most distinguished manufacturers of handguns, rifles and shotguns, H&R had suspended operations in 1986. In 1987, New England Firearms was established as an independent company. In 1991, H&R 1871, Inc. was formed from the residual of the parent company.

> **NOTE**
>
> For ease in finding a particular firearm, H&R handguns are grouped into Automatic/Single-Shot Pistols, followed by Revolvers. For a complete listing, please refer to the Index.

AUTOMATIC/SINGLE-SHOT PISTOLS

Harrington & Richardson SL 25 Pistol **$350**
Modified Webley & Scott design. Caliber: 25 Auto. 6-shot magazine. 2-inch bbl. 4.5 inches overall. Weight: 12 oz. Fixed sights. Blued finish. Black hard rubber grips. Made 1912-16.

Harrington & Richardson SL 32 Pistol **$295**
Modified Webley & Scott design. Caliber: 32 Auto. 8-shot magazine. 3.5-inch bbl. 6.5 inches overall. Weight: about 20 oz. Fixed sights. Blued finish. Black hard rubber grips. Made 1916-24.

Harrington & Richardson USRA Model Single-Shot Target Pistol **$425**
Hinged frame. Caliber: 22 LR. Bbl. lengths: 7-, 8- and 10-inch. Weight: 31 oz. w/10-inch bbl. Adj. target sights. Blued finish. Checkered walnut grips. Made 1928-41.

Hämmerli-Walther
Olympia Model 203
1958-Type

Hämmerli-Walther
Olympia Model 205

SIG-Hämmerli Model P240
Target

Harrington & Richardson
SL 32

**Harrington & Richardson
USRA Model
Single-Shot Target Pistol**

**Harrinoton & Richardson
Model 5**

REVOLVERS

Harrinoton & Richardson Model 4 (1904) DA **$75**
Solid frame. Calibers: 32 S&W Long, 38 S&W. 6-shot cylinder (32 cal.), 5-shot (38 cal.). Bbl. lengths: 2.5-, 4.5- and 6-inch. Weight: about 16 oz., 32 cal. Fixed sights. Blued or nickel finish. Hard rubber grips. Discontinued prior to 1942.

**Harrinoton & Richardson
Model 4 Double-Action Revolver**

**Harrinoton & Richardson
Model 6**

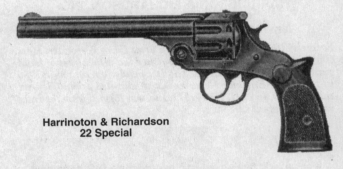

**Harrinoton & Richardson
22 Special**

Harrington & Richardson Model 5 (1905) DA **$80**
Solid frame. Caliber: 32 S&W. 5-shot cylinder. Bbl. lengths: 2.5-,4.5- and 6-inch. Weight: about 11 oz. Fixed sights. Blued or nickel finish. Hard rubber grips. Discontinued prior to 1942.

Harrington & Richardson Model 6 (1906) DA **$85**
Solid frame. Caliber: 22 LR. 7-shot cylinder. Bbl. lengths: 2.5- 4.5- and 6-inch. Weight: about 10 oz. Fixed sights. Blued or nickel finish. Hard rubber grips. Discontinued prior to 1942.

Harrington & Richardson 22 Special DA **$140**
Heavy hinged frame. Calibers: 22 LR, 22 Mag. 9-shot cylinder. 6-inch bbl. Weight: 23 oz. Fixed sights, front gold-plated. Blued finish. Checkered walnut grips. Recessed safety cylinder on later models for high-speed ammunition. Discontinued prior to 1942.

**Harrington & Richardson Model 199
Sportsman SA Revolver** **$225**
Hinged frame. Caliber: 22 LR. 9-shot cylinder. 6-inch bbl. 11 inches overall. Weight: 30 oz. Adj. target sights. Blued finish. Checkered walnut grips. Discontinued 1951.

**Harrinoton & Richardson
Model 199 Sportsman**

Harrington & Richardson Model 504 DA **$145**
Caliber: 32 H&R Magnum. 5-shot cylinder. 4- or 6-inch bbl. (square butt), 3- or 4-inch bbl. (round butt). Made 1984-86.

Harrington & Richardson Model 532 DA **$85**
Caliber: 32 H&R Magnum. 5-shot cylinder. 2.5- or 4-inch bbl. Weight: approx. 20 and 25 oz. respectively. Fixed sights. American walnut grips. Lustre blued finish. Made 1984-86.

Harrington & Richardson Model 586 DA **$145**
Caliber: 32 H&R Magnum. 5-shot cylinder. Bbl. lengths: 4.5, 5.5, 7.5, 10 inches. Weight: 30 oz. average. Adj. rear sight, blade front. Walnut finished hardwood grips. Made 1984-86.

**Harrington & Richardson
Model 622**

**Harrington & Richardson
Model 632**

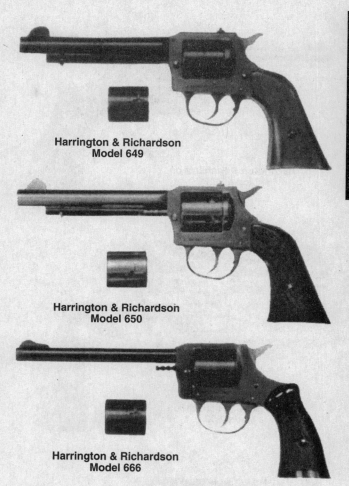

**Harrington & Richardson
Model 649**

**Harrington & Richardson
Model 650**

**Harrington & Richardson
Model 666**

Harrington & Richardson Model 603 Target $125
Similar to Model 903, except in 22 WMR. 6-shot capacity w/un-fluted cylinder. Made 1980-83.

Harrington & Richardson Model 604 Target $135
Similar to Model 603, except w/6-inch bull bbl. Weight: 38 oz. Made 1980-83.

Harrington & Richardson Model 622/623 DA $85
Solid frame. Caliber: 22 Short, Long, LR, 6-shot cylinder. Bbl. lengths: 2.5-, 4-, 6-inch. Weight: 26 oz. (4-inch bbl.). Fixed sights. Blued finish. Plastic grips. Made 1957-86. *Note:* Model 623 is same, except chrome or nickel finish.

Harrington & Richardson Model 632/633 Guardsman DA Revolver $85
Solid Frame. Caliber: 32 S&W Long. 6-shot cylinder. Bbl. lengths: 2.5- or 4-inch. Weight: 19 oz. (2.5-inch bbl.). Fixed sights. Blued or chrome finish. Checkered Tenite grips (round butt on 2.5-inch, square butt on 4-inch). Made 1953-86. *Note:* Model 633 is the same, except for chrome or nickel finish.

Harrington & Richardson Model 649/650 DA $120
Solid frame. Side loading and ejection. Convertible model w/two 6-shot cylinders. Calibers: 22 LR, 22 WMR. 5.5-inch bbl. Weight: 32 oz. Adj. rear sight, blade front. Blued finish. One-piece, Western-style walnut grip. Made 1976-86. *Note:* **Model 650** is same, except nickel finish.

Harrington & Richardson Model 666 DA $85
Solid frame. Convertible model w/two 6-shot cylinders. Calibers: 22 LR, 22 WMR. 6-inch bbl. Weight: 28 oz. Fixed sights. Blued finish. Plastic grips. Made 1976-78.

Harrington & Richardson Model 676 DA $120
Solid frame. Side loading and ejection. Convertible model w/two 6-shot cylinders. Calibers: 22 LR, 22 WMR. Bbl. lengths: 4.5-, 5.5-, 7.5-, 12-inch. Weight: 32 oz. (5.5-inch bbl.). Adj. rear sight, blade front. Blued finish, color-casehardened frame. One-piece, Western-style walnut grip. Made 1976-1980.

**Harrington & Richardson
Model 676**

**Harrington & Richardson
Model 686**

**Harrington & Richardson
Model 733**

**Harrington & Richardson
Model 830**

**Harrington & Richardson
Model 900**

**Harrington & Richardson
Model 903**

Harrington & Richardson Model 686 DA $150
Caliber: 22 LR and 22 WMR. 4.5-, 5.5-, 7.5-, 10- or 12-inch bbl. 6-shot magazine. Adj. rear sight, ramp and blade front. Blued, color-casehardened frame. Weight: 31 oz. (4.5-inch bbl.). Made 1980-86.

Harrington & Richardson Model 732/733 DA $115
Solid frame, swing-out 6-shot cylinder. Calibers: 32 S&W, 32 S&W Long. Bbl. lengths: 2.5-, 4-inch. Weight: 26 oz.(4-inch bbl.). Fixed sights (windage adj. rear on 4-inch bbl. model). Blued finish. Plastic grips. Made 1958-86. *Note:* Model 733 is same, except nickel finish.

Harrington & Richardson Model 826 DA $110
Caliber: 22 WMR. 6-shot magazine. 3-inch bull bbl. Ramp and blade front sight, adj. rear. American walnut grips. Weight: 28 oz. Made 1981-83.

Harrington & Richardson Model 829/830 DA
Same as Model 826, except in 22 LR caliber. 9-shot capacity. Made 1981-83.
Model 829, Blued $115
Model 830, Nickel 120

Harrington & Richardson Model 832,1833 DA
Same as Model 826, except in 32 S W Long. Blued or nickel finish. Made 1981-83.
Model 832, Blued $125
Model 833, Nickel 130

Harrington & Richardson Model 900/901 DA $75
Solid frame, snap-out cylinder. Calibers: 22 Short, Long, LR. 9-shot cylinder. Bbl. lengths: 2.5-, 4-, 6-inch. Weight: 26 oz. (6-inch bbl.). Fixed sights. Blued finish. Blade Cycolac grips. Made 1962-73. *Note:* **Model 901** (discontinued in 1963) is the same, except has chrome finish and white Tenite grips.

**Harrington & Richardson
Model 905**

Harrington & Richardson Model 903 Target $125
Caliber: 22 LR. 9-shot capacity. SA/DA. 6-inch targetweight flat-side bbl. Swing-out cylinder. Weight: 35 oz. Blade front sight, adj. rear. American walnut grips. Made 1980-83.

Harrington & Richardson Model 904 Target $130
Similar to Model 903, except 4- or 6-inch bull bbl. 4-inch bbl. weighs 32 oz. Made 1980-86.

Harrington & Richardson Model 905 Target $135
Same as Model 904, except w/4-inch bbl. only. Nickel finish. Made 1981-83.

**Harrington & Richardson
Model 922, First Issue**

Harrington & Richardson Model 922 DA Revolver, First Issue $150
Solid frame. Caliber: 22 LR. 9-shot cylinder. 10-inch, octagon bbl. (early model) or 6-inch, round bbl. (later production). Weight: 26 oz. (6-inch bbl.). Fixed sights. Blued finish. Checkered walnut grips. Safety cylinder on later models. Discontinued prior to 1942.

**Harrington & Richardson
Model 922, Second Issue**

Harrington & Richardson Model 922/923 DA Revolver, Second Issue $75
Solid frame. Caliber: 22 LR. 9-shot cylinder. Bbl. lengths: 2.5-, 4-, 6-inch. Weight: 24 oz. (4-inch bbl.). Fixed sights. Blued finish. Plastic grips. Made 1950-86. *Note:* Second Issue Model 922 has a different frame from that of the First Issue. **Model 923** is same as Model 922, Second Issue, except for nickel finish.

Harrington & Richardson Model 925 Defender $115
DA. Hinged frame. Caliber: 38 S&W. 5-shot cylinder. 2.5-inch bbl. Weight: 22 oz. Adj. rear sight, fixed front. Blued finish. One-piece wraparound grip. Made 1964-78.

Harrington & Richardson Model 926 DA $115
Hinged frame. Calibers: 22 LR, 38 S&W. 9-shot (22 LR) or 5-shot (38) cylinder. 4-inch bbl. Weight: 31 oz. Adj. rear sight, fixed front. Blued finish. Checkered walnut grips. Made 1968-78.

Harrington & Richardson Model 929/930 Sidekick DA Revolver $85
Caliber: 22 LR. Solid frame, swing-out 9-shot cylinder. Bbl. lengths: 2.5-, 4-, 6-inch. Weight: 24 oz. (4-inch bbl.). Fixed sights. Blued finish. Checkered plastic grips. Made 1956-86. *Note:* **Model 930** is same, except nickel finish.

Harrington & Richardson Model 939/940 Ultra Sidekick DA Revolver $110
Solid frame, swing-out 9-shot cylinder. Safety lock. Calibers: 22 Short, Long, LR. Flat-side 6-inch bbl. w/vent rib. Weight: 33 oz. Adj. rear sight, ramp front. Blued finish. Checkered walnut grips. Made 1958-86, reintroduced by H&R 1871 in 1992. *Note:* **Model 940** is same, except has round bbl.

**Harrington & Richardson
Model 925**

**Harrington & Richardson
Model 926**

**Harrington & Richardson
Model 929**

**Harrington & Richardson
Model 939**

Harrington & Richardson Model 949

Harrington & Richardson Model 950

Harrington & Richardson Model 999, First Issue

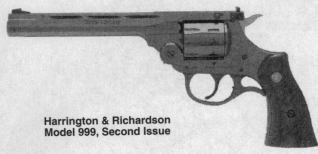

Harrington & Richardson Model 999, Second Issue

Harrington & Richardson Automatic Ejecting

Harrington & Richardson Model 949/950 Forty-Niner DA Revolver $110
Solid frame. Side loading and ejection. Calibers: 22 Short, Long, LR. 9-shot cylinder. 5.5-inch bbl. Weight: 31 oz. Adj. rear sight, blade front. Blued or nickel finish. One-piece, Western-style walnut grip. Made 1960-86, reintroduced by H&R 1871 in 1992. *Note:* **Model 950** is same, except nickel finish.

Harrington & Richardson Model 976 DA $90
Same as Model 949, except has color-casehardened frame, 7.5-inch bbl. weighs 36 oz. Intro. 1977. Discontinued.

Harrington & Richardson Model 999 Sportsman DA Revolver, First Issue $165
Hinged frame. Calibers: 22 LR, 22 Mag. Same specifications as Model 199 Sportsman Single Action. Discontinued before 1942.

Harrington & Richardson Model 999 Sportsman DA Revolver, Second Issue $175
Hinged frame. Caliber: 22 LR. 9-shot cylinder. 6-inch bbl. w/vent rib. Weight: 30 oz. Adj. sights. Blued finish. Checkered walnut grips. Made 1950-86.

Harrington & Richardson (New) Model 999 Sportsman DA Revolver $170
Hinged frame. Caliber: 22 Short, Long, LR. 9-shot cylinder. 6-inch bbl. w/vent rib. Weight: 30 oz. Blade front sight adj. for elevation, square-notched rear adj. for windage. Blued finish. Checkered hardwood grips. Reintroduced by H&R 1871 in 1992.

Harrington & Richardson American DA $75
Solid frame. Calibers: 32 S&W Long, 38 S&W. 6-shot (32 cal.) or 5-shot (38 car.) cylinder. Bbl. lengths: 2.5-,4.5- and 6-inch. Weight: about 16 oz. Fixed sights. Blued or nickel finish. Hard rubber grips. Discontinued prior to 1942.

Harrington & Richardson Automatic Ejecting DA Revolver $145
Hinged frame. Calibers: 32 S&W Long, 38 S&W. 6-shot cylinder (32 cal.), 5-shot (38 cal.). Bbl. lengths: 3.25-, 4-, 5- and 6-inch. Weight: 16 oz. (32 cal.), 15 oz. (38 cal.). Fixed sights. Blued or nickel finish. Black hard rubber grips. Discontinued prior to 1942.

Harrington & Richardson Bobby

Harrington & Richardson Bobby DA $250
Hinged frame. Calibers: 32 S&W, 38 S&W. 6-shot cylinder (32 cal.), 5-shot (38 cal.). 4-inch bbl. 9 inches overall. Weight: 23 oz. Fixed sights. Blued finish. Checkered walnut grips. Discontinued 1946. *Note:* Originally designed and produced for use by London's bobbies.

**Harrington & Richardson
Defender 38**

**Harrington & Richardson
Hammerless, Small Frame**

**Harrington & Richardson
Premier**

**Harrington & Richardson
Target**

Harrington & Richardson Hammerless DA, Large Frame $105
Hinged frame. Calibers: 32 S&W Long 38 S&W. 6-shot cylinder (32 cal.), 5-shot (38 cal.). Bbl. lengths: 3.25-, 4-, and 6-inch. Weight: about 17 oz. Fixed sights. Blued or nickel finish. Hard rubber grips. Discontinued prior to 1942.

Harrington & Richardson Hammerless DA, Small Frame $100
Hinged frame. Calibers: 22 LR, 32 S&W. 7-shot (22 cal.) 5-shot (32 cal.) cylinder. Bbl. lengths: 2-, 3-, 4-, 5- and 6-inch. Weight: about 13 oz. Fixed sights. Blued or nickel finish. Hard rubber grips. Discontinued. prior to 1942.

Harrington & Richardson Hunter Model DA $115
Solid frame. Caliber: 22 LR. 9-shot cylinder. 10-inch octagon bbl. Weight: 26 oz. Fixed sights. Blued finish. Checkered walnut grips. Safety cylinder on later models. *Note:* An earlier Hunter Model was built on the smaller 7-shot frame. Discontinued prior to 1942.

Harrington & Richardson New Defender DA $215
Hinged frame. Caliber: 22 LR. 9-shot cylinder. 2-inch bbl. 6.25 inches overall. Weight: 23 oz. Adj. sights. Blued finish. Checkered walnut grips, round butt. *Note:* Basically, this is the Sportsman DA w/a short bbl. Discontinued prior to 1942.

Harrington & Richardson Premier DA $90
Small hinged frame. Calibers: 22 LR, 32 S&W. 7-shot (22 LR) or 5-shot (32) cylinder. Bbl. lengths: 2-, 3-, 4-, 5 and 6-inch. Weight: 13 oz. (22 LR), 12 oz. (32 S&W). Fixed sights. Blued or nickel finish. Black hard rubber grips. Discontinued prior to 1942.

Harrington & Richardson Model STR 022 Blank Revolver $65
Caliber: 22 RF blanks. 9-shot cylinder. 2.5-inch bbl. Weight: 19 oz. Satin blued finish.

Harrington & Richardson Model STR 032 Blank Revolver $75
Same general specifications as STR 022 except chambered for 32 S&W blank cartridges.

Harrington & Richardson Target Model DA $130
Small hinged frame. Calibers: 22 LR, 22 W.R.F. 7-shot cylinder. 6-inch bbl. Weight: 16 oz. Fixed sights. Blued finish. Checkered walnut grips. Discontinued prior to 1942.

**Harrington & Richardson
Trapper**

Harrington & Richardson Defender 38 DA $120
Hinged frame. Based on the Sportsman design. Caliber: 38 S&W. Bbl. lengths: 4- and 6-inch. 9 inches overall (4-inch bbl.). Weight: 25 oz., 4-inch bbl. Fixed sights. Blued finish. Black plastic grips. Discontinued 1946. *Note:* This model was manufactured during WW II as an arm for plant guards, auxiliary police, etc.

Harrington & Richardson Expert Model DA $150
Same specifications as 22 Special, except has 10-inch bbl. Weight: 28 oz. Discontinued prior to 1942.

Harrington & Richardson Trapper Model DA $115
Solid frame. Caliber: 22 LR. 7-shot cylinder. 6-inch octagon bbl. Weight: 12.5 oz. Fixed sights. Blued finish. Checkered walnut grips. Safety cylinder on later models. Discontinued prior to 1942.

**Harrington & Richardson
Ultra Sportsman**

**Harrington & Richardson
Vest Pocket**

**Harrington & Richardson
Young American**

Harrington & Richardson Ultra Sportsman **$195**
SA. Hinged frame. Caliber: 22 LR. 9-shot cylinder. 6-inch bbl.
Weight: 30 oz. Adj. target sights. Blued finish. Checkered walnut
grips. This model has short action, wide hammer spur, cylinder is
length of a 22 LR cartridge. Discontinued prior to 1942.

Harrington & Richardson Vest Pocket DA **$75**
Solid frame. Spurless hammer. Calibers: 22 Rimfire, 32 S&W. 7-
shot (22 cal.) or 5-shot (32 cal.) cylinder. 1.13-inch bbl. Weight:
about 9 oz. Blued or nickel finish. Hard rubber grips. Discontinued
prior to 1942.

Harrington & Richardson Young America DA **$75**
Solid frame. Calibers: 22 Long, 32 S&W. 7-shot (22 cal.) or 5-shot
(32 cal.) cylinder. Bbl. lengths: 2-, 4.5- and 6-inch. Weight: about 9
oz. Fixed sights. Blued or nickel finish. Hard rubber grips. Discont.
prior to 1942.

HARTFORD ARMS & EQUIPMENT CO.
Hartford, Connecticut

*Hartford pistols were the forebears of the original High Stan-
dard line, since High Standard Mfg. Corp. acquired Hartford
Arms & Equipment Co. in 1932. The High Standard Model B is
essentially the same as the Hartford Automatic.*

**Hartford
Automatic Target Pistol**

Hartford Automatic Target Pistol **$550**
Caliber. 22 LR. 10-shot magazine. 6.75-inch bbl. 10.75 inches overall. Weight: 31 oz. Target sights. Blued finish. Black rubber grips.
This arm closely resembles the early Colt Woodsman and High Standard pistols. Made 1929-30.

Hartford Repeating Pistol **$425**
Same general design as the automatic pistol of this manufacture, but
this model is a hand-operated repeating pistol on the order of the Fiala. Made 1929-30.

Hartford Single-Shot Target Pistol **$395**
Similar in appearance to the Hartford Automatic. Caliber: 22 LR.
6.75-inch bbl. 10.75 inches overall. Weight: 38 oz. Target sights.
Mottled frame and slide, blued bbl. Black rubber or walnut grips.
Made 1929-30.

HASKELL MANUFACTURING
Lima, Ohio

See listings under Hi-Point.

HAWES FIREARMS
Van Nuys, California

**Hawes
Deputy Denver Marshal**

Hawes Deputy Denver Marshal
Same as Deputy Marshal SA, except has brass frame. Imported from
1973-81.
22 LR (plastic grips) **$75**
Combination, 22 LR/22 WMR (plastic) 75
Extra for walnut grips 10

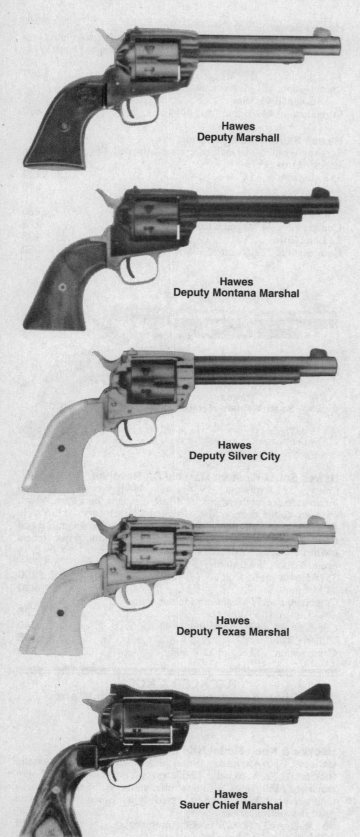

Hawes Deputy Marshall

Hawes Deputy Montana Marshal

Hawes Deputy Silver City

Hawes Deputy Texas Marshal

Hawes Sauer Chief Marshal

Hawes Deputy Marshal SA Revolver

Calibers: 22 LR, also 22 WMR in two-cylinder combination. 6-shot cylinder. 5.5-inch bbl. 11 inches overall. Weight: 34 oz. Adj. rear sight, blade front. Blued finish. Plastic or walnut grips. Imported from 1973-81.

22 Long Rifle (plastic grips)	$70
Combination, 22 LR/22 WMR (plastic)	75
Extra for walnut grips	10

Hawes Deputy Montana Marshal

Same as Deputy Marshal, except has brass grip frame. Walnut grips only. Imported from 1973-81.

22 Long Rifle	$95
Combination, 22 LR/22 WMR	120

Hawes Deputy Silver City Marshal

Same as Deputy Marshal, except has chrome-plated frame, brass grip frame, blued cylinder and bbl. Imported from 1973-81.

22 Long Rifle (plastic grips)	$85
Combination, 22 LR/22 WMR (plastic)	120
Extra for walnut grips	10

Hawes Deputy Texas Marshal

Same as Deputy Marshal, except has chrome finish. Imported from 1973-81.

22 Long Rifle (plastic grips)	$95
Combination, 22 LR/22 WMR (plastic)	120
Extra for walnut grips	10

Hawes Favorite Single-Shot Target Pistol $150

Replica of Stevens No. 35. Tip-up action. Caliber: 22 LR. 8-inch bbl. 12 inches overall. Weight: 24 oz. Target sights. Chrome-plated frame. Blued bbl. Plastic or rosewood grips (**add** $5). Imported 1972-76.

Hawes Sauer Chief Marshal SA Target Revolver

Same as Western Marshal, except has adj. rear sight and front sight, oversized rosewood grips. Not made in 22 caliber. Imported from 1973-81.

357 Magnum or 45 Colt	$200
44 Magnum	225
Combination, 357 Magnum and 9mm Para. 45 Colt and 45 Auto	250
Combination, 44 Magnum and 44-40	245

Hawes Sauer Federal Marshal

Hawes Sauer Federal Marshal

Same as Western Marshal, except has color-casehardened frame, brass grip frame, one-piece walnut grip. Not made in 22 caliber. Imported from 1973-81.

357 Magnum or 45 Colt	$200
44 Magnum	230
Combination, 357 Magnum and 9mm Para., 45 Colt and 45 Auto	250
Combination, 44 Magnum and 44-40	245

**Hawes
Sauer Montana Marshal**

**Hawes
Sauer Montana Marshal-22**

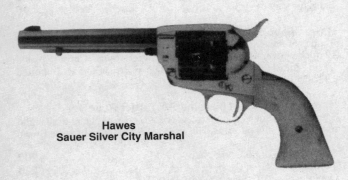

**Hawes
Sauer Silver City Marshal**

**Hawes
Sauer Texas Marshal**

Hawes Sauer Montana Marshal
Same as Western Marshal, except has brass grip frame. Imported from 1973-81.

357 Magnum or 45 Colt	**$200**
44 Magnum	**225**
Combination, 357 Magnum and 9mm Para., 45 Colt and 45 Auto	**240**
Combination, 44 Magnum and 44-40	**250**
22 LR	**195**
Combination, 22 LR and 22 WMR	**210**

Hawes Sauer Silver City Marshal
Same as Western Marshal except has nickel-plated frame, brass grip frame, blued cylinder and bbl., pearlite grips. Imported from 1973-81.

44 Magnum	**$240**
Combination, 357 Magnum and 9mm Para. 45 Colt and 45 Auto	**225**
Combination, 44 Magnum and 44-40	**260**

Hawes Sauer Texas Marshal
Same as Western Marshal, except nickel-plated, has pearlite grips. Imported from 1973-81.

357 Magnum or 45 Colt	**$225**
44 Magnum	**240**
Combination, 357 Magnum and 9mm Para., 45 Colt and 45 Auto	**260**
Combination, 44 Magnum and 44-40	**275**
22 Long Rifle	**195**
Combination, 22 LR and 22 WMR	**230**

**Hawes
Sauer Western Marshal**

Hawes Sauer Western Marshal SA Revolver
Calibers: 22 LR (discont.), 357 Magnum, 44 Magnum, 45 Auto. Also in two-cylinder combinations: 22 WMR (discont.), 9mm Para., 44-40, 45 Auto. 6-shot cylinder. Bbl. lengths: 5.5-inch (discont.), 6-inch. 11.75 inches overall (6-inch bbl.). Weight: 46 oz. Fixed sights. Blued finish. Originally furnished w/simulated stag plastic grips. Recent production has smooth rosewood grips. Made from 1968 by J. P. Sauer & Sohn, Eckernforde, Germany. Imported from 1973-81.

357 Magnum or 45 Colt	**$200**
44 Magnum	**230**
Combination, 357 Magnum and 9mm Para., 45 Colt and 45 Auto	**225**
Combination, 44 Magnum and 44-40	**240**
22 Long Rifle	**195**
Combination, 22 LR and 22 WMR	**210**

HECKLER & KOCH
Oberndorf/Neckar, West Germany, and Chantilly, Virginia

Heckler & Koch Model HK4 DA Auto Pistol
Calibers: 380 Automatic (9mm Short), 22 LR, 25 Automatic (6.35mm), 32 Automatic (7.65mm) w/conversion kits. 7-shot magazine (380 Auto), 8-shot in other calibers. 3.4-inch bbl. 6.19 inches overall. Weight: 18 oz. Fixed sights. Blued finish. Plastic grip. Discontinued 1984.

22 LR, 25 ACP, 32 ACP or 380 Automatic	**$325**
380 Automatic w/22 conversion unit	**425**
380 Automatic w/22, 25, 32 conversion units	**595**

**Heckler & Koch
Model HK4**

**Heckler & Koch
Model P7K3**

**Heckler & Koch
Model P7M8**

**Heckler & Koch
Model P7M13**

Heckler & Koch Model Mark 23 DA Auto Pistol . . . $1525
Short-recoil semiautomatic pistol w/polymer frame and steel slide. Caliber: .45 ACP. 10-shot magazine. 5.87-inch bbl. 9.65 inches overall. Weight: 43 oz. Seven interchangeable rear sight adjustment units w/3-Dot system. Developed primarily in response to specifications by the Special Operations Command (SOCOM) for a Special Operations Forces Offensive Handgun Weapon System. Imported 1996 to date.

Heckler & Koch Model P7K3 DA Auto Pistol
Caliber: 380 ACP. 8-round magazine. 3.8 inch-bbl. 6.3 inches overall. Weight: about 26 oz. Adj. rear sight. Imported 1988-94.
P7K3 in 380 Cal. **$725**
22 LR Conversion Kit. **495**

Heckler & Koch Model P7M8 . . . . $750
Squeeze-cock SA semiautomatic pistol. Caliber: 9mm Para. 8-shot magazine. 4.13-inch bbl. 6.73 inches overall. Weight: 29.9 oz. Matte black or nickel finish. Adj. rear sight. Imported 1985 to date.

Heckler & Koch Model P7M10
Caliber: 40 S&W. 9-shot magazine. 4.2-inch bbl. 6.9 inches overall. Weight: 43 oz. Fixed front sight blade, adj. rear w/3-dot system. Imported 1992-94.
Blued Finish **$795**
Nickel Finish. **850**

Heckler & Koch Model P7M13 . . . . $825
Caliber: 9mm. 13-shot magazine. 4.13-inch bbl. 6.65 inches overall. Weight: 34.42 oz. Matte black finish. Adj. rear sight. Imported 1985-94.

Heckler & Koch Model P7(PSP) Auto Pistol $695
Caliber: 9mm Para. 8-shot magazine. DA. 4.13-inch bbl. 6.54 inches overall. Weight: about 33.5 oz. Blued finish. Imported 1983-85 and again in 1990 with limited availability.

Heckler & Koch Model P9S DA Automatic Pistol
Calibers: 9mm Para., 45 Automatic. 9-shot (9mm) or 7-shot (45 Auto) magazine. 4-inch bbl. 7.63 inches overall. Weight: 32 ounces. Fixed sights. Blued finish. Contoured plastic grips. Discontinued 1986.
9mm **$595**
45 Automatic. **625**

Heckler & Koch Model P9S 9mm Target Competition Kit
Same as Model P9S 9mm Target, except comes w/extra 5.5-inch bbl. and bbl. weight. Also available w/walnut competition grip.
W/Standard grip **$725**
W/Competition grip **950**

**Heckler & Koch
Model P7 (PSP)**

**Heckler & Koch
Model P9S DA**

**Heckler & Koch
Model P9S Target**

**Heckler & Koch
Model vp'70Z**

**Heckler & Koch
Model USP Compact**

**Heckler & Koch
Model USP Stainless**

**Heckler & Koch
Model USPTactical**

Heckler & Koch Model SP89. . . . . **$2195**
Semiautomatic, recoil-operated, delayed roller-locked bolt system.
Caliber: 9mm Para. 15-shot magazine. 4.5-inch bbl. 13 inches overall. Weight: 68 oz. Hooded front sight, adj. rotary-aperture rear. Imported 1989-93.

Heckler & Koch Model USP Auto Pistol
Polymer integral grip/frame design w/recoil reduction system.
Calibers: 9mm Para., 40 S&W or 45 ACP. 15-shot (9mm) or 13-
shot (40 S&W and 45ACP) magazine. 4.13- or 4.25-inch bbl.
6.88 to 7.87 inches overall. Weight: 26.5-30.4 oz. Blade front
sight, adj. rear w/3-dot system. Matte black or stainless finish.
Stippled black polymer grip. Available in SA/DA or DAO. Im-
ported 1993 to date.
Matte Black Finish **$465**
Stainless 475
W/Tritium Sights, **add** 95

Heckler & Koch Model USP Compact
Similar to USP Standard Model, except w/3.58- or 3.8-inch bbl. and
shorter slide. Weight: 25.5-28 oz. Imported 1997 to date.
Matte Black Finish **$475**
Stainless 480

Heckler & Koch Model USP Tactical **$485**
SOCOM Enhanced version of the USP Standard Model, w/4.92-inch
threaded bbl. chambered 45 ACP only. Imported 1998 to date.

Heckler & Koch Model VP'70Z Auto Pistol **$335**
Caliber: 9mm Para. 18-shot magazine. DA. 4.5-inch bbl. 8 inches
overall. Weight: 32.5 oz. Fixed sights. Blued slide, plastic receiver
and grip. Discontinued 1986.

HELWAN PISTOLS

See listings under Interarms.

HERITAGE MANUFACTURING
Opa Locka, Florida

Heritage Model HA25 Auto Pistol

Caliber: 25 ACP. 6-shot magazine. 2.5-inch bbl. 4.63 inches overall. Weight: 12 oz. Fixed sights. Blued or chrome finish. Made 1993 to date.

Blued **$85**
Chrome **90**

**Heritage
Rough Rider**

Heritage Rough Rider SA Revolver

Calibers: 22 LR, 22 Mag. 6-shot cylinder. Bbl. lengths: 2.75, 3.75, 4.75, 6.5 or 9 inches. Weight: 31-38 oz. Blade front sight, fixed rear. High-polished blued finish w/gold accents. Smooth walnut grips. Made 1993 to date.

22 Long Rifle . . . . **$ 75**
22 LR/22 WRF Combo **95**

Heritage Sentry

Heritage Sentry DA Revolver

Calibers: 22 LR, 22 Mag., 32 Mag., 9mm or 38 Special. 6- or 8-shot (rimfire) cylinder. 2- or 4-inch bbl. 6.25 inches overall (2-inch bbl.). Ramp front sight, fixed rear. Blued or nickel finish. Checkered polymer grips. Made from 1993-97.

Blued **$85**
Nickel **95**

Heritage Stealth DA Auto Pistol **$195**

Calibers: 9mm, 40 S&W. 10-shot magazine. 3.9-inch bbl. Weight: 20.2 oz. Gas-delayed blow back, double action only. Ambidextrous trigger safety. Blade front sight, drift-adj. rear. Black chrome or stainless slide. Black polymer grip frame. Made from 1996 to date.

HI-POINT FIREARMS
Mansfield, Ohio

**Hi-Point
Model JS-9mm**

Hi-Point Model JS-9mm Auto Pistol **$90**

Caliber: 9mm Para. 8-shot magazine. 4.5-inch bbl. 7.75 inches overall. Weight: 39 oz. Fixed low-profile sights w/3-dot system. Matte blue, matte black or chrome finish. Checkered synthetic grips. Made from 1990-98.

Hi-Point Model JS-9mm Comp Pistol (Stallard) **$110**

Similar to standard JS-9, except w/4-inch compensated bbl. w/ shortened slide and adj. sights. 10-shot magazine. 7.25 inches overall. Weight: 30 oz. Made from 1998 to date.

Hi-Point Model JS-9mm/C-9mm
Compact Pistol (Beemiller) **$85**

Similar to standard JS-9, except w/3.5-inch bbl. and shortened slide w/alloy or polymer frame. 6.72 inches overall. Weight: 29 oz. or 32 oz. 3-dot style sights. Made from 1993 to date.

Hi-Point Model CF-380 Polymer **$75**

Caliber: 380 ACP, 8-shot magazine. 3.5-inch bbl. 6.72 inches overall. Weight: 32 oz. 3-dot sights. Made from 1994 to date.

Hi-Point Model JS-40/JC-40
Auto Pistol (Iberia) **$110**

Similar to Model JS-9mm, except in caliber 40 S&W.

Hi-Point Model JS-45/JH-45
Auto Pistol (Haskell) **$115**

Similar to Model JS-9mm, except in caliber 45 ACP w/7-shot magazine and two-tone Polymer finish.

J. C. HIGGINS HANDGUNS

See Sears, Roebuck & Company

HIGH STANDARD SPORTING FIREARMS
East Hartford, Connecticut
Formerly High Standard Mfg. Co., Hamden, Connecticut

A long-standing producer of sporting arms, High Standard discontinued its operations in 1984. See new High Standard models under separate entry, HIGH STANDARD MFG. CO., INC.

NOTE

For ease in finding a particular firearm, High Standard handguns are grouped into three sections: Automatic Pistols (below), Derringers, and Revolvers which follow. For a complete listing, please refer to the index.

AUTOMATIC PISTOLS

High Standard Model A Automatic Pistol **$565**
Hammerless. Caliber: 22 LR. 10-shot magazine. Bbl. lengths: 4.5-, 6.75-inch. 11.5 inches overall (6.75-inch bbl.). Weight: 36 oz. (6.75-inch bbl.). Adj. target sights. Blued finish. Checkered walnut grips. Made 1938-42.

High Standard Model B Automatic Pistol **$495**
Original Standard pistol. Hammerless. Caliber: 22 LR. 10-shot magazine. Bbl. lengths: 4.5-, 6.75-inch. 10.75 inches overall (6.75-inch bbl.). Weight: 33 oz. (6.75-inch bbl.). Fixed sights. Blued finish. Hard rubber grips. Made 1932-42.

High Standard Model C Automatic Pistol **$635**
Same as Model B, except in 22 Short. Made 1935-42.

High Standard Model D Automatic Pistol **$650**
Same general specifications as Model A, but heavier bbl. Weight: 40 oz. (6.75-inch bbl.). Made 1937-42.

High Standard Dura-Matic Automatic Pistol **$285**
Takedown. Caliber: 22 LR. 10-shot magazine. 4.5 or 6.5 inch interchangeable bbl. 10.88 inches overall (6.5-inch bbl.). Weight: 35 oz. (6.5-inch bbl.). Fixed sights. Blued finish. Checkered grips. Made 1952-70.

High Standard Model E Automatic Pistol **$895**
Same general specifications as Model A, but w/extra heavy bbl. and thumbrest grips. Weight: 42 oz. (6.75inch bbl.). Made 1937-42.

High Standard Field-King Automatic Pistol
Same general specifications as Sport-King but w/heavier bbl. and target sights. Late model 6.75-inch bbls. have recoil stabilizer feature. Weight: 43 oz. (6.75 inch bbl.). Made 1951-58.
W/One Bbl. **$495**
W/Both Bbls. **575**

High Standard Flite-King Automatic Pistol — First Model
Same general specifications as Sport-King, except in caliber 22 Short w/aluminum alloy frame and slide. Weighs 26 oz. (6.5-inch bbl.). Made 1953-58.
W/One Bbl. **$395**
W/Both Bbls. **475**

High Standard Flite-King Automatic Pistol — Second Model **$495**
Same as Sport-King—Second Model, except caliber 22 Short and weighs 2 oz. lighter. Made 1958-66.

High Standard
Model A

High Standard
Model B

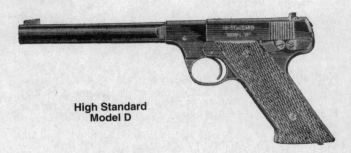

High Standard
Model D

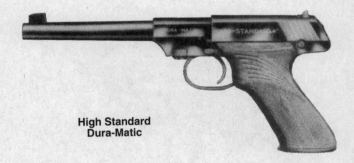

High Standard
Dura-Matic

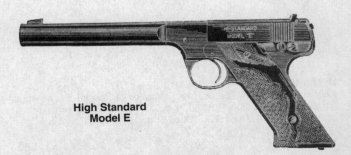

High Standard
Model E

**High Standard
G-380**

**High Standard
Model G-B**

**High Standard
Model G--E**

**High Standard
Model H-A**

**High Standard
Model H-B**

High Standard Model G-380 Automatic Pistol $535
Lever takedown. Visible hammer. Thumb safety. Caliber: 380 Automatic. 6-shot magazine. 5-inch bbl. Weight: 40 oz. Fixed sights. Blued finish. Checkered plastic grips. Made 1943-50.

High Standard Model G-B Automatic Pistol
Lever takedown. Hammerless. Interchangeable bbls. Caliber: 22 LR. 10-shot magazine. Bbl. lengths: 4.5, 6.75 inches. 10.75 inches overall (6.75-inch bbl.). Weight: 36 oz. (6.75-inch bbl.). Fixed sights. Blued finish. Checkered plastic grips. Made 1948-51.
W/One Bbl. **$465**
W/Both Bbls.. . . . . **535**

High Standard Model G-D Automatic Pistol
Lever takedown. Hammerless. Interchangeable bbls. Caliber: 22 LR. 10-shot magazine. Bbl. lengths: 4.5, 6.75 inches. 11.5 inches overall (6.75-inch bbl.). Weight: 41 oz. (6.75-inch bbl.). Target sights. Blued finish. Checkered walnut grips. Made 1948-51.
W/One Bbl. **$645**
W/Both Bbls.. . . . . **750**

High Standard Model G-E Automatic Pistol
Same general specifications as Model G-D, but w/extra heavy bbl. and thumbrest grips. Weight: 44 oz. (6.75inch bbl.). Made 1949-51.
W/One Bbl. **$885**
W/Both Bbls.. . . . . **995**

High Standard Model H-A Automatic Pistol **$765**
Same as Model A, but w/visible hammer, no thumb safety. Made 1939-42.

High Standard Model H-B Automatic Pistol **$545**
Same as Model B, but w/visible hammer, no thumb safety. Made 1940-42.

High Standard Model H-D Automatic Pistol **$1325**
Same as Model D, but w/visible hammer, no thumb safety. Made 1939-42.

High Standard Model H-DM Automatic Pistol. **$450**
Also called H-D Military. Same as Model H-D, but w/thumb safety. Made 1941-51.

High Standard Model H-E Automatic Pistol **$1475**
Same as Model E, but w/visible hammer, no thumb safety. Made 1939-42.

High Standard Olympic Automatic — First Model
Same general specifications as Model G-E, but in 22 Short w/light alloy slide. Made 1950-51.
W/one bbl. **$1050**
W/both bbls. **1225**

High Standard Olympic Automatic — Second Model
Same general specifications as Supermatic, but in 22 Short w/light alloy slide. Weight: 39 oz. (6.75-inch bbl.). Made 1951-58.
W/one bbl.. . . . . **$725**
W/both bbls. **875**

High Standard Olympic Automatic Pistol — Third Model **$695**
Same as Supermatic Trophy w/bull bbl., except in caliber 22 Short. Made 1963-66.

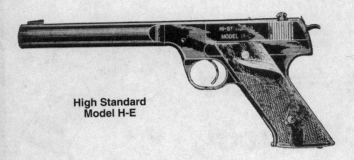

**High Standard
Model H-E**

**High Standard
Olympic Automatic — First Model**

**High Standard
Olympic Automatic — Second Model**

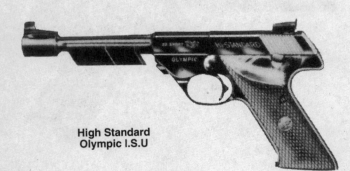

**High Standard
Olympic I.S.U**

High Standard Olympic Commemorative
Limited edition of Supermatic Trophy Military issued to commemorate the only American-made rimfire target pistol ever to win an Olympic Gold Medal. Highly engraved w/Olympic rings inlaid in gold. Deluxe presentation case. Two versions issued: in 1972 (22 LR) and 1980 (22 Short). *Note:* Value shown is for pistol in new, unfired condition.
1972 Issue. **$3595**
1980 Issue. **1295**

High Standard Olympic I.S.U $695
Same as Supermatic Citation, except caliber 22 Short. 6.75- or 8--inch tapered bbl. w/stabilizer, detachable weights. Made from 1958-77. 8-inch bbl. discontinued in 1966.

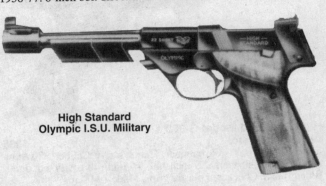

**High Standard
Olympic I.S.U. Military**

High Standard Olympic I.S.U. Military $725
Same as Olympic I.S.U., except has military grip and bracket rear sight. Intro. in 1965. Discontinued.

**High Standard
Olympic Military**

High Standard Olympic Military $625
Same as Olympic — Third Model, except has military grip and bracket rear sight. Made in 1965.

High Standard Plinker $275
Similar to Dura-Matic w/same general specifications. Made 1971-73.

High Standard Sharpshooter Automatic Pistol $375
Takedown. Hammerless. Caliber: 22 LR. 10-shot magazine. 5.5-inch bull bbl. 9 inches overall. Weight: 42 oz. Micrometer rear sight, blade front sight. Blued finish. Plastic grips. Made 1971-83.

High Standard Sport-King Automatic — First Model
Takedown. Hammerless. Interchangeable bbls. Caliber: 22 LR. 10-shot magazine. Bbl. lengths: 4.5-, 6.75-inch. 11.5 inches overall (6.75-inch bbl.). Weight: 39 oz. (6.75-inch bbl.). Fixed sights. Blued finish. Checkered plastic thumbrest grips. Made 1951-58. *Note:* 1951-54 production has lever takedown as in "G" series. Later version (illustrated) has push-button takedown.
W/One Bbl. **$345**
W/Both Bbls. **425**

High Standard Sport-King Automatic Pistol
Second Model $295
Caliber: 22 LR. 10-shot magazine. 4.5- or 6.75 inch interchangeable bbl.. 11.25 inches overall (6.75-inch bbl.). Weight: 42 oz. (6.75-inch bbl.). Fixed sights. Blued finish. Checkered grips. Made 1958-70.

High Standard
Plinker

High Standard
Sport-King Automatic — First Model

High Standard — Sharpshooter

High Standard
Sport-King Automatic — Second Model

High Standard Sport-King Automatic Pistol
Third Model **$265**
Similar to Sport-King — Second Model, w/same general specifications. Blued or nickel finish. Intro. in 1974. Discontinued.

High Standard Sport-King Lightweight
Same as standard Sport-King, except has forged aluminum alloy frame, weighs 30 oz. w/6.75-inch bbl. Made 1954-65.
W/One Bbl. **$350**
W/Both Bbls. **425**

High Standard Supermatic Automatic Pistol
Takedown. Hammerless. Interchangeable bbls. Caliber: 22 LR. 10-shot magazine. Bbl. lengths: 4.5-, 6.75-inch. Late model 6.75-inch bbls. have recoil stabilizer feature. Weight: 43 oz. (6.75-inch bbl.). 11.5 inches overall (6.75-inch bbl.). Target sights. Elevated serrated rib between sights. Adj. bbl. weights add 2 or 3 oz. Blued finish. Checkered plastic thumbrest grips. Made 1951-58.
W/One Bbl. **$550**
W/Both Bbls. **675**

High Standard
Sport-King Automatic — Third Model

High Standard Supermatic Citation
Same as Supermatic Tournament, except 6.75-, 8- or 10-inch tapered bbl. w/stabilizer and two removable weights. Also furnished w/Tournament's 5.5-inch bull bbl., adj. trigger pull, recoil-proof click-adj. rear sight (bbl.-mounted on 8- and 10-inch bbls.), checkered walnut thumbrest grips on bull bbl. model. Currently mfd. w/only bull bbl. Made 1958-66.
W/Bull Bbl. **$575**
W/Tapered Bbl. **695**

High Standard Supermatic Citation Military
Same as Supermatic Citation, except has military grip and bracket rear sight as in Supermatic Trophy. Made from 1965-73.
W/Bull Bbl. **$535**
W/Fluted Bbl. **575**

High Standard
Supermatic

High Standard Supermatic Tournament **$550**
Takedown. Caliber: 22 LR. 10-shot magazine. Interchangeable 5.5-inch bull or 6.75-inch heavy tapered bbl., notched and drilled for stabilizer and weights. 10 inches overall (5.5-inch bbl.). Weight: 44 oz. (5.5-inch bbl.). Click adj. rear sight, undercut ramp front. Blued finish. Checkered grips. Made 1958-66.

**High Standard
Supermatic Citation — Bull Barrel**

**High Standard
Supermatic Citation Military — Fluted Barrel**

**High Standard
Supermatic Tournament — Bull Barrel**

**High Standard
Supermatic Tournament Military
Tapered Barrel**

**High Standard
Supermatic Trophy — Bull Barrel**

High Standard Supermatic Tournament
Military **$450**
Same as Supermatic Tournament, except has military grip. Made 1965-71.

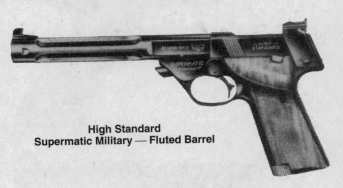

**High Standard
Supermatic Military — Fluted Barrel**

High Standard Supermatic Trophy
Same as Supermatic Citation except 5.5-inch bull bbl. or 7.25-inch fluted bbl. w/detachable stabilizer and weights, extra magazine. High-luster blued finish, checkered walnut thumbrest grips. Made 1963-66.
W/Bull Bbl. **$650**
W/Fluted Bbl. **695**

High Standard Supermatic Trophy Military
Same as Supermatic Trophy, except has military grip and bracket rear sight. Made 1965-84.
W/Bull Bbl. **$625**
W/Fluted Bbl. **695**

**High Standard
Victor — Solid Rib Barrel**

High Standard Victor Automatic Pistol **$550**
Takedown. Caliber: 22 LR. 10-shot magazine. 4.5-inch solid or vent rib and 5.5-inch vent rib, interchangeable bbl. 9.75 inches overall (5.5-inch bbl.). Weight: 52 oz. (5.5-inch bbl.). Rib-mounted target sights. Blued finish. Checkered walnut thumbrest grips. Standard or military grip configuration. Made from 1972-84 (standard-grip model, 1974-75).

> **NOTE**
>
> High Standard Automatic Pistols can be found in the preceding section, while Revolvers immediately follow this Derringer listing.

DERRINGER

**High Standard
Derringer**

High Standard Derringer

Hammerless, double action, 2-shot, double bbl. (over/under). Calibers: 22 Short, Long, LR or 22 Magnum Rimfire. 3.5-inch bbls. 5 inches overall. Weight: 11 oz. Standard model has blued or nickel finish w/plastic grips. Presentation model is gold-plated in walnut case. Standard model made from 1963 (22 S-L-LR) and 1964 (22 MRF) to 1984. Gold model, 1965-83.

Gold Presentation, one derringer	**$375**
Gold Presentation, matched pair, consecutive numbers	**695**
Standard Model	**195**

> **NOTE**
>
> Only High Standard Revolvers can be found in this section. For Automatic Pistols and Derringers see the preceding sections. For a complete listing of High Standard handguns, please refer to the Index.

REVOLVERS

High Standard Camp Gun......................**$200**
Same as Sentinel Mark I/Mark IV, except has 6-inch bbl., adj. rear sight, target-style checkered walnut grips. Caliber: 22 LR or 22 WMR. Made 1976-83.

High Standard Double-Nine DA Revolve —Aluminum Frame...........................**$215**
Western-style version of Sentinel. Blued or nickel finish w/simulated ivory, ebony or stag grips. 5.5-inch bbl. 11 inches overall. Weight: 27.25 oz. Made 1959-71.

High Standard Double-Nine Deluxe..............**$225**
Same as Double-Nine — Steel Frame, except has adj. target rear sight. Intro. in 1971. Discontinued.

High Standard Double-Nine—Steel Frame........**$250**
Similar to Double-Nine—Aluminum Frame, w/same general specifications, except has extra cylinder for 22 WMR, walnut grips. Intro. in 1971. Discontinued.

High Standard Durango......................**$145**
Similar to Double-Nine—Steel Frame, except 22 LR only, available w/4.5- or 5.5-inch bbl. Made 1971-73.

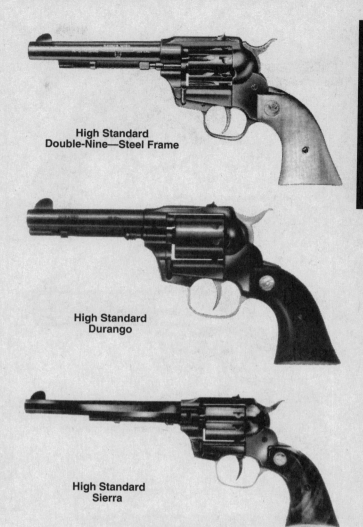

**High Standard
Double-Nine—Steel Frame**

**High Standard
Durango**

**High Standard
Sierra**

High Standard High Sierra DA Revolver
Similar to Double-Nine—Steel Frame, except has 7-inch octagon bbl. w/gold-plated grip frame, fixed or adj. sights. Made 1973-1983.
W/fixed sights	**$190**
W/adj. sights	**250**

High Standard Hombre........................**$195**
Similar to Double-Nine—Steel Frame, except 22 LR only, lacks single-action type ejector rod and tube, has 4.5-inch bbl. Made 1971-73.

High Standard Kit Gun DA Revolver.............**$175**
Solid frame, swing-out cylinder. Caliber: 22 LR. 9-shot cylinder. 4-inch bbl. 9 inches overall. Weight: 19 oz. Adj. rear sight, ramp front. Blued finish. Checkered walnut grips. Made 1970-73.

High Standard Longhorn — Aluminum Frame
Similar to Double-Nine—Aluminum Frame, except has Longhorn hammer spur. 4.5-, 5.5- or 9.5-inch bbl. Walnut, simulated pearl or simulated stag grips. Blued finish. Made 1960-1971.
W/4.5- or 5.5-inch bbl.	**$160**
W/9.5-inch bbl.	**225**

**High Standard
Kit Gun**

**High Standard
Longhorn — Aluminum Frame**

**High Standard
Longhorn — Steel Frame**

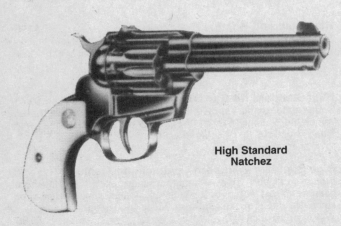

**High Standard
Natchez**

High Standard Longhorn — Steel Frame
Similar to Double-Nine — Steel Frame, except has 9.5-inch bbl. w/fixed or adj. sights. Made 1971-1983
W/Fixed Sights **$180**
W/Adj. Sights **295**

High Standard Natchez **$130**
Similar to Double-Nine — Aluminum Frame, except 4.5-inch bbl. (10 inches overall, weighs 25.25 oz.), blued finish, simulated ivory bird'shead grips. Made 1961-66.

High Standard Posse **$135**
Similar to Double-Nine — Aluminum Frame, except 3.5-inch bbl. (9 inches overall, weighs 23.25 oz.), blued finish, brass-grip frame and trigger guard, walnut grips. Made 1961-66.

High Standard Sentinel DA Revolver **$295**
Solid frame, swing-out cylinder. Caliber: 22 LR. 9-shot cylinder. 3-, 4- or 6-inch bbl. 9 inches overall (4inch-bbl.). Weight: 19 oz. (4-inch bbl.). Fixed sights. Aluminum frame. Blued or nickel finish. Checkered grips. Made 1955-56.

**High Standard
Posse**

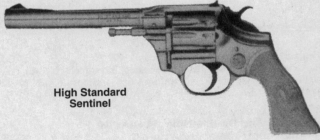

**High Standard
Sentinel**

High Standard Sentinel Deluxe **$245**
Same as Sentinel, except w/4- or 6-inch bbl., wide trigger, drift-adj. rear sight, two-piece square-butt grips. Made 1957-74. *Note:* Designated Sentinel after 1971.

High Standard Sentinel Imperial **$220**
Same as Sentinel, except has onyx-black or nickel finish two-piece checkered walnut grips, ramp front sight. Made 1962-65.

High Standard Sentinel I DA Revolver
Steel frame. Caliber: 22 LR. 9-shot cylinder. Bbl. lengths: 2-, 3-, 4-inch. 6.88 inches overall (2-inch bbl.). Weight: 21.5 oz. (2-inch bbl.). Ramp front sight, fixed or adj. rear. Blued or nickel finish. Smooth walnut grips. Made 1974-83.
W/Fixed Sights **$240**
W/Adj. Sights **255**

**High Standard
Sentinel I**

**High Standard
Sentinel Mark II**

**High Standard
Sentinel Mark III**

**High Standard
Sentinel Snub**

High Standard Sentinel Mark II DA Revolver **$275**
Caliber: 357 Magnum. 6-shot cylinder. Bbl. lengths: 2.5-, 4-, 6-inch.
9 inches overall w/4-inch bbl. Weight: 38 oz. (4-inch bbl.). Fixed
sights. Blued finish. Walnut service or combat-style grips. Made
1974-76.

High Standard Sentinel Mark III **$295**
Same as Sentinel Mark II, except has ramp front and adj. rear sights.
Weight: 40 oz. (4-inch bbl.). Blued finish. Made 1974-76.

High Standard Sentinel Mark IV
Same as Sentinel Mark I, except in caliber 22 WMR. Made 1974-83.
W/Fixed Sights **$225**
W/Adj. Sights **245**

High Standard Sentinel Snub
Same as Sentinel Deluxe, except w/2.75-inch bbl. (7.25 inches over-
all, weighs 15 oz.), checkered bird'shead-type grips. Made 1957-74.
Blued Finish **$155**
Nickel Finish. **175**

HIGH STANDARD MFG. CO., INC.
Houston, Texas
Distributed from Hartford, Connecticut

High Standard 10X Automatic Pistol **$595**
Caliber: 22 LR. 10-shot magazine. 5.5-inch bbl. 9.5 inches overall.
Weight: 45 oz. Checkered walnut grips. Blued finish. Made from
1994 to date.

High Standard Citation MS Auto Pistol **$475**
Similar to the Supermatic Citation, except has 10-inch bbl. 14 inches
overall. Weight: 49 oz. Made 1994 to date.

High Standard Olympic I.S.U. Automatic Pistol
Same specifications as the 1958 I.S.U. issue. *See* listing under pre-
vious High Standard Section.
Olympic I.S.U. Model **$425**
Olympic I.S.U. Military Model. **335**

High Standard Sport King Auto Pistol **$260**
Caliber: 22 LR.10-shot magazine. 4.5- or 6.75-inch bbl. 8.5 or 10.75
inches overall. Weight: 44 oz. (4.5-inch bbl.), 46 oz. (6.75-inch bbl.).
Fixed sights, slide mounted. Checkered walnut grips. Parkerized finish. Made 1994 to date.

High Standard Supermatic Citation Auto Pistol
Caliber: 22 LR.10-shot magazine. 5.5- or 7.75-inch bbl. 9.5 or
11.75 inches overall. Weight: 44 oz. (5.5-inch bbl.), 46 oz. (7.75-
inch bbl.). Frame-mounted, micro-adj. rear sight, undercut ramp
front sight. Blued or Parkerized finish. Made 1994 to date.
Supermatic Citation Model. **$295**
22 Short Conversion **275**

High Standard Supermatic Tournament **$295**
Caliber: 22 LR. 10-shot magazine. Bbl. length: 4.5, 5.5, or 6.75
inches. Overall length: 8.5, 9.5 or 10.75 inches. Weight: 43, 44 or 45
oz. depending on bbl. length. Micro-adj. rear sight, undercut ramp
front sight. Checkered walnut grips. Parkerized finish. Made 1994 to
date.

High Standard Supermatic Trophy
Caliber: 22 LR. 10-shot magazine. 5.5 or 7.25-inch bbl. 9.5 or 11.25 inches overall. Weight: 44 oz. (5.5-inch bbl.). Micro-adj. rear sight, undercut ramp front sight. Checkered walnut grips w/thumbrest. Blued or Parkerized finish. Made 1994 to date.
Supermatic Trophy **$345**
22 Short Conversion **275**

High Standard Victor Automatic
Caliber: 22 LR. 10-shot magazine. 4.5- or 5.5-inch ribbed bbl. 8.5 or 9.5 inches overall. Weight: 45 oz. (4.5-inch bbl.), 46 oz. (5.5-inch bbl.). Micro-adj. rear sight, post front. Checkered walnut grips. Blued or Parkerized finish. Made 1994 to date.
Victor Model. **$355**
22 Short Conversion **275**

HOPKINS & ALLEN ARMS CO.
Norwich, Connecticut

Hopkins & Allen Revolvers
See listings of comparable Harrington & Richardson and Iver Johnson models for values.

INGRAM
Mfd. by Military Armament Corp.

See listings under M.A.C. (Military Armament Corp.)
Note: Military Armament Corp. cease production of the select fire automatic, M10 (9mm & 45 ACP) and M11 (380 ACP) in 1977. Commercial production resumed on semiautomatic versions under the M.A.C. banner until 1982.

INTERARMS
Alexandria, Virginia

See also Bersa Pistol.

Interarms/Helwan Brigadier Auto Pistol **$165**
Caliber: 9mm Para. 8-shot magazine. 4.25-inch bbl. 8 inches overall. Weight: 32 oz. Blade front sight, dovetailed rear. Blued finish. Grooved plastic grips. Imported 1987-95.

Interarms Virginian Dragoon SA Revolver
Calibers: 357 Magnum, 44 Magnum, 45 Colt. 6-shot cylinder. Lbs.: 5- (not available in 44 Magnum), 6-, 7.5-, 8.38-inch (latter only in 44 Magnum w/adj. sights). 11.88 inches overall (6-inch bbl.). Weight: 48 oz. (6-inch bbl.). Fixed sights or micrometer rear and ramp front sights. Blued finish w/color-casetreated frame. Smooth walnut grips. SWIS-SAFE base pin safety system. Manufactured by Interarms Industries Inc., Midland, VA. 1977-84.
Standard Dragoon. **$235**
Engraved Dragoon **495**
Deputy Model **225**
Stainless **235**

Interarms Virginian Revolver Silhouette Model **$325**
Same general specifications as regular model except designed in stainless steel w/untapered bull bbl., lengths of 7.5, 8.38 and 10.5 inches. Made 1985-86.

Interarms Virginian SA Revolver **$395**
Similar to Colt Single-Action Army, except has base pin safety system. Imported 1973-76. See also listing under Hämmerli (manufacturer).

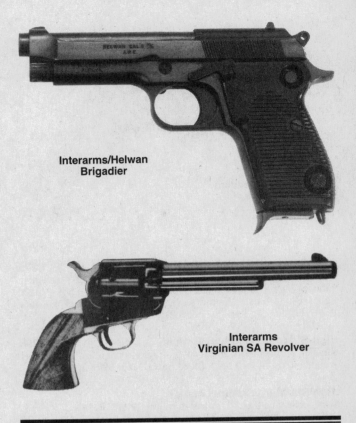

Interarms/Helwan Brigadier

Interarms Virginian SA Revolver

INTRATEC U.S.A. INC.
Miami, Florida

Intratec Category 9 DAO Semiautomatic **$155**
Blowback action w/polymer frame. Caliber: 9mm Para. 8-shot magazine. 3-inch bbl. 7.7 inches overall. Weight: 18 oz. Textured black polymer grips. Matte black finish. Made 1993 to date.

Intratec Category 40 DAO Semiautomatic **$165**
Locking-breech action w/polymer frame. Caliber: 40 S&W. 7-shot magazine. 3.25-inch bbl. 8 inches overall. Weight: 21 oz. Textured black polymer grips. Matte black finish. Made 1994 to date.

Intratec Category 45 DAO Semiautomatic **$175**
Locking-breech action w/polymer frame. Caliber: 45 ACP. 6-shot magazine. 3.25-inch bbl. 8 inches overall. Weight: 21 oz. Textured black polymer grips. Matte black finish. Made 1994 to date.

Intratec Model ProTec 22 DA Semiautomatic
Caliber: 25 ACP. 10-shot magazine. 2.5-inch bbl. 5 inches overall. Weight: 14 oz. Wraparound composition grips. Black Teflon, satin grey or Tec-Kote finish. Advertised in 1992.
ProTec 22 Standard. . . . . **SRP $112**
ProTec 22 w/Satin or Tec-Kote. . . . . **SRP 119**

Intratec Model ProTec 25 DA Semiautomatic
Caliber: 25 ACP. 8-shot magazine. 2.5-inch bbl. 5 inches overall. Weight: 14 oz. Fixed sights. Wraparound composition grips. Black Teflon, satin grey or Tec-Kote finish. Made 1991 to date. *Note:* Formerly Model Tec-25.
ProTec 25 Standard. **$65**
ProTec 25 w/Satin or Tec-Kote **70**

Intratec Model Tec-9 Semiautomatic
Caliber: 9mm Luge. 20- or 36-round magazine. 5-inch bbl. Weight: 50-51 oz. Open fixed front sight, adj. rear. Military nonglare blued or stainless finish.
Tec-9 w/Blued Finish $185
Tec-9 w/Electroless Nickel Finish 195
Tec 9S w/Stainless Finish 250

Intratec Model Tec-9M Semiautomatic
Same specifications as Model Tec-9, except has 3-inch bbl. without shroud and 20-round magazine. Blued or stainless finish.
Tec-9M w/Blued Finish $175
Tec-9MS w/Stainless Finish 235

Intratec Model Tec-22T Semiautomatic
Caliber: 22 LR. 10/22-type 30-shot magazine. 4-inch bbl. 11.19 inches overall. Weight: 30 oz. Protected post front sight, adj. rear sight. Matte black or Tec-Kote finish. Made 1989 to date.
Tec-22T Standard $125
Tec-22TK Tec-Kote 155

Intratec Model Tec Double Derringer $90
Calibers: 22 WRF, 32 H&R Mag., 38 Special. 2-shot capacity. 3-inch bbl. 4.63 inches overall. Weight: 13 oz. Fixed sights. Matte black finish. Made 1986-88.

ISRAEL ARMS
Kfar Sabs, Israel

Imported by Israel Arms International, Houston TX

Israel Arms Bul-M5 Locked Breech SA $780
Similar to the M1911 U.S. Government Model. Caliber: .45 ACP. 10-shot magazine. 5-inch bbl. 8.5 inches overall. Weight: 36 oz. Blade front and fixed low-profile rear sights.

Israel Arms Kareen MK II (1500) Auto Pistol
Single-action only. Caliber: 9mm Para. 10-shot magazine. 4.75-inch bbl. 8.0 inches overall. Weight: 33.6 oz. Blade front sight, rear adjustable for windage. Textured black composition or rubberized grips. Blued, two-tone, matte black finish. Imported 1996 to date.
Blued or Matte Black Finish $275
Two-Tone Finish 425
Meprolite Sights add 40

Israel Arms Kareen MK II Compact Auto Pistol $325
Similar to standard Kareen MKII, except w/3.85-inch bbl. 7.1 inches overall. Weight: 32 oz. Imported 1997 to date.

Israel Arms Golan Model (2500) Auto Pistol
Single or double action. Caliber: 9mm Para., S&W 40. 10-shot magazine. 3.85-inch bbl. 7.1 inches overall. Weight: 34 oz. Steel slide and alloy frame w/ambidextrous safety and decocking lever. Matte black finish. Imported 1997 to date.
9mm Para $375
40 S&W 395

Israel Arms GAL Model (5000) Auto Pistol $275
Caliber: 45 ACP. 8-shot magazine. 4.25-inch bbl. 7.25 inches overall. Weight: 42 oz. Low profile 3-dot sights. Combat-style black rubber grips. Imported 1997 to date.

JAPANESE MILITARY PISTOLS
Tokyo, Japan
Manufactured by Government Plant

**Japanese
Model 14 (1925)**

Japanese Model 14 (1925) Automatic Pistol $495
Modification of the Nambu Model 1914, changes chiefly intended to simplify mass production. Standard round trigger guard or oversized guard for use w/gloves. Caliber: 8mm Nambu. 8-shot magazine. 4.75-inch bbl. 9 inches overall. Weight: about 29 oz. Fixed sights. Blued finish. Grooved wood grips. Intro. 1925 and mfd. through WW II.

**Japanese
Model 26 DAO Revolver**

Japanese Model 26 DAO Revolver $355
Top-break frame. Caliber: 9mm, 6-shot cylinder w/automatic extractor/ejector. 4.7-inch bbl. Adopted by the Japanese Army from 1893 to 1914, when replaced by the Model 14 Automatic Pistol, but remained in service through World War II.

**Japanese
Model 94 (1934)**

Japanese Model 94 (1934) Automatic Pistol **$225**
Poorly design and constructed, this pistol can be fired merely by applying pressure on the sear, which is exposed on the left side. Caliber: 8mm Nambu. 6-shot magazine. 3.13-inch bbl. 7.13 inches overall. Weight: about 27 oz. Fixed sights. Blued finish. Hard rubber or wood grips. Intro. in 1934, principally for export to Latin American countries, production continued thru WW II.

Japanese Nambu Model 1914 Automatic Pistol **$465**
Original Japanese service pistol, resembles Luger in appearance and Glisenti in operation. Caliber: 8mm Nambu. 7-shot magazine. 4.5-inch bbl. 9 inches overall. Weight: about 30 oz. Fixed front sight, adj. rear sight. Blued finish. Checkered wood grips. Made 1914-1925.

JENNINGS FIREARMS INC.
Currently Mfd. by Bryco Arms, Irvine, California
Previously by Calwestco, Inc. & B.L. Jennings

See additional listings under Bryco Arms.

**Jennings
Model J Auto Pistol**

Jennings Model J Auto Pistol
Calibers: 22 LR, 25 ACP. 6-shot magazine. 2.5-inch bbl. About 5 inches overall. Weight: 13 oz. Fixed sights. Chrome, satin nickel or black Teflon finish. Walnut, grooved black Cycolac or resin-impregnated wood grips. Made from 1981-85 under Jennings and Calwestco logos; from 1985 to date by Bryco Arms.
Model J-22 . . . . **$75**
Model J-25 . . . . **60**

IVER JOHNSON'S ARMS, INC.
Jacksonville, Arkansas

Operation of this company dates back to 1871, when Iver Johnson and Martin Bye partnered to manufacture metallic cartridge revolvers. Johnson became the sole owner and changed the name to Iver Johnson's Arms & Cycle Works which it was known as for almost 100 years. Modern management shortened the name, and after several owner changes, the firm was moved from Massachusetts, its original base, to Jacksonville, Arkansas. In 1987, the American Military Arms Corporation (AMAC) acquired the operation, which subsequently ceased in 1993.

> **NOTE**
>
> For ease in finding a particular firearm, Iver Johnson handguns are divided into two sections: Automatic Pistols (below) and Revolvers, which follows. For the complete handgun listing, please refer to the Index.

AUTOMATIC PISTOLS

Iver Johnson 9mm DA Automatic **$295**
Caliber: 9mm. 6-round magazine. 3-inch bbl. 6.5 inches overall. Weight: 26 oz. Blade front sight, adj. rear. Smooth hardwood grip. Blued or matte blued finish. Intro. in 1986.

Iver Johnson Compact 25 ACP **$150**
Bernardelli V/P design. Caliber: 25 ACP. 5-shot magazine. 2.13-inch bbl. 4.13 inches overall. Weight: 9.3 oz. Fixed sights. Checkered composition grips. Blued slide, matte blued frame and color-casehardened trigger. Made 1991 to 93.

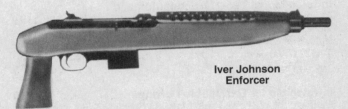

**Iver Johnson
Enforcer**

Iver Johnson Enforcer **$325**
Caliber: 30 US Carbine. 5-, 15-, or 30-round magazine. Semiautomatic. 9.5- inch bbl. Weight: 5.5 lbs. Adj. sights. Walnut stock. Made mid-1980s to 93.

Iver Johnson PP30 Super Enforcer Automatic **$375**
Caliber: 30 US Carbine. 15- or 30-shot magazine. 9.5-inch bbl. 17 inches overall. Weight: 4 pounds. Adj. peep rear sight, blade front. American walnut stock. Made 1984-86.

Iver Johnson Pony Automatic Pistol **$250**
Caliber: 380 Auto. 6-shot magazine. 3.1-inch bbl. 6.1 inches overall. Blued, matte or nickel finish. Weight: 20 oz. Wooden grips. Smallest of the locked breech automatics. Made 1982-88. Reintroduced 1990-93.

**Iver Johnson
Model TP**

Iver Johnson Model TP-22 DA Automatic **$125**
Calibers: 22 LR, 7-shot magazine. 2.85-inch bbl. 5.39 inches overall. Blued finish. Weight: 14.46 oz. Made 1982-93.

Iver Johnson Model TP25 DA Pocket Pistol **$140**
Double-action automatic. Caliber: 25 ACP. 7-shot magazine. 3-inch bbl. 5.5 inches overall. Weight: 12 oz. Black plastic grips and blued finish. Made 1982-93.

Iver Johnson Trailsman Automatic Pistol
Caliber: 22 LR. 10-shot magazine. 4.5 or 6-inch bbl. 8.75 inches overall (4.5-inch bbl.). Weight: 46 oz. Fixed target-type sights. Checkered composition grips. Made 1984-90.
Standard Model $165
Deluxe Model **195**

NOTE

Only Iver Johnson Revolvers can be found in the section below. For Pistols, please see the preceding pages. For a complete listing, please refere to the Index.

REVOLVERS

Iver Johnson Model 55 Target DA Revolver $115
Solid frame. Caliber: 22 LR. 8-shot cylinder. Bbl. lengths: 4.5-, 6-inch. 10.75 inches overall (6-inch bbl.). Weight: 30.5 oz. (6-inch bbl.). Fixed sights. Blued finish. Walnut grips. *Note:* Original model designation was 55 changed to 55A when loading gate was added in 1961. Made 1955-77.

Iver Johnson Model 55-S Revolver $125
Same general specifications as the Model 55 except for 2.5-inch bbl. and small molded pocket-size grip.

Iver Johnson Model 56 Blank Revolver $65
Solid frame. Caliber: 22 blanks only. 8-shot cylinder. 2.5-inch solid bbl., 6.75 inches overall. Weight: 10 oz.

Iver Johnson Model 57A Target DA Revolver $135
Solid frame. Caliber: 22 LR. 8-shot cylinder. Bbl. lengths: 4.5-, 6-inch. 10.75 bbl.). Weight: 30.5 oz., 6-inch bbl. Adj. sights. Blued finish. Walnut grips. *Note:* Original model designation was 57, changed to 57A when loading gate was added in 1961. Made 1956-75.

Iver Johnson Model 66 Trailsman DA Revolver $95
Hinged frame. Rebounding hammer. Caliber: 22 LR. 8-shot cylinder. 6-inch bbl. 11 inches overall. Weight: 34 oz. Adj. sights. Blued finish. Walnut grips. Made 1958-75.

Iver Johnson Model 67 Viking DA Revolver $110
Hinged frame. Caliber: 22 LR. 8-shot cylinder. Bbl. lengths: 4.5- and 6-inch. 11 inches overall (6-inch bbl.). Weight: 34 oz. (6-inch bbl.). Adj. sights. Walnut grips w/thumbrest. Made 1964-75.

Iver Johnson Model 67S Viking Snub Revolver $125
DA. Hinged frame. Calibers: 22 LR, 32 S&W Short and Long, 38 S&W. 8-shot cylinder in 22, 5-shot in 32 and 38 calibers. 2.75-inch bbl. Weight: 25 oz. Adj. sights. Tenite grips. Made 1964-75.

Iver Johnson Model 1900 DA Revolver $110
Solid frame. Calibers: 22 LR, 32 S&W, 32 S&W Long, 38 S&W. 7-shot cylinder (22 cal.), 6-shot (32 S&W), 5-shot (32 S&W Long, 38 S&W). Bbl. lengths: 2.5-, 4.5- and 6-inch. Weight: 12 oz. (32 S&W w/2.5-inch bbl.). Fixed sights. Blued or nickel finish. Hard rubber grips. Made 1900-47.

Iver Johnson Model 1900 Target DA Revolver $150
Solid frame. Caliber: 22 LR. 7-shot cylinder. Bbl. lengths: 6- and 9.5-inch. Fixed sights. Blued finish. Checkered walnut grips. This earlier model does not have counterbored chambers as in the Target Sealed 8. Made 1925-42.

**Iver Johnson
Model 55 — Target**

**Iver Johnson
Model 55-S**

**Iver Johnson
Model 56 — Blank Revolver**

**Iver Johnson
Model 57A — Target**

Iver Johnson American Bulldog DA Revolver
Solid frame. Calibers: 22 LR, 22 WMR, 38 Special. 6-shot cylinder in 22, 5-shot in 38. Bbl. lengths: 2.5-, 4-inch. 9 inches overall (4-inch bbl.). Weight: 30 oz. (4-inch bbl.). Adj. sights. Blued or nickel finish. Plastic grips. Made 1974-76.
38 Special $145
Other calibers **125**

Iver Johnson
Model 66 — Trailsman

Iver Johnson
Model 67 — Viking

Iver Johnson
Model 67S — Viking Snub

Iver Johnson
Model 1900 — Target

Iver Johnson
Cadet

Iver Johnson Armsworth Model 855 SA **$135**
Hinged frame. Caliber: 22 LR. 8-shot cylinder. 6-inch bbl. 10.75 inches overall. Weight: 30 oz. Adj. sights. Blued finish. Checkered walnut one-piece grip. Adj. finger rest. Made 1955-57.

Iver Johnson Cadet DA Revolver. **$135**
Solid frame. Calibers: 22 LR, 22 WMR, 32 S&W Long, 38 S&W, 38 Special. 6- or 8-shot cylinder in 22, 5-shot in other calibers. 2.5-inch bbl. 7 inches overall. Weight: 22 oz. Fixed sights. Blued finish or nickel finish. Plastic grips. *Note:* Loading gate added in 1961, 22 cylinder capacity changed from 8 to 6 rounds in 1975. Made 1955-77.

Iver Johnson Cattleman SA Revolver
Patterned after the Colt Army SA Revolver. Calibers: 357 Magnum, 44 Magnum, 45 Colt. 6-shot cylinder. Bbl. lengths: 4.75-, 5.5- (not available in 44), 6- (44 only), 7.25-inch. Weight: about 41 oz. Fixed sights. Blued bbl. and cylinder color-casehardened frame, brass grip frame. One-piece walnut grip. Made by Aldo Uberti, Brescia, Italy, 1973-78.
44 Magnum. **$260**
Other calibers **195**

Iver Johnson Cattleman Buckhorn SA Revolver
Same as standard Cattleman except has adj. rear and ramp front sights. 4.75- (44 only), 5.75- (not available in 44), 6- (44 only), 7.5- or 12-inch bbl. Weight: almost 44 oz. Made 1973-78.
357 Magnum or 45 Colt w/12-inch bbl. **$300**
357 Magnum or 45 Colt w/5.75- or 7.5-inch bbl. **330**
44 Magnum , w/12-inch bbl. **300**
44 Magnum, other bbls. **330**

Iver Johnson Cattleman Buntline SA Revolver
Same as Cattleman Buckhorn, except has 18-inch bbl., walnut shoulder stock w/brass fittings. Weight: about 56 oz. Made 1973-78.
44 Magnum **$385**
Other calibers **345**

Iver Johnson Cattleman Trail Blazer **$170**
Similar to Cattleman Buckhorn, except 22 caliber has interchangeable 22 LR and 22 WMR cylinders, 5.5- or 6.5-inch bbl. Weight: about 40 oz. Made 1973-78.

Iver Johnson Champion 22 Target SA **$185**
Hinged frame. Caliber: 22 LR. 8-shot cylinder. Single action. Counterbored chambers as in Sealed 8 models. 6-inch bbl. 10.75 inches overall. Weight: 28 oz. Adj. target sights. Blued finish. Checkered walnut grips, adj. finger rest. Made 1938-48.

Iver Johnson Deluxe Target **$170**
Same as Sportsman, except has adj. sights. Made 1975-76.

Iver Johnson Protector Sealed 8 DA Revolver **$150**
Hinged frame. Caliber: 22 LR. 8-shot cylinder. 2.5-inch bbl. 7.25 inches overall. Weight: 20 oz. Fixed sights. Blued finish. Checkered walnut grips. Made 1933-49.

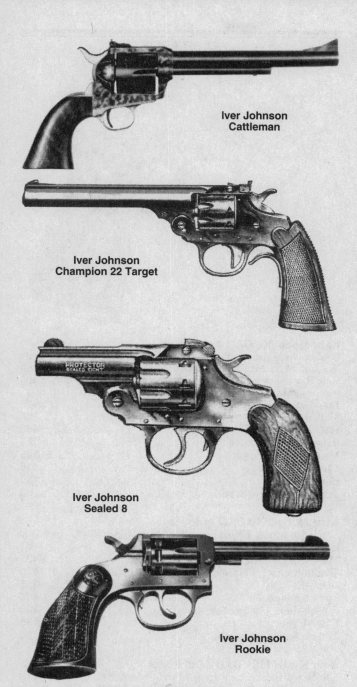

**Iver Johnson
Cattleman**

**Iver Johnson
Champion 22 Target**

**Iver Johnson
Sealed 8**

**Iver Johnson
Rookie**

Iver Johnson Rookie DA Revolver. **$95**
Solid frame. Caliber: 38 Special. 5-shot cylinder. 4-inch bbl. 9-inches overall. Weight: 30 oz. Fixed sights. Blued or nickel finish. Plastic grips. Made 1975-77.

Iver Johnson Safety Hammer DA Revolver **$125**
Hinged frame. Calibers: 22 LR, 32 S&W, 32 S&W Long, 38 S&W. 7-shot cylinder (22 cal.), 6-shot (32 S&W Long), 5-shot (32 S&W, 38 S&W). Bbl. lengths: 2, 3, 3.25, 4, 5 or 6 inches. Weight w/4-inch bbl.: 15 oz. (22, 32 S&W), 19.5 oz. (32 S&W Long) or 19 oz. (38 S&W). Fixed sights. Blued or nickel finish. Hard rubber, round butt grips or square butt, rubber or walnut grips available. *Note:* 32 S&W Long and 38 S&W models built on heavy frame. Made 1892-1950.

Iver Johnson Safety Hammerless DA Revolver **$135**
Similar to the Safety Hammer Model, except w/shrouded hammerless frame. Made 1895-1950.

Iver Johnson Sidewinder DA Revolver **$135**
Solid frame. Caliber: 22 LR. 6- or 8-shot cylinder. Bbl. lengths: 4.75-, 6-inch. 11.25 inches overall (6-inch bbl.). Weight: 31 oz. (6-inch bbl.). Fixed sights. Blued or nickel finish w/plastic "staghorn" grips or color-casehardened frame w/walnut grips. *Note:* Cylinder capacity changed from 8 to 6 rounds in 1975. Intro. 1961. Discontinued.

Iver Johnson Sidewinder "S". **$145**
Same as Sidewinder, except has interchangeable cylinders in 22 LR and 22 WMR, adj. sights. Intro. 1974. Discontinued.

Iver Johnson Sportsman DA Revolver **$95**
Solid frame. Caliber: 22 LR. 6-shot cylinder. Bbl. lengths: 4.75-, 6-inch. 10.75 inches overall (6-inch bbl.). Weight: 30.5 oz. (6-inch bbl.). Fixed sights. Blued finish. Plastic grips. Made 1974-76.

Iver Johnson Supershot 9-Shot DA Revolver. **$135**
Same as Supershot Sealed 8, except has nine non-counterbored chambers. Made 1929-49.

Iver Johnson Supershot 22 DA Revolver **$105**
Hinged frame. Caliber: 22 LR. 7-shot cylinder. 6-inch bbl. Fixed sights. Blued finish. Checkered walnut grips. This earlier model does not have counterbored chambers as in the Supershot Sealed 8. Made 1929-49.

Iver Johnson Supershot Model 844 DA. **$195**
Hinged frame. Caliber: 22 LR. 8-shot cylinder. Bbl. lengths: 4.5- or 6-inch. 9.25 inches overall (4.5-inch bbl.). Weight: 27 oz. (4.5-inch bbl.). Adj. sights. Blued finish. Checkered walnut one-piece grip. Made 1955-56.

**Iver Johnson
Safety Hammer**

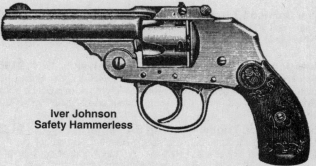

**Iver Johnson
Safety Hammerless**

Iver Johnson
Sidewinder

Iver Johnson
Supershot — Sealed 8

Iver Johnson
Target — Sealed 8

Iver Johnson Supershot Sealed 8 DA Revolver **$175**
Hinged frame. Caliber: 22 LR. 8-shot cylinder. 6-inch bbl. 10.75 inches overall. Weight: 24 oz. Adj. target sights. Blued finish. Checkered walnut grips. Postwar model does not have adj. finger rest as earlier version. Made 1931-57.

Iver Johnson Swing Out DA Revolver
Calibers: 22 LR, 22 WMR, 32 S&W Long, 38 Special. 6-shot cylinder in 22, 5-shot in 32 and 38. 2-, 3-, 4-inch plain bbl. or 4- 6-inch vent rib bbl. 8.75 inches overall (4-inch bbl.). Fixed or adj. sights. Blued or nickel finish. Walnut grips. Made in 1977.
W/Plain Barrel, Fixed Sights $135
W/Vent rib, Adj. Sights 195

Iver Johnson Target 9-Shot DA Revolver **$145**
Same as Target Sealed 8, except has nine non-counterbored chambers. Made 1929-46.

Iver Johnson Target Sealed 8 DA Revolver **$155**
Solid frame. Caliber: 22 LR. 8-shot cylinder. Bbl. lengths: 6- and 10-inch. 10.75 inches overall (6-inch bbl.). Weight: 24 oz. (6-inch bbl.). Fixed sights. Blued finish. Checkered walnut grips. Made 1931-57.

Iver Johnson Trigger-Cocking SA Target **$165**
Hinged frame. First pull on trigger cocks hammer, second pull releases hammer. Caliber: 22 LR. 8-shot cylinder, counterbored chambers. 6-inch bbl. 10.75 inches overall. Weight: 24 oz. Adj. target sights. Blued finish. Checkered walnut grips. Made 1940-47.

KAHR ARMS
Blauvelt, New York

Kahr
Model K9

Kahr Model K9 DAO Auto Pistol
Caliber: 9mm Para. 7-shot magazine. 3.5-inch bbl. 6 inches overall. Weight: 24 oz. Fixed sights. Matte black, electroless nickel, Birdsong Black-T or matte stainless finish. Wraparound textured polymer or hardwood grips. Made 1994 to date.
Duo-Tone Finish **$535**
Matte Black Finish 395
Electroless Nickel Finish 460
Black-T Finish 525
Matte Stainless Finish 425
Lady K9 Model 410
Elite Model 435
Tritium Night Sights, **add** 90

Kahr Model K40 DAO Auto Pistol
Similar to Model K9, except chambered 40 S&W w/5- or 6-shot magazine. Weight: 26 oz. Made 1997 to date.
Matte Black Finish **$405**
Electroless Nickel Finish 475
Black-T Finish 535
Matte Stainless Finish 445
Covert Model (Shorter Grip-frame) 425
Elite Model 435
Tritium Night Sights, **add** 90

Kahr Model MK9 DAO Auto Pistol
Similar to Model K9, except w/Micro-Compact frame. 6- or 7- shot magazine. 3- inch bbl. 5.5 inches overall. Weight: 22 oz. Stainless or Duo-Tone finish. Made 1998 to date.
Duo-Tone Finish **$535**
Matte Stainless Finish 435
Elite Model 445
Tritium Night Sights, **add** 90

KBI, INC
Harrisburg, Pennsylvania

KBI Model PSP-25 Auto Pistol **$195**
Caliber: 25 ACP, 6-shot magazine. 2.13-inch bbl., 4.13 inches overall. Weight: 9.5 oz. All-steel construction w/dual safety system. Made 1994 to date.

KEL-TEC CNC INDUSTRIES, INC.
Cocoa, Florida

**Kel-Tec
Model P-11**

Kel-Tec Model P-11 DAO Pistol
Caliber: 9mm Parabellum or 40 S&W. 10-shot magazine. 3.1- inch bbl. 5.6 inches overall. Weight: 14 oz. Blade front sight, drift adjustable rear. Aluminum frame w/steel slide. Checkered black, gray, or green polymer grips. Matte blue, electroless nickel, stainless steel or parkerized finish. Made 1995 to date.

9mm	**$195**
40 S&W	225
Parkerized Finish, **add**	45
Nickel Finish, **add** (Disc. 1995)	35
Stainless Finish, **add** (1996 to date)	55
Tritium Night Sights, **add**	50
40 Cal. Conversion Kit, **add**	150

KIMBER MANUFACTURING, INC.
Yonkers, New York
(Formerly Kimber of America, Inc.)

Kimber Model Classic .45
Similar to Government 1911 built on steel, polymer or alloy full-size or compact frame. Caliber: 45 ACP. 7-, 8-, 10- or 14-shot magazine. 4- or 5-inch bbl., 7.7 or 8.75 inches overall. Weight: 28 oz. (Compact LW), 34 oz. (Compact or Polymer) or 38 oz. (Custom FS). McCormick low-profile combat or Kimber adj. target sights. Blued, matte black oxide or stainless finish. Checkered custom wood or black synthetic grips. Made from 1994 to date.

Custom (matte black)	**$430**
Custom Royal (polished blue)	525
Custom Stainless (satin stainless)	495
Custom Target (matte black)	505
Target Gold Match (polished blue)	695
Target Stainless Match (polished stainless)	775
Polymer (matte black)	565
Polymer Stainless (satin stainless slide)	665
Polymer Target (matte black slide)	635
Compact (matte black)	445
Compact Stainless (satin stainless)	495
Compact LW (matte black w/alloy frame)	465

KORTH PISTOLS
Ratzeburg, Germany
***Currently imported by Keng's Firearms Specialty, Inc.
Previously by Beeman Precision Arms; Osborne's and
Mandall Shooting Supply***

**Kimber
Model Classic .45**

Korth Revolvers Combat, Sport, Target
Calibers: 357 Mag. and 22 LR w/interchangeable combo cylinders of 357 Mag./9mm Para. or 22 LR/22 WMR also 22 Jet, 32 S&W and 32 H&R Mag. Bbls: 2.5-, 3-, 4-inch (combat) and 5.25- or 6-inch (target). Weight: 33 to 42 oz. Blued, stainless, matte silver or polished silver finish. Checkered walnut grips. Imported 1967 to date.

Standard Rimfire Model	**$2295**
Standard Centerfire Model	2350
ISU Match Target Model	2995
Custom Stainless Finish, **add**	450
Matte Silver Finish, **add**	650
Polished Silver Finish, **add**	950

Korth Semiautomatic Pistol
Calibers: 30 Luger, 9mm Para., 357 SIG, 40 S&W, 9x21mm. 10- or 14-shot magazine. 4- or 5-inch bbl. All-steel construction, recoil-operated. Ramp front sight, adj. rear. Blued, stainless, matte silver or polished silver finish. Checkered walnut grips. Imported 1988 to date.

Standard Model	**$3750**
Matte Silver Finish, **add**	650
Polished Silver Finish, **add**	950

LAHTI PISTOLS
Mfd. by Husqvarna Vapenfabriks A. B. Huskvarna,
Sweden, and Valtion Kivaar Tedhas ("VKT")
Jyväskyla, Finland

**Lahti
Automatic Pistol**

Lahti Automatic Pistol
Caliber: 9mm Para. 8-shot magazine. 4.75-inch bbl. Weight: about 46 oz. Fixed sights. Blued finish. Plastic grips. Specifications given are those of the Swedish Model 40 but also apply in general to the Finnish Model L-35, which differs only slightly. A considerable number of Swedish Lahti pistols were imported and sold in the U.S., the Finnish Model, somewhat better made, is a rather rare pistol. Finnish Model L-35 adopted 1935. Swedish Model 40 adopted 1940 and mfd. through 1944.

Finnish L-35 Model	**$1695**
Swedish 40 Model	395

L.A.R. MANUFACTURING, INC.
West Jordan, Utah

**L.A.R.
Mark I Grizzly**

L.A.R. Mark I Grizzly Win. Mag. Automatic Pistol

Calibers: 357 Mag., 45 ACP, 45 Win. Mag. 7-shot magazine. 6.5-inch bbl. 10.5 inches overall. Weight: 48 oz. Fully adj. sights. Checkered rubber combat-style grips. Blued finish. Made from 1983 to date. 8- or 10-inch bbl. Made 1987 to date.

357 Mag. (6.5" barrel)	$650
45 Win. Mag.(6.5" barrel)	635
8-inch barrel	925
10-inch barrel	995

L.A.R. Mark IV Grizzly Automatic Pistol $725

Same general specifications as the L.A.R. Mark I, except chambered for 44 Magnum- has 5.5- or 6.5-inch bbl. beavertail grip safety, matte blued finish. Made 1991 to date.

L.A.R. Mark V Auto Pistol $795

Similar to the Mark I, except chambered in 50 Action Express. 6-shot magazine, 5.4- or 6.5-inch bbl. 10.6 inches overall (5.4-inch bbl.). Weight: 56 oz. Checkered walnut grips. Made 1993 to date.

LASERAIM TECHNOLOGIES INC.
Little Rock, Arkansas

Laseraim Series I SA Auto Pistol

Calibers: 40 S&W, 45 ACP, 10mm. 7- or 8- shot magazine. 3.875- or 5.5-inch dual-port compensated bbl. 8.75 or 10.5 inches overall. Weight: 46 or 52 oz. Fixed sights w/Laseraim or adjustable Millet sights. Textured black composition grips. Extended slide release, ambidextrous safety and beveled magazine well. Stainless or matte black Teflon finish. Made 1993 to date.

Series I W/Adjustable Sights	$325
Series I W/Fixed Sights	295
Series I W/Fixed Sights (HotDot)	425
Series I Dream Team (RedDot)	625
Series I Illusion (Laseraim)	675

Laseraim Series II SA Auto Pistol

Similar to Series I, except w/stainless finish and no bbl. compensator. Made 1993-96.

Series II W/Adjustable Sights	$275
Series II W/Fixed Sights	250
Series II Dream Team	495
Series II Illusion	525

Laseraim Series III SA Auto Pistol

Similar to Series II, except w/serrated slide and 5-inch compensated bbl.only. Made 1994 to date.

Series III W/Adjustable Sights	$435
Series III W/Fixed Sights	395

Laseraim Velocity Series SA Auto Pistol

Similar to Series I, except chambered 357 Sig. or 400 Cor-Bon. 3.875-inch unported bbl. (compact) or 5.5-inch dual-port compensated bbl. Made 1997 to date.

Compact Model (unported)	$275
Government Model (ported)	325
W/Wireless Laser (HotDot), **add**	150

**Laseraim
Series I**

**Laseraim® Velocity 400™
Series shown w/ Laser Sight
(Optional Accessory)**

**Laseraim® Series III
w/LA93 Illusion III ™ Scope**

LIGNOSE PISTOLS
Suhl, Germany
Aktien-Gesellschaft "Lignose" Abteilung

The following Lignose pistols were manufactured from 1920 to the mid-1930s. They were also marketed under the Bergmann name.

Lignose Model 2 Pocket Auto Pistol **$195**
Conventional Browning type. Same general specifications as Einhand Model 2A, but lacks the one-hand operation.

**Lignose Einhand
Model 2A — Pocket**

Lignose Einhand Model 2A Pocket Auto Pistol **$225**
As the name implies, this pistol is designed for one-hand operation, pressure on a "trigger" at the front of the guard retracts the slide. Caliber: 25 Automatic (6.35 mm). 6-shot magazine. 2-inch bbl. 4.75 inches overall. Weight: about 14 oz. Blued finish. Hard rubber grips.

**Lignose Einhand
Model 3A — Pocket**

Lignose Einhand Model 3A Pocket Auto Pistol **$275**
Same as the Model 2A except has longer grip, 9-shot magazine, weighs about 16 oz.

LLAMA HANDGUNS
Mfd. by Gabilondo y Cia, Vitoria, Spain
(*Imported by S.G.S., Wanamassa, New Jersey*)

> **NOTE**
>
> For ease in finding a particular Llama handgun, the listings are divided into two groupings: Automatic Pistols (below) and Revolvers, which follows. For a complete listing of Llama handguns, please refer to the Index.

AUTOMATIC PISTOLS

Llama Model IIIA Automatic Pistol **$225**
Caliber: 380 Auto. 7-shot magazine. 3.69-inch bbl. 6.5 inches overall. Weight: 23 oz. Adj. target sights. Blued finish. Plastic grips. Intro. 1951. Discontinued.

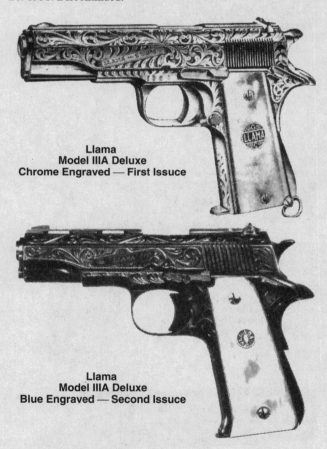

**Llama
Model IIIA Deluxe
Chrome Engraved — First Issue**

**Llama
Model IIIA Deluxe
Blue Engraved — Second Issue**

Llama Models IIIA, XA, XV Deluxe
Same as standard Model IIIA, XA and XV, except engraved w/blued or chrome finish and simulated pearl grips. Disc. 1984.
Chrome-engraved Finish **$250**
Blue-engraved Finish **225**

Llama Model VIII Automatic Pistol **$295**
Caliber: 38 Super. 9-shot magazine. 5-inch bbl. 8.5 inches overall. Weight: 40 oz. Fixed sights. Blued finish. Wood grips. Intro. in 1952. Disc.

**Llama
Model XA — First Issuce**

**Llama
Model C-XI**

**Llama
Model CE-IIIA**

**Llama
Model Compact**

Llama Models VIII, IXA, XI Deluxe
Same as standard Models VIII, IXA and XI, except finish—chrome engraved or blued engraved—and simulated pearl grips. Discontinued 1984.
Chrome-engraved Finish **$325**
Blue-engraved Finish **350**

Llama Model IXA Automatic Pistol **$275**
Same as Model VIII, except caliber 45 Automatic, 7-shot magazine.

Llama Model XA Automatic Pistol **$220**
Same as Model IIIA, except caliber 32 Automatic, 8-shot magazine.

Llama Model XI Automatic Pistol **$245**
Same as Model VIII, except caliber 9mm Para.

Llama Model XV Automatic Pistol **$220**
Same as Model XA, except caliber 22 LR.

Llama Models BE-IIIA, BE-XA, BE-XV **$250**
Same as Models IIIA, XA and XV, except w/blue-engraved finish. Made 1977-84.

Llama Models BE-VIII, BE-IXA, BE-XI Deluxe **$320**
Same as Models VIII, IXA and XI, except w/blue-engraved finish. Made 1977-84.

Llama Models C-IIIA, C-XA, C-XV **$320**
Same as Models IIIA, XA and XV, except in satin-chrome.

Llama Models C-VIII, C-IXA, C-XI **$320**
Same as Models VIII, IXA and XI, except in satin-chrome.

Llama Models CE-IIIA, CE-XA, CE-XV **$315**
Same as Models IIIA, XA and XV, except w/chrome-engraved finish. Made 1977-84.

Llama Models CE-VIII, CE-IXA, CE-XI **$395**
Same as Models VIII, IXA and XI, w/except chrome-engraved finish. Made 1977-84.

Llama Compact Frame Auto Pistol **$335**
Calibers: 9mm Para., 38 Super, 45 Auto. 7-, 8- or 9-shot. 5-inch bbl. 7.88 inches overall. Weight: 34 oz. Blued, satin-chrome or Duo-Tone finishes. Made 1990 to date. Duo-Tone discontinued 1993.

Llama Duo-Tone Large Frame Auto Pistol **$295**
Caliber: 45 ACP. 7-shot magazine. 5-inch bbl. 8.5 inches overall. Weight: 36 oz. Adj. rear sight. Blued finished w/satin chrome. Polymer black grips. Made 1990-93.

Llama Duo-Tone Small Frame Auto Pistol **$285**
Calibers: 22 LR, 32 and 380 Auto. 7- or 8-shot magazine 3.69 inch bbl. 6.5 inches overall. Weight: 23 oz. Square-notch rear sight, Patridge-type front. Blued finish w/chrome. Made 1990-93.

Llama Model G-IIIA Deluxe **$850**
Same as Model IIIA, except is gold-damascened w/simulated pearl grips. Discontinued 1982.

Llama Large-Frame Automatic Pistol (IXA)
Caliber: 45 Auto. 7-shot magazine. 5-inch bbl. Weight: 2 lbs. 8 oz. Adj. rear sight, Patridge-type front. Walnut grips or teakwood on satin chrome model. Later models w/polymer grips. Made from 1984 to date.
Blued Finish **$295**
Satin Chrome Finish **325**

Llama M-82 DA Automatic Pistol $575
Caliber: 9mm Para. 15-shot magazine. 4.25-inch bbl. 8 inches overall. Weight: 39 oz. Drift-adj. rear sight. Matte blued finish. Matte black polymer grips. Made from 1988-93.

Llama M-87 Comp Pistol. $875
Caliber: 9mm Para. 15-shot magazine. 5.5-inch bbl. 9.5 inches overall. Weight: 40 oz. Low-profile combat sights. Satin nickel finish. Matte black grip panels. Built-in ported compensator to minimize recoil and muzzle rise. Made 1989-93.

Llama MICRO-MAX SA Automatic Pistol
Caliber: 380 ACP. 7-shot magazine. 3.125-inch bbl. Weight: 23 oz. Blade front sight, drift adjustable rear w/3-Dot system. Matte blue or satin chrome finish. Checkered polymer grips. Imported 1997 to date.
Matte Blue Finish $185
Satin Chrome Finish 215

Llama MINI-MAX SA Automatic Pistol
Calibers: 9mm, 40 S&W or 45 ACP. 6- or 8-shot magazine. 3.5-inch bbl. 8.3 inches overall. Weight: 35 oz. Blade front sight, drift adjustable rear w/3-Dot system. Matte blue, Duo-Tone or satin chrome finish. Checkered polymer grips. Imported 1996 to date.
Duo-Tone Finish $225
Matte Blue Finish 195
Satin Chrome Finish 235
Stainless (disc.). 250

Llama MINI-MAX II SA Automatic Pistol
Caliber: 45 ACP only. 10- shot magazine 3.625-inch bbl. 7.375 inches overall. Weight: 37 oz. Blade front sight, drift adjustable rear w/3-Dot system. Shortened barrel and grip. Matte and Satin Chrome Finish. Imported 1998.
Matte Blue Finish $300
Satin Chrome Finish 325

Llama MAX-I SA Automatic Pistol
Calibers: 9mm or 45 ACP. 7- or 9-shot magazine. 4.25- to 5.125-inch bbl. Weight: 34 or 36 oz. Blade front sight, drift adjustable rear w/3-Dot system. Matte blue, Duo-Tone or satin chrome finish. Checkered black rubber grips. Imported 1995 to date.
Duo-Tone Finish $230
Matte Blue Finish 215
Satin Chrome Finish 250

Llama MAX-II SA Automatic Pistol
Same as the MAX-I Compact (4.25 bbl.), except w/10-shot magazine. Weight: 39 oz. Matte blue or satin chrome finish. Imported 1996 to date.
Matte Blue Finish $215
Satin Chrome Finish 245

Llama Omni 9mm Double-Action Automatic $350
Same general specifications as 45 Omni, except chambered for 9mm w/13-shot magazine. Made 1983-86.

Llama Omni 45 Double-Action Automatic Pistol . . . $375
Caliber: 45 Auto. 7-shot magazine. 4.25-inch bbl. 7.75 inches overall. Weight: 40 oz. Adj. rear sight, ramp front. Highly polished deep blued finish. Made 1984-86.

Llama Single-Action Automatic Pistol $345
Calibers: 38 Super, 9mm, 45 Auto. 9-shot magazine (7-shot for 45 Auto). 5-inch bbl. 8.5 inches overall. Weight: 2 lhs. 8 oz. Intro. in 1981.

Llama
Duo-Tone Large Frame

Llama
M-82 DA Auto

Llama
MINI-MAX II

Llama MAX-I

Llama Small-Frame Automatic Pistol

Calibers: 380 Auto (7-shot magazine), 22 RF (8 shot magazine). 3.69-inch bbl. Weight: 23 oz. Patridge-blade front sight, adj. rear. Blued or satin-chrome finish.

Blued Finish	**$265**
Satin-Chrome Finish	**325**

**Llama
Martial Deluxe — Gold Damascened**

> **NOTE**
>
> This section contains only Llama Revolvers. Pistols may be found in the preceding section. For a complete listing of Llama handguns, please refer to the Index.

REVOLVERS

Llama Comanche I

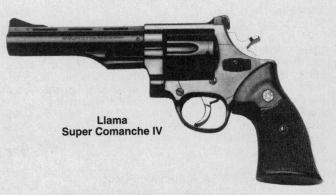

**Llama
Super Comanche IV**

**Llama
Comanche III Chrome**

Llama Comanche III Double-Action Revolver $250

Caliber: 357 Magnum. 6-shot cylinder. 4-inch bbl. 9.25 inches overall. Weight: 36 oz. Adj. rear sight, ramp front. Blued finish. Checkered walnut grips. Made from 1975-95. *Note:* Prior to 1977, this model was designated "Comanche."

Llama Comanche III Chrome $295

Same gen. specifications as Comanche III, except has satin chrome finish, 4- or 6-inch bbl. Made 1979-92.

Llama Martial Double-Action Revolver $215

Calibers: 22 LR, 38 Special. 6-shot cylinder. Bbl. lengths: 4-inch (38 Special only), 6-inch. 11.25 inches overall w/6-inch bbl. Weight: about 36 oz. w/6-inch bbl. Target sights. Blued finish. Checkered walnut grips. Made 1969-76.

**Llama
Martial Double-Action Revolver**

Llama Martial Double-Action Deluxe

Same as standard Martial, except w/satin chrome, chrome engraved, blued engraved or gold damascened finish. Simulated pearl grips. Made 1969-78.

Satin-chrome Finish	**$ 275**
Chrome-engraved Finish	**325**
Blue-engraved Finish	**300**
Gold-damascened Finish	**995**

Llama Super Comanche IV DA Revolver $295

Caliber: 44 Magnum. 6-shot cylinder. 6-inch bbl. 11.75 inches overall. Weight: 50 oz. Adj. rear sight, ramp front. Polished deep blued finish. Checkered walnut grips. Made from 1980-93.

Llama Comanche I Double-Action Revolver $195

Same general specifications as Martial 22. Made 1977-83.

Llama Comanche II $215

Same general specifications as Martial 38. Made 1977-83.

Llama Super Comanche V DA Revolver $275

Caliber: 357 Mag. 6-shot cylinder. 4-, 6- or 8.5-inch bbl. Weight: 48 ozs. Ramped front blade sight, click-adj. Rear. Made from 1980-89.

LORCIN Engineering Co., Inc.
Mira Loma, California

Lorcin Model L-22 Semiautomatic Pistol $65
Caliber: 22LR. 9-shot magazine. 2.5-inch bbl. 5.25 inches overall. Weight: 16 oz. Blade front sight, fixed notch rear w/3-dot system. Black Teflon or chrome finish. Black, pink or pearl composition grips. Made 1990 to date.

Lorcin Model L-25, LT-25 Semiautomatic Pistol
Caliber: 25ACP. 7-shot magazine. 2.4-inch bbl. 4.8 inches overall. Weight: 12 oz. (LT-25) or 14.5 oz. (L-25). Blade front sight, fixed rear. Black Teflon or chrome finish. Black, pink or pearl composition grips. Made 1989 to date.
Model L-25 . . . . **$50**
Model LT-25 . . . . **60**
Model Lady Lorcin **65**

Lorcin Model L-32 Semiautomatic Pistol $65
Caliber: 32ACP. 7-shot magazine. 3.5-inch bbl. 6.6 inches overall. Weight: 27 oz. Blade front sight, fixed notch rear. Black Teflon or chrome finish. Black composition grips. Made 1992 to date.

Lorcin Model L-380 Semiautomatic Pistol
Caliber: 380ACP. 7- or 10-shot magazine. 3.5-inch bbl. 6.6 inches overall. Weight: 23 oz. Blade front sight, fixed notch rear. Matte Black finish. Grooved black composition grips. Made 1994 to date.
Model L9MM (7-shot) **$70**
Model L9MM (10-shot) **95**

Lorcin Model L9MM Semiautomatic Pistol
Caliber: 9mm Parabellum. 10- or 13-shot* magazine. 4.5-inch bbl. 7.5 inches overall. Weight: 31 oz. Blade front sight, fixed notch rearw/3-dot system. Black Teflon or chrome finish. Black composition grips. Made 1992 to date.
Model L-380 (10-shot) **$110**
Model L-380 (13-shot) **135**

Lorcin Over/Under Derringer $65
Caliber: 38 Special/357 Mag., 45LC. 2-shot Derringer. 3.5-inch bbls. 6.5 inches overall. Weight: 12 oz. Blade front sight, fixed notch rear. Stainless finish. Black composition grips. Made 1996 to date.

LUGER PISTOLS

Mfd. by Deutsche Waffen und Munitionsfabriken (DWM), Berlin, Germany. Previously by Koniglich Gewehrfabrik Erfurt, Heinrich Krieghoff Waffenfabrik, Mauser-Werke, Simson & Co., Vickers Ltd., Waffenfabrik, Bern.

Luger 1900 American Eagle $2400
Caliber: 7.65 mm. 8-shot magazine. Thin, 4.75-inch long, tapered bbl. 9.5 inches overall. Weight: 32 oz. Fixed rear sight, dovetailed front sight. Grip safety. Checkered walnut grips. Early-style toggle, narrow trigger, wide guard, no stock lug. American Eagle over chamber. Estimated 8000 production.

Luger 1900 Commercial $2625
Same specifications as Luger 1900 American Eagle, except DWM on early-style toggle, no chamber markings. Estimated 8000 production.

Luger 1900 Swiss $2950
Same specifications as Luger 1900 American Eagle, except Swiss cross in sunburst over chamber. Estimated 9000 production.

**Luger
1900 American Eagle**

Luger 1902 American Eagle $6350
Caliber: 9mm Para. 8-shot magazine. 4-inch, heavy tapered bbl. 8.75 inches overall. Weight: 30 oz. Fixed rear sight, dovetailed front sight. Grip safety. Checkered walnut grips. American Eagle over chamber, DWM on early-style toggle, narrow trigger, wide guard, no stock lug. Estimated 700 production.

Luger 1902 Carbine $9500
Caliber: 7.65mm. 8-shot magazine. 11.75-inch tapered bbl. 16.5 inches overall. Weight: 46 oz. Adj. 4-position rear sight, long ramp front sight. Grip safety. Checkered walnut grips and forearm. DWM on early-style toggle, narrow trigger, wide guard, no chamber markings, stock lug. Estimated 3200 production.

Luger 1902 Cartridge Counter $16,250
Caliber: 9mm Para. 8-shot magazine. Heavy, tapered 4-inch bbl. 8.75 inches overall. Weight: 30 oz. Fixed rear sight, dovetailed front sight. Grip safety. Checkered walnut grips. DWM on dished toggle w/lock, American Eagle over chamber when marked. No stock lug. Estimated production unknown.

Luger 1902 Commercial $5850
Same basic specifications as Luger 1902 Cartridge Counter, except DWM on early-style toggle, narrow trigger, wide guard, no chamber markings, no stock lug. Estimated 400 production.

Luger 1903 American Eagle $7500
Same basic specifications as Luger 1902 Cartridge Counter, except American Eagle over chamber, DWM on early-style toggle, narrow trigger, wide guard, no stock lug. Estimated 700 production.

Luger 1904 GL "Baby" $150,000
Caliber: 9mm Para. 7-shot magazine. 3.25-inch bbl. 7.75 inches overall. Weight: approx. 20 oz. Serial number 10077B. "GL" marked on rear of toggle. Georg Luger's personal sidearm. Only one made in 1904.

Luger 1904 Naval (Reworked) $9750
Caliber: 9mm Para. 8-shot magazine. Bbl. length altered to 4 inches. 8.75 inches overall. Weight: 30 oz. Adj. two-position rear sight, dovetailed front sight. Thumb lever safety. Checkered walnut grips. Heavy tapered bbl. DWM on new-style toggle w/lock, 1902 over chamber. W/or without grip safety and stock lug. Estimated 800 production.

Luger 1906 (11.35) $90,000
Caliber: 45 ACP. 6-shot magazine. 5-inch bbl. 9.75 inches overall. Weight: 36 oz. Fixed rear sight, dovetailed front sight. Grip safety. Checkered walnut grips. GL monogram on rear toggle link, larger frame w/altered trigger guard and trigger, no proofs, no markings over chamber. No stock lug. Estimated production is only 2. *Note:* this version of the Luger pistol is the most valuable next to the "GL" Baby Luger.

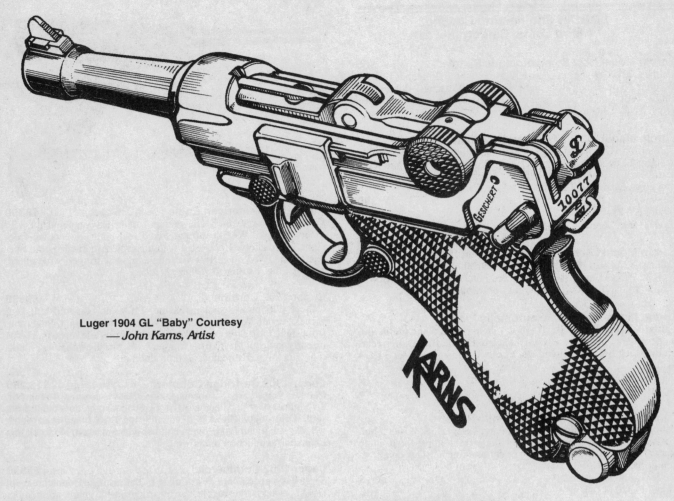

Luger 1904 GL "Baby" Courtesy
— John Karns, Artist

Luger 1906 American Eagle (7.65) **$1650**
Caliber: 7.65mm.8-shot magazine. Thin, 4.75-inch, tapered bbl. 9.5 inches overall. Weight: 32 oz. Fixed rear sight dovetailed front sight. Grip safety. Checkered walnut grips. DWM on new-style toggle, American Eagle over chamber. No stock lug. Estimated 8000 production.

Luger 1906 American Eagle (9mm) **$1950**
Same basic specifications as the 7.65mm 1906, except in 9mm Para. w/4-inch barrel, 8.75 inches overall and weight of 30 ounces. Estimated 3500 production.

Luger 1906 Bern (7.65mm) **$1850**
Same basic specifications as the 7.65mm 1906 American Eagle, except checkered walnut grips w/.38-inch borders, Swiss Cross on new-style toggle, Swiss proofs, no markings over chamber, no stock lug. Estimated 17,874 production.

Luger 1906 Brazilian (7.65mm) **$1850**
Same general specifications as the 7.65mm 1906 American Eagle, except w/Brazilian proofs, no markings over chamber, no stock lug. Estimated 4500 production.

Luger 1906 Brazilian (9mm) **$1950**
Same basic specifications as the 9mm 1906 American Eagle, except w/Brazilian proofs, no markings over chamber, no stock lug. Estimated production unknown, but less than 4000 is estimated by collectors.

Luger 1906 Commercial
Calibers: 7.65mm, 9mm. Same specifications as the 1906 American Eagle versions, above, except no chamber markings and no stock lug. Estimated production: 6000 (7.65mm) and 3500 (9mm).
7.65mm **$1425**
9mm **1950**

Luger 1906 Dutch **$1450**
Caliber: 9mm Para. Same specifications as the 9mm 1906 American Eagle, except tapered bbl. w/proofs, no markings over chamber, no stock lug. Estimated 3000 production.

Luger 1906 Loewe and Company **$3550**
Caliber: 7.65mm. 8-shot magazine. 6-inch tapered bbl. 10.75 inches overall. Weight: 35 oz. Adj. two-position rear sight, dovetailed front sight. Grip safety. Checkered walnut grips. Loewe & Company over chamber, Naval proofs, DWM on new-style toggle, no stock lug. Estimated production unknown.

Luger 1906 Naval **$2650**
Caliber: 9mm Para. 8-shot magazine. 6-inch tapered bbl. 10.75 inches overall. Weight: 35 oz. Adj. two-position rear sight, dovetailed front sight. Grip safety and thumb safety w/lower marking (1st issue), higher marking (2nd issue). Checkered walnut grips. No chamber markings, DWM on new-style toggle w/o lock, but w/stock lug. Est. production: 8000 (1st issue); 12,000 (2nd issue).

Luger 1906 Naval Commercial **$2850**
Same as the 1906 Naval, except lower marking on thumb safety, but no chamber markings. DWM on new-style toggle, w/stock lug and commercial proofs. Estimated 3000 production.

Luger 1906 Portuguese Army **$1250**
Same specifications as the 7.65mm 1906 American Eagle except w/Portuguese proofs, crown and crest over chamber. No stock lug. Estimated 3500 production.

Luger 1906 Portuguese Naval **$6500**
Same as the 9mm 1906 American Eagle except w/Portuguese proofs, crown and anchor over chamber, but no stock lug.

Luger 1906 Russian **$9750**
Same general specifications as the 9mm 1906 American Eagle, except thumb safety has markings concealed in up position, DWM on new-style toggle, DWM bbl. proofs, crossed rifles over chamber. Estimated production unknown.

Luger 1906 Swiss **$2200**
Same general specifications as the 7.65mm 1906 American Eagle Luger, except Swiss Cross in sunburst over chamber, no stock lug. Estimated 10,300 production.

Luger 1906 Swiss (Rework) **$3000**
Same basic specifications as the 7.65mm 1906 Swiss, except in bbl. lengths of 3.63, 4 and 4.75 inches, overall length 8.38 inches (4-inch bbl.). Weight 32 oz. (4-inch bbl.). DWM on new-style toggle, bbl. w/serial number and proof marks, Swiss Cross in sunburst or shield over chamber, no stock lug. Estimated production unknown.

Luger 1906 Swiss Police **$2400**
Same general specifications as the 7.65mm 1906 Swiss except DWM on new-style toggle, Swiss Cross in matted field over chamber, no stock lug. Estimated 10,300 production.

Luger 1908 Bulgarian **$2150**
Caliber: 9mm Para. 8-shot magazine. 4-inch tapered bbl. 8.75 inches overall. Weight: 30 oz. Fixed rear sight dovetailed front sight. Thumb safety w/lower marking concealed. Checkered walnut grips. DWM chamber marking, no proofs, crown over shield on new-style toggle lanyard loop, no stock lug. Estimated production unknown.

Luger 1908 Commercial **$995**
Same basic specifications as the 1908 Bulgarian, except higher marking on thumb safety. No chamber markings, commercial proofs, DWM on new-style toggle, but no stock lug. Estimated 4000 production.

Luger 1908 Erfurt Military **$995**
Caliber: 9mm Para. 8-shot magazine. 4-inch tapered bbl. 8.75 inches overall. Weight: 30 oz. Fixed rear sight dovetailed front sight. Thumb safety w/higher marking concealed. Checkered walnut grips. Serial number and proof marks on barrel, crown and Erfurt on new-style toggle, dated chamber, but no stock lug. Estimated production unknown.

Luger 1908 Military
Same general specifications as the 9mm 1908 Erfurt Military Luger, except *1st and 2nd Issue* have thumb safety w/higher marking concealed, serial number on bbl. no chamber markings, proofs on frame, DWM on new style toggle, but no stock lug. Estimated production: 10,000 (lst issue) and 5000 (2nd issue). *3rd Issue* has serial number and proof marks on barrel, dates over chamber, DWM on new-style toggle, but no stock lug. Estimated 3000 production.
1st Issue **$1350**
2nd Issue 1500
3rd Issue 995

Luger 1908 Naval **$2950**
Same basic specifications as the 9mm 1908 Military Lugers, except w/6-inch bbl, adj. two-position rear sight, no chamber markings, DWM on new-style toggle, w/stock lug. Estimated 26,000 production.

Luger 1908 Naval (Commercial) **$3250**
Same specifications as the 1908 Naval Luger, except no chamber markings or date. Commercial proofs, DWM on new-style toggle, w/stock lug. Estimated 1900 produced.

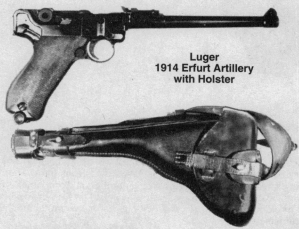

Luger
**1914 Erfurt Artillery
with Holster**

Luger 1914 Erfurt Artillery **$1850**
Caliber: 9mm Para. 8-shot magazine. 8-inch tapered bbl. 12.75 inches overall. Weight: 40 oz. Artillery rear sight, Dovetailed front sight. Thumb safety w/higher marking concealed. Checkered walnut grips. Serial number and proof marks on barrel, crown and Erfurt on new-style toggle, dated chamber, w/stock lug. Estimated production unknown.

Luger 1914 Erfurt Military **$695**
Same specifications as the 1914 Erfurt Artillery, except w/4-inch bbl. and corresponding length, weight, etc., and fixed rear sight. Estimated 3000 production.

Luger 1914 Naval **$2395**
Same specifications as 9mm 1914 Lugers, except has 6-inch bbl. w/corresponding length and weight, and adj. two-position rear sight. Dated chamber, DWM on new-style toggle, w/stock lug. Estimated 40,000 produced.

Luger 1914-1918 DWM Artillery **$1295**
Caliber: 9mm Para. 8-shot magazine. 8-inch tapered bbl. 12.75 inches overall. Weight: 40 oz. Artillery rear sight, dovetailed front sight. Thumb safety w/higher marking concealed. Checkered walnut grips. Serial number and proof marks on barrel, DWM on new-style toggle, dated chamber, w/stock lug. Estimated 3000 production.

Luger 1914-1918 DWM Military **$795**
Same specifications as the 9mm 1914-1918 DWM Artillery, except w/4-inch tapered bbl. and corresponding length, weight, etc., and fixed rear sight. Estimated production unknown.

Luger 1920 Carbine
Caliber: 7.65mm. 8-shot magazine. 11.75-inch tapered bbl. 15.75 inches overall. Weight: 44 oz. Four-position rear sight, long ramp front sight. Grip (or thumb) safety. Checkered walnut grips and forearm. Serial numbers and proof marks on barrel, no chamber markings, various proofs, DWM on new-style toggle, w/stock lug. Estimated production unknown.
Carbine W/Forearm **$7500**
Carbine Less Forearm 4950

**Luger
1923 Stoeger**

Luger 1920 Commercial $595
Calibers: 7.65mm, 9mm Para. 8-shot magazine. 3.63-, 3.75-, 4-, 4.75-, 6-, 8-, 10-, 12-, 16-, 18- or 20-inch tapered bbl. Overall length: 8.375 to 24.75 inches. Weight: 30 oz. (3.63-inch bbl.). Varying rear sight configurations, dovetailed front sight. Thumb safety. Checkered walnut grips. Serial numbers and proof marks on barrel, no chamber markings, various proofs, DWM or crown over Erfurt on new-style toggle, w/stock lug. Estimated production not documented.

Luger 1920 DWM and Erfurt Military $695
Cariber: 9mm Para. 8-shot magazine. 4-inch tapered barrel. 8.75 inches overall. Weight: 30 oz. Fixed rear sight dovetailed front sight. Thumb safety. Checkered walnut grips. Serial numbers and proof marks on barrel, dated chamber, various proofs, DWM or crown over Erfurt on new-style toggle, w/stock lug. Esimated 3000 production.

Luger 1920 Police $900
Same specifications as 9mm 1920 DWM w/some dated chambers, various proofs, DWM or crown over Erfurt on new-style toggle, identifying marks on grip frame, w/stock lug. Estimated 3000 production.

Luger 1923 Commerical $725
Calibers: 7.65mm and 9mm Para. 8-shot magazine. 3.63-, 3.75-, 4-, 6-, 8-, 12- or 16-inch tapered bbl. Overall length: 8.38 inches (3.63-inch bbl.). Weight: 30 oz. (3.63-inch bbl.). Various rear sight configurations, dovetailed front sight. Thumb lever safety. Checkered walnut grips. DWM on new-style toggle, serial number and proofs on barrel, no chamber markings, w/stock lug. Estimated 15,000 production.

Luger 1923 Dutch Commerical. $2350
Same basic specifications as 1923 Commercial Luger, w/same caliber offerings, but only 3.63- or 4-inch bbl. Fixed rear sight, thumb lever safety w/arrow markings. Estimated production unknown.

Luger 1923 Krieghoff Commercial. $1950
Same specifications as 1923 Commercial Luger, w/same caliber offerings but bbl. lengths of 3.63, 4, 6, and 8 inches. "K" marked on new-style toggle. Serial number, proofs and Germany on barrel. No chamber markings, but w/ stock lug. Estimated production unknown.

Luger 1923 Safe and Loaded $1150
Same caliber offerings, bbl. lengths and specifications as the 1923 Commercial, except thumb lever safety w/ safe markings. Other markings the same. Estimated 10,000 production.

Luger 1923 Stoeger $2900
Same general specifications as the 1923 Commercial Luger, w/the same caliber offerings and bbl. lengths of 3.75, 4, 6, and 8 inches. Thumb lever safety. DWM on new-style toggle, serial number and/or proof marks on barrel. American Eagle over chamber, but no stock lug. Estimated production unknown (also see Stoeger).

Luger 1926 "Baby" Prototype $85,000
Calibers: 7.65mm Browning and 9mm Browning (short). 5-shot magazine. 2.31-inch bbl. About 6.25 inches overall. Small-sized frame and toggle assembly. Prototype for a Luger "pocket pistol," but never manufactured commercially. Checkered walnut grips, slotted for safety. Only four known to exist, but possibly as many as a dozen could have been made.

Luger 1929 Swiss $1550
Caliber: 7.65mm. 8-shot magazine. 4.75-inch tapered bbl. 9.5 inches overall. Weight: 32 oz. Fixed rear sight, dovetailed front sight. Long grip safety and thumb lever w/S markings. Stepped receiver and straight grip frame. Checkered plastic grips. Swiss Cross in shield on new-style toggle. Serial numbers and proofs on barrel, no markings over chamber and no stock lug. Estimated 1900 produuction.

Luger 1934 Krieghoff Commercial (Side Frame). $2950
Caliber: 7.65mm or 9mm Para. 8-shot magazine. Bbl. lengths: 4, 6, and 8 inches. Overall length: 8.75 (4-inch bbl.). Weight: 30 oz. (4-inch bbl.). Various rear sight configurations w/dovetailed front sight. Thumb lever safety. Checkered brown plastic grips. Anchor w/H K Krieghoff Suhl on new-style toggle, but no chamber markings. Tapered bbl. w/serial number and proofs; w/stock lug. Estimated 1700 production.

Luger 1934 Krieghoff S
Caliber: 9mm Para. 8-shot magazine. 4-inch tapered bbl. 8.75 inches overall. Weight: 30 oz. Fixed rear sight, dovetailed front sight. Thumb lever safety. Anchor w/H K Krieghoff Suhl on new-style toggle, S dated chamber, bbl. proofs and stock lug. *Early Model:* Checkered walnut or plastic grips. Estimated 2500 production. *Late Model:* Checkered brown plastic grips. Est. 1200 production.
Early Model **$2495**
Late Model **1950**

Luger 1934 byf. $895
Caliber: 9mm Para. 8-shot magazine. 4-inch tapered bbl. 8.75 inches overall. Weight: 30 oz. Fixed rear sight, dovetailed front sight. Thumb lever safety. Checkered walnut or plastic grips. byf on new-style toggle, serial number and proofs on bbl. 41-42 dated chamber and w/stock lug. Estimated 3000 productlon.

Luger 1934 Mauser S/42 K. $2550
Caliber: 9mm Para. 8-shot magazine. 4-inch tapered bbl. 8.75 inches overall. Weight: 30 oz. Fixed rear sight dovetailed front sight. Thumb lever safety. Checkered walnut or plastic grips. 42 on new-style toggle, serial number and proofs on barrel, 1939-40 dated chamber markings and w/stock lug. Estimated 10,000 production.

Luger 1934 Mauser S/42 (Dated) $650
Same specifications as Luger 1934 Mauser 42, above, except 41 dated chamber markings and w/stock lug. Estimated production unknown.

Luger 1934 Mauser Banner (Military) $1695
Same specifications as Luger 1934 Mauser 42, except Mauser in banner on new-style toggle, tapered bbl. w/serial number and proofs usually, dated chamber markings and w/stock lug. Estimated production unknown.

Luger S/42

Luger 1934 Mauser Commercial $2800
Same specifications as Luger 1934 Mauser 42, except checkered walnut grips. Mauser in banner on new-style toggle, tapered bbl. w/serial number and proofs usually, no chamber markings, but w/stock lug. Estimated production unknown.

Luger 1934 Mauser Dutch. $1595
Same specifications as Luger 1934 Mauser 42, except checkered walnut grips. Mauser in banner on new-style toggle, tapered bbl. w/caliber, 1940 dated chamber markings and w/stock lug. Estimated production unknown.

Luger 1934 Mauser Latvian $2850
Caliber: 7.65mm. 8-shot magazine. 4-inch tapered bbl. 8.75 inches overall. Weight: 30 oz. Fixed square-notched rear sight, dovetailed Patridge front sight. Thumb lever safety. Checkered walnut stosks. Mauser in banner on new-style toggle,1937 dated chamber markings and w/stock lug. Estimated production unknown.

Luger 1934 Mauser (Oberndorf). $2200
Same as 1934 Mauser 42, except checkered walnut grips. Oberndorf 1934 on new-style toggle, tapered bbl. w/proofs and caliber, Mause banner over chamber and w/stock lug (also see Mauser).

Luger 1934 Simson-S Toggle $1850
Same as 1934 Mauser 42, except checkered walnut grips. S on new-style toggle, tapered bbl. w/serial number and proofs, no chamber markings; w/stock lug. Estimated 10,000 production.

Luger 42 Mauser Banner (byf) $1350
Same specifications as Luger 1934 Mauser 42, except weight 32 oz. Mauser in banner on new-style toggle, tapered bbl. w/serial number and proofs usually, dated chamber markings and w/stock lug. Estimated 3,500 production.

Luger Abercrombie and Fitch $4895
Calibers: 7.65mm and 9mm Para. 8-shot magazine. 4.75-inch tapered bbl. 9.5 inches overall. Weight: 32 oz. Fixed rear sight, dovetailed front sight. Grip safety. Checkered walnut grips. DWM on new-style toggle Abercrombie & Fitch markings on barrel, Swiss Cross in sunburst over chamber, but no stock lug. Est.100 production.

Luger Dutch Royal Air Force $1550
Caliber: 9mm Para. 8-shot magazine. 4-inch tapered bbl. 8.75 inches overall. Weight: 30 oz. Fixed rear sight dovetailed front sight. Grip safety and thumb safety w/markings and arrow. Checkered walnut grips. DWM on new-style toggle, bbl. dated w/serial number and proofs, no markings over chamber, no stock lug. Estimated 4000 production.

Luger DWM (G Date) $850
Caliber: 9mm Para. 8-shot magazine. 4-inch tapered bbl. 8.75 inches overall. Weight: 30 oz. Fixed rear sight, dovetailed front sight. Thumb lever safety. Checkered walnut grips. DWM on new-style toggle, serial number and proofs on barrel, G (1935 date) over chamber and w/stock lug. Estimated production unknown.

Luger DWM and Erfurt $895
Caliber: 9mm Para. 8-shot magazine. 4- or 6-inch tapered bbl. Overall length: 8.75, 10.75 inches. Weight: 30 or 38 oz. Fixed rear sight, dovetailed front sight. Thumb safety. Checkered walnut grips. Serial numbers and proof marks on barrel, double dated chamber, various proofs, DWM or crown over Erfurt on new-style toggle and w/stock lug. Estimated production unknown.

Luger Krieghoff 36 . . . . $2350
Caliber: 9mm Para. 8-shot magazine. 4-inch tapered bbl. 8.75 inches overall. Weight: 30 oz. Fixed rear sight dovetailed front sight. Thumb lever safety. Checkered brown plastic grips. Anchor w/H K Krieghoff Suhl on new-style toggle, 36 dated chamber, serial number and proofs on barrel and w/stock lug. Estimated 700 production.

Luger Krieghoff-Dated 1936-1945 . . . . $1950
Same specifications as Luger Krieghoff 36, except 1936-45 dated chamber, bbl. proofs. Est. 8600 production.

Luger Krieghoff (Grip Safety). $3645
Same specifications as Luger Krieghoff 36, except grip safety and thumb lever safety. No chamber markings, tapered bbl. w/serial number, proofs and caliber, no stock lug. Estimated production unknown.

Luger Mauser Banner (Grip Safety). $2850
Caliber: 7.65mm. 8-shot magazine. 4.75-inch tapered bbl. 9.5 inches overall. Weight: 30 oz. Fixed rear sight, dovetailed front sight. Grip safety and thumb lever safety. Checkered walnut grips. Mauser in banner on new-style toggle, serial number and proofs on barrel, 1939 dated chamber markings, but no stock lug. Estimated production unknown.

Luger Mauser Banner 42 (Dated) $1350
Caliber: 9mm Para. 8-shot magazine. 4-inch tapered bbl. 8.75 inches overall. Weight: 30 oz. Fixed rear sight dovetailed front sight. Thumb lever safety. Checkered walnut or plastic grips. Mauser in banner on new-style toggle serial number and proofs on bbl. usually, 42 dated chamber markings and stock lug. Estimated production unknown.

Luger Mauser Banner (Swiss Proof) $1850
Same specifications as Luger Mauser Banner 42, above except checkered walnut grips and 1939 dated chamber.

Luger Mauser Freise $3550
Same specifications as Mauser Banner 42, except checkered walnut grips tapered bbl. w/proofs on sight block and Freise above chamber. Estimated production unknown.

Luger S/42
Caliber: 9mm Para. 8-shot magazine. 4-inch tapered barrel. 8.75 inches overall. Weight: 30 oz. Fixed rear sight, dovetailed front sight. Thumb lever safety. Checkered walnut grips. S/42 on new-style toggle, serial number and proofs on barrel and w/stock lug. *Dated Model:* has dated chamber; estimated 3000 production. *G Date:* has G (1935 date) over chamber; estimated 3000 production. *K Date:* has K (1934 date) over chamber; prod. figures unknown.
Dated Model . . . . $ 895
G Date Model 875
K Date Model 1895

Luger Russian Commercial **$2150**
Caliber: 7.65mm. 8-shot magazine. 3.63-inch tapered bbl. 8.38 inches overall. Weight: 30 oz. Fixed rear sight, dovetailed front sight. Thumb lever safety. Checkered walnut grips. DWM on new-style toggle, Russian proofs on barrel, no chamber markings, but w/stock lug. Estimated production unknown.

Luger Simson and Company **$1295**
Calibers: 7.65mm and 9mm Para. 8-shot magazine. Weight: 32 oz. Fixed rear sight, dovetailed front sight. Thumb lever safety. Checkered walnut grips. Simson & Company Suhl on new-style toggle, serial number and proofs on barrel, date over chamber and w/stock lug. Estimated 10,000 production.

Luger Vickers-Dutch **$2550**
Caliber: 9mm Para. 8-shot magazine. 4-inch tapered bbl. 8.75 inches overall. Weight: 30 oz. Fixed rear sight, dovetailed front sight. Grip safety and thumb lever w/arrow markings. Checkered walnut grips (coarse). Vickers LTD on new-style toggle, no chamber markings, dated barrel, but no stock lug. Estimated 10,000 production.

LUNA FREE PISTOL
Zella-Mehlis, Germany
Originally mfd. by Ernst Friedr. Buchel and later
by Udo Anschutz

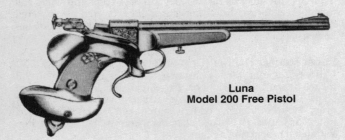

**Luna
Model 200 Free Pistol**

Luna Model 200 Free Pistol **$1025**
Single-shot. System Aydt action. Set trigger. Caliber: 22 LR. 11-inch bbl. Weight: 40 oz. Target sights. Blued finish. Checkered and carved walnut grip and forearm improved design w/adj. hand base on later models of Udo Anschutz manufacture. Made prior to WWII.

M.A.C. (Military Armament Corp.)
Stephensville, Texas
Dist. by Defense Systems International
Marietta, Georgia
Previously by Cobray, SWD and RPB Industries

M.A.C. Ingram Model 10 Auto Pistol
Select fire (NFA-Title II-Class III) SMG based on Ingram M10 blowback system using an open bolt design with or without telescoping stock. Calibers: 9mm or 45 ACP. Cyclic rate: 750 RPM (9mm) or 900 RPM (45ACP). 32- or 30-round magazine. 5.75-inch threaded bbl. to accept muzzle break, bbl. extension or suppressor. 10.5 inches overall w/o stock or 10.6 (w/telescoped stock) and 21.5 (w/extended stock). Weight: 6.25 pounds. Front protected post sight, fixed aperture rear sight. Garand-style safety in trigger guard.
9mm Model **$895**
45 ACP Model 950
W/Bbl. Extension, **add** 195
W/Suppressor, **add** 495

M.A.C. Ingram Model 10A1S Semiautomatic
Similar to the MAC 10, except (Class I) semiautomatic w/closed bolt design to implement an interchangable component system to easily convert to fire 9mm and 45 ACP.
9mm Model **$395**
45 ACP Model 450
W/Bbl. Extension, **add** 150
W/Fake Suppressor, **add** 195

M.A.C. Ingram Model 11 Semiautomatic
Similar to the MAC 10A1, except (Class I) semiautomatic chambered 380 ACP.
380 ACP Model **$375**
W/Bbl. Extension, **add** 150
W/Fake Suppressor, **add** 195

MAGNUM RESEARCH INC.
Minneapolis, Minnesota

Magnum Research Baby Eagle Semiautomatic **$295**
DA. Calibers: 9mm, 40 S&W, 41 AE. 15-shot magazine (9mm), 9-shot magazine (40 S&W), 10-shot magazine (41 AE). 4.75-inch bbl. 8.15 inches overall. Weight: 35.4 oz. Combat sights. Matte blued finish. Imported 1991-96 and 1999 to date.

**Magnum Research
Model Desert Eagle Mark XIX**

Magnum Research Desert Eagle Mk VII Semiautomatic
Calibers: 357 Mag., 41 Mag., 44 Mag., 50 Action Express (AE). 8- or 9-shot magazine. Gas-operated. 6-inch w/standard bbl. or 10- and 14-inch w/polygonal bbl. 10.6 inches overall (6-inch bbl.). Weight: 52 oz. (w/alum. alloy frame) to 67 oz. (w/steel frame). Fixed or adj. combat sights. Combat-type trigger guard. Finish: Military black oxide, nickel, chrome, stainless or blued. Wraparound rubber grips. Made by Israel Military Industries 1984-95..
357 Standard (steel) or Alloy (6-inch bbl.) . . . . **$ 595**
357 Stainless Steel (6-inch bbl.) 725
41 Mag. Standard (steel) or Alloy (6-inch bbl.) 675
41 Mag. Stainless Steel (6-inch bbl.) 650
44 Mag. Standard (steel) or Alloy (6-inch bbl.) 635
44 Mag. Stainless Steel (6-inch bbl.) 695
50 AE Magnum Standard 750
Add for 10-inch Bbl. 100
Add for 14-inch Bbl. 125

Magnum Research Model Desert Eagle Mark XIX Semi-Automatic Pistol

Interchangeable component system based on .50 caliber frame. Calibers: .357 Mag., 44 Mag., 50 AE. 9-, 8-, 7-shot magazine. 6- or 10-inch bbl w/dovetail design and cross slots to accept scope rings. Weight: 70.5 oz. (6-inch bbl.) or 79 oz. 10.75 or 14.75 inches overall. Sights: post front and adjustable rear. Blue, chrome or nickel finish available brushed, matte or polished. Hogue soft rubber grips. Made 1995 to date.

357 Mag. (W/6-inch bbl.)	**$ 725**
.44 Mag. (W/6-inch bbl.)	**750**
.50 AE (W/6-inch bbl.)	**775**
W/10-inch bbl., **add**	**50**
Caliber Conversion (bbl., bolt & magazine), **add**	**395**
XIX Platform System (3 caliber-conversion w/6 bbls.)	**2195**
XIX6 System (2 caliber-conversion w/2 6-inch bbls.)	**1450**
XIX10 System (2 caliber-conversion w/2 10-inch bbls.)	**1575**
Custom Shop Finish, **add**	**15%**
24K Gold Finish, **add**	**35%**

Magnum Research (ASIA) Model One Pro 45 Pistol

Calibers: 45 ACP or 400 COR-BON. 3.75- inch bbl. 7.04 or 7.83 (IPSC Model) inches overall. Weight: 23.5 (alloy frame) or 31.1 oz. 10-shot magazine. Short recoil action. SA or DA mode w/decocking lever. Steel or alloy grip-frame. Textured black polymer grips. Imported 1998 to date.

Model 1P45	**$425**
Model 1C45/400 (Compensator Kit), **add**	**150**
Model 1C400NC (400 Conversion Kit), **add**	**100**

**Magnum Research
Model One Pro 45**

**Magnum Research
Mountain Eagle**

Magnum Research Mountain Eagle
Semiautomatic **$155**

Caliber: 22 LR.15-shot polycarbonate resin magazine. 6.5-inch injection-molded polymer and steel bbl. 10.6 inches overall. Weight: 21 oz. Ramp blade front sight, adj. rear. Injection-molded, checkered and textured grip. Matte black finish. Made 1992-96.

Magnum Research SSP-91 Lone Eagle Pistol

Single-shot action w/interchangeable rotating breech bbl. assembly. Calibers: 22 LR, 22 Mag., 22 Hornet, 22-250, 223 Rem., 243 Rem., 6mm BR, 7mm-08, 7mm BR, 30-06, 30-30, 308 Win., 35 Rem., 357 Mag., 44 Mag., 444 Marlin. 14-inch interchangeable bbl. assembly. 15 inches overall. Weight: 4.5 lbs. Made 1991 to date.

SSP-91 S/S Pistol (Complete)	**$295**
14-inch Bbl. Assembly	**250**
Stock Assembly	**75**

**Magnum Research
SSP-91 Lone Eagle Pistol
(Shown w/Optional Leupold® Scope)**

MAUSER PISTOLS
Oberndorf, Germany
Waffenfabrik Mauser of Mauser-Werke A.G.

Mauser Model 80-SA Automatic. **$325**
Caliber: 9mm Para. 13-shot magazine. 4.66-inch bbl. 8 inches overall. Weight: 31.5 oz. Blued finish. Hardwood grips. Made 1991-94.

Mauser Model 90 DA Automatic. **$335**
Caliber: 9mm Para. 14-shot magazine. 4.66-inch bbl. 8 inches overall. Weight: 35 oz. Blued finish. Hardwood grips. Made 1991-94.

**Mauser
Model 80-SA**

**Mauser
Model 90-DA**

**Mauser
Model 1898 (1896) Military**

**Mauser
Model HSC**

Mauser Model 90 DAC Compact **$345**
Caliber: 9mm Para. 14-shot magazine. 4.13-inch bbl. 7.4 inches
overall. Weight: 33.25 oz. Blued finish. Hardwood grips. Made
1991-94.

Mauser Model 1898 (1896) Military Auto Pistol
Caliber: 7.63mm Mauser, but also chambered for 9mm Mauser
and 9mm Para. w/the latter being identified by a large red "9" in
the grips. Box magazine, 10-shot. 5.25-inch bbl. 12 inches over-
all. Weight: 45 oz. Adj. rear sight. Blued finish. Walnut grips.
Made 1897-1939. *Note:* Specialist collectors recognize a number
of variations at significantly higher values. Price here is for more
common commercial and military types with original finish.
Commercial Model (Pre-War) **$1595**
Commercial Model (Wartime) 1250
Red 9 Commercial Model (Fixed Sight) 795
Red 9 WWI Contract (Tangent Sight) 1795
W/Stock Assembly (Matching SN), **add** 500

Mauser Model HSC DA Auto Pistol **$450**
Calibers: 32 Auto (7.65mm), 380 Auto (9mm Short). 8-shot (32) or
7-shot (380) magazine. 3.4-inch bbl. 6.4 inches overall. Weight:
23.6 oz. Fixed sights. Blued or nickel finish. Checkered walnut
grips. Made 1938-45 and from 1968-96.

Mauser Luger Lange Pistole 08 . . . . **$1995**
Caliber: 9mm Para. 8-inch bbl. Checkered grips. Blued finish. Accessorized w/walnut shoulder stock, front sight tool, spare magazine, leather case. Currently in production. Commemorative version
made in limited quantities w/ivory grips and 14-carat gold monogram plate.

Mauser Parabellum Luger Auto Pistol **$1650**
Current commercial model. Swiss pattern w/grip safety. Calibers:
7.65mm Luger, 9mm Para. 8-shot magazine. Bbl. lengths: 4-,6-
inch. 8.75 inches overall (4-inch bbl.). Weight: 30 oz. (4-inch
bbl.). Fixed sights. Blued finish. Checkered walnut grips. Made
from 1970 to date. *Note:* Pistols of this model sold in the U.S. have
the American Eagle stamped on the receiver.

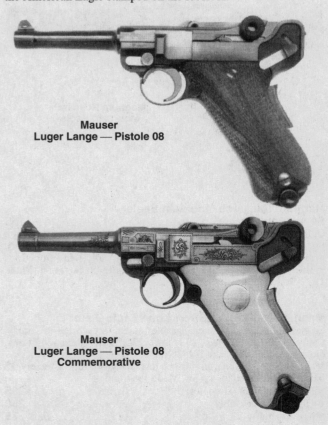

**Mauser
Luger Lange — Pistole 08**

**Mauser
Luger Lange — Pistole 08
Commemorative**

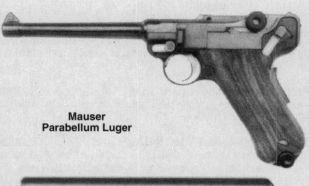

**Mauser
Parabellum Luger**

**Mauser
WTP Model I**

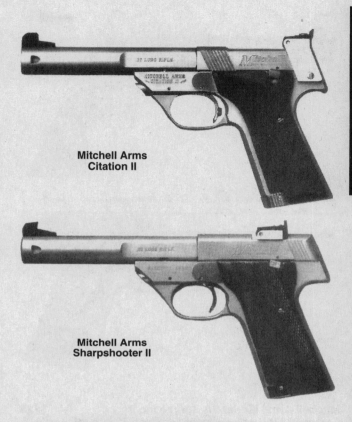

**Mitchell Arms
Citation II**

**Mitchell Arms
Sharpshooter II**

Mauser Pocket Model 1910 Auto Pistol **$335**
Caliber: 25 Auto (6.35mm). 9-shot magazine. 3.1-inch bbl. 5.4 inches overall. Weight: 15 oz. Fixed sights. Blued finish. Checkered walnut or hard rubber grips. Made 1910-34.

Mauser Pocket Model 1914 Automatic **$325**
Similar to Pocket Model 1910. Caliber: 32 Auto (7.65mm). 8-shot magazine. 3.4-inch bbl. 6 inches overall. Weight: 21 oz. Fixed sights. Blued finish. Checkered walnut or hard rubber grips. Made 1914-34.

Mauser Pocket Model 1934. . . . . **$365**
Similar to Pocket Models 1910 and 1914 in the respective calibers. Chief difference is in the more streamlined one-piece grips. Made c. 1934-39.

Mauser WTP Model I Auto Pistol **$395**
"Westentaschen-Pistole" (Vest Pocket Pistol). Caliber: 25 Automatic (6.35mm). 6-shot magazine. 2.5-inch bbl. inches overall. Weight: 11.5 oz. Blued finish. Hard rubber grips. Made 1922-37.

Mauser WTP Model II Auto Pistol **$550**
Similar to Model I, but smaller and lighter. Caliber: 25 Automatic (6.35mm). 6-shot magazine. 2-inch bbl. 4 inches overall. Weight: 9.5 oz. Blued finish. Hard rubber grips. Made 1938-40.

MITCHELL ARMS, INC.
Santa Ana, California

Mitchell Arms Model 1911 Gold Signature
Caliber: 45 ACP. 8-shot magazine. 5-inch bbl. 8.75 inches overall. Weight: 39 oz. Interchangeable blade front sight, drift-adj. combat or fully adj. rear. Smooth or checkered walnut grips. Blued or stainless finish. Made 1994-96.
Blued Model w/Fixed Sights **$365**
Blued Model w/Adj. Sights **400**
Stainless Model w/Fixed Sights **390**
Stainless Model w/Adj. Sights **425**

Mitchell Arms Alpha Model Auto Pistol
Dual action w/interchangeable trigger modules. Caliber: 45 ACP. 8-shot magazine. 5-inch bbl. 8.75 inches overall. Weight: 39 oz. Interchangeable blade front sight, drift-adj. rear. Smooth or checkered walnut grips. Blued or stainless finish. Made from 1994 to date. Advertised 1995, but not manufactured.
Blued Model w/Fixed Sights . . . . **(SRP) $698**
Blued Model w/Adj. Sights . . . . **(SRP) 725**
Stainless Model w/Fixed Sights . . . . **(SRP) 725**
Stainless Model w/Adj. Sights . . . . **(SRP) 749**

Mitchell Arms American Eagle Pistol **$475**
Stainless-steel re-creation of the American Eagle Parabellum auto pistol. Caliber: 9mm Para. 7-shot magazine. 4-inch bbl. 9.6 inches overall. Weight: 26.6 oz. Blade front sight, fixed rear. Stainless finish. Checkered walnut grips. Made 1993-94.

Mitchell Arms Citation II Auto Pistol **$295**
Re-creation of the High Standard Supermatic Citation Military. Caliber: 22 LR. 10-shot magazine. 5.5-inch bull bbl. or 7.25 fluted bbl. 9.75 inches overall (5.5-inch bbl.). Weight: 44.5 oz. Ramp front sight, slide-mounted micro-adj. rear. Satin blued or stainless finish. Checkered walnut grips w/thumbrest. Made 1992-96.

Mitchell Arms Olympic I.S.U. Auto Pistol **$445**
Similar to the Citation II Model, except chambered in 22 Short. 6.75-inch round tapered bbl. w/stabilizer and removable counterweights. Made 1992-96.

Mitchell Arms Sharpshooter II Auto Pistol **$275**
Re-creation of the High Standard Sharpshooter. Caliber: 22 LR. 10-shot magazine. 5-inch bull bbl. 10.25 inches overall. Weight: 42 oz. Ramp front sight, slide-mounted micro-adj. rear. Satin blued or stainless finish. Checkered walnut grips w/thumbrest. Made 1992-96.

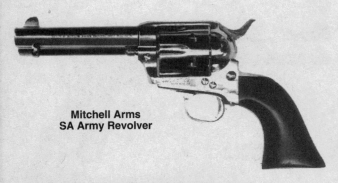

**Mitchell Arms
SA Army Revolver**

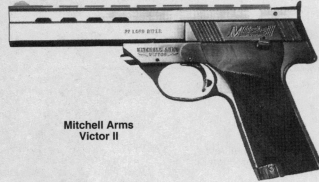

**Mitchell Arms
Victor II**

Mitchell Arms Model SA Sport King II **$225**
Caliber: 22 LR. 10-shot magazine. 4.5- or 6.75-inch bbl. 9 or 11.25 inches overall. Weight: 39 or 42 oz. Checkered walnut or black plastic grips. Blade front sight and drift adjustable rear. Made 1993-94.

Mitchell Arms SA Army Revolver
Calibers: 357 Mag., 44 Mag., 45 Colt/45 ACP. 6-shot cylinder. Bbl. lengths: 4.75, 5.5, 7.5 inches. Weight: 40-43 oz. Blade front sight, grooved topstrap or adj. rear. Blued or nickel finish w/color-casehardened frame. Brass or steel backstrap/trigger guard. Smooth one-piece walnut grips. Imported 1987-94 and 1997 to date.
Standard Model w/Blued Finish **$295**
Standard Model w/Nickel Finish   325
Standard Model w/Stainless Backstrap   350
45 Combo w/Blued Finish   395
45 Combo w/Nickel Finish   425

Mitchell Arms Trophy II Auto Pistol **$325**
Similar to the Citation II Model, except w/gold-plated trigger and gold-filled markings. Made 1992-96.

Mitchell Arms Victor II Auto Pistol **$395**
Re-creation of the High Standard Victor w/full-length vent rib. Caliber: 22 LR. 10-shot magazine. 4.5- or 5.5-inch bbl. 9.75 inches overall (5.5-inch bbl.). Weight: 52 oz., (5.5-inch bbl.). Rib-mounted target sights. Satin blued or stainless finish. Checkered walnut grips w/thumbrest. Made 1992-96.

**Mitchell Arms Model Guardian
Angel Derringer** **$100**
Hammerless, double-action O/U derringer w/interchangeable drop-in breech block. Calibers: 22 LR, 22 WRM. 2-shot capacity. 2-inch bbl. 5 inches overall. Weight: 12 oz. Blue, nickel or gold finish. Blade front and fixed rear sights. Checkered black grips. Made 1996-97.

Mitchell Arms Model Guardian II **$190**
Caliber: 38 Special, 6-shot cylinder 2-, 4- or 6-inch bbl. 8.5 inches overall (4-inch bbl.). Weight: 32 oz (4-inch bbl.). Blade ramp front and fixed rear sights. Checkered combat or target grips. Blued finish. Made 1995.

Mitchell Arms Model Guardian III **$200**
Same specifications as Guardian II model except w/adjustable rear sights. Made 1995.

Mitchell Arms Model Titan II DA **$255**
Caliber: 357 Mag. 6-shot cylinder. 2-, 4- or 6-inch bbl. 7.75 inches overall (4-inch bbl.). Weight: 38 oz (4-inch bbl.). Blade front and fixed rear sights. Crane mounted cylinder release. Blued or stainless finish. Made 1995.

Mitchell Arms Model Titan III DA **$295**
Same specification as the Titan II, except w/adjustable rear sight. Made 1995.

MKE PISTOL
Ankara, Turkey
Mfd. by Makina ve Kimya Endüstrisi Kurumu

**MKE
Kirikkale**

MKE Kirikkale DA Automatic Pistol **$295**
Similar to Walther PP. Calibers: 32 Auto (7.65mm), 380 Auto (9mm Short). 7-shot magazine. 3.9-inch bbl. 6.7 inches overall. Weight: 24 oz. Fixed sights. Blued finish. Checkered plastic grips. Made 1948-88. *Note:* This is a Turkish Army standard service pistol.

MOA CORPORATION
Dayton, Ohio

MOA Maximum Single-Shot Pistol
Calibers: 22 Hornet to 454 Casull Mag. Armoloy, Chromoloy or stainless falling block action fitted w/blued or stainless 8.75-, 10- or 14-inch Douglas bbl. Weight: 60 - 68 oz. Smooth walnut grips. Made 1986 to date.
Chromoloy Receiver (Blued bbl.). **$445**
Armoloy Receiver (Blued bbl.)   475
Stainless Receiver (Blued bbl.).   525
W/Stainless Bbl., **add**   95
W/Extra Bbl., **add** 250

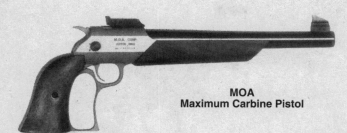

**MOA
Maximum Carbine Pistol**

MOA Maximum Carbine Pistol **$575**
Similar to Maximum Pistol, except w/18-inch bbl. Made 1986-88 and 1994 to date.

O.F. MOSSBERG & SONS, INC.
North Haven, Connecticut

**Mossberg
Brownie "Pepperbox" Pistol**

Mossberg Brownie "Pepperbox" Pistol **$425**
Hammerless, top-break, double-action, four bbls. w/revolving firing pin. Caliber: 22 LR. 2.5-inch bbls.. Weight: 14 oz. Blued finish. Serated grips. Approximately 37,000 made 1919-32.

NAMBU PISTOLS

See Listings under Japanese Military Pistols.

NAVY ARMS COMPANY
Martinsburg, West Virginia

Navy Arms Model 1873 SA Revolver
Calibers: 44-40, 45 Colt. 6-sfiot cylinder. Bbl. lengths: 3, 4.75, 5.5, 7.5 inches. 10.75 inches overall (5.5-inch bbl.). Weight: 36 oz. Blade front sight, grooved topstrap rear. Blued w/color-casehardened frame or nickel finish. Smooth walnut grips. Made 1991 to date.
Blued Finish w/Brass Backstrap **$275**
U.S. Artillery Model w/5-inch bbl. **325**
U.S. Cavalry Model w/7-inch bbl **350**
Bisley Model **295**
Sheriff's Model (Disc. 1998) **345**

Navy Arms Model 1875 Schofield Revolver
Replica of S&W Model 3, Top-break single-action w/auto ejector. Calibers: 44-40 or 45 LC. 6-shot cylinder, 5- or 7-inch bbl. 10.75 or 12.75 inches overall. Weight: 39 oz. Blade front sight, square-notched rear. Polished blued finish. Smooth walnut grips. Made 1994 to date.
Cavalry Model (7-inch bbl.) **$525**
Wells Fargo Model (5-inch bbl.) **495**

**Navy Arms
Model 1875 Schofield**

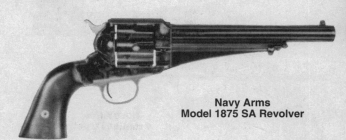

**Navy Arms
Model 1875 SA Revolver**

Navy Arms Model 1875 SA Revolver **$295**
Replica of Remington Model 1875. Calibers: 357 Magnum, 44-40, 45 Colt. 6-shot cylinder. 7.5-inch bbl. 13.5 inches overall. Weight: about 48 oz. Fixed sights. Blued or nickel finish. Smooth walnut grips. Made in Italy c.1955-1980. *Note:* Originally marketed in the U.S. as Replica Arms Model 1875 (that firm was acquired by Navy Arms Co).

Navy Arms Buntline Frontier **$375**
Same as Target Frontier, except has detachable shoulder stock and 16.5-inch bbl. Calibers: 357 Magnum and 45 Colt only. Made 1975-79.

Navy Arms Frontier SA Revolver **$230**
Calibers: 22 LR, 22 WMR, 357 Mag., 45 Colt. 6-shot cylinder. Bbl. lengths: 4 .5-, 5.5-, 7.5-inch. 10.25 inches overall (4.5-inch bbl.). Weight: about 36 oz. (4.5-inch bbl.). Fixed sights. Blued bbl. and cylinder, color-casehardened frame, brass grip frame. One-piece smooth walnut grip. Imported 1975-79.

Navy Arms Frontier Target Model **$250**
Same as Standard Frontier, except has adj. rear sight and ramp front sight. Imported 1975-79.

Navy Arms Luger (Standard) Automatic **$125**
Caliber: 22 LR, standard or high velocity. 10-shot magazine. Bbl.: 4.5 inches. 8.9 inches overall. Weight: 1 lb. 13.5 oz. Square blade front sight w/square notch, stationary rear sight. Walnut checkered grips. Non-reflecting black finish. Made 1986-88.

Navy Arms Rolling Block Single-Shot Pistol **$180**
Calibers: 22 LR, 22 Hornet, 357 Magnum. 8-inch bbl. 12 inches overall. Weight: about 40 oz. Adjustable sights. Blued bbl., color-casehardened frame, brass trigger guard. Smooth walnut grip and forearm. Imported 1965-80.

Navy Arms TT-Olympia Pistol **$250**
Reproduction of the Walther Olympia Target Pistol. Caliber: 22 LR. 4.6-inch bbl. 8 inches overall. Weight: 28 oz. Blade front sight, adj. rear. Blued finish. Checkered hardwood grips. Imported 1992-94.

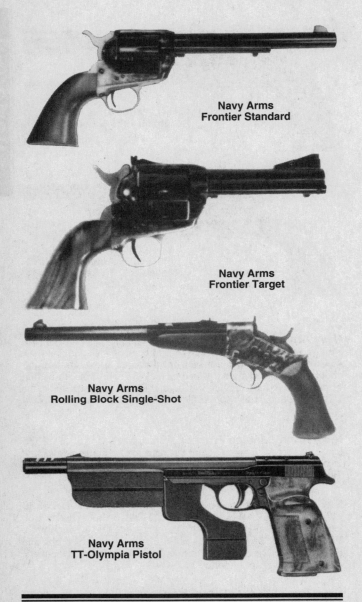

**Navy Arms
Frontier Standard**

**Navy Arms
Frontier Target**

**Navy Arms
Rolling Block Single-Shot**

**Navy Arms
TT-Olympia Pistol**

**New England Firearms
Model R73 Revolver**

**New England Firearms
Model 832 Starter Pistol**

NEW ENGLAND FIREARMS
Gardner, Massachusetts

New England Firearms Model R73 Revolver **$105**
Caliber: 32 H&R Mag. 5-shot cylinder. 2.5- or 4-inch bbl. 8.5 inches overall (4-inch bbl.). Weight: 26 oz. (4 inch bbl.). Fixed or adjustable sights. Blued or nickel finish. Walnut-finish hardwood grips. Made from 1988 to date.

New England Firearms Model R92 Revolver **$95**
Same general specifications as Model R73, except chambered for 22 LR. 9-shot cylinder. Weight: 28 oz. w/4 inch bbl. Made from 1988 to date.

**New England Firearms Model 832
Starter Pistol** **$75**
Calibers: 22 Blank, 32 Blank. 9- and 5-shot cylinders, respectively. Push-pin swing-out cylinder. Solid wood grips w/NEF medallion insert.

New England Firearms Ultra Revolver **$125**
Calibers: 22 LR, 22 Mag. 9-shot cylinder in 22 LR, 6-shot cylinder in 22 Mag. 4- or 6-inch ribbed bull bbl. 10.75 inches overall (6-inch bbl.). Weight: 36 oz. (6-inch bbl.). Blade front sight, adj. square-notched rear. Blued or nickel finish. Walnut-finish hardwood grips. Made from 1989 to date.

New England Firearms Lady Ultra Revolver **$135**
Same basic specifications as the Ultra, except in 32 H&R Mag. w/5-shot cylinder and 3-inch ribbed bull bbl. 7.5 inches overall. Weight: 31 oz. Made 1992 to date.

NORTH AMERICAN ARMS
Provo, Utah

North American Arms Model 22LR **$125**
Same as Model 22S, except chambered for 22 LR., is 3.88-inches overall, weighs 4.5 oz. made 1976 to date.

North American Arms Model 22S Mini Revolver ... **$135**
SA. Caliber: 22 Short .5-shot cylinder. 1.13-inch bbl. 3.5-inches overall. Weight: 4 oz. Fixed sights. Stainless steel. Plastic grips. Made from 1975 to date.

**North American Arms Model 450 Magnum
Express** **$850**
SA. Calibers: 450 Magnum Express, 45 Win. Mag. 7.5-inch bbl. Matte stailess steel finish. Cased. Discontinued 1986.

North American Arms Model 454C SA Revolver . . . **$755**
Caliber: 454 Casull. 5-shot cylinder. 7.5-inch bbl, 14-inches overall. Weight: 50 oz. Fixed sights. Stainless steel. Smooth hardwood grips. Intro. 1977.

North American Arms Black Widow Revolver
SA. Calibers: 22 LR., 22 WMR. 5-shot cylinder. 2-inch heavy vent bbl. 5.88-inches overall. Weight: 8.8 oz. Fixed or adj. sights. Full-size black rubber grips. Stainless steel brush finish. Made from 1990 to date.
Adj. Sight Model $210
Adj. Sight Combo Model 220
Fixed Sight Model. 175
Fixed Sight Combo Model 210

North American Arms Guardian DAO Pistol. **$320**
Caliber: 32 ACP. 6-shot magazine. 2-inch bbl. 4.4 inches overall. Weight: 13.5 oz. Fixed sights. Black synthetic grips. Stainless steel. Made 1997 to date.

North American Arms Mini-Master Revolver
SA. Calibers: 22 LR., 22 WMR. 5-shot cylinder. 4-inch heavy rib bbl. 7.75-inches overall. Weight: 10.75 oz. Fixed or adj. sights. Black rubber grips. Stainless steel brush finish. Made from 1990 to date.
Adj Sight Model $215
Adj, Sight Combo Model 250
Fixed Sight Combo Model 220
Fixed Sight Combo model 255

NORWEGIAN MILITARY PISTOLS
Mfd. by Kongsberg Vaapenfabrikk,
the government arsenal at Kongsberg, Norway

Norwegian Model 1914 Automatic Pistol **$325**
Similar to Colt Model 1911 45 Automatic w/same general specifications, except has lengthened slide stop. Made 1919-46.

Norwegian Model 1912 . . . . **$2100**
Same as the model 1914 except has conventional slide stop. Only 500 were made.

ORTGIES PISTOLS
Erfurt, Germany
Manufactured by Deutsche Werke A.G.

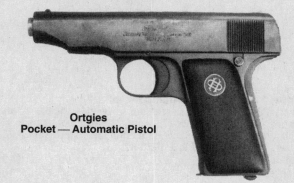

Ortgies
Pocket — Automatic Pistol

Ortgies Pocket Automatic Pistol **$205**
Calibers: 32 Automatic (7.65mm), 380 Automatic (9mm) 7-shot magazine (380 cal.), 8-shot (32 cal.). 3.25-inch bbl. 6.5 inches overall. Weight: 22 oz. Fixed sights. Blued finish. Plain walnut grips. Made in 1920's.

Ortgies Vest Pocket Automatic Pistol **$225**
Caliber: 25 Automatic (6.35mm). 6-shot magazine. 2.75-inch bbl. 5.19-inches overall. Weight: 13.5 oz. Fixed sights. Blued finish. Plain walnut grips. Made in 1920's.

OLYMPIC ARMS, INC.
Olympic, Washington

Olympic OA-93 AR Semiautomatic Pistol
AR-15 style receiver with no buffer tube or charging handle. Caliber: 223 Rem. or 7.62x39mm. 5-, 20- or 30-shot detachable magazine. 6-, 9- or 14-inch stainless steel bbl. 15.75 inches overall w/6-inch bbl. Weight: 4 lbs. 3 oz. Flat top upper with no open sights. Vortex flash suppressor. A2 stoaway pistol grip and forward pistol grip. Made 1993-94. Note: All Post-ban versions of OA-93 style weapons are classified by BATFas "Any Other Weapon" and must be transferred by a Class III dealer. Values listed here are for limited production Pre-ban guns.
Model OA-93 (223 Rem.) $2995
Model OA-93 (7.62x39mm). 3850

Olympic OA-96 AR Semiautomatic Pistol **$695**
Similar to Model OA-93 AR, except w/6.5-inch bbl.only chambered 223 Rem. Additional compliance modifications include a fixed (nonremovable) well-style magazine and no forward pistol grip. Made 1996 to date.

PHOENIX ARMS
Ontario, Canada

**Phoenix Arms Model HP22/HP25 SA
Auto Pistols** **$65**
Caliber: 22 LR, 25 ACP. 10-shot magazine. 3- inch bbl. Weight: 20 oz. 5.5 inches overall. Checkered synthetic grips. Blade front sight, adjustable rear. Blue, chrome or nickel finish. Made 1994 to date.

**Phoenix Arms Model HP Rangemaster Target
SA Auto Pistol** **$95**
Similar to Model HP22, except w/5.5-inch target bbl. and extended magazine. Ramp front sight, adjustable notch rear on vent rib. Checkered synthetic grips. Blue or satin nickel finish. Made 1998 to date.

**Phoenix Arms Model HP Rangemaster Deluxe Target
SA Auto Pistol** **$95**
Similar to Model HP Rangemaster Target Model, except w/dual-2000 lazer sight and custom wood grips. Made 1998 to date.

Phoenix Arms Model "Raven" SA Auto Pistol **$50**
Caliber: 25 ACP. 6-shot magazine. 2.5inch bbl. 4.75 inches overall. Weight: 15 oz. Ivory, pink pearl, or black slotted stocks. Fixed sights. Blue, chrome or nickel finish. Made 1993 to date.

PARA-ORDNANCE MFG. INC.
Scarborough, Ontario, Canada

Para-Ordnance Limited Edition Series
Custom tuned and fully accessorized "Limited Edition" versions of standard "P" Models. Enhanced gripframe and serrated slide fitted w/match grade bbl. and full length recoil spring guide system. Beavertail grip safety and skeletonized hammer. Ambidextrous safety and trigger-stop adjustment. Fully adjustable or contoured low-mount sights. For pricing see individual models.

**Para-Ordnance
P-12 Compact**

**Para-Ordnance
P-14 Auto Pistol**

Para-Ordnance Model P-10 SA Auto Pistol
Super compact. Calibers: 40 S&W, 45 ACP. 10-shot magazine. 3-inch bbl. Weight: 24 oz. (alloy frame) or 31 oz. (steel frame). Ramp front sight and drift adjustable rear w/3-Dot system. Steel or alloy frame. Matte black, Duo-Tone or stainless finish. Made 1997 to date.
Alloy Model **$455**
Duo-Tone Model. **495**
Stainless Steel Model **525**
Steel Model. **475**
Limited Model (Tuned & Accessorized), **add** **100**

Para-Ordnance P-12 Compact Auto Pistol $450
Calibers: 45 ACP. 11-shot magazine. 3.5-inch bbl. 7-inches overall. Weight: 24 oz. (alloy frame). Blade front sight, adj. rear w/3 Dot system. Textured composition grips. Matte black alloy or steel finish. Made 1990 to date.

Para-Ordnance P-13 Auto Pistol
Same general specifications as Model P-12, except w/12-shot magazine. 4.5-inch bbl. 8-inches overall. Weight: 25 oz. (alloy frame). Blade front sight, adj. rear w/3-dot system. Textured composition grips. Matte balck alloy or steel finish. Made 1990 to date.
Model P1345 (Alloy) **$475**
Model P1345C (Steel). **565**

Para-Ordnance P-14 Auto Pistol
Caliber: 45 ACP. 13-shot magazine. 5-inch bbl. 8.5-inches overall. Weight: 28 oz. Alloy frame. Blade front sight, adj. rear w/3-dot system. Textured composition grips. matte black alloy or steel finish. Made from 1990 to date.
Model P1445 (Alloy) **$455**
Model P1445C (Steel). **545**

Para-Ordnance Model P-16 SA Auto Pistol
Caliber: 40 S&W. 10- or 16-shot magazine. 5-inch bbl. 8.5 inches overall. Weight: 40 oz. Ramp front sight and drift adjustable rear w/3-Dot system. Carbon steel or stainless frame. Matte black, Duo-Tone or stainless finish. Made 1997 to date.
Blue Steel Model **$485**
Duotone Model. **500**
Stainless Model. **535**
Limited Model (Tuned & Accessorized), **add** **100**

Para-Ordnance Model P-18 SA Auto Pistol
Caliber: 9mm Parabellum. 10- or 18-shot magazine. 5-inch bbl. 8.5 inches overall. Weight: 40 oz. Dovetailed front sight and fully adjustable rear. Bright stainless finish. Made 1998 to date.
Stainless Model. **$535**
Limited Model (Tuned & Accessorized), **add** **100**

PLAINFIELD MACHINE COMPANY
Dunellen, New Jersey
This firm discontinued operation about 1982.

**Plainfield
Model 71**

**Plainfield
Model 72**

Plainfield Model 71 Automatic Pistol
Calibers: 22 LR, 25 Automatic w/conversion kit available. 10-shot magazine (22 LR) or 8-shot (25 Auto). 2.5-inch bbl. 5.13 inches overall. Weight: 25 oz. Fixed sights. Stainless steel frame/slide. Checkered walnut grips. Made 1970-82.
22 LR or 25 Auto only **$135**
W/Conversion Kit. **160**

Plainfield Model 72 Automatic Pistol
Same as Model 71, except has aluminum slide, 3.5-inch bbl., is 6 inches overall. Made 1970-82
22 LR or 25 Auto only **$145**
W/Conversion Kit. **170**

PROFESSIONAL ORDNANCE, INC.
Lake Havasu City, Arizona

Professional Ordnance Model Carbon-15 Type 20 Semiautomatic Pistol **$995**
AR-15 operating system w/recoil reduction system. Caliber: 223 Rem. 10-shot magazine. 7,25-inch bbl. 20 inches overall. Weight: 40 oz. Ghost ring sights. Carbon-fiber upper and lower receivers w/chromoly bolt carrier. Matte black finish. Checkered composition grip. Made 1996 to date.

RADOM PISTOL
Radom, Poland
Manufactured by the Polish Arsenal

Radom P-35

Radom P-35 Automatic Pistol **$795**
Variation of the Colt Government Model 45 Auto. Caliber: 9mm Para. 8-shot magazine. 4.75-inch bbl. 7.75 inches overall. Weight: 29 oz. Fixed sights. Blued finish. Plastic grips. Made 1935 thru WWII.

RECORD-MATCH PISTOLS
Zella-Mehlis, Germany
Manufactured by Udo Anschütz

**Record-Match
Model 200 Free Pistol**

Record-Match Model 200 Free Pistol **$895**
Basically the same as Model 210 except plainer, w/different stock design and conventional set trigger, spur trigger guard. Made prior to WW II.

Record-Match Model 210 Free Pistol **$1295**
System Martini action, set trigger w/button release. Caliber: 22 LR. Single-shot. 11-inch bbl. Weight: 46 oz. Target sights micrometer rear. Blued finish. Carved and checkered walnut forearm and stock w/adj. hand base. Also made w/dural action (Model 210A); weight of this model, 35 oz. Made prior to WWII.

REISING ARMS CO.
Hartford, Connecticut

Reising Target Automatic Pistol **$425**
Hinged frame. Outside hammer. Caliber: 22 LR. 12-shot magazine. 6.5-inch bbl. Fixed sights. Blued finish. Hard rubber grips. Made 1921-24.

REMINGTON ARMS COMPANY
Ilion, New York

**Remington
Model 51**

Remington Model 51 Automatic Pistol **$625**
Calibers: 32 Auto, 380 Auto. 7-shot magazine. 3.5-inch bbl. 6.63 inches overall. Weight: 21 oz. Fixed sights. Blued finish. Hard rubber grips. Made 1918-34.

**Remington
Model 95 Double Derringer**

Remington Model 95 Double Derringer
SA. Caliber: 41 Short Rimfire. 3-inch double bbls. (superposed). 4.88 inches overall. Early models have long hammer spur and two-armed extractor, but later production have short hammer spur and sliding extractor (a few have no extractor). Fixed front blade sight and grooved rear. Finish: blued, blued w/nickel-plated frame or fully nickel-plated; also w/factory engraving. Grips: walnut, checkered hard rubber, pearl, ivory. Weight: 11 oz. Made 1866-1935. Approximately 150,000 were manufactured. *Note:* During the 70 years of its production, serial numbering of this model was repeated two or three times. Therefore, aside from hammer and extractor differences between the earlier and later models, the best clue to the age of a Double Derringer is the stamping of the company's name on the top of the bbl. or side rib. Prior to 1888, derringers were stamped "E. Remington & Sons"; 1888-1910, "Remington Arms Co."; 1910-35, "Remington Arms-U.M.C. Co."
Standard Model (Early Type I) **$1295**
Standard Model (Late Type II) 850
Factory-engraved Model w/ivory or pearl grips 1950

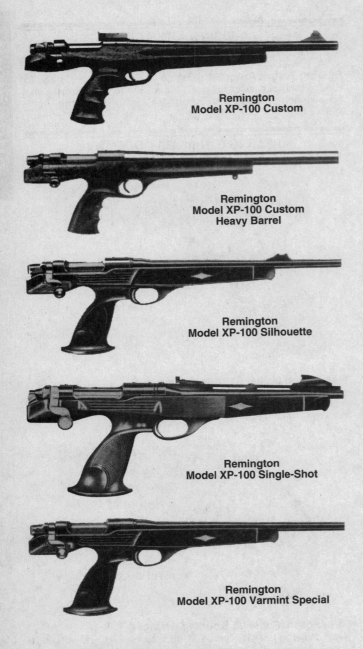

Remington
Model XP-100 Custom

Remington
Model XP-100 Custom
Heavy Barrel

Remington
Model XP-100 Silhouette

Remington
Model XP-100 Single-Shot

Remington
Model XP-100 Varmint Special

Remington New Model Single-Shot Target Pistol $1850
Also called Model 1901 Target. Rolling-block action. Calibers: 22 Short, 22 LR, 44 S&W Russian. 10-inch bbl., half-octagon. 14 inches overall. Weight: 45 oz. (22 cal.). Target sights. Blued finish. Checkered walnut grips and forearm. Made 1901-09.

Remington Model XP-100 Custom Pistol $725
Bolt-action, single-shot, long-range pistol. Calibers: 223 Rem., 7mm-08 or 35 Rem. 14.5-inch bbl., standard contour or heavy. Weight: about 4.25 lbs. Currently in production.

Remington Model XP-100 Silhouette $425
Same general specifications as Model XP-100, except chambered for 7mm BR Rem. And 35 Rem. 14.75-inch bbl. Weighs 4.13 lbs. Made 1987-92.

Remington Model XP-100 Single-Shot Pistol $335
Bolt action. Caliber: 221 Rem. "Fire Ball." 10.5-inch vent rib bbl. 16.75 inches overall. Weight: 3.75 lbs. Adj. rear sight, blade front, receiver drilled and tapped for scope mounts. Blued finish. One-piece brown nylon stock. Made 1963-88.

Remington Model XP-100 Varmint Special $345
Bolt-action, single-shot, long-range pistol. Calibers: 223 Rem., 7mm BR. 14.5-inch bbl. 21.25 inches overall. Weight: about 4.25 lbs. One-piece Du Pont nylon stock w/universal grips. Made 1986-92.

Remington Model XP-100R Custom Repeater
Same general specifications as Model XP-100 Custom, except 4- or 5- shot repeater chambered for 22-250, 223 Rem., 250 Savage, 7mm-08 Rem., 308 Win., 35 Rem. and 350 Rem. Mag. Kevlar-reinforced synthetic or fiberglass stock w/blind magazine and sling swivel studs. Made 1992-94 and 1998 to date.
Model XP-100R (Fiberglass Stock) **$525**
Model XP-100R KS (Kevlar Stock) **695**

Remington XP-22R Rimfire Repeater . . . . (SRP) $432
Bolt-action clip repeater built on Model 541-style action. Calibers: 22 Short, Long, LR. 5-shot magazine. 14.5-inch bbl. Weight: 4.25 lbs. Rem. synthetic stock. Announced 1991, but not produced.

RG REVOLVERS
Mfg. By Rohm Gmbh, Germany
(Imported by R.G. Industries, Miami, Florida)

RG Model 23
SA/DA. 6-shot magazine, swing-out cylinder. Caliber: 22 LR. 1.75- or 3.38-inch bbl. Overall length: 5.13 and 7.5 inches. Weight: 16-17 oz. Fixed sights. Blued or nickel finish. Discontinued 1986.
Blued Finish $80
Nickel Finish 85

RG Model 38S
SA/DA. 6-shot magazine, swing-out cylinder. Caliber: 38 Special. 3- or 4-inch bbl. Overall length: 8.25 and 9.25 inches. Weight: 32-34 oz. Windage-adj. rear sight. Blued finish. Discontinued 1986.
W/Plastic Grips $105
W/Wood Grips 120

ROSSI REVOLVERS
Sáo Leopoldo, Brazil
Manufactured by Amadeo Rossi S.A.
(Imported by Interarms, Alexanderia, Virginia)

Rossi Model 31 DA Revolver $105
Caliber: 38 Special. 5-shot cylinder. 4-inch bbl. Weight: 20 oz. Blued or nickel finish. Discontinued 1985.

Rossi Model 51 DA Revolver $115
Caliber: 22 LR. 6-shot cylinder. 6-inch bbl. Weight: 28 oz. Blued finish. Discontinued 1985.

Rossi Model 68 . . . . $135
Caliber: 38 Special. 5-round magazine. 2- or 3-inch bbl. Overall length: 6.5 and 7.5 inches. Weight: 21-23 oz. Blued finish. Nickel finish available w/3-inch bbl.

Rossi Model 84 DA Revolver $145
Caliber: 38 Special. 6-shot. 3-inch bbl. 8 inches overall. Weight: 27.5 oz. Stainless steel finish. Imported 1984-86.

**Rossi
Model 88 DA Revolver**

**Rossi
Model 851 DA Revolver**

Rossi DA Revolver

Rossi Model 85 DA Revolver **$175**
Same as Model 84 except has vent rib. Imported 1985-86.

Rossi Model 88 DA Revolver
Caliber: 38 Special. 5-shot cylinder. 2- or 3-inch bbl. Weight: 21 oz. Stainless steel finish.
Model 88 (Disc.). **$135**
Model 88 Lady Rossi (Round Butt) **175**

Rossi Model 88/2 DA Revolver **$145**
Caliber: 38 Special. 5-shot cylinder. 2- or 3-inch bbl. 6.5 inches overall. Weight: 21 oz. Stainless steel finish. Imported 1985-87.

Rossi Model 89 DA Revolver **$145**
Caliber: 32 S&W. 6-shot cylinder. 3-inch bbl. 7.5 inches overall. Weight: 17 oz. Stainless steel finish. Imported 1989-90.

Rossi Model 94 DA Revolver **$135**
Caliber: 38 Special. 6-shot cylinder. 3-inch bbl. 8 inches overall. Weight: 29 oz. Imported 1985-88.

Rossi Model 95 (951) Revolver **$150**
Caliber: 38 Special. 6-round magazine. 3-inch bbl. 8 inches overall. Weight: 27.5 oz. Vent rib. Blued finish. Imported 1985-90.

Rossi Model 511 DA Revolver **$145**
Similar to the Model 51, except in stainless steel. Imported 1986-1990.

Rossi Model 515 DA Revolver **$165**
Calibers: 22 LR, 22 Mag. 6-shot cylinder. 4-inch bbl. 9 inches overall. Weight: 30 oz. Red ramp front sight, adj. square-notched rear. Stainless finish. Checkered hardwood grips. Imported from 1992 to date.

Rossi Model 518 DA Revolver **$155**
Similar to the Model 515, except in caliber 22 LR. Imported 1993 to date.

Rossi Model 677 DA Revolver **$165**
Caliber: 357 Mag. 6-shot cylinder. 2-inch bbl. 6.87 inches overall. Weight: 26 oz. Serrated front ramp sight, channel rear. Matte blue finish. Contoured rubber grips. Made 1997 to date.

Rossi Model 720 DA Revolver **$175**
Caliber: 44 Special. 5-shot cylinder. 3-inch bbl. 8 inches overall. Weight: 27.5 oz. Red ramp front sight, adj. square-notched rear. Stainless finish. Checkered Neoprene combat-style grips. Imported from 1992 to date.

Rossi Model 841 DA Revolver **$165**
Same general specifications as Model 84, except has 4-inch bbl. (9 inches overall), weighs 30 oz. Imported 1985-86.

Rossi Model 851 DA Revolver **$155**
Same general specifications as Model 85, except w/3-or 4-inch bbl. 8 inches overall (3-inch bbl.). Weight: 27.5 oz. (3-inch bbl.). Red ramp front sight, adj. square-notched rear. Stainless finish. Checkered hardwood grips. Imported 1991 to date.

Rossi Model 877 DA Revolver **$165**
Same general specifications as Model 677, except stainless steel. Made 1996 to date.

Rossi Model 941 DA Revolver **$150**
Caliber: 38 Special. 6-shot cylinder. 4-inch bbl. 9 inches overall. Weight: 30 oz. Blued finish. Imported 1985-86.

Rossi Model 951 DA Revolver **$160**
Previous designation M95 w/same general specifications.

Rossi Model 971 DA Revolver
Caliber: 357 Magnum. 6-shot cylinder. 2.5-, 4- or 6-inch bbl. 9 inches overall (4-inch bbl.). Weight: 36 oz. (4-inch bbl.). Blade front sight, adj. square-notched rear. Blued or stainless finish. Checkered hardwood grips. Imported 1990 to date.
Blued Finish **$155**
Stainless Finish **165**
W/Compensated Bbl., **add** **15**

Rossi Model 971 VRC DA Revolver **$225**
Same general specifications as Model 971 stainless, except w/ventilated rib and compensated bbl. Made 1996 to date.

**Rossi
Model 971 DA Revolver**

**Ruger Mark I Target
W/6.88-inch Heavy Tappered Barrel**

**Ruger Mark I Target
W/5.5-inch Untappered Bull Barrel**

Ruger Mark II

Rossi Cyclops DA Revolver **$275**
Caliber: 357 Mag. 6-shot cylinder. 6- or 8-inch compensated slab-sided bbl. 11.75 or 13.75 inches overall. Weight: 44 oz. or 51oz. Undercut blade front sight, fully adjustable rear. B-Square mount and rings. Stainless steel finish. Checkered rubber grips. Made 1997 to date.

Rossi DA Revolver **$125**
Calibers: 22 LR, 32 S&W Long, 38 Special. 5-shot (38) or 6-shot cylinder (other calibers). Bbl. lengths: 3-, 6-inch. Weight: 22 oz. (3-inch bbl.). Adj. Rear sight, ramp front. Blued or nickel finish. Wood or plastic grips. Imported 1965-91.

Rossi Sportsman's 22 . . . . **$175**
Caliber: 22 LR. 6-round magazine. 4-inch bbl. 9 inches overall. Weight: 30 oz. Stainless steel finish. Disc. 1991.

RUBY PISTOL
Manufactured by Gabilondo y Urresti, Eibar, Spain, and others

Ruby 7.65mm Automatic Pistol **$165**
Secondary standard service pistol of the French Army in World Wars I and II. Essentially the same as the Alkartasuna (*see* separate listing). Other manufacturers: Armenia Elgoibarresa y Cia., Eceolaza y Vicinai y Cia., Hijos de Angel Echeverria y Cia., Bruno Salaverria y Cia., Zulaika y Cia., all of Eibar, Spain-Gabilondo y Cia., Elgoibar Spain; Ruby Arms Company, Guernica, Spain. Made 1914-22.

RUGER HANDGUNS
Southport, Connecticut
Manufactured by Sturm, Ruger & Co.

Rugers made in 1976 are designated "Liberty" in honor of the U.S. Bicentennial and bring a premium of approximately 25 percent in value over regular models.

AUTOMATIC/SINGLE-SHOT PISTOLS

> **NOTE**
>
> For ease in finding a particular Ruger handgun, the listings are divided into two groupings: Automatic/Single-Shot Pistols (below) and Revolvers, which follow. For a complete listing, please refer to the Index.

Ruger Hawkeye Single-Shot Pistol **$1095**
Built on a SA revolver frame w/cylinder replaced by a swing-out breechblock and fitted w/a bbl. w/integral chamber. Caliber: 256 Magnum. 8.5-inch bbl. 14.5 inches overall. Weight: 45 oz. Blued finish. Ramp front sight, click adj. rear. Smooth walnut grips. Made 1963-65 (3300 produced).

Ruger Mark I Target Model Automatic Pistol
Caliber: 22 LR. 10-shot magazine. 5.25- and 6.88-inch heavy tapered or 5.5-inch untapered bull bbl. 10.88 inches overall (6.88-inch bbl.). Weight: 42 oz. (5.5- or 6.88-inch bbl.). Under-cut target front sight, adj. rear. Blued finish. Hard rubber grips or checkered walnut thumbrest grips. Made 1951-81.
Standard $185
W/Red Medallion 495
Walnut Grips, **add** 15

Ruger Mark II Automatic Pistol
Caliber: 22 LR, standard or high velocity. 4.75- or 6-inch tapered bbl. 10-shot magazine. 8.31 inches overall (4.75-inch bbl.). Weight: 36 oz. Fixed front sight, square notch rear. Blued or stainless finish. Made 1982 to date.
Blued $155
Stainless 185
Bright Stainless (Ltd. prod. 5,000 in 1982) 350

**Ruger
Mark II Bull Barrel
Stainless W/10-inch Barrel**

**Ruger Mark II
Target Model**

**Ruger
Model P-85**

**Ruger
Model P-89**

**Ruger
Model P-90**

Ruger Mark II 22/45 Automatic Pistol

Same general specifications as Ruger Mark II 22 LR, except w/blued or stainless receiver and bbl. in four lengths: 4- inch tappered w/ addition. sights(P4), 4.75-inch tapered w/fixed sights (KP4), 5.25-inch tapered w/adj. sights (KP 514) and 5.5-inch bull (KP 512). Fitted w/Zytel grip frame of the same design as the Model 191145 ACP. Made 1993 to date.

Model KP4 (4.75-inch bbl.) **$165**
Model KP512, KP514 (w/5.5- or 5.25-inch bbl.). **195**
Model P4, P512 (Blued w/4- or 5.5-inch bbl.) **155**

Ruger Mark II Bull Barrel Model

Same as standard Mark II, except for bull bbl. (5.5- or 10-inch). Weight: about 2.75 lbs.

Blued Finish **$185**
Stainless Finish (intro. 1985) **250**

Ruger Mark II Government Model Auto Pistol

Civilian version of the Mark II used by U.S. Armed Forces. Caliber: 22 LR rimfire. 10-shot magazine. 6.88-inch bull bbl. 11.13 inches overall. Weight: 44 oz. Blued or stainless finish. Made from 1986 to date.

Blued Model (MK687G commercial) **$225**
Stainless Steel Model (KMK678G commercial) **260**
W/US Markings (military model). **495**

Ruger Mark II Target Model

Caliber: 22 LR. 10-shot magazine. 4-, 5.5- and 10-inch bull bbl. or 5.25- and 6.88-inch heavy tappered bbl. Weight: 38oz. to 52oz. 11.13 inches overall (6.88-inch bbl.). Made from 1982 to date.

Blued. **$200**
Stainless Steel. **225**

Ruger Model P-85 Automatic Pistol

Caliber: 9mm. DA, recoil-operated. 15-shot capacity. 4.5 inch bbl. 7.84 inches overall. Weight: 32 oz. Fixed rear sight, square-post front. Available w/decocking levers, ambidextrous safety or in DA only. Blued or stainless finish. Made 1987-92.

Blued Finish **$275**
Stainless Steel Finish **295**

Ruger Model P-89 Automatic Pistol

Caliber: 9mm. DA w/slide-mounted safety levers. 15-shot magazine. 4.5-inch bbl. 7.84 inches overall. Weight: 32 oz. Square-post front sight, adj. rear w/3-dot system. Blued or stainless steel finish. Grooved black Xenoy grips. Made from 1992 to date.

P-89 Blued **$285**
P-89 Stainless **325**

**Ruger
Model P-91**

**Ruger
Model P-94**

Ruger Model P-89 DAC/DAO Auto Pistols
Similar to the standard Model P-89, except the P-89 DAC has ambidextrous decocking levers. The P-89 DAO operates in double-action-only mode. Made1991 to date.

P-89 DAC Blued. $295
P-89 DAC Stainless 335
P-89 DAO Stainless 330

Ruger Model P-90, KP90 DA Automatic Pistol
Caliber: 45 ACP. 7-shot magazine. 4.5-inch bbl. 7.88 inches overall. Weight: 33.5 oz. Square-post front sight adj. square-notched rear w/3-dot system. Grooved black Xenoy composition grips. Blued or stainless finish. DAC model has ambidextrous decocking levers. Made 1991 to date.

Model P-90 Blued. $305
Model P-90 DAC (Decockers) 310
Model KP-90 DAC Stainless 315
Model KP-90 DAC (Decockers) 320

Ruger Model P-91 DA Automatic Pistol
Same general specifications as the Model P-90, except chambered for 40 S&W w/12-shot double-column magazine. Made 1992-94.

Model P-91 Standard $315
Model P-91 DAC (Decockers) 320
Model P-91 DAO (DA Only) 325

Ruger Model P-93 Compact Auto Pistol
Similar to the standard Model P-89, except w/3.9-inch bbl. (7.3 inches overall) and weighs 31 oz. Stainless steel finish. Made 1993 to date.

Model P-93 DAC (Decocker) (Disc. 1994) $365
Model P-93 DAO (DA Only) 345

Ruger Model P-94 Automatic Pistol
Similar to the Model P-91, except w/4.25-inch bbl. Calibers: 9mm or 40 S&W. Blued or stainless steel finish. Made 1994 to date.

Model KP-94 DAC (S/S Decocker) $325
Model KP-94 DAO (S/S Double Action Only) 315
Model P-94 DAC (Blued Decocker). 295
Model P-94 DAO (Blued Double Action Only). 285

Ruger Model P-95 Automatic Pistol
Caliber: 9mm Parabellum. 10-shot magazine. 3.9-inch bbl. 7.3 inches overall. Weight: 27 oz. Square-post front sight, drift adjustable rear w/3-Dot system. Molded polymer grip-frame fitted w/stainless or chrome-moly slide. Ambidextrous decocking levers (P-95D) or double action only (DAO). Matte black or stainless finish. Made 1997 to date.

P-95 Blued $235
P-95 Stainless 255

Ruger Standard Model Automatic Pistol
Caliber: 22 LR. 9-shot magazine. 4.75- or 6-inch bbl. 8.75 inches overall (4.75-inch bbl.). Weight: 36 oz. (4.75 inch bbl.). Fixed sights. Blued finish. Hard rubber or checkered walnut grips. Made from 1949 to date. *Note:* In 1951, after the death of Alexander Sturm, the color of the eagle on the grip medallion was changed from red to black as a memorial. Known as the "Red Eagle Automatic," this early type is now a collector's item. Made 1951-81.

W/Red Eagle Medallion $495
W/Black Eagle Medallion. 150
Walnut Grips, **add** 15

> ### NOTE
>
> This section contains only Ruger Revolvers. Automatic and Single-Shot Pistols may be found on the preceding pages. For a complete listing of Ruger handguns, please refer to the Index.

REVOLVERS

**Ruger
Bearcat SA (Old Model)**

Ruger Bearcat SA (Old Model) $295
Aluminum frame. Caliber: 22 LR. 6-shot cylinder. 4-inch bbl. 8.88 inches overall. Weight: 17 oz. Fixed sights. Blued finish. Smooth walnut grips. Made 1958-73.

Ruger Bearcat, Super $325
Same general specifications as Bearcat, except has steel frame. Weight: 25 oz. Made 1971-73.

**Ruger Bisley
Colt 45 Long — (New Model)**

**Ruger Bisley
Single-Six — Small Frame**

Ruger Blackhawk

**Ruger
Blackhawk SA 44**

Ruger Blackhawk SA Convertible

Same as Blackhawk, except has extra cylinder. Caliber combinations: 357 Magnum and 9mm Para., 45 Colt and 45 Automatic. Made 1967-72.

357/9mm Combo (early w/o prefix SN)	**$395**
357/9mm Combo (late w/prefix SN)	**315**
45 LC/45 ACP Combo	**435**

Ruger Blackhawk SA Revolver $255

Calibers: 30 Carbine, 357 Magnum, 41 Magnum, 45 Colt. 6-shot cylinder. Bbl. lengths: 4.63-inch (357, 41, 45 caliber), 6.5-inch (357, 41 caliber), 7.5-inch (30, 45 caliber). 10.13 inches overall (357 Mag. w/4.63-inch bbl.). Weight: 38 oz. (357 w/4.63-inch bbl.). Ramp front sight, adj. rear. Blued finish. Checkered hard rubber or smooth walnut grips. Made 1956-73.

Ruger Blackhawk SA 44 Magnum Revolver

SA w/heavy frame and cylinder. Caliber: 44 Magnum. 6-shot cylinder. 6.5-inch bbl. 12.13 inches overall. Weight: 40 oz. Adj. rear sight, ramp front. Blued finish. Smooth walnut grips. Made 1956-73.

Standard	**$350**
Flat Top	**625**

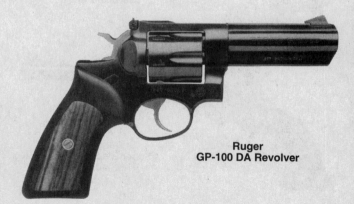

**Ruger
GP-100 DA Revolver**

Ruger Bisley SA Revolver, Large Frame

Calibers: 357 Mag., 41 Mag. 44 Mag., 45 Long Colt. 7.5-inch bbl. 13 inches overall. Weight: 48 oz. Non-fluted or fluted cylinder, no engraving. Ramp front sight, adj. rear. Satin blued or stainless. Made 1986 to date.

Blued Finish	**$265**
Vaquero/Bisley (Blued w/case colored frame)	**275**
Vaquero/Bisley (Stainless Steel)	**260**

Ruger Bisley Single-Six Revolver,
Small Frame **$225**

Calibers: 22 LR and 32 Mag. 6-shot cylinder. 6.5-inch bbl. 11.5 inches overall. Weight: 41 oz. Fixed rear sight, blade front. Made 1986 to date.

Ruger GP-100 DA Revolver

Caliber: 357 Magnum. 4-inch heavy bbl., or 6-inch standard or heavy bbl. Overall length: 9.38 or 11.38 inches. Cushioned grip panels. Made from 1986 to date.

Blued Finish	**$265**
Stainless Steel Finish	**285**

Ruger New Model Bearcat Revolver

Same general specifications as Super Bearcat, except all steel frame and trigger guard. Interlocked mechanism and transfer bar. Calibers: 22 LR and 22WMR. Interchangeable 6-shot cylinders (disc. 1996). Smooth walnut stocks w/Ruger medallion. Made 1994 to date.

Convertible Model (disc. 1996)	**$525**
Standard Model (22 LR only)	**195**

**Ruger Blackhawk
High-Gloss Stainless (New Model)
357 Magnum**

**Ruger Blackhawk
Blued Finish (New Model)**

**Ruger
Single-Six SSM
(New Model)**

**Ruger
Super Blackhawk
(New Model)**

**Ruger
Super Single-Six — Convertable
(New Model)**

**Ruger
Police Service-Six
Stainless Steel**

Ruger New Model Blackhawk Convertible
Same as New Model Blackhawk, except has extra cylinder. Blued finish only. Caliber combinations: 357 Magnum/9mm Para., 44 Magnum/44-40, 45 Colt/45 ACP. Made 1973 to date.

357/9mm Combo **$225**
44/44-40 Combo (Disc.1982) **285**
45 LC/45 ACP Combo (Disc. 1985) **295**

Ruger New Model Blackhawk SA Revolver
Interlocked mechanism. Calibers: 30 Carbine, 357 Magnum, 357 Maximum, 41 Magnum, 44 Magnum, 44 Special, 45 Colt. 6-shot cylinder. Bbl. lengths: 4.63-inch (357, 41, 45 Colt); 5.5 inch (44 Mag., 44 Spec.); 6.5-inch (357, 41, 45 Long Colt); 7.5-inch (30, 45, 44 Special, 44 Mag.); 10.5-inch in 44 Mag. 10.38 inches overall (357 Mag. w/4.63-inch bbl.). Weight: 40 oz. (357 w/4.63-inch bbl.). Adj. rear sight, ramp front. Blued finish or stainless steel; latter only in 357 or 45 LC. Smooth walnut grips. Made 1973 to date.

Blued finish. **$185**
High-Gloss Stainless (357Mag., 45 LC). **295**
Satin Stainless (357Mag., 45 LC). **285**
357 Maximum (1984-85) **425**
44 Magnum (1987 to date in combo only) **295**

Ruger New Model Single-Six SSM Revolver **$275**
Same general specifications as standard Single-Six, except chambered for 32 H&R Magnum cartridge. Bbl. lengths: 4.63, 5.5, 6.5 or 9.5 inches.

Ruger New Model Super Blackhawk SA Revolver
Interlocked mechanism. Caliber: 44 Magnum. 6-shot cylinder. 5.5-inch, 7.5-inch and 10.5-inch bull bbl. 13.38 inches overall. Weight: 48 oz. Adj. rear sight, ramp front. Blued and stainless steel finish. Smooth walnut grips. Made 1973 to date. 5.5-inch bbl. made 1987 to date.

Blued Finish **$245**
High-Gloss Stainless (1994-96) **295**
Satin Stainless steel. **285**

Ruger New Model Super Single-Six Convertible Revolver
SA w/interlocked mechanism. Calibers: 22 LR and 22 WMR. Interchangeable 6-shot cylinders. Bbl. lengths: 4.63, 5.5, 6.5, 9.5 inches. 10.81 inches overall (4.63 inch bbl.). Weight: 33 oz. (4.63-inch bbl.). Adj. rear sight, ramp front. Blued finish or stainless steel; latter only w/5.5- or 6.5-inch bbl. Smooth walnut grips. Made 1972 to date.

Blued Finish **$165**
High-Gloss Stainless (1994-96) **185**
Stainless Steel **205**

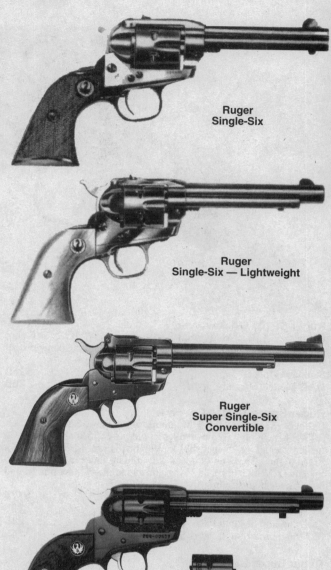

**Ruger
Super Redhawk Stainless DA
Scope-Ring**

**Ruger
Single-Six**

**Ruger
Single-Six — Lightweight**

**Ruger
Super Single-Six
Convertible**

**Ruger Single-Six
Fixed Sight — New Model**

**Ruger
Ruger Vaquero — Stainless Steel**

Ruger Police Service-Six

Same general specifications as Speed-Six, except has square butt. Stainless steel models and 9mm Para. caliber available w/only 4-inch bbl. Made 1971-88.

38 Special, Blued Finish. **$195**
38 Special, Stainless Steel **205**
357 Magnum or 9mm Para., Blued Finish **205**
357 Magnum, Stainless Steel **215**

Ruger Redhawk DA Revolver

Calibers: 357 Mag., 41 Mag., 45 LC, 44 Mag. 6-shot cylinder. 5.5- and 7.5-inch bbl. 11 and 13 inches overall, respectively. Weight: about 52 oz. Adj. rear sight, interchangeable front sights. Stainless finish. Made 1979 to date; 357 Mag. discontinued 1986. Alloy steel model w/blued finish intro. in 1986 in 41 Mag. and 44 Mag. calibers.

Blued Finish **$295**
Stainless Steel **325**

Ruger Super Redhawk Stainless DA Scope-Ring
Revolver **$355**

Same general specifications as standard Redhawk, except w/7.5- or 9-inch bbl. chambered for 44 Mag. only. Integral scope mounting system w/stainless rings.

Ruger Security-Six DA Revolver

Caliber: 357 Magnum, handles 38 Special. 6-shot cylinder. Bbl. lengths: 2.25-, 4-, 6-inch. 9.25 inches overall (4-inch bbl.). Weight: 33.5 oz. (4-inch bbl.). Adj. rear sight, ramp front. Blued finish or stainless steel. Square butt. Checkered walnut grips. Made 1971-85.

Blued Finish **$200**
Stainless Steel **235**

Ruger Single-Six SA Revolver

Calibers: 22 LR, 22 WMR. 6-shot cylinder. Bbl. lengths: 4.63, 5.5, 6.5, 9.5 inches. 10.88 inches overall (5.5-inch bbl.). Weight: about 35 oz. Fixed sights. Blued finish. Checkered hard rubber or smooth walnut grips. Made 1953-73. *Note:* Pre-1956 model w/flat loading gate is worth about twice as much as later version.

Standard **$195**
Convertible (w/2 cylinders, 22 LR/22 WMR) **225**

Ruger Single-Six — Lightweight **$325**

Same general specifications as Single-Six, except has 4.75-inch bbl., lightweight alloy cylinder and frame, 10 inches overall length, weighs 23 oz. Made in 1956.

**Ruger
Ruger Vaquero — Blued**

Ruger SP101 DA Revolver
Calibers: 22 LR, 32 Mag., 9mm, 38 Special+P, 357 Mag. 5- or 6-shot cylinder. 2.25-, 3.06- or 4-inch bbl. Weight: 25-34 oz. Stainless steel finish. Cushioned grips. Made 1988 to date.
Standard Model. **$250**
DAO Model (DA only, spurless hammer) **265**

Ruger Speed-Six DA Revolver
Calibers: 38 Special, 357 Magnum, 9mm Para. 6-shot cylinder. 2.75-, 4-inch bbl. (9mm available only w/2.75-inch bbl.). 7.75 inches overall (2.75-inch bbl.). Weight: 31 oz. (2.75-inch bbl.). Fixed sights. Blued or stainless steel finish; latter available in 38 Special (2.75 inch bbl.), 357 Magnum and 9mm w/either bbl. Round butt. Checkered walnut grips. Made 1973-87.
38 Special, Blued Finish. **$175**
38 Special, Stainless Steel **195**
357 Magnum or 9mm Para., Blued Finish **215**
357 Magnum or 9mm Para., Stainless Steel. **245**

Ruger Super Bearcat
See Ruger Bearcat, Super.

Ruger Super Blackhawk SA Revolver **$275**
Caliber: 44 Magnum. 6-shot cylinder. 7.5-inch bbl. 13.38 inches overall. Weight: 48 oz. Click adj. rear sight, ramp front. Blued finish. Steel or brass grip frame. Plain walnut grips. Made 1959-73.

Ruger Super Single-Six Convertible Revolver
Same general specifications as Single-Six, except has ramp front sight, click-adj. rear. 5.5- or 6.5-inch bbl. only. Two interchangeable cylinders, 22 LR and 22 WMR. Made 1973 to date.
Blued. **$185**
Stainless Steel **255**

Ruger Vaquero SA Revolver
Calibers: 44-40, 44 Magnum, 45 Colt. 6-shot cylinder. Bbl. lengths: 4.63, 5.5, 7.5 inches. 13.63 inches overall (7.5-inch bbl.). Weight: 41 oz. (7.5-inch bbl.). Blade front sight, grooved topstrap rear. Blued w/color casehardened frame or polished stainless finish. Smooth rosewood grips w/Ruger medallion. Made 1993 to date.
Blued w/Color-casehardened Frame. **$250**
Stainless Finish **265**

RUSSIAN SERVICE PISTOLS
Mfd. by Government plants at Tula and elsewhere

Tokarev-type pistols have also been made in Hungary, Poland, Yugoslavia, People's Republic of China, N. Korea.

Russian Model 30 Tokarev Service Automatic **$295**
Modified Colt-Browning type. Caliber: 7.62mm Russian Automatic (also uses 7.63mm Mauser Automatic cartridge). 8-shot magazine. 4.5-inch bbl. 7.75 inches overall. Weight: about 29 oz. Fixed sights. Made 1930 mid-1950s. *Note:* A slightly modified version w/improved locking system and different disconnector was adopted in 1933.

Russian Model PM Makarov Auto Pistol **$145**
Double-action, blowback design. Caliber: 9mm Makarov. 8-shot magazine. 3.8-inch bbl. 6.4 inches overall. Weight: 26 oz. Blade front sight, square-notched rear. Checkered composition grips.

SAKO HANDGUNS
Riihimaki, Finland
Manufactured by Oy Sako Ab

Sako 22-32 Olympic Pistol
Calibers: 22 LR, 22 Short, 32 S&W Long. 5-round magazine. 6- or 8.85- (22 Short) inch bbl. Weight: about 46 oz. (22 LR); 44 oz. (22 Short); 48 oz. (32). Steel frame. ABS plastic, anatomically designed grip. Non-reflecting matte black upper surface and chromium-plated slide. Equipped w/carrying case and tool set. Made 1983-89.
Sako 22-32 Single Pistol **$1050**
Sako Triace, triple-barrel set w/wooden grip **1995**

SAUER HANDGUNS
Mfd. through WW II by J. P. Sauer & Sohn, Suhl, Germany. Now mfd. by J. P. Sauer & Sohn, GmbH, Eckernförde, West Germany

See **also listings under Sig Sauer.**

Sako
Model 22-32 Olympic

SAVAGE ARMS CO.
Utica, New York

Savage Model 101

Sauer Model 1913 Pocket Automatic Pistol **$235**
Caliber: 32 Automatic (7.65mm). 7-shot magazine. 3-inch bbl. 5.88 inches overall. Weight: 22 oz. Fixed sights. Blued finish. Black hard rubber grips. Made 1913-30.

Sauer 1930
Pocket

Sauer Model 1930 Pocket Automatic Pistol **$250**
Authority Model (Behorden Modell). Successor to Model 1913, has improved grip and safety. Caliber: 32 Auto (7.65mm). 7-shot magazine. 3-inch bbl. 5.75 inches overall. Weight: 22 oz. Fixed sights. Blued finish. Black hard rubber grips. Made 1930-38. *Note:* Some pistols made w/indicator pin showing when cocked. Also mfd. w/dural slide and receiver; this type weighs about 7 oz. less than the standard model.

Sauer Model 38H DA Automatic Pistol **$395**
Calibers: 25 Auto (6.35mm), 32 Auto (7.65mm), 380 Auto (9mm). Specifications shown are for 32 Auto model. 7-shot magazine. 3.25-inch bbl. 6.25 inches overall. Weight: 20 oz. Fixed sights. Blued finish. Black plastic grips. Also made in dural model weighing about 6 oz. less. Made 1938-1945. *Note:* This pistol, designated Model 38, was mfd. during WW II for military use. Wartime models are inferior to earlier production, as some lack safety lever.

Sauer Pocket 25 Automatic Pistol **$250**
Smaller version of Model 1913, issued about same time as 32 caliber model. Caliber: 25 Auto (6.35mm). 7-shot magazine. 2.5-inch bbl. 4.25 inches overall. Weight: 14.5 oz. Fixed sights. Blued finish. Black hard rubber grips. Made 1913-30.

Sauer Single-Action Revolvers
See listings under Hawes.

Savage Model 101 SA Single-Shot Pistol **$150**
Barrel integral w/swing-out cylinder. Calibers: 22 Short, Long, LR. 5.5-inch bbl. Weight: 20 oz. Blade front sight, slotted rear, adj. for windage. Blued finish. Grips of compressed, impregnated wood. Made 1960-68.

Savage Model 1907 Automatic Pistol **$295**
Caliber: 32 ACP, 10-shot magazine. 3.25-inch bbl., 6.5 inches overall. Weight: 19 oz. Checkered steel grips marked "Savage Quality," circling an Indian-head logo. Made 1908-20.

Savage Model 1910 Automatic Pistol
Calibers: 32 Auto, 380 Auto. 10-shot magazine (32 cal.), 9-shot (380 cal.). 3.75-inch bbl. (32 cal.), 4.25-inch (380 cal.). 6.5 inches overall (32 cal.), 7 inches (380 cal.). Weight: about 23 oz. Fixed sights. Blued finish. Hard rubber grips. Made in hammerless type w/grip safety or w/exposed hammer spur. Made 1910-17.
32 ACP **$225**
380 ACP **375**

Savage Model 1915 Automatic Pistol
Same general specifications as the Savage Model 1907 except the Model 1915 is hammerless and has a grip safety. It is also chambered for both the 32 and 380 ACP. Made 1915-17.
32 ACP **$395**
380 ACP **450**

Savage U.S. Army Test Model **$5000**
Caliber: 45 ACP, 7-shot magazine w/exposed hammer. An enlarged version of the Model 1910 manufactured for military trials between 1907 and 1911.

Savage Model 1917 Automatic Pistol
Same specifications as 1910 Model, except has spur-type hammer and redesigned heavier grip. Made 1917-28.
32 ACP **$185**
380 ACP **375**

SEARS, ROEBUCK & COMPANY
Chicago, Illinois

Sears/J.C. Higgins Model 80 Auto Pistol **$150**
Caliber: 22 LR. 10-shot magazine. 4.5- or 6.5-inch interchangeable bbl. 10.88 inches overall (6.5-inch bbl.). Weight: 41 oz. (6.5-inch bbl.). Fixed Patridge sights. Blued finish. Checkered grips w/thumbrest.

Sears/J.C. Higgins Model 88 DA Revolver

Sears/J.C. Higgins Model 88 DA Revolver **$95**
Caliber: 22 LR. 9-shot cylinder. 4- or 6-inch bbl. 9.5 inches (4-inch bbl.). Weight: 23 oz. (4-inch bbl.). Fixed sights. Blued or nickel finish. Checkered plastic grips.

Sears/J.C. Higgins Ranger DA Revolver **$115**
Caliber: 22 LR. 9-shot cylinder. 5.5-inch bbl. 10.75 inches overall. Weight: 28 oz. Fixed sights. Blued or chrome finish. Checkered plastic grips.

SECURITY INDUSTRIES OF AMERICA
Little Ferry, New Jersey

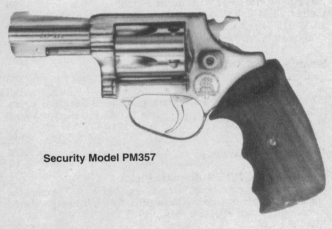

Security Model PM357

Security Model PPM357

Security Model PM357 DA Revolver **$185**
Caliber: 357 Magnum. 5-shot cylinder. 2.5-inch bbl. 7.5 inches overall. Weight: 21 oz. Fixed sights. Stainless steel. Walnut grips. Made 1975-78.

Security Model PPM357 DA Revolver **$175**
Caliber: 357 Magnum. 5-shot cylinder. 2-inch bbl. 6.13 inches overall. Weight: 18 oz. Fixed sights. Stainless steel. Walnut grips. Made from 1976-78. *Note:* Spurless hammer (illustrated) was discontinued in 1977; this model has the same conventional hammer as other Security revolvers.

Security Model PSS38

Security Model PSS38 DA Revolver **$150**
Caliber: 38 Special. 5-shot cylinder. 2-inch bbl. 6.5 inches overall. Weight: 18 oz. Fixed sights. Stainless steel. Walnut grips. Intro. 1973. Discontinued.

R. F. SEDGLEY. INC.
Philadelphia, Pennsylvania

Sedgley Baby Hammerless Ejector Revolver **$425**
DA. Solid frame. Folding trigger. Caliber: 22 Long. 6-shot cylinder. 4 inches overall. Weight: 6 oz. Fixed sights. Blued or nickel finish. Rubber grips. Made c. 1930-39.

L. W. SEECAMP, INC.
Milford, Connecticut

Seecamp Model LWS 25 DAO Pistol **$295**
Caliber: 25 ACP. 7-shot magazine. 2-inch bbl. 4.125 inches overall. Weight: 12 oz. Checkered black polycarbonate grips. Matte stainless finish. No sights. Made 1981-85.

Seecamp Model LWS 32 DAO Pistol
Caliber: 32 ACP. 6-shot magazine. 2-inch bbl. 4.25 inches overall. Weight: 12.9 oz. Ribbed sighting plane with no sights. Checkered black Lexon grips. Stainless steel. Made 1985 to date. Limited production results in inflated resale values.
Matte Stainless Finish. **$495**
Polished Stainless Finish **525**

SHERIDAN PRODUCTS, INC.
Racine, Wisconsin

Sheridan Knocabout Single-Shot Pistol **$140**
Tip-up type. Caliber: 22 LR, Long, Short. 5-inch bbl. 6.75 inches overall. Weight: 24 oz. Fixed sights. Checkered plastic grips. Blued finish. Made 1953-60.

SIG PISTOLS
Neuhausen am Rheinfall, Switzerland
Mfd. by SIG Schweizerische Industrie-Gesellschaft

See also listings under SIG-Sauer.

SIG Model P210-1

**SIG Model P210-6
Target Pistol**

SIG Model P210-1 Automatic Pistol **$1525**
Calibers: 22 LR, 7.65mm Luger, 9mm Para. 8-shot magazine. 4.75-inch bbl. 8.5 inches overall. Weight: 33 oz. (22 cal.) or 35 oz. (7.65mm, 9mm). Fixed sights. Polished blued finish. Checkered wood grips. Made 1949-86.

SIG Model P210-2 . . . . **$1195**
Same as Model P210-1, except has sandblasted finish, plastic grips. Not avail. in 22 LR. Discont. 1987.

SIG Model P210-5 Target Pistol **$1495**
Same as Model P210-2, except has 6-inch bbl., micrometer adj. rear sight, target front sight, adj. trigger stop. 9.7 inches overall. Weight: about 38.3 oz. Discontinued 1997.

SIG Model P210-6 Target Pistol **$1250**
Same as Model P210-2, except has micrometer adj. rear sight, target front sight, adj. trigger stop. Weight: about 37 oz. Discontinued 1987.

SIG P210 22 Conversion Unit **$595**
Converts P210 pistol to 22 LR. Consists of bbl. w/recoil spring, slide and magazine.

SIG SAUER HANDGUNS
Mfd. by J. P. Sauer & Sohn of West Germany, SIG of Switzerland, and other manufacturers

SIG Sauer Model P220 DA Automatic Pistol **$545**
Calibers: 9mm Para., 38 Super, 45 Automatic. 7-shot in 45, 9-shot in other calibers. 4.4-inch bbl. 8 inches overall. Weight: 9mm, 26.5 oz. Fixed sights. Blued finish Checkered plastic grips. Imported 1976 to date. *Note:* Also sold in U.S. as Browning BDA.

SIG Sauer Model P225 DA Automatic **$525**
Caliber: 9mm Para. 8-shot magazine. 3.85-inch bbl. 7 inches overall. Weight: 26.1 oz. Blued finish.

SIG Sauer Model P226 DA Automatic **$550**
Caliber: 9mm Para. 15-shot magazine. 4.4-inch bbl. 7.75 inches overall. Weight: 26.5 oz. Blued finish. Imported 1985 to date.

SIG Sauer Model P228 DA Automatic
Same general specifications as Model P226, except w/3.86-inch bbl. 7.13 inches overall. Blued or K-Kote finish. Imported 1990-97.
Blued Finish **$545**
K-Kote Finish **575**
For Siglite Nite Sights, **add** **80**

SIG Sauer Model P229 DA Automatic
Same general specifications as Model P226, except chambered in 40 S&W w/12-shot magazine. 3.86-inch bbl. 7.13 inches overall. Weight: 30.5 oz. Blued finish. Imported 1991 to date.
Blued Finish **$575**
Blued Finish DAO (double action only) **580**
For Siglite Nite Sights, **add** **80**

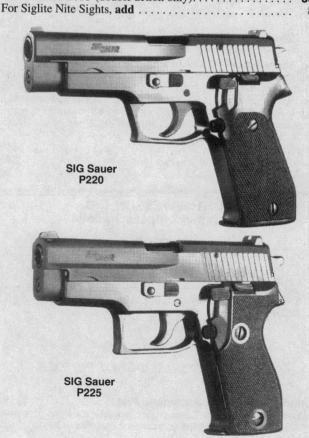

**SIG Sauer
P220**

**SIG Sauer
P225**

SIG Sauer
P230

SIG Sauer Model P230 DA Automatic Pistol
Calibers: 22 LR, 32 Automatic (7.65mm), 380 Automatic (9mm Short), 9mm Police.10-shot magazine in 22, 8-shot in 32, 7-shot in 9mm. 3.6-inch bbl. 6.6 inches overall. Weight: 32 Auto, 18.2 oz. Fixed sights. Blued or stainless finish. Plastic grips. Imported 1976-96.
Blued Finish **$350**
Stainless Finish **395**

SIG Sauer Model P232 Automatic Pistol
Caliber: 380 ACP. 7-shot magazine. 3.6-inch bbl. 6.6 inches overall. Weight: 16.2 oz. or 22.4 oz. (steel frame). Double/singleaction or double action only. Blade front and notch rear drift adjustable sights. Alloy or steel frame. Automatic firing pin lock and heel mounted magazine release. Blue or stainless finish. Stippled black composite stocks. Imported 1997 to date.
Blued Finish **$335**
Stainless Finish **375**
W/Nickel Slide, **add** **35**

Sig Sauer Model P239 Automatic Pistol
Caliber: 357 SIG or 9mm Parabellum. 7- or 8-shot magazine. 3.6-inch bbl. 6.6 inches overall. Weight: 25.2 oz. Double/single action or double action only. Blade front and notch rear adjustable sights. Alloy frame w/stainless slide. Ambidextrous frame mounted magazine release. Blued finish. Stippled black composite stocks. Made 1996 to date.
Blued Finish **$365**
Blued Finish DAO (double action only) **375**
For Siglite Nite Sights, **add** **80**

SMITH & WESSON, INC.
Springfield, Massachusetts

NOTE

For ease in locating a particular S&W handgun, the listings are divided into two groupings: Automatic/Single- Shot Pistols (below) and Revolvers (page 153). For a complete handgun listing, please refer to the Index.

AUTOMATIC/SINGLE-SHOT PISTOLS

Smith & Wesson 32 Automatic Pistol **$2195**
Caliber: 32 Automatic. Same general specifications as 35 caliber model, but barrel is fastened to the receiver instead of hinged. Made 1924-37.

Smith & Wesson Model 22 Sport Series
Caliber: 22 LR. 10-shot magazine. 4-, 5.5- or 7-inch standard (A-series) or bull bbl. (S-series). Single action. 8, 9.5 or 11 inches overall. Weight: 28 oz. to 33 oz. Partridge front sight, fully adjustable rear. Alloy frame w/stainless slide. Blued finish. Black polymer or Dymondwood grips. Made 1997 to date.
Model 22A (w/4-inch bbl.) **$150**
Model 22A (w/5.5-inch bbl.) **165**
Model 22A (w/7-inch bbl.) **175**
Model 22S (w/5.5-inch bbl.) **185**
Model 22S (w/7-inch bbl.) **195**
W/Bull bbl., **add** **40**

Smith & Wesson Model 22A Semi-automatic Target Pistol **$165**
Caliber: 22 LR, 10-shot magazine. 5.5 inch bbl., 9.31 inches overall. Weight: 40 oz. Click-adjustable rear sight, post front sight. Full-length Weaver-style sight-mounting rail. Laminated grips. Made 1997 to date.

Smith & Wesson 35 (1913) Automatic Pistol **$625**
Caliber: 35 S&W Automatic. 7-shot magazine. 3.5-inch bbl. (hinged to frame). 6.5 inches overall. Weight: 25 oz. Fixed sights. Blued or nickel finish. Plain walnut grips. Made 1913-21.

Smith & Wesson Model 39 9mm DA Auto Pistol
Caliber: 9mm Para. 8-shot magazine. 4-inch barrel. Overall length: 7.44- inches. Weight: 26.5 oz. Click adjustable rear sight, ramp front. Blued or nickel finish. Checkered walnut grips. Made 1954-82. *Note:* Between 1954 and 1966, 927 pistols of this model were made w/steel instead of alloy frames.
W/Steel Frame **$895**
W/Alloy Frame **325**

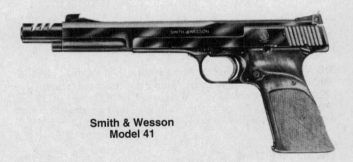

Smith & Wesson
Model 41

Smith & Wesson Model 41 22 Automatic Pistol
Caliber: 22 LR, 22 Short (not interchangeably). 10-shot magazine. Bbl. lengths: 5-, 5.5-, 7.75-inch; latter has detachable muzzle brake. 12 inches overall (7.75-inch bbl.). Weight: 43.5 oz. (7.75-inch bbl.). Click adj. rear sight, undercut Patridge front. Blued finish. Checkered walnut grips w/thumbrest. 1957 to date.
22 LR Model **$425**
22 Short Model **875**
W/Muzzle Brake, **add** **50**

Smith & Wesson Model 46 22 Auto Pistol **$425**
Caliber: 22 LR. 10-shot magazine. Bbl. lengths: 5, 5.5, 7 inches. 10.56 inches overall (7-inch bbl.). Weight: 42 oz. (7-inch bbl.). Click adj. rear sight, undercut Patridge front. Blued finish. Molded nylon grips w/thumbrest. Only 4,000 produced. Made 1957-66.

**Smith & Wesson
Model 46**

**Smith & Wesson
Model 59**

**Smith & Wesson
Model 52**

**Smith & Wesson
Model 422**

**Smith & Wesson
Model 439**

Smith & Wesson Model 52 38 Master Auto

Caliber: 38 Special (midrange wadcutter only). 5-shot magazine. 5-inch bbl. Overall length: 8.63 inches. Weight: 41 oz. Micrometer click rear sight, Patridge front on ramp base. Blued finish. Checkered walnut grips. Made 1961-94.

Model 52 (1961-63)	**$750**
Model 52-1 (1963-71)	**575**
Model 52-2 (1971-93)	**550**
Model 52-A USA Marksman (Less than 100 mfg.)	**2550**

Smith & Wesson Model 59 9mm DA Auto $325

Similar specifications as Model 39, except has 14-shot magazine, checkered nylon grips. Made 1971-81.

Smith & Wesson Model 61 Escort Pocket
Automatic Pistol $230

Caliber: 22 LR. 5-shot magazine. 2.13-inch bbl. 4.69 inches overall. Weight: 14 oz. Fixed sights. Blued or nickel finish. Checkered plastic grips. Made 1970-74.

Smith & Wesson Model 410 Auto Pistol $245

Caliber: 40 S&W. Double action. 10-shot magazine. 4-inch bbl. 7.5 inches overall. Weight: 28.5 oz. Alloy frame w/steel slide. Post front sight, fixed rear w/3-Dot system. Matte blue finish. Checkered synthetic grips w/straight backstrap. Made 1996 to date.

Smith & Wesson Model 411 Auto Pistol $350

Similar to S&W Model 915, except in caliber 40 S&W. 11-shot magazine. Made 1994-96.

Smith & Wesson Model 422 SA Auto Pistol

Caliber: 22 LR. 10-shot magazine. 4.5- or 6-inch bbl. 7.5 inches overall (4.5-inch bbl.). Weight: 22-23.5 oz. Fixed or adjustable sights. Checkered plastic or walnut grips. Blued finish. Made from 1987-96.

Standard Model	**$135**
Target Model	**165**

Smith & Wesson Model 439 9mm Automatic $275

DA. Caliber: 9mm Para. Two 8-round magazines. 4-inch bbl. 7.44 inches overall. Weight: 30 oz. Serrated ramp square front sight, square notch rear. Checkered walnut grips. Blued or nickel finish. Made 1980-88.

**Smith & Wesson
Model 459**

**Smith & Wesson
Model 469**

**Smith & Wesson
Model 639**

**Smith & Wesson
Model 645**

Smith & Wesson Model 457
Compact Auto Pistol **$245**
Caliber: 45 ACP. Double action. 7-shot magazine. 3.75-inch bbl. 7.25 inches overall. Weight: 29 oz. Alloy frame w/steel slide. Post front sight, fixed rear w/3-Dot system. Bobbed hammer. Matte blue finish. Wrap-arround synthetic grip w/straight backstrap. Made 1996 to date.

Smith & Wesson Model 459 DA Automatic
Caliber: 9mm Para. Two 14-round magazines. 4-inch bbl. 7.44 inches overall. Weight: 28 oz. Blued or nickel finish. Made 1981-88.
Standard Model **$325**
FBI Model (Brushed Finish) 575

Smith & Wesson Model 469 9mm Automatic **$315**
DA. Caliber: 9mm Para. Two 12-round magazines. 3.5-inch bbl. 6.88 inches overall. Weight: 26 oz. Yellow ramp front sight, dovetail mounted square-notch rear. Sandblasted blued finish. Optional ambidextrous safety. Made 1982-88.

Smith & Wesson Model 622 SA Auto Pistol
Same general specifications as Model 422, except w/stainless finish. Made from 1990-96.
Standard Model **$175**
Target Model 215

Smith & Wesson Model 639 Automatic **$305**
Caliber: 9mm Para. Two 12-round magazines. 3.5-inch bbl. 6.9 inches overall. Weight: 36 oz. Stainless. Made 1982-88.

Smith & Wesson Model 645 DA Automatic **$355**
Caliber: 45 ACP. 8-shot. 5-inch bbl. Overall length: 8.5 inches. Weight: Approx. 38 oz. Red ramp front, fixed rear sights. Stainless. Made 1986-88.

**Smith & Wesson
Model 659**

Smith & Wesson Model 659 9mm Automatic **$335**
DA. Similar to S&W Model 459, except weight is 39.5 oz. and finish is satin stainless steel. Made 1983-88.

Smith & Wesson Model 669 Automatic **$305**
Caliber: 9mm. 12-shot magazine. Bbl.: 3.5 inches. 6.9 inches overall. Weight: 26 oz. Serrated ramp front sight w/red bar, fixed rear. Nonglare stainless-steel finish. Made 1986-88.

Smith & Wesson Model 745 Automatic Pistol
Caliber: 45 ACP. 8-shot magazine. 5-inch bbl. 8.63 inches overall. Weight: 38.75 oz. Fixed sights. Blued slide, stainless frame. Checkered walnut grips. Similar to the model 645, but w/o DA capability. Made 1987-90.
W/Standard Competition Features **$395**
IPSC Commemorative (first 5,000) 545

Smith & Wesson
Model 669

Smith & Wesson
Model 745

Smith & Wesson
Model 1026

Smith & Wesson
Model 3906

Smith & Wesson Model 908/909/910 Auto Pistols
Caliber: 9mm Parabellum. Double action. 8-shot (Model 908), 9-shot (Model 909) or 10-shot (Model 910) magazine. 3.5- or 4-inch bbl. 6.83 or 7.38 inches overall. Weight: 26oz. to 28.5 oz. Post front sight, fixed rear w/3-Dot system. Matte blue steel slide w/alloy frame. Delrin synthetic wrap-arround grip w/straight backstrap. Made 1994 to date.

Model 908	**$255**
Model 909 (disc 1996)	**265**
Model 910	**255**

Smith & Wesson Model 909/910 Auto Pistols
Caliber: 9mm Para. 9-shot (Model 909) or 10-shot (Model 910) magazine. 4-inch bbl. 7.38 inches overall. Weight: 28 oz. Post front sight, fixed rear. Delrin straight backstrap grips. Blued steel slide w/alloy frame. Made 1994-96.

Model 909	**$265**
Model 910	**255**

Smith & Wesson Model 915 Auto Pistol $265
DA. Caliber: 9mm Para. 15-shot magazine. 4-inch bbl. 7.5 inches overall. Weight: 28.5 oz. Post front sight, fixed square-notched rear w/3-dot system. Xenoy® wraparound grip. Blued steel slide and alloy frame. Made 1992-94.

Smith & Wesson Model 1000 Series DA Auto
Caliber: 10mm. 9-shot magazine. 4.25- or 5-inch bbl. 7.88 or 8.63 inches overall. Weight: about 38 oz. Post front sight, adj. or fixed square-notched rear w/3-dot system. One-piece Xenoy® wraparound grips. Stainless slide and frame. Made 1990-94.

Model 1006 (Fixed Sights, 5" bbl.)	**$425**
Model 1006 (Adj. Sights, 5" bbl.)	**450**
Model 1026 (Fixed Sights, 5" bbl., Decocking Lever)	**435**
Model 1066 (Fixed Sights, 4.25" bbl.)	**425**
Model 1076 (Fixed Sights, 4.25" bbl., Frame-mounted Decocking Lever, Straight Backstrap)	**445**
Model 1076 (same as above w/Tritium Night Sight)	**475**
Model 1086 (same as Model 1076 in DA only)	**435**

Smith & Wesson Model 2206 SA Automatic Pistol
Similar to Model 422, except w/stainless-steel slide and frame, weighs 35-39 oz. Patridge front sight on adj. sight model; post w/white dot on fixed sight model. Plastic grips. Made 1990-96.

Standard Model	**$195**
Target Model	**250**

Smith & Wesson Model 2213 Sportsman Auto $175
Caliber: 22 LR. 8-shot magazine. 3-inch bbl. 6.13 inches overall. Weight: 18 oz. Patridge front sight, fixed square-notched rear w/3-Dot system. Black synthetic molded grips. Stainless steel slide w/alloy frame. Made 1992 to date.

Smith & Wesson Model 2214 Sportsman Auto..... $170
Same general specifications as Model 2214, except w/blued slide and matte black alloy frame. Made from 1990 to date.

Smith & Wesson Model 3904/3906 DA Auto Pistol
Caliber: 9mm. 8-shot magazine. 4-inch bbl. 7.5 inches overall. Weight: 25.5 oz. (Model 3904) or 34 oz. (Model 3906). Fixed or adj. sights. Delrin one-piece wraparound, checkered grips. Alloy frame w/blued carbon steel slide (Model 3904) or satin stainless (Model 3906). Made 1989-91.

Model 3904 w/Adjustable Sights	**$385**
Model 3904 w/Fixed Sights	**355**
Model 3904 w/Novak LC Sight	**360**
Model 3906 w/Adjustable Sights	**415**
Model 3906 w/Novak LC Sight	**395**

**Smith & Wesson
Model 3914**

**Smith & Wesson
Model 4046**

**Smith & Wesson
Model 4506**

Smith & Wesson Model 3913/3914 DA Automatic
Caliber: 9mm Parabellum (Luger). 8-shot magazine. 3.5-inch bbl. 6.88 inches overall. Weight: 25 oz. Post front sight, fixed or adj. square-notched rear. One-piece Xenoy® wraparound grips w/straight backstrap. Alloy frame w/stainless or blued slide. Made 1990 to date.
Model 3913 Stainless **$385**
Model 3913LS Lady Smith Stainless w/contoured
   Trigger Guard 415
Model 3914 Blued Compact (Disc 1995) 370

Smith & Wesson Model 3953/3954 DA Auto Pistol
Same general specifications as Model 3913/3914, except double action only. Made from 1991 to date.
Model 3953 Stainless, Double Action Only **$395**
Model 3954 Blued, Double Action Only (Disc. 1992) 365

Smith & Wesson Model 4000 Series DA Auto
Caliber: 40 S&W. 11-shot magazine. 4-inch bbl. 7.88 inches overall. Weight: 28-30 oz. w/alloy frame or 36 oz. w/stainless frame. Post front sight, adj. or fixed square-notched rear w/2 white dots. Straight backstrap. One-piece Xenoy wraparound grips. Blued or stainless finish. Made between 1990-93.
Model 4003 Stainless w/Alloy Frame **$445**
Model 4004 Blued w/Alloy Frame 425
Model 4006 Stainless Frame, Fixed Sights 465
Model 4006 Stainless Frame, Adj. Sights 495
Model 4026 w/Decocking Lever (Disc. 1994) 475
Model 4043 DA only, Stainless w/Alloy Frame 465
Model 4044 DA only, Blued w/Alloy Frame 445
Model 4046 DA only, Stainless Frame, Fixed Sights 425
Model 4046 DA only, Stainless Frame, Tritium
   Night Sight 565

Smith & Wesson Model 4013/4014 DA Automatic
Caliber: 40 S&W .8-shot capacity. 3.5-inch bbl. 7 inches overall. Weight: 26 oz. Post front sight, fixed Novak LC rear w/3-dot system. One-piece Xenoy wraparound grips. Stainless or blued slide w/alloy frame. Made 1991 to date.
Model 4013 w/Stainless Slide (Disc. 1996) **$425**
Model 4013 Tactial w/Stainless Slide. 450
Model 4014 w/Blued Slide (Disc. 1993) 395

Smith & Wesson Model 4053/4054 DA Auto Pistol
Same general specifications as Model 4013/4014, except double action only. Alloy frame fitted w/blued steel slide. Made 1991-97.
Model 4053 DA only w/Stainless Slide. **$435**
Model 4054 DA only w/Blued Slide (Disc. 1992) 395

Smith & Wesson Model 4500 Series DA Automatic
Caliber: 45 ACP. 6-, 7- or 8-shot magazine. Bbl. lengths: 3.75, 4.25 or 5 inches. 7.13 to 8.63 inches overall. Weight: 34.5 to 38.5 oz. Post front sight, fixed Novak LC rear w/3-Dot system or adj. One-piece Xenoy wraparound grips. Satin stainless finish. Made 1990 to date.
Model 4505 w/Fixed Sights, 5-inch bbl **$465**
Model 4505 w/Novak LC Sight, 5-inch bbl 485
Model 4506 w/Fixed Sights, 5-inch bbl 450
Model 4506 w/Novak LC Sight, 5-inch bbl 465
Model 4513T w/3.75-inch bbl, Tactical Combat 445
Model 4516 w/3.75-inch bbl. 465
Model 4526 w/5-inch bbl., Alloy Frame, Decocking
   Lever, Fixed Sights 445
Model 4553T w/3.75-inch bbl, Tactical Combat. 450
Model 4566 w/4.25-inch bbl., Ambidextrous Safety,
   Fixed Sights 435
Model 4576 w/4.25-inch bbl., Decocking Lever 445
Model 4586 w/4.25-inch bbl., DA only 450

Smith & Wesson Model 5900 Series DA Automatic
Caliber: 9mm. 15-shot magazine. 4-inch bbl. 7.5 inches overall. Weight: 26-38 oz. Fixed or adj. sights. One-piece Xenoy wraparound grips. Alloy frame w/stainless-steel slide (Model 5903) or blued slide (Model 5904) stainless-steel frame and slide (Model 5906). Made 1989/90 to date.
Model 5903 w/Adjustable Sights **$435**
Model 5903 w/Novak LC Rear Sight 425
Model 5904 w/Adjustable Sights 430

**Smith & Wesson
Model 5904
w/Adjustable Sights**

Smith & Wesson Model 5900 Series *(Con't)*

Model 5904 w/Novak LC Rear Sight	405
Model 5905 w/Adjustable Sights	495
Model 5905 w/Novak LC Rear Sight	445
Model 5906 w/Adjustable Sights	440
Model 5906 w/Novak LC Rear Sight	405
Model 5906 w/Tritium Night Sight	550
Model 5924 Anodized frame, Blued Slide	395
Model 5926 Stainless frame, Decocking Lever.	445
Model 5943 Alloy Frame/Stainless Slide, DA only	440
Model 5944 Alloy Frame/Blued Slide, DA only	385
Model 5946 Stainless Frame/Slide, DA only	425

Smith & Wesson Model 6900 Compact Series

Double action. Caliber: 9mm. 12-shot magazine. 3.5-inch bbl. 6.88 inches overall. Weight: 26.5 oz. Ambidextrous safety. Post front sight, fixed Novak LC rear w/3-Dot system. Alloy frame w/blued carbon steel slide (Model 6904) or stainless steel slide (Model 6906). Made 1989 to date.

Model 6904	$365
Model 6906 w/Fixed Sights	395
Model 6906 w/Tritium Night Sight	495
Model 6926 Same as Model 6906 w/Decocking Lever.	425
Model 6944 Same as Model 6904 in DA only	395
Model 6946 Same as Model 6906 in DA only, Fixed Sights	425
Model 6946 w/Tritium Night Sight	495

Smith & Wesson Sigma SW380
Automatic Pistol $225
Caliber: .380 ACP. Double action only. 6-shot magazine. 3inch bbl. Weight: 14 oz. Black integral polymer gripframe w/checkered back and front straps. Fixed channel sights. Polymer frame w/hammerless steel slide. Made 1996 to date.

Smith & Wesson Sigma SW9 Series Automatic Pistol

Caliber: 9mm Parabellum. Double action only. 10-shot magazine. 3.25-, 4- or 4.5-inch bbl. Weight: 17.9 oz. to 24.7 oz. Polymer frame w/hammerless steel slide. Post front sight and drift adjustable rear w/3-Dot system. Gray or black integral polymer gripframe w/checkered back and front straps. Made 1994 to date.

Model SW9C (compact w/3.25-inch bbl.)	$335
Model SW9F (blue slide w/4.5inch bbl.)	345
Model SW9M (compact w/3.25inch bbl.)	255
Model SW9V (stainless slide w/4inch bbl.)	265
Tritium night sight, **add**	100

Smith & Wesson SW40 Series Automatic Pistol
Same general specifications as SW9 Series, except chambered 40 S&W w/4- or 4.5-inch bbl. Weight: 24.4 to 26 oz. Made 1994 to date.

Model SW40C (compact w/4inch bbl.)	$335
Model SW40F (blue slide w/4.5inch bbl.)	345
Model SW40V (stainless slide w/4inch bbl.)	265
Tritium night sight, **add**	100

Smith & Wesson Model 1891 Single-Shot
Target Pistol, First Model
Hinged frame. Calibers: 22 LR, 32 S&W, 38 S&W. Bbl. lengths: 6-, 8- and 10-inch. Approx. 13.5 inches overall (10-inch bbl.). Weight: about 25 oz. Target sights, barrel catch rear adj. for windage and elevation. Blued finish. Square butt, hard rubber grips. Made 1893-1905. *Note:* This model was available also as a combination arm w/accessory 38 revolver bbl. and cylinder enabling conversion to a pocket revolver. It has the frame of the 38 SA Revolver Model 1891 w/side flanges, hand and cylinder stop slots.

Single-shot Pistol, 22 LR	$ 650
Single-shot Pistol, 32 S&W or 38 S&W	1095
Combination Set, Revolver and Single-shot Barrel	1395

Smith & Wesson Model 1891 Single-Shot
Target Pistol, Second Model $595
Similar to the First Model, except side flanges, hand and stop slots eliminated, cannot be converted to revolver, redesigned rear sight. Caliber: 22 LR only. 10-inch bbl. only. Made 1905-09.

Smith & Wesson Perfected Single-Shot Target Pistol
Similar to the Second Model, except has double-action lockwork. Caliber: 22 LR only. 10-inch bbl. Checkered walnut grips, extended square-butt target type. Made 1909-23. *Note:* In 1920 and thereafter, this model was made w/barrels having bore diameter of .223 instead of .226 and tight, short chambering. The first group of these pistols was produced for the U.S. Olympic Team of 1920, thus the designation Olympic Model.

Pre-1920 Type	$625
Olympic Model.	895

Smith & Wesson Straight Line Single-Shot
Target Pistol $1095
Frame shaped like that of an automatic pistol, barrel swings to the left on pivot for extracting and loading, straight line trigger and hammer movement. Caliber: 22 LR. 10-inch bbl. 11.25 inches overall. Weight: 34 oz. Target sights. Blued finish. Smooth walnut grips. Supplied in metal case w/screwdriver and cleaning rod. Made 1925-36.

> **NOTE**
>
> The following section contains only S&W Revolvers. Both Automatic and Single-Shot Pistols may be found in the preceding section. For a complete listing of S&W handguns, please refer to the index.

REVOLVERS

Smith & Wesson Model 1 Hand Ejector
DA Revolver $525
First Model. Forerunner of the 32 Hand Ejector and Regulation Police models, this was the first S&W revolver of the solid-frame, swing-out cylinder type. Top strap of this model is longer than those of later models, and it lacks the usual S&W cylinder latch. Caliber: 32 S&W Long. Bbl. lengths: 3.25-, 4.25-, and 6-inch. Fixed sights. Blued or nickel finish. Round butt, hard rubber stocks. Made 1896-1903.

**Smith & Wesson
Model 1**

**Smith & Wesson
Model 10 (Two-inch Barrel)**

**Smith & Wesson
Model 12 (Two-inch Barrel)**

**Smith & Wesson
Model 13 (Heavy Barrel)**

Smith & Wesson No. 3 SA Frontier **$3595**
Caliber: 44-40 WCF. Bbl. lengths: 4-, 5- and 6.5-inch. Fixed or target sights. Blued or nickel finish. Round, hard rubber or checkered walnut grips. Made 1885-1908.

Smith & Wesson No. 3 SA (New Model) **$2995**
Hinged frame. 6-shot cylinder. Caliber: 44 S&W Russian. Bbl. lengths: 4-, 5-, 6-, 6.5-, 7.5- and 8-inch. Fixed or target sights. Blued or nickel finish. Round, hard rubber or checkered walnut grips. Made 1878-1908. *Note:* Value shown is for standard model. Specialist collectors recognize numerous variations w/a range of higher values.

Smith & Wesson No. 3 SA Target **$2450**
Hinged frame. 6-shot cylinder. Calibers: 32/44 S&W, 38/44 S&W Gallery & Target. 6.5-inch bbl. only. Fixed or target sights. Blued or nickel finish. Round, hard rubber or checkered walnut grips. Made 1887-1910.

Smith & Wesson Model 10 38 Military & Police DA
Also called Hand Ejector Model of 1902, Hand Ejector Model of 1905, Model K. Manufactured substantially in its present form since 1902, this model has undergone numerous changes, most of them minor. Round- or square-butt models, the latter intro. in 1904. Caliber: 38 Special. 6-shot cylinder. Bbl. lengths: 2-(intro. 1933), 4-, 5-, 6- and 6.5-inch (latter discontinued 1915) also 4-inch heavy bbl. (intro. 1957). 11.13 inches overall (square-butt model w/6-inch bbl.). Round-butt model is ¼-inch shorter, weighs about ½ oz. less. Fixed sights. Blued or nickel finish. Checkered walnut grips, hard rubber available in round-butt style. Current Model 10 has short action. Made 1902 to date. *Note:* S&W Victory Model, wartime version of the M & P 38, was produced for the U.S. Government from 1940 to the end of the war. A similar revolver, designated 38/200 British Service Revolver, was produced for the British Government during the same period. These arms have either brush-polish or sandblast blued finish, and most of them have plain, smooth walnut grips, lanyard swivels.
Model of 1902 (1902-05) **$475**
Model of 1905 (1905-40) **400**
38/200 British Service (1940-45) **325**
Victory Model (1942-45) **285**
Model of 1944 (1945-48) **175**
Model 10 (1948 - date) **180**

**Smith & Wesson Model 10 38 Military & Police
Heavy Barrel** **$195**
Same as standard Model 10, except has heavy 4-inch bbl., weighs 34 oz. Made 1957 to date.

Smith & Wesson Model 12 38 M & P Airweight **$225**
Same as standard Military & Police, except has light alloy frame, furnished w/2- or 4-inch bbl. only, weighs 18 oz. (w/2-inch bbl.). Made 1952-86.

**Smith & Wesson Model 12/13 (Air Force Model)
DA Revolver** **$695**
Special "Air Force" Model designed with alloy cylinder and frame to be used as a "Survival Weapon" for air crews. Athough this weapon was actually a first series Model 12, the Air Force stamped "M13" on the top strap. Issued 1953 but recalled for function problems in 1954.

Smith & Wesson Model 13 357 Military/Police **$195**
Same as Model 10 38 Military & Police Heavy Barrel except chambered for 357 Magnum and 38 Special w/3- or 4-inch bbl. Round or squar buttconfiguration. Made 1974-98.

**Smith & Wesson
Model 14**

**Smith & Wesson
Model 17 K-22**

Smith & Wesson Models 14 (K38) and 16 (K32) Masterpiece Revolvers

Calibers: 22 LR, 22 Magnum Rimfire, 32 S&W Long, 38 Special. 6-shot cylinder. DA/SA. Bbl. lengths: 4- (22 WMR only), 6-, 8.38-inch (latter not available in K32). 11.13 inches overall (6-inch bbl.). Weight: 38.5 oz. (6-inch bbl.). Click adj. rear sight, Patridge front. Blued finish. Checkered walnut grips. Made 1947 to date. (Model 16 discontinued 1974, w/only 3,630 produced; reissued 1990-93.)

Model 14 (K-38 Double-Action)	**$235**
Model 14 (K-38 Single Action, 6-inch bbl.)	**255**
Model 14 (K-38 Single Action, 8.38-inch bbl.)	**285**
Model 16 (K-32 Double-Action) 1st Issue	**1025**
Model 16 (K-32 Double-Action)	**250**

**Smith & Wesson
Model 18
(*See* Model 15 for**

**Smith & Wesson
Model 15**

**Smith & Wesson
Model 19 (Round Butt)**

Smith & Wesson Models 15 (38) and 18 (22) Combat Masterpiece DA Revolvers

Same as K-22 and K-38 Masterpiece but w/2- (38) or 4-inch bbl. and Baughman quick-draw front sight. 9.13 inches overall w/4-inch bbl. Weight: 34 oz. (38 cal.). Made 1950 to date.

Model 15	**$255**
Model 18 (Disc. 1985)	**275**

Smith & Wesson Model 17 K-22 Masterpiece
DA Revolver **$265**
Caliber: 22 LR. 6-shot cylinder. Bbl. lengths: 4, 6 or 8.38 inches. 11.13 inches overall (6-inch bbl.). Weight: 38.5 oz. (6-inch bbl.). Patridge-type front sight, S&W micrometer click rear. Checkered walnut Service grips w/S&W monogram. S&W blued finish. Made 1947-93 and 1996 to date.

Smith & Wesson Model 19 357 Combat Magnum
DA Revolver **$245**
Caliber: 357 Magnum. 6-shot cylinder. Bbl lengths: 2.5 (round butt), 4, or 6 inches. 9.5 inches overall (4-inch bbl.). Weight: 35 oz. (4-inch bbl.). Click adj. rear sight, ramp front. Blued or nickel finish. Target grips of checkered Goncalo Alves. Made 1956 to date (2.5- and 6-inch bbls. were discontinued in 1991).

**Smith & Wesson
Model 19 (Square Butt)**

Smith & Wesson Model 20

Smith & Wesson Model 21

Smith & Wesson Model 22

Smith & Wesson Model 22/32 Kit Gun

Smith & Wesson Model 22/32 Target Revolver

Smith & Wesson Model 20 38/44 Heavy Duty DA

Caliber: 38 Special. 6-shot cylinder. Bbl. lengths: 4, 5 and 6.5 inches.10.38 inches overall (5-inch bbl.). Weight: 40 oz. (5-inch bbl.). Fixed sights. Blued or nickel finish. Checkered walnut grips. Short action after 1948. Made 1930-56 and 1957-67.
Pre-World War II $495
Postwar 275

Smith & Wesson Model 21 1950 44 Military
DA Revolver $1425
Postwar version of the 1926 Model 44 Military. Caliber: .44 Special, 6-shot cylinder. Bbl.: 4-, 5- and 6.5-inch, 11.75 inches overall w/6.5-inch bbl. Weight: 39.5 oz. w/6.5-inch bbl. Fixed front sight w/square-notch rear sight; target model has micrometer click rear sight adj. for windage and elevation. Checkered walnut grips w/S&W monogram. Blued or nickel finish. Made 1950-67.

Smith & Wesson Model 22 1950 Army DA $1125
Postwar version of the 1917 Army w/same general specifications, except redesigned hammer. Made 1950-67.

Smith & Wesson 22/32 Kit Gun $595
Same as 22/32 Target, except has 4-inch bbl., round grips, 8 inches overall, weighs 21 oz. Made 1935-53.

Smith & Wesson 22/32 Target DA Revolver $795
Also known as the Bekeart Model. Design based upon "32 Hand Ejector." Caliber: 22 LR (recessed head cylinder for high-speed cartridges intro. 1935). 6-shot cylinder. 6-inch bbl. 10.5 inches overall. Weight: 23 oz. Adj. target sights. Blued finish. Checkered walnut grips. Made 1911-53. *Note:* In 1911, San Francisco gun dealer Phil Bekeart, who suggested this model, received 292 pieces. These are the true "Bekeart Model" revolvers worth about double the value shown for the standard 22/32 Target.

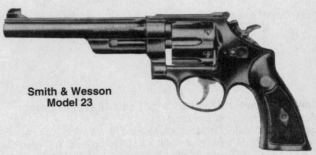

Smith & Wesson Model 23

Smith & Wesson Model 23 38/44 Outdoorsman DA Revolver

Target version of the 38/44 Heavy Duty. 6-inch bbl. only. Weight: 41.75 oz. Target sights, micrometer-click rear on postwar models. Blued finish only. 1950 model has ribbed barrel, redesigned hammer. Made 1930-67.
Prewar.. $595
Postwar 495

Smith & Wesson Model 24 1950 44 Target
DA Revolver $625
Postwar version of the 1926 Model 44 Target w/same specifications, except has redesigned hammer, ribbed bbl., micrometer click rear sight. Made 1950-67.

Smith & Wesson Model 25 1955 45 Target
DA Revolver $395
Same as 1950 Model 44 Target but chambered for 45 ACP cartridge, and later, the 45 LC cartridge. Made 1950, in several variations, to 1991.

Smith & Wesson Model 27 357 Magnum DA

Caliber: 357 Magnum. 6-shot cylinder. Bbl. lengths: 3.5-, 5-, 6-, 6.5- and 8.38-inch. 11.38 inches overall (6-inch bbl.). Weight: 44 oz. (6-inch bbl.). Adj. target sights, Baughman quick-draw ramp front sight on 3.5-inch bbl. Blued or nickel finish. Checkered walnut grips. Made 1935-94. *Note:* Until 1938, the 357 Magnum was custom made in any barrel length from 3.5-inch to 8.75-inch, each of these revolvers was accompanied by a registration certificate and has its registration number stamped on the inside of the yoke. Postwar Magnums have a redesigned hammer w/shortened fall and the new S&W micrometer click rear sight.

Prewar Registered Model	**$995**
Prewar Model without Registration Number	**695**
Late Model w/8.38-inch bbl	**325**
Late Model, other bbl. lengths	**295**

Smith & Wesson Model 28 Highway Patrolman $235

Caliber: 357 Magnum. 6-shot cylinder. Bbl. lengths: 4- or 6-inch. 11.25 inches overall (6-inch bbl.). Weight: 44 oz. (6-inch bbl.). Adj.rear sight, ramp front. Blued finish. Checkered walnut grips, Magna or target type. Made 1954-86.

Smith & Wesson Model 29 44 Magnum DA Revolver

Caliber: 44 Magnum. 6-shot cylinder. Bbl. lengths: 4-, 6.5-, 8.38-inch. 11.88 inches overall (6.5-inch bbl.). Weight: 47 oz. (6.5-inch bbl.). Click adj. rear sight, ramp front. Blued or nickel finish. Checkered Goncalo Alves target grips. Made 1956 to date.

3-Screw Model (1962-83)	**$525**
4-Screw Model (1957-61)	**650**
5-Screw Model (1956-57)	**725**
Late Series Model (Disc. 1998)	**425**

Smith & Wesson Model 30 32 Hand Ejector
DA Revolver $295

Caliber: 32 S&W Long. 6-shot cylinder. Bbl. lengths: 2- (intro. 1949), 3-, 4- and 6-inch. 8 inches overall (4-inch bbl.). Weight: 18 oz. (4-inch bbl.). Fixed sights. Blued or nickel finish. Round, checkered walnut or hard rubber grips. Made 1903-76.

Smith & Wesson Models 31 & 33 Regulation
Police DA Revolver

Same basic type as 32 Hand Ejector, except has square buttgrips. Calibers: 32 S&W Long (Model 31) 38 S&W (Model 33). 6-shot cylinder in 32 cal., 5-shot in 38 caliber. Bbl. lengths: 2- (intro. 1949), 3-, 4- and 6-inch in 32 cal., 4-inch only in 38 cal. 8.5 inches overall (4-inch bbl.). Weight: 18 oz. (38 cal. w/4-inch bbl.), 18.75 oz. (32 cal. w/4-inch bbl.). Fixed sights. Blued or nickel finish. Checkered walnut grips. Made from 1917. Model 33 discontinued in 1974; Model 31 discontinued in 1992.

Model 31	**$215**
Model 33	**250**

Smith & Wesson 32 Double-Action Revolver $195

Hinged frame. Caliber: 32 S&W. 5-shot cylinder. Bbl. lengths: 3-, 3.5- and 6-inch. Fixed sights. Blued or nickel finish. Hard rubber grips. Made from 1880-1919. *Note:* Value shown applies generally to the several varieties. Exception is the rare first issue of 1880 (identified by squared sideplate and serial no. 1 to 30) valued at **$2,500.**

Smith & Wesson Model 32 Terrier DA $225

Caliber: 38 S&W. 5-shot cylinder. 2-inch bbl., 6.25 inches overall. Weight: 17 oz. Fixed sights. Blued or nickel finish. Checkered walnut or hard rubber grips. Built on 32 Hand Ejector frame. Made 1936-74.

Smith & Wesson 32-20 Military & Police DA

Same as M & P 38, except chambered for 32-20 Winchester cartridge. First intro. in the 1899 model, M & P Revolvers were produced in this caliber until about 1940. Values same as M & P 38 models.

Smith & Wesson
Model 24 Target

Smith & Wesson
Model 25 Target

Smith & Wesson
Model 27

Smith & Wesson
Model 28

Smith & Wesson Model 34 1953 22/32 Kit Gun..... $230

Same general specifications as previous 22/32 Kit Gun, except w/2-inch or 4-inch bbl. and round or square buttgrips, blued or nickel finish. Made 1936-91.

Smith & Wesson Model 35 1953 22/32 Target...... $295

Same general specifications as previous model 22/32 Target, except has micrometer-click rear sight. Magna type target grips. Weight: 25 oz. Made 1953-74.

**Smith & Wesson
Model 29**

**Smith & Wesson
Model 31**

**Smith & Wesson
Model 32**

**Smith & Wesson
Model 34**

**Smith & Wesson
Model 36**

**Smith & Wesson
Model 37**

**Smith & Wesson
Model 38
Bodyguard Airweight**

Smith & Wesson Model 36 Chiefs Special DA **$205**
Based on 32 Hand Ejector w/frame lengthened to permit longer cylinder for 38 Special cartridge. Caliber: 38 Special. 5-shot cylinder. Bbl. lengths: 2- or 3-inch. 6.5 inches overall (2-inch bbl.). Weight: 19 oz. Fixed sights. Blued or nickel finish. Checkered walnut grips, round or square butt. Made 1952 to date.

**Smith & Wesson Model 37 Airweight Chiefs
Special** **$225**
Same general specifications as standard Chiefs Special except has light alloy frame, weighs 12.5 oz. w/2-inch bbl., blued finish only. Made 1954 to date.

**Smith & Wesson
Model 48**

Smith & Wesson Model 38 Bodyguard Airweight DA Revolver

"Shrouded" hammer. Light alloy frame. Caliber: 38 Special. 5-shot cylinder. 2- or 3-inch bbl. 6.38 inches overall w/2-inch bbl. Weight: 14.5 oz. Fixed sights. Blued or nickel finish. Checkered walnut grips. Made 1955-98.

Blued Model	$215
Nickel Model	235
Early Model (Pinned & Recessed, Pre-1981)	250

Smith & Wesson 38 DA Revolver $860

Hinged frame. Caliber: 38 S&W. 5-shot cylinder. Bbl. lengths: 4-, 4.25-, 5-, 6-, 8- and 10- inch. Fixed sights. Blued or nickel finish. Hard rubber grips. Made 1880-1911. *Note:* Value shown applies generally to the several varieties. Exceptions are the first issue of 1880 (identified by squared sideplate and serial no. 1 to 4,000) and the 8- and 10-inch bbl. models of the third issue (1884-95).

Smith & Wesson Model 38 Hand Ejector DA

Military & Police — First Model. Resembles Colt New Navy in general appearance, lacks bbl. lug and locking bolt common to all later S&W hand ejector models. Caliber: 38 Long Colt. 6-shot cylinder. Bbl. lengths: 4-, 5-, 6- and 6.5-inch. 11.5 inches overall (6.5-inch bbl.). Fixed sights. Blued or nickel finish. Round, checkered walnut or hard rubber grips. Made 1899-1902.

Standard Model (Civilian Issue)	$215
Navy Model (Marked USN, 1000 issued)	1495

Smith & Wesson 38 Military & Police Target DA $960

Target version of the Military & Police w/standard features of that model. Caliber: 38 Special. 6-inch bbl. Weight: 32.25 oz. Adj. target sights. Blued finish. Checkered walnut grips. Made 1899-1940. For values, **add $175** to those shown for corresponding M&P 38 models.

Smith & Wesson Model 38 Perfected DA $540

Hinged frame. Similar to earlier 38 DA Model, but heavier frame, side latch as in solid-frame models improved lockwork. Caliber: 38 S&W. 5-shot cylinder. Bbl. lengths: 3.25 , 4, 5 and 6 inches. Fixed sights. Blued or nickel finish. Hard rubber grips. Made 1909-20.

Smith & Wesson Model 40 Centennial DA Hammerless Revolver $375

Similar to Chiefs Special, but has Safety Hammerless-type mechanism w/grip safety. 2-inch bbl.. Weight: 19 oz. Made 1953-74.

Smith & Wesson Model 42 Centennial Airweight $425

Same as standard Centennial model, except has light alloy frame, weighs 13 oz. Made 1954-74.

Smith & Wesson Model 43 1955 22/32 Kit Gun Airweight $350

Same as Model 34 Kit Gun, except has light alloy frame, furnished w/3.5-inch bbl. only, weighs 14.25 oz., square grip. Made 1954-74.

Smith & Wesson Model 44 1926 Military DA Revolver

Basically the same as the early New Century model, having the extractor rod casing but lacking the "Triple Lock" feature. Caliber: 44 S&W Special. 6-shot cylinder. Bbl. lengths: 4, 5 and 6.5 inches. 11.75 inches overall (6.5-inch bbl.). Weight: 39.5 oz. (6.5-inch bbl.). Blued or nickel finish. Checkered walnut grips. Made 1926-41.

Standard Model	$ 725
Target Model w/6.5-inch bbl., Target Sights, Blued	2950

Smith & Wesson 44 and 38 DA Revolvers

Also called Wesson Favorite (lightweight model), Frontier (caliber 44-40). Hinged frame. 6-shot cylinder. Calibers: 44 S&W Russian, 38-40, 44-40. Bbl. lengths: 4-, 5-, 6- and 6.5-inch. Weight: 37.5 oz. (6.5-inch bbl.). Fixed sights. Blued or nickel finish. Hard rubber grips. Made 1881-1913, Frontier discontinued 1910.

Standard Model, 44 Russian	$1495
Standard Model, 38-40	2995
Frontier Model	1595
Favorite Model	6295

Smith & Wesson 44 Hand Ejector, Second Model DA Revolver $595

Basically the same as New Century, except crane lock ("Triple Lock" feature) and extractor rod casing eliminated. Calibers: 44 S&W Special 44-40 Win. 45 Colt. Bbl. lengths: 4-, 5-, 6.5- and 7.5-inch. 11.75 inches overall (6.5-inch bbl.). Weight: 38 oz. (6.5-inch bbl.). Fixed sights. Blued or nickel finish. Checkered walnut grips. Made 1915-37.

Smith & Wesson Model 48 (K-22) Masterpiece M.R.F. DA Revolver $265

Caliber: 22 Mag. and 22 RF. 6-shot cylinder. Bbl. lengths: 4 and 6 inches. 11.13 inches overall. Weight: 39 oz. Adj. rear sight, ramp front. Made 1959-86.

**Smith & Wesson
Model 49 Bodyguard**

Smith & Wesson Model 49 Bodyguard $215

Same as Model 38 Bodyguard Airweight, except has steel frame, weighs 20.5 oz. Made 1959-96.

**Smith & Wesson
Model 57**

**Smith & Wesson
Model 60**

**Smith & Wesson
Model 63**

**Smith & Wesson
Model 64**

Smith & Wesson Model 51 1960 22/32 Kit Gun $355
Same as Model 34 Kit Gun, except chambered for 22 WMR. 3.5-inch bbl. Weighs 24 oz. Made 1960-74.

Smith & Wesson Model 53 22 Magnum DA $625
Caliber: 22 Rem. Jet C.F. Magnum. 6-shot cylinder (inserts permit use of 22 Short, Long, or LR cartridges). Bbl. lengths: 4-, 6-, 8.38-inches. 11.25 inches (6-inch bbl.). Weight: 40 oz. (6-inch bbl.). Micrometer-click rear sight ramp front. Checkered walnut grips. Made 1960-74.

Smith & Wesson Model 57 41 Magnum DA Revolver
Caliber: 41 Magnum. 6-shot cylinder. Bbl. lengths: 4-, 6-, 8.38-inches. Weight: 40 oz. (6-inch bbl.). Micrometer click rear sight, ramp front. Target grips of checkered Goncalo Alves. Made 1964-93.
W/8.38-inch bbl. $375
Other bbl. lengths 295

Smith & Wesson Model 58 41 Military & Police DA Revolver $395
Caliber: 41 Magnum. 6-shot cylinder. 4-inch bbl.. 9.25 inches overall. Weight: 41 oz. Fixed sights. Checkered walnut grips. Made 1964-82.

Smith & Wesson Model 60 Stainless DA
Caliber: 38 Special or 357 Magnum. 5-shot cylinder. Bbl. lengths: 2-, 2.1 or 3 inches. 6.5 or 7.5 inches overall. Weight: 19 to 23 oz. Square-notch rear sight, ramp front. Satin finish stainless steel. Made 1965-96 (38 Special) and 1996 to date 357/38).
38 Special (Disc. 1996) $235
357 Mag. 265
Lady Smith (W/Smaller Grip) 250

Smith & Wesson Model 63 (1977) Kit Gun DA $250
Caliber: 22 LR. 6-shot cylinder. 2- or 4-inch bbl. 6.5 or 8.5 inches overall. Weight: 19 to 24.5 oz. Adj. rear sight, ramp front. Stainless steel. Checkered walnut or synthetic grips.

Smith & Wesson Model 64 38 M&P Stainless $205
Same as standard Model 10, except satin-finished stainless steel, square butt w/4-inch heavy bbl. or round butt w/2-inch bbl. Made 1970 to date.

Smith & Wesson Model 65 357 Military/Police Stainless
Same as Model 13, except satin-finished stainless steel. Made 1974 to date.
Model 65 M&P $215
Model 65 Lady Smith (W/Smaller Grip) 225

**Smith & Wesson
Model 66 Combat Magnum**

HANDGUNS

Smith & Wesson Model 66 357 Combat Magnum Stainless $265
Same as Model 19, except satin-finished stainless steel. Made 1971 to date.

Smith & Wesson Model 67 38 Combat Masterpiece Stainless $225
Same as Model 15, except satin-finished stainless steel available only w/4-inch bbl. Made 1972-88 and 1991 to date.

Smith & Wesson 125th Anniversary Commemorative
Issued to celebrate the 125th anniversary of the 1852 partnership of Horace Smith and Daniel Baird Wesson. Standard Edition is Model 25 revolver, caliber 45 Colt w/6.5-inch bbl., bright blued finish, gold-filled bbl. roll mark "Smith & Wesson 125th Anniversary," sideplate marked w/gold-filled Anniversary seal, smooth Goncalo Alves grips, in presentation case w/nickel silver Anniversary medallion and book, *125 Years w/Smith & Wesson,* by Roy Jinks. Deluxe Edition is same, except revolver is Class A engraved w/gold-filled seal on sideplate, ivory grips, Anniversary medallion is sterling silver and book is leather bound. Limited to 50 units. Total issue is 10,000 units, of which 50 are Deluxe Edition and two are a Custom Deluxe Edition not for sale. Made in 1977. Values are for revolvers in new condition.
Standard Edition .... $ 595
Deluxe Edition 1695

Smith & Wesson 317 Airlite DA Revolver
Caliber: 22 LR. 8-shot cylinder. 1.88- or 3-inch bbl. 6.3 or 7.2 inches overall. Weight: 9.9 oz. or 11 oz. Ramp front sight, notched frame rear. Aluminum, carbon fiber, stainless and titanium construction. Brushed aluminum finish. Synthetic or Dymondwood grips. Made 1997 to date.
Model 317 (w/1.88-inch bbl.)....................... $205
Model 317 (w/3-inch bbl.) 235
Model 317 (w/Dymondwood grips), **add**................ 30

Smith & Wesson Model 547 DA Revolver $270
Caliber: 9mm. 6-shot cylinder. Bbl. lengths: 3 or 4 inches. 7.31 inches overall. Weight: 32 oz. Square-notch rear sight, ramp front. Discontinued 1986.

Smith & Wesson Model 581 Revolver
Caliber: 357 Magnum. Bbl.: 4 inches. Weight: 34 oz. Serrated ramp front sight, square notch rear. Checkered walnut grips. Made 1985-92.
Blued Finish $195
Nickel Finish 230

Smith & Wesson Model 586 Distinguished Combat Magnum $255
Caliber: 357 Magnum. 6-shot. Bbl. lengths: 4, 6 and 8.38 inches. Overall length: 9.75 inches (4-inch bbl.). Weight: 42, 46, 53 oz., respectively. Red ramp front sight, micrometer-click adj. rear. Checkered grip. Blued or nickel finish. Made 1980 to date.

Smith & Wesson Model 610 DA Revolver $395
Similar to Model 625, except in caliber 10mm. Magna classic grips. Made 1990-91 and 1998 to date.

Smith & Wesson Model 617 DA Revolver
Similar to Model 17, except in stainless. Made 1990 to date.
Semi-Target Model w/4- or 6-inch bbl. $245
Target Model w/6-inch bbl. 295
Target Model w/8.38-inch bbl. 325
W/10-Shot Cylinder, **add** 50

Smith & Wesson
Model 67

Smith & Wesson
125th Commemorative
Deluxe Edition

Smith & Wesson
Model 586
Distinguished Combat Magnum

Smith & Wesson Model 624 Double-Action Revolver
Same general specifications as Model 24, except satin-finished stainless steel. Limited production of 10,000. Made 1985-86.
Model 624 w/4-inch bbl. $295
Model 624 w/6.5-inch bbl. 325
Model 624 w/3-inch bbl., rd. butt (Lew Horton Spec.)..... 350

Smith & Wesson Model 625 DA Revolver $365
Same general specifications as Model 25, except 3-, 4- or 5-inch bbl., round-butt Pachmayr grips and satin stainless steel finish. Made 1989 to date.

Smith & Wesson Model 627 DA Revolver $375
Same general specifications as Model 27, except satin stainless steel finish. Made 1989-91.

Smith & Wesson Model 629 DA Revolver
Same as Model 29 in 44 Magnum, except in stainless steel. Classic made 1990 to date.
Model 629 (3-inch bbl, Backpacker) $435
Model 629 (4- and 6-inch bbl.) 385
Model 629 (8.38-inch bbl.) 385
Model 629 Classic (5- and 6.5-inch bbl.)............... 395
Model 629 Classic (8.38-inch bbl.) 425

**Smith & Wesson
Model 625**

**Smith & Wesson
Model 629**

**Smith & Wesson
Model 642
Centennial Airweight**

Smith & Wesson Model 631 DA Revolver
Similar to Model 31, except chambered for 32 H&R Mag. Goncalo Alves combat grips. Made 1991-92.

Fixed Sights, 2-inch bbl. **$250**
Adjustable Sights, 4-inch bbl. **275**

Smith & Wesson Model 632 Centennial DA **$265**
Same general specifications as Model 640, except chambered for 32 H&R Mag. 2-inch bbl. Weight: 15.5 oz. Stainless slide w/alloy frame. Santoprene combat grips. Made in 1991.

Smith & Wesson Model 640 Centennial DA **$275**
Caliber: 38 Special. 5-shot cylinder. 2-, 2.1 or 3-inch bbl. 6.31 inches overall. Weight: 20-22 oz. Fixed sights. Stainless finish. Smooth hardwood service grips. Made 1990 to date.

Smith & Wesson Model 642 Centennial Airweight DA Revolver
Same general specifications as Model 640, except w/stainless steel/aluminum alloy frame and finish. Weight of 15.8 oz. Santoprene combat grips. Made 1990-93.

Model 640 Centennial **$275**
Model 640 Lady Smith (W/Smaller Grip) **295**

Smith & Wesson Model 648 DA Revolver **$250**
Same general specifications as Models 17/617, except in stainless and chambered for 22 Magnum. Made 1990-93.

Smith & Wesson Model 649 Bodyguard DA **$245**
Caliber: 38 Special. 5-shot cylinder. Bbl. length: 2 inches. 6.25 inches overall. Weight: 20 oz. Square-notch rear sight ramp front. Stainless frame and finish. Made 1986 to date.

Smith & Wesson Model 650 Revolver **$195**
Caliber: 22 Mag. 6-shot cylinder. 3-inch bbl. 7 inches overall. Weight: 23.5 oz. Serrated ramp front sight, fixed square-notch rear. Round butt, checkered walnut monogrammed grips. Stainless steel finish. Made 1983-86.

Smith & Wesson Model 651 Stainless DA **$225**
Caliber: 22 Mag. Rimfire. 6-shot cylinder. Bbl. length: 3 and 4 inches. 7 and 8.63 inches, respectively, overall. Weight: 24.5 oz. Adj. rear sight, ramp front. Made 1983-87 and 1990 to date..

Smith & Wesson Model 657 Revolver
Caliber: 41 Mag. 6-shot cylinder. Bbl.: 4, 6 or 8.4 inches. 9.6, 11.4, and 13.9 inches overall. Weight: 44.2, 48 and 52.5 oz. Serrated black ramp front sight on ramp base click rear, adj. for windage and elevation. Satin finished stainless steel. Made 1986 to date.

W/4- or 6-inch bbl. **$295**
W/8.4-inch bbl. **350**

Smith & Wesson Model 681 . . . . **$220**
Same as S&W Model 581, except in stainless finish only. Made 1991-93.

Smith & Wesson Model 686 . . . . **$275**
Same as S&W Model 586 Distinguished Combat Magnum, except in stainless finish w/additional 2.5-inch bbl. Made 1991 to date.

Smith & Wesson Model 686 Plus
Same as standard Model 686 Magnum, except w/7-shot cylinder and 2.5-, 4- or 6-inch bbl. Made 1996 to date.

686 Plus Model (w/2.5-inch bbl.) **$275**
686 Plus Model (w/4-inch bbl.) **285**
686 Plus Model (w/6-inch bbl.) **295**

**Smith & Wesson
Model K-22
Outdoorsman**

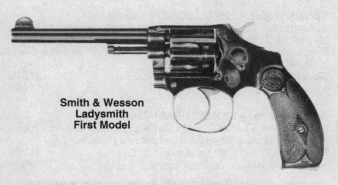

**Smith & Wesson
Ladysmith
First Model**

**Smith & Wesson
Ladysmith
Second Model**

**Smith & Wesson
Regulation Police Target**

**Smith & Wesson
Safety Hammerless**

Smith & Wesson Model 696 .... $275

Caliber: 44 S&W Special. L-Frame w/5-shot cylinder. 3-inch shrouded bbl. 8.38 inches overall. Weight: 48 oz. Red ramp front sight, micrometer-click adj. rear. Checkered synthetic grip. Satin stainless steel. Made 1997 to date.

Smith & Wesson Model 940 Centennial DA $265

Same general specifications as Model 640, except chambered for 9mm. 2- or 3-inch bbl. Weight: 23-25 oz. Santoprene combat grips. Made 1991 to date.

Smith & Wesson Model 1891 SA Revolver

Hinged frame. Caliber: 38 S&W. 5-shot cylinder. Bbl. lengths: 3.25, 4-, 5- and 6-inch. Fixed sights. Blued or nickel finish. Hard rubber grips. Made 1891-1911. *Note:* Until 1906, an accessory single-shot target bbl. *(see* Model 1891 Single-Shot Target Pistol) was available for this revolver.

Revolver only $1595
Set w/22 single-shot bbl. 2450

Smith & Wesson 1917 Army DA Revolver

Caliber: 45 Automatic, using 3-cartridge half-moon clip or 45 Auto Rim, without clip. 6-shot cylinder. 5.5-inch bbl. 10.75 inches overall. Weight: 36.25 oz. Fixed sights. Blued finish (blue-black finish on commercial model, brush polish on military). Checkered walnut grips (commercial model, smooth on military). Made under U.S. Government contract 1917-19 and produced commercially 1919-1941. *Note:* About 175,000 of these revolvers were produced during WW I. The DCM sold these to NRA members during the 1930s at $16.15 each.

Commercial Model $425
Military Model 495

Smith & Wesson K-22 Masterpiece DA $950

Improved version of K-22 Outdoorsman w/same specifications but w/micrometer-click rear sight, short action and antibacklash trigger. Less than 1100 manufactured in 1940.

Smith & Wesson K-22 Outdoorsman DA $495

Design based on the 38 Military & Police Target. Caliber: 22 LR. 6-shot cylinder Bbl..11.13 inches overall. Weight: 35 oz. Adj. target sights. Blued finish. Checkered walnut grip. Made 1931-40.

Smith & Wesson K32 and K38 Heavy Masterpieces

Same as K32 and K38 Masterpiece, but w/heavy bbl. Weight: 38.5 oz. Made 1950-53. *Note:* All K32 and K38 revolvers made after September 1953 have heavy bbls. and the "Heavy Masterpiece" designation was discontinued. Values for Heavy Masterpiece models are the same as shown for Models 14 and 16. *(See* separate listing).

Smith & Wesson K-32 Target DA Revolver $1050

Same as 38 Military & Police Target, except chambered for 32 S&W Long cartridge, slightly heavier bbl. Weight: 34 oz. Only 98 produced. Made 1938-40.

Smith & Wesson Ladysmith (Model M Hand Ejector) DA Revolver

Caliber: 22 LR. 7-shot cylinder. Bbl. lengths: 2.25-, 3-, 3.5- and 6-inch (Third Model only). Approximately 7 inches overall w/3.5-inch bbl. Weight: about 9.5 oz. Fixed sights, adj. target sights available on Third Model. Blued or nickel finish. Round butt, hard rubber grips on First and Second Model; checkered walnut or hard rubber square buttgrips on Third Model. First Model —1902-06: cylinder locking bolt operated by button on left side of frame, no bbl. lug and front locking bolt. Second Model —1906-11: rear cylinder latch eliminated, has bbl. llug, forward cylinder lock w/draw-bolt fastening. Third Model —1911-21: same as Second Model except has square buttgrips, target sights and 6-inch bbl. available. *Note:* Legend has it that a straight-laced D.B. Wesson ordered discontinuance of the Ladysmith when he learned of the little revolver's reputed popularity w/ladies of the evening. The story, which undoubtedly has enhanced the appeal of this model to collectors, is not true: Wesson Ladysmith was discontinued because of difficulty of manufacture and high frequency of repairs.

First Model $1650
Second Model 1350
Third Model, fixed sights, 2.25- or 3.5-inch bbl......... 1295
Third Model, fixed sights, 6-inch bbl. 1995
Third Model, adj. sights, 6-inch bbl. 2550

Smith & Wesson New Century Model Hand Ejector DA Revolver $1950
Also called "Triple Lock" because of its third cylinder lock at the crane. 6-shot cylinder. Calibers: 44 S&W Special, 450 Eley, 455 Mark II. Bbl. lengths: 4-, 5-, 6.5- and 7.5-inch. Weight: 39 oz. (6.5-inch bbl.). Fixed sights. Blued or nickel finish. Checkered walnut grips. Made 1907-15.

Smith & Wesson Regulation Police Target DA $475
Target version of the Regulation Police w/standard features of that model. Caliber: 32 S&W Long. 6-inch bbl. 10.25 inches overall. Weight: 20 oz. Adj. tar get sights. Blued finish. Checkered walnut grips. Made about 1917-40.

Smith & Wesson Safety Hammerless Revolver $650
Also called New Departure Double Action. Hinged frame. Calibers: 32 S&W, 38 S&W. 5-shot cylinder. Bbl. lengths: 2-, 3- and 3.5-inch (32 cal.) or 2-, 3.25-, 4-, 5- and 6-inch (38 cal.). 6.75 inches overall (32 cal. w/3-inch bbl.) or 7.5 inches (38 cal. w/3.25-inch bbl.). Weight: 14.25 oz. (32 cal. w/3-inch bbl.) or 18.25 oz. (38 cal. .25-inch bbl.). Fixed sights. Blued or nickel finish. Hard rubber grips. Made 1888-1937 (32 cal.); 1887-1941 (38 cal. w/various minor changes.)

Smith & Wesson Texas Ranger Commemorative $895
Issued to honor the 150th anniversary of the Texas Rangers. Model 19 357 Combat Magnum w/4-inch bbl., sideplate stamped w/Texas Ranger Commemorative Seal, smooth Goncalo Alves grips. Special Bowie knife. In presentation case. 8,000 sets made in 1973. Value is for set in new condition.

SPHINX ENGINEERING SA.
Porrentruy, Switzerland

**Sphinx
Model AT2000S
Scope Optional**

Sphinx Model AT2000S DA Automatic Pistol
Calibers: 9mm Parabellum, .40 S&W. 15- or 11-shot magazine respectively. 4.53-inch bbl. (S-standard), 3.66-inch bbl. (P-compact), 3.34-inch bbl. (H-subcompact). 8.25inches overall. Weight: 36.5 oz. Fixed sights w/3 Dot system. Stainless frame w/blued slide or Palladium finish. Ambidextrous safety. Checkered walnut or neoprene grips. Imported 1993-96.
Model AT2000S (standard) $650
Model AT2000P (compact) 565
Model AT2000H (subcompact) 545
Caliber 40 S&W, **add** 60
For Palladium Finish (disc. 1994). 90

Sphinx Model AT2000C/2000CS Competitor
Similar to the Model AT2000S, except also chambered 9x21mm. 10-shot magazine. 5.31-inch compensated bbl. 9.84 inches overall. Weight: 40.56 oz. Fully adjustable BoMar or ProPoint sights. Made 1993-96.
Model 2000C (w/BoMar sight) $1095
Model 2000CS (w/ProPoint sight) 1295

Sphinx Model AT2000GM/GMS Grand Master
Similar to the AT2000C, except single action only w/square trigger guard and extended beavertail grip. Imported 1993-96.
Model 2000GM (w/BoMar sight) $1795
Model 2000GMS (w/ProPoint sight) 1895

SPRINGFIELD, INC.
Geneseo, Illinois
(Formerly Springfield Armory)

Springfield Armory Model M1911 Series Auto Pistol
Springfield builds the "PDP" (*Personal Defense Pistol*) Series based on the self-loading M 1911-A1 pistol (Mil-Spec Model) as adopted for a standard service weapon by the U.S. Army. With enhancements and modifications they produce a full line of firearms including Ultra-Compacts, Lightweights, Match Grade and Competition Models. For values see specific models.

Springfield Armory Model 1911-A1 Government
Calibers: 9mm Para., 38 Super, 40 S&W, 10mm or 45 ACP., 7-, 8-, 9- or 10-shot magazine. 4- or 5-inch bbl. 8.5 inches overall. Weight: 36 oz. Fixed combat sights. Blued, Parkerized or Duo-Tone finish. Checkered walnut grips. *Note:* This is an exact duplicate of the Colt M1911-A1 that was used by the U.S. Armed Forces as a service weapon.
Blued Finish $395
Parkerized Finish 360

Springfield Armory Model 1911-A1 (Pre '90 Series)
Calibers: 9mm Parabellum, 38 Super, 10mm, 45 ACP., 7-, 8-, 9- or 10-shot magazine. Bbl. Length: 3.63, 4, 4.25 or 5 inches. 8.5 inches overall. Weight: 36 oz. Fixed combat sights. Blued, Duo-Tone or parkerized finish. Checkered walnut stocks. Made 1985-90.
Government Model (Blued) . . . . $ 395
Government Model (Parkerized) 365
Bullseye Model (Wadcutter) 995
Combat Commander Model (Blued) 365
Combat Commander Model (Parkerized) 345
Commander Model (Blued) 385
Commander Model (Duo-Tone) 425
Commander Model (Parkerized). 355
Compact Model (Blued). 375
Compact Model (Duo-Tone). 415
Compact Model (Parkerized) 350
Defender Model (Blued). 400
Defender Model (Parkerized) 370
Defender Model (Custom Carry) 675
National Match Model (Hardball) 595
Trophy Master (Competition). 995
Trophy Master (Distinguished) 1495
Trophy Master (Competition). 1295

Springfield Armory Model 1911-A1 (Post '90 Series)
SA linkless operating system w/steel or alloy frame. Calibers: 9mm Parabellum, 38 Super, 40 S&W, 10mm, 45 ACP. 7-, 8-, 9- or 10-shot magazine. Bbl. Length: 3.63, 4, 4.25 or 5 inches. 8.5 inches overall. Weight: 28 oz. to 36 oz. Fixed combat sights. Blued, Duo-Tone, parkerized or stainless finish. Checkered composition or walnut stocks. Made 1990 to date.

**Springfield Armory
1911-A1 Post '90 Series
Trophy Model**

**Springfield Armory
1911-A1 Ultra Compact
Parkerized**

**Springfield Armory
1911-A1 PDP Series
Defender**

**Springfield Armory
1911-A1 Champion**

Springfield Armory Model 1911-A1 PDP Series *(Con't)*

system. Blued, Duo-Tone, parkerized or stainless finish. Checkered composition or walnut stocks. Made 1991 to date.

Defender Model (Blued)	**$650**
Defender Model (Duo-Tone)	**675**
Defender Model (Parkerized)	**595**
45 ACP Champion Comp Model (Blued)	**595**
45 ACP Compact Comp HC Model (Blued)	**635**
38 Sup Factory Comp Model (Blued)	**615**
45 ACP Factory Comp Model (Blued)	**585**
38 Sup Factory Comp HC Model (Blued)	**695**
45 ACP Factory Comp HC Model (Blued)	**665**

Springfield Armory Model M1911-A1 Champion

Calibers: 38 ACP, 45 ACP. 6- or 7-shot magazine. 4-inch bbl. Weight: 26.5 to 33.4 oz. Low profile post front sight and drift adjustable rear w/3-Dot sighting system. Commander style hammer and slide. Checkered walnut grips. Blue, Bi-Tone, parkerized or stainless finish. Made 1992 to date.

380 ACP Standard (disc. 1995)	**$295**
45 ACP Parkerized	**325**
45 ACP Blued	**360**
45 ACP Bi-Tone (B/H Model)	**365**
45 ACP Stainless	**375**
45 Super Tuned	**575**

Springfield Armory Model M1911-A1 Compact

Similar to the standard M1911 w/champion length slide on a steel or alloy frame w/a shortened grip. Caliber: 45 ACP. 6- or 7-shot magazine (10+ law enforcement only). 4-inch bbl. Weight: 26.5 to 32 oz. Low profile sights w/3-Dot system. Checkered walnut grips. Matte blue, DuoTone or parkerized finish. Made 1991-96.

Compact Parkerized	**$305**
Compact Blued	**350**
Compact DuoTone	**325**
Compact Stainless	**365**
Compact Comp (ported)	**625**
High Capacity Blue	**435**
High Capacity Stainless	**475**

Springfield Armory Model M1911-A1 Ultra Compact

Similar to M1911 Compact, except chambered 380 ACP, or 45 ACP w/6- or 7-shot magazine. 3.5-inch bbl. Weight: 22 oz. to 30 oz. Matte Blue, Bi-Tone, parkerized (MilSpec) or stainless finish. Made 1995 to date.

380 ACP Ultra (disc. 1996)	**$285**
45 ACP Ultra Parkerized	**325**
45 ACP Ultra Blued	**350**
45 ACP Ultra Bi-Tone	**425**

Springfield Armory Model 1911-A1 *(Con't)*

Mil-Spec Model (Blued)	**$ 355**
Mil-Spec Model (Parkerized)	**325**
Standard Model (Blued)	**365**
Standard Model (Parkerized)	**335**
Standard Model (Stainless)	**375**
Trophy Model (Blued)	**575**
Trophy Model (Duo-Tone)	**585**
Trophy Model (Stainless)	**595**

Springfield Armory Model 1911-A1 PDP Series

PDP Series (*Personal Defense Pistol*). Calibers: 38 Super, 40 S&W, 45 ACP. 7-, 8-, 9-, 10-, 13- or 17-shot magazine. Bbl. Length: 4, 5, 5.5 or 5.63 inches. 9 to 11 inches overall w/compensated bbl. Weight: 34.5 oz. to 42.8 oz. Post front sight, adjustable rear w/3-Dot

Springfield Armory Model 1911-A1 *(Con't)*

45 ACP Ultra Stainless	**465**
45 ACP Ultra High Capacity Parkerized	**425**
45 ACP Ultra High Capacity Blue	**455**
45 ACP Ultra High Capacity Stainless	**495**
45 ACP V10 Ultra Comp Parkerized	**395**
45 ACP V10 Ultra Comp Blue	**435**
45 ACP V10 Ultra Comp Stainless	**475**
45 ACP V10 Ultra Super Tuned	**595**

Springfield Armory Bobcat Auto Pistol . . . . . (SRP) $449
Caliber: 380 ACP. 13-shot magazine. 3.5-inch bbl. 6.6 inches overall. Weight: 21.95 oz. Blade front sight, rear adj. for windage. Textured composition grip. Matte blued finish. Announced 1992, but not produced.

Springfield Armory Firecat Automatic Pistol
Calibers: 9mm, 40 S&W. 8-shot magazine (9mm) or 7-shot magazine (40 S&W). 3.5-inch bbl. 6.5 inches overall. Weight: 25.75 oz. Fixed sights w/3-Dot system. Checkered walnut grip. Matte blued finish. Made 1991-93.

9mm	**$385**
40 S&W	**395**

Springfield Armory Lynx Auto Pistol . . . . . . (SRP) $269
Caliber: 25 ACP. 7-shot magazine. 2.25-inch bbl. 4.45 inches overall. Weight: 10.55 oz. Blade front sight, rear adj. for windage w/3-Dot system. Checkered composition grip. Matte blued finish. Announced 1992, but not produced.

Springfield Armory Panther Automatic Pistol $395
Calibers: 9mm, 40 S&W. 15-shot magazine (9mm) or 11-shot magazine (40 S&W). 3.8-inch bbl. 7.5 inches overall. Weight: 28.95 oz. Blade front sight, rear adj. for windage w/3-Dot system. Checkered walnut grip. Matte blued finish. Made 1991-93.

Springfield Armory Model P9 DA Combat Series
Calibers: 9mm, 40 S&W, 45 ACP. Magazine capacity: 15-shot (9mm), 11-shot (40 S&W) or 10-shot (45 ACP). 3.66-inch bbl. (Compact and Sub-Compact), or 4.75-inch bbl. (Standard). 7.25 or 8.1 inches overall. Weight: 32 to 35oz. Fixed sights w/3-Dot system. Checkered walnut grip. Matte blued, Parkerized, stainless or Duo-Tone finish. Made 1990-94.

Compact Model (9mm,Parkerized)	**$325**
Sub-Compact Model (9mm, Parkerized)	**354**
Standard Model (9mm, Parkerized)	**395**
W/Blued Finish, **add**	**25**
W/Duo-Tone Finish, **add**	**55**
W/Stainless Finish, **add**	**20**
40 S&W , **add**	**35**
45 ACP, **add**	**45**

Springfield Armory Model P9 Competition Series
Same general specifications as Model P9, except in target configuration w/5-inch bbl. (LSP Ultra) or 5,25-inch bbl. (Factory Comp) w/dual port compensator system, extended safety and magazine release.

Factory Comp Model (9mm Bi-Tone)	**$475**
Factory Comp Model (9mm Stainless)	**565**
LSP Ultra Model (9mm Bi-Tone)	**435**
LSP Ultra Model (9mm Stainless)	**465**
40 S&W, 45 ACP, **add**	**40**

STALLARD ARMS
Mansfield, Ohio

See listings under Hi-Point.

**Springfield Armory
PD9 Series**

STAR PISTOLS
Eibar, Spain
Star, Bonifacio Echeverria, S.A.

Star Model 30M DA Auto Pistol $325
Caliber: 9mm Para. 15-shot magazine. 4.38-inch bbl. 8 inches overall. Weight: 40 oz. Steel frame w/combat features. Adj. sights. Checkered composition grips. Blued finish. Made 1984-91.

Star Model 30PK DA Auto Pistol $325
Same gen. specifications as Star Model 30M, except 3.8-inch bbl. 30-oz. weight and alloy frame. Made 1984-89.

Star Model 31P DA Auto Pistol
Same general specifications as Model 30M, except removable back-strap houses complete firing mechanism. Weight: 39.4 oz. Made 1990-94.

Blued Fininh	**$285**
Starvel Finish	**375**

Star Model 31 PK DA Auto Pistol $295
Same general specifications as Model 31P, except w/alloy frame. Weight: 30 oz. Made 1990-97.

Star Model A Automatic Pistol $225
Modification of the Colt Government Model 45 Auto, which it closely resembles, but lacks grip safety. Caliber: 38 Super. 8-shot magazine. 5-inch bbl. 8-inches overall. Weight: 35 oz. Fixed sights. Blued finish. Checkered grips. Made 1934-97. (No longer imported.)

Star Models AS, BS, PS $295
Same as Models A, B and P, except have magazine safety. Made in 1975.

Star Model B $235
Same as Model A, except in 9mm Para. Made 1934-75.

Star Model BKM $245
Similar to Model BM, except has aluminum frame and weighs 25.6 oz. Made 1976-92.

Star Model BKS Starlight Automatic Pistol $235
Light alloy frame. Caliber: 9mm Para. 8-shot magazine. 4.25-inch bbl. 7 inches overall. Weight: 25 oz. Fixed sights. Blued or chrome finish. Plastic grips. Made 1970-81.

Star Model 30M

Star Model 30PK

Star Modedl AS

Star Model BKS

Star Model BM Automatic Pistol
Caliber: 9mm. 8-shot magazine. 3.9-inch bbl. 6.95 inches overall. Weight: 34.5 oz. Fixed sights. Checkered walnut grips. Blued or Starvel finish. Made 1976-92.

Blued Finish **$245**
Starvel Finish 260

Star Model CO Pocket Automatic Pistol **$185**
Caliber: 25 Automatic (6.35mm). 2.75-inch bbl. 4.5 inches overall. Weight: 13 oz. Fixed sights. Blued finish. Plastic grips. Made 1941-57.

Star Model CU Starlet Pocket Pistol **$195**
Light alloy frame. Caliber: 25 Automatic (6.35mm). 8-shot magazine. 2.38-inch bbl. 4.75 inches overall. Weight: 10.5 oz. Fixed sights. Blued or chrome-plated slide w/frame anodized in black, blue, green, gray or gold. Plastic grips. Made 1957-97. (U.S. importation Disc. 1968.)

Star Model Model F

Star Model F Automatic Pistol **$175**
Caliber: 22 LR. 10-shot magazine. 4.5-inch bbl. 7.5 inches overall. Weight: 25 oz. Fixed sights. Blued finish. Plastic grips. Made 1942-67.

Star
Model F
Olympic

Star Model F Olympic Rapid-Fire **$325**
Caliber: 22 Short. 9-shot magazine. 7-inch bbl. 11.06 inches overall. Weight: 52 oz. w/weights. Adj. target sight. Adj. 3-piece bbl. weight. Aluminum alloy slide. Muzzle brake. Plastic grips. Made 1942-67.

Star Model FM **$175**
Similar to Model FR, except has heavier frame w/web in front of trigger guard, 4.25-inch heavy bbl. Weighs 32 oz. Made 1972-91.

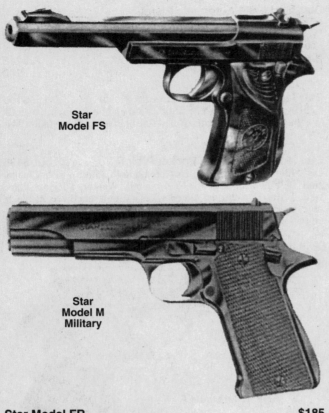

Star
Model FS

Star
Model M
Military

Star Model FR **$185**
Similar to Model F w/same general specifications, but restyled, has slide stop and adj. rear sight. Made 1967-72.

Star Model FRS **$175**
Same as Model FR, except has 6-inch bbl. Weighs 28 oz. Also avail. in chrome finish. Made 1967-91.

Star Model FS **$185**
Same as regular Model F, but w/6-inch bbl. and adj. sights. Weight: 27 oz. Made 1942-67.

Star Model H **$195**
Same as Model HN except caliber 32 Automatic (7.65mm). 7-shot magazine. Weight: 20 oz. Made 1934-41.

Star Model HK Lancer Automatic Pistol **$220**
Similar to Starfire w/same general specifications, except caliber 22 LR. Made 1955-68.

Star Model HN Automatic Pistol **$185**
Caliber: 380 Automatic (9mm Short). 6-shot magazine. 2.75-inch bbl. 5.56 inches overall. Weight: 20 oz. Fixed sights. Blued finish. Plastic grips. Made 1934-41.

Star Model I Automatic Pistol. **$195**
Caliber: 32 Automatic (7.65mm). 9-shot magazine. 4.81-inch bbl. 7.5 inches overall. Weight: 24 oz. Fixed sights. Blued finish. Plastic grips. Made 1934-36.

Star Model IN **$200**
Same as Model I, except caliber 380 Automatic (9mm Short), 8-shot magazine, weighs 24.5 oz. Made 1934-36.

Star Model M Military Automatic Pistol **$265**
Modification of the Colt Government Model 45 Auto, which it closely resembles, but lacks grip safety. Calibers: 9mm Bergmann, 38 Super, 9mm Para. 8-shot magazine, except 7-shot in 45 caliber. 5-inch bbl. 8.5 inches overall. Weight: 36 oz. Fixed sights. Blued finish. Checkered grips. Made 1934-39.

Star Models M40, M43, M45 Firestar Auto Pistols
Calibers: 9mm, 40 S&W, 45 ACP. 7-shot magazine (9mm) or 6-shot (other calibers). 3.4-inch bbl., 6.5 inches overall. Weight: 30.35 oz. Blade front sight, adj. rear w/3-Dot system. Checkered rubber grips. Blued or Starvel finish. Made 1990-97.
M40 Blued (40 S&W) **$235**
M40 Starvel (40 S&W)............................. 255
M43 Blued (9mm)................................. 230
M43 Starvel (9mm)............................... 245
M45 Blued (45 ACP) 265
M45 Starvel (45 ACP) 295

Star Megastar Automatic Pistol
Calibers: 10mm, 45 ACP. 12-shot magazine. 4.6-inch bbl. 8.44 inches overall. Weight: 47.6 oz. Blade front sight, adj. rear. Checkered composition grip. Finishes: Blued or Starvel. Made from 1992-97.
Blued Finish, 10mm or 45 ACP **$375**
Starvel Finish, 10mm or 45 ACP..................... 395

Star Model P. **$265**
Same as Model A, except caliber 45 Automatic, has 7-shot magazine. Made 1934-75.

Star
Model PD

Star Model PD Automatic Pistol
Caliber: 45 Automatic. 6-shot magazine. 3.75-inch bbl. 7 inches overall. Weight: 25 oz. Adj. rear sight, ramp front. Blued or Starvel finish. Checkered walnut grips. Made 1975-92.
Blued Finish **$255**
Starvel Finish 275

Star Model S. **$175**
Same as Model SI except caliber 380 Automatic (9mm), 7-shot magazine, weighs 19 oz. Made 1941-65.

Star Model SI Automatic Pistol **$195**
Reduced-size modification of the Colt Government Model 45 Auto, lacks grip safety. Caliber: 32 Automatic (7.65mm). 8-shot magazine. 4-inch bbl. 6.5 inches overall. Weight: 20 oz. Fixed sights. Blued finish. Plastic grips. Made 1941-65.

HANDGUNS

Star Starfire Automatic Pistol **$365**
Light alloy frame. Caliber: 380 Automatic (9mm Short). 7-shot magazine. 3.13-inch bbl.. 5.5 inches overall. Weight: 14.5 oz. Fixed sights. Blued or chrome-plated slide w/frame anodized in black, blue, green, gray or gold. Plastic grips. Made 1957-97. U.S. importation discontinued 1968.

Star Models Super A, Super B, Super **$250**
Same as Models A, B and P, except w/improvements described under Super Star. Made c. 1946-89/90.

Star Models Super SI, Super S **$260**
Same general specifications as the regular Model SI and S, except w/improvements described under Super Star. Made c. 1946-72.

Star Model Super SM **$235**
Similar to Model Super S, except has adj. rear sight, wood grips. Made 1973-81.

Star Super Automatic Pistol **$295**
Improved version of the Model M w/same general specifications, but has disarming bolt permitting easier takedown, indicator of cartridge in chamber, magazine safety, take-down magazine, improved sights w/luminous spots for aiming in darkness. Calibers: 38 Super, 9mm Para. This is the standard service pistol of the Spanish Armed Forces adopted 1946.

Star Super Target Model **$395**
Same as regular Super Star, except w/adj. target rear sight. Discontinued.

Star Ultrestar DA Automatic Pistol **$235**
Calibers: 9mm Parabellum or 40 S&W. 9-shot magazine. 3.57-inch bbl. 7 inches overall. Weight: 26 oz. Blade front, adjustable rear w/3 Dot system. Polymer frame. Blue metal finish. Checkered black polymer grips. Imported 1994-97.

STENDA-WERKE PISTOL
Suhl, Germany

Stenda Pocket Automatic Pistol **$230**
Essentially the same as the Beholla, *see* listing of that pistol for specifications. Made c. 1920-25. *Note:* This pistol may be marked "Beholla" along w/the Stenda name and address.

STERLING ARMS CORPORATION
Gasport, New York

Sterling
Model 283 Target 300

Sterling Model 283 Target 300 Auto Pistol **$150**
Caliber: 22 LR. 10-shot magazine. Bbl. lengths: 4.5-, 6- 8-inch. 9 inches overall w/4.5-inch bbl. Weight: 36 oz. w/4.52-inch bbl. Adj. sights. Blued finish. Plastic grips. Made 1970-71.

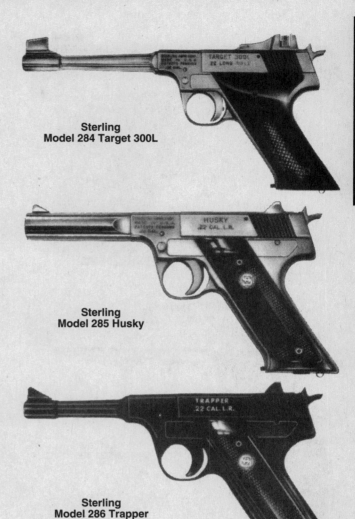

Sterling
Model 284 Target 300L

Sterling
Model 285 Husky

Sterling
Model 286 Trapper

Sterling Model 284 Target 300L **$165**
Same as Model 283, except has 4.5- or 6-inch "Luger"-type bbl. Made 1970-71.

Sterling Model 285 Husky **$165**
Same as Model 283, except has fixed sights, 4.5-inch bbl only. Made 1970-71.

Sterling Model 286 Trapper **$165**
Same as Model 284, except w/fixed sights. Made 1970-71.

Sterling Model 287 PPL-380 Automatic Pistol **$100**
Caliber: 380 Automatic. 6-shot magazine. 1-inch bbl. 5.38 inches overall. Weight: 22.5 oz. Fixed sights. Blued finish. Plastic grips. Made 1971-72.

Sterling Model 300 Automatic Pistol **$125**
Caliber: 25 Automatic. 6-shot magazine. 2.33-inch bbl. 4.5 inches overall. Weight: 13 oz. Fixed sights. Blued or nickel finish. Plastic grips. Made 1972-83.

Sterling Model 300S **$115**
Same as Model 300, except in stainless steel. Made 1976-83.

**Sterling
Model 300**

**Sterling
Model 400**

Sterling Model 302. . . . . **$125**
Same as Model 300, except in 22 LR. Made 1973-83.

Sterling Model 302S **$125**
Same as Model 302 except in stainless steel. Made 1976-83.

Sterling Model 400 DA Automatic Pistol **$195**
Caliber: 380 Automatic. 7-shot magazine. 3.5-inch bbl. 6.5 inches overall. Weight: 24 oz. Adj. rear sight. Blued or nickel finish. Checkered
walnut grips. Made 1975-83.

Sterling Model 400S **$245**
Same as Model 400, except stainless steel. Made 1977-83.

Sterling Model 450 DA Auto Pistol **$295**
Caliber: 45 Automatic. 8-shot magazine. 4-inch bbl. 7.5 inches overall. Weight: 36 oz. Adj. rear sight. Blued finish. Smooth walnut
grips. Made 1977-83.

Sterling Model PPL-22 Automatic Pistol **$175**
Caliber: 22 LR. 10-shot magazine. 1-inch bbl. 5.5 inches overall.
Weight: about 24 oz. Fixed sights. Blued finish. Wood grips. *Note:*
Only 382 made 1970-71.

J. STEVENS ARMS & TOOL CO.
Chicopee Falls, Mass.

*This firm was established in Civil War days by Joshua Stevens, for
whom the company was named. In 1936 it became a subsidiary of
Savage Arms.*

Stevens No. 10 Single-Shot Target Pistol **$250**
Caliber: 22 LR. 8-inch bbl. 11.5 inches overall. Weight: 37 oz. Target sights. Blued finish. Hard rubber grips. In external appearance
this arm resembles an automatic pistol, but it has a tip-up action.
Made 1919-39.

**Stevens No. 35 Offhand Model Single-Shot
Target Pistol** **$350**
Tip-up action. Caliber: 22 LR. Bbl. lengths: 6, 8, 10, 12.25 inches.
Weight: 24 oz. w/6-inch bbl. Target sights. Blued finish. Walnut
grips. *Note:* This pistol is similar to the earlier "Gould" model. Made
1907-39.

**Stevens Offhand No. 35 Single-Shot
Pistol/Shotgun** **$275**
Same general specifications as the standard No. 35 pistol except cham-
bered for the 410 shotshell. 6-, 8-, 10-, or 12-inch half-ocatagonal bbl.
Iron frame either blued, nickel plated, or casehardened. BATF Class 3
license required to purchase or dispose of. Made 1923-42.

Stevens No. 10

Stevens No. 35

Stevens No. 38

Stevens-Lord No. 36 Single-Shot Pistol **$575**
Tip-up action. Calibers: 22 Short and LR, 22 WRF, 25 Stevens, 32 Short Colt, 38 Long Colt, 44 Russian. 10- or 12-inch half-octagonal bbl. Iron or brass frame w/nickel plated finish. Blued bbl. Checkered walnut grips. Made 1880-1911.

Stevens-Gould No. 37 Single-Shot Pistol **$750**
Similar specifications to the No. 38 except the finger spur on the trigger guard has been omitted. Made 1889-1903.

Stevens-Conlin No. 38 Single-Shot Pistol **$450**
Tip-up action. Calibers: 22 Short and LR, 22 WRF, 25 Stevens, 32 Stevens, 32 Short Colt. Iron or brass frame. Checkered grips. Made 1884-1903.

Stevens No. 41 Tip-Up Single-Shot Pistol **$235**
Tip-up action. Caliber: 22 Short. 3.5-inch half-octagonal bbl. Blued metal parts w/optional nickel frame. Made 1896-1915.

STEYR PISTOLS
Steyr, Austria

Steyr GB

Steyr GB Semiautomatic Pistol **$395**
Caliber: 9mm Para. 18-round magazine. 5.4-inch bbl. 8.9 inches overall. Weight: 2.9 lbs. Post front sight, fixed, notched rear. Double, gas-delayed, blow-back action. Made 1981-88.

Steyr-Hahn M12 Automatic Pistol **$375**
Caliber: 9mm Steyr. 8-shot fixed magazine, charger loaded. 5.1-inch bbl. 8.5 inches overall. Weight: 35 oz. Fixed sights. Blued finish. Checkered wood grips. Made 1911-19. Adopted by the Austro-Hungarian Army in 1912. *Note:* Confiscated by the Germans in 1938, an estimated 250,000 of these pistols were converted to 9mm Para. and stamped w/an identifying "08" on the left side of the slide. Mfd. by Osterreichische Waffenfabrik-Gesellschaft.

STOEGER LUGERS
Formerly mfd. by Stoeger Industries, So. Hackensack, N.J.; latter by Classic Arms, Union City, N.J.

Stoeger American Eagle Luger
Caliber: 9mm Para. 7-shot magazine. 4- or 6-inch bbl. 8.25 inches overall (4-inch bbl.). or 10.25 inches (6-inch bbl.). Weight: 30 or 32 oz. Checkered walnut grips. Stainless steel w/brushed or matte black finish. Made from 1994 to date.
P-08 Model (4-inch bbl.) **$525**
Navy Model (6-inch bbl.) **535**

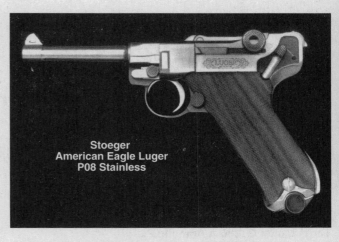

**Stoeger
American Eagle Luger
P08 Stainless**

Stoeger Standard Luger 22 Automatic Pistol **$150**
Caliber: 22 LR. 10-shot magazine. 4.5- or 5.5-inch bbl. 8.88 inches overall (4.5-inch bbl.). Weight: 29.5 oz. (4.5-inch bbl.). Fixed sights. Black finish. Smooth wood grips. Made 1969-86.

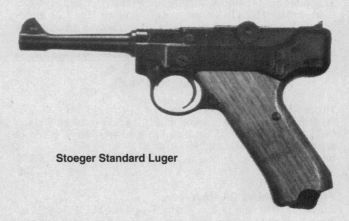

Stoeger Standard Luger

Stoeger Steel Frame Luger 22 Auto Pistol **$165**
Caliber: 22 LR. 10-shot magazine. 4.5-inch bbl. 8.88 inches overall. Blued finish. Checkered wood grips. Features one-piece solidly forged and machined steel frame. Made 1980-86.

Stoeger Target Luger 22 Automatic Pistol **$195**
Same as Standard Luger 22, except has target sights. 9.38 inches overall w/4.5-inch bbl. Checkered wood grips. Made 1975-86.

TARGA PISTOLS
Italy
Manufactured by Armi Tanfoglio Guiseppe

Targa Model GT26S Auto Pistol **$75**
Caliber: 25 ACP. 6-shot magazine. 2.5-inch bbl. 4.63 inches overall. Weight: 15 oz. fixed sights. Checkered composition grips. Blued or chrome finish. Discontinued 1990.

Targa Model GT32 Auto Pistol
Caliber: 32 ACP. 6-shot magazine. 4.88-inch bbl. 7.38 inches overall. Weight: 26 oz. fixed sights. Checkered composition or walnut grips. Blued or chrome finish.
Blued finished . . . . **$ 95**
Chrome finish **110**

**Targa
Model GT380XE**

Targa Model GT380 Automatic Pistol

Same as the Targa GT32, except chambered for 380 ACP.

Blued finish.	$115
Chrome finish	125

Targa Model GT380XE Automatic Pistol $135

Caliber: 380 ACP.11-shot magazine. 3.75-inch bbl. 7.38 inches overall. Weight: 28 oz. Fixed sights. Blued or satin nickel finish. Smooth wooden grips. Made 1980-90.

FORJAS TAURUS S.A.
Porto Alegre, Brazil

Taurus Model 44 DA Revolver

Caliber: 44 Mag. 6-shot cylinder. 4-, 6.5-, or 8.38-inch bbl. Weight: 44.75 oz., 52.5 or 57.25 oz. Brazilian hardwood grips. Blued or stainless steel finish. Made 1994 to date.

Blued Finish	$275
Stainless	325

Taurus Model 65 DA Revolver

Caliber: 357 Magnum. 6-shot cylinder. 3- or 4-inch bbl. Weight: 32 oz. Front ramp sight, square notch rear. Checkered walnut target grip. Royal blued or satin nickel finish. Discontinued 1997.

Blue.	$165
Stainless	215

Taurus Model 66 DA Revolver

Calibers: 357 Magnum, 38 Special. 6-shot cylinder. 3-, 4- and 6-inch bbl. Weight: 35 oz. Serrated ramp front sight, rear click adj. Checkered walnut grips. Royal blued or nickel finish. Discontinued 1997.

Blue.	$175
Stainless	225

Taurus Model 73 DA Revolver $145

Caliber: 32 Long. 6-shot cylinder. 3-inch heavy bbl. Weight: 20 oz. Checkered grips. Blued or satin finish. Discontinued 1993.

Taurus Model 74 Target Grade DA Revolver $150

Caliber: 32 S&W Long. 6-shot cylinder. 3-inch bbl. 8.25 inches overall. Weight: 20 oz. Adj. rear sight, ramp front. Blued or nickel finish. Checkered walnut grips. Made 1971-90.

Taurus Model 80 DA Revolver

Caliber: 38 Special. 6-shot cylinder. Bbl. lengths: 3, 4, inches. 9.25 inches overall (4-inch bbl.). Weight: 30 oz. (4-inch bbl.), Fixed sights. Blued or nickel finish. Checkered walnut grips. Made 1996-97.

Blued.	$135
Stainless	165

Taurus Model 44

Taurus Model 66

**Taurus Model 74
Target Grade**

Taurus Model 80

Taurus Model 82 Heavy Barrel

Same as Model 80, except has heavy bbl. Weight: w/4-inch bbl., 33 oz. Made 1971 to date.

Blued.	$145
Stainless	175

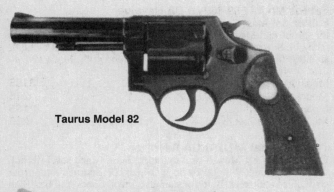

Taurus Model 82

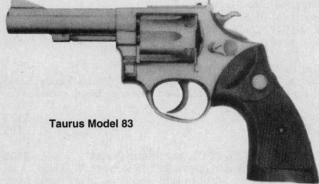

Taurus Model 83

Taurus Model 84

**Taurus Model 85
Concealed Hammer**

Taurus Model 83 Heavy Barrel Target Grade
Same as Model 84, except has heavy bbl., weighs 34.5 oz. Made 1977 to date.
Blued. **$150**
Stainless 180

Taurus Model 84 Target Grade Revolver $185
Caliber: 38 Special. 6-shot cylinder. 4-inch bbl. 9.25 inches overall. Weight: 31 oz. Adj. rear sight, ramp front. Blued or nickel finish. Checkered walnut grips. Made 1971-89.

**Taurus Model 85
w/Spur Hammer**

Taurus Model 85 DA Revolver
Caliber: 38 Special. 5-shot cylinder. Bbl.. 2- or 3-inch. Weight: 21 oz. Fixed sights. Checkered walnut grips. Blued, satin nickel or stainless-steel finish. Currently in production. **Model 85CH** is the same as the standard version except for concealed hammer.
Blued or Satin Nickel **$145**
Stainless Steel 165

Taurus Model 86

Taurus Model 86 Target Master DA Revolver $185
Caliber: 38 Special. 6-shot cylinder. 6-inch bbl. 11.25 inches overall. Weight: 34 oz. Adj. rear sight, Patridge-type front. Blued finish. Checkered walnut grips. Made 1971-94.

Taurus Model 94 Target Grade
Same as Model 74, except caliber 22 LR, w/9-shot cylinder. 3- or 4-inch bbl. Weight: 25 oz. Blued or stainless finish. Made 1971 to date.
Blued Finish **$165**
Stainless Finish 215

Taurus Model 96 Target Master $185
Same as Model 86, except in 22 LR. Made 1971 to date.

Taurus Model 431 DA Revolver
Caliber: 44 Spec. 5-shot cylinder. 3- or 4-inch solid-rib bbl. w/ejector shroud. Weight: 35 oz.w/4-inch bbl. Serrated ramp front sight, notched topstrap rear. Blued or stainless finish. Made 1992-97.
Blued Finish **$175**
Stainless Finish 225

Taurus Model 669

Taurus Model PT-22

Taurus Model PT-58

Taurus Model 441 DA Revolver
Similar to the Model 431, except w/6-inch bbl. and fully adj. target sights. Weight: 40 oz. Made 1991-97.
Blued Finish **$185**
Stainless Finish **235**

Taurus Model 445 DA Revolver
Caliber: 44 Special. 5-shot cylinder. 2-inch bbl. 6.75 inches overall. 6.75 inches overall. Weight: 28.25 oz. Serrated ramp front sight, notched frame rear. Standard or concealed hammer. Santoprene I grips. Blue or stainless finish. Imported 1997 to date.
Blue Model **$175**
Stainless Model **195**

Taurus Model 454 DA Raging Bull Revolver
Caliber: 454 Casull. 5-shot cylinder. Ported 6.5- or 8.4-inch vent rib bbl. 12 inches overall w/6.5-inch bbl. Weight: 53 or 63 oz. Partridge front sight, micrometer adj. rear. Santoprene I or walnut grips. Blue or stainless finish. Imported 1997 to date.
Blue Model **$595**
Stainless Model **615**

Taurus Model 669/669VR DA Revolver
Caliber: 357 Mag. 6-shot cylinder. 4- or 6-inch solid-rib bbl. w/ejector shroud and Model 669VR has vent rib bbl. Weight: 37 oz. w/4-inch bbl. Serrated ramp front sight, micro-adj. rear. Royal blued or stainless finish. Checkered Brazilian hardwood grips. Made 1989 to date.
Model 669 Blued **$165**
Model 669 Stainless **215**
Model 669VR Blued **185**
Model 669VR Stainless **235**

Taurus Model 741/761 DA Revolver
Caliber: 32 H&R Mag. 6-shot cylinder. 3- or 4-inch solid-rib bbl. w/ejector shroud. Weight: 20 oz. w/3-inch bbl. Serrated ramp front sight, micro-adj. rear. Blued or stainless finish. Checkered Brazilian hardwood grips. Made 1991-97.
Model 741 Blued **$145**
Model 741 Stainless **205**
Model 761 (6-inch bbl., blued, 34 oz.) **165**

Taurus Model 941 Target Revolver
Caliber: 22 Magnum. 8-shot cylinder. Solid-rib bbl. w/ejector shroud. Micro-adj. rear sight. Brazilian hardwood grips. Blued or stainless finish.
Blued Finish **$175**
Stainless Finish **195**

Taurus Model PT 22 DA Automatic Pistol **$135**
Caliber: 22 LR. 9-shot magazine. 2.75-inch bbl. Weight: 12.3 oz. Fixed open sights. Brazilian hardwood grips. Blued finish. Made 1991 to date.

Taurus Model PT 25 DA Automatic Pistol **$145**
Same general specifications as Model PT 22, except in 25 ACP w/8-shot magazine. Made from 1992 to date.

Taurus Model PT-58 Semiautomatic Pistol **$250**
Caliber: 380 ACP. 12-shot magazine. 4-inch bbl. 7.2 inches overall. Weight: 30 oz. Blade front sight, rear adj. for windage w/3-Dot sighting system. Blued, satin nickel or stainless finish. Made 1988-96.

Taurus Model PT92

Taurus Model PT 92AF Semiautomatic Pistol
Caliber: 9mm Para. 15-round magazine. Double action. 5-inch bbl. 8.5 inches overall. Weight: 24 oz. Blade front sight, notched bar rear. Smooth Brazilian walnut grips. Blued, satin nickel or stainless finish. Made 1991 to date.
Blued Finish **$295**
Satin Nickel Finish **335**
Stainless Finish **345**

Taurus Model PT-92AFC Compact Pistol

Same general specifications as Model PT-92AF, except w/13-shot magazine. 4.25-inch bbl. 7.5 inches overall. Weight: 31 oz. Made 1991-96

Blued Finish	**$275**
Satin Nickel Finish	315
Stainless Finish	325

Taurus Model PT 99AF Semiautomatic Pistol $295

Same general specifications as Model PT-92AF, except rear sight is adj. for elevation and windage, and finish is blued or satin nickel.

Taurus Model PT 100 DA Automatic Pistol

Caliber: 40 S&W. 11-shot magazine. 5-inch bbl. Weight: 34 oz. Fixed front sight, adj. rear w/3-dot system. Smooth hardwood grip. Blued, satin nickel or stainless finish. Made 1991-97.

Blued Finish	**$285**
Satin Finish	325
Stainless Finish	335

Taurus Model PT 101 DA Automatic Pistol

Same general specifications as Model 100, except w/micrometer click adj. sights. Made 1992-96.

Blued Finish	**$310**
Satin Nickel Finish	350
Stainless Finish	365

Taurus Model PT 111 Millennium DAO Pistol

Caliber: 9mm Parabellum. 10-shot magazine. 3.12-inch bbl. 6.0 inches overall. Weight: 19.1 oz. Fixed low-profile sights w/3-dot system. Black polymer grip/frame. Blue or stainless slide. Imported 1998 to date.

Blue Model	**$235**
Stainless Model	250

Taurus Model PT 908 Semiautomatic Pistol

Caliber: 9mm Para. 8-shot magazine. 3.8-inch bbl. 7 inches overall. Weight: 30 oz. Post front sight, drift-adj. combat rear w/3-Dot system. Blued, satin nickel or stainless finish. Made 1993-97.

Blued Finish	**$275**
Satin Nickel Finish	315
Stainless Finish	325

Taurus Model PT 911 Compact Semiautomatic Pistol

Caliber: 9mm Parabellum. 10-shot magazine. 3.75-inch bbl. 7.05 inches overall. Weight: 28.2 oz. Fixed low-profile sights w/3-dot system. Santoprene II grips. Blue or stainless finish. Imported 1997 to date.

Blue Model	**$260**
Stainless Model	275

Taurus Model PT 938 Compact Semiautomatic Pistol

Caliber: 380ACP. 10-shot magazine. 3.72-inch bbl. 6.75 inches overall. Weight: 27 oz. Fixed low-profile sights w/3-dot system. Santoprene II grips. Blue or stainless finish. Imported 1997 to date.

Blue Model	**$255**
Stainless Model	270

Taurus Model PT 940 Compact Semiautomatic Pistol

Caliber: 40 S&W. 10-shot magazine. 3.75-inch bbl. 7.05 inches overall. Weight: 28.2 oz. Fixed low-profile sights w/3-dot system. Santoprene II grips. Blue or stainless finish. Imported 1997 to date.

Blue Model	**$265**
Stainless Model	280

Taurus Model PT-99AF

Taurus Model PT-908

Taurus Model PT 945 Compact Semiautomatic Pistol

Caliber: 45ACP. 8-shot magazine. 4.25-inch bbl. 7.48 inches overall. Weight: 29.5 oz. Fixed low-profile sights w/3-dot system. Santoprene II grips. Blue or stainless finish. Imported 1995 to date.

Blue Model	**$275**
Stainless Model	290

TEXAS ARMS
Waco, Texas

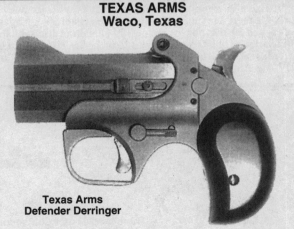

**Texas Arms
Defender Derringer**

Texas Arms Defender Derringer $225

Calibers: 9mm, 357 Mag., 44 mag., 45 ACP, 45 Colt/.410. 3-inch bbl. 5 inches overall. Weight: 21 oz. Blade front sight, fixed rear. Matte gun-metal grey finish. Smooth grips. Made 1993 to date.

TEXAS LONGHORN ARMS
Richmond, Texas

Texas Longhorn Arms "The Jezebel" Pistol **$200**
Top-break, single-shot. Caliber: 22 Short, Long or LR. 6-inch half-round bbl. 8 inches overall. Weight: 15 oz. Bead front sight, adj. rear. One-piece walnut grip. Stainless finish. Intro. in 1987.

Texas Longhorn Arms Sesquicentennial SA Revolver **$1895**
Same as South Texas Army Limited Edition, except engraved and nickel-plated w/one-piece ivory grip. Intro. in 1986.

Texas Longhorn Arms SA Revolver Cased Set
Set contains one each of the Texas Longhorn Single Actions. Each chambered in the same caliber and w/the same serial number. Intro. in 1984.
Standard Set **$4400**
Engraved Set 5750

Texas Longhorn Arms South Texas Army Limited Edition SA Revolver **$1295**
Calibers: all popular centerfire pistol calibers. 6-shot cylinder. 4.75-inch bbl. 10.25 inches overall. Weight: 40 oz. Fixed sights. Color casehardened frame. One-piece deluxe walnut grips. Blued bbl. Intro. in 1984.

Texas Longhorn Arms Texas Border Special SA Revolver **$1195**
Same as South Texas Army Limited Edition, except w/3.5-inch bbl. and bird's-head grips. Intro. in 1984.

Texas Longhorn Arms West Texas Flat Top Target SA Revolver **$1525**
Same as South Texas Army Limited Edition, except w/choice of bbl. lengths from 7.5 to 15 inches. Same special features w/flat-top style frame and adj. rear sight. Intro. in 1984.

THOMPSON PISTOL
West Hurley, New York
Mfd. by Auto-Ordnance Corporation

Thompson Model 27A-5 Semiautomatic Pistol
Similar to Thompson Model 1928A submachine gun, except has no provision for automatic firing, does not have detachable buttstock. Caliber: 45 Automatic. 20-shot detachable box magazine (5-, 15- and 30-shot box magazines, 39-shot drum also available). 13-inch finned bbl. Overall length: 26 inches. Weight: about 6.75 lbs. Adj. rear sight, blade front. Blued finish. Walnut grips. Intro. 1977. *See* Auto-Ordnance in Handgun Section.

THOMPSON/CENTER ARMS
Rochester, New Hampshire

Thompson/Center Contender
Single-Shot Pistol

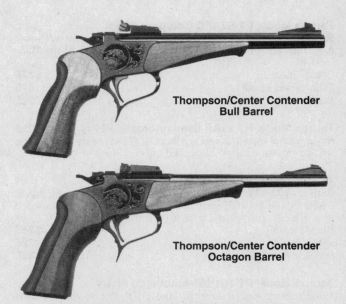

Thompson/Center Contender
Bull Barrel

Thompson/Center Contender
Octagon Barrel

Thompson/Center Contender Single-Shot Pistol
Break frame, underlever action. Calibers: (rimfire) 22 LR 22 WMR, 5mm RRM; (standard centerfire), 218 Bee, 22 Hornet, 22 Rem. Jet, 221 Fireball, 222 Rem., 25-35, 256 Win. Mag., 30 M1 Carbine, 30-30, 38 Auto, 38 Special 357 Mag./Hot Shot, 9mm Para., 45 Auto, 45 Colt, 44 Magnum/Hot Shot; (wildcat centerfire) 17 Ackley Bee, 17 Bumblebee, 17 Hornet, 17 K Hornet, 17 Mach IV, 17-222, 17-223, 22 K Hornet, 30 Herrett, 357 Herrett, 357-4 B&D. Interchangeable bbls.: 8.75- or 10-inch standard octagon (357 Mag., 44 Mag. and 45 Colt available w/detachable choke for use w/Hot Shot cartridges); 10-inch w/vent rib and detachable internal choke tube for Hot Shots, 357 and 44 Magnum only; 10-inch bull bbl., 30 or 357 Herrett only. 13.5 inches overall w/10-inch bbl. Weight: 43 oz. (w/standard 10-inch bbl.). Adj. rear sight, ramp front; vent rib model has folding rear sight, adj. front; bull bbl. available w/or w/o sights. Lobo 1½ X scope and mount (**add** $40 to value). Blued finish. Receiver photoengraved. Checkered walnut thumbrest grip and forearm (pre-1972 model has different grip w/silver grip cap). Made 1967 to date, w/the following revisions and variations.
Standard Model. **$220**
Vent rib Model 250
Bull Bbl. Model, w/sights 255
Bull Bbl. Model, without sights 225
Extra standard bbl. 85
Extra vent rib or bull bbl. 90

Thompson/Center Contender — Bull Barrel **$260**
Caliber offerings of the bull bbl. version expanded in 1973, w/another bump in 1978, making it the Contender model w/the widest range of caliber options: 22 LR, 22 Win. Mag., 22 Hornet, 223 Rem., 7mm T.C.U., 7×30 Waters, 30 M1 Carbine, 30-30 Win., 32 H&R Mag., 32-20 Win., 357 Rem. Max., 357 Mag., 10mm Auto, 44 Magnum, 445 Super Magnum. 10-inch heavy bbl. Patridge-style iron sights. Contoured Competitor grip. Blued finish.

Thompson/Center Contender — Internal Choke Model
Originally made in 1968-69 w/octagonal bbl., this Internal Choke version in 45 Colt/.410 caliber only was reintroduced in 1986 w/10-inch bull bbl. Vent rib also available. Fixed iron rear sight, bead front. Detachable choke screws into muzzle. Blued finish. Contoured American black walnut Competitor Grip, also intro. in 1986, has nonslip rubber insert permanently bonded to back of grip.
W/Bull Bbl. **$285**
W/Vent Rib 340

Thompson/Center Contender — Octagon Barrel $245

The original Contender design, this octagonal bbl. version began to see the discontinuance of caliber offerings in 1980, so that now it is available in 22 LR only. 10-inch octagonal bbl. Patridge-style iron sights. Contoured Competitor Grip. Blued finish.

Thompson/Center Contender — Stainless

Similar to the standard Contender Models, except stainless steel w/blued sights. Black Rynite forearm and ambidextrous finger-groove grip. Made from 1993 to date.

Standard SS Model (10-inch bbl.)	**$295**
SS Super 14.	315
SS Super 16	325

Thompson/Center Contender Super 14 . . . . $275

Calibers: 22 LR, 222 Rem., 223 Rem., 6mm T.C.U., 6.5mm T.C.U., 7mm T.C.U., 7× 30 Waters, 30 Herrett, 30-30 Win., 357 Herrett, 357 Rem. Max., 35 Rem., 10mm Auto, 44 Mag., 445 Super Mag. 14-inch bull bbl. 18 inches overall. Weight: 56 oz. Patridge-style ramp front sight, adj. target rear. Blued finish. Made 1978 to date.

Thompson/Center Contender TC Alloy II

Calibers: 22 LR, 223 Rem., 357 Magnum, 357 Rem. Max., 44 Magnum, 7mm T.C.U., 30-30 Win., 45 Colt/.410 (w/internal choke), 35 Rem. and 7×30 Waters (14-inch bbl.). 10- or 14-inch bull bbl. or 10-inch vent rib bbl. (w/internal choke). All metal parts permanently electroplated w/T/C Alloy II, which is harder than stainless steel, ensuring smoother action, 30 percent longer bbl. life. Other design specifications the same as late model Contenders. Made 1986-89.

T/C Alloy II 10-inch Bull Bbl.	**$295**
T/C Alloy II Vent rib Bbl. w/Choke	325
T/C Alloy II Super 14	350

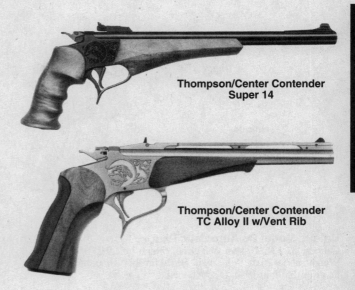

Thompson/Center Contender Super 14

Thompson/Center Contender TC Alloy II w/Vent Rib

UBERTI HANDGUNS
Mfd. by Aldo Uberti, Ponte Zanano, Italy
(Imported by Uberti USA, Inc.)

Uberti Model 1871 Rolling Block Target Pistol. $250

Single shot. Calibers: 22 LR, 22 Magnum, 22 Hornet and 357 Magnum. 9.5-inch bbl. 14 inches overall. Weight: 44 oz. Ramp front sight, fully adjustable rear. Smooth walnut grip and forearm. Color casehardened frame w/brass trigger guard. Blued halfoctagon or full round barrel.

Uberti Model 1873 Cattleman SA Revolver

Calibers: 357 Magnum, 38-40, 44-40, 44 Special, 45 Long Colt, 45 ACP. 6-shot cylinder. Bbl length: 3.5, 4.5, 4.75, 5.5, 7.5 or 18 inches. 10.75 inches overall (5.5-inch bbl.). Weight: 38 oz. (5.5-inch bbl.). Color casehardened steel frame w/steel or brass back strap and trigger guard. Nickel plated or blued barrel and cylinder. Imported 1997 to date.

First Issue	**$255**
Bisley	265
Bisley (Flattop)	275
Buntline (reintroduced 1992)	285
Quick Draw	225
Sabre (Bird Head)	250
Sheriff's Model	230
Convertible Cylinder, **add**	75
Stainless Steel, **add**	65
Steel Backstrap and Triggerguard, **add**	30
Target Sights, **add**	35

Uberti Model 1875 Remington Outlaw

Replica of Model 1875 Remington. Calibers: 357 Mag., 44-40, 45 ACP, 45 Long Colt. 6-shot cylinder. 5.5- to 7.5-inch bbl. 11.75 to 13.75 inches overall. Weight: 44 oz. (7.5 inch bbl). Color casehardened steel frame w/steel or brass back strap and trigger guard. Blue or nickel finish.

Blue Model	**$235**
Nickel Model (disc. 1995)	285
Convertible Cylinder (45 LC/45 ACP), **add**	50

Uberti Model 1890 Remington Police

Similar to Model 1875 Remington, except without the web under the ejector housing.

Blue Model	**$235**
Nickel Model (disc. 1995)	285
Convertible Cylinder (45 LC/45 ACP), **add**	50

ULTRA LIGHT ARMS, INC
Granville, WV.

Ultra Light Arms Model 20 Series Pistols

Calibers: 22-250 thru 308 Win. 5-shot magazine. 14-inch bbl. Weight: 4 lbs. Composite Kevlar, graphite reinforced stock. Benchrest grade action available in right- or left-hand models. Timney adjustable trigger w/three function safety. Bright or matte finish. Made 1987 to date.

Model 20 Hunter's Pistol	**$995**
Model 20 Reb Pistol (disc. 1989)	950

UNIQUE PISTOLS
Hendaye, France
Mfd. by Manufacture d'Armes des Pyrénées
Currently imported by Nygord Precision Products
(Previously by Beeman Precision Arms)

Unique Model B/cf Automatic Pistol $195

Calibers: 32 ACP, 380 ACP, 9-shot (32) or 8-shot (38) magazine. 4-inch bbl. 6.6 inches overall. Weight: 24.3 oz. Blued finish. Plain or thumbrest plastic grips. Intro. 1954. Disc.

Unique Model D2 . . . . $235

Same as Model D6, except has 4.5-inch bbl., 7.5 inches overall, weighs 24.5 oz. Made 1954 to date.

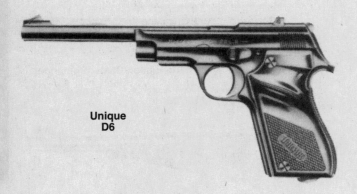

**Unique
D6**

Unique Model D6 Automatic Pistol $225
Caliber: 22 LR. 10-shot magazine. 6-inch bbl. 9.25 inches overall. Weight: about 26 oz. Adj. sights. Blued finish. Plain or thumbrest plastic grips. Intro. in 1954. Discontinued.

Unique Model DES/32U Rapid Fire Pistol
Caliber: 32 S&W Long (wadcutter). 5- or 6-shot magazine. 5.9-inch bbl. Weight: 40.2 oz. Blade front sight, micro-adj. rear. Trigger adj. for weight and position. Blued finish. Stippled handrest grips. Imported 1990 to date.
Right-Hand Model . . . . $ 995
Left-Hand Model 1025

**Unique
DES/69 Standard Match**

Unique Model DES/69-U Target Pistol
Caliber: 22 LR. 5-shot magazine. 5.9-inch bbl. w/250 gm counterweight. 10.6 inches overall. Trigger adjusts for position and pull. Weight: 35.3 oz. Blade front sight, micro-adj. rear. Checkered walnut thumbrest grips w/adj. handrest. Blued finish. Imported 1969 to date.
Right-Hand Model $950
Left-Hand Model 975

**Unique Model
DES/VO Rapid Fire Match**

Unique Model DES/VO Rapid Fire Match Automatic Pistol $750
Caliber: 22 Short. 5-shot magazine. 5.9-inch bbl. 10.4 inches overall. Weight: 43 oz. Click adj. rear sight blade front. Checkered walnut thumbrest grips w/adj. handrest. Trigger adj. for length of pull. Made 1974 to date.

Unique Kriegs Model L Automatic Pistol $250
Caliber: 32 Automatic (7.65mm). 9-shot magazine. 3.2-inch bbl. 5.8 inches overall. Weight: 26.5 oz. Fixed sights. Blued finish. Plastic grips. Mfd. during German occupation of France 1940-45. *Note:* Bears the German military acceptance marks and may have grips marked "7.65m/m 9 SCHUSS."

Unique Model L Automatic Pistol $195
Calibers: 22 LR, 32 Auto (7.65mm), 380 Auto (9mm Short). 10-shot magazine in 22, 7 in 32, 6 in 380. 3.3-inch bbl. 5.8 inches overall. Weight: 16.5 oz. (380 Auto w/light alloy frame), 23 oz. (w/steel frame). Fixed sights. Blued finish. Plastic grips. Intro. 1955. Discontinued.

**Unique
Mikros Pocket**

Unique Model Mikros Pocket Automatic Pistol $160
Calibers: 22 Short, 25 Auto (6.35mm). 6-shot magazine. 2.25-inch bbl. 4.44 inches overall. Weight: 9.5 oz. (light alloy frame), 12.5 oz. (steel frame.). Fixed sights. Blued finish. Plastic grips. Intro. 1957. Discontinued.

**Unique
Rr**

Unique Model Rr Automatic Pistol $185
Postwar commercial version of WWII Kriegsmodell w/same general specifications. Intro. 1951. Discontinued.

Unique Model 2000-U Match Pistol

Caliber: 22 Short. Designed for U.I.T. rapid fire competition. 5-shot top-inserted magazine. 5.5-inch bbl. w/five vents for recoil reduction. 11.4 inches overall. Weight: 43.4 oz. Special light alloy frame, solid steel slide and shock absorber. Stippled French walnut w/adj. handrest. Imported 1990-96.
Right-Hand Model $950
Left-Hand Model 995

UNITED STATES ARMS CORPORATION
Riverhead, New York

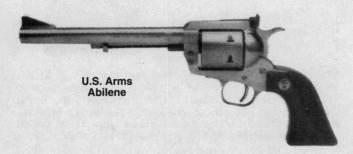

U.S. Arms Abilene

U.S. Arms Abilene SA Revolver

Safety Bar action. Calibers: 357 Mag., 41 Mag., 44 Mag., 45 Colt and 357/9mm convertible model w/two cylinders. 6-shot cylinder. Bbl. lengths: 4.63-, 5.5-, 6.5-inch, 7.5- and 8.5-inch in 44 Mag. only. Weight: about 48 oz. Adj. rear sight, ramp front. Blued finish or stainless steel. Smooth walnut grips. Made 1976-83.
44 Magnum, blued finish $250
44 Magnum, stainless steel 295
Other calibers, blued finish 225
357 Magnum, stainless steel 275
Convertible, 357 Mag./9mm Para., blued finish 250

UNIVERSAL FIREARMS CORPORATION
Hialeah, Florida

This company was purchased by Iver Johnson's Arms in the mid-1980s, when the Enforcer listed below was discontinued. An improved version was issued under the Iver Johnson name (see separate listing).

**Universal
Enforcer (3000)**

Universal Enforcer (3000) Semiautomatic Pistol . . . $275

M-1 Carbine-type action. Caliber: 30 Carbine. 5-, 15- or 30-shot clip magazine. 10.25-inch bbl. 17.75 inches overall. Weight: 4.5 lbs. (30-shot magazine). Adj. rear sight, blade front. Blued finish. Walnut stock w/pistol grip and handguard. Made 1964-83.

UZI PISTOLS
Mfd. by Israel Military Industries, Israel
(*Currently imported by UZI America*)

**Uzi
Semiautomatic Pistol**

Uzi Semiautomatic Pistol $995

Caliber: 9mm Para. 20-round magazine. 4.5-inch bbl. About 9.5 inches overall. Weight: 3.8 lbs. Front post-type sight, rear open-type, both adj. Discontinued 1993.

Uzi "Eagle" Series Semiautomatic DA Pistol

Caliber: 9mm Parabellum, 40 S&W, 45 ACP (Short Slide). 3.5-, 3.7- and 4.4-inch bbl. 10-shot magazine. Weight: 32 oz. to 35 oz. Blade front sight, drift adjustable tritium rear. Matte blue finish. Black synthetic grips. Imported 1997 to date.
Compact Model (DA or DAO) $415
Polymer Compact Model 395
Full Size Model 400
Short Slide Model 425

WALTHER PISTOLS
Manufactured by German, French and Swiss firms

The following Walther pistols were made before and during World War II by Waffenfabrik Walther, Zella-Mehlis (Thür.), Germany.

Walther Model 1 Automatic Pistol $425

Caliber: 25 Auto (6.35mm). 6-shot. 2.1-inch bbl. 4.4 inches overall. Weight: 12.8 oz. Fixed sights. Blued finish. Checkered hard rubber grips. Intro. 1908.

Walther Model 2 Automatic Pistol

Caliber: 25 Auto (6.35mm). 6-shot magazine. 2.1-inch bbl. 4.2 inches overall. Weight: 9.8 oz. Fixed sights. Blued finish. Checkered hard rubber grips. Intro. 1909.
Standard Model. $375
Pop-Up Sight Model 925

Walther Model 3 Automatic Pistol $1225

Caliber: 32 Auto (7.65mm). 6-shot magazine. 2.6-inch bbl. 5 inches overall. Weight: 16.6 oz. Fixed sights. Blued finish. Checkered hard rubber grips. Intro. 1910.

Walther Model 4 Automatic Pistol $275

Caliber: 32 Automatic (7.65mm). 8-shot magazine. 3.5-inch bbl. 5.9 inches overall. Weight: 18.6 oz. Fixed sights. Blued finish. Checkered hard rubber grips. Made 1910-18.

Walther Model 5

Walther Model 9

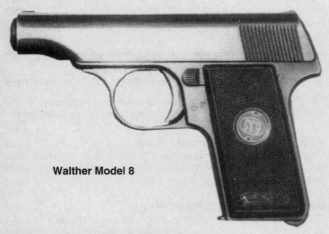

Walther Model 8

Walther Model 5 Automatic Pistol **$325**
Improved version of Model 2 w/same general specifications, distinguished chiefly by better workmanship and appearance. Intro. in 1913.

Walther Model 6 Automatic Pistol **$3250**
Caliber: 9mm Para. 8-shot magazine. 4.75-inch bbl. 8.25 inches overall. Weight: 34 oz. Fixed sights. Blued finish. Checkered hard rubber grips. Made 1915-17. *Note:* Since the powerful 9mm Para. cartridge is too much for the simple blow-back system of this pistol, firing is not recommended.

Walther Model 7 Automatic Pistol **$475**
Caliber: 25 Auto. (6.35mm). 8-shot magazine. 3-inch bbl. 5.3 inches overall. Weight: 11.8 oz. Fixed sights. Blued finish. Checkered hard rubber grips. Made 1917-18.

Walther Model 8 Automatic Pistol **$375**
Caliber: 25 Auto. (6.35mm). 8-shot magazine. 2.88-inch bbl. 5.13 inches overall. Weight: 12.38 oz. Fixed sights. Blued finish. Checkered plastic grips. Made 1920-45.

Walther Model 8 Lightweight Automatic Pistol **$395**
Same as standard Model 8 except about 25 percent lighter due to use of aluminum alloys.

Walther Model 9 Vest Pocket Automatic Pistol **$395**
Caliber: 25 Automatic (6.35mm). 6-shot magazine. 2-inch bbl. 3.94 inches overall. Weight: 9 oz. Fixed sights. Blued finish. Checkered plastic grips. Made 1921-45.

Walther Model HP Double-Action Automatic **$1050**
Prewar commercial version of the P38. "HP" is abbreviation of "Heeres Pistole" (Army Pistol). Caliber: 9mm Para. 8-shot magazine. 5-inch bbl. 8.38 inches overall. Weight: about 34.5 oz. Fixed sights. Blued finish. Checkered wood or plastic grips. The Model HP is distinguished by its notably fine material and workmanship. Made 1937-44.

Walther Olympia Funfkampf Model Automatic **$925**
Caliber: 22 LR. 10-shot magazine. 9.6-inch bbl. 13 inches overall. Weight: 33 oz., less weight. Set of 4 detachable weights. Adj. target sights. Blued finish. Checkered grips. Intro. 1936.

Walther Olympia Hunting Model Automatic **$950**
Same general specifications as Olympia Sport Model, but w/4-inch bbl. Weight: 28.5 oz.

Walther Olympia Rapid Fire Model Automatic . . . **$1050**
Caliber: 22 Short. 6-shot magazine. 7.4-inch bbl. 10.7 inches overall. Weight: without 12.38 oz. detachable muzzle weight, 27.5 oz. Adj. target sights. Blued finish. Checkered grips. Made about 1936-40.

Walther Olympia Sport Model Automatic **$795**
Caliber: 22 LR. 10-shot magazine. 7.4-inch bbl. 10.7 inches overall. Weight: 30.5 oz., less weight. Adj. target sights. Blued finish. Checkered grips. Set of four detachable weights was supplied at extra cost. Made about 1936-40.

Walther P38 Military DA Automatic **$625**
Modification of the Model HP adopted as an official German Service arm in 1938 and produced throughout WW II by Walther (code "ac"), Mauser (code "byf") and a few other manufacturers. General specifications and appearance same as Model HP, but w/a vast difference in quality, the P38 being a mass-produced military pistol. Some of the late wartime models were very roughly finished and tolerances were quite loose.

Walther Model PP DA Automatic Pistol
Polizeipistole (Police Pistol). Calibers: 22 LR (5.6mm), 25 Auto (6.35mm), 32 Auto (7.65mm), 380 Auto (9mm). 8-shot magazine, 7-shot in 380. 3.88-inch bbl. 6.94 inches overall. Weight: 23 oz. Fixed sights. Blued finish. Checkered plastic grips. *Note:* Wartime models are inferior in quality to prewar commercial pistols. Made 1929-45.
22 caliber, commercial model. **$750**
25 caliber, commercial model 2795
32 and 380 caliber, commercial model 425
Wartime model 550
380 Caliber, Commercial Model. 795

Walther
PP (Prewar)

Walther Model PP Lightweight
Same as standard Model PP, except about 25 percent lighter due to use of aluminum alloys. Values 50 percent higher.

Walther Model PP 7.65mm Presentation $1400
Made of soft aluminum alloy in green-gold color, these pistols were not intended to be fired.

Walther
PPK (WW II)

Walther Model PPK Double-Action Automatic Pistol
Polizeipistole Kriminal (Detective Pistol). Calibers: 22 LR (5.6mm), 25 Auto (6.35mm), 32 Auto (7.65mm), 380 Auto (9mm). 7-shot magazine, 6-shot in 380. 3.25-inch bbl. 5.88 inches overall. Weight: 19 oz. Fixed sights. Blued finish. Checkered plastic grips. *Note:* Wartime models are inferior in workmanship to prewar commercial pistols. Made 1931-45.

22 LR	$895
25, Commercial model	3950
32, Commercial model	595
380 (Pre-War Commercial model)	1895

Walther Model PPK Lightweight
Same as standard Model PPK, except about 25 percent lighter due to aluminum alloys. Values 50 percent higher.

Walther Model PPK 7.65mm Presentation $1095
Made of soft aluminum alloy in green-gold color, these pistols were not intended to be fired.

NOTE

The following Walther pistols are now manufactured by Carl Walther Waffenfabrik, Ulm/Donau, Germany.

Walther Self-Loading Sport Pistol $610
Caliber: 22 LR. 10-shot magazine. Bbl. lengths: 6- and 9-inch. 9.88 inches overall w/6-inch bbl. Target sights. Blued finish. One-piece, wood or plastic grips, checkered. Intro. in 1932.

Walther Free Pistol

Walther Model Free Pistol $1295
Single-Shot. Caliber: 22 LR. 11.7-inch heavy bbl. Weight: 48 oz. Adj. grips and target sights w/electronic trigger. Importation discontinued 1991.

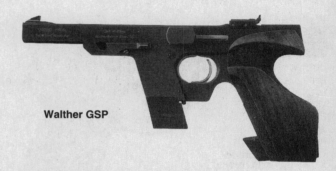

Walther GSP

Walther Model GSP Target Automatic Pistol
Calibers: 22 LR, 32 S&W Long Wadcutter. 5-shot magazine. 4.5-inch bbl. 11.8 inches overall. Weights: 44.8 oz. (22 cal.) or 49.4 oz. (32 cal.). Adj. target sights. Black finish. Walnut thumbrest grips w/adj. handrest. Made 1969-94.

22 Long Rifle	$ 995
32 S&W Long Wadcutter	1195
Conversion unit. 22 Short or 22 LR extra	1095

Walther Model OSP Rapid Fire Target Pistol $925
Caliber: 22 Short. 5-shot magazine. 4.5-inch bbl. 11.8 inches overall. Weight: 42.3 oz. Adj. target sights. Black finish. Walnut thumbrest grips w/adj. handrest. 22 LR conversion unit available (**add** $275). Made 1968-94.

Walther Model P4 (P38-IV) DA Pistol $550
Similar to P38, except has an uncocking device instead of a manual safety. Caliber: 9mm Para. only. 4.3-inch bbl. 7.9 inches overall. Other general specifications same as for current model P38. Made 1974-82.

Walther Model P5 DA Pistol $595
Alloy frame w/frame-mounted decocking levers. Caliber: 9mm Para. 8-shot magazine. 3.5-inch bbl. 7 inches overall. Weight: 28 oz. blued finish. Checkered walnut or synthetic grips. Made 1988 to date.

Walther Model P5 Compact Pistol $585
Similar to model P5, except w/3.1-inch bbl. and weighs 26 oz. Made 1988-96.

Walther P5

Walther P38 (WW II)

Walther Model P38 (P1) DA Automatic
Postwar commercial version of the P38, has light alloy frame. Calibers: 22 LR, 7.65mm Luger, 9mm Para. 8-shot magazine. Bbl. lengths: 5.1-inch in 22 caliber, 4.9- inch in 7.65mm and 9mm. 8.5 inches overall. Weight: 28.2 oz. Fixed sights. Nonreflective black finish. Checkered plastic grips. Made 1957-89. *Note:* The "P1" is W. German Armed Forces official pistol.
22 Long Rifle **$575**
Other calibers **445**

Walther Model P38 Deluxe Engraved Pistol
Elaborately engraved. Available in blued, chrome-, silver- or gold-plated finish.
Blued finish **$1150**
Chrome-plated **1095**
Silver-plated **1250**
Gold-plated **1650**

Walther Model P38K **$750**
Short-barreled version of current P38, the "K" standing for kurz (short). Same general specifications as standard model, except 2.8-inch bbl., 6.3 inches overall, weighs 27.2 oz., front sight is slide mounted. Caliber: 9mm Para. only. Made 1974-80.

Walther Model P 88 DA Automatic Pistol **$825**
Caliber: 9mm Para. 15-shot magazine. 4-inch bbl. 7.38 inches overall. Weight: 31.5 oz. Blade front sight, adj. rear. Checkered black synthetic grips. External hammer w/ambidextrous decocking levers. Alloy frame w/matte blued steel slide. Made 1987-93.

Walther Model P 88 DA Compact
Similar to the standard P 88 Model, except w/10- or 13-shot magazine. 3.8-inch bbl. 7.1 inches overall. Weight: 29 oz. Imported 1993 to date.
Model P88 (early importation) **$795**
Model P88 (post 94 importation) **$625**

Walther Model P99 DA Automatic Pistol **$575**
Caliber: 9mm Para. 10-shot magazine. 4-inch bbl. 7.2 inches overall. Weight: 22 oz. Ambidextrous magazine release, decocking lever and 3-function safety. Interchangeable front post sight, micro-adj. rear. Polymer gripframe w/blued slide. Imported 1997 to date..

Walther Model PP DA Automatic Pistol
Calibers: 22 LR, 32 Auto (7.65mm), 380 Auto (9mm Short). 8-shot magazine in 22 and 32, 7-shot in 380. 3.9-inch bbl. 6.7 inches overall. Weight: 32 cal., 23.3 oz. Fixed sights. Blued finish. Checkered plastic grips. Made 1963 to date.
22 Long Rifle **$495**
Other calibers **475**

Walther P38K

Walther P88

Walther PPK (Current)

Walther TPH (Current)

Walther Model PP Super DA Pistol $575
Caliber: 9×18mm. 7-shot magazine. 3.6-inch bbl. 6.9 inches overall. Weight: 30 oz. Fixed sights. Blued finish. Checkered plastic grips. Made 1974-81.

Walther Model PPK DA Automatic Pistol
Steel or dural frame. Calibers: 22 LR, 32 Auto (7.65mm) 380 Auto (9mm Short); latter caliber not available in model w/dural frame. 3.3-inch bbl. 6.1 inches overall. Weight (32 caliber): w/steel frame, 20.8 oz., w/dural frame,16.6 oz. Fixed sights. Blued finish. Checkered plastic grips. German-made 1963 to date, U.S. importation discontinued in 1968. U.S. version made by Interarms since 1986 including a stainless-steel model.
22 Long Rifle $595
Other calibers 550

Walther Model PPK/S DA Automatic Pistol
Designed to meet the requirements of the U.S. Gun Control Act of 1968, this model has the frame of the PP and the shorter slide and bbl. of the PPK. Overall length: 6.1 inches. Weight: 23 oz. Other specifications are the same as those of standard PPK, except steel frame only. German-made 1971 to date. U.S. version made by Interarms 1979 to date.
22 Long Rifle $575
Other calibers 525

> **NOTE**
>
> The following Mark IIs have been made in France since 1950 by Manufacture de Machines du Haut-Rhin (MANURHIN) at Mulhouse-Bourtzwiller. The designation "Mark II" is used here to distinguish between these and the prewar models. Early (1950-54) production bears MANURHIN trademark on slide and grips. Later models are marked "Walther Mark II." U.S. importation discontinued.

Walther Models PP, PPK, PPK/S Deluxe Engraved
These elaborately engraved models are available in blued finish, chrome-, silver- or gold-plated.
Blued finish . . . . $ 995
Chrome-plated 1025
Silver-plated 1150
Gold-plated. 1295
Add for 22 Long Rifle 50

Walther Model TPH DA Pocket Pistol
Light alloy frame. Calibers: 22 LR, 25 ACP (6.35mm). 6-shot magazine. 2.25-inch bbl. 5.38 inches overall. Weight: 14 oz. Fixed sights. Blued finish. Checkered plastic grips. Made 1968 to date. *Note:* Few Walther-made models reached the U.S. because of import restrictions; a U.S.-made version has been mfd. by Interarms since 1986.
German Model $575
U.S. Model 295

Walther Mark II Model PP Auto Pistol $325
Same general specifications as prewar Model PP.

Walther Mark II Model PPK Auto Pistol $350
Same general specifications as prewar Model PPK.

Walther Mark II Model PPK Lightweight $310
Same as standard PPK except has dural receiver. Calibers: 22 LR and 32 Auto.

Hammerelli-Walther Model 200, 1952 Type

Hämmerli-Walther Olympia Model 200 Auto Pistol, 1952 Type $575
Similar to 1936 Walther Olympia Funfkampf Model.

> **NOTE**
>
> The Walther Olympia Model pistols were manufactured 1952-1963 by Hämmerli AG Jagd-und Sportwaffenfabrik, Lenzburg, Switzerland, and marketed as "Hämmerli-Walther." See Hämmerli listings for specific data.

For the following Hammerelli-Walther Models 200, 201, 202, 203, 204, and 205

See Listings under Hammerli Section.

WARNER PISTOL
Norwich, Connecticut
Warner Arms Corp. (or Davis-Warner Arms Co.)

Warner Infallible Pocket Automatic Pistol $275
Caliber: 32 Automatic. 7-shot magazine. 3-inch bbl. 6.5 inches overall. Weight: about 24 oz. Fixed sights. Blued finish. Hard rubber grips. Made 1917-19.

WEBLEY & SCOTT LTD.
London and Birmingham, England

Webley Model 9mm Military & Police Automatic . . . $795
Caliber: 9mm Browning Long. 8-shot magazine. 8 inches overall. Weight: 32 oz. Fixed sights. Blued finish. Checkered vulcanite grips. Made 1909-30.

Webley Model 25 Hammer Automatic **$225**
Caliber: 25 Automatic. 6-shot magazine. Overall length: 4.75 inches. Weight: 11.75 oz. No sights. Blued finish. Checkered vulcanite grips. Made 1906-40.

Webley Model 25 Hammerless Automatic **$220**
Caliber: 25 Automatic. 6-shot magazine. Overall length: 4.25 inches. Weight: 9.75 oz. Fixed sights. Blued finish. Checkered vulcanite grips. Made 1909-40.

Webley Mark I 455 Automatic Pistol **$1150**
Caliber: 455 Webley Automatic. 7-shot magazine. 5-inch bbl. 8.5 inches overall. Weight: about 39 oz. Fixed sights. Blued finish. Checkered vulcanite grips. Made 1913-31. Reissued during WWII. Total production about 9,300. *Note:* **Mark I No. 2** is same pistol w/adj. rear sight and modified manual safety.

**Webley Mark III
38 Military & Police Revolver**

Webley Mark III 38 Military & Police Revolver **$275**
Hinged frame. DA. Caliber: 38 S&W. 6-shot cylinder. Bbl. lengths: 3- and 4-inch. 9.5 inches overall (4-inch bbl.). Weight: 21 oz. (4-inch bbl.). Fixed sights. Blued finish. Checkered walnut or vulcanite grips. Made 1897-45.

**Webley Mark IV
22 Caliber Target Revolver**

Webley Mark IV 22 Caliber Target Revolver **$345**
Same frame and general appearance as Mark IV 38. Caliber: 22 LR. 6-shot cylinder. 6-inch bbl. 10.13 inches overall. Weight: 34 oz. Target sights. Blued finish. Checkered grips. Discontinued 1945.

Webley Mark IV 38 Military & Police Revolver **$275**
Identical in appearance to the Mark IV 22 on the preceding page. Hinged frame. DA. Caliber: 38 S&W. 6-shot cylinder. Bbl. lengths: 3-, 4- and 5-inch. 9.13 inches overall (5-inch bbl.). Weight: 27 oz. (5-inch bbl.). Fixed sights. Blued finish. Checkered grips. Made 1929-c. 1957.

**Webley Mark VI
No. 1 British Service Revolver**

Webley Mark VI No. 1 British Service Revolver **$275**
DA. Hinged frame. Caliber: 455 Webley. 6-shot cylinder. Bbl. lengths: 4-, 6- and 7.5-inch. 11.25 inches overall (6-inch bbl.). Weight: 38 oz. (6-inch bbl.). Fixed sights. Blued finish. Checkered walnut or vulcanite grips. Made 1915-1947.

Webley Mark VI 22 Target Revolver **$295**
Same frame and general appearance as the Mark VI 455. Caliber: 22 LR. 6-shot cylinder. 6-inch bbl. 11.25 inches overall. Weight: 40 oz. Target sights. Blued finish. Checkered walnut or vulcanite grips. Discontinued 1945.

Webley Metropolitan Police Automatic Pistol **$275**
Calibers: 32 Auto, 380 Auto. 8-shot (32) or 7-shot (380) magazine. 3.5-inch bbl. 6.25 inches overall. Weight: 20 oz. Fixed sights. Blued finish. Checkered vulcanite grips. Made 1906-40 (32) and 1908-20 (380).

Webley RIC Model DA Revolver **$295**
Royal Irish Constabulary or Bulldog Model. Solid frame. Caliber: 455 Webley. 5-shot cylinder. 2.25-inch bbl. Weight: 21 oz. Fixed sights. Blued finish. Checkered walnut or vulcanite grips. Discontinued.

Webley "Semiautomatic" Single-Shot Pistol **$695**
Similar in appearance to the Webley Metropolitan Police Automatic, this pistol is "semiautomatic" in the sense that the fired case is extracted and ejected and the hammer cocked as in a blow-back automatic pistol; it is loaded singly and the slide manually operated in loading. Caliber: 22 Long. 4.5- or 9-inch bbl. 10.75 inches overall (9-inch bbl.). Weight: 24 oz. (9-inch bbl.). Adj. sights. Blued finish. Checkered vulcanite grips. Made 1911-27.

Webley Single-Shot Target Pistol **$195**
Hinge frame. Caliber: 22 LR. 10-inch bbl. 15 inches overall. Weight: 37 oz. Fixed sights on earlier models, current production has adj. rear sight. Blued finish. Checkered walnut or vulcanite grips. Made 1909 to date.

Webley-Fosbery Automatic Revolver
Hinged frame. Recoil action revolves cylinder and cocks hammer. Caliber: 455 Webley. 6-shot cylinder. 6-inch bbl. 12 inches overall. Weight: 42 oz. Fixed sights. Blued finish. Checkered walnut grips. Made 1901-1939. *Note:* A few were produced in caliber 38 Colt Auto w/an 8-shot cylinder (very rare).

1901 Model	**$3,295**
1902 Model	2,950
1904 Model	2,850
38 Colt (8-shot), **add**	1,295

WESSON FIREARMS CO., INC.
Palmer, Massachusetts
Formerly Dan Wesson Arms, Inc.

Dan Wesson Model 8 Service
Same general specifications as Model 14, except caliber 38 Special. Made 1971-75. Values same as for Model 14.

Dan Wesson Model 8-2 Service
Same general specifications as Model 14-2, except caliber 38 Special. Made 1975 to date. Values same as for Model 14-2.

Dan Wesson Model 9 Target
Same as Model 15, except caliber 38 Special. Made 1971-75. Values same as for Model 15.

Dan Wesson Model 9-2 Target
Same as Model 15-2, except caliber 38 Special. Made 1975 to date. Values same as for Model 15-2.

Dan Wesson Model 9-2H Heavy Barrel
Same general specifications as Model 15-2H, except caliber 38 Special. Made 1975 to date. Values same as for Model 15-2H. Discontinued 1983.

Dan Wesson Model 9-2HV Vent Rib Heavy Barrel
Same as Model 15-2HV, except caliber 38 Special. Made 1975 to date. Values same as for Model 15-2HV.

Dan Wesson Model 9-2V Vent rib
Same as Model 15-2V, except caliber 38 Special. Made 1975 to date. Values same as for Model 15-2H.

Dan Wesson Model 11 Service DA Revolver
Caliber: 357 Magnum. 6-shot cylinder. Bbl. lengths: 2.5-, 4-, 6-inch- interchangeable bbl. assemblies. 9 inches overall (4-inch bbl.). Weight: 38 oz. (4-inch bbl.). Fixed sights. Blued finish. Interchangeable grips. Made 1970-71. *Note:* The Model 11 has an external bbl. nut.

W/one bbl. assembly and grip	$185
Extra bbl. assembly	60
Extra grip	45

Dan Wesson Model 12 Target
Same general specifications as Model 11, except has adj. sights. Made 1970-71.

W/one-bbl. assembly and grip	$215
Extra bbl. assembly	50
Extra grip	25

Dan Wesson Model 14 Service DA Revolver
Caliber: 357 Magnum. 6-shot cylinder. Bbl. lengths: 2.25-, 3.75, 5.75-inch; interchangeable bbl. assemblies. 9 inches overall (3.75-inch bbl.). Weight: 36 oz. (3.75-inch bbl.). Fixed sights. Blued or nickel finish. Interchangeable grips. Made 1971-75. *Note:* Model 14 has recessed bbl. nut.

W/one-bbl. assembly and grip	$225
Extra bbl. assembly	50
Extra grip	25

**Dan Wesson
Model 14 Service**

**Dan Wesson
Model 14-2 Service**

Dan Wesson Model 14-2 Service DA Revolver
Caliber: 357 Magnum. 6-shot cylinder. Bbl. lengths: 2.5-, 4-, 6-, 8-inch; interchangeable bbl. assemblies. 9.25 inches overall (4-inch bbl.) Weight: 34 oz. (4-inch bbl.).. Fixed sights. Blued finish. Interchangeable grips. Made 1975 to date. *Note:* Model 14-2 has recessed bbl. nut.

W/one-bbl. assembly (8") and grip	$150
W/one-bbl. assembly (other lengths) and grip	135
Extra bbl. assembly, 8"	65
Extra bbl. assembly, other lengths	50
Extra grip	25

Dan Wesson Model 15 Target
Same general specifications as Model 14, except has adj. sights. Made 1971-75.

W/one-bbl. assembly and grip	$165
Extra bbl. assembly	50
Extra grip	25

Dan Wesson Model 15-2 Target
Same general specifications as Model 14-2 except has adj. rear sight and interchangeable blade front; also avail. w/10-, 12- or 15-inch bbl. Made 1975 to date.

W/one-bbl. assembly (8") and grip	$150
W/one-bbl. assembly (10") and grip	165
W/one-bbl. assembly (12")/grip. Disc	175
W/one-bbl. assembly (15")/grip. Disc	185
W/one-bbl. assembly (other lengths)/grip.	140
Extra bbl. assembly (8")	75
Extra bbl. assembly (10")	95
Extra bbl. assembly (12"). Discontinued	100
Extra bbl. assembly (15"). Discontinued	115
Extra bbl. assembly (other lengths).	50
Extra grip.	20

Dan Wesson Model 15-2H Heavy Barrel
Same as Model 15-2, except has heavy bbl. assembly weight, w/4-inch bbl., 38 oz. Made 1975-83.

W/one-bbl. assembly (8") and grip	$155
W/one-bbl. assembly (10") and grip	175
W/one-bbl. assembly (12") and grip	185
W/one-bbl. assembly (15") and grip	195
W/one-bbl. assembly (other lengths)/grip	145
Extra bbl. assembly (8")	80
Extra bbl. assembly (10")	100
Extra bbl. assembly (12")	110
Extra bbl. assembly (15")	135
Extra bbl. assembly (other lengths).	60
Extra grip.	20

Dan Wesson
Model 15-2H
Interchangeable Heavy

Dan Wesson Model 15-2HV Vent Rib Heavy Barrel
Same as Model 15-2, except has vent rib heavy bbl. assembly; weight, w/4-inch bbl., 37 oz. Made 1975 to date.

W/one-bbl. assembly (8") and grip	$175
W/one-bbl. assembly (10") and grip	190
W/one-bbl. assembly (12") and grip	195
W/one-bbl. assembly (15") and grip	235
W/one-bbl. assembly (other lengths) and grip	170
Extra bbl. assembly (8")	90
Extra bbl. assembly (10")	110
Extra bbl. assembly (12")	125
Extra bbl. assembly (15")	140
Extra bbl. assembly (other lengths)	70
Extra grip	20

Dan Wesson Model 15-2V Vent rib
Same as Model 15-2, except has vent rib bbl. assembly weighs 35 oz. (4-inch bbl.). Made 1975 to date. Values same as for 15-2H.

Dan Wesson Hunter Pacs
Dan Wesson Hunter Pacs are offered in all Magnum calibers and include heavy vent 8-inch shroud bbl. revolver Burris scope mounts, bbl. changing tool in a case.

HP22M-V	$525
HP22M-2	485
HP722M-V	565
HP722M-2	545
HP32-V	525
HP32-2	455
HP732-V	550
HP732-2	515
HP15-V	515
HP15-2	475
HP715-V	555
HP715-2	515
HP41-V	450
HP741-V	595
HP741-2	515
HP44-V	585
HP44-2	515
HP744-V	645
HP744-2	625
HP40-V	395
HP40-2	565
HP740-V	675
HP740-2	625
HP375-V	385
HP375-2	565
HP45-V	495

WHITNEY FIREARMS COMPANY
Hartford, Connecticut

Whitney Wolverine Automatic Pistol **$235**
Dural frame/shell contains all operating components. Caliber: 22 LR. 10-shot magazine. 4.63-inch bbl. 9 inches overall. Weight: 23 oz. Patridge-type sights. Blued or nickel finish. Plastic grips. Made 1955-62.

WICHITA ARMS
Wichita, Kansas

Wichita Classic Pistol
Caliber: Chambered to order. Bolt-action, single-shot. 11.25-inch octagonal bbl. 18 inches overall. Weight: 78 oz. Open micro sights. Custom-grade checkered walnut stock. Blued finish. Made 1980-97.

Standard	**$2295**
Presentation Grade (engraved)	3695

Wichita Hunter Pistol **$470**
Bolt-action, single-shot. Calibers: 22 LR, 22 WRF, 7mm Super Mag., 7-30 Waters, 30-30 Win., 32 H&R Mag., 357 Mag., 357 Super Mag. 10.5-inch bbl. 16.5 inches overall. Weight: 60 oz. No sights (scope mount only). Stainless steel finish. Walnut stock. Made 1983-94.

Wichita International Pistol **$495**
Top-break, single-shot. SA. Calibers: 7-30 Waters, 7mm Super Mag., 7R (30-30 Win. necked to 7mm), 30-30 Win. 357 Mag., 357 Super Mag., 32 H&R Mag., 22 Mag., 22 LR. 10- and 14-inch bbl. (10.5" for centerfire calibers). Weight: 50-71 oz. Patridge front sight, adj. rear. Walnut forend and grips.

Wichita MK-40 Silhouette Pistol **$820**
Calibers: 22-250, 7mm IHMSA, 308 Win. Bolt-action, single-shot. 13-inch bbl. 19.5 inches overall. Weight: 72 oz. Wichita Multi-Range sight system. Aluminum receiver w/blued bbl. Gray fiberthane glass stock. Made 1981 to date.

Wichita Silhouette Pistol **$840**
Calibers: 22-250, 7mm IHMSA 308 Win. Bolt-action, single-shot. 14.94-inch bbl. 21.38 inches overall. Weight: 72 oz. Wichita Multi-Range sight system. Blued finish. Walnut or gray fiberthane glass stock. Walnut center or rear grip. Made 1979 to date.

WILKINSON ARMS
Parma, Idaho

Wilkinson "Linda" Semiautomatic Pistol **$295**
Caliber: 9mm Para. Luger. 31-shot magazine. 8.25-inch bbl. 12.25 inches overall. Weight: 77 oz., empty. Rear peep sight w/blade front. Blued finish. Checkered composition grips.

Wilkinson "Sherry" Semiautomatic Pistol **$150**
Caliber: 22 LR. 8-shot magazine. 2.13-inch bbl. 4.38 inches overall. Weight: 9.25 oz., empty. Crossbolt safety. Fixed sights. Blued or blue-gold finish. Checkered composition grips.

Section II
RIFLES

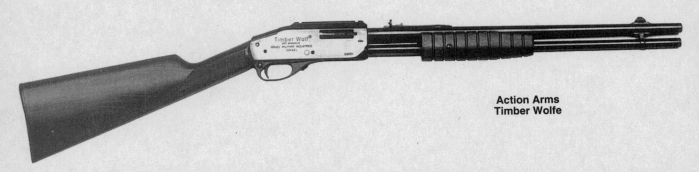

**Action Arms
Timber Wolfe**

AA ARMS
Monroe, North Carolina

AA Arms AR-9 Semiautomatic Carbine........... **$625**
Semiautomatic recoil-operated rifle w/side-folding metal stock design. Fires from a closed bolt. Caliber: 9mm Parabellum. 20-shot magazine. 16.25-inch bbl. 33 inches overall. Weight: 6.5 lbs. Fixed blade, protected post front sight adjustable for elevation, winged square notched rear. Matte phosphate/blue or electroless nickel finish. Checkered polymer grip/frame. Made 1991-94.

ACTION ARMS
Philadelphia, Pennsylvania

**Action Arms Model B Sporter Semiautomatic
Carbine** **$525**
Similar to Uzi Carbine (*see* separate listing), except w/thumbhole stock. Caliber: 9mm Parabellum, 10-shot magazine. 16-inch bbl. Weight: 8.75 lbs. Post front sight; adj. rear. Imported 1993-94.

Action Arms Timber Wolfe Repeating Rifle
Calibers: 357 Mag./38 Special and 44 Mag. slide-action. Tubular magazine holds 10 and 8 rounds, respectively. 18.5-inch bbl. 36.5 inches overall. Weight: 5.5 lbs. Fixed blade front sight; adj. rear. Receiver w/integral scope mounts. Checkered walnut stock. Imported 1989-94.

Action Arms Timber Wolfe Repeating Rifle *(Con't)*
Blued Model **$225**
Chrome Model, **add** **85**
44 Mag., **add**. **150**

ALPHA ARMS, INC.
Dallas, Texas

Alpha Arms Alaskan Bolt-Action Rifle **$1125**
Similar to Custom model, except w/stainless-steel bbl. and receiver w/all other parts coated w/Nitex. Weight: 6.75 to 7.25 lbs. Open sights w/bbl-band sling swivel. Classic-style Alphawood stock w/Niedner-style steel grip cap and solid recoil pad. Made 1984-89.

Alpha Arms Custom Bolt-Action Rifle........... **$1075**
Calibers: 17 Rem. thru 338 Win. Mag. Right or left-hand action in three action lengths w/three-lug locking system and 60-degree bolt rotation. 20- to 24-inch round or octagonal bbl. Weight: 6 to 7 lbs. No sights. Presentation-grade California claro walnut stock w/hand-rubbed oil finish, custom inleted sling swivels and Ebony forend tip. Made 1984-89.

Alpha Arms Grand Slam Bolt-Action Rifle **$950**
Same as Custom model, except w/Alphawood (fiberglass and wood) classic-style stock featuring Niedner-style grip cap. Weight: 6.5 lbs. Left-hand models. Made 1984-89.

Alpha Arms Alaskan

Alpha Arms Custom

**Alpha Arms
Jaguar**

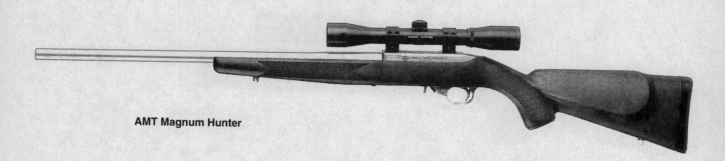

AMT Magnum Hunter

Alpha Arms Jaguar Bolt-Action Rifle
Same as Custom Rifle, except designed on Mauser-style action w/claw extractor, drilled and tapped and for scope. Originally designated Alpha Model 1. Calibers: 243, 7mm-08, 308 original chambering (1984-85) up to 338 Win. Mag. in standard model; 338 thru 458 Win. Mag. in Big Five model (1984-85). Teflon coated trigger guard/floorplate assembly. Made 1984-88.

Jaguar Grade I $795
Jaguar Grade II 895
Jaguar Grade III 950
Jaguar Grade IV 995
Big Five Model 1195

AMERICAN ARMS
N. Kansas City, Missouri

American Arms 1860 Henry $595
Replica of 1860 Henry rifle. Calibers: 44-40 or 45 LC. 24.25-inch half-octagonal bbl. 43.75 inches overall. Weight: 9.25 lbs. Brass frame and appointments. Straight-grip walnut buttstock.

American Arms 1866 Winchester
Replica of 1866 Winchester. Calibers: 44-40 or 45 LC. 19-inch round tapered bbl. (carbine) or 24.25-inch tapered octagonal bbl. (rifle). 38 to 43.25 inches overall. Weight: 7.75 or 8.15 lbs. Brass frame, elevator and buttplate. Walnut buttstock and forend.

Carbine $495
Rifle 525

American Arms 1873 Winchester
Replica of 1873 Winchester rifle. Calibers: 44-40 or 45 LC. 24.25-inch tapered octagonal bbl. w/tubular magazine. Color casehardened steel frame w/brass elevator and ejection port cover. Walnut buttstock w/steel buttplate.

Standard Model $550
Deluxe Model 675

AMT (ARCADIA MACHINE & TOOL)
Irwindale, California

AMT Bolt Action Repeating Rifle
Winchester type push-feed or Mauser type controlled-feed short, medium or long action. Calibers: 223 Rem., 22-250 Rem., 243 A, 243 Win., 6mm PPC, 25-06 Rem., 6.5x08, 270 Win., 7x57 Mauser, 7mm-08 Rem., 7mm Rem Mag., 7.62x39mm, 308 Win., 30-06, 300 Win. Mag., 338 Win. Mag., 375 H&H, 416 Rem., 416 Rigby, 458 Win. Mag. 22- to 28-inch #3 contour bbl. Weight: 7.75 to 8.5 lbs. Sights: None furnished, drilled and tapped for scope mounts. Classic composite or Kevlar stock. Made 1996-97.

Standard Model $795
Deluxe Model 995

AMT Bolt Action Single Shot Rifle
Winchester type cone breech push-feed or Mauser type controlled-feed action. Calibers: 22 Hornet, 22 PPC, 222 Rem., 223 Rem., 22-250 Rem., 243 A, 243 Win., 6mm PPC, 6.5x08, 270 Win., 7mm-08 Rem., 308 Win. 22- to 28-inch #3 contour bbl. Weight: 7.75 to 8.5 lbs. Sights: None furnished, drilled and tapped for scope mounts. Classic composite or Kevlar stock. Made 1996-97.

Standard Model $875
Deluxe Model 1550

AMT Challenger AutoloadingTarget Rifle Series I, II & III
Similar to Small Game Hunter, except w/McMillan target fiberglass stock. Caliber: 22 LR. 10-shot magazine. 16.5-, 18-, 20- or 22-inch bull bbl. Drilled and tapped for scope mount; no sights. Stainless steel finish. Made 1994-98.

Challenger I Standard $475
Challenger II w/Muzzle Break 625
Challenger III w/Bbl Extension 550
W/Jewell Trigger, **add** 200

AMT Lightning 25/22 Autoloading Rifle $165
Caliber: 22 LR. 25-shot magazine. 18-inch tapered or bull bbl. Weight: 6 lbs. 37 inches overall. Sights: adj. rear; ramp front. Folding stainless-steel stock w/matte finish. Made 1984-94.

AMT Lightning Small Game Hunter Series
Similar to AMT 25/22, except w/conventional matte black fiberglass/nylon stock. 10-shot rotary magazine. 22-inch bbl. 40.5 inches overall. Weight: 6 lbs. Grooved for scope; no sights. Made 1987-94 (Series I), 1992-93 (Series II).

Hunter I $175
Hunter II (w/22-inch heavy target bbl.) 185

AMT Magnum Hunter Auto Rifle $295
Similar to Lightning Small Game Hunter II Model, except chambered in 22 WRF w/22-inch match-grade bbl. Made 1993 to date.

ANSCHUTZ RIFLES
Ulm, Germany
Mfd. by J.G. Anschutz GmbH Jagd und Sportwaffenfabrik
Currently Imported by AcuSport Corp.; Accuracy International; Champion's Choice; Champion Shooter's Supply Go Sportsmen's Supply, Inc.; Gunsmithing Inc. (Previously by Precision Sales Int'l., Inc.)

Anschutz Models 1407 ISU, 1408-ED, 1411, 1413, 1418, 1432, 1433, 1518 and 1533 were marketed in the U.S by Savage Arms. Further, Anschutz Models 1403, 1416, 1422D, 1441, 1516 and 1522D were sold as Savage/Anschutz with Savage model designations (see also listings under Savage Arms.)

RIFLES

**Anschutz
Model 54.18MS-REP**

Anschutz Model 54.18MS **$950**
Bolt-action, single-shot. Caliber: 22 LR. 22-inch bbl. European hardwood stock w/cheekpiece. Stipple checkered Forend and Wundhammer swell pistol grip. Receiver grooved, drilled and tapped for scope blocks. Weight: 8.4 lbs. Imported 1982-97.

Anschutz Model 54.18MS-REP Repeating Rifle
Same as Model 54.18MS, except w/repeating action and 5-shot magazine. 22- to 30-inch bbl. 41-49 inches overall. Avg. weight: 7 lbs. 12 oz. Hardwood or synthetic gray thumbhole stock. Imported 1989-97.
Standard MS-REP Model **$1095**
MS-REP Deluxe w/Fibergrain Stock **995**

Anschutz Model 64S Bolt-Action Single-Shot Rifle
Bolt-action, single-shot. Caliber: 22 LR. 26-inch bbl. Checkered European hardwood stock w/Wundhammer swell pistol grip and adj. buttplate. Single-stage trigger. Aperture sights. Weight: 8.25 lbs. Imported 1963-82.
Standard **$395**
Left-Hand Model **425**

Anschutz Model 64MS Bolt-Action Single-Shot Rifle
Bolt-action, single-shot. Caliber: 22 LR. 21.25-inch bbl. European hardwood silhouette-style stock w/cheekpiece. Forend base and Wundhammer swell pistol grip, stipple checkered. Adj. two-stage trigger. Receiver grooved, drilled and tapped for scope blocks. Weight: 8 lbs. Imported 1982-96.
Standard or Featherweight (Disc. 1988) **$595**
Left-Hand Model **650**

Anschutz Model 64MSR Bolt-Action Repeater
Similar to Anschutz Model 64MS, except repeater w/5-shot magazine. Imported 1996 to date.
Standard Model **$625**
Left-Hand Model **695**

Anschutz Model 520/61 Semiautomatic **$195**
Caliber: 22 LR. 10-shot magazine. 24-inch bbl. Sights: folding leaf rear, hooded ramp front. Receiver grooved for scope mounting. Rotary-style safety. Monte Carlo stock and beavertail forend, checkered. Weight: 6.5 lbs. Imported 1982-83.

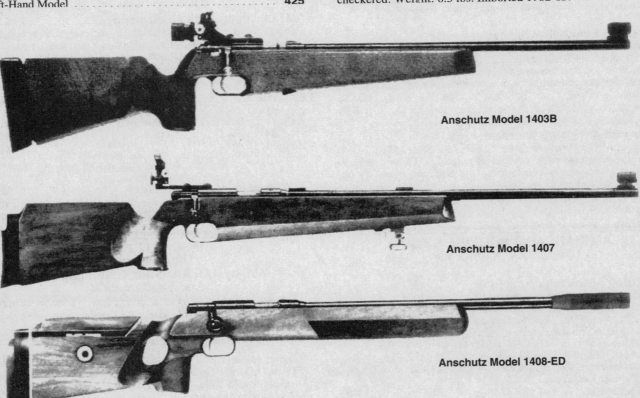

Anschutz Model 1403B

Anschutz Model 1407

Anschutz Model 1408-ED

Anschutz Model 1411

Anschutz Model 525 Autoloading
Rifle. **$375**
Caliber: 22 LR. 10-shot magazine. 24-inch bbl. 43 inches overall. Weight: 6.5 lbs. Adj. folding rear sight; hooded ramp front.Checkered European hardwood Monte Carlo style buttstock and beavertail forend. Sling swivel studs. Imported since 1982.

Anschutz Model 1403B **$695**
A lighter weight model designed for Biathlon competition. Caliber: 22 LR. 21.5-inch bbl. Adj. two-stage trigger. Adj. grooved wood buttplate, stipple-checkered deep thumbrest flute and straight pistol grip. Weight: 9 lbs. w/sights. Imported 1982-1992.

Anschutz Model 1403D Match Single-Shot
Target Rifle
Caliber: 22 LR. 25-inch bbl. 43 inches overall. Weight: 8.6 lbs. No sights, receiver grooved for Anschutz target sights. Walnut-finished hardwood target stock w/adj. buttplate. Importation disc. 1992.
Standard Model. **$495**
W/Match Sights **695**

Anschutz Model 1407 ISU Match 54 Rifle
Bolt-action, single-shot. Caliber: 22 LR. 26.88-inch bbl. Scope bases. Receiver grooved for Anschutz sights. Single-stage adj. trigger. Select walnut target stock w/deep forearm for position shooting, adj. buttplate, hand stop and swivel. Weight: 10 lbs. Imported 1970-81.
Standard Model. **$335**
Left-Hand Model **375**
W/International Sights, **add** **65**

Anschutz Model 1408-ED Super Running Boar **$475**
Bolt-action, single-shot. Caliber: 22 LR. 23.5-inch bbl. w/sliding weights. No metallic sights. Receiver drilled and tapped for scope-sight bases. Single-stage adj. trigger. Oversize bolt knob. Select walnut stock w/thumbhole, adj. comb and buttplate. Weight: 9.5 lbs. Intro. 1976. Disc.

Anschutz Model 1411 Match 54 Rifle
Bolt-action, single-shot. Caliber: 22 LR. 27.5-inch extra heavy bbl. w/mounted scope bases. Receiver grooved for Anschutz sights. Single-stage adj. trigger. Select walnut target stock w/cheekpiece (adj. in 1973 and later production), full pistol grip, beavertail forearm, adj. buttplate, hand stop and swivel. Model 1411-L has left-hand stock. Weight: 11 lbs. Disc.
W/Non-Adj. Cheekpiece. **$325**
W/adj. Cheekpiece **450**
Extra for Anschutz International Sight Set **125**

Anschutz Model 1413 Super Match 54 Rifle
Freestyle international target rifle w/specifications similar to those of Model 1411, except w/special stock w/thumbhole, adj. pistol grip, adj. cheekpiece in 1973 and later production, adj. hook buttplate, adj. palmrest. Model 1413-L has left-hand stock. Weight: 15.5 lbs. Disc.
W/Non-Adj. Cheekpiece. **$450**
W/Adj. Cheekpiece. **485**
Extra for Anschutz International Sight Set **125**

Anschutz Model 1416D **$445**
Bolt-action sporter. Caliber: 22 LR. 22.5-inch bbl. Sights: folding leaf rear; hooded ramp front. Receiver grooved for scope mounting. Select European stock w/cheekpiece, skip-checkered pistol grip and forearm. Weight: 6 lbs. Imported 1982 to date.

Anschutz Model 1416D Classic/Custom
Sporters
Same as Model 1416D, except w/American classic-style stock (Classic) or modified European-style stock w/Monte Carlo roll-over cheekpiece and Schnabel forend (Custom). Weight: 5.5 lbs. (Classic); 6 lbs. (Custom). Imported 1986 to date.
Model 1416D Classic **$460**
Model 1416D Classic, "True" Left-Hand. **495**
Model 1416D Custom. **445**
Model 1416D Fiberglass (1991-92) **545**

RIFLES

Anschutz Model 1413

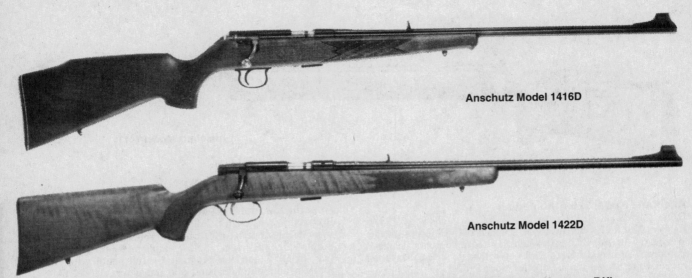

Anschutz Model 1416D

Anschutz Model 1422D

Anschutz Model 1418 Bolt-Action
Sporter **$275**
Caliber: 22 LR. 5- or 10-shot magazine. 19.75-inch bbl. Sights: folding leaf rear; hooded ramp front. Receiver grooved for scope mounting. Select walnut stock, Mannlicher type, w/cheekpiece, pistol grip and forearm skip checkered. Weight: 5.5 lbs. Intro. 1976. Disc.

Anschutz Model 1418D Bolt-Action
Sporter **$695**
Caliber: 22 LR. 5- or 10-shot magazine. 19.75-inch bbl. European walnut Monte Carlo stock, Mannlicher type w/cheekpiece, pistol grip and forend skip-line checkered buffalo horn Schnabel tip. Weight: 5.5 lbs. Imported from 1982-95 and 1998 to date.

Anschutz Model 1422D Classic/Custom Rifle
Bolt-action sporter. Caliber: 22 LR. 5-shot removable straight-feed clip magazine. 24-inch bbl. Sights: folding leaf rear; hooded ramp front. Select European walnut stock, classic type (Classic); Monte Carlo w/hand-carved rollover cheekpiece (Custom). Weight: 7.25 lbs. (Classic) 6.5 lbs. (Custom). Imported 1982-89.
Model 1422D Classic **$575**
Model 1422D Custom 625

Anschutz Model 1427B Biathlon Rifle
Bolt-action clip repeater. Caliber: 22 LR. 21.5-inch bbl. Two-stage trigger w/wing-type safety. Hardwood stock w/deep fluting, pistol grip and deep forestock with adj. hand stop rail. Target sights w/adjustable weights. Advertized in 1981 but imported from 1982 to date as Model 1827B. (See that model designation for current values)

Anschutz Model 1430D Match **$525**
Improved version of Model 64S. Bolt-action, single-shot. Caliber: 22 LR. 26-inch medium-heavy bbl. Walnut Monte Carlo stock w/cheekpiece, adj. buttplate, deep midstock tapered to forend. Pistol grip and contoured thumb groove w/stipple checkering. Single-stage adj. trigger. Target sights. Weight: 8.38 lbs. Imported 1982-83.

Anschutz Model 1432 Bolt-Action Sporter
Caliber: 22 Hornet. 5-shot box magazine. 24-inch bbl. Sights: folding leaf rear, hooded ramp front. Receiver grooved for scope mounting. Select walnut stock w/Monte Carlo comb and cheekpiece, pistol grip and forearm skip checkered. Weight: 6.75 lbs. Imported 1974-87. (Reintroduced as 1700/1730 series)
Early Model (1974-85) **$895**
Late Model (1985-87) 815

Anschutz Model 1432D Classic/Custom Rifle
Bolt-action sporter similar to Model 1422D except chambered for 22 Hornet. 4-shot magazine. 23.5-inch bbl. Weight: 7.75 lbs. (Classic); 6.5 lbs. (Custom). Classic stock on Classic model; fancy-grade Monte Carlo w/hand-carved rollover cheekpiece (Custom). Imported 1982-87. (Reintroduced as 1700/1730 series)
Model 1432D Classic **$895**
Model 1432D Custom 930

Anschutz Model 1433 Bolt-Action Sporter **$795**
Caliber: 22 Hornet. 5-shot box magazine. 19.75-inch bbl. Sights: folding leaf rear, hooded ramp front. Receiver grooved for scope mounting. Single-stage or double-set trigger. Select walnut Mannlicher stock; cheekpiece, pistol grip and forearm skip checkered. Weight: 6.5 lbs. Imported 1976-86.

Anschutz Model 1449 Sporter **$185**
Bolt-action sporter version of Model 2000. Caliber: 22 LR. 5-shot box magazine. 16.25-inch bbl. Weight: 3.5 lbs. Hooded ramp front sight, addition. Walnut-finished hardwood stock. Imported 1989-92.

Anschutz Model 1450B Target Rifle **$475**
Biathlon rifle developed on 2000 Series action. 19.5-inch bbl. Weight: 5.5 lbs. Adj. buttplate. Target sights. Imported 1993-94.

Anschutz Model 1516D Classic/Custom Rifle
Same as Model 1416D, except chambered for 22 Magnum RF, with w/American classic-style stock (Classic) or modified European-style stock w/Monte Carlo rollover cheekpiece and Schnabel forend (Custom). Weight: 5.5 lbs. (Classic), 6 lbs. (Custom). Imported 1986 to date.
Model 1516D Classic **$495**
Model 1516D Custom 515

Anschutz Models 1516D/1518D Luxus Rifles
The alpha designation for these Models was changed from Custom to Luxus in 1996-98. (See current Custon listings for Luxus values)

Anschutz Models 1518/1518D Sporting Rifles
Same as Model 1418, except chambered for 22 Magnum RF, 4-shot box magazine. Model 1518 intro. 1976. Disc. Model 1518D has full Mannlicher-type stock. Imported 1982 to date.
Model 1518 . . . . **$550**
Model 1518D 695

Anschutz Model 1427B

Anschutz Model 1432

Anschutz Model 1432D

RIFLES

Anschutz Model 1522D Classic/Custom Rifle
Same as Model 1422D, except chambered for 22 Magnum RF, 4-shot magazine. Weight: 6.5 lbs. (Custom). Fancy-grade Classic or Monte Carlo stock w/hand-carved rollover cheekpiece. Imported 1982-89. (Reintroduced as 1700D/1730D series)
Model 1522D Classic) $795
Model 1522D Custom) 725

Anschutz Model 1532D Classic/Custom Rifle
Same as Model 1432D except chambered for 222 Rem. 3-shot mag. Weight: 6.5 lbs. (Custom). Classic stock on Classic Model; fancy-grade Monte Carlo stock w/handcarved rollover cheekpiece (Custom). Imported 1982-89. (Reintroduced as 1700D/174 D0 series)
Model 1532D Classic $695
Model 1532D Custom. 895

Anschutz Model 1533 . . . . $725
Same as Model 1433 except chambered for 222 Rem. 3-shot box magazine. Imported 1976-94.

Anschutz Model 1700 Series Bolt-Action Repeater
Match 54 Sporter. Calibers: 22 LR, 22 Magnum, 22 Hornet, 222 Rem. 5-shot removable magazine.24-inch bbl. 43 inches overall. Weight: 7.5 lbs. Folding leaf rear sight, hooded ramp front. Select European walnut stock w/cheekpiece and Schnabel forend tip. Imported 1989 to date.
Standard Model 1700 Bavarian — RF calibers $825
Standard Model 1700 Bavarian — CF calibers 985
Model 1700D Classic (Classic stock, 6.75 lbs.)
   Rimfire calibers. 875
Model 1700D Classic — Centerfire calibers 950
Model 1700D Custom — Rimfire calibers 795
Model 1700D Custom — Centerfire calibers 975
Model 1700D Graphite Custom (McMillian graphite
   reinforced stock, 22" bbl., intro. 1991). 875
Model 1700 FWT Featherweight (6.5 lbs.)
   Rimfire calibers. 825
Model 1700 FWT — Centerfire calibers 1025

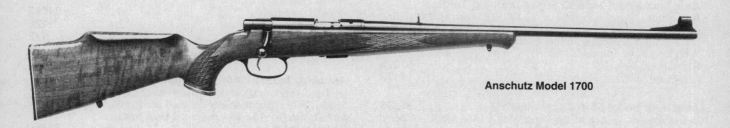

Anschutz Model 1700

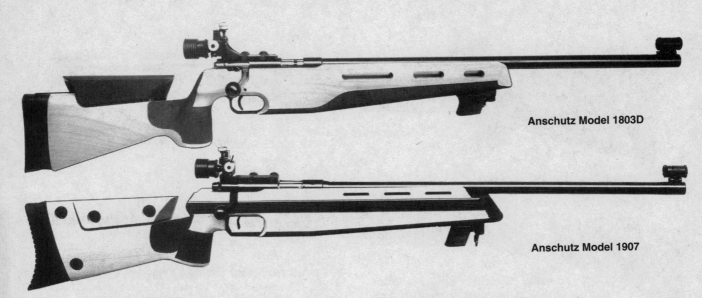

Anschutz Model 1803D

Anschutz Model 1907

Anschutz Model 1733D Mannlicher $925
Same as Model 1700D, except w/19-inch bbl. and Mannlicher-style stock. 39 inches overall. Weight: 6.25 lbs. Imported 1993 to date.

Anschutz Model 1740 Monte Carlo Sporter
Caliber: 22 Hornet or 222 Rem. 3- and 5-shot magazines respectively. 24-inch bbl. 43.25 inches overall. Weight: 6.5 lbs. Hooded ramp front, folding leaf rear. Drilled and tapped for scope mounts. Select European walnut stock w/roll-over cheekpiece, checkered grip and forend. Imported 1997 to date.
Model 1740 Custom **$895**
Model 1740 Classic (Meistergrade) **1095**

Anschutz Model 1743 Monte Carlo Sporter $925
Similar to Model 1740, except w/Mannlicher full stock. Imported 1997 to date.

Anschutz Model 1803D Match Single-Shot Target Rifle
Caliber: 22 LR. 25.5-inch bbl. 43.75 inches overall. Weight: 8.5 lbs. No sights, receiver grooved, drilled and tapped for scope mounts. Blonde or walnut-finished hardwood stock w/adj. cheekpiece, stippled grip and forend. Left-hand version. Imported 1987-92.
Right-Hand Model (Reintroduced as 1903D) **$695**
Left-Hand Model **775**

Anschutz Model 1807 ISU Standard Match $895
Bolt-action, single-shot. Caliber: 22 LR. 26-inch bbl. Improved Super Match 54 action. Two-stage match trigger. Removable cheekpiece, adj. buttplate, thumbpiece and forestock w/stipple checkered. Weight: 10 lbs. Imported 1982-88. (Reintroduced as 1907 ISU)

Anschutz Model 1808ED Super-Running Target
Bolt-action, single-shot. Caliber: 22 LR. 23.5-inch bbl. w/sliding weights. Improved Super Match 54 action. Heavy beavertail forend, adj. cheekpiece and buttplate. Adj. single-stage trigger. Weight: 9.5 lbs. Imported from 1982 to date.
Right-Hand Model **$1095**
Left-Hand Model **1175**

Anschutz Model 1810 Super Match II $1295
A less detailed version of the Super Match 1813 Model. Tapered forend w/deep receiver area. Select European hardwood stock. Weight: 13.5 lbs. Imported 1982-88. (Reintroduced as 1910 series)

Anschutz Model 1811 Prone Match $1095
Bolt-action, single-shot. Caliber: 22 LR. 27.5-inch bbl. Improved Super Match 54 action. Select European hardwood stock w/beavertail forend, adj. cheekpiece, and deep thumb flute. Thumb groove and pistol grip w/stipple checkering. Adj. buttplate. Weight: 11.5 lbs. Imported 1982-88. (Reintroduced as 1911 Prone Match)

Anschutz Model 1813 Super Match $1250
Bolt-action, single-shot. Caliber: 22 LR. 27.5-inch bbl. Improved Super Match 54 action w/light firing pin, one-point adj. trigger. European walnut thumbhole stock, adj. palmrest, forend and pistol grip stipple checkered. Adj. cheekpiece and hook buttplate. Weight: 15.5 lbs. Imported 1982-88. (Reintroduced as 1913 Super Match)

Anschutz Model 1827 Biathlon Rifle
Bolt-action clip repeater. Caliber: 22 LR. 21.5-inch bbl. 42.5 inches overall. Weight: 8.5 to 9 lbs. Slide safety. Adj. target sight set w/snow caps. European walnut stock w/cheekpiece, stippled pistol grip and forearm w/adj. weights. Fortner straight-pull bolt option offered in 1986. Imported 1982 to date.
Model 1827B w/Super Match 54 action **$1350**
Model 1827B, Left-Hand **1495**
Model 1827BT w/Fortner Option, Right-Hand **1595**
Model 1827BT, Left-Hand **1895**
Model 1827BT w/Laminated Stock, **add** . . . . /175

Anschutz Model 1907 ISU International Match Rifle
Updated version of Model 1807 w/same general specifications Model 1913, except w/26-inch bbl. 44.5 inches overall. Weight: 11 lbs. Designed for ISU 3-position competition. Fitted w/vented beechwood, walnut, blonde finished or color laminated stock. Imported 1989 to date.
Right-Hand Model **$1095**
Left-Hand Model **1225**
W/Laminated Stock, **add** **175**
W/Walnut Stock, **add** **100**

Anschutz Model 1910 International Super Match Rifle
Updated version of Model 1810 w/same general specifications Model 1913, except w/less-detailed hardwood stock w/tapered forend. Weight: 13.5 lbs. Imported 1989 to date.
Right-Hand Model **$1650**
Left-Hand Model **1795**

Anschutz Model 2013

Anschutz Model 1911 Prone Match Rifle
Updated version of Model 1811 w/same general specifications Model 1913, except w/specialized prone match hardwood stock w/beavertail forend. Weight: 11.5 lbs. Imported 1989 to date.
Right-Hand Model **$1295**
Left-Hand Model **1395**

Anschutz Model 1913 Standard Rifle $1095
Similar to 1913 Super Match w/economized appointments. Imported 1997 to date.

Anschutz Model 1913 Super Match Rifle
Bolt-action, single-shot Super Match (updated version of Model 1813). Caliber: 22 LR. 27.5-inch bbl. Weight: 14.2 lbs. Adj. two-stage trigger. Vented International thumbhole stock w/adj. cheekpiece, hand and palmrest, fitted w/10-way butthook. Imported 1989 to date.
Right-Hand Model **$1595**
Left-Hand Model **1750**

Anschutz Model 2007 ISU Standard Rifle $1395
Bolt-action, single-shot. Caliber: 22 LR. 19.75-inch bbl. 43.5 to 44.5 inches overall. Weight: 10.8 lbs. Two-stage trigger. Standard ISU stock w/adj. cheekpiece. Imported 1992 to date.

Anschutz Model 2013 Silhouette Rifle
Bolt-action, single-shot. Caliber: 22 LR. 20-inch bbl. 43 to 45.5 inches overall. Weight: 11.5 lbs. Two-stage trigger. International thumbhole black synthetic or laminated stock w/adj. cheekpiece hand and palmrest; fitted w/10-way butthook. Imported 1994 to date.
Model 2013 Silhouette **$1495**
Model 2013/BR-50 Laminated (Reintroduced 1997)...... **1525**

Anschutz Model 2013 Super Match Rifle
Bolt-action, single-shot. Caliber: 22 LR. 19.75-inch bbl. 43 to 45.5 inches overall. Weight: 12.5 lbs. Two-stage trigger. International thumbhole black synthetic or laminated stock w/adj. cheekpiece hand and palmrest; fitted w/10-way butthook. Imported 1992 to date.
Right-Hand Model **$1650**
Left-Hand Model **1795**

Anschutz Achiever Bolt-Action Rifle............ $250
Caliber: 22 LR. 5-shot magazine. Mark 2000-type repeating action. 19.5-inch bbl. 36.5 inches overall. Weight: 5 lbs. Adj. open rear sight; hooded ramp front. Plain European hardwood target-style stock w/vented forend and adj. buttplate. Imported since 1987.

Anschutz Achiever ST-Super Target $335
Same as Achiever, except single-shot w/22-inch bbl. and adj. stock. 38.75 inches overall. Weight: 6.5 lbs. Target sights. Imported since 1994.

Anschutz Model BR-50 Bench Rest Rifle $1495
Single-shot. Caliber: 22 LR. 19.75-inch bbl. (23 inches w/muzzle weight). 37.75-42.5 inches overall. Weight: 11 lbs. Grooved receiver, no sights. Walnut-finished hardwood or synthetic benchrest stock w/adj. cheekpiece. Imported 1994-97. Reintroduced as Model 2013 BR-50.

Anschutz Kadett Bolt-Action Repeating Rifle...... $195
Caliber: 22 LR. 5-shot detachable box magazine. 22-inch bbl. 40 inches overall. Weight: 5.5 lbs. Adj. folding leaf rear sight; hooded ramp front. Checkered European hardwood stock w/walnut-finish. Imported 1987-88.

Anschutz Mark 2000 Match $265
Takedown. Bolt-action, single-shot. Caliber: 22 LR. 26-inch heavy bbl. Walnut stock w/deep-fluted thumb-groove, Wundhammer swell pistol grip, beavertail-style forend. Adj. buttplate, single-stage adj. trigger. Weight: 8.5 lbs. Imported 1982-89.

ARMALITE INC.
Geneseo, Illinois
(Formerly Costa Mesa, California)

Armalite AR-7 Explorer Survival Rifle............. $95
Takedown. Semiautomatic. Caliber: 22 LR. 8-shot box magazine. 16-inch cast aluminum bbl. w/steel liner. Sights: peep rear; blade front. Brown plastic stock, recessed to stow barrel, action, and magazine. Weight: 2.75 lbs. Will float stowed or assembled. Made 1959-1973 by Armalite; 1974-90 by Charter Arms; 1990-97 by Survival Arms, Cocoa, FL.; 1997 to date by Henry Repeating Arms Co., Brooklyn, NY.

Anschutz Achiever

Armalite AR-10

Armalite AR-10 (T)
Target Carbine

Armalite M-15A2
National Match

Armalite M-15A2
HBAR

Armalite AR-7 Explorer Custom Rifle **$135**
Same as AR-7 Survival Rifle except w/deluxe walnut stock
w/cheekpiece and pistol grip. Weight: 3.5 lbs. Made 1964-70.

Armalite AR-10 (A) Semiautomatic Series
Gas-operated semiautomatic action. Calibers: 243 Win. or 308 Win.
(7.62 × 51mm). 10-shot magazine. 16- or 20-inch bbl. 35.5 or 39.5
inches overall. Weight: 9 to 9.75 lbs. Post front sight, adj. aperature
rear. Black or green composition stock. Made 1995 to date.
AR-10 A2 (Std. Carbine) **$875**
AR-10 A2 (Std. Rifle). **895**
AR-10 A4 (S.P. Carbine) **850**
AR-10 A4 (S.P. Rifle). **865**
W/Stainless Steel Bbl., **add** **120**

Armalite AR-10 (T) Target
Similar to Armalite Model AR-10A, except in National Match configu-
ration w/three-slot short Picatinny rail system and case deflector. 16- or
24-inch bbl. Weight: 8.25 to 10.4 lbs. Composite stock and handguard.
No sights. Optional National Match carry handle and detachable front
sight. Made 1995 to date.
AR-10 T (Rifle) **$1395**
AR-10 T (Carbine) **1350**

Armalite AR-180 Semiautomatic Rifle
Commercial version of full automatic AR-18 Combat Rifle. Gas-
operated semiautomatic. Caliber: 223 Rem. (5.56mm). 5-, 20-, 30--
round magazines. 18.25-inch bbl. w/flash hider/muzzle brake.
Sights: flip-up "L" type rear, adj. for windage; post front, adj. for elevation. Accessory 3X scope and mount (**add** $60 to value). Folding butt-
stock of black nylon, rubber buttplate, pistol grip, heat-dissipating
fiberglass forend (hand guard), swivels, sling. 38 inches overall, 28.75
inches folded. Weight: 6.5 lbs. *Note:* Made by Armalite Inc. 1969-72,
manufactured for Armalite by Howa Machinery Ltd., Nagoya, Japan,
1972-73; by Sterling Armament Co. Ltd., Dagenham, Essex, England,
1976 to 94. Importation disc. due to federal restrictions.
Armalite AR-180 (Mfg. By Armalite- Costa Mesa) . . . . . . **$ 895**
Armalite AR-180 (Mfg. By Howa) **1095**
Armalite AR-180 (Mfg. By Sterling) **795**
W/3X scope and mount, **add** **75**

Armalite M15 Series
Gas operated semiautomatic w/A2-style forward assist mechanism
and push-type pivot pin for easy takedown. Caliber: 223. 7-shot
magazine. 16-, 20- or 24-inch bbl. Weight: 7-9.2 lbs. Composite
or retractable stock. Fully adj. sights. Black anodized finish.
Made 1995 to date.

Armi Jager
AP-74 Wood Stock

Armalite M15 Series *(Con't)*

M-15A2 (Carbine) $875
M-15A2 (Service Rifle) 895
M-15A2 (National Match) 950
M-15A2 (Golden Eagle Heavy Bbl.) 925
M-15A2 M4C (Retractable Stock, Disc. 1997) 850
M-15A4 (Action Master, Disc 1997) 825
M-15A4 (Predator) 895
M-15A4 (S.P. Rifle) 650
M-15A4 (S.P. Carbine) 625
M-15A4T (Eagle Eye Carbine) 850
M-15A4T (Eagle Eye Rifle) 895

ARMI JAGER
Turin, Italy

Armi Jager AP-74 Commando $185
Similar to standard AP-74, but styled to resemble original version of
Uzi 9mm submachine gun w/wood buttstock. Lacks carrying handle
and flash suppressor. Has different type front sight mount and guards,
wood stock, pistol grip and forearm. Intro. 1976. Disc.

Armi Jager AP-74 Semiautomatic Rifle
Styled after U.S. M16 military rifle. Calibers: 22 LR, 32 Auto (pistol
cartridge). Detachable clip magazine; capacity: 14 rounds 22 LR, 9
rounds 32 ACP. 20-inch bbl. w/flash suppressor. Weight: 6.5 lbs.
M16 type sights. Stock, pistol grip and forearm of black plastic,
swivels and sling. Intro. 1974. Disc.
22 LR $225
32 Automatic 250

Armi Jager AP-74 Wood Stock Model
Same as standard AP-74, except w/wood stock, pistol grip and forearm, weighs 7 lbs. Disc.
22 LR $275
32 Automatic 295

ARMSCOR (Arms Corp.)
Manila, Philippines
Imported until 1991 by Armscor Precision, San Mateo,
CA; 1991-95 by Ruko Products, Inc., Buffalo NY: Cur-
rently improted by K.B.I., Harrisburg, PA

Armscor Model 20 Auto Rifle
Caliber: 22 LR. 15-shot magazine. 21-inch bbl. 39.75 inches overall.
Weight: 6.5 lbs. Sights: hooded front; adj. rear. Checkered or plain
walnut-finished mahogany stock. Blued finish. Imported 1990-91.
Reinstated by Ruko in the M series.
Model 20 (Checkered Stock) $95
Model 20C (Carbine-style Stock) 85
Model 20P (Plain Stock) 75

Armscor Model 1600 Auto Rifle
Caliber: 22 LR. 15-shot magazine. 19.5-inch bbl. 38 inches overall.
Weight: 6 lbs. Sights: post front; aperture rear. Plain mahogany
stock. Matte black finish. Imported 1987-91. Reinstated by Ruko in
the M series.
Standard Model $110
Retractable Stock Model 115

Armscor Model AK22 Auto Rifle
Caliber: 22 LR. 15- or 30-shot magazine. 18.5-inch bbl. 36 inches
overall. Weight: 7 lbs. Sights: post front; adj. rear. Plain mahogany
stock. Matte black finish. Imported 1987-91.
Standard Model $145
Folding Stock Model 195

Armscor Model M14 Series Bolt-Action Rifle
Caliber: 22 LR. 10-shot magazine. 23-inch bbl. Weight: 6.25 lbs. Open
sights. Walnut or walnut-finished mahogany stock. Imported 1991-97.
M14P Standard Model $70
M14D Deluxe Model (Checkered Stock, Disc. 1995) 85

Armscor Model M20 Series Semiautomatic Rifle
Caliber: 22 LR. 10- or 15-shot magazine. 18.25- or 20.75-inch bbl.
Weight:5.5 to 6.5 lbs. 38 to 40.5 inches overall. Hooded front sight
w/windage adj. rear. Walnut-finished mahogany stock. Imported
1990-97.
M20C Carbine Model . . . . $ 85
M20P Standard Model 75
M20S Sporter Deluxe (Checkered Mahogany Stock) 115
M20SC Super Classic (Checkered Walnut Stock) 195

Armscor Model M1400 Bolt-Action Rifle
Similar to Model 14P, except w/checkered stock w/Schnabel forend.
Weight: 6 lbs. Imported 1990-97.
M1400LW (Lightweight, Disc 1992) $155
M1400S (Sporter) $110
M1400SC (Super Classic) 195

Armscor Model M1500 Bolt-Action Rifle
Caliber: 22 Mag. 5-shot magazine. 21.5-inch bbl. Weight: 6.5 lbs. Open
sights. Checkered mahogany stock. Imported 1991-97.
M1500 (Standard) $110
M1500LW (Euro-style Walnut Stock, Disc. 1992) 165
M1500SC (Monte Carlo Stock) 185

Armscor Model M1600 Auto Rifle
Rimfire replica of Armalite Model AR 180 (M16), except cham-
bered for 22 LR. 15-shot magazine. 18-inch bbl. Weight: 5.25 lbs.
Composite or retractable buttstock w/composite handguard and
pistol grip. Carrying handle w/adj. aperature rear sight and pro-
tected post front. Black anodized finish. Imported 1991-97.
M-1600 (Standard w/fixed stock) $115
M-1600R (Retractable stock) 125

RIFLES

A-Square — Hannibal

Armscor Model M1800 Bolt-Action Rifle
Caliber: 22 Hornet. 5-shot magazine. 22-inch bbl. Weight: 6.6 lbs. Checkered hardwood or walnut stock. Sights: post front; adj. rear. Imported 1995 to date.
M-1800 (Standard) **$195**
M-1800SC (Checkered Walnut Stock) **295**

Armscor Model M2000 Auto Rifle
Similar to Model 20P, except w/checkered mahogany stock and adj. sights. Imported 1991 to date.
M2000S (Standard) **$100**
M2000SC (Checkered Walnut Stock) **185**

ARNOLD ARMS
Arlington, Washington

Arnold Arms African Safari
Calibers: 243 to 458 Win. Magnum. 22- to 26-inch bbl. Weight: 7-9 lbs. Scope mount standard or w/optional M70 Express sights. Chrome-moly in four finishes. "A" and "AA" Fancy Grade English walnut stock with #5 standard wraparound checkering pattern. Ebony forend tip. Made 1996 to date.
With "A" Grade English Walnut: Matte Blue **$3495**
Std. Polish **3695**
Hi-Luster **3850**
Stainless Steel Matte **3495**
With "AA" Grade English Walnut: C-M Matte Blue **3475**
Std. Polish **3695**
Hi-Luster **3850**
Stainless Steel Matte **3495**

Arnold Arms Alaskan Trophy
Calibers: 300 Magnum to 458 Win. Magnum. 24- to 26-inch bbl. Weight: 7-9 lbs. Scope mount w/Express sights standard. Stainless steel or chrome-moly Apollo action w/fibergrain or black synthetic stock. Barrel band on 357 H&H and larger magnums. Made 1996 to date.
Matte finish **$2395**
Std. Polish **2595**
Stainless steel **2450**

A-SQUARE COMPANY INC.
Louisville, Kentucky
(Formerly Bedford, KY)

A-Square Caesar Bolt-Action Rifle
Custom rifle built on Remington's 700 receiver. Calibers: Same as Hannibal, Groups I, II and III. 20- to 26-inch bbl. Weight: 8.5 to 11 lbs. Express 3-leaf rear sight, ramp front. Synthetic or classic Claro oil-finished walnut stock w/flush detachable swivels and Coil-Check recoil system. Three-way adj. target trigger; 3-position safety. Right- or left-hand. Made 1984 to date.
Synthetic Stock Model **$2495**
Walnut Stock Model **2195**

A-Square Genghis Khan Bolt-Action Rifle
Custom varmint rifle developed on Winchester's 70 receiver; fitted w/heavy tapered bbl. and Coil-Chek stock. Calibers: 22-250 Rem., 243 Win., 25-06 Rem., 6mm Rem. Weight: 8-8.5 lbs. Made 1994 to date.
Synthetic Stock Model **$2995**
Walnut Stock Model **2550**

A-Square Hamilcar Bolt-Action Rifle
Similar to Hannibal Model except lighter. Calibers: 25-06, 257 Wby., 6.5×55 Swedish, 270 Wby., 7×57, 7mm Rem., 7mm STW, 7mm Wby., 280 Rem., 30-06, 300 Win., 300 Wby., 338-06, 9.3×62. Weight: 8-8.5 lbs. Made 1994 to date.
Synthetic Stock Model **$2895**
Walnut Stock Model **2550**

A-Square Hannibal Bolt-Action Rifle
Custom rifle built on reinforced P-17 Enfield receiver. Calibers: *Group I:* 30-06, *Group II:* 7mm Rem. Mag., 300 Win. Mag., 416 Taylor, 425 Express, 458 Win. Mag.; *Group III:* 300 H&H, 300 Wby. Mag., 8mm Rem. Mag., 340 Wby. Mag., 375 H&H, 375 Wby. Mag., 404 Jeffery, 416 Hoffman, 416 Rem Mag., 450 Ackley, 458 Lott; *Group IV:* 338 A-Square Mag., 375 A-Square Mag., 378 Wby. Mag., 416 Rigby, 416 Wby. Mag., 460 Short Square Mag., 500 A-Square Mag. 20- to 26-inch bbl. Weight: 9 to 11.75 lbs. Express 3-leaf rear sight, ramp front. Classic Claro oil-finished walnut stock, or synthetic stock w/flush detachable swivels and Coil-Check recoil system. Adj. trigger w/2-position safety. Made 1983 to date.
Synthetic Stock Model **$2850**
Walnut Stock Model **2695**

AUSTRIAN MILITARY RIFLES
Steyr, Austria
Manufactured at Steyr Armory

Austrian Model 90 Steyr-Mannlicher Rifle **$175**
Straight-pull bolt action. Caliber: 8mm. 5-shot magazine. Open sights. 10-inch bayonet. Cartridge clip forms part of the magazine mechanism. Some of these rifles were provided with a laced, canvas hand guard, others were of wood.

Austrian Model 90 Steyr-Mannlicher Carbine **$190**
Same general specifications as Model 90 Rifle, except w/19.5-inch bbl., weighs 7 lbs. No bayonet stud or supplemental forend grip.

Austrian Model 95 Steyr-Mannlicher Carbine **$140**
Same general specifications as Model 95 Rifle, except w/19.5-inch bbl., weighs 7 lbs. Post front sight; adj. rear carbine sight.

Austrian Model 95 Steyr-Mannlicher
Service Rifle **$130**
Straight-pull bolt action. Caliber: 8×50R Mannlicher (many of these rifles were altered during World War II to use the 7.9mm German service ammunition). 5-shot Mannlicher-type box magazine. 30-inch bbl. Weight: 8.5 lbs. Sights: blade front; rear adj. for elevation. Military-type full stock.

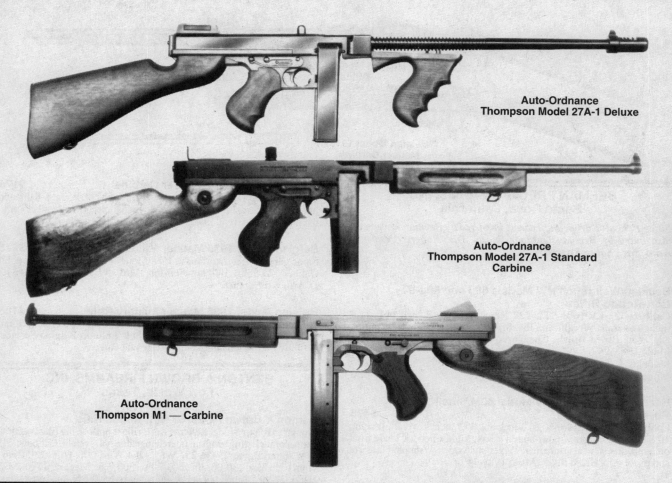

**Auto-Ordnance
Thompson Model 27A-1 Deluxe**

**Auto-Ordnance
Thompson Model 27A-1 Standard
Carbine**

**Auto-Ordnance
Thompson M1 — Carbine**

AUTO-ORDNANCE CORPORATION
West Hurley, New York
(Manufacturing rights acquired by Kahr Arms in 1999)

Auto-Ordnance Thompson Model 22-27A-3 . . . . . . **$375**
Same-bore version of Delux Model 27A-1. Same general specifications, except caliber 22 LR w/lightweight alloy receiver, weighs 6.5 lbs. Magazines include 5-, 20-, 30- and 50-shot box types, 80-shot drum. Made 1977-94.

Auto-Ordnance Thompson Model 27A-1 Deluxe
Same as Standard Model 27A-1, except w/finned bbl. w/compensator, adj. rear sight, pistol-grip forestock. Caliber: 22 LR, 10mm (1991-93) or 45 ACP. Weight: 11.5 lbs. Made 1976-99.
22 LR (Limited Production) **$995**
10mm or 45 ACP **695**
50-Round Drum Magazine, **add** **250**
100-Round Drum Magazine, **add** **450**
Violin Carrying Case, **add** **100**

**Auto-Ordnance Thompson Model 27A-1 Standard
Semiauto Carbine** **$550**
Similar to Thompson submachine gun ("Tommy Gun"), except has no provision for automatic firing. Caliber: 45 Automatic. 20-shot detachable box magazine (5-,15- and 30-shot box magazines, 39-shot drum also available). 16-inch plain bbl. Weight: 14 lbs. Sights: aperture rear; blade front. Walnut buttstock, pistol grip and grooved forearm, sling swivels. Made 1976-86.

**Auto-Ordnance Thompson 27A-1C Lightweight
Carbine** **$575**
Similar to Model 27A-1, except w/lightweight alloy receiver. Weight: 9.25 lbs. Made 1984 to date.

**Auto-Ordnance Thompson M1 Semiautomatic
Carbine** **$525**
Similar to Model 27A-1, except in M-1 configuration w/side cocking lever and horizontal forearm. Weight: 11.5 lbs. Made 1986 to date.

BARRETT FIREARMS MFG., INC.
Murfreesboro, Tennessee

Barrett Model 82 A-1 Semiautomatic Rifle **$4795**
Caliber: 50 BMG. 10-shot detachable box magazine. 29-inch recoiling bbl. w/muzzlebrake. 57 inches overall. Weight: 28.5 lbs. Open iron sights and 10X scope. Composition stock w/Sorbothance recoil pad and self-leveling bipod. Blued finish. Made 1985 to date.

Barrett Model 90 Bolt-Action Rifle **$2695**
Caliber: 50 BMG. 5-shot magazine. 29-inch match bbl. 45 inches overall. Weight: 22 lbs. Composition stock w/retractable bipod. Made 1990-95.

Barrett Model 95 Bolt Action **$2795**
Similar to Model 90 Bullpup design chambered 50 BMG, except w/improved muzzlebrake and extendible bipod. Made 1995 to date.

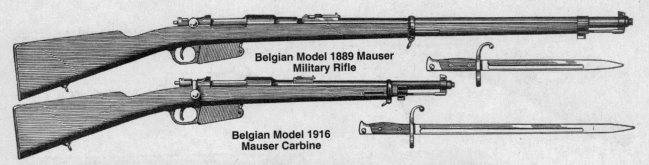

Belgian Model 1889 Mauser Military Rifle

Belgian Model 1916 Mauser Carbine

BEEMAN PRECISION ARMS, INC.
Santa Rosa, California

Since 1993 all European firearms imported by Beeman have been distributed by Beeman Outdoor Sports, Div. Roberts Precision Arms, Inc., Santa Rosa, CA.

Beeman/Weihrauch HW Models 60J and 60J-ST Bolt-Action Rifles
Calibers: 22 LR (60J-ST), 222 Rem. (60J).22.8-inch bbl. 41.7 inches overall. Weight: 6.5 lbs. Sights: hooded blade front; open adj. rear. Blued finish. Checkered walnut stock w/cheekpiece. Made 1988-94.
Model 60J $795
Model 60J-ST 450

Beeman/Weihrauch HW Model 60M Small Bore Rifle $475
Caliber: 22 LR. Single-shot. 26.8-inch bbl. 45.7 inches overall. Weight: 10.8 lbs. Adj. trigger w/push-button safety. Sights: hooded blade front on ramp, precision aperture rear. Target-style stock w/stippled forearm and pistol grip. Blued finish. Made 1988-94.

Beeman/Weihrauch HW Model 660 Match Rifle $650
Caliber: 22 LR.26-inch bbl.45.3 inches overall. Weight: 10.7 lbs. Adj. match trigger. Sights: globe front; precision aperture rear. Match-style walnut stock w/adj. cheekpiece and buttplate. Made 1988-94.

Beeman/Feinwerkbau Model 2600 Series Target Rifle
Caliber: 22 LR. Single-shot. 26.3-inch bbl. 43.7 inches overall. Weight: 10.6 lbs. Match trigger w/fingertip weight adjustment dial. Sights: globe front; micrometer match aperture rear. Laminated hardwood stock w/adj. cheekpiece. Made 1988-94.
Standard Model 2600 (left-hand) $1250
Standard Model 2600 (right-hand) 1095
Free Rifle Model 2602 (left-hand) 1595
Free Rifle Model 2602 (right-hand) 1450

BELGIAN MILITARY RIFLES
Mfd. by Fabrique Nationale D'Armes de Guerre, Herstal, Belgium; Fabrique D'Armes de L'Etat, Lunich, Belgium

Hopkins & Allen Arms Co. of Norwich, Conn., as well as contractors in Birminaham, England, also produced these arms during World War 1.

Belgian Model 1889 Mauser Military Rifle $125
Caliber: 7.65mm Belgian Service (7.65mm Mauser). 5-shot projecting box magazine. 30.75-inch bbl. w/jacket. Weight: 8.5 lbs. Adj. rear sight, blade front. Straight-grip military stock. This and the carbine version were the principal weapons of the Belgian Army at the start of WWII. Made 1889 to c.1935.

Belgian Model 1916 Mauser Carbine $175
Same as Model 1889 Rifle, except w/20.75-inch bbl. Weighs 8 lbs. and has minor differences in the rear sight graduations, lower band closer to the muzzle and swivel plate found on side of buttstock.

Belgian Model 1935 Mauser Military Rifle $225
Same general specifications as F.N. Model 1924; minor differences. Caliber: 7.65mm Belgian Service. Mfd. by Fabrique Nationale D'Armes de Guerre.

Belgian Model 1936 Mauser Military Rifle $185
An adaptation of Model 1889 w/German M/98-type bolt, Belgian M/89 protruding box magazine. Caliber: 7.65mm Belgian Service. Mfd. by Fabrique Nationale D'Armes de Guerre.

BENTON & BROWN FIREARMS, INC.
Fort Worth, Texas

Benton & Brown Model 93 Bolt-Action Rifle
Similar to Blaser Model R84 (the B&B rifle is built on the Blaser action, *see* separate listing) with an interchangeable bbl. system. Calibers: 243 Win., 6mm Rem., 25-06, 257 Wby., 264 Win., 270 Win., 280 Rem., 7mm Rem Mag., 30-06, 308, 300 Wby., 300 Win. Mag., 338 Win., 375 H&H. 22- or 24-inch bbl. 41 or 43 inches overall. Bbl.-mounted scope rings and one-piece base; no sights. Two-piece walnut or fiberglass stock. Made 1993 to date.
Walnut Stock Model. $1295
Fiberglass Stock Model 1195
Extra Bbl. Assembly, **add** 475
Extra Bolt Assembly, **add** 425

BERETTA U.S.A. CORP.
Accokeek, Maryland
Manufactured by Fabbrica D'Armi Pietro Beretta S.p.A. in the Gardone Val Trompia (Brescia), Italy

Beretta 455 SxS Express Double Rifle
Sidelock action w/removable sideplates. Calibers: 375 H&H, 458 Win. Mag., 470 NE, 500 NE (3 inches), 416 Rigby. Bbls.: 23.5 or 25.5-inch. Weight: 11 lbs. Double triggers. Sights: blade front; V-notch folding leaf rear. Checkered European walnut forearm and buttstock w/recoil pad. Color casehardened receiver w/blued bbls. Made 1990 to date.
Model 455 . . . . $29,750
Model 455EELL 39,500

Beretta 500 Bolt-Action Sporter
Centerfire bolt-action rifle w/Sako AI short action. Calibers: 222 Rem., 223 Rem. 5-shot magazine. 23.63-inch bbl. Weight: 6.5 lbs. No sights. Tapered dovetailed receiver. European walnut stock. Disc. 1986.
Standard . . . . $ 495
DL Model 1125
500 EELL (Engraved) 1295

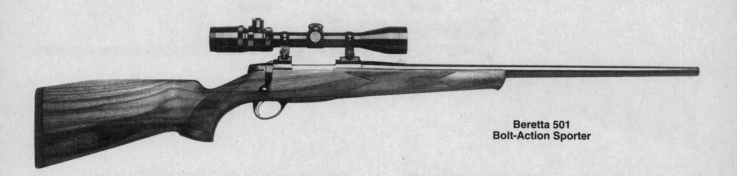

Beretta 501
Bolt-Action Sporter

Beretta 501 Bolt-Action Sporter

Same as Model 500, except w/Sako AII medium action. Calibers: 243 Win., 308 Win. Weight: 7.5 lbs. Disc. 1986.

Standard .... $ 495
Standard w/Iron Sights 525
DL Model 1195
501 EELL (Engraved) 1350

Beretta 502 Bolt-Action Sporter

Same as Model 500, except w/Sako AIII long action. Calibers: 270 Win., 7mm Rem. Mag., 30.06, 375 H&H. Weight: 8.5 lbs. Disc. 1986.

Standard .... $ 545
DL Model 1225
502 EELL (Engraved) 1395

Beretta AR-70 Semiautomatic Rifle $1395

Caliber: 223 Rem. (5.56mm). 30-shot magazine. 17.75-inch bbl. Weight: 8.25 lbs. Sights: rear peep adj. for windage and elevation; blade front. High-impact synthetic buttstock. Imported 1984-89.

Beretta Mato

Calibers: 270 Win., 280 Rem., 7mm Rem. Mag, 300 Win. Mag., 338 Win. Mag., 375 H&H. 23.6-inch bbl. Weight: 8 lbs. Adjustable trigger. Drop-out box magazine. Drilled and tapped for scope w/ or w/o adj. sights. Walnut or synthetic stock. Manufactured based on Mauser 98 action. Made in1997 to date.

Standard Model $1025
Deluxe Model 1295

Beretta Small Bore Sporting Carbine $295

Semiautomatic w/bolt handle raised, conventional bolt-action repeater w/handle in lowered position. Caliber: 22 LR. 4-, 8-, or 20-shot magazines.20.5-inch bbl. Sights: 3-leaf folding rear- partridge front. Stock w/checkered pistol grip and forend, sling swivels. Weight: 5.5 lbs.

Beretta Express SSO O/U Express Double Rifle

Sidelock. Calibers: 375 H&H Mag., 458 Win. Mag., 9.3 X 74R. 23-, 24- or 25.5-inch blued bbls. Weight: 11 lbs. Double triggers. Express sights w/blade front and V-notch folding leaf rear. Optional Zeiss scope w/claw mounts. Color casehardened receiver w/scroll engraving, game scenes and gold inlays on higher grades. Checkered European walnut forearm and buttstock w/cheekpiece and recoil pad. Imported 1985 to date.

Model SS0 (Disc. 1989) .... $ 8,795
Model SS05 (Disc. 1990) 9,200
Model SS06 Custom 16,500
Model SS06 EELL Gold Custom 19,500
Extra Bbl. Assembly, **add** 5,800
Claw Mounts, **add** 550

BLASER USA, INC.
Fort Worth, Texas
Mfd. by Blaser Jagdwaffen GmbH, Germany
(Imported by Sigarms, Exeter, NH; Autumn Sales, Inc.,
Fort Worth, TX)

Blaser Model R84 Bolt-Action Rifle

Calibers: 22-250, 243, 6mm Rem., 25-06, 270, 280 Rem., 30-06 257 Wby. Mag., 264 Win. Mag., 7mm Rem Mag., 300 Win. Mag., 300 Wby. Mag., 338 Win. Mag., 375 H&H. Interchangeable bbls. w/standard or Magnum bolt assemblies. Bbl. length: 23 inches (standard); 24 inches (Magnum). 41 to 42 inches overall. Weight: 7 to 7.25 lbs. No sights. Bbl.-mounted scope system. Two-piece Turkish walnut stock w/solid black recoil pad. Imported 1989-94.

Model R84 Standard $1395
Model R84 Deluxe (Game Scene) 1695
Model R84 Super Deluxe (Gold and Silver Inlays) 1895
Left-Hand Model, **add** 125
Extra Bbl. Assembly, **add** 650

Beretta AR-70

Blaser Model R84

Blaser Model R93 Safari Series Bolt-Action Repeater
Similar to Model R84, except restyled action w/straight-pull bolt w/
unique safety and searless trigger mechanism. Additional chamber-
ings: 6.5×55, 7×57, 308, 416 Rem. Optional open sights. Imported
1994-98.

Model R93 Safari **$2595**
Model R93 Safari Deluxe 2950
Model R84 Safari Super Deluxe 3295
Extra Bbl. Assembly, **add** 595

**Blaser Model R93 Classic Series
Bolt-Action Repeater**
Similar to Model R93 Safari, except w/expanded model variations.
Imported 1998 to date.

Model R93 Attache (Premium Wood, Fluted Bbl.) **$3795**
Model R93 Classic (22-250 to 375 H&H) 2595
Model R93 Classic Safari (416 Rem.) 2895
Model R93 LX (22-250 to 416 Rem.) 1295
Model R93 Synthetic (22-250 to 375 H&H) 995
Extra Bbl. Assembly, **add** 550

BRITISH MILITARY RIFLES
Mfd. at Royal Small Arms Factory, Enfield Lock, Middlesex, England, as well as private contractors

British Army Rifle No. 1 Mark III* **$165**
Short Magazine Lee-Enfield (S.M.L.E.). Bolt action. Caliber: 303 British Service. 10-shot box magazine. 25.25-inch bbl. Weight: 8.75 lbs.
Sights: adj. rear; blade front w/guards. Two-piece, full-length military
stock. *Note:* The earlier Mark III (approved 1907) is virtually the same
as Mark III* (adopted 1918) except for sights and different magazine
cut-off that was eliminated on the latter.

British Army Rifle No. 3 Mark I* (Pattern '14) **$175**
Modified Mauser-type bolt action. Except for caliber, 303 British Service, and long-range sights, this rifle is the same as U.S. Model 1917 Enfield. *See* listing of the latter for general specifications.

British Army Rifle No. 4 Mark I* **$150**
Post-World War I modification of the S.M.L.E. intended to simplify
mass production. General specifications same as Rifle No. 1 Mark
III*, except w/aperture rear sight and minor differences in construction and weighs 9.25 lbs.

British Army Light Rifle No. 4 Mark I* **$135**
Modification of the S.M.L.E. Caliber: 303 British Service. 10-
shot box magazine. 23-inch bbl. Weight: 6.75 lbs. Sights: micrometer click rear peep; blade front. One-piece military-type
stock w/recoil pad. Made during WWII.

British Army Rifle No. 5 Mark I* **$225**
Jungle Carbine. Modification of the S.M.L.E. similar to Light Rifle No.
4 Mark I* except w/20.5-inch bbl. w/flash hider, carbine-type stock.
Made during WWII, originally designed for use in the Pacific Theater.

BRNO SPORTING RIFLES
Brno, Czech Republic
Manufactured by Ceska Zbrojovka
Imported by Euro-Imports, El Cajon, CA
(Previously by Bohemia Arms & Magnum Research)

See also CZ Rifles.

Brno Model I Bolt-Action Sporting Rifle **$495**
Caliber: 22 LR. 5-shot detachable magazine. 22.75-inch bbl.
Weight: 6 lbs. Sights: three-leaf open rear; hooded ramp front. Sport-
ing stock w/checkered pistol grip, swivels. Made 1946-58.

British Army Rifle No. 1 Mark III

British Army Rifle No. 3 Mark I

Brno Model II

Brno Model 21H
Bolt-Action Sporting Rifle

Brno Model 22F

Brno Hornet
Bolt-Action Sporting Rifle

RIFLES

Brno Model II. **$535**
Same as Model I except w/deluxe grade stock. Disc.

Brno Model 21H Bolt-Action Sporting Rifle **$625**
Mauser-type action. Calibers: 6.5×57mm, 7×57mm 8×57mm. 5-shot box magazine. 20.5-inch bbl. Double set trigger. Weight: 6.75 lbs. Sights: two-leaf open rear-hooded ramp front. Half-length sporting stock w/cheekpiece, checkered pistol grip and forearm, swivels. Made 1946-55.

Brno Model 22F **$795**
Same as Model 21H except w/full-length Mannlicher-type stock, weighs 6 lbs., 14 oz. Disc.

Brno Model 98 Standard
Calibers: 243 Win., 270 Win., 30-06, 308 Win., 300 Win. Mag., 7×57mm, 7×64mm, or 9.3×62mm. 23.8-inch bbl. Overall 34.5 inches. Weight: 7.25 lbs. Checkered walnut stock w/Bavarian cheekpiece. Imported 1998 to date.
Standard Calibers **$345**
Calibers 300 Win., Mag., 9.3×62mm 385
W/Set Triggers, **add** 100

Brno Model 98 Mannlicher
Similar to Model 98 Standard, except full length stock and set triggers. Imported 1998 to date.
Standard Calibers **$475**
Calibers 300 Win. Mag., 9.3×62mm 525

Brno ZBK-110 Single-Shot
Calibers: 22 Hornet, 222 Rem., 5.6×52R, 5.6×50 Mag., 6.5×57R, 7×57R, and 8×57JRS. 23.8-inch bbl. Weight: 6.1 lbs. Walnut checkered buttstock and forearm W/Bavarian cheekpiece. Imported 1998 to date.
Standard Model. **$165**
Lux Model. 245
Calibers 7×57R and 8×57 JRS, **add** 25
W/Interchangeable 12 ga. Shotgun Bbl., **add** 100

Brno Hornet Bolt-Action Sporting Rifle **$725**
Miniature Mauser action. Caliber: 22 Hornet. 5-shot detachable box magazine. 23-inch bbl. Double set trigger. Weight: 6.25 lbs. Sights: three-leaf open rear hooded ramp front. Sporting stock w/checkered pistol grip and forearm, swivels. Made 1949-74. *Note:* This was also marketed in U.S. as "Z-B Mauser Varmint Rifle." No longer imported.

Brno Model ZKB 680 Bolt-Action Rifle **$350**
Calibers: 22 Hornet, 222 Rem. 5-shot detachable box magazine. 23.5-inch bbl. Weight: 5.75 lbs. Double-set triggers. Adj. open rear sight, hooded ramp front. Walnut stock. Imported 1985-92.

Brno Model ZKM 611 Semiautomatic Rifle **$345**
Caliber: 22 WMR. 6-round magazine. 20-inch bbl. 37 inches overall. Weight: 6 lbs. 2oz. Hooded front sight; mid-mounted rear sight. Checkered walnut stock; beechwood stock available for $375. Single thumb-screw takedown. Grooved receiver for scope mounting. Imported 1992 to date. *See* Illustration next page.

Brno Model ZKM 611

BROWN PRECISION COMPANY
Los Molinos, California

Brown Precision Model 7 Super Light Sporter **$795**
Lightweight sporter built on a Remington Model 7 barreled action
w/18-inch factory bbl. Weight: 5.25 lbs. Kevlar stock. Made
1984-92.

Brown Precision High Country Bolt-Action Sporter
Custom sporting rilfes built on Blaser, Remington 700, Ruger 77 and
Winchester 70 actions. Calibers: 243 Win., 25-06, 270 Win., 7mm
Rem. Mag., 308 Win., 30-06. 5-shot magazine (4-shot in 7mm Mag.).
22- or 24-inch bbl. Weight: 6.5 lbs. Fiberglass stock w/recoil pad, sling
swivels. No sights. Made 1975 to date.

Standard High Country	$ 795
Custom High Country	1295
Left-Hand Action, **add**	100
Stainless Bbl., **add**	100
70, 77 or Blaser Actions, **add**	125
70 SG Action, **add**	350

Brown Precision High Country Youth Rifle **$850**
Similar to standard Model 7 Super Light, except w/Kevlar or
graphite stock scaled-down to youth dimensions. Calibers: 223,
243, 6mm, 7mm-08, 308. Made 1992 to date.

Brown Precision Pro-Hunter Bolt-Action Rifle
Custom sporting rifle built on Remington 700 or Winchester 70
SG action, fitted w/match-grade Shilen bbl. chambered in customer's choice of caliber. Matte blued electroless nickel or Teflon
finish. Express-style rear sight hooded ramp front. Synthetic
stock. Made 1989 to date

Standard Pro-Hunter	$1750
Pro-Hunter Elite (1993 to date)	2450

Brown Precision Pro-Varminter Bolt-Action Rifle
Custom varminter built on a Remington 700 or 40X action fitted
w/Shilen stainless steel benchrest bbl. Varmint or benchrest-style
stock. Made 1993 to date.

Standard Pro-Varminter	$1295
Pro-Hunter w/Rem 40X Action	1695

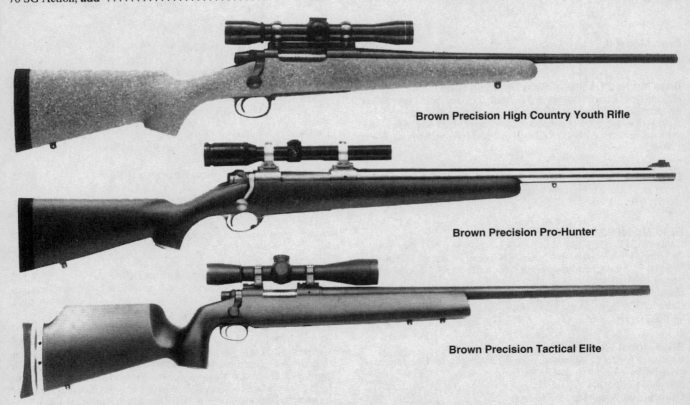

Brown Precision High Country Youth Rifle

Brown Precision Pro-Hunter

Brown Precision Tactical Elite

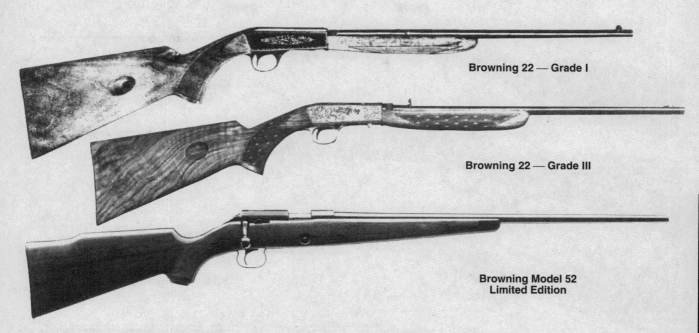

Browning 22 — Grade I

Browning 22 — Grade III

**Browning Model 52
Limited Edition**

Brown Precision Selective Target Model **$850**
Tactical law-enforcement rifle built on a Remington 700V action. Caliber: 308 Win. 20-, 22- or 24-inch bbl. Synthetic stock. Made 1989-92.

Brown Precision Tactical Elite Rifle **$1695**
Similar to Selective Target Model, except fitted w/select match-grade Shilen benchrest heavy stainless bbl. Calibers: 223, 308, 300 Win. Mag. Black or camo Kevlar/graphite composite fiberglass stock w/adj. buttplate. Non-reflective black Teflon metal finish. Made 1993 to date.

BROWNING RIFLES
Morgan, Utah
Mfd. for Browning by Fabrique Nationale d'Armes de Guerre (now Fabrique Nationale Herstal), Herstal, Belgium; Miroku Firearms Mfg. Co., Tokyo, Japan; A.T.I., Salt Lake City, Ut; Oy Sako Ab, Riihimaki, Finland.

Browning 22 Automatic Rifle, Grade I
Similar to discontinued Remington Model 241A. Autoloading. Takedown. Calibers: 22 LR, 22 Short (not interchangeably). Tubular magazine in buttstock holds 11 LR, 16 Short. Bbl. lengths: 19.25 inches (22 LR), 22.25 inches (22 Short). Weight: 4.75 lbs. (22 LR); 5 lbs. (22 Short). Receiver scroll engraved. Open rear sight, bead front. Checkered pistol-grip buttstock, semibeavertail forearm. Made 1965-72 by FN; 1972 to date by Miroku. *Note:*Illustrations are of rifles manufactured by FN.
FN Manufacture **$375**
Miroku Manufacture **225**

Browning 22 Automatic Rifle, Grade II
Same as Grade I, except satin chrome-plated receiver engraved w/small game animal scenes, gold-plated trigger select walnut stock and forearm. 22 LR only. Made 1972-84.
FN manufacture **$695**
Miroku manufacture **325**

Browning 22 Automatic Rifle, Grade III
Same as Grade I, except satin chrome-plated receiver elaborately hand-carved and engraved w/dog and game-bird scenes, scrolls and leaf clusters: gold-plated trigger, extra-fancy wal-nut stockand forearm, skip-checkered. 22 LR only. Made 1972-84.
FN manufacture **$1550**
Miroku manufacture **595**

**Browning 22
Automatic, Grade VI**

Browning 22 Automatic, Grade VI **$495**
Same general specifications as standard 22 Automatic, except for engraving, high-grade stock w/checkering and glossy finish. Made 1986 to date.

Browning Model 52 Bolt-Action Rifle **$495**
Limited Edition of Winchester Model 52C Sporter. Caliber: 22 LR. 5-shot magazine. 24-inch bbl. Weight: 7 lbs. Micro-Motion trigger. No sights. Checkered select walnut stock w/rosewood forend and metal grip cap. Blued finish. Only 5000 made in 1991-92.

Browning Model 53 Lever-Action Rifle **$475**
Limited Edition of Winchester Model 53. Caliber: 32-20. 7-shot tubular half-magazine. 22-inch bbl. Weight: 6.5 lbs. Adj. rear sight, bead front. Select walnut checkered pistol-grip stock w/high-gloss finish. Classic-style forearm. Blued finish. Only 5000 made in 1990.

RIFLES

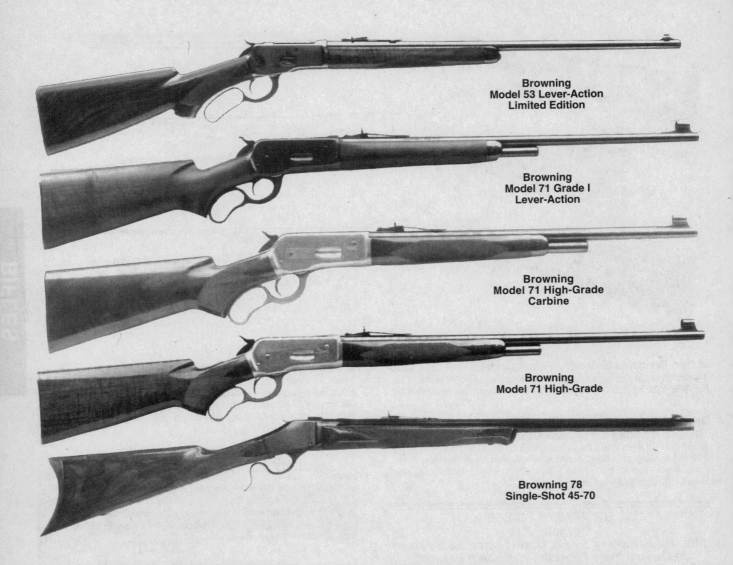

Browning
Model 53 Lever-Action
Limited Edition

Browning
Model 71 Grade I
Lever-Action

Browning
Model 71 High-Grade
Carbine

Browning
Model 71 High-Grade

Browning 78
Single-Shot 45-70

Browning Model 65 Grade I Lever-Action Rifle ... **$450**
Caliber: 218 Bee. 7-shot tubular half-magazine. 24-inch bbl. Weight: 6.75 lbs. Sights: adj. buckhorn-style rear, hooded bead front. Select walnut pistol-grip stock w/high-gloss finish. Semibeavertail forearm. Limited edition of 3500 made in 1989.

Browning Model 65 High Grade Rifle **$795**
Same general specifications as Model 65 Grade I, except w/engraving and gold-plated animals on grayed receiver. Cut-checkered on pistol grip and forearm. Limited edition of 1500 made in 1989.

Browning Model 71 Grade I Carbine **$425**
Same general specifications as Model 71 Grade I Rifle, except carbine w/20-inch round bbl. and weighs 8 lbs. Limited edition of 4000 made in 1986-87.

Browning Model 71 Grade I Lever-Action Rifle ... **$450**
Caliber: 348 Win. 4-shot magazine. 24-inch round bbl. Weight: 8 lbs. 2 oz. Open buckhorn sights. Select walnut straight grip stock w/satin finish. Classic-style forearm, flat metal buttplate. Limited edition of 3000 made in 1986-87.

Browning Model 71 High-Grade Carbine **$650**
Same general specifications as Model 71 High Grade Rifle, except carbine w/20-inch round bbl. Limited edition of 3000 made in 1986-88.

Browning Model 71 High-Grade Rifle **$685**
Caliber: 348 Win. 4-shot magazine. 24-inch round bbl. Weight: 8 lbs. 2 oz. Engraved receiver. Open buckhorn sights. Select walnut checkered pistol-grip stock w/high-gloss finish. Classic-style forearm, flat metal buttplate. Limited edition of 3000 made in 1987.

Browning 78 Bicentennial Set **$2250**
Special Model 78 45-70 w/same specifications as standard type, except sides of receiver engraved w/bison and eagle, scroll engraving on top of receiver, lever, both ends of bbl. and buttplate; high-grade walnut stock and forearm. Accompanied by an engraved hunting knife and stainless steel commemorative medallion all in an alder wood presentation case. Each item in set has matching serial number beginning with "1776" and ending with numbers 1 to 1,000. Edition limited to 1,000 sets. Made in 1976. Value is for set in new condition.

**Browning Model 1885 High-Wall
Single-Shot**

Browning 78 Single-Shot Rifle

Falling-block lever-action similar to Winchester 1885 High Wall single-shot rifle. Calibers: 22-250, 6mm Rem., 243 Win., 25-06, 7mm Rem. Mag., 30-06, 45-70 Govt. 26-inch octagon or heavy round bbl.; 24-inch octagon bull bbl. on 45-70 model. Weight: 7.75 lbs.w/octagon bbl.; w/round bbl., 8.5 lbs.; 45-70, 8.75 lbs. Furnished w/o sights, except 45-70 model w/open rear sight, blade front. Checkered fancy walnut stock and forearm. 45-70 model w/straight-grip stock and curved buttplate; others have stock w/Monte Carlo comb and cheekpiece, pistol grip w/cap, recoil pad. Made 1973-83 by Miroku. Reintroduced in 1985 as Model 1885.
All calibers except 45-70 . . . . **$495**
45-70 . . . . **525**

Browning Model 1885 Single-Shot Rifle

Calibers: 22 Hornet, 223, 243, (Low Wall); 357 Mag., 44 Mag., 45 LC (L/W Traditional Hunter); 22-250, 223 Rem., 270 Win., 7mm Rem. Mag., 30-06, 454 Casull Mag., 45.70 (High Wall); 30.30 Win., 38-55 WCF, 45 Govt. (H/W Traditional Hunter); 40-65, 45 Govt. and 45.90 (BPCR). 24-, 28-, 30 or 34-inch round, octagonal or octagonal and round bbl. 39.5, 43.5, 44.25 or 46.125 inches overall. Weight: 6.25, 8.75, 9, 11, or 11.75 lbs. respectively. Blued or color casehardened receiver. Gold-colored adj. trigger. Drilled and tapped for scope mounts w/no sights or vernier tang rear sight w/globe front and open sights on 45-70 Govt. Walnut straight-grip stock and Schnabel forearm w/cut checkering and high-gloss or oil finish. Made 1985 to date.
Low Wall Model w/o sights (Intro. 1995) . . . . **$ 525**
   Traditional Hunter Model (Intro. 1998) **695**
High Wall Model w/o sights (Intro. 1985) **525**
   Traditional Hunter Model (Intro. 1997) **675**
   BPCR Model w/no ejector (Intro. 1996) **1150**
   BPCR Creedmoor Model 45.90 (Intro. 1998) **1195**

Browning Model 1886 Montana Centennial Rifle **$1325**

Same general specifications as Model 1886 High Grade Lever-Action, except w/specially engraved receiver designating Montana Centennial; also different stock design. Made in 1986 in limited issue 1 of 2,000 by Miroku.

Browning Model 1886 Grade I Lever-Action Rifle **$795**

Caliber: 45-70 Govt. 8-round magazine. 26-inch octagonal bbl. 45 inches overall. Weight: 9 lbs. 5 oz. Deep blued finish on receiver. Open buckhorn sights. Straight-grip walnut stock. Classic-style forearm. Metal buttplate. Satin finish. Made in 1986 in limited issue 7,000 by Miroku.

Browning Model 1886 High-Grade Lever-Action Rifle **$1095**

Same general specifications as theModel 1886 Grade I, except receiver is grayed steel embellished w/scroll; game scenes of elk and American bison engraving. High-gloss stock. Made in 1986 in limited issue 3,000 by Miroku.

Browning Model 1895 Grade I Lever-Action Rifle **$485**

Caliber: 30-06, 30-40 Krag. 4-shot magazine. 24-inch round bbl. 42 inches overall. Weight: 8 lbs. French walnut stock and Schnabel forend. Sights: rear buckhorn; gold bead on elevated ramp front. Made in 1984 in limited issue of 8,000 (2,000 chambered 30-40 Krag and 6,000 chambered 30.06). Mfd. by Miroku.

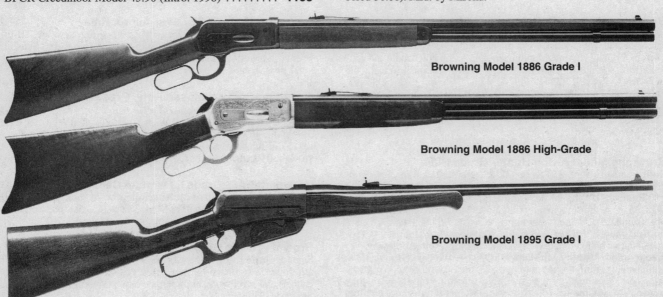

Browning Model 1886 Grade I

Browning Model 1886 High-Grade

Browning Model 1895 Grade I

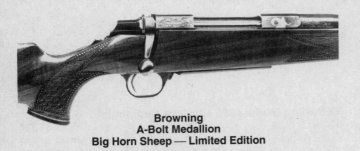

**Browning
A-Bolt Medallion
Big Horn Sheep — Limited Edition**

Browning Model 1895 High-Grade Lever-Action Rifle $950
Same general specifications as Model 1895 Grade I except engraved receiver and Grade III French walnut stock and forend w/fine checkering. Made in 1985 in limited issue of 1000 in each caliber by Miroku.

Browning Model A-Bolt 22 Rifle
Calibers: 22 LR, 22 Magnum. 5- and 15-shot magazines. 22-inch round bbl. 40.25 inches overall. Weight: 5 lbs. 9 oz. Gold-colored adj. trigger. Laminated walnut stock w/checkering. Rosewood forend grip cap; pistol grip. With or w/o sights. Ramp front and adj. folding leaf rear on open sight model. 22 LR made 1985-96; 22 Magnum, 1990-96.
Grade I 22 LR $225
Grade I 22 Magnum 265
Deluxe Grade Gold Medallion 350

Browning Model A-Bolt Eclipse Bolt Rifle
Same general specifications as Hunter Grade, except fitted w/gray and black laminated thumbhole stock. Available in both short and long action w/two bbl. configurations w/BOSS. Mfd. by Miroku 1996 to date.
Eclipse w/standard bbl. $595
Eclipse Varmint w/heavy bbl. 650
Eclipse M-1000 (Target 300 Win. Mag.) 675

Browning Model A-Bolt Euro-Bolt Rifle
Same general specifications as Hunter Grade, except w/checkered satin-finished walnut stock w/continental-style cheekpiece, palm-swell grip ans Schnabel forend. Mannlicher-style spoon bolt handle and contoured bolt shroud. 22- or 26-inch bbl. w/satin blued finish. Weight: 6.8 to 7.4 lbs. Calibers: 22-250 Rem., 243 Win., 270 Win., 30.06, 308 Win., 7mm Rem. Mag. Mfd. by Miroku 1993-94; 1994-96 (Euro-Bolt II).
Euro-Bolt $425
Euro-Bolt II 450
BOSS Option, **add** 90

Browning Model A-Bolt Hunter Grade Rifle
Calibers: 22 Hornet, 223 Rem., 22-250 Rem., 243 Win., 257 Roberts, 7mm-08 Rem., 308 Win., (short action) 25-06 Rem., 270 Win., 280 Rem., 284 Win., 30-06, 7mm Rem. Mag., 300 Win. Mag., 338 Win. Mag. 4-shot magazine (standard), 3-shot (magnum) 22-inch bbl. (standard), 24-inch (magnum). Weight: 7.5 lbs. (standard), 8.5 lbs. (magnum). With or w/o sights. Classic-style walnut stock. Produced in two action lengths w/nine locking lugs, fluted bolt w/60 degree rotation. Mfd. by Miroku 1985-93; 1994 to date (Hunter II).
Hunter $325
Hunter II 350
BOSS Option, **add** 90
Open Sights, **add** 50

Browning Model A-Bolt Medallion Grade Rifle
Same as Hunter Grade, except w/high-gloss deluxe stock rosewood grip cap and forend; high-luster blued finish. Also in 375 H&H w/open sights. Left-Hand Models available in long action only. Mfd. by Miroku 1988-93; 1994 to date (Medallion II).
Big Horn Sheep Ltd. Ed.
 (600 made 1986, 270 Win.) $895
Gold Medallion Deluxe Grade 525
Gold Medallion II Deluxe Grade 545
Medallion, Standard Grade 395
Medallion II, Standard Grade 425
Medallion, 375 H&H 575
Medallion II, 375 H&H 595
Micro Medallion 375
Micro Medallion II 385
Pronghorn Antelope Ltd. Ed.
 (500 made 1987, 243 Win.) 850
BOSS Option, **add** 90
Open Sights, **add** 50

Browning Model A-Bolt Stalker Rifle
Same general specifications as Model A-Bolt Hunter Rifle except w/checkered graphite-fiberglass composite stock and matte blued or stainless metal. Nonglare matte finish off all exposed metal surfaces. 3 models: Camo Stalker orig. w/multi-colored laminated wood stock, matte blued metal; Composite Stalker w/graphite-fiberglass stock, matte blued metal; w/composite stock, stainless metal. Mfd. by Miroku 1987-93; 1994 to date. (Stalkert II).
Camo Stalker (orig. Laminated Stock) $335
Composite Stalker 345
Composite Stalker II 345
Stainless Stalker 395
Stainless Stalker II 415
Stainless Stalker, 375 H&H 545
BOSS Option, **add** 90
Left-Hand Model, **add** 90

Browning Model A-Bolt Varmint II Rifle $495
Same general specifications as Stalker Model, except w/22-inch heavy bbl. w/BOSS system and varmint-style black laminated wood stock. Calibers: 22-250, 223 or 308. No sights. Bright blued or satin finish. Mfd. by Miroku 1994 to date.

Browning Model B-92 Lever-Action Rifle $325
Calibers: 357 Mag. and 44 Rem. Mag. 11-shot magazine. 20-inch round bbl. 37.5 inches overall. Weight: 5.5 to 6.4 lbs. Seasoned French walnut stock w/high gloss finish. Cloverleaf rear sight; steel post front. Made 1979-89 by Miroku.

Browning BAR Automatic Rifle, Grade I, Standard Calibers $435
Gas-operated semiautomatic. Calibers: 243 Win., 270 Win., 280 Rem., 308 Win., 30-06. 4-round box magazine. 22-inch bbl. Weight: 7.5 lbs. Folding leaf rear sight, hooded ramp front. French walnut stock and forearm checkered, QD swivels. Made 1967-92 by FN.

Browning BAR, Grade I, Magnum Calibers $475
Same as BAR in standard calibers, except w/24-inch bbl. chambered 7mm Rem. Mag. or 300 Win. Mag. 338 Win. Mag. w/3-round box magazine and recoil pad. Weight: 8.5 lbs. Made 1969-92 by FN.

Browning BAR, Grade II
Same as Grade I, except receiver engraved w/big-game heads (deer and antelope on standard-caliber rifles, ram and grizzly on Magnum-caliber) and scrollwork, higher grade wood. Made 1967-74 by FN.
Standard Calibers $575
Magnum Calibers 625

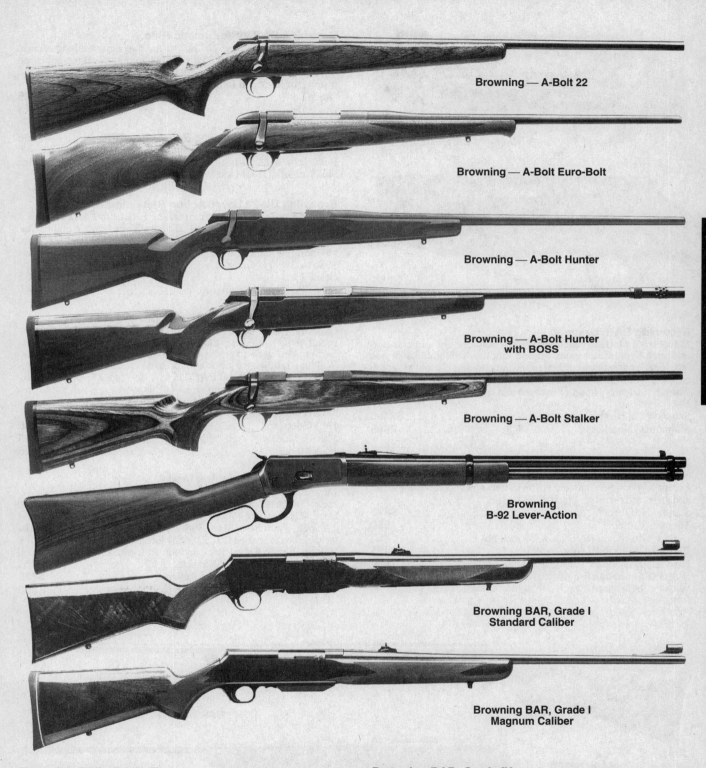

Browning — A-Bolt 22

Browning — A-Bolt Euro-Bolt

Browning — A-Bolt Hunter

Browning — A-Bolt Hunter with BOSS

Browning — A-Bolt Stalker

Browning B-92 Lever-Action

Browning BAR, Grade I Standard Caliber

Browning BAR, Grade I Magnum Caliber

RIFLES

Browning BAR, Grade III

Same as Grade I, except receiver of grayed steel engraved w/big-game heads (deer and antelope on standard-caliber rifles, moose and elk on Magnum-caliber) framed in fine-line scrollwork, gold-plated trigger, stock and forearm of highly figured French walnut, hand-checkered and carved. Made 1971-74 by FN.

Standard Calibers **$725**
Magnum Calibers 850

Browning BAR, Grade IV

Same as Grade I, except receiver of grayed steel engraved w/full-detailed rendition of running deer and antelope on standard-caliber rifles, moose and elk on Magnum-caliber gold-plated trigger, stock and forearm of highly figured French walnut, hand checkered and carved. Made 1971-86 by FN.

Standard Calibers **$1250**
Magnum Calibers 1450

Browning — BAR, Grade IV

Browning — BAR, Grade V

Browning BAR, Grade V
Same as Grade I, except receiver w/complete big-game scenes executed by a master engraver and inlaid w/18K gold (deer and antelope on standard-caliber rifles, moose and elk on Magnum caliber), gold-plated trigger, stock and forearm of finest French walnut, intricately hand-checkered and carved. Made 1971-74 by FN.

Standard Calibers	$2450
Magnum Calibers	2695

Browning Model BAR Mark II Safari Automatic Rifle
Same general specifications as standard BAR semiauotmatic rifle, except w/redesigned gas and buffer systems, new bolt release lever, and engraved receiver. Made 1993 to date.

Standard Calibers	$475
Magnum Calibers	550
Lightweight (Alloy Receiver w/20-inch bbl.)	450
BAR Mk II Grade III (Intro. 1996)	1495
BAR Mk II Grade IV (Intro. 1996)	1550
W/BOSS Option, **add**	60
W/Open Sights, **add**	15

Browning BAR-22 Automatic Rifle
Semiautomatic. Caliber: 22 LR. Tubular magazine holds 15 rounds. 20.25-inch bbl. Weight: 6.25 lbs. Sights: folding-leaf rear, gold bead front on ramp. Receiver grooved for scope mounting. French walnut pistol-grip stock and forearm checkered. Made 1977-8.5

Grade I	$225
Grade II	315

Browning BBR Lightning Bolt-Action Rifle $350
Bolt-action rifle w/short throw bolt of 60 degrees. Calibers: 25-06 Rem., 270 Win., 30-06, 7mm Rem. Mag., 300 Win. Mag. 24-inch bbl. Weight: 8 lbs. Made 1979-84.

Browning BL-22 Lever-Action Repeating Rifle
Short-throw lever-action. Caliber: 22 LR, Long, Short. Tubular magazine holds 15 LR, 17 Long 22 Short. 20-inch bbl. Weight: 5 lbs. Sights: folding leaf rear; bead front. Receiver grooved for scope mounting. Walnut straight-grip stock and forearm, bbl. band. Made 1970 to date by Miroku.

Grade I	$195
Grade II w/Scroll Engraving	245

Browning BLR Lever-Action Repeating Rifle
Calibers: (short action only) 243 Win., 308 Win., 358 Win. 4-round detachable box magazine. 20-inch bbl. Weight: 7 lbs. Sights: windage and elevation adj. open rear; hooded ramp front. Walnut straight-grip stock and forearm, checkered, bbl. band, recoil pad. Made 1966 by BAC/USA; 1969-73 by FN; 1974-80 by Miroku. *Note:* USA manufacture of this model was limited to prototypes and pre-production guns only and may be identified the "MADE IN USA" roll stamp on the bbl.

FN Model	$450
Miroku Model	325
USA Model	825

Browning BLR Lightning Model
Lightweight version of the Browning BLR '81 w/forged alloy receiver and redesigned trigger group. Calibers: Short Action— 22250 Rem., 223 Rem., 243 Win., 7mm-08 Rem., 308 Win.; Long Action— 270 Win., 7mm Rem. Mag., 30-06, 300 Win. Mag. 3- or 4-round detachable box magazine. 20-, 22- or 24-inch bbl. Weight: 6.5 to 7.75 lbs. Pistol-grip style walnut stock and forearm, cut checkering and recoil pad. Made by Miroku 1995 to date.

BLR Lightning Model Short Action	$325
BLR Lightning Model Long Action	350

**Browning
BAR Mark II Safari**

**Browning
BAR Mark II Safari
with BOSS**

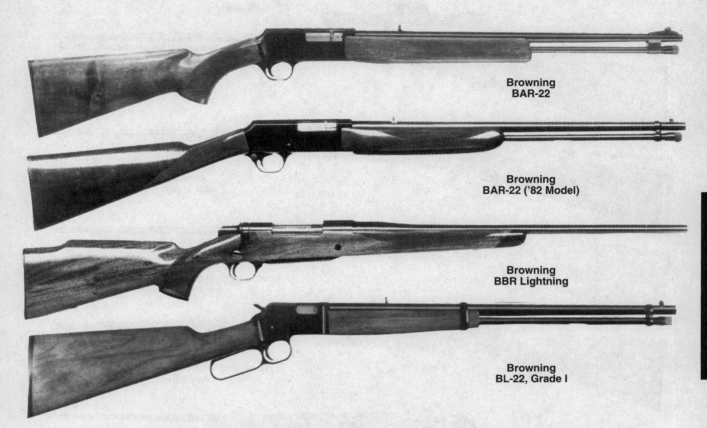

Browning
BAR-22

Browning
BAR-22 ('82 Model)

Browning
BBR Lightning

Browning
BL-22, Grade I

RIFLES

Browning BLR Model '81

Redesigned version of the Browning BLR. Calibers: 222-50 Rem., 243 Win., 308 Win., 358 Win; Long Action— 270 Win., 7mm Rem. Mag., 30-06. 4-round detachable box magazine. 20-inch bbl. Weight: 7 lbs. Walnut straight-grip stock and forearm, cut checkering, recoil pad. Made by Miroku 1981-95; Long Action intro. 1991.

BLR Model '81 Short Action **$345**
BLR Model '81 Long Action 355

Browning BPR-22 Pump Rifle

Hammerless slide-action repeater. Specifications same as for BAR-22, except also available chambered for 22 Magnum RF; magazine capacity, 11 rounds. Made 1977-82 by Miroku.

Model I **$185**
Model II 325

Browning BPR Pump Rifle

Slide-action repeater based on proven BAR designs w/forged alloy receiver and slide that cams down to clear bbl. and receiver. Calibers: 243 Win., 308 Win., 270 Win., 30-06, 7mm Rem. Mag. 300 Win. Mag. 3- or 4-round detachable box magazine. 22- or 24-inch bbl. w/ramped front sight and open adj. rear. Weight: 7.2 to 7.4 lbs. Made 1997 to date by Miroku.

BPR Model Standard Calibers **$525**
BPR Model Magnum Calibers 595

Browning High-Power Bolt-Action Rifle, Medallion Grade **$1195**

Same as Safari Grade, except receiver and bbl. scroll engraved, ram's head engraved on floorplate; select walnut stock w/rosewood forearm tip, grip cap. Made 1961-74.

Browning High-Power Bolt-Action Rifle, Olympian Grade **$2395**

Same as Safari Grade, except bbl. engraved; receiver, trigger guard and floorplate satin chrome-plated and engraved w/game scenes appropriate to caliber; finest figured walnut stock w/rosewood forearm tip and grip cap, latter w/18K-gold medallion. Made 1961-74.

Browning High-Power Bolt-Action Rifle, Safari Grade, Medium Action **$695**

Same as Standard, except medium action. Calibers: 22/250, 243 Win., 264 Win. Mag., 284 Win. Mag., 308 Win. Bbl.: 22-inch lightweight bbl.; 22/250 and 243 also available w/24-inch heavy bbl. Weight: 6 lbs. 12 oz. w/lightweight bbl.; 7 lbs. 13 oz. w/heavy bbl. Made 1963-74 by Sako.

Browning High-Power Bolt-Action Rifle, Safari Grade, Short Action **$695**

Same as Standard, except short action. Calibers: 222 Rem., 222 Rem. Mag. 22-inch lightweight or 24-inch heavy bbl. No sights. Weight: 6 lbs. 2 oz. w/lightweight bbl.; 7.5 lbs. w/heavy bbl. Made 1963-74 by Sako.

Browning High-Power Bolt-Action Rifle, Safari Grade, Standard Action **$895**

Mauser-type action. Calibers: 270 Win., 30-06, 7mm Rem. Mag., 300 H&H Mag., 300 Win. Mag., 308 Norma Mag. 338 Win. Mag., 375 H&H Mag., 458 Win. Mag. Cartridge capacity: 6 rounds in 270, 30-06; 4 in Magnum calibers. Bbl. length: 22 in., 270, 30-06, 24 in., Magnum calibers. Weight: 7 lbs. 2 oz., 270, 30-06; 8.25 lbs., Mag. calibers. Folding leaf rear sight, hooded ramp front. Checkered stock w/pistol grip, Monte Carlo cheekpiece, QD swivels; recoil pad on Magnum models. Made 1959-74 by FN.

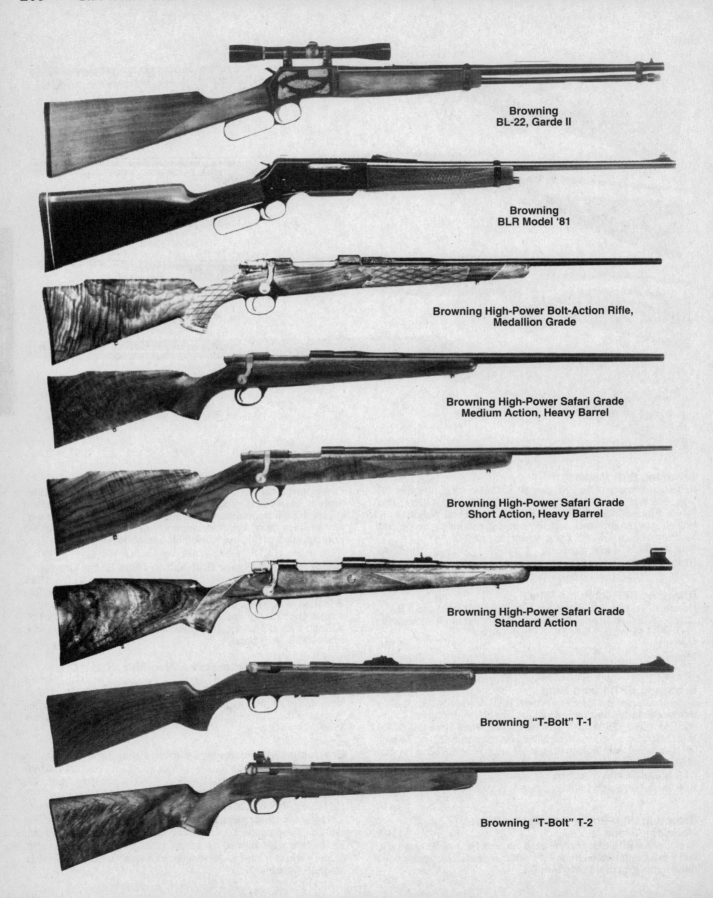

**Browning
BL-22, Garde II**

**Browning
BLR Model '81**

**Browning High-Power Bolt-Action Rifle,
Medallion Grade**

**Browning High-Power Safari Grade
Medium Action, Heavy Barrel**

**Browning High-Power Safari Grade
Short Action, Heavy Barrel**

**Browning High-Power Safari Grade
Standard Action**

Browning "T-Bolt" T-1

Browning "T-Bolt" T-2

Browning "T-Bolt" T-1 22 Repeating Rifle
Straight-pull bolt action. Caliber: 22 LR. 5-shot clip magazine. 24-inch bbl. Peep rear sight w/ramped blade front. Plain walnut stock w/pistol grip and laquered finish. Weight: 6 lbs. Also left-hand model. Made 1965-74 by FN.
Right-Hand Model $350
Left-Hand Model 395

Browning "T-Bolt" T-2
Same as T-1 Model, except w/checkered fancy-figured walnut stock. Made 1966-74 by FN. *Note:* Reintroduced briefly during the late 1980's with oil-finished stock.
Original Model $395
Reintroduced Model 295

F.N. Browning FAL Semiautomatic Rifle
Same as F.N. FAL Semiautomatic Rifle. *See* F.N. listing for specifications. Sold by Browning for a brief period c. 1960.
F.N. FAL Standard Model (G-series). $2595
F.N. FAL Lightweight Model (G-series) $3325
F.N. FAL Heavy Bbl. Model (G-series). $4975
BAC FAL Model $1995

BSA GUNS LTD.
Birmingham, England
(Previously Imported by Samco Glogal Arms, BSA Guns Ltd and Precision Sports)

BSA No. 12 Martini Single-Shot Target Rifle $325
Caliber: 22 LR. 29-inch bbl. Weight: 8.75 lbs. Parker-Hale Model 7 rear sight and Model 2 front sight. Straight-grip stock, checkered forearm. *Note:* This model was also available w/open sights or w/BSA No. 20 and 30 sights. Made before WWII.

BSA Model 12/15 Martini Heavy $425
Same as Standard Model 12/15 except w/extra heavy bbl., weighs 11 lbs.

BSA Model 12/15 Martini Single-Shot Target Rifle $365
Caliber: 22 LR. 29-inch bbl. Weight: 9 lbs. Parker-Hale No. PH-7A rear sight and No. FS-22 front sight. Target stock w/high comb and cheekpiece, beavertail forearm. *Note:* This is a post-WWII model; however, a similar rifle, the BSA-Parker Model 12/15, was produced c. 1938.

BSA No. 13 Martini Single-Shot Target Rifle $375
Caliber: 22 LR. Lighter version of the No.12 w/same general specifications, except w/25-inch bbl., weighs 6.5lbs. Made before WWII.

BSA No. 13 Sporting Rifle
Same as No. 13 Target except fitted w/Parker-Hale "Sportarget" rear sight and bead front sight. Also available in 22 Hornet. Made before WWII.
22 Long Rifle $375
22 Hornet 525

BSA Model 15 Martini Single-Shot Target Rifle $375
Caliber: 22 LR. 29-inch bbl. Weight: 9.5 lbs. BSA No.30 rear sight and No. 20 front sight. Target stock w/cheekpiece and pistol grip, long semibeavertail forearm. Made before WWII.

BSA Centurion Model Match Rifle $350
Same general specifications as Model 15 except w/"Centurion" match bbl. Made before WWII.

BSA CF-2 Bolt-Action Hunting Rifle $295
Mauser-type action. Calibers: 7mm Rem. Mag., 300 Win. Mag. 3-shot magazine. 23.6-inch bbl. Weight: 8 lbs. Sights: adj. rear; hooded ramp front. Checkered walnut stock w/Monte Carlo comb, rollover cheekpiece, rosewood forend tip, recoil pad, sling swivels. Made 1975-87. *See* Ithaca-BSA CF-2.

BSA CF-2 Stutzen Rifle. $350
Calibers: 222 Rem., 22/250, 243 Win., 270 Win., 30-06. 4-round capacity (5 in 222 Rem.). 20.6-inch bbl. 41.5 inches (approx.) overall length. Weight: 7.5 to 8 lbs. Williams front and rear sights. Hand-finished European walnut stock. Monte Carlo cheekpiece and Wundhammer palmswell. Double-set triggers. Importation disc. 1987.

BSA CFT Target Rifle $595
Single-shot bolt action. Caliber: 7.62mm. 26.5-inch bbl. About 47.5 inches overall. Weight: 11 lbs., incl. accessories. Bbl. and action weight: 6 lbs. 12 oz. Importation disc. 1987.

BSA Majestic Deluxe Featherweight Bolt-Action Hunting Rifle
Mauser-type action. Calibers: 243 Win., 270 Win., 308 Win., 30-06, 458 Win. Mag. 4-shot magazine. 22-inch bbl. w/BESA re-coil reducer. Weight: 6.25 lbs.; 8.75 lbs. in 458. Folding leaf rear sight, hooded ramp front. Checkered European-style walnut stock w/cheekpiece, pistol grip, Schnabel forend, swivels, recoil pad. Made 1959-65.
458 Win. Mag. Caliber $395
Other Calibers 245

BSA Majestic Deluxe Standard Weight $245
Same as Featherweight Model, except heavier bbl. w/o recoil reducer. Calibers: 22 Hornet, 222 Rem., 243 Win., 7×57mm, 308 Win., 30-06. Weight: 7.25 to 7.75 lbs. Disc.

BSA Martini-International ISU Match Rifle $675
Similar to MK III, but modified to meet International Shooting Union "Standard Rifle" specifications. 28-inch standard weight bbl. Weight: 10.75 lbs. Redesigned stock and forearm, latter attached to bbl. w/"V" section alloy strut. Intro. 1968. Disc.

BSA Martini-International Mark V Match Rifle $685
Same as ISU model, except w/heavier bbl. Weight: 12.25 lbs. Intro. 1976. Disc.

BSA Martini-International Match Rifle Single-Shot Heavy Pattern $475
Caliber: 22 LR. 29-inch heavy bbl. Weight: 14 lbs. Parker-Hale "International" front and rear sights. Target stock w/full cheekpiece and pistol grip, broad beavertail forearm, handstop, swivels. Right- or left-hand models. Mfd. 1950-53.

BSA Martini-International Match Rifle — Light Pattern $465
Same general specifications as Heavy Pattern, except w/26-inch lighter weight bbl. Weight: 11 lbs. Disc.

BSA Martini-International MK II Match Rifle $575
Same general specifications as original model. Heavy and Light Pattern. Improved trigger mechanism and ejection system. Redesigned stock and forearm. Made 1953-59.

BSA Martini-International MK III Match Rifle $665
Same general specifications as MK II Heavy Pattern. Longer action frame w/I-section alloy strut to which forearm is attached; bbl. is fully floating. Redesigned stock and forearm. Made 1959-67.

RIFLES

BSA Model 12/15 Martini

BSA Model 15 Martini

BSA CFT Target

BSA Martini-International ISU Match

BSA Martini-International
Mark V Match

BSA Martini-International MK III Match

BSA Monarch Deluxe Varmint

BSA Monarch Deluxe Bolt-Action Hunting Rifle . . . **$225**
Same as Majestic Deluxe Standard Weight Model, except w/re-designed stock of U.S. style w/contrasting hardwood forend tip and grip cap. Calibers: 222 Rem., 243 Win., 270 Win., 7mm Rem. Mag., 308 Win., 30-06. 22-inch bbl. Weight: 7 to 7.25 lbs. Made 1965-74.

BSA Monarch Deluxe Varmint Rifle **$245**
Same as Monarch Deluxe except w/24-inch heavy bbl. and weighs 9 lbs. Calibers: 222 Rem., 243 Win.

BUSHMASTER FIREARMS
(Quality Parts Company)
Windham, Maine

Bushmaster M17S Bullpup **$595**
Caliber: 223. 21.5-inch bbl. Weight: 8.2 lbs. Polymer stocks. Handle w/fixed open sights w/Weavertype rail for any optics. Semi-auto, self-compensating short stroke gas piston. Forward trigger/grip w/rear chamber Bullpup style. Alloy receiver. Synthetic lower receiver is hinged to upper w/hinged takedown system. Accepts M-16 type magazines. Made 1992 to date.

Bushmaster Model XM15 E2S Series
Caliber: 223. 16-, 20-, 24- or 26-inch bbl. Weight: 7 to 8.6 lbs. Polymer stocks. Adjustable sights w/dual flip-up aperture; optional flattop rail accepts scope. Direct gas operated w/rotating bolt. Forged alloy receiver. All steel coated w/manganese phosphate. Accepts M-16 type magazines. Made 1989 to date.
XM15 E2S Carbine **$750**
XM15 E2S Target Rifle 795

CALICO LIGHT WEAPONS SYSTEMS
Bakersville, California

Calico Liberty 50/100 Semiautomatic Rifle
Retarded blowback action. Caliber: 9mm. 50- or 100-shot helical-feed magazine. 16.1-inch bbl. 34.5 inches overall. Weight: 7 lbs. Adjustable post front sight and aperture rear. Ambidextrous rotating safety. Glass-filled Polymer or thumbhole-style wood stock. Made 1995 to date.
Model Liberty 50 . . . . **$525**
Model Liberty 100 595

Calico Model M-100 Semiautomatic Series
Similar to the Liberty 100 Model, except chambered 22 LR. Weight: 5 lbs. 34.5 inches overall. Folding or glass-filled Polymer stock and forearm. Made 1986 to date.
Model M-100 w/folding stock (Disc. 1994) **$250**
Model M-100 FS w/fixed stock (Inrto. 1996) 395

Calico Model M-105 Semiautomatic Sporter **$225**
Similar to the Liberty 100 Model, except fitted w/walnut buttstock and forearm. Made 1986 to date.

Calico Model M-900 Semiautomatic Carbine
Caliber: 9mm Parabellum. 50- or 100-shot magazine. 16.1-inch bbl. 28.5 inches overall. Weight: 3.7 lbs. Post front sight adj. for windage and elevation, fixed notch rear. Collapsible steel buttstock and glass-filled polymer grip. Matte black finish. Made 1989-94.
Model M-100 w/folding stock (Disc. 1994) **$395**
Model M-100 FS w/fixed stock (Inrto. 1996) 375

Calico Model M-951 Tactical Carbine
Similar to Model 900, except w/long compensator and adj. forward grip. Made 1990-94.
Model 951 . . . . **$375**
Model 951-S 385

CANADIAN MILITARY RIFLES
Quebec, Canada
Manufactured by Ross Rifle Co.

Canadian Model 1907 Mark II Ross
Military Rifle **$185**
Straight-pull bolt action. Caliber: 303 British. 5-shot box magazine. 28-inch bbl. Weight: 8.5 lbs. Sights: adj. rear; blade front. Military-type full stock. *Note:* The Ross was originally issued as a Canadian service rifle in 1907. There were several variations; it was the official weapon at the start of WWI, but has been obsolete for many years. For Ross sporting rifle, *see* listing under Ross Rifle company.

CENTURY INTERNATIONAL ARMS, INC.
Boca Raton, Florida
(Formerly St. Albans, Vermont)

Century International Centurion M38/M96
Bolt-Action Sporter
Sporterized Swedish M38/96 Mauser action. Caliber: 6.5×55mm. 5-shot magazine. 24-inch bbl. 44 inches overall. Adj. rear sight. Blade front. Black synthetic or checkered European hardwood Monte Carlo stock. Holden Ironsighter see-through scope mount. Imported 1987 to date.
W/Hardwood Stock **$125**
W/Synthetic Stock 150

Century International Centurion M98
Bolt-Action Sporter
Sporterized VZ24 or 98 Mauser action. Calibers: 270 Win., 7.62×39mm, 308 Win., 30-06. 5-shot magazine. 22-inch bbl. 44 inches overall. Weight: 7.5 lbs. W/Millet or Weaver Scope base(s), rings and no iron sights. Classic or Monte Carlo laminated hardwood, black synthetic or checkered European hardwood stock. Imported 1992 to date.
M98 Action W/Black Synthetic Stock (w/o rings) **$175**
M98 Action W/Hardwood Stock (w/o rings) 140
VZ24 Action W/Laminated Hardwood Stock (Elite) 235
VZ24 Action W/Black Synthetic Stock 225
W/Millet Base and Rings, **add** 25

Century International Centurion P-14 Sporter
Sporterized P-14 action. Caliber: 7mm Rem Mag., 300 Win. Mag. 5-shot magazine. 24-inch bbl. 43.4 inches overall. Weight: 8.25 lbs. Weaver-type scope base. Walnut stained hardwood or fiberglass stock. Imported 1987 to date.
W/Hardwood Stock **$150**
W/Fiberglass Stock 185

Century International Enfield Sporter #4
Bolt-Action Rifle
Sporterized Lee-Enfield action. Caliber: 303 British. 10-shot magazine. 25.25-inch bbl. 44.5 inches overall. Blade front sight, adj. aperture rear. Sporterized beechwood military stock or checkered walnut Monte Carlo stock. Blued finish. Imported 1987 to date.
W/Sporterized Military Stock . . . . **$ 85**
W/Checkered Walnut Stock 135

RIFLES

Chipmunk Bolt-Action Single-Shot

Century International L1A1 FAL Sporter. **$595**
Sporterized L1A1 FAL semiautomatic. Caliber: 308 Win. 20.75-inch bbl. 41 inches overall. Weight: 9.75 lbs. Protected front post sight, adj. aperture rear. Matte blued finish. Black or camo Bell & Carlson thumbhole sporter stock w/rubber buttpad. Imported 1988-98.

Century International M-14 Sporter. **$225**
Sporterized M-14 gas operated semiautomatic action. Caliber: 308 Win. 10-shot magazine. 22-inch bbl. 41 inches overall. Weight: 8.25 lbs. Blade front sight, adj. aperture rear sight. Parkerized finish. Walnut stock w/rubber recoil pad. Forged receiver. Imported 1991 to date.

Century International Tiger Dragunov. **$1695**
Russian SVD semiautomatic sniper rifle. Caliber: 7.62×54R. 5-shot magazine. 21-inch bbl. 43 inches overall. Weight: 8.5 lbs. Blade front sight, open rear adj. for elevation. Blued finish. Europeon laminated hardwood thumbhole stock. 4x range finding scope w/lighted reticle and sunshade. Quick detachable scope mount. Imported 1994-95.

CHARTER ARMS CORPORATION
Stratford, Connecticut

Charter AR-7 Explorer Survival Rifle. **$100**
Same as Armalite AR-7, except w/black, instead of brown, "wood grain" plastic stock. *See* listing of that rifle for specifications. Made 1973-90.

**Charter AR-7 Explorer
Survival Rifle
(Disassembled and Stowed in Stock)**

CHIPMUNK RIFLES
Prospect, Oregon
MFD. by Rogue Rifle Company
(Formerly Oregon Arms Company and
Chipmunk Manufacturing, Inc.)

Chipmunk Bolt-Action Single-Shot Rifle
Calibers: 22 LR or 22 WMR. Bbl.: 16.13 inches. Weight: 2.5 lbs. Sights: peep rear; ramp front. Plain American walnut stock. Made 1982 to date.
Standard Model (Disc. 1987) **$105**
Camouflage Model **130**
Deluxe Grade **155**

CHURCHILL RIFLES
Mfd. in England.
Imported by Ellett Brothers, Inc.,Chapin, SC
(Previously by Kassnar Imports, Inc.)

Churchill Highlander Bolt-Action Rifle **$350**
Calibers: 243 Win., 25-06 Rem., 270 Win., 308 Win., 30-06, 7mm Rem. Mag., 300 Win. Mag. 4-shot magazine (standard); 3-shot (magnum). Bbl. length: 22-inch (standard); 24-inch (magnum). 42.5 to 44.5 inches overall. Weight: 7.5 lbs. Adj. rear sight, blade front. Checkered European walnut pistol-grip stock. Imported 1986-91.

Churchill "One of One Thousand" Rifle **$995**
Made for Interarms to commemorate that firm's 20th anniversary. Mauser-type action. Calibers: 270, 7mm Rem. Mag., 308, 30-06, 300 Win. Mag., 375 H&H Mag., 458 Win. Mag. 5-shot magagine (3-shot in Magnum calibers). 24-inch bbl. Weight: 8 lbs. Classic-style French walnut stock w/cheekpiece, black forend tip, checkered pistol grip and forearm, swivel-mounted recoil pad w/cartridge trap, pistol-grip cap w/trap for extra front sight, barrel-mounted sling swivel. Limited issue of 1,000 rifles made in 1973.

Churchill Regent Bolt-Action Rifle **$425**
Calibers: 243 Win., 25-06 Rem., 270 Win., 308 Win., 30-06, 7mm Rem. Mag.,300 Win. Mag. 4-shot magazine. 22-inch round bbl. 42.5 inches overall. Weight: 7.5 lbs. Ramp front sight w/gold bead; adj. rear. Hand-checkered Monte Carlo-style stock of select European walnut; recoil pad. Made 1986-90.

CIMARRON ARMS
Fredericksburg, Texas

Cimarron 1860 Henry Lever-Action
Replica of 1860 Henry w/original Henry loading system.Calibers: 44-40, 44 Special, 45 Colt. 13-shot magazine. 22-inch bbl. (carbine) or 24.25-inch bbl. (rifle). 43 inches overall (rifle). Weight: 9.5 lbs. (rifle). Bead front sight, open adj. rear. Brass receiver and buttplate. Smooth European walnut buttstock. Imported 1991 to date.
Carbine Model **$575**
Rifle Model **595**
Civil War Model (U.S. issue martially marked) **625**
W/A-Engraving, **add** **395**
W/B-Engraving, **add** **495**
W/C-Engraving, **add** **695**

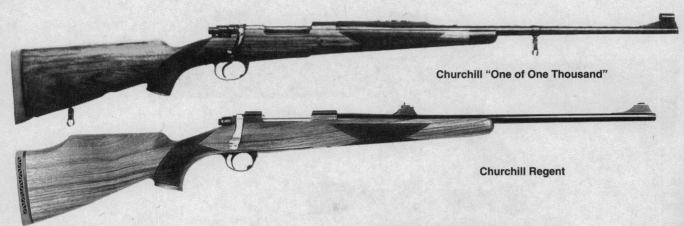

Churchill "One of One Thousand"

Churchill Regent

Cimarron 1866 Yellowboy Lever-Action
Replica of 1866 Winchester. Calibers: 22 LR, 22WMR, 38 Special, 44-40, 45 Colt. 16-inch round bbl. (trapper), 19-inch round bbl. (carbine) or 24.25-inch ocatagonal bbl. (rifle). 43 inches overall (rifle). Weight: 9 lbs. (rifle). Bead front sight, open adj. rear. Brass receiver, buttplate and forend cap. Smooth European walnut stock. Imported 1991 to date.

Carbine Model	$475
Rifle Model	495
Indain Model (Disc.)	425
Trapper Model (44-40 WCF only, Disc.)	400
W/A-Engraving, **add**	425
W/B-Engraving, **add**	595
W/C-Engraving, **add**	995

Cimarron 1873 Lever-Action
Replica of 1873 Winchester. Calibers: 22 LR, 22WMR, 357 Magnun, 44-40 or 45 Colt. 16-inch round bbl. (Trapper), 19-inch round bbl. (SRC), 20-inch ocatagonal bbl. (short rifle), 24.25-inch ocatagonal bbl. (sporting rifle) and 30-inch octagonal bbl. (express rifle). 43 inches overall (sporting rifle). Weight: 8 lbs. Fixed blade front sight, adj. semi-buckhorn rear or tang peep sight. Walnut stock and forend. Color case-hardened receiver. Imported 1989 to date.

Express Rifle Model	$595
Short Rifle Model (Disc.)	575
Sporting Rifle Model	550
SRC Carbine Model	565
Trapper Model (Disc.)	425
One of 1000 Engraved Model	1550

CLERKE RECREATION PRODUCTS
Santa Monica, California

Clerke Deluxe Hi-Wall **$275**
Same as standard model, except w/adj. trigger, half-octagon bbl., select wood, stock w/cheekpiece and recoil pad. Made 1972-74.

Clerke Hi-Wall Single-Shot Rifle **$225**
Falling-block lever-action similar to Winchester 1885 High Wall S.S. Color casehardened investment-cast receiver. Calibers: 222 Rem., 22-250, 243 Rem., 6mm Rem., 25-06, 270 Win., 7mm Rem. Mag., 30-06, 45-70 Govt. 26-inch medium-weight bbl. Weight: 8 lbs. Furnished w/o sights. Checkered walnut pistol-grip stock and Schnabel forearm. Made 1972-74.

CLIFTON ARMS
Medina, Texas

Clifton Arms Scout Bolt-Action Rifle
Custom rifle built on the Dakota 76, Ruger 77 or Winchester 70 action. Shilen match grade barrel cut and chambered to customer's specification. Clifton composite stock fitted and finished to customer's preference. Made 1992-97.

African Scout	$1995
Pseudo Scout	1750
Standard Scout	1695
Super Scout	1850

COLT INDUSTRIES, FIREARMS DIVISION
Hartford, Connecticut

Colt AR-15 A2 Delta Match H-BAR Rifle **$1350**
Similar to AR-15A2 Government Model except w/standard stock and heavy refined bbl. Furnished w/3-9× rubber armored scope and removeable cheekpiece.

Colt AR-15 A2 Government Model Carbine **$895**
Caliber: 223 Rem., 5- shot magazine. 16-inch bbl. w/flash suppressor. 35 inches overall. Weight: 5.8 lbs. Telescoping aluminum buttstock; sling swivels. Made 1985-91.

Colt AR-15 A2 Sporter II **$850**
Same general specifications as standard AR-15 Sporter except heavier bbl., improved pistol grip, weight of 7.5 lbs. and optional 3X or 4X scope. Made 1985-89.

Colt AR-15 Compact 9mm Carbine **$1095**
Semiautomatic. Caliber: 9mm NATO. 20-round detachable magazine. Bbl.: 16-inch round. Weight: 6.3 lbs. Adj. rear and front sights. Adj. buttstock. Ribbed round handguard. Made 1985-86.

Colt AR-15 Semiautomatic Sporter
Commercial version of U.S. M16 rifle. Gas-operated. Takedown. Caliber: 223 Rem. (5.56mm). 20-round magazine w/spacer to reduce capacity to 5 rounds. 20-inch bbl. w/flash suppressor. Sights: rear peep w/windage adjustment in carrying handle; front adj. for windage. 3X scope and mount optional (**add** $70 to value). Black molded buttstock of high-impact synthetic material, rubber buttplate. Barrel surrounded by handguards of black fiberglass w/heat-reflecting inner shield. Swivels, black web sling strap. Weight: w/o accessories, 6.3 lbs. Made 1964-94.

Standard Sporter	$ 995
w/Adj. Stock, Redesigned Forearm (Disc. 1988)	1195

Colt AR-15 A2

Colt AR-15 A2
Delta Match H-BAR

Colt AR-15 A2
Government Model

Coit AR-15
Sporter Competition H-BAR

**Colt AR-15 Sporter Competition
H-BAR Rifle** **$875**
Similar to AR-15 Sporter Target Model, except w/integral Weaver-
type mounting system on a flat-top receiver. 20-inch bbl. w/counter-
bored muzzle and 1:9" rifling twist. Made 1991 to date.

NOTE

On Colt AR-15 Sporter models currently produced (i.e., Competition H-BAR, Sporter Match Target Lightweight, and Sporter
Target Rifle, **add $200** to pre-ban models made prior to 10-13-
94.

Colt AR-15 Sporter Competition H-BAR (RS) $1275
Similar to AR-15 Sporter Competition H-BAR Model, except "Range Selected" for accuracy w/3×9 rubber-clad scope w/mount. Carrying handle w/iron sights. Made 1992-94.

Colt AR-15 Sporter Match Target Lightweight
Calibers: 223 Rem., 7.62×39mm, 9mm. 5-shot magazine. 16-inch bbl. (non-threaded after 1994). 34.5-35.5 inches overall. Weight: 7.1 lbs. Redesigned stock and shorter handguard. Made 1991 to date.

Standard LW Sporter (except 9mm)	$750
Standard LW Sporter, 9mm	685
22 LR Conversion (Disc. 1994), **add**	175

Colt AR-15 Sporter Target Rifle
Caliber: 223 Rem. 5-shot magazine. 20-inch bbl. w/flash suppressor (non-threaded after 1994). 39 inches overall. Weight: 7.5 lbs. Black composition stock, grip and handguard. Sights: post front; adj. aperture rear. Matte black finish. Made 1993 to date.

Sporter Target Rifle	$850
22 LR Conversion (Disc. 1994), **add**	200

Colt Lightning Carbine
Same as Lightning Magazine Rifle, except w/12-shot magazine, 20-inch bbl., weighs 6.25 lbs.

Carbine	$3550
Baby Carbine (5.5 lbs.)	7950

Colt Lightning Magazine Rifle $2595
Slide-action. Calibers: 32-20, 38-40, 44-40. 15-shot tubular magazine. 26-inch bbl., round or octagon. Weight: 6.75 lbs. (round barrel). Sights: open rear; bead or blade front. Walnut stock and forearm. Made 1884-1902.

Colt Lightning Magazine Rifle
22 Small Frame $1450
Slide-action. Caliber: 22 Rimfire (Short or Long). Tubular magazine holding 15 long or 16 short. 24-inch bbl., round or octagon. Weight: 5.75 lbs. (round bbl). Sights: open rear, bead front. Walnut stock and forearm. Available in small, medium and large frames. Made 1887-1904.

Colt Stagecoach 22 Autoloader $275
Same as Colteer 22 Autoloader, except w/engraved receiver, saddle ring, 16.5-inch bbl. Weight: 4 lbs. 10 oz. Made 1965-75.

Colteer 1-22 Single-Shot Bolt-Action Rifle $225
Caliber: 22 LR, Long, Short. 20- or 22-inch bbl. Sights: open rear; ramp front. Pistol-grip stock w/Monte Carlo comb. Weight: 5 lbs. Made 1957-67.

Colteer 22 Autoloader $230
Caliber: 22 LR. 15-round tubular magazine. 19.38-inch bbl. Sights: open rear; hooded ramp front. Straight-grip stock, Western carbine-style forearm w/bbl. band. Weight: 4.75 lbs. Made 1964-75.

RIFLES

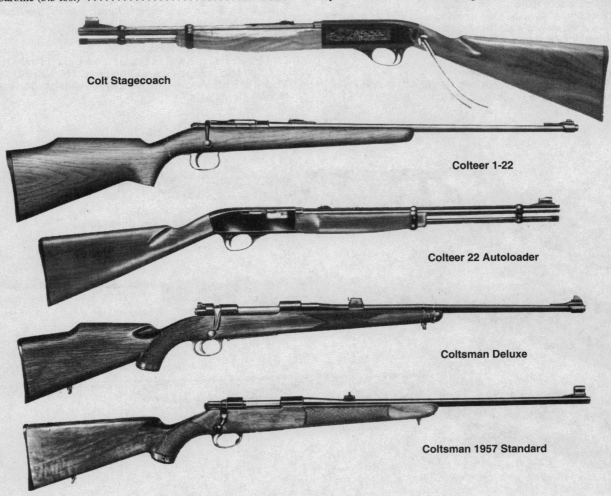

Colt Stagecoach

Colteer 1-22

Colteer 22 Autoloader

Coltsman Deluxe

Coltsman 1957 Standard

Coltsman 1961 Custom

Coltsman 1961 Standard

Coltsman Custom Bolt-Action Sporting Rifle **$395**
FN Mauser action, side safety, engraved floorplate. Calibers: 30-06, 300 H&H Mag. 5-shot box magazine. 24-inch bbl., rampfront sight. Fancy walnut stock. Monte Carlo comb, cheekpiece, pistol grip, checkered, QD swivels. Weight: 7.25 lbs. Made 1957-61.

Coltsman Deluxe Rifle **$595**
FN Mauser action. Same as Custom model except plain floorplate, plainer wood and checkering. Made 1957-61. Value shown is for rifle, as furnished by manufacturer, w/o rear sight.

Coltsman Models of 1957 Rifles
Sako Medium action. Calibers: 243, 308. Weight: 6.75 lbs. Other specifications similar to those of models w/FN actions. Made 1957-61.
Custom **$595**
Deluxe 575
Standard 495

Coltsman Model of 1961, Custom Rifle **$475**
Sako action. Calibers: 222, 222 Mag., 223, 243, 264, 270, 308, 30-06, 300 H&H. 23-, 24-inch bbl. Sights: folding leaf rear; hooded ramp front. Fancy French walnut stock w/Monte Carlo comb, rosewood forend tip and grip cap skip checkering, recoil pad, sling swivels. Weight: 6.5 - 7.5 lbs. Made 1963-65.

Coltsman Model of 1961, Standard Rifle **$495**
Same as Custom model, except plainer, American walnut stock. Made 1963-65.

Coltsman Standard Rifle **$450**
FN Mauser action. Same as Deluxe model, except in 243, 30-06, 308, 300 Mag. and stock w/o cheekpiece, bbl. length 22 inches. Made 1957-61. Value shown is for rifle, as furnished by manufacturer, w/o rear sight.

Colt-Sauer Drillings
See Colt shotgun listings.

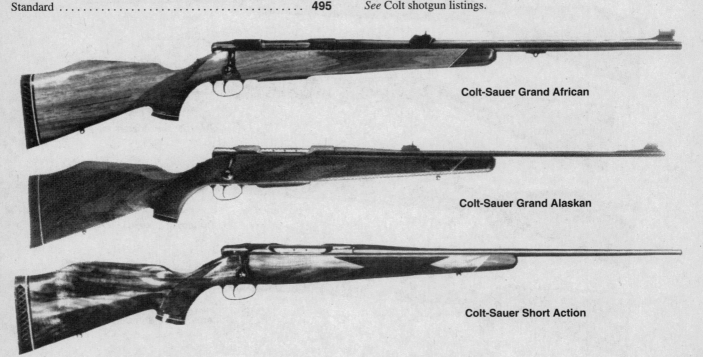

Colt-Sauer Grand African

Colt-Sauer Grand Alaskan

Colt-Sauer Short Action

**Cooper Arms — Model 22
Pro Varmint**

Colt-Sauer Grand African **$1200**
Same specifications as standard model, except caliber 458 Win. Mag., weighs 9.5 lbs. Sights: adj. leaf rear; hooded ramp front. Magnum-style stock of Bubinga. Made 1973-85.

Colt-Sauer Grand Alaskan **$1195**
Same specifications as standard model, except caliber 375 H&H, weighs 8.5 lbs. Sights: adj. leaf rear; hooded ramp front. Magnum-style stock of walnut.

Colt-Sauer Magnum **$965**
Same specifications as standard model, except calibers 7mm Rem. Mag., 300 Win. Mag., 300 Weatherby. Weight: 8.5 lbs. Made 1973-85.

Colt-Sauer Short Action **$895**
Same specifications as standard model, except shorter action chambered for the following calibers: 22-250, 243 Win., 308 Win. and similar length cartridges. Weight: 7.5 lbs.; 8.25 lbs. (22-250). Drilled and tapped for scope mount. No front or rear open sights. Made 1973-88.

Colt-Sauer Sporting Rifle,Standard Model **$925**
Sauer 80 nonrotating bolt action. Calibers: 25-06, 270 Win., 30-06. 3-round detachable box magazine. 24-inch bbl. Weight: 7.75 lbs., 8.5 lbs. (25-06). Furnished w/o sights. American walnut stock w/Monte Carlo cheekpiece, checkered pistol grip and forearm, rosewood forend tip and pistol-grip cap, recoil pad. Made 1973-88.

COMMANDO CARBINES
Knoxville, Tennessee
(Formerly Volunteer Enterprises, Inc.)

Commando Mark III Semiautomatic Carbine
Blow-back action, fires from closed bolt. Caliber: 45 ACP. l5- or 30-shot magazine. 16.5-inch bbl. w/cooling sleeve and muzzle brake. Weight: 8 lbs. Sights: peep rear; blade front. "Tommy Gun" style stock and forearm or grip. Made I969-76.
W/Horizontal Forearm **$335**

Commando Mark 9
Same specifications as Mark III and Mark 45, except caliber 9mm Luger. Made 1976-81.
W/Horizontal Forearm **$360**
W/Vertical Foregrip 375

Commando Mark 45
Same specifications as Mark III. Has redesigned trigger housing and magazines. Made 1976-88.
W/Horizontal Forearm **$375**
W/Vertical Foregrip 435

CONTINENTAL RIFLES
Manufactured in Belgium for Continental Arms Corp., New York, N.Y.

Continental Double Rifle **$4750**
Calibers: 270, 303 Sav., 30-40, 348 Win., 30-06, 375 H&H, 400 Jeffrey, 465, 470, 475 No. 2, 500, 600. Side-by-side. Anson-Deeley reinforced boxlock action w/triple bolting lever-work. Two triggers. Nonautomatic safety.24- or 26-inch bbls. Sights: express rear; bead front. Checkered cheekpiece stock and forend. Weight: from 7 lbs., depending on caliber. Imported 1956-75.

COOPER ARMS
Stevensville, Montana

Cooper Arms Model 21
Similar to Model 36CF, except in calibers 17 Rem., 17 Mach IV, 221 Fireball, 222, 223, 6×45, 6×47. 24-inch stainless or chrome-moly bbl. 43.5 inches overall. Weight: 8.75 lbs. Made 1994 to date.
Benchrest **$1195**
Varmint Extreme 895

Cooper Arms Model 22 Pro Varmint
Bolt-action, single-shot. Calibers: 22 BR, 22-250 Rem., 220 Swift, 243 Win., 6mm PPC, 6.5×55mm, 25-06 Rem., 7.62×39mm 26-inch bbl, 45.63 inches overall. Weight: 8 lbs., 12 oz. Single-stage trigger. AAA claro walnut stock. Made 1996 to date.
Pro Varmint **$995**
Black Jack. 950

Cooper Arms Model 22 RF/BR 50 . . . . **$1050**
Caliber: 22 LR. Bolt action. Single-shot. 22-inch bbl. 40.5 inches overall. Weight: 6.8 lbs. No sights. Fully-adj. match grade trigger. Stainless barrel. McMillan benchrest stock. Three mid-bolt locking lugs. Made 1994 to date.

Cooper Arms Model 36CF Bolt-Action Rifle
Calibers: 17 CCM, 22 CCM, 22 Hornet. 4-shot magazine. 23.75-inch bbl. 42.5 inches overall. Weight: 7 lbs. Walnut or synthetic stock. Made 1992-94.
Marksman **$675**
Sportsman 625
Classic Grade 895
Custom Grade 795
Custom Classic Grade 1095

Cooper Arms Model 36RF Bolt-Action Rifle
Similar to Model 36CF, except in caliber 22 LR. 5-shot magazine. Weight: 6.5-7 lbs. Made 1992 to date.
BR-50 (22-inch stainless bbl.) **$995**
Custom Grade 775
Custom Classic Grade 875
Featherweight 850

RIFLES

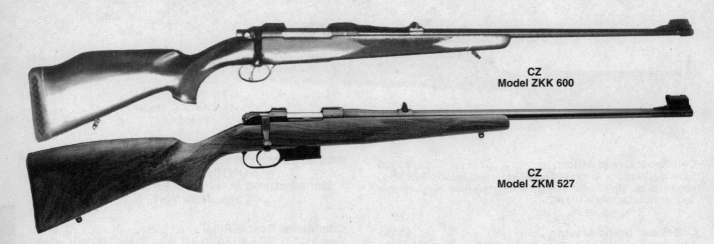

**CZ
Model ZKK 600**

**CZ
Model ZKM 527**

Cooper Arms Model 36 TRP-1 Series
Similar to Model 36RF, except in target configuration w/ ISU or silhouette style stock. Made 1991-93.
TRP-1 (ISU single-shot) **$695**
TRP-1S (Silhouette) 650

Cooper Arms Model 38 Single Shot
Similar to Model 36CF, except in calibers 17 or 22 CCM w/3-shot magazine. Weight: 8 lbs. Walnut or synthetic stock. Made 1992-93.
Sporter Standard **$650**
Classic Grade 795
Custom Grade 875
Custom Classic Grade 995

Cooper Arms Model 40 Classic Bolt-Action Rifle
Calibers: 17 CCM, 17 Ackley Hornet, 22 Hornet, 22K Hornet, 22 CCM, 4- or 5-shot magazine. 23.75-inch bbl. Checkered oil-finished AAA Claro walnut stock. Made1995-97.
Classic. **$995**
Custom Classic 1150
Classic Varminter 1195

CUMBERLAND MOUNTAIN ARMS
Winchester, Tennessee

Cumberland Mountain Arms Plateau Rifle
Falling block action w/underlever. Calibers: 40-65, and 45-70. 32-inch round bbl. 48 inches overall. Weight: 10.5 lbs. American walnut stock. Bead front sight, adj. buckhorn rear. Blued finish. Lacquer finish walnut stock w/crescent buttplate. Made 1995 to date.
Standard Model. **$695**
Deluxe Model 995

CZ RIFLES
Strankonice, Czechoslovakia
(Currently Uhersky Brod and Brno, Czecho.)
Mfd. by Ceska Zbrojovka-Nardoni Podnik
(Formerly Bohmische Waffenfabrik A.G.)

See also listings under Brno Sporting Rifles and Springfield, Inc.

CZ ZKK 600 Bolt-Action Rifle
Calibers: 270 Win., 7x57, 7x64, 30-06. 5-shot magazine. 23.5- inch bbl. Weight: 7.5 lbs. Adj. folding-leaf rear sight, hooded ramp front. Pistol-grip walnut stock. Imported 1990 to date.
Standard Model **$425**
Deluxe Model 495

CZ ZKK 601 Bolt-Action Rifle
Similar to Model ZKK 600, except w/short action in calibers 223 Rem., 243 Win., 308 Win. 43 inches overall. Weight: 6 lbs. 13 oz. Checkered walnut pistol-grip stock w/Monte Carlo cheekpiece. Imported 1990 to date.
Standard Model **$395**
Deluxe Model 425

CZ ZKK 602 Bolt-Action Rifle
Similar to Model ZKK 600, except w/Magnum action in calibers 300 Win. Mag., 8×68S, 375 H&H, 458 Win. Mag. 25-inch bbl. 45.5 inches overall. Weight: 9.25 lbs. Imported 1990 to date.
Standard Model **$525**
Deluxe Model 625

CZ ZKM 452 Bolt-Action Repeating Rifle
Calibers: 22 LR or 22 WMR. 5-, 6- or 10-shot magazine. 25-inch bbl. 43.5 inches overall. Weight: 6 lbs. Adj. rear sight, hooded bead front. Oil-finished beechwood or checkered walnut stock w/Schnabel forend. Imported 1995 to date.
Standard Model (22 LR) **$185**
Deluxe Model (22 LR) 225
Varmint Model (22 LR) 250
22 WMR, **add** 60

CZ ZKM 527 Bolt-Action Rifle
Calibers: 22 Hornet, 222 Rem., 223 Rem. 5-shot magazine. 23.5-inch bbl. 42.5 inches overall. Weight: 6.75 lbs. Adj. rear sight, hooded ramp front. Grooved receiver. Adj. double-set triggers. Oil-finished beechwood or checkered walnut stock . Imported 1995 to date.
Standard Model **$375**
Deluxe Model 450

CZ ZKM 537 Sporter Bolt-Action Rifle
Calibers: 243 Win., 270 Win., 7x57mm, 308 Win., 30-06. 4- or 5-shot magazine. 19- or 23.5-inch bbl. 40.25 or 44.75 inches overall. Weight: 7 to 7.5 lbs. Adj. folding leaf rear sight, hooded ramp front. Shrouded bolt. Standard or Mannlicher-style checkered walnut stock. Imported 1992-94.
Standard Stock Model **$395**
Mannlicher Stock Model 495

CZ 550 Bolt-Action Series
Calibers: 243 Win., 6.5x55mm, 270 Win., 7mm Mag., 7x57, 7x64, 30-06, 300 Win Mag., 375 H&H, 416 Rem., 416 Rigby, 458 Win. Mag., 9.3x62. 4- or 5-shot detachable magazine. 20.5- or 23.6-inch bbl. Weight: 7.25 to 8 lbs. No sights or Express sights on magnum models.

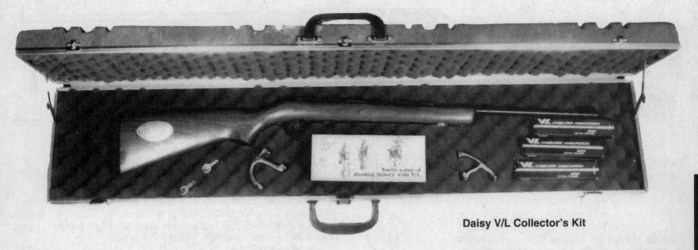

Daisy V/L Collector's Kit

CZ 550 Bolt-Action Series *(Con't)*

Receiver drilled and tapped for scope mount. Standard or Mann-licher-style checkered walnut stock w/buttpad. Imported 1995 to date.

Standard Model	**$350**
Magnum Model	425
Lux Model	375
Mannlicher Model	425
Calibers 416 Rem., 416 Rigby, 458 Win. Mag., **add**	50

CZECHOSLOVAKIAN MILITARY RIFLES
Brno, Czechoslovakia
Manufactured by Ceska Zbrojovka

Czech Model 1924 (VZ24) Mauser Military Rifle **$185**

Basically same as German Kar.,98k and F.N. (Belgian Model 1924.) Caliber: 7.9mm Mauser. 5-shot box magazine. 23.25-inch bbl. Weight: 8.5 lbs. Sights: adj. rear; blade front w/guards. Military stock of Belgian-type, full handguard. Mfd. 1924 thru WWII. Many of these rifles were made for export. As produced during the German occupation, this model was known as Gewehr 24t.

**Czech Model 1933 (VZ33) Mauser Military
Carbine** **$220**

Modification of German M/98 action w/smaller receiver ring. Caliber: 7.9mm Mauser. 19.25-inch bbl. Weight: 7.5 lbs. Sights: adj. rear; blade front w/guards. Military-type full stock. Mfd. 1933 thru WWII a similar model, produced during the German occupation, was designated Gew. 33/40.

DAEWOO PRECISION INDUSTRIES
Manufactured in Korea
(Previously Imported by Kimber of America; Daewoo Percision Industries; Nationwide Sports and KBI, Inc.)

Daewoo DR200 SA Semiautomatic Sporter

Caliber: 223 Rem. (5.56mm). 6- or 10-shot magazine. 18.4-inch bbl. 39.25 inches overall. Weight: 9 lbs. Protected post front sight, fully-adj. aperture rear. Forged aluminum receiver w/rotating locking bolt assembly. Synthetic sporterized thumbhole stock. Imported 1994-96.

Sporter Model	**$425**
Varmint Model	395

Daewoo DR300 SA Semiautomatic Sporter **$450**

Similar to Model Daewoo DR200, except chambered 7.62x39mm. Imported 1994-96.

DAISY RIFLES
Rogers, Arkansas

Daisy V/L rifles carry the first and only commercial caseless cartridge system. These rifles are expected to appreciate considerably in future years. The cartridge, no longer made is also a collector's item.

Daisy V/L Collector's Kit **$325**

Presentation-grade rifle w/gold plate inscribed w/owner's name and gun serial number mounted on the stock. Also includes a special gun case, pair of brass gun cradles for wall-hanging, 300 rounds of 22 V/L ammunition and a certificate signed by Daisy president, Cass S. Hough. Approx. 1,000 manufactured 1968-69.

Daisy V/L Presentation Grade **$295**

Same specifications as standard model, except w/walnut stock. Approx. 4,000 manufactured 1968-69.

Daisy V/L Standard Rifle **$195**

Single-shot, under-lever action. Caliber: 22 V/L (caseless cartridge, propellant ignited by jet of hot air). 18-inch bbl. Weight: 5 lbs. Sights: adj. open rear, ramp w/blade front. Wood-grained Lustran stock (foam-filled). Approx. 19,000 manufactured 1968-69.

DAKOTA ARMS, INC.
Sturgis, South Dakota

Dakota Model 10 Single-Shot Rifle **$1695**

Chambered for most commercially loaded calibers. 23-inch bbl. 39.5 inches overall. Weight: 5.5 lbs. Top tang safety. No sights. Checkered pistol-grip buttstock and semi-beavertail forearm, QD swivels, rubber recoil pad. Made 1992 to date.

Dakota Model 22 Bolt-Action Sporter Rifle **$975**

Calibers: 22 LR 22 Hornet. 5-shot magazine. 22-inch bbl. Weight: 6.5 lbs. Adj. trigger. Checkered classic-style Clara or English walnut stock w/black recoil pad. Made 1992 to date.

Daisy V/L Standard

Dakota Model 76 African Bolt-Action Rifle **$2995**
Same general specifications as Model 76 Safari. Calibers: 404 Jeffery, 416 Rigby, 416 Dakota, 450 Dakota. 24-inch bbl. Weight: 8 lbs. Checkered select walnut stock w/two crossbolts. Made 1989 to date.

Dakota Model 76 Alpine Bolt-Action Rifle **$1495**
Same general specifications as Model 76 Classic, except short action w/blind magazine. Calibers: 22-250, 243, 6mm Rem., 250-3000, 7mm-08, 308. 21-inch bbl. Weight: 7.5 lbs. Made 1989-93.

Dakota Model 76 Classic Bolt-Action Rifle **$1695**
Calibers: 257 Roberts, 270 Win., 280 Rem., 30-06, 7mm Rem. Mag., 300 Win. Mag., 338 Win. Mag., 375 H&H Mag., 458 Win. Mag. 21- or 23-inch bbl. Weight: 7.5 lbs. Receiver drilled and tapped for sights. Adj. trigger. Classic-style checkered walnut stock w/steel grip cap and solid recoil pad. Right- and left-hand models. Made 1988 to date.

Dakota Model 76 Longbow Tactical Bolt-Action Rifle **$2995**
Calibers: 300 Dakota Mag., 330 Dakota Mag., 338 Capua Mag. Blind magazine. Ported 28-inch bbl. 50 to 51 inches overall. Weight: 13.7 lbs. Black or oliver green fiberglass stock w/adj. cheekpiece and buttplate. Receiver drilled and tapped w/one-piece rail mount and no sights. Made 1997 to date.

Dakota Model 76 Safari Bolt-Action Rifle **$2595**
Calibers: 300 Win. Mag., 338 Win. Mag.,375 H&H Mag. 458 Win. Mag. 23-inch bbl. w/bbl. band swivel. Weight: 8.5 lbs. Ramp front sight, standing leaf rear. Checkered fancy walnut stock w/ebony forend tip and solid recoil pad. Made 1988 to date.

Dakota Model 76 Varmint Bolt-Action Rifle **$1595**
Similar to Model 76 Classic, except single-shot action w/ heavy bbl. chambered 17 Rem. to 6mm PPC. Weight: 13.7 lbs. Checkered walnut or synthetic stock. Receiver drilled and tapped for scope mounts and no sights. Made 1994-98.

Dakota Model 97 Hunter Bolt-Action Series
Calibers: 22-250 Rem. to 330 Dakota Mag.(Lightweight), 25-06 to 375 Dakota Mag. (Long Range). 22-, 24- or 26-inch bbl. 43 to 46 inches overall. Weight: 6.16 lbs. to 7.7 lbs. Black composite fiberglass stock w/recoil pad. Fully adj. match trigger. Made 1997 to date.
Lightweight. **$1195**
Long Range. **1250**

Dakota Model 97 Varmint Hunter Bolt-Action Rifle **$1095**
Similar to Model 97 Hunter, except single-shot action w/ heavy bbl. chambered 22-250 Rem. to 308 Win. Checkered walnut stock. Receiver drilled and tapped for scope mounts and no sights. Made 1998 to date.

CHARLES DALY RIFLE
Harrisburg, Pennsylvania
Imported by K.B.I., Inc., Harrisburg, PA
(Previously by Outdoor Sports Headquarters, Inc.)

Charles Daly Empire Grade Bolt-Action Rifle
Similar to Superior Grade, except w/checkered California walnut stock w/rosewood grip cap and forearm cap. High polished blued finish and damascened bolt. Made 1998 to date.
Empire Grade (22 LR) **$245**
Empire Grade (22WMR) **265**
Empire Grade (22 Hornet) **350**

Charles Daly Field Grade Bolt-Action Rifle
Caliber: 22 LR. 16.25-, 17.5- or 22.63-inch bbl. 32 to 41 inches overall. Single-shot (True Youth) and 6- or 10-shot magazine. Plain walnut-finished hardwood stock. Imported 1998 to date.
Field Grade .... **$ 90**
Field Grade (True Youth) **119**

Charles Daly Hammerless Drilling
See listing under Charles Daly shotguns.

Charles Daly Hornet Rifle **$895**
Same as Herold Rifle. See listing of that rifle for specifications. imported during the 1930s

Charles Daly Mauser 98
Calibers: 243 Win., 270 Win., 7mm Rem. Mag. 308 Win., 30-06, 300 Win. Mag., 375 H&H, or 458 Win. Mag. 3-, 4-, or 5-shot magazine. 23-inch bbl. 44.5 inches overall. Weight: 7.5 lbs. Checkered European walnut (Superior) or fiberglass/graphic (field) stock w/recoil pad. Ramped front sight, adj. rear. Receiver drilled and tapped w/side saftey. Imported 1998 to date.
Field Grade (standard calibers). **$265**
Field Grade (375 H&H and 458 Win. Mag.) **425**
Superior Grade (standard calibers) **315**
Superior Grade (magnum calibers). **495**

Charles Daly Mini-Mauser 98
Similar to Mauser 98, except w/19.25-inch bbl. chambered 22 Hornet, 22-250 Rem., 223 Rem., or 7.62×39mm. 5-shot magazine. Imported 1998 to date.
Field Grade **$295**
Superior Grade **325**

Charles Daly Superior Grade Bolt-Action Rifle
Calibers: 22 LR, 22 WMR, 22 Hornet. 20.25- to 22.63-inch bbl. 40.5 to 41.25 inches overall. 5-, 6-, or 10-shot magazine. Ramped front sight, adj. rear w/grooved receiver. Checkered walnut stock. Made 1998 to date.
Superior Grade (22 LR) **$119**
Superior Grade (22WMR) **135**
Superior Grade (22 Hornet) **285**

Dakota
Model 10

Dakota
22 LR Sporter

Dakota
76 Classic Grade

Dakota
76 Safari Grade

Charles Daly Semiautomatic Rifle
Caliber: 22 LR. 20.75-inch bbl. 40.5 inches overall. 10-shot magazine. Ramped front sight, adj. rear w/grooved receiver. Plain walnut-finished hardwood stock (Field), checkered walnut (Superior) or checkered California walnut stock w/rosewood grip cap and forearm cap (Empire). Imported 1998 to date.

Field Grade	$ 90
Superior Grade	125
Empire Grade	225

EAGLE ARMS INC.
Geneseo, Illinois
(Previously Coal Valley, Illinois)

In 1995, Eagle Arms Inc. became a division of ArmaLite and reintroduced that logo. For current production see models under that listing.

Eagle Arms Model EA-15 Carbine
Caliber: 223 Rem. (5.56mm). 30-shot magazine. 16-inch bbl. and collapsible buttstock. Weight: 5.75 lbs. (E1); 6.25 lbs. (E2 w/heavy bbl. & NM sights). Made 1989 to 94.

Eagle Arms Model EA-15 Carbine
E1 Carbine	$625
E2 Carbine	675

Eagle Arms Model EA-15 Golden Eagle
Match Rifle $825
Same general specifications as EA-15 Standard, except w/E2-style National Match sights. 20-inch Douglas Heavy Match bbl. NM trigger and bolt-carrier group. Weight: 12.75 lbs. Made 1991 to 94.

Eagle Arms Model EA-15 Semiautomatic Rifle..... $650
Same general as EA-15 Carbine except 20-inch bbl., 39 inches overall, and weighs 7 lbs. Made 1989 to date.

EMF COMPANY, INC.
Santa Ana, California

EMF Model AP-74 Semiautomatic Carbine $235
Calibers: 22 LR or 32 ACP, 15-shot magazine. 20-inch bbl. w/flash reducer. 38 inches overall. Weight: 6.75 lbs. Protected pin front sight; protected rear peep sight. Lightweight plastic buttstock; ventilated snap-out forend. Importation. Disc. 1989.

EMF
AP-74 Semiautomatic

**EMF
AP74-W Sporter**

**EMF
AP-74 Paratrooper**

EMF Model AP74-W Sporter Carbine **$265**
Sporterized version of AP-74 w/wood buttstock and forend. Importation. Disc. 1988.

EMF Model AP74 Paratrooper **$275**
Same general specifications as Model AP74-W except w/folding tubular buttstock. Made in 22 LR only. Importation. Disc. 1987.

EMF Model 1860 Henry Rifle
Calibers: 44-40 and 45 LC. 24.25-inch bbl.; upper-half octagonal w/magazine tube in one-piece steel. 43.75-inches overall. Weight: 9.25 lbs. Varnished American walnut wood stock. Polished brass frame and brass buttplate. Original rifle was patented by B. Tyler Henry and produced by the New Haven Arms Company, where Oliver Winchester was then president. Imported 1987 to date.
Deluxe Model **$495**
Engraved Model **750**

EMF Model 1866 Yellow Boy Rifle **$395**
Calibers: 44-40, 45LC and 38 Special. Lever-action. Bbl: 24 inches, 43 inches overall. Bead front sight. Exact reproduction. Offered w/blued finish, walnut stock and brass frame.

EMF Model 1866 Yellow Boy Carbine
Same features as 1866 Yellow Boy Rifle, except carbine.
Standard Carbine **$375**
Engraved Carbine **495**

EMF Model 1873 Sporting Rifle
Calibers: 22 LR, 22 WMR, 357 Mag., 44-40 and 45 LC. 24.25-inch octagonal bbl. 43.25 inches overall. Weight: 8.16 lbs. Color casehardened frame w/blued steel magazine tube . Walnut stock and forend.
Standard Rifle **$475**
Engraved Rifle **550**
Boy's Rifle (Youth Model, 22 LR) **450**

EMF Model 1873 Sporting Rifle Carbine
Same features as 1873 sporting rifle, except w/19-inch bbl. Overall length: 38.25 inches. Weight: 7.38 lbs. Color casehardened or blued frame.
Standard Carbine **$475**
Boy's Carbine (Youth Model, 22 LR) **450**

ERMA-WERKE
Dachau, Germany
(Previously imported by Precision Sales International; Nygord Precision Products; Mandall's Shooting Supplies)

Erma Model EG72 Pump Action Repeater **$115**
Visible hammer. Caliber: 22 LR. 15-shot magazine. 18.5-inch bbl. Weight: 5.25 lbs. Sights: open rear; hooded ramp front. Receiver grooved for scope mounting. Straight-grip stock, grooved slide handle. Imported 1970-76.

Erma Model EG73 . . . . **$215**
Same as Model EG712, except chambered for 22 WMR, w/12-shot tubular magazine, 19.3-inch bbl. Imported 1973-97.

**Erma Model EG712 Lever-Action Repeating
Carbine** **$205**
Styled after Winchester Model 94. Caliber: 22 LR, Long, Short. Tubular magazine holds 15 LR, 17 Long, 21 Short. 18.5-inch bbl. Weight: 5.5 lbs. Sights: open rear; hooded ramp front. Receiver grooved for scope mounting. Western carbine-style stock and forearm w/ bbl. band. Imported 1976-97. *Note:* A similar carbine of Erma manufacture is marketed in U.S. as Ithaca Model 72 Saddlegun.

Erma Model EGM1 . . . . **$215**
Same as Model EM1, except w/unslotted buttstock, ramp front sight, 5-shot magazine standard. Imported 1970-95.

Erma Model EM1 22 Semiautomatic Carbine **$265**
Styled after U.S. Carbine Cal. 30 M1. Caliber: 22 LR. 10- or 15-round magazine. 18-inch bbl. Weight: 5.5 lbs. Carbine-type sights. Receiver grooved for scope mounting. Military stock/handguard. Imported 1966-97.

EUROPEAN AMERICAN ARMORY
Sharpes, Florida

**EAA Model HW 660 Bolt-Action
Single-Shot Rifle** **$625**
Caliber: 22 LR. 26.8-inch bbl., 45.7 inches overall. Weight: 10.8 lbs. Match-type aperture rear sight; hooded ramp front. Stippled walnut stock. Imported 1992-96.

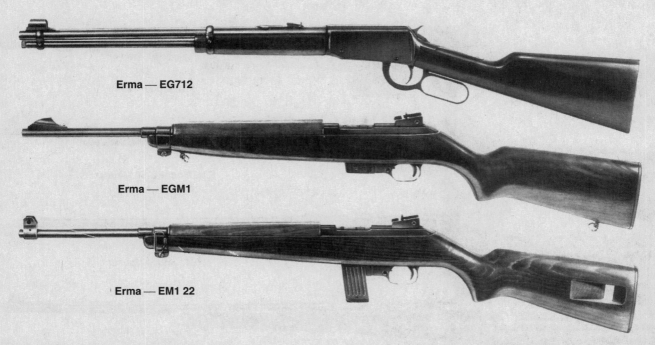

Erma — EG712

Erma — EGM1

Erma — EM1 22

EAA Model HW Bolt-Action Single-Shot
Target Rifle **$525**
Same general specification as Model HW 660 except equipped w/a target stock. Imported 1995-96.

EAA Model Sabitti SP1822
Caliber: 22 LR. 10-round detachable magazine. 18.5 inch bbl. 37.5 inches overall. Weight: 5.25 to 7.15 lbs. No sights. Hammer-forged heavy non-tapered bbl. Scope-mounted rail. Flush-mounted magazine release. Alloy receiver w/non-glare finish. Manual bolt lock. Wide claw extractor. Blowback action. Cross-trigger safety. Imported 1994-96.
Traditional Sporter Model **$155**
Thumbhole Sporter Model (synthetic stock) **250**

FABRIQUE NATIONALE HERSTAL
Herstal & Liege, Belgium
(Formerly Fabrique Nationale d'Armes de Guerre)

F.N. Models 1924, 1934/30 and 1930 Mauser
Military Rifles **$275**
Similar to German Kar.98k w/straight bolt handle. Calibers: 7mm, 7.65mm and 7.9mm Mauser. 5-shot box magazine. 23.5-inch bbl. Weight: 8.5 lbs. Sights: adj. rear; blade front. Military stock of M/98 pattern w/slight modification. Model differences are minor. Also produced in a short carbine model w/17.25-inch bbl. *Note:* These rifles were manufactured under contract for Abyssinia, Argentina, Belgium, Bolivia, Brazil, Chile, China, Colombia, Ecuador, Iran, Luxembourg, Mexico, Peru, Turkey, Uruguay and Yugoslavia. Such arms usually bear the coat of arms of the country for which they were made, together with the contractor's name and date of manufacture. Also sold commercially and exported to all parts of the world.

F.N. Model 1949 Semiautomatic Military Rifle $350
Gas-operated. Calibers: 7mm, 7.65mm, 7.92mm, 30-06. 10-round box magazine, clip fed or loaded singly. 23.2-inch bbl. Weight: 9.5 lbs. Sights: tangent rear-shielded post front. Pistol-grip stock, handguard. *Note:* Adopted by Belgium in 1949; also by Belgian Congo, Brazil, Colombia, Luxembourg, Netherlands, East Indies; and Venezuela. Approx. 160,000 were made.

F.N. Model 1950 Mauser Military Rifle $295
Same as previous F.N. models of Kar. 98k type, except chambered for 30-06.

F.N. Deluxe Mauser Bolt-Action Sporting Rifle $550
American calibers: 220 Swift, 243 Win., 244 Rem., 250/ 3000, 257 Roberts, 270 Win., 7mm, 300 Sav., 308 Win. 30-06. *Europearn calibers:* 7×57, 8×57JS, 8×60S, 9.3×62, 9.5×57, 10.75×68mm. 5-shot box magazine. 24-inch bbl. Weight: 7.5-8.25 lbs. American model is standard w/hooded ramp front sight and Tri-Range rear; Continental model w/two-leaf rear. Checkered stock w/cheekpiece, pistol grip, swivels. Made 1947-63.

F.N. Deluxe Mauser — Presentation Grade $995
Same as regular model, except w/select grade stock; engraving on receiver, trigger guard, floorplate and bbl. breech. Disc. 1963.

F.N. FAL/FNC/LAR Semiautomatic
Same as standard FAL military rifle except w/o provision for automatic firing. Gas-operated. Calibers: 7.62mm NATO (308 Win.) or 5.56mm (223 Rem.). 10- or 20-round box magazine. 25.5-inch bbl. (including flash hider). Weight: 9 lbs. Sights: post front; aperture rear. Fixed wood or folding buttstock, pistol grip, forearm/handguard w/carrying handle and sling swivels. Disc. 1988 .
F.N. FAL/LAR Model (Light Automatic Rifle). **$1895**
F.N. FAL/HB Model (Heavy Bbl.) **$2150**
F.N. FAL/PARA (Paratrooper) **$2295**
F.N. FNC Carbine Model (223 cal.) **$1650**
F.N. FNC Carbine Model w/flash suppresser (223 cal.) . . . **$1795**

F.N. Supreme Mauser Bolt-Action Sporting Rifle $575
Calibers: 243, 270, 7mm, 308, 30-06. 4-shot magazine in 243 and 308; 5-shot in other calibers. 22-inch bbl. in 308; 24-inch in other calibers. Sights: hooded ramp front, Tri-Range peep rear. Checkered stock w/ Monte Carlo checkpiece, pistol grip, swivels. Weight: 7.75 lbs. Made 1957-75.

F.N. Supreme Magnum Mauser $595
Calibers: 264 Mag., 7mm Mag., 300 Win. Mag. Specifications same as for standard-caliber model except 3-shot magazine capacity.

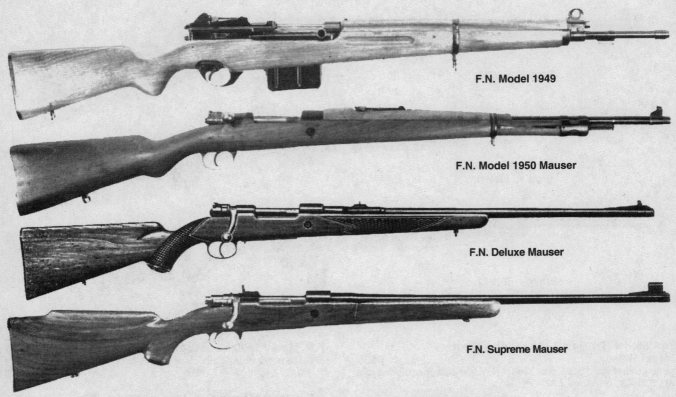

F.N. Model 1949

F.N. Model 1950 Mauser

F.N. Deluxe Mauser

F.N. Supreme Mauser

FEATHER INDUSTRIES, INC.
Boulder, Colorado

See MITCHHELL ARMS. For current production.

Feather Model AT-9 Semiautomatic Rifle
Caliber: 9mm Parabellum. 25-shot magazine. 17-inch bbl. 35 inches overall (extended). Hooded post front sight, adj. aperture rear. Weight: 5 lbs. Telescoping wire stock w/composition pistol-grip and barrel-shroud handguard. Matte black finish. Made 1988-95.
Model AT-9 . . . . $395

Feather Model AT-22 . . . . $175
Caliber: 22 LR. 20-shot magazine. 17-inch bbl. 35 inches overall (extended). Hooded post front sight; adj. aperture rear. Weight: 3.25 lbs. Telescoping wire stock w/composition pistol-grip and barrel-shroud handguard. Matte black finish.

Feather Model F2 SA Carbine $185
Similar to AT-22, except w/fixed Polymer stock and pistol grip. Made 1992-95.

Feather Model F9 SA Carbine $395
Similar to AT-9, except w/fixed Polymer stock and pistol grip. Made 1992-95.

FINNISH LION RIFLES
Jyväkylylä, Finland
Manufactured by Valmet Oy, Tourula Works

Finnish Lion Champion Free Rifle $475
Bolt-action, single-shot target rifle. Double-set trigger. Caliber: 22 LR. 28.75-inch heavy bbl. Weight: 16 lbs. Sights: extension rear peep; aperture front. Walnut free-rifle stock w/full pistol grip, thumbhole, beavertail forend, hook buttplate, palmrest, hand stop, swivel. Made 1965-72.

**F.N. FAL
Semiautomatic**

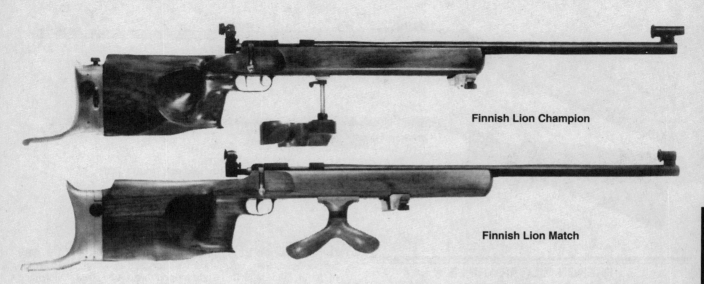

Finnish Lion Champion

Finnish Lion Match

Finnish Lion Standard ISU Target Rifle $275
Bolt-action, single-shot. Caliber: 22 LR. 27.5-inch bbl. Weight: 10.5 lbs. Sights: extension rear peep; aperture front. Walnut target stock w/full pistol grip, checkered beavertail forearm, adj. buttplate, sling swivel. Made 1966-77.

Finnish Lion Match Rifle $395
Bolt-action, single-shot. Caliber: 22 LR. 28.75-inch heavy bbl. Weight: 14.5 lbs. Sights: extension rear peep; aperture front. Walnut free-rifle stock w/full pistol grip, thumbhole, beavertail forearm, hook buttplate, palmrest, hand stop, swivel. Made 1937-72.

Finnish Lion Standard Target Rifle
Bolt-action, single-shot. Caliber: 22 LR. 27.5-inch bbl. 44.5 inches overall. Weight: 10.5 lbs. No sights; micrometer rear and globe front International-style sights available. Select walnut stock in target configuration. Currently in production.

Standard Model	$550
Thumbhole Stock Model	625
Standard Model	250
Deluxe Model	295

LUIGI FRANCHI, S.P.A.
Brescia, Italy

Franchi Centennial Automatic Rifle
Commemorates Franchi's 100th anniversary (1868-1968). Centennial seal engraved on receiver. Semiautomatic. Takedown. Caliber: 22 LR. 11-shot magazine in buttstock. 21-inch bbl. Weight: 5.13 lbs. Sights: open rear; goldbead front, on ramp. Checkered walnut stock and forend. Deluxe model w/fully engraved receiver, premium grade wood. Made 1968.

Standard Model...............................	$250
Engraved Model	325

FRANCOTTE RIFLES
Leige, Belgium
Imported by Armes de Chasse, Herford, NC
(Previously by Abercrombie & Fitch)

Francotte Bolt-Action Rifle
Custom rifle built on Mauser style bolt action. Available in three action lengths. Cal;ibers: 17 Bee to 505 Gibbs. Barrel length: 21- to 24.5-

Francotte Bolt-Action Rifle *(Con't)*
inches. Weight: 8 to 12 lbs. Stock dimensions, wood type and style to customer's specifications. Engraving, appointments and finish to customer's preference. *Note:* **Deduct 25%** for models w/o engraving.

Short Action	**$6250**
Standard Action	5250
Magnum Francotte Action	9750

Francotte Boxlock Mountain Rifle
w/claw mouths and scope

Francotte Boxlock Mountain Rifle
Custom single-shot rifle built on Anson & Deeley style boxlock or Holland & Holland style sidelock action. 23- to 26-inch barrels chambered to customer's specification. Stock dimensions, wood type and style to customer's specifications. Engraving, appointments and finish to customer's preference. *Note:* **Deduct 30%** for models w/o engraving.

Boxlock	**$9,750**
Sidelock....................................	17,550

Francotte Double Rifle
Custom side-by-side rifle. Built on Francotte system boxlock or back-action sidelock. 23.5- to 26-inch barrels chambered to customer's specification. Stock dimensions, wood type and style to customer's specifications. Engraving, appointments and finish to customer's preference. *Note:* **Deduct 30%** for models w/o engraving. *See* Illustration next page.

Boxlock	**$12,750**
Sidelock....................................	21,500

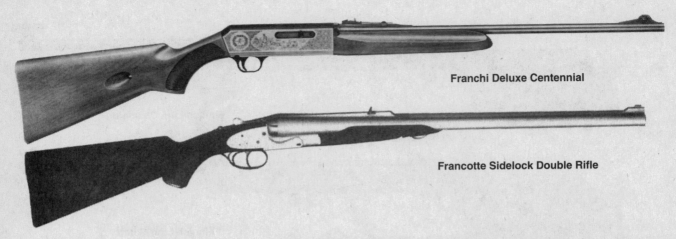

Franchi Deluxe Centennial

Francotte Sidelock Double Rifle

FRENCH MILITARY RIFLE
Saint Etienne, France

French Model 1936 MAS Military Rifle **$110**
Bolt action. Caliber: 7.5mm MAS. 5-shot box magazine. 22.5-inch bbl. Weight: 8.25 lbs. Sights: adj. rear; blade front. Two-piece military-type stock. Bayonet carried in forend tube. Made 1936-1940 by Manufacture Francaise d'Armes et de Cycles de St. Etienne (MAS).

GALIL RIFLES
Manufactured by Israel Military Industries, Israel
Imported by UZI America Inc., North Haven, CT
(Previously by Action Arms, Springfield Armory and Magnum Research, Inc.)

Galil AR Semiautomatic Rifle
Calibers: 308 Win. (7.62 NATO), 223 Rem. (5.56mm). 25-shot (308) or 35-shot (223) magazine. 16-inch (223) or 18.5-inch (308) bbl. w/flash suppressor. Weight: 9.5 lbs. Folding aperture rear sight, post front. Folding metal stock w/carrying handle. Imported 1982-94. Currently select fire models available to law enforcement only.
Model 223 AR **$1395**
Model 308 AR **1425**
Model 223 ARM **1695**
Model 308 ARM **1750**

Galil Sporter Semiautomatic Rifle **$995**
Same general specifications as AR Model, except w/hardwood thumb-hole stock and 5-shot magazine. Weight: 8.5 lbs. Imported 1991-94.

GARCIA CORPORATION
Teaneck, New Jersey

Garcia Bronco 22 Single-Shot Rifle **$85**
Swing-out action. Takedown. Caliber: 22 LR, Long, Short. 16.5-inch bbl. Weight: 3 lbs. Sights: open rear-blade front. One-piece stock and receiver, crackle finish. Intro. 1967. Discontinued.

GERMAN MILITARY RIFLES
Mfd. by Ludwig Loewe & Co., Berlin, other contractors and by German arsenals and various plants under German Government control

German Model 24T (Gew. 24T) Mauser Rifle **$365**
Same general specifications as Czech Model 24 (VZ24) Mauser Rifle w/minor modification and laminated woodstock. Weight: 9.25 lbs. Made in Czechoslovakia during German occupation; adopted 1940.

German Model 29/40 (Gew. 29/40) Mauser Rifle . . . **$225**
Same general specifications as Kar. 98K, w/minor differences. Made in Poland during German occupation; adopted 1940.

German Model 33/40 (Gew. 33/40) Mauser Rifle . . . **$695**
Same general specifications as Czech Model 33 (VZ33) Mauser Carbine w/minor modifications and laminated wood stock as found in war-time Model 98K carbines. Made in Czechoslovakia during German occupation; adopted 1940.

German Models 41 and 41-W (Gew. 41, Gew. 41-W) Semiautomatic Military Rifles
Gas-operated, muzzle cone system. Caliber: 7.9mm Mauser. 10-shot box magazine. 22.5-inch bbl. Weight: 10.25 lbs. Sights: adj. leaf rear; blade front. Military-type stock w/semi-pistol grip, plastic handguard. *Note:* Model 41 lacks bolt release found on Model 41-W; otherwise, the models are the same. These early models were mfd. in Walther's Zella-Mehlis plant. Made c.1941-43.
Model 41 . . . . **$3250**
Model 41-W **2495**

**French Model 1936
MAS Military Rifle**

**Galil 223 AR
Semiautomatic Rifle**

German Model 43 (Gew. 43, Kar. 43) Semiauto Military Rifles $950

Gas-operated, bbl. vented as in Russian Tokarev. Caliber: 7.9mm Mauser.10-shot detachable box magazine. 22- or 24-inch bbl. Weight: 9 lbs. Sights: adj. rear; hooded front. Military-type stock w/semipistol grip, wooden handguard. *Note:* These rifles are alike except for minor details, have characteristic late WWII mfg. short cuts: cast receiver and bolt cover, stamped steel parts, etc. Gew. 43 may have either 22- or 24-inch bbl. The former length was standardized in late 1944, when weapon designation was changed to "Kar. 43." Made 1943-45.

German Model 1888 (Gew. 88) Mauser-Mannlicher Service Rifle $195

Bolt action w/straight bolt handle. Caliber: 7.9mm Mauser (8×57mm). 5-shot Mannlicher box magazine. 29-inch bbl. w/jacket. Weight: 8.5 lbs. Fixed front sight, adj. rear. Military-type full stock. Mfd. by Ludwig Loewe & Co., Haenel, Schilling and other contractors.

German Model 1888 (Kar. 88) Mauser-Mannlicher Carbine $215

Same general specifications as Gew. 88, except w/18-inch bbl., w/o jacket, flat turned-down bolt handle, weighs 6.75 lbs. Mfd. by Ludwig Loewe & Co., Haenel, Schilling and other contractors.*See* Illustration next page.

German Model 1898 (Gew. 98) Mauser Military Rifle $250

Bolt action with straight bolt handle. Caliber: 7.9mm Mauser (8×57mm). 5-shot box magazine. 29-inch steeped bbl. Weight: 9 lbs. Sights: blade front; adj. rear. Military-type full stock w/rounded bottom pistol grip. Adopted 1898.

German Model 1898A (Kar. 98A) Mauser Carbine $275

Same general specifications as Model 1898 (Gew.98) Rifle, except has turned down bolt handle, smaller receiver ring, light 23.5-inchstraight taper bbl., front sight guards, sling is attached to left side of stock, weighs 8 lbs. *Note:* Some of these carbines are marked "Kar. 98"; the true Kar. 98 is the earlier original M/98 carbine w/17-inch bbl. and is rarely encountered.

German Model 1898B (Kar. 98B) Mauser Carbine $350

Same general specifications as Model 1898 (Gew.98) Rifle, except has turned-down bolt handle and sling attached to left side of stock. This is post-WWI model.

German Model 1898K (Kar. 98K) Mauser Carbine $265

Same general specifications as Model 1898 (Gew.98) Rifle, except has turned-down bolt handle, 23.5-inch bbl., may have hooded front sight, sling is attached to left side of stock, weighsabout 8.5 lbs. Adopted in 1935, this was the standard German Service rifle of WWII. *Note:* Late-war models had stamped sheet steel trigger guards and many of the Model 98K carbines made during WWII had laminated wood stocks, these weigh ½ to ¾ pound more than the previous Model 98K. Value shown is for earlier type.

German Model VK 98 People's Rifle ("Volksaewehr") $175

Kar. 98K-type action. Caliber: 7.9mm. Single-shot or repeater (latter w/rough hole-in-the-stock 5-shot "magazine" or fitted w/10-shot clip of German Model 43 semiauto rifle). 20.9-inch bbl. Weight: 7 lbs. Fixed V-notch rear sight dovetailed into front receiver ring; front blade

German Gew. 33/40

German Gew. 43

German Kar. 88

German Model VK 98 People's Rifle *(Con't)*
welded to bbl. Crude, unfinished, half-length stock w/o buttplate. Last ditch weapon made in 1945 for issue to German civilians. *Note:* Of value only as a military arms collector's item, this hastily made rifle should be regarded as *unsafe* to shoot.

GÉVARM RIFLE
Saint Etienne, France
Manufactured by Gevelot

Gévarm E-1 Autoloading Rifle **$175**
Caliber: 22 LR. 8-shot clip magazine. 19.5-inch bbl. Sights: open rear; post front. Pistol-grip stock and forearm of French walnut.

GOLDEN EAGLE RIFLES
Houston, Texas
Mfd. by Nikko Firearms Ltd., Tochigi, Japan

Golden Eagle Model 7000 Grade I African **$595**
Same as Grade I Big Game, except calibers 375 H&H Mag. and 458 Win. Mag., 2-shot magazine in 458, weighs 8.75 lbs. in 375 and 10.5 lbs. in 458, furnished w/sights. Imported 1976-81.

Golden Eagle Model 7000 Big Game Series
Bolt action. Calibers: 22-250, 243 Win., 25-06, 270 Win., Weatherby Mag., 7mm Rem. Mag., 30-06, 300 Weatherby Mag., 300 Win. Mag., 338 Win. Mag. Magazine capacity: 4 rounds in 22-250, 3 rounds in other calibers. 24- or 26-inch bbl. (26-inch only in 338). Weight: 7 lbs., 22-250; 8.75, lbs., other calibers. Furnished w/o sights. Fancy American walnut stock, skip checkered, contrasting wood forend tip and grip cap w/gold eagle head, recoil pad. Imported 1976-81.
Model 7000 Grade I **$525**
Model 7000 Grade II. **595**

GREIFELT & CO.
Suhl, Germany

Greifelt Sport Model 22 Hornet Bolt-Action Rifle . **$1695**
Caliber: 22 Hornet. 5-shot box magazine. 22-inch Krupp steel bbl. Weight: 6 lbs. Sights: two-leaf rear; ramp front. Walnut stock, checkered pistol grip and forearm. Made before WWII.

CARL GUSTAF RIFLES
Eskilstuna, Sweden
Mfd. by Carl Gustafs Stads Gevärsfaktori

Carl Gustaf Model 2000 Bolt-Action Rifle
Calibers: 243, 6.5×55, 7×64, 270, 308 Win., 30-06, 7mm Rem. Mag., 300 Win. Mag. 3-shot magazine. 24-inch bbl. 44 inches overall. Weight: 7.5 lbs. Receiver drilled and tapped. Hooded ramp front sight, open rear. Adj. trigger. Checkered European walnut stock w/Monte Carlo cheekpiece and Wundhammer palmswell grip. Imported 1991-95
Model 2000 w/o Sights **$1095**
Model 2000 w/Sights **1250**
Model 2000 LUXE. **1295**

Carl Gustaf Deluxe **$575**
Same specifications as Monte Carlo Standard. Calibers: 6.5×55, 308 Win., 30-06, 9.3×62. 4-shot magazine in 9.3×62. Jeweled bolt. Engraved floorplate and trigger guard. Deluxe French walnut stock w/rosewood forend tip. Imported 1970-77.

Carl Gustaf Grand Prix Single-Shot Target Rifle **$495**
Special bolt action with "world's shortest lock time." Single-stage trigger adjusts down to 18 oz. Caliber: 22 LR. 26.75-inch heavy bbl. w/adj. trim weight. Weight: 9.75 lbs. Furnished w/o sights. Target-type Monte Carlo stock of French walnut, adj. cork buttplate. Imported 1970-77

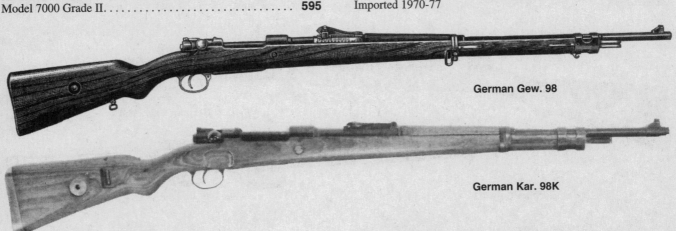

German Gew. 98

German Kar. 98K

Golden Eagle Model 7000

Carl Gustaf Monte Carlo Standard
Bolt-Action Sporting Rifle $350
Carl Gustaf 1900 action. Calibers: 6.5×55, 7×64, 270 Win., 7mm Rem. Mag., 308 Win., 30-06, 9.3×62. 5-shot magazine, except 4-shot in 9.3×62 and 3-shot in 7mm Rem. Mag. 23.5-inch bbl. Weight: 7 lbs. Sights: folding leaf rear; hooded ramp front. French walnut Monte Carlo stock w/cheekpiece, checkered forearm and pistol grip, sling swivels. Also available in left-hand model. Imported 1970-77.

Carl Gustaf Special $395
Also designated "Grade II" in U.S. and "Model 9000" in Canada. Same specifications as Monte Carlo Standard. Calibers: 22-250, 243 Win., 25-06, 270 Win., 7mm Rem. Mag., 308 Win., 30-06, 300 Win. Mag. 3-shot magazine in Magnum calibers. Select wood stock w/rosewood forend tip. Left-hand model avail. Imported 1970-77.

Carl Gustaf Sporter $475
Also designated "Varmint-Target" in U.S. Fast bolt action w/large bakelite bolt knob. Trigger pull adjusts down to 18 oz. Calibers: 222 Rem., 22-250, 243 Win., 6.5×55. 5-shot magazine, except 6-shot in 222 Rem. 26.75-inch heavy bbl. Weight: 9.5 lbs. Furnished w/o sights. Target-type Monte Carlo stock of French walnut. Imported 1970 to date.

Carl Gustaf Standard $425
Same specifications as Monte Carlo Standard. Calibers: 6.5×55, 7×64, 270 Win., 308 Win., 30-06, 9.3×62. Classic-style stock w/o Monte Carlo. Imported 1970-77.

Carl Gustaf Trofé $595
Also designated "Grade III" in U.S. and "Model 8000" in Canada. Same specifications as Monte Carlo Standard. Calibers: 22-250, 25-06, 6.5×55, 270 Win., 7mm Rem. Mag., 308 Win., 30-06, 300 Win. Mag. 3-shot magazine in Magnum calibers. Furnished w/o sights. Fancy wood stock w/rosewood forend tip, high-gloss lacquer finish. Imported 1970-77.

C.G. HAENEL
Suhl, Germany

RIFLES

Carl Gustaf Model 2000

Carl Gustaf Deluxe

Carl Gustaf Grand Prix

Carl Gustaf Sporter

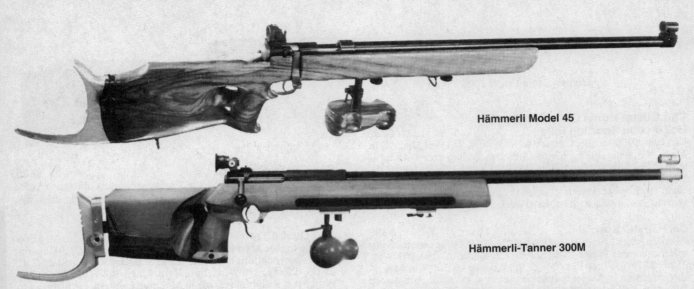

Hämmerli Model 45

Hämmerli-Tanner 300M

Haenel '88 Mauser Sporter **$395**
Same general specifications as Haenel Mauser-Mannlicher, except
w/Mauser 5-shot box magazine.

**Haenel Mauser-Mannlicher Bolt-Action
Sporting Rifle** **$350**
Mauser M/88-type action. Calibers: 7×57, 8×57, 9×57mm. Mannlicher
clip-loading box magazine, 5-shot. 22- or 24-inch half or full octagon
bbl. w/raised matted rib. Double-set trigger. Weight: 7.5 lbs. Sights:
leaf-type open rear; ramp front. Sporting stock w/cheekpiece, checkered pistol grip, raised side-panels, Schnabel tip, swivels.

HÄMMERLI AG JAGD-UND
SPORTWAFFENFABRIK
Lenzburg, Switzerland
Imported by Sigarms, Exetre, NH
(Previously by Hammerli USA; Mandall Shooting Sup-
plies, Inc. & Beeman Precision Arms)

**Hämmerli Model 45 Smallbore Bolt-Action
Single-Shot Match Rifle** **$525**
Calibers: 22 LR, 22 Extra Long. 27.5-inch heavy bbl. Weight:
15.5 lbs. Sights: micrometer peep rear; globe front. Free-rifle
stock w/cheekpiece, full pistol grip, thumbhole, beavertail forearm, palmrest, Swiss-type buttplate, swivels. Made 1945-57.

Hämmerli Model 54 Smallbore Match Rifle **$495**
Bolt-action, single-shot. Caliber: 22 LR. 27.5-inch heavy bbl. Weight:
15 lbs. Sights: micrometer peep rear; globe front. Free-rifle stock
w/cheekpiece, thumbhole, adj. hook buttplate, palmrest, swivel. Made
1954-57.

Hämmerli Model 503 Free Rifle **$485**
Bolt-action, single-shot. Caliber: 22 LR. 27.5-inch heavy bbl. Weight: 15
lbs. Sights: micrometer peep rear; globe front. Free-rifle stock w/cheek-
piece, thumbhole, adj. hook buttplate, palmrest, swivel. Made 1957-62.

Hämmerli Model 506 Smallbore Match Rifle **$525**
Bolt-action, single-shot. Caliber: 22 LR. 26.75-inch heavy bbl. Weight:
16.5 lbs. Sights: micrometer peep rear; globe front. Free-rifle stock
w/cheekpiece, thumbhole adj. hook buttplate, palmrest, swivel. Made
1963-66.

**Hämmerli Model Olympia 300 Meter Bolt-Action
Single-Shot Free Rifle** **$695**
Calibers: 30-06, 300 H&H Magnum for U.S.A.; ordinarily produced in
7.5mm, other calibers available on special order. 29.5-inch heavy bbl.
Double-pull or double-set trigger. Sights: micrometer peep rear, globe
front. Free-rifle stock w/cheekpiece, full pistol grip, thumbhole, beavertail forend, palmrest, Swiss-type buttplate, swivels. Made 1945-59.

Hämmerli-Tanner 300 Meter Free Rifle **$795**
Bolt-action, single-shot. Caliber: 7.5mm standard, available in
most popular centerfire calibers. 29.5-inch heavy bbl. Weight:
16.75 lbs. Sights: micrometer peep rear; globe front. Free-rifle
stock w/cheekpiece, thumbhole, adj. hook buttplate, palmrest,
swivel. Intro. 1962. Disc.

HARRINGTON & RICHARDSON, INC.
Gardner, Massachusetts
(Now H&R 1871, INC., Gardner, Mass.)

Formerly Harrington & Richardson Arms Co. of Worcester Mass.
After a long and distinguished career in gunmaking, this firm sus-
pended operation in 1986. However, in 1992, the firm was bought
by New England Firearms of Gardner, Mass. Some models are
mfd. under that banner as well as H&R 1871, Inc.

**Harrington & Richardson Model 60 Reising
Semiautomatic Rifle** **$350**
Caliber: 45 Automatic. 12- and 20-shot detachable box magazines.
18.25-inch bbl. Weight: 7.5 lbs. Sights: open rear; blade front. Plain
pistol-grip stock. Made 1944-46.

**Harrington & Richardson Model 65 Military Autoloading
Rifle** **$245**
Also called "General." Caliber: 22 LR. 10-shot detachable box
magazine. 23-inch heavy bbl. Weight: 9 lbs. Sights: Redfield 70
rear peep, blade front w/protecting "ears." Plain pistol-grip stock,
"Garand" dimensions. Made 1944-46. *Note:* This model was used
as a training rifle by the U.S. Marine Corps.

**Harrington & Richardson Model 150
Leatherneck Autoloader** **$110**
Caliber: 22 LR only. 5-shot detachable box magazine. 22-inch bbl.
Weight: 7.25 lbs. Sights: open rear; blade front, on ramp. Plain pistol-
grip stock. Made 1949-53.

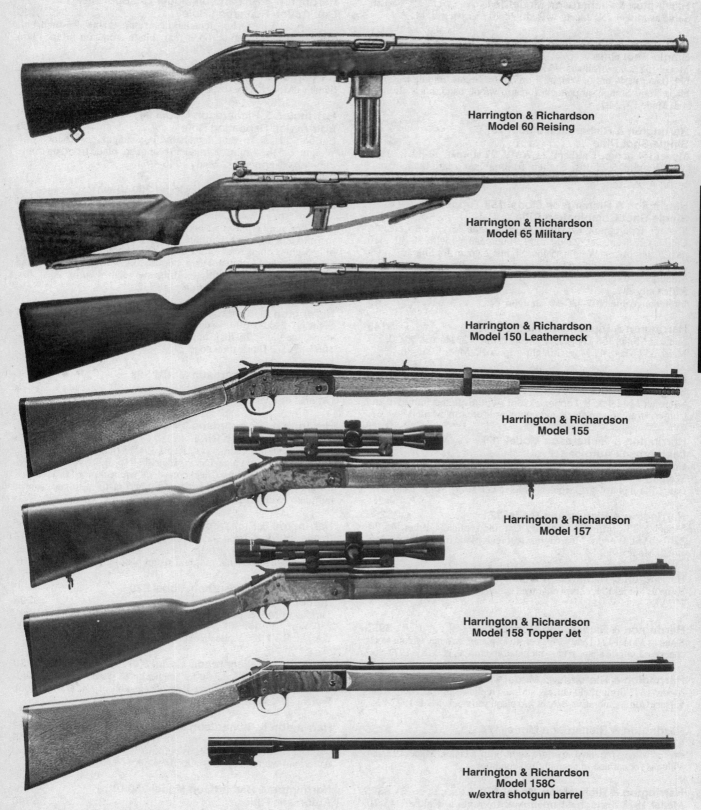

**Harrington & Richardson
Model 60 Reising**

**Harrington & Richardson
Model 65 Military**

**Harrington & Richardson
Model 150 Leatherneck**

**Harrington & Richardson
Model 155**

**Harrington & Richardson
Model 157**

**Harrington & Richardson
Model 158 Topper Jet**

**Harrington & Richardson
Model 158C
w/extra shotgun barrel**

RIFLES

Harrington & Richardson Model 151 . . . . **$125**
Same as Model 150 except w/Redfield 70 rear peep sight.

**Harrington & Richardson Model 155
Single-Shot Rifle** **$110**
Model 158 action. Calibers: 44 Rem. Mag., 45-70 Govt. 24- or 28-inch
bbl. (latter in 44 only). Weight: 7 or 7.5 lbs. Sights: folding leaf rear;
blade front. Straight-grip stock, forearm w/bbl. band, brass cleaning
rod. Made 1972-82.

**Harrington & Richardson Model 157
Single-Shot Rifle** **$100**
Model 158 action. Calibers: 22 WMR, 22 Hornet, 30-30. 22-inch
bbl. Weight: 6.25 lbs. Sights: folding leaf rear; blade front.
Pistol-grip stock, full-length forearm, swivels. Made 1976-86.

**Harrington & Richardson Model 158 Topper Jet
Single-Shot Combination Rifle**
Shotgun-type action w/visible hammer, side lever, auto ejector.
Caliber: 22 Rem. Jet. 22-inch bbl. (interchanges with 30-30, .410
ga., 20 ga. bbls.). Weight: 5 lbs. Sights: Lyman folding adj. open
rear; ramp front. Plain pistol-grip stock and forearm, recoil pad.
Made 1963-67.
Rifle Only **$130**
Interchangeable bbl.—30-30, shotgun 45

Harrington & Richardson Model 158C **$145**
Same as Model 158 Topper Jet, except calibers 22 Hornet, 30-30, 357
Mag., 357 Max., 44 Mag. Straight-grip stock. Made 1963-86.

**Harrington & Richardson Model 163 Mustang
Single-Shot Rifle** **$115**
Same as Model 158 Topper except w/gold-plated hammer and
trigger, straight-grip stock and contoured forearm. Made 1964-67.

**Harrington & Richardson Model 165
Leatherneck Autoloader** **$125**
Caliber: 22 LR. 10-shot detachable box magazine. 23-inch bbl.
Weight: 7.5 lbs. Sights: Redfield 70 rear peep; blade front, on
ramp. Plain pistol-grip stock, swivels, web sling. Made 1945-61.

Harrington & Richardson Model 171 . . . . **$295**
Model 1873 Springfield Cavalry Carbine replica. Caliber: 45-70.
22-inch bbl. Weight: 7 lbs. Sights: leaf rear; blade front. Plain walnut
stock. Made 1972-81.

Harrington & Richardson Model 171 Deluxe **$345**
Same as Model 171, except engraved action and different sights. Made
1972-86.

Harrington & Richardson Model 172 . . . . **$995**
Same as Model 171 Deluxe, except silver-plated, w/fancy walnut stock,
checkered, w/grip adapter; tang-mounted aperture sight. Made 1972-86.

Harrington & Richardson Model 173 . . . . **$325**
Model 1873 Springfield Officer's Model replica, same as 100th Anniversary Commemorative, except w/o plaque on stock. Made 1972-86.

Harrington & Richardson Model 174 . . . . **$350**
Little Big Horn Commemorative Carbine. Same as Model 171 Deluxe,
except w/tang-mounted aperture sight, grip adapter. Made 1972-84.
Value is for carbine in new, unfired condition.

Harrington & Richardson Model 178 . . . . **$325**
Model 1873 Springfield Infantry Rifle replica. Caliber: 45-70.
32-inch bbl. Weight: 8 lbs. 10 oz. Sights: leaf rear; blade front.
Full-length stock w/bbl. bands, swivels, ramrod. Made 1973-86.

**Harrington & Richardson Model 250 Sportster
Bolt-Action Repeating Rifle** **$80**
Caliber: 22 LR. 5-shot detachable box magazine. 23-inch bbl.
Weight: 6.5 lbs. Sights: open rear; blade front, on ramp. Plain
pistol-grip stock. Made 1948-61.

Harrington & Richardson Model 251 . . . . **$85**
Same as Model 250 except w/Lyman 55H rear sight.

**Harrington & Richardson Model 265 "Reg'lar"
Bolt-Action Repeating Rifle** **$80**
Caliber: 22 LR. 10-shot detachable box magazine. 22-inch bbl.
Weight: 6.5 lbs. Sights: Lyman 55 rear peep; blade front, on ramp.
Plain pistol-grip stock. Made 1946-49.

**Harrington & Richardson Model 300 Ultra
Bolt-Action Rifle** **$435**
Mauser-type action. Calibers: 22-250, 243 Win., 270 Win., 30-
06, 308 Win., 7mm Rem. Mag., 300 Win. Mag. 3-round magazine
in 7mm and 300 Mag. calibers, 5-round in others. 22- or 24-inch
bbl. Sights: open rear; ramp front. Checkered stock w/rollover
cheekpiece and full pistol grip, contrasting wood forearm tip and
pistol grip, rubber buttplate, sling swivels. Weight: 7.25 lbs.
Made 1965-82. *See* Illustration page 232.

Harrington & Richardson Model 301 Carbine **$395**
Same as Model 300, except w/18-inch bbl., Mannlicher-style
stock, weighs 7.25 lbs.; not available in caliber 22-250. Made
1967-82. *See* Illustration page 232.

**Harrington & Richardson Model 308
Automatic Rifle** **$325**
Original designation of the Model 360 Ultra. Made 1965-67.

**Harrington & Richardson Model 317 Ultra
Wildcat Bolt-Action Rifle** **$525**
Sako short action. Calibers: 17 Rem., 17/223 (handload), 222 Rem.,
223 Rem. 6-round magazine. 20-inch bbl. No sights, receiver dovetailed for scope mounts. Checkered stock w/cheekpiece and full pistol grip, contrasting wood forearm tip and pistol-grip cap, rubber
buttplate. Weight: 5.25 lbs. Made 1968-76.

**Harrington & Richardson Model 317P
Presentation Grade** **$595**
Same as Model 317, except w/select grade fancy walnut stock
w/basketweave carving on forearm and pistol grip. Made 1968-76.

**Harrington & Richardson Model 330
Hunter's Rifle** **$295**
Similar to Model 300, but w/plainer stock. Calibers: 243 Win.,
270 Win., 30-06, 308 Win., 7mm Rem. Mag., 300 Win. Mag.
Weight: 7.13 lbs. Made 1967-72. *See* Illustration page 232.

Harrington & Richardson Model 333 . . . . **$230**
Plainer version of Model 300 w/uncheckered walnut-finished hard-
wood stock. Calibers: 7mm Rem. Mag. and 30-06. 22-inch bbl.
Weight: 7.25 lbs. No sights. Made in 1974.

Harrington & Richardson Model 340 . . . . **$295**
Mauser-type action. Calibers: 243 Win., 308 Win., 270 Win., 30-
06, 7×57mm. 22-inch bbl. Weight: 7.25 lbs. Hand-checkered,
American walnut stock. Made 1982-84.

**Harrington & Richardson Model 360 Ultra
Automatic Rifle** **$345**
Gas-operated semiautomatic. Calibers: 243 Win., 308 Win. 3-round de-
tachable box magazine. 22-inch bbl. Sights: open rear; ramp front.

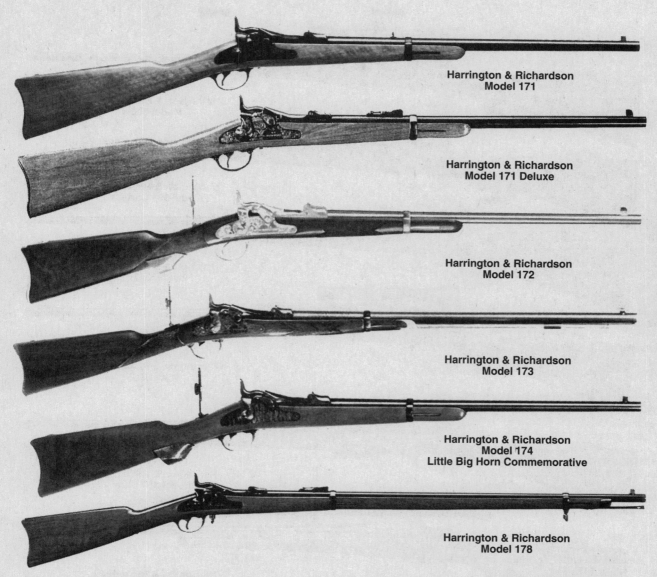

Harrington & Richardson
Model 171

Harrington & Richardson
Model 171 Deluxe

Harrington & Richardson
Model 172

Harrington & Richardson
Model 173

Harrington & Richardson
Model 174
Little Big Horn Commemorative

Harrington & Richardson
Model 178

RIFLES

Harrington & Richardson Model 360 Ultra *(Con't)*
Checkered stock w/rollover cheekpiece, full pistol grip, contrasting wood forearm tip and pistol-grip cap, rubber buttplate, sling swivels. Weight: 7.25 lbs. Made 1967-78.

Harrington & Richardson Model 361 . . . . $375
Same as Model 360, except w/full rollover cheekpiece for right- or left-hand shooters. Made 1970-73.

Harrington & Richardson Model 365 Ace
Bolt-Action Single-Shot Rifle $100
Caliber: 22 LR.22-inch bbl. Weight: 6.5 lbs. Sights: Lyman 55 rear peep, blade front, on ramp. Plain pistol-grip stock. Made 1946-47.

Harrington & Richardson Model 370 Ultra
Medalist $395
Varmint and target rifle based on Model 300. Calibers: 22-250, 243 Win., 6mm Rem. 5-round magazine. 24-inch varmint weight bbl. No sights. Target-style stock w/semibeavertail forearm. Weight: 9.5 lbs. Made 1968-73.

Harrington & Richardson Model 422
Slide-Action Repeater $125
Caliber: 22 LR, Long, Short. Tubular magazine holds 21 Short, 17 Long, 15 LR. 24-inch bbl. Weight: 6 lbs. Sights: open rear; ramp front. Plain pistol-grip stock grooved slide handle. Made 1956-58.

Harrington & Richardson Model 450 . . . . $130
Same as Model 451, except w/o front and rear sights.

Harrington & Richardson Model 451 Medalist
Bolt-Action Target Rifle $155
Caliber: 22 LR. 5-shot detachable box magazine. 26-inch bbl. Weight: 10.5 lbs. Sights: Lyman 524F extension rear; Lyman 77 front, scope bases. Target stock w/full pistol grip and forearm, swivels and sling. Made 1948-61.

Harrington & Richardson Model 465 Targeteer Special
Bolt-Action Repeater $145
Caliber: 22 LR.10-shot detachable box magazine. 25-inch bbl. Weight: 9 lbs. Sights: Lyman 57 rear peep; blade front, on ramp. Plain pistol-grip stock, swivels, web sling strap. Made 1946-47.

Harrington & Richardson
Model 300

Harrington & Richardson
Model 301 Carbine

Harrington & Richardson
Model 317P

Harrington & Richardson
Model 330

Harrington & Richardson
Model 360 Ultra

Harrington & Richardson
Model 370 Ultra Medalist

**Harrington & Richardson Model 700
Autoloader** **$225**
Caliber: 22 WMR. 5-shot magazine. 22-inch bbl. Weight: 6.5 lbs.
Sights: folding leaf rear; blade front, on ramp. Monte Carlo-style
stock of American walnut. Made 1977-86.

Harrington & Richardson Model 700 Deluxe **$345**
Same as Model 700 Standard except w/select custom polished
and blued finish, select walnut stock, hand ceckering, and no iron
sights. Fitted w/H&R Model 432 4× scope. Made 1980-86.

**Harrington & Richardson Model 750 Pioneer
Bolt-Action Single-Shot Rifle** **$90**
Caliber: 22 LR, Long, Short. 22- or 24-inch bbl. Weight: 5 lbs.
Sights: open rear; bead front. Plain pistol-grip stock. Made 1954-81;
redesigned 1982; discont. 1985.

**Harrington & Richardson Model 751
Single-Shot Rifle** **$75**
Same as Model 750, except w/Mannlicher-style stock. Made
1971.

**Harrington & Richardson Model 755 Sahara
Single-Shot Rifle** **$75**
Blow-back action, automatic ejection. Caliber: 22 LR, Long, Short.
18-inch bbl. Weight: 4 lbs. Sights: open rear; military-type front.
Mannlicher-style stock. Made 1963-71.

Harrington & Richardson Model 760 Single-Shot **$85**
Same as Model 755, except w/conventional sporter stock. Made
1965-70.

**Harrington & Richardson Model 765 Pioneer
Bolt-Action Single-Shot Rifle** **$60**
Caliber: 22 LR, Long, Short. 24-inch bbl. Weight: 5 lbs. Sights: open
rear; hooded bead front. Plain pistol-grip stock. Made 1948-54.

Harrington & Richardson
Model 700 Deluxe

Harrington & Richardson
Model 750 Pioneer

Harrington & Richardson
"New" Model 750

Harrington & Richardson
Model 755

Harrington & Richardson
Model 760

Harrington & Richardson
Model 866

Harrington & Richardson
Model 1873 — 100th Anniversary

Harrington & Richardson
Model 5200 Sporter

Harrington & Richardson
Ultra Varmint

RIFLES

Harrington & Richardson Model 800 Lynx Autoloading Rifle **$110**
Caliber: 22 LR. 5- or 10-shot clip magazine. 22-inch bbl. Open sights. Weight: 6 lbs. Plain pistol-grip stock. Made 1958-60.

Harrington & Richardson Model 852 Fieldsman Bolt-Action Repeater **$90**
Caliber: 22 LR, Long, Short. Tubular magazine holds 21 Short, 17 Long, 15 LR. 24-inch bbl. Weight: 5.5 lbs. Sights: open rear; bead front. Plain pistol-grip stock. Made 1952-53.

Harrington & Richardson Model 865 Plainsman Bolt-Action Repeater **$85**
Caliber 22 LR, Long, Short. 5-shot detachable box magazine. 22- or 24-inch bbl. Weight: 5.25 lbs. Sights: open rear, bead front. Plain pistol-grip stock. Made 1949-86.

Harrington & Richardson Model 866 Bolt-Action Repeater **$85**
Same as Model 865, except w/Mannlicher-style stock. Made 1971.

Harrington & Richardson Model 1873 100th Anniversary (1871-1971) Commemorative Officer's Springfield Replica **$595**
Model 1873 "trap door" single-shot action. Engraved breech block, receiver, hammer, lock, band and buttplate. Caliber: 45-70. 26-inch bbl. Sights: peep rear; blade front. Checkered walnut stock w/anniversary plaque. Ramrod. Weight: 8 lbs.10,000 made 1971. Value is for rifle in new, unfired condition.

Harrington & Richardson Model 5200 Sporter **$495**
Turn-bolt repeater. Caliber: 22 LR. 24-inch bbl. Classic-style American walnut stock. Adj. trigger. Sights: peep receiver; hooded ramp front. Weight: 6.5 lbs. Disc. 1983.

Harrington & Richardson Model 5200 Match Rifle **$525**
Same action as 5200 Sporter. Caliber: 22 LR. 28-inch target weight bbl. Target stock of American walnut. Weight: 11 lbs. Made 1982-86.

Harrington & Richardson Custer Memorial Issue
Limited Edition Model 1873 Springfield Carbine replica, richly engraved and inlaid w/gold, fancy walnut stock, in mahogany display case. Made 1973. Value is for carbine in new, unfired condition.
Officers' Model, limited to 25 pieces **$4500**
Enlisted Men's Model, limited to 243 pieces **2500**

Harrington & Richardson Targeteer Jr. Bolt-Action Rifle **$110**
Caliber: 22 LR 5-shot detachable box magazine. 20-inch bbl. Weight: 7 lbs. Sights: Redfield 70 rear peep; Lyman 17A front. Target stock, junior size w/pistol grip, swivels and sling. Made 1948-51.

Harrington & Richardson Ultra Single-Shot Rifle
Side-lever single-shot. Calibers: 22-250 Rem., 223 Rem., 25-06 Rem., 308 Win. 22- to 26-inch bbl. Weight: 7-8 lbs. Curly maple or laminated stock. Barrel-mounted scope mount, no sights. Made 1993 to date.
Ultra Hunter (25-06, 308) **$150**
Ultra Varmint **190**

HARRIS GUNWORKS
Phoenix, Arizona
(Formerly McMillan Gun Works)

Harris Signature Alaskan Bolt-Action Rifle **$2550**
Same general specifications as Classic Sporter, except w/match-grade bbl. Rings and mounts. Sights: single-leaf rear, bbl. band front. Checkered Monte Carlo stock w/palmswell and solid recoil pad. Electroless nickel finish. Calibers: LA (long): 270 Win., 280 Rem., 30-06, MA (Magnum): 7mm Rem. Mag., 300 Win. Mag., 300 Wby. Mag., 340 Wby. Mag., 358 Win., 375 H&H Mag. Made 1990 to date.

Harris Signature Classic Sporter
The "prototype" for Harris's Signature Series, this bolt action is available in three lengths: SA (standard/ short) — from 22-250 to 350 Rem Mag.; LA (long) — 25-06 to 30-06; MA (Magnum) — 7mm STW to 416 Rem. Mag. Four-shot or 3-shot (Magnum) magazine. Bbl. lengths: 22, 24 or 26 inches. Weight: 7 lbs. (short action). No sights; rings and bases provided. Harris fiberglass stock, Fibergrain or wood stock optional. Stainless, matte black or black chrome sulfide finish. Available in right- and left-hand models. Made 1987 to date. Has pre-64 Model 70-style action for dangerous game.
Classic Sporter Standard......................... **$1750**
Classic Sporter Stainless........................ **1795**
Talon Sporter **1850**

Harris Signature Mountain Rifle............... **$2395**
Same general specifications as Harris (McMillan) Classic Sporter, except w/titanium action and graphite-reinforced fiberglass stock. Weight: 5.5 lbs. Calibers: 270 Win., 280 Rem., 30-06, 7mm Mag., 300 Win. Mag. Other calibers on special order. Made 1995 to date.

Harris Signature Super Varminter **$1750**
Same general specifications as Harris (McMillan) Classic Sporter, except w/heavy, contoured bbl., adj. trigger, fiberglass stock and field bipod. Calibers: 223, 22-250, 220 Swift, 244 Win., 6mm Rem., 25-06, 7mm-08, 308 Win., 350 Win. Mag. Made 1995 to date.

Harris Talon Safari Rifle
Same general specifications as Harris (McMillan) Classic Sporter, except w/Harris Safari-grade action, match-grade bbl. and "Safari" fiberglass stock. Calibers: Magnum — 300 H&H Mag., 300 Win Mag., 300 Wby. Mag., 338 Win. Mag., 340 Wby. Mag., 375 H&H Mag., 404 Jeffrey, 416 Rem. Mag., 458 Win., Super Mag. — 300 Phoenix, 338 Lapua, 378 Wby.Mag., 416 Rigby, 416 Wby. Mag., 460 Wby. Mag. Matte black finish. Other calibers available on special order, and at a premium, but the "used gun" value remains the same. Imported 1989 to date.
Safari Magnum **$2695**
Safari Super Magnum **2995**

HECKLER & KOCH, GMBH
Oberndorf/Neckar, Germany
Imported by Heckler & Koch, Inc., Sterling, VA

Heckler & Koch Model 911 Semiauto Rifle **$1395**
Caliber: 308 (7.62mm). 5-shot magazine. 19.7-inch bull bbl. 42.4 inches overall. Sights: hooded post front; adj. aperture rear. Weight: 11 lbs. Kevlar reinforced fiberglass thumbhole-stock. Imported 1989-93.

Heckler & Koch Model HK91 A-2 Semiauto **$1350**
Delayed roller-locked blow-back action. Caliber: 7.62mm × 51 NATO (308 Win.) 5- or 20-round box magazine. 19-inch bbl. Weight: w/o magazine, 9.37 lbs. SightsL "V" and aperture rear, post front. Plastic buttstock and forearm. Disc. 1991.

Heckler & Koch Model HK91 A-3 .... **$1695**
Same as Model HK91 A-2, except w/retractable metal buttstock, weighs 10.56 lbs. Disc. 1991.

**Heckler & Koch
Model HK91 A-2**

**Heckler & Koch
Model HK91 A-3**

**Heckler & Koch
Model HK93 A-2**

**Heckler & Koch
Model HK94 Carbine**

**Heckler & Koch
Model HK PSG-1**

RIFLES

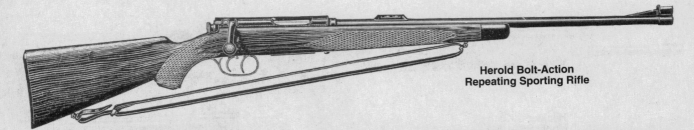

**Herold Bolt-Action
Repeating Sporting Rifle**

Heckler & Koch Model HK93 Semiautomatic

Delayed roller-locked blow-back action. Caliber: 5.56mm × 45 (223 Rem.). 5- or 20-round magazine. 16.13-inch bbl. Weight: w/o magazine, 7.6 lbs. Sights: "V" and aperture rear; post front. Plastic buttstock and forearm. Disc. 1991.

HK93 A-2 . . . . **$1295**
HK93 A-3 w/retractable stock **1550**

Heckler & Koch Model HK94 Semiautomatic Carbine

Caliber: 9mm Para. 15-shot magazine. 16-inch bbl. Weight: 6.75 lbs. Aperture rear sight, front post. Plastic buttstock and forend or retractable metal stock. Imported 1983-91.

HK94-A2 w/Standard Stock **$1995**
HK94-A3 w/Retractable Stock **2395**

Heckler & Koch Model HK300 Semiautomatic **$575**

Caliber: 22 WMR. 5- or 15-round box magazine.19.7-inch bbl. w/polygonal rifling. Weight: 5.75 lbs. Sights: V-notch rear; ramp front. High-luster polishing and bluing. European walnut stock w/cheekpiece, checkered forearm and pistol grip. Disc. 1989.

Heckler & Koch Model HK630 Semiautomatic **$895**

Caliber: 223 Rem. 4- or 10-round magazine. 24-inch bbl. Overall length: 42 inches. Weight: 7 lbs. Sights: open rear; ramp front. European walnut stock w/Monte Carlo cheekpiece. Imported 1983-90.

Heckler & Koch Model HK770 Semiautomatic **$1125**

Caliber: 308 Win. 3- or 10-round magazine. Overall length: 44.5 inches. Weight: 8 lbs. Sights: open rear; ramp front. European walnut stock w/Monte Carlo cheekpiece. Imported 1983-90.

Heckler & Koch Model HK940 Semiautomatic **$1595**

Caliber: 30-06 Springfield. 3- or 10-round magazine. Overall length: 47 inches. Weight: 8.8 lbs. Sights: open rear; ramp front. European walnut stock w/Monte Carlo cheekpiece. Imported 1983-90.

Heckler & Koch Model HK PSG-1
Marksman's Rifle **$8250**

Caliber: 308 (7.62mm). 5- and 20-shot magazine. 25.6-inch bbl. 47.5 inches overall. Hensoldt 6×42 telescopic sight. Weight: 17.8 lbs. Matte black composition stock w/pistol grip. Imported 1988 to date.

Heckler & Koch Model SR-9 Semiauto Rifle **$1350**

Caliber: 308 (7.62mm). 5-shot magazine. 19.7-inch bull bbl. 42.4 inches overall. Hooded post front sight; adj. aperture rear. Weight: 11 lbs. Kevlar reinforced fiberglass thumbhole-stock w/wood grain finish. Imported 1989-93.

Heckler & Koch Model SR-9 Target Rifle **$1950**

Same general specifications as standard SR-9, except w/ PSG-1 trigger group and adj. buttstock. Imported 1992-94.

HERCULES RIFLES

See listings under "W" for Montgomery Ward.

HEROLD RIFLE
Suhl,Germany
Made by Franz Jaeger & Company

Herold Bolt-Action Repeating Sporting Rifle **$950**

"Herold-Repetierbüchse." Miniature Mauser-type action w/unique 5-shot box magazine on hinged floorplate. Double-set triggers. Caliber: 22 Hornet. 24-inch bbl. Sights: leaf rear; ramp front. Weight: 7.75 lbs. Fancy checkered stock. Made before WWII. *Note:* These rifles were imported by Charles Daly and A.F. Stoeger Inc. of New York City and sold under their own names.

HEYM RIFLES AMERICA, INC.
Mfd. By Hyem, GmbH & Co JAGWAFFEN KD.
Gleichamberg, Germany
(Previously imported by Heym America, Inc.; Heckler & Koch; JagerSport, Ltd.)

Heym Model 55B O/U Double Rifle

Kersten boxlock action w/double cross bolt and cocking indicators. Calibers: 308 Win., 30-06, 375 H&H, 458 Win. Mag., 470 N.E. 25-inch bbl. 42 inches overall. Weight: 8.25 lbs. Sights: fixed V-type rear; front ramp w/silver bead. Engraved receiver w/optional sidelocks, interchangeable bbls. and claw mounts. Checkered European walnut stock. Imported from Germany.

Model 55 (boxlock) **$7850**
Model 55 (sidelock) **9550**
W\Extra Rifle Bbls., **add** **5500**
W/Extra Shotgun Bbls., **add**. **2800**

Heym Model 88B Double Rifle

Modified Anson & Deeley boxlock action w/standing sears, double underlocking lugs and Greener extension w/crossbolt. Calibers: 8×57 JRS, 9.3×74R, 30-06, 375 H&H, 458 Win. Mag., 470 Nitro Express, 500 Nitro Express. Other calibers available on special order. Weight: 8 to 10 lbs. Top tang safety and cocking indicators. Double triggers w/front set. Fixed or 3-leaf express rear sight, front ramp w/silver bead. Engraved receiver w/optional sidelocks. Checkered French walnut stock. Imported from Germany.

Model 88B Boxlock . . . . **from $ 8,775**
Model 88B/SS Sidelock . . . . **from 11,500**
Model 88B Safari (Magnum) **11,950**

Heym Express Bolt-Action Rifle

Same general specifications as Model SR-20 Safari, except w/modified Magnum Mauser action. Checkered AAA-grade European walnut stock w/cheekpiece, solid rubber recoil pad, rosewood forend tip and grip cap. Calibers: 338 Lapua Magnum, 375 H&H, 378 Wby. Mag., 416 Rigby 450 Ackley, 460 Wby. Mag., 500 A-Square, 500 Nitro Express, 600 Nitro Express. Other calibers available on special order, but no change in "used gun value." Imported from Germany 1989-95.

Standard Express Magnum **$4150**
600 Nitro Express **7650**
Left-Hand Models, **add** **600**

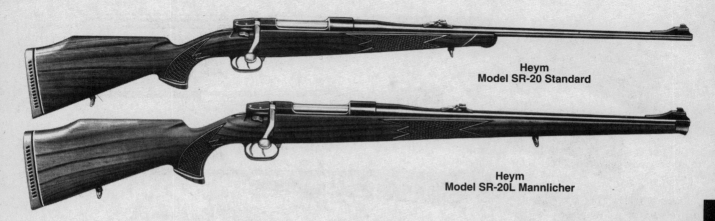

**Heym
Model SR-20 Standard**

**Heym
Model SR-20L Mannlicher**

Heym SR-20 Bolt-Action Rifle

Calibers: 243 Win., 270 Win., 308 Win., 30-06, 7mm Rem. Mag., 300 Win. Mag., 375 H&H. 5-shot (standard) or 3-shot (Magnum) magazine. Bbl. length: 20.5-inch (SR-20L); 24-inch (SR-20N); 26-inch (SR-20G). Weight: 7.75 lbs. Adj. rear sight, blade front. Checkered French walnut stock in Monte Carlo style (N&G Series) or full Mannlicher (L Series). Imported from Germany. Disc. 1992.

SR-20L	$1095
SR-20N	1295
SR-20G	1495

Heym SR-20 Classic Bolt-Action Rifles

Same as SR-20, except w/22-250 and 338 Win. Mag., plus metric calibers on request. 24-inch (standard) or 25-inch (Magnum) bbl. Checkered French walnut stock. Left-hand models. Imported from Germany since 1985; Sporter version 1989-93.

Classic (Standard)	$1395
Classic (Magnum)	1550
Left-Hand Models, **add**	300
Classic Sporter (Std. w/22-inch bbl.)	1495
Classic Sporter (Mag. w/24-inch bbl.)	1650

Heym SR-20 Alpine, Safari and Trophy Series

Same general specifications as Model SR-20 Classic Sporter, except **Alpine Series** w/20-inch bbl., Mannlicher stock, chambered in standard calibers only; **Safari Series** w/24-inch bbl., 3-leaf express sights and magnum action in calibers 375 H & H, 404 Jeffrey, 425 Express, 458 Win. Mag.; **Trophy Series** w/Krupp-Special tapered octagon bbl. w/quarter rib and open sights, standard and Magnum calibers. Imported from Germany 1989-93.

Alpine Series	$1495
Safari Series	1525
Trophy Series (Standard Calibers)	1950
Trophy Series (Magnum Calibers)	2125

J.C. HIGGINS RIFLES

See Sears, Roebuck & Company.

HI-POINT FIREARMS
Dayton, Ohio

Hi-Point Model 995 Carbine

Semiautomatic recoil-operated carbine. Calibers: 9mm Parabellum or 40 S&W. 10-shot magazine. 16.5-inch bbl. 31.5 inches overall. Protected post front sight, aperture rear w/integral scope mount. Matte blue, chrome or pakerized finish. Checkered polymer grip/frame. Made 1996 to date.

Model 995, 9mm (blue or parkerized)	$125
Model 995, 40 S&W (blue or parkerized)	150
W/Lazer Sights, **add**	35
W/Chrome Finish, **add**	15

HIGH STANDARD SPORTING FIREARMS
East Hartford, Connecticut
(Formerly High Standard Mfg. Co., Hamden, CT)

A long-standing producer of sporting arms, High Standard discontinued its operations in 1984.

High Standard Flite-King Pump Rifle **$115**
Hammerless slide-action. Caliber: 22 LR, 22 Long, 22 Short. Tubular mag. holds 17 LR, 19 Long, or 24 Short. 24-inch bbl. Weight: 5.5 lbs. Sights: Patridge rear; bead front. Monte Carlo stock w/pistol grip, serrated semibeavertail forearm. Made 1962-75.

High Standard Hi-Power Deluxe Rifle **$295**
Mauser-type bolt action, sliding safety. Calibers: 270, 30-06. 4-shot magazine. 22-inch bbl. Weight: 7 lbs. Sights: folding open rear; ramp front. Walnut stock w/checkered pistol grip and forearm, Monte Carlo comb, QD swivels. Made 1962-66.

High Standard Hi-Power Field Bolt-Action Rifle **$225**
Same as Hi-Power Deluxe, except w/plain field style stock. Made 1962-66.

High Standard Sport-King Autoloading Carbine **$135**
Same as Sport-King Field Autoloader, except w/18.25-inch bbl., Western-style straight-grip stock w/bbl. band, sling and swivels. Made 1964-73.

High Standard Sport-King Deluxe Autoloader **$155**
Same as Sport-King Special Autoloader, except w/checkered stock. Made 1966-75.

High Standard Sport-King Field Autoloader **$110**
Calibers: 22 LR, 22 Long, 22 Short (high speed). Tubular magazine holds 15 LR, 17 Long, or 21 Short. 22.25-inch bbl. Weight: 5.5 lbs. Sights: open rear; beaded post front. Plain pistol-grip stock. Made 1960-66.

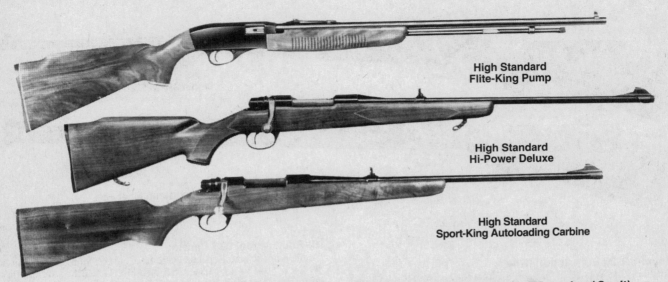

High Standard
Flite-King Pump

High Standard
Hi-Power Deluxe

High Standard
Sport-King Autoloading Carbine

High Standard Sport-King Special Autoloader **$125**
Same as Sport-King Field, except stock w/Monte Carlo comb and
semibeavertail forearm. Made 1960-66.

HOLLAND & HOLLAND, LTD.
London, England
Imported by Holland & Holland, NY, NY

**Holland & Holland No. 2 Model Hammerless
Ejector Double Rifle** **$11,950**
Same general specifications as Royal Model, except plainer finish.
Disc. 1960.

**Holland & Holland Best Quality Magazine
Rifle** **$11,250**
Mauser or Enfield action. Calibers: 240 Apex, 300 H&H Mag., 375
H&H Magnum. 4-shot box magazine. 24-inch bbl. Weight: 7.25 lbs.,
240 Apex; 8.25 lbs., 300 Mag. and 375 Mag. Sights: folding leaf rear;
hooded ramp front. Detachable French walnut stock w/cheekpiece,

Holland & Holland Best Quality Magazine (*Con't*)
checkered pistol grip and forearm, swivels. Currently mfd. Specifications given apply to most models.

Holland & Holland Deluxe Magazine Rifle **$11,950**
Same specifications as Best Quality, except w/exhibition grade
stock and special engraving. Currently mfd.

Holland & Holland Royal Deluxe Double Rifle **$45,500**
Formerly designated "Modele Deluxe." Same specifications as Royal
Model, except w/exhibition grade stock and special engraving. Cur-
rently mfd.

**Holland & Holland Royal Hammerless
Ejector Rifle** **$32,500**
Sidelock. Calibers: 240 Apex, 7mm H&H Mag., 300 H&H Mag.,
300 Win. Mag., 30-06, 375 H&H Mag., 458 Win. Mag., 465 H&H
Mag. 24- to 28-inch bbls. Weight: from 7.5 lbs. Sights: folding leaf
rear, ramp front. Cheekpiece stock of select French walnut, checkered pistol grip and forearm. Currently mfd. Same general specifications apply to prewar model.

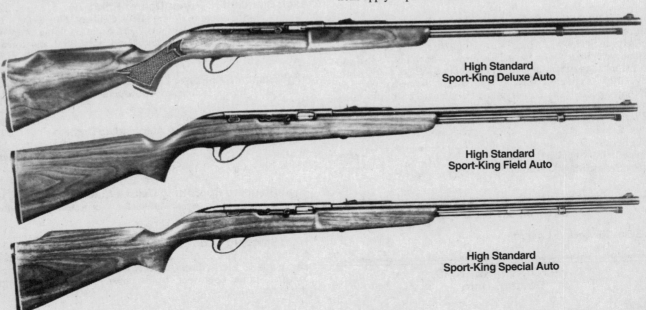

High Standard
Sport-King Deluxe Auto

High Standard
Sport-King Field Auto

High Standard
Sport-King Special Auto

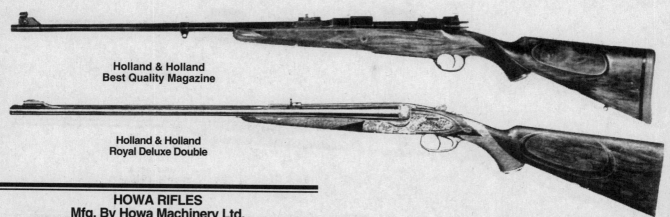

**Holland & Holland
Best Quality Magazine**

**Holland & Holland
Royal Deluxe Double**

HOWA RIFLES
Mfg. By Howa Machinery Ltd.
Shinkawa-Chonear, Nagoya 452, Japan
Imported by Interarms, Alexandria, VA

See also Mossberg (1500) Smith & Wesson (1500 & 1700) and Weatherby (Vanguard).

Howa Model 1500 Hunter
Similar to Trophy Model, except w/standard walnut stock. No Monte Carlo cheekpiece or grip cap. Imported 1988-89.
Standard Calibers $300
Magnum Calibers 325

Howa Model 1500 Lightning Bolt-Action Rifle
Similar to Hunter Model; except fitted w/black Bell & Carlson Carbelite stock w/checkered grip and forend. Weight: 7.5 lbs. Imported 1988-89.
Standard Calibers $250
Magnum Calibers 275

Howa Model 1500 Realtree Camo Rifle $375
Similar to Trophy Model, except fitted w/Camo Bell & Carlson Carbelite stock w/checkered grip and forend. Weight: 8 lbs. Stock, action and barrel finished in Realtree camo. Available in standard calibers only. Imported 1993 to date.

Howa Model 1500 Trophy/Varmint Bolt-Action Rifle
Calibers: 22-250, 223, 243 Win., 270 Win., 308 Win., 30-06, 7mm Mag., 300 Win. Mag., 338 Win. Mag. 22-inch bbl. (standard); 24-inch bbl. (Magnum). 42.5 inches overall (standard). Weight: 7.5 lbs. Adj. rear sight hooded ramp front. Checkered walnut stock w/Monte

Howa Model 1500 Trophy/Varmint *(Con't)*
Carlo cheekpiece. Varmint Model w/24-inch heavy bbl., weight of 9.5 lbs. in calibers 22-250, 223 and 308 only. Imported 1979-93.
Trophy Standard $350
Trophy Magnum 375
Varmint (Parkerized Finish) 325

Howa Model 1500 Woodgrain Lightning Rifle
Calibers: 243, 270, 7mm Rem. Mag., 30-06. Mag. 5-shot magazine 22-inch. 42-inches overall. Weight: 7.5 lbs. Receiver drilled and tapped for scope mount, no sights. Checkered woodgrain synthetic polymer stock. Imported 1993-94.
Standard Calibers $375
Magnum Calibers 395

H-S PRECISION
Rapid City, South Dakota

H-S Precision Pro-Series
Custom rifle built on Remington 700 bolt action. Calibers: 22 to 416 Rem 24- or 26-inch bbl. w/fluted option Aluminum bedding block system w/take-down option. Kevlar/carbon fiber stock to customer's specifications. Appointments and options to customer's preference. Made 1990 to date.
Sporter Model. $1295
Pro-Hunter Model (PHR) 1325
Long Range Model 1275
Long Range Takedown Model 1750
Marksman Model 1295

**Howa
Model 1500 Hunter**

**Howa
Model 1500 Lightning**

Husqvarna Series
1100 Deluxe

Husqvarna
1951 Hi-Power

Husqvarna
3000 Crown Grade

Husqvarna
4100 Lightweight

Husqvarna
6000 Imperial Custom

H-S Precision Pro-Series

Marksman Takedown Model 1725
Varmint Takedown Model (VTD) 1250
Left-Hand Model, **add** 200

HUNGARIAN MILITARY RIFLES
Budapest, Hungary
Manufactured at Government Arsenal

**Hungarian Model 1935M Mannlicher
Military Rifle** **$200**
Caliber: 8×52mm Hungarian. Bolt action, straight handle. 5-shot projecting box magazine. 24-inch bbl. Weight: 9 lbs. Adj. leaf rear sight, hooded front blade. Two-piece military-type stock. Made 1935-40.

**Hungarian Model 1943M (German Gew. 98/40)
Mannlicher Military Rifle** **$250**
Modification, during German occupation, of Model 1935M. Caliber: 7.9mm Mauser. Turned-down bolt handle and Mauser M/98-type box magazine; other differences are minor. Made 1940 to end of war in Europe.

HUSQVARNA VAPENFABRIK A.B.
Huskvarna, Sweden

**Husqvarna Model 456 Lightweight
Full-Stock Sporter** **$425**
Same as Series 4000/4100 except w/sporting-style full stock w/slope-away cheekrest. Weight: 6.5 lbs. Made 1959-70.

Husqvarna Series 1000 Super Grade **$395**
Same as 1951 Hi-Power, except w/European walnut sporter stock w/Monte Carlo comb and cheekpiece. Made 1952-56.

**Husqvarna Series 1100 Deluxe Model Hi-Power
Bolt-Action Sporting Rifle** **$400**
Same as 1951 Hi-Power, except w/"jeweled" bolt, European walnut stock. Made 1952-56.

Husqvarna 1950 Hi-Power Sporting Rifle **$350**
Mauser-type bolt action. Calibers: 220 Swift, 270 Win. 30-06 (*see* note below), 5-shot box magazine. 23.75-inch bbl. Weight: 7.75 lbs. Sights: open rear; hooded ramp front. Sporting stock of Arctic beech, checkered pistol grip and forearm, swivels. *Note:* Husqvarna sporters were first intro. in U.S. about 1948; earlier models were also available in calibers 6.5×55, 8×57 and 9.3×57. Made 1946-1951.

**Interarms
Mark X Lightweight Sporter**

**Interarms
Mini-Mark X**

Husqvarna 1951 Hi-Power Rifle **$375**
Same as 1950 Hi-Power, except w/high-comb stock, low safety.
See Illustration previous page.

Husqvarna Series 3000 Crown Grade **$425**
Same as Series 3100, except w/Monte Carlo comb stock. *See* Illustration previous page.

Husqvarna Series 3100 Crown Grade **$435**
HVA improved Mauser action. Calibers: 243, 270, 7mm, 30-06, 308
Win. 5-shot box magazine. 23.75-inch bbl. Weight: 7.75 lbs. Sights:
open rear; hooded ramp front. European walnut stock, checkered,
cheekpiece, pistol-grip cap, black foretip, swivels. Made 1954-72.

Husqvarna Series 4000 Lightweight Rifle **$450**
Same as Series 4100, except w/Monte Carlo comb stock and no rear
sight.

Husqvarna Series 4100 Lightweight Rifle **$425**
HVA improved Mauser action. Calibers: 243, 270, 7mm, 30-06, 308
Win. 5-shot box magazine. 20.5-inch bbl. Weight: 6.25 lbs.
Sights: open rear; hooded ramp front. Lightweight walnut stock
w/cheekpiece, pistol grip, Schnabel foretip, checkered, swivels.
Made 1954-72. *See* Illustration previous page.

**Husqvarna Series 6000 Imperial Custom
Grade** **$525**
Same as Series 3100, except fancy-grade stock, 3-leaf folding rear
sight, adj. trigger. Calibers: 243, 270, 7mm Rem. Mag. 308, 30-06.
Made 1968-70. *See* Illustration previous page.

**Husqvarna Series 7000 Imperial Monte Carlo
Lightweight** **$530**
Same as Series 4000 Lightweight, except fancy-grade stock, 3-leaf
folding rear sight, adj. trigger. Calibers: 243, 270, 308, 30-06. Made
1968-70.

Husqvarna Model 8000 Imperial Grade Rifle **$550**
Same as Model 9000, except w/jeweled bolt, engraved floorplate, deluxe French walnut checkered stock, no sights. Made
1971-72.

Husqvarna Model 9000 Crown Grade Rifle **$425**
New design Husqvarna bolt action. Adj. trigger. Calibers: 270, 7mm
Rem. Mag., 30-06, 300 Win. Mag. 5-shot box magazine, hinged
floorplate. 23.75-inch bbl. Sights: folding leaf rear; hooded ramp
front. Checkered walnut stock w/Monte Carlo cheekpiece, rosewood forearm tip and pistol-grip cap. Weight: 7 lbs. 3 oz. Made
1971-72.

Husqvarna Series P-3000 Presentation Rifle **$695**
Same as Crown Grade Series 3000, except w/selected stock, engraved action, adj. trigger. Calibers: 243, 270, 7mm Rem. Mag.,
30-06. Made 1968-70.

INTERARMS RIFLES
Alexandria, Virginia

*The following Mark X rifles are manufactured by Zavodi Crvena
Zastava, Belgrade, Yugoslavia.*

Interarms Mark X Alaskan **$395**
Same specifications as Mark X Sporter, except chambered 375 H&H
Mag. and 458 Win. Mag. w/3-round magazine. Stock w/recoil-
absorbing cross bolt and heavy duty recoil pad. Weighs 8.25 lbs.
Made 1976-84.

Interarms Mark X Bolt-Action Sporter Series
Mauser-type action. Calibers: 22-250, 243, 25-06, 270, 7×57, 7mm
Rem. Mag., 308, 30-06, 300 Win. Mag. 5-shot magazine (3-shot in
Magnum calibers). 24-inch bbl. Weight: 7.5 lbs. Sights: adj. leaf rear;
ramp front, w/hood. Classic-style stock of European walnut w/Monte
Carlo comb and cheekpiece, checkered pistol grip and forearm, black
forend tip, QD swivels. Made 1972-97.
Mark X Standard Model **$295**
Mark X Camo Model (Realtree) 375
American Field Model, Std. (Rubber Recoil Pad) 465
American Field, Magnum (Rubber Recoil Pad) 495

Interarms Mark X Cavalier **$325**
Same specifications as Mark X Sporter, except w/contemporary-style
stock w/rollover cheekpiece, rosewood forend tip/grip cap, recoil pad.
Intro. 1974; Disc.

**Interarms Mark X Continental Mannlicher
Style Carbine** **$350**
Same specifications as Mark X Sporter, except straight European-
style comb stock w/sculptured cheekpiece. Precise double-set trig-
gers and classic "butter-knife" bolt handle. French checkering.
Weight: 7.25 lbs. Disc.

Interarms Mark X Lightweight Sporter **$285**
Calibers: 270, 7mm Rem. Mag., 30-06, 5-round magazine. 20-inch
bbl. Weight: 7 lbs. Imported 1988-84.

**Interarms Mark X Marquis Mannlicher
Style Carbine** **$345**
Same specifications as Mark X Sporter, except w/20-inch bbl., full-
length Mannlicher-type stock w/metal forend/muzzle cap. Calibers:
270, 7×57, 308, 30-06. Imported 1976-84.

RIFLES

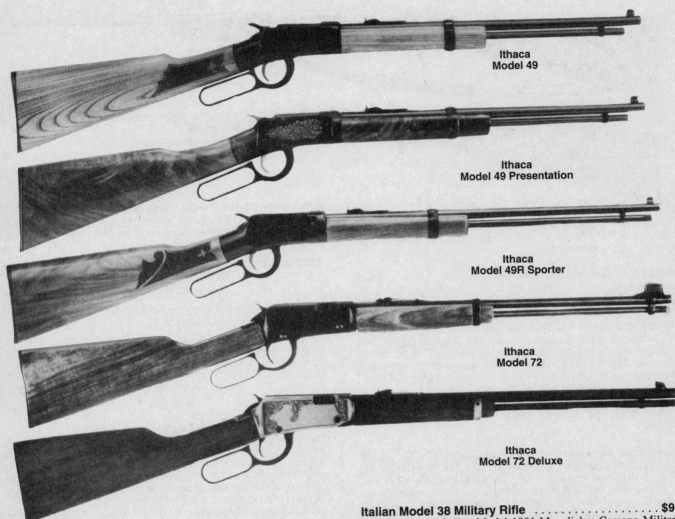

Ithaca
Model 49

Ithaca
Model 49 Presentation

Ithaca
Model 49R Sporter

Ithaca
Model 72

Ithaca
Model 72 Deluxe

Interarms Mini-Mark X Bolt-Action Rifle **$325**
Miniature M98 Mauser action. Caliber: 223 Rem. 5-shot magazine. 20-inch bbl. 39.75 inches overall. Weight: 6.25 lbs. Adj. rear sight, hooded ramp front. Checkered hardwood stock. Imported 1987-94.

Interarms Mark X Viscount **$275**
Same specifications as Mark X Sporter, except w/plainer field grade stock. Imported 1974-87.

**Interarms Whitworth Express Rifle,
African Series** **$525**
Mauser-type bolt action. Calibers: 375 H&H Mag., 458 Win. Mag. 3-shot magazine. 24-inch bbl. Weight: 8 lbs. Sights: 3-leaf express open rear, ramp front w/hood. English-style stock of European walnut, w/cheekpiece, black forend tip, checkered pistol grip and forearm, recoil pad, QD swivels. Imported 1974-96 by Whitworth Rifle Co., England.

ITALIAN MILITARY RIFLES
Manufactured by Government plants at Brescia, Gardone, Terni and Turin, Italy

Italian Model 38 Military Rifle **$95**
Modification of Italian Model 1891 Mannlicher-Carcano Military Rifle w/turned-down bolt handle, detachable folding bayonet. Caliber: 7.35mm Italian Service (many arms of this model were later converted to the old 6.5mm caliber). 6-shot box magazine. 21.25-inch bbl. Weight: 7.5 lbs. Sights: adj. rear-blade front. Military straight-grip stock. Adopted 1938.

ITHACA GUN COMPANY, INC.
King Ferry, New York
(Formerly Ithaca, NY)

**Ithaca Model 49 Saddlegun Lever-Action
Single-Shot Rifle** **$95**
Martini-type action. Hand-operated rebounding hammer. Caliber: 22 LR, Long, Short. 18-inch bbl. Open sights. Western carbine-style stock. Weight: 5.5 lbs. Made 1961-78.

Ithaca Model 49 Saddlegun — Deluxe **$115**
Same as standard Model 49, except w/gold-plated hammer and trigger, figured walnut stock, sling swivels. Made 1962-75.

Ithaca Model 49 Saddlegun — Magnum **$125**
Same as standard Model 49, except chambered for 22 WMR cartridge. Made 1962-78.

Ithaca
Model LSA-65 Standard

Ithaca
Model X5-T

Ithaca
Model X-15

Ithaca
BSA CF-2

RIFLES

Ithaca Model 49 Saddlegun — Presentation **$165**
Same as standard Model 49 Saddlegun, except w/gold-plated hammer and trigger, engraved receiver, full fancy-figured walnut stock w/gold nameplate. Available in 22 LR or 22 WMR. Made 1962-74.

**Ithaca Model 49 Saddlegun — St. Louis
Bicentennial** **$225**
Same as Model 49 Deluxe, except w/commemorative inscription. 200 made in 1964. Value is for rifle in new, unfired condition.

Ithaca Model 49 Youth Saddlegun **$100**
Same as standard Model 49, except shorter stock for young shooters. Made 1961-78.

Ithaca Model 49R Saddlegun Repeating Rifle **$145**
Similar in appearance to Model 49 Single-Shot. Caliber: 22 LR, Long, Short. Tubular magazine holds 15 LR, 17 Long, 21 Short. 20-inch bbl. Weight: 5.5 lbs. Sights: open rear-bead front. Western-style stock, checkered grip. Made 1968-71.

**Ithaca Model 72 Saddlegun Lever-Action
Repeating Carbine** **$245**
Caliber: 22 LR, Long, Short. Tubular magazine holds 15 LR, 17 Long, 21 Short. 18.5-inch bbl. Weight: 5.5 lbs. Sights: open rear; hooded ramp front. Receiver grooved for scope mounting. Western carbine stock and forearm of American walnut. Made 1973-78.

Ithaca Model 72 Saddlegun — Deluxe **$275**
Same as standard Model 72, except w/silver-finished and engraved receiver, octagon bbl., higher grade walnut stock and forearm. Made 1974-76.

> **NOTE**
>
> Barrel and action for the Model 72s are manufactured by Erma-Werke, Dachau, W. Germany; wood installed by Ithaca.

**Ithaca Model LSA-65 Bolt Action
Standard Grade** **$380**
Same as Model LSA-55 Standard Grade, except calibers 25-06, 270, 30-60; 4-shot magazine, 23-inch bbl., weighs 7 lbs. Made 1969-77.

Ithaca Model LSA-65 Deluxe. **$435**
Same as Model LSA-65 Standard Grade, except w/special features of Model LSA-55 Deluxe. Made 1969-77.

Ithaca Model X5-C Lightning Autoloader. **$135**
Takedown. Caliber: 22 LR. 7-shot clip magazine. 22-inch bbl. Weight: 6 lbs. Sights: open rear; Raybar front. Pistol-grip stock, grooved forearm. Made 1958-64.

Japanese
Model 38

Japanese
Model 99

Ithaca Model X5-T Lightning Autoloader Tubular Repeating Rifle **$125**
Same as Model as model X5-C, except w/16-shot tubular magazine, stock w/plain forearm.

Ithaca Model X-15 Lightning Autoloader **$130**
Same general specifications as Model X5-C, except forend is not grooved. Made 1964-67.

Ithaca-BSA CF-2 Bolt-Action Repeating Rifle **$250**
Mauser-type action. Calibers: 7mm Rem. Mag., 300 Win. Mag. 3-shot magazine. 23.6-inch bbl. Weight: 8 lbs. Sights: adj. rear; hooded ramp front. Checkered walnut stock w/Monte Carlo comb, rollover cheekpiece, rosewood forend tip, recoil pad, sling swivels. Imported 1976-77. Mfd. by BSA Guns Ltd., Birmingham, England.

JAPANESE MILITARY RIFLES
Tokyo, Japan
Manufactured by Government Plant

Japanese Model 38 Arisaka Carbine **$130**
Same general specifications as Model 38 Rifle, except w/19-inch bbl., heavy folding bayonet, weighs 7.25 lbs.

Japanese Model 38 Arisaka Service Rifle **$135**
Mauser-type bolt action. Caliber: 6.5mm Japanese. 5-shot box magazine. Bbl. lengths: 25.38 and 31.25 inches. Weight: 9.25 lbs. w/long bbl. Sights: fixed front, adj. rear. Military-type full stock. Adopted in 1905, the 38th year of the Meiji reign; hence, the designation "Model 38."

Japanese Model 44 Cavalry Carbine **$195**
Same general specifications as Model 38 Rifle except w/19-inch bbl., heavy folding bayonet, weighs 8.5 lbs. Adopted in 1911, the 44th year of the Meiji reign; hence, the designation "Model 44."

Japanese Model 99 Service Rifle **$165**
Modified Model 38. Caliber: 7.7mm Japanese. 5-shot box magazine. 25.75-inch bbl. Weight: 8.75 lbs. Sights: fixed front; adj. aperture rear; anti-aircraft sighting bars on some early models; fixed rear sight on some late WWII rifles. Military-type full stock, may have bipod. Takedown paratroop model was also made during WWII. Adopted in 1939 Japanese year 2599 from which the designation "Model 99" is taken. *Note:* The last Model 99 rifles made were of poor quality; some have cast steel receivers. Value shown is for earlier type.

JARRETT CUSTOM RIFLES
Jackson, South Carolina

Jarrett Model No. 2 Walkabout Bolt-Action Rifle **$2150**
Custom lightweight rifle built on Remington 7 action. Jarrett match grade barrel cut and chambered to customer's specification in short action calibers only. Macmillan fiberglass stock pillar-bedded to action and finished to customer's preference.

Jarrett Model No. 3 Custom Bolt-Action Rifle **$2095**
Custom rifle built on Remington 700 action. Jarrett match grade barrel cut and chambered to customer's specification. Macmillan classic fiberglass stock pillar-bedded to action and finished to customer's preference. made 1989 to date.

Jarrett Model No. 4 Professional Hunter Bolt-Action Rifle **$4595**
Custom magnum rifle built on Winchester 70 "controlled feed" action. Jarrett match grade barrel cut and chambered to customer's specification in magnum calibers only. Quarter rib w/iron sights and two Leupold scopes w/Q-D rings and mounts. Macmillan classic fiberglass stock fitted and finished to customer's preference.

JOHNSON AUTOMATICS, INC.
Providence, Rhode Island

Johnson Model 1941 Semiauto Military Rifle **$1395**
Short-recoil-operated. Removable, air-cooled, 22-inch bbl. Caliber: 30-06, 7mm Mauser. 10-shot rotary magazine. Two-piece, wood stock, pistol grip, perforated metal radiator sleeve over rear half of bbl. Sights: receiver peep; protected post front. Weight: 9.5 lbs. *Note:* The Johnson M/1941 was adopted by the Netherlands Government in 1940-41 and the major portion of the production of this rifle, 1941-43, was on Dutch orders. A quantity was also bought by the U.S. Government for use by Marine Corps parachute troops (1943) and for Lend Lease. All these rifles were caliber 30-06; the 7mm Johnson rifles were made for the South American government.

Johnson Sporting Rifle Prototype **$11,000**
Same general specifications as military rifle except fitted w/sporting stock, checkered grip and forend. Blade front sight; receiver peep sight. Less than a dozen made prior to World War II.

IVER JOHNSON'S ARMS, INC.
Jacksonville, Arkansas
(Formerly of Fitchburg, Massachusetts, and Middlesex, New Jersey)

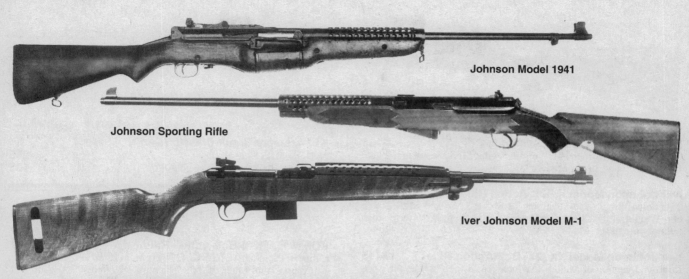

Johnson Model 1941

Johnson Sporting Rifle

Iver Johnson Model M-1

Iver Johnson Li'l Champ Bolt-Action Rifle $70
Caliber: 22 S, L, LR. Single-shot. 16.25-inch bbl. 32.5 inches
overall. Weight: 3.25 lbs. Adj. rear sight, blade front. Synthetic
composition stock. Made 1986-88.

Iver Johnson Model M-1 Semiautomatic Carbine
Similar to U.S. M-1 Carbine. Calibers: 9mm Parabellum 30 U.S.
Carbine. 15- or 30-shot magazine. 18-inch bbl. 35.5 inches over-
all. Weight: 6.5 lbs. Sights: blade front, w/guards; adj. peep rear.
Walnut, hardwood or collapsible wire stock. Parkerized finish.
Model M-1 (30 cal. w/hardwood) $265
Model M-1 (30 cal. w/walnut) 290
Model M-1 (30 cal. w/wire) 350
Model M-1 (9mm w/hardwood) 295
Model M-1 (9mm w/walnut) 335
Model M-1 (9mm w/wire) 395

**Iver Johnson Model PM.30 Semiautomatic
Carbine** $275
Similar to U.S. Carbine, Cal. 30 M1. 18-inch bbl. Weight: 5.5 lbs.
15- or 30-round detachable magazine. Both hardwood and walnut
stock.

**Iver Johnson Model SC30FS Semiautomatic
Carbine** $350
Similar to Survival Carbine except w/folding stock. Made 1983 to
date.

Iver Johnson Survival Semiautomatic Carbine $325
Similar to Model PM.30 except in stainless steel w/high-impact
plastic, one-piece stock. Made 1983 to date. W/folding high-impact
plastic stock **add $35.**

Iver Johnson Model SC30FS

Iver Johnson
Survival Semiautomatic Carbine

Iver Johnson Trailblazer

Iver Johnson Model XX (2X) Bolt-Action Rifle

KDF Model K15

Iver Johnson Trailblazer Semiauto Rifle **$135**
Caliber: 22 LR. 18-inch bbl. Weight: 5.5 lbs. Sights: open rear: blade front. Hardwood stock. Made 1983-85

Iver Johnson Model X Bolt-Action Rifle **$110**
Takedown. Single-shot. Caliber: 22 Short, Long and LR. 22-inch bbl. Weight: 4 lbs. Sights: open rear; blade front. Pistol grip stock w/knob forend tip. Made 1928-32.

Iver Johnson Model XX (2X) Bolt-Action Rifle **$115**
Improved version of Model X, w/heavier 24-inch bbl. larger stock (w/o knob tip), weighs 4.5 lbs. Made 1932-55.

Iver Johnson Model 5100A1 Bolt-Action Rifle . . . **$3995**
Single-shot long-range rifle w/removable bolt for breech loading. Caliber: 50BMG. Fluted 29-inch bbl. w/muzzle break. 51.5 inches overall. Adj. composition stock w/folding bipod. Scope rings; no sights.

K.B.I., INC.
Harrisburg, Pennsylvania
See listing under Armscor; Charles Daly; FEG; Liberty and I.M.I.

KBI Armscor Super Classic
Calibers: 22 LR, 22 Mag., RF, 22 Hornet. 5- or 10-round capacity. Bolt and semiauto action. 22.6- or 20.75-inch bbl. 41.25 or 40.5 inches overall. Weight: 6.4 to 6.7 lbs. Blue finish. Oil-finished American walnut stock w/hardwood grip cap and forend tip. Checkered Monte Carlo comb and cheekpiece. High polish blued barreled action w/damascened bolt. Dovetailed receiver and iron sights. Recoil pad. QD swivel posts.
22 Long Rifle (M-1500 SC) $165
22 Magnum Rimfire (M-1500 SC) 185

KBI Armscor Super Classic *(Con't)*
22 Hornet (M-1800-S) 295
22 Long Rifle Semiauto (M-2000 SC) 185

K.D.F. INC.
Sequin, Texas

KDF Model K15 Bolt-Action Rifle
Calibers: (Standard) 22-250, 243 Win., 6mm Rem., 25-06, 270 Win., 280 Rem., 7mm Mag., 30-60; (Magnum) 300 Wby., 300 Win., 338 Win., 340 Wby., 375 H&H, 411 KDF, 416 Rem., 458 Win. 4-shot magazine (standard), 3-shot (magnum). 22-inch (standard) or 24-inch (magnum) bbl. 44.5 to 46.5 inches overall. Weight: 8 lbs. Sights optional. Kevlar composite or checkered walnut stock in Classic, European or thumbhole-style. *Note:* U.S. Manufacture limited to 25 prototypes and preproduction variations.
Standard Model **$1225**
Magnum Model 1325

KEL-TEC CNC INDUSTRIES, INC.
Cocoa, Florida

Kel-Tec Sub-Series Semiautomatic Rifles
Semiautomatic blowback action w/pivioting bbl. takedown. 9mm Parabellum or 40 S&W. Interchangeable grip assembly accepts most double column high capacity handgun magazines. 16.1-inch bbl. 31.5 inches overall. Weight: 4.6 lbs. Hooded post front sight, flip-up rear. Matte black finish. Tubular buttstock w/grooved polymer buttplate and vented handguard. Made 1997 to date.
Sub-9 Model (9mm) **$550**
Sub-40 Model (40 S&W) 575

**Kel-Tec Sub-Series
Semiautomatic Rifles**

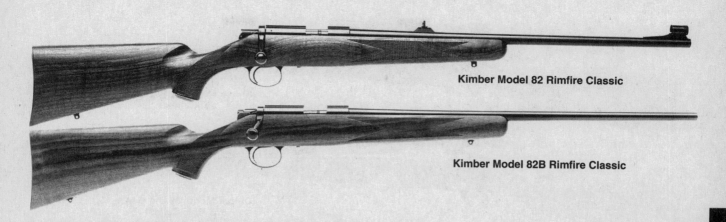

Kimber Model 82 Rimfire Classic

Kimber Model 82B Rimfire Classic

KIMBER RIFLES
Mfd. By Kimber Manufacturing, Inc., Younkers, NY
(Formerly Kimber of America, Inc.;
Kimber of Oregon, Inc.)

Note: From 1980-91 Kimber of Oregon produced Kimber firearms. A redesigned action designated by serialization with a "B" suffix was introduced 1986. Pre-1986 production is recognized as the "A" series, but is not so marked. These early models in rare configurations and limited-run calibers command premium prices from collectors. Kimber of America, located in Clackamas, Oregon, acquired the Kimber trademark and resumed manufactured of Kimber rifles. During this transition Nationwide Sports Distributors, Inc. located in Pennsylvania and Nevada became exclusive distributors of Kimber products. In 1997 Kimber Manufacturing acquired the trademark with manufacturing rights and expanded production to include a 1911-A1-style semiautomatic pistol, the Kimber Classic 45.

Rifle production is projected to resume in late 1998 with the announcement of an all-new Kimber .22 rifle and a refined Model 84 in both single shot and repeater configurations which is expected to be available in the spring of 1999. All initial sales are from residual stock of a few remaining Model 98 Sporters chambered 30-06, 7mm Rem. Mag. & .338 Win. Mag. Additional there are a few remaining 22 rifles (82C & 84C Models) available from Master Dealers' stock.

Kimber Model 82 Bolt-Action Rifle
Small action based on Kimber's "A" Model 82 rimfire receiver w/twin rear locking lugs. Calibers: 22 LR, 22 WRF, 22 Hornet, 218 Bee, 25-20. 5- or 10-shot magazine (22 LR); 5-shot magazine (22WRF); 3-shot magazine (22 Hornet). 218 Bee and 25-20 are single-shot. 18- to 25-inch bbl. 37.63 to 42.5 inches overall. Weight: 6 lbs. (Light Sporter), 6.5 lbs. (Sporter), 7.5 lbs. (Varmint); 10.75 lbs. (Target). Right- and left-hand actions are available in distinctive stock styles.

Cascade (disc. 1987)	$ 685
Classic (disc. 1988)	650
Continental	1050
Custom Classic (disc. 1988)	725

Kimber Model 82 Bolt-Action Rifle *(Con't)*

Mini Classic	450
Super America	825
Super Continental	1100
1990 Classifications	
All-American Match	595
Deluxe Grade (disc. 1990)	850
Hunter (Laminated Stock)	575
Super America	825
Target (Government Match)	595

Kimber Model 82C Classic Bolt-Action Rifle
Caliber: 22 LR. 4-shot or 10-shot magazine. 21-inch air-gauged bbl. 40.5 inches overall. Weight: 6.5 lbs. Receiver drilled and tapped for Warne scope mounts; no sights. Single-set trigger. Checkered Claro walnut stock w/red buttpad and polished steel grip cap. Reintroduced 1993.

Classic Model	$575
Left-Hand Model, **add**	75

Kimber Model 84 Bolt-Action Rifle
Classic (disc. 1988) Compact medium action based on a "scaled down" Mauser-type receiver, designed to accept small base centerfire cartridges. Calibers: 17 Rem., 221 Fireball, 222 Rem., 223 Rem. 5-shot magazine. Same general barrel and stock specifications as Model 82.

Classic (discontinued 1988)	$ 695
Continental	950
Custom Classic (disc. 1988)	850
Super America (disc. 1988)	975
Super Continental (disc. 1988)	1025
l990 Classifications	
Deluxe Grade (disc. 1990)	895
Hunter/Sporter (Laminated Stock)	795
Super America (disc. 1991)	995
Super Varmint (disc. 1991)	1025
Ultra Varmint (disc. 1991)	950

Kimber Model 84 Classic

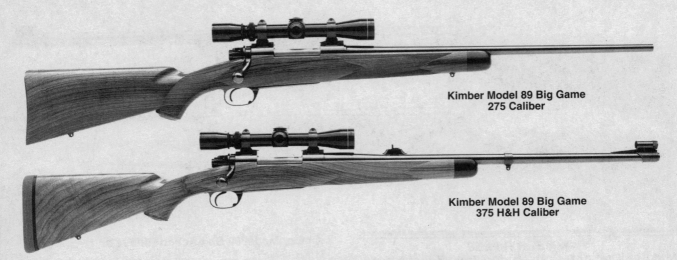

**Kimber Model 89 Big Game
275 Caliber**

**Kimber Model 89 Big Game
375 H&H Caliber**

Kimber Model 89 Big-Game Rifle

Large action combining the best features of the pre-64 Model 70 Winchester and the Mauser 98. Three action lengths are offered in three stock styles. Calibers: 257 Roberts, 25-06, 7×57, 270 Win., 280 Win., 30-06, 7mm Rem. Mag., 300 Win. Mag., 300 H&H, 338 Win., 35 Whelen, 375 H&H, 404 Jeffrey, 416 Rigby, 460 Wby., 505 Gibbs (308 cartridge family to follow). 5-shot magazine (standard calibers); 3-shot magazine (Magnum calibers). 22- to 24-inch bbl. 42 to 44 inches overall. Weight: 7.5 to 10.5 lbs. Model 89 African features express sights on contoured quarter rib, banded front sight. Barrel-mounted recoil lug w/integral receiver lug and twin recoil crosspins in stock.

BGR Long Action Classic (Disc. 1988)	$ 695
Custom Classic (disc. 1988)	895
Super America	1095

1990 Classifications
Deluxe Grade

Featherweight	1295
Medium	1325
375 H&H	1395

Hunter Grade (Laminated Stock)

270 and 30-06	950
375 H&H	1195

Super America

Featherweight	1495
Medium	1570
375 H&H	2000
African — All calibers	3595

KNIGHT'S MANUFACTURING COMPANY
Vero Beach, Florida

Knight's Stoner SR-15 Semiautomatic
Match Rifle **$1295**
AR-15 configuration. Caliber: 223 Rem. (5.56mm). 5- or 10-shot magazine. 20-inch w/free floating, match-grade bbl. 38 inches overall. Weight: 7.9 lbs. Integral Weaver-style rail. Two-stage target trigger. Matte black oxide finish. Black synthetic AR-15A2-style stock and forearm. Made 1997 to date.

Knight's Stoner SR-15 M-4 Semiautomatic Carbine
Similar to SR-15 Rifle, except w/16-inch bbl. Sights and mounts optional. Fixed synthetic or collapsible buttstock. Made 1997 to date.

Model SR-15 Carbine (w/collapsible stock)	**$950**
Model SR-15 Carbine (w/fixed stock)	1025

Knight's Stoner SR-15 M-5
Semiautomatic Rifle **$975**
Caliber: 223 Rem. (5.56mm). 5- or 10-shot magazine. 20-inch bbl. 38 inches overall. Weight: 7.6 lbs. Integral Weaver-style rail. Two-stage target trigger. Matte black oxide finish. Black synthetic AR-15A2-style stock and forearm. Made 1997 to date.

Knight's Stoner SR-25 Match Rifle
Similar to SR-25 Sporter, except w/free floating 20- or 24-inch match bbl. 39.5-43.5 inches overall. Weight: 9.25 and 10.75 lbs. Respectively. Integral Weaver-style rail. Sights and mounts optional. 1 MOA guaranteed. Made 1993 to date.

Model SR-25 LW Match (w/20-inch bbl.)	**$2050**
Model SR-25 Match (w/24-inch bbl.)	**$2095**
W/RAS (Rail Adapter System), **add**	**300**

Knight's Stoner SR-25 Semiautomatic Carbine
Similar to SR-25 Sporter, except w/free floating 16-inch bbl. 35.75 inches overall. Weight: 7.75 lbs. Integral Weaver-style rail. Sights and mounts optional. Made 1995 to date.

Model SR-25 Carbine (w/o sights)	**$1995**
W/RAS (Rail Adapter System), **add**	**300**

Knight's Stoner SR-25 Semiautomatic
Sporter Rifle **$2195**
AR-15 configuration. Caliber: 308 Win. (7.62 NATO). 5-, 10- or 20-shot magazine. 20-inch bbl. 39.5 inches overall. Weight: 8.75 lbs. Integral Weaver-style rail. Protected post front sight adjustable for elevation, detachable rear adjustable for windage. Two-stage target trigger. Matte black oxide finish. Black synthetic AR-15A2-style stock and forearm. Made 1993-97.

Knight's Stoner SR-50 Semiautomatic
Long Range Precision Rifle **$4695**
Gas-operated semiautomatic action. Caliber: 50 BMG. 10-shot magazine. 35.5-inch bbl. 58.5 inches overall. Weight: 31.75 lbs. Integral Weaver-style rail. Two-stage target trigger. Matte black oxide finish. Tublar-style stock. Made 1996 to date.

KONGSBERG RIFLES
Kongsberg, Norway
(Imported by Kongsberg America L.L.C., Fairfield, CT)

Kongsberg Model 393 Classic Sporter
Calibers: 22-250 Rem., 243 Win., 6.5x55, 270 Win., 7mm Rem. Mag., 30-06, 308 Win. 300 Win. Mag., 338 Win. Mag. 3- or 4-shot

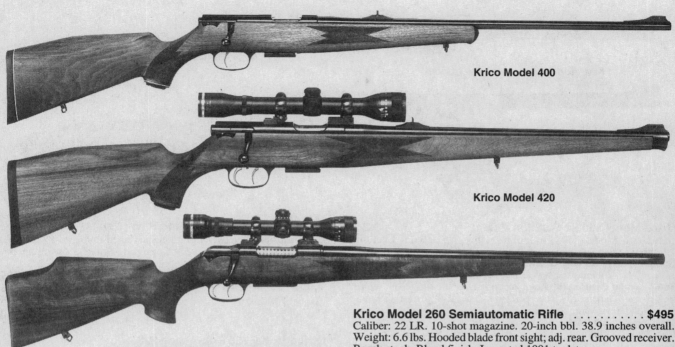

Krico Model 400

Krico Model 420

Krico 640 Varmint

Kongsberg Model 393 Classic Sporter *(Con't)*

rotary magazine. 23-inch bbl. (Standard) or 26-inch bbl. (magnum). Weight: 7.5 to 8 lbs. 44 to 47 inches overall. No sights w/ dovetailed receiver or optional hooded blade front sight, adjustable rear. Blue finish. Checkered European walnut stock w/rubber buttplate. Imported 1994-98.

Standard Calibers	$695
Magnum Calibers	825
Left-Handed Model, **add**	125
W/Optional Sights, **add**	50

Kongsberg Model 393 Deluxe Sporter

Similar to Classic Model, except w/deluxe European walnut stock. Imported 1994-98.

Standard Calibers	$725
Magnum Calibers	850
Left-Handed Model, **add**	125
W/Optional Sights, **add**	50

Kongsberg Model 393 Thumbhole Sporter

Calibers: 22-250 Rem. or 308 Win. 4-shot rotary magazine. 23-inch heavy bbl. Weight: 8.5 lbs. 44 inches overall. No sights w/ dovetailed receiver. Blue finish. Stippled American walnut thumbhole stock w/adjustable cheekpiece. Imported 1993-98.

Right-Handed Model	$1050
Left-Handed Model	1195

KRICO RIFLES
Stuttgart-Hedelfingen, Germany
Mfd. by Sportwaffenfabrik Kriegeskorte GmbH
Imported by Mandall Shooting Supplies, Scottsdale, AZ
(Previously by Beeman Precision Arms, Inc.)

Krico Model 260 Semiautomatic Rifle $495
Caliber: 22 LR. 10-shot magazine. 20-inch bbl. 38.9 inches overall. Weight: 6.6 lbs. Hooded blade front sight; adj. rear. Grooved receiver. Beech stock. Blued finish. Imported 1991 to date.

Krico Model 300 Bolt-Action Rifle
Calibers: 22 LR, 22 WMR, 22 Hornet. 19.6-inch bbl. (22 LR), 23.6-inch (22 Hornet). 38.5 inches overall. Weight: 6.3 lbs. Double-set triggers. Sights: ramped blade front, adj. open rear. Checkered walnut-finished hardwood stock. Blued finish. Imported 1993 to date.

Model 300 Standard	$495
Model 300 Deluxe	525
Model 300 SA (Monte-Carlo walnut stock)	565
Model 300 Stutzen (full-length walnut stock)	620

Krico Model 311 Small-Bore Rifle
Bolt action. Caliber: 22 LR. 5- or 10-shot clip magazine. 22-inch bbl. Weight: 6 lbs. Single- or double-set trigger. Sights: open rear; hooded ramp front; available w/factory-fitted Kaps 2.5× scope. Checkered stock w/cheekpiece, pistol grip and swivels. Disc.

W/Scope Sight	$350
W/Iron Sights Only	295

Krico Model 320 Bolt-Action Sporter $550
Caliber: 22 LR. 5-shot detachable box magazine. 19.5-inch bbl. 38.5 inches overall. Weight: 6 lbs. Adj. rear sight, blade ramp front. Checkered European walnut Mannlicher-style stock w/low comb and cheekpiece. Single or double-set triggers. Imported 1986-91.

Krico Model 340 Metallic Silhouette
Bolt-Action Rifle $595
Caliber: 22 LR. 5-shot magazine. 21-inch heavy, bull bbl. 39.5 inches overall. Weight: 7.5 lbs. No sights. Grooved receiver for scope mounts. European walnut stock in off-hand match-style configuration. Match or double-set triggers. Imported 1983-86.

Krico Model 360S Biathlon Rifle $1025
Caliber: 22 LR. Five 5-shot magazines. 21.25-inch bbl. w/snow cap. 40.5 inches overall. Weight: 9.25 lbs. Straight-pull action. Match trigger w/17½-oz. pull. Sights: globe front, adj. match peep rear. Biathlon-style walnut stock w/high comb and adj. butt-plate. Imported 1991 to date.

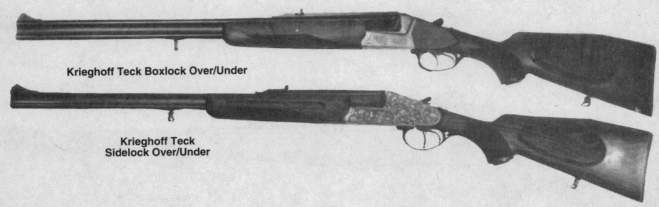

Krieghoff Teck Boxlock Over/Under

**Krieghoff Teck
Sidelock Over/Under**

Krico Model 360 S2 Biathlon Rifle $950
Similar to Model 360S, except w/pistol-grip activated action. Biathlon-style walnut stock w/black epoxy finish. Imported 1991 to date.

Krico Model 400 Bolt-Action Rifle $675
Caliber: 22 Hornet. 5-shot detachable box magazine. 23.5-inch bbl. Weight: 6.75 lbs. Adj. open rear sight, ramp front. European walnut stock. Disc. 1990.

Krico Model 420 Bolt-Action Rifle $695
Same as Model 400, except w/full-length Mannlicher-style stock and double-set triggers. Scope optional, extra. Disc. 1989.

Krico Model 440 S Bolt-Action Rifle $650
Caliber: 22 Hornet. Detachable box magazine. 20-inch bbl. 36.5 inches overall. Weight: 7.5 lbs. No sights. French walnut stock w/ventilated forend. Disc. 1988.

Krico Model 500 Match Rifle $2650
Caliber: 22 LR. Single-shot. 23.6-inch bbl. 42 inches overall. Weight: 9.4 lbs. Kricotronic electronic ignition system. Sights: globe front; match micrometer aperture rear. Match-style European walnut stock w/adj. butt.

Krico Model 600 Bolt-Action Rifle $895
Same general specifications as Model 700, except w/short action. Calibers: 17 Rem., 222, 223, 22-250, 243, 5.6×50 Mag. and 308.

Krico Model 620 Bolt-Action Rifle $925
Same as Model 600, except w/short action chambered 308 Win. only and full-length Mannlicher-style stock w/Schnabel forend tip. 20.75-inch bbl. Weight: 6.5 lbs. No longer imported.

Krico Model 640 Super Sniper Bolt-Action
Repeating Rifle $1095
Calibers: 223 Rem., 308 Win. 3-shot magazine. 26-inch bbl. 44.25 inches overall. Weight: 9.5 lbs. No sights drilled and tapped for scope mounts. Single or double-set triggers. Select walnut stock w/adj. cheekpiece and recoil pad. Disc. 1989.

Krico Model 640 Varmint Rifle $695
Caliber: 222 Rem. 4-shot magazine. 23.75-inch bbl. Weight: 9.5 lbs. No sights. European walnut stock. No longer imported.

Krico Model 700 Bolt-Action Rifle
Calibers: 17 Rem., 222, 222 Rem. Mag., 223, 22-250, 5.6×50 Mag., 243, 5.6×57 RSW, 6×62, 6.5×55, 6.5×57, 6.5×68 270 Win., 7×64, 7.5 Swiss, 7mm Mag., 30-06, 300 Win., 8×68S, 9.3×64. 24-inch (standard) or 26-inch (magnum) bbl. 44 inches overall (standard). Weight: 7.5 lbs.

Krico Model 700 Bolt-Action Rifle *(Con't)*
Adj. rear sight; hooded ramp front. Checkered European-style walnut stock w/Bavarian cheekpiece and rosewood Schnabel forend tip. Imported 1983 to date.
Model 700 . . . . $ 795
Model 700 Deluxe 895
Model 700 Deluxe S 1095
Model 700 Stutzen 950

Krico Model 720 Bolt-Action Rifle
Same general specifications as Model 700, except in calibers 270 Win. and 30-06 w/full-length Mannlicher-style stock and Schnabel forend tip. 20.75-inch bbl. Weight: 6.75 lbs. Disc. importing 1990.
Sporter Model . . . . $ 895
Ltd. Edition 1795

Krico Bolt-Action Sporting Rifle $535
Miniature Mauser action. Single- or double-set trigger. Calibers: 22 Hornet, 222 Rem. 4-shot clip magazine. 22-24- or 26-inch bbl. Weight: 6.25 lbs. Sights: open rear; hooded ramp front. Checkered stock w/cheekpiece, pistol grip, black forend tip, sling swivels. Imported 1956-62.

Krico Carbine $525
Same as Krico Sporting Rifle, except w/20- or 22-inch bbl., full-length Mannlicher-type stock.

Krico Special Varmint Rifle $535
Same as Krico Rifle, except w/heavy bbl., no sights weighs 7.25 lbs. Caliber: 222 Rem. only.

KRIEGHOFF RIFLES
Ulm (Donau), Germany
Mfd. by H. Krieghoff Jagd und Sportwaffenfabrik

See also combination guns under Krieghoff shotgun listings.

Krieghoff Teck Over/Under Rifle
Kersten action, double crossbolt, double underlugs. Boxlock. Calibers: 7×57r5, 7×64, 7×65r5, 30-30, 308 Win. 30-06, 300 Win. Mag., 9.3×74r5, 375 H&H Mag. 458 Win. Mag. 25-inch bbls. Weight: 8 to 9.5 lbs. Sights: express rear; ramp front. Checkered walnut stock and forearm. Made 1967 to date.
Standard calibers $5795
375 H&H Mag. (Disc. 1988),458 Win. Mag 6550

Krieghoff Ulm Over/Under Rifle $8950
Same genmeral specifications as Teck model, except w/sidelocks w/leaf arabesque engraving. Made 1963 to date.

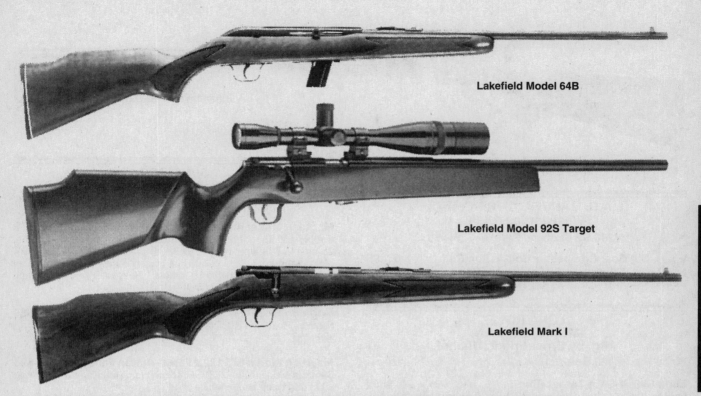

Lakefield Model 64B

Lakefield Model 92S Target

Lakefield Mark I

Krieghoff ULM-Primus Over/Under Rifle **$11,250**
Delux version of Ulm model, w/detachable sidelocks, higher grade engraving and stock wood. Made 1963 to date.

LAKEFIELD ARMS LTD.
Ontario, Canada

See also listing under Savage for production since 1994.

Lakefield Model 64B Semiautomatic Rifle **$120**
Caliber: 22 LR. 10-shot magazine. 20-inch bbl. Weight: 5.5 lbs. 40 inches overall. Bead front sight, adj. rear. Grooved receiver for scope mounts. Stamped checkering on walnut-finished hardwood stock w/Monte Carlo cheekpiece. Imported 1990-94.

Lakefield Model 90B Bolt-Action Target Rifle **$295**
Caliber: 22 LR. 5-shot magazine. 21-inch bbl. w/snow cap. 39.63 inches overall. Weight: 8.25 lbs. Adj. receiver peep sight; globe front w/colored inserts. Receiver drilled and tapped for scope mounts. Biathlon-style natural finished hardwood stock w/shooting rails, hand stop and butthook. Made 1991-94.

Lakefield Model 91T/91TR Bolt-Action Target Rifle
Calibers: 22 Short, Long, LR. 25-inch bbl. 43.63 inches overall. Weight: 8 lbs. Adj. rear peep sight; globe front w/inserts. Receiver drilled and tapped for scope mounts. Walnut-finished hardwood stock w/shooting rails and hand stop. **Model 91TR** is a 5-shot clip-fed repeater. Made 1991-94.
Model 91T Single-Shot **$245**
Model 91TR Repeater (22 LR only) **265**

Lakefield Model 92S Target Rifle **$225**
Same general specifications as Model 90B, except w/conventional target-style stock. 8 lbs. No sights, but drilled and tapped for scope mounts. Made 1993-95.

Lakefield Arms Model 93M Bolt Action **$110**
Caliber: 22 WMR. 5-shot magazine. 20.75-inch bbl. 39.5 inches overall. Weight: 5.75 lbs. Bead front sight, adj. open rear. Receiver grooved for scope mount. Thumb operated rotary safety. Checkered walnut-finished hardwood stock. Blued finish. Made 1995.

Lakefield Mark I Bolt-Action Rifle **$70**
Calibers: 22 Short, Long, LR. Single-shot. 20.5-inch bbl. (19-inch Youth Model); available in smoothbore. Weight: 5.5 lbs. 39.5 inches overall. Bead front sight; adj. rear. Grooved receiver for scope mounts. Checkered walnut-finished hardwood stock w/Monte Carlo and pistol grip. Blued finish. Made 1990-94.

Lakefield Mark II Bolt-Action Rifle
Same general specifications as Mark I, except has repeating action w/10-shot detachable box magazine. 22 LR only. Made 1992-94.
Mark II Standard . . . . **$ 75**
Mark II Youth (19-inch barrel) **85**
Mark II Left-Hand **105**

LAURONA RIFLES
Mfg. in Eibar, Spain
Imported by Galaxy Imports, Victoria, TX

Laurona Model 2000X O/U Express Rifle
Calibers: 30-06, 8×57 JRS, 8×75 JR, 375 H&H, 9.3×74R 5-shot magazine. 24-inch separated bbls. Weight: 8.5 lbs. Quarter rib drilled and tapped for scope mount. Open sights. Matte black chrome finish. Monte Carlo-style checkered walnut buttstock; tulip forearm. Imported 1993 to date.
Standard Calibers **$2250**
Magnum Calibers **2795**

Luna Single-Shot Target

L.A.R. MANUFACTURING, INC.
West Jordan, Utah

L.A.R. Big Boar Competitor Bolt-Action Rifle **$1895**
Single-shot, bull-pup action. Caliber: 50 BMG. 36-inch bbl. 45.5 inches overall. Weight: 28.4 lbs. Made 1994 to date.

LUNA RIFLE
Mehlis, Germany
Mfg. by Ernst Friedr. Büchel

Luna Single-Shot Target Rifle **$850**
Falling block action. Calibers: 22 LR, 22 Hornet. 29-inch bbl. Weight: 8.25 lbs. Sights: micrometer peep rear tang; open rear; ramp front. Cheekpiece stock w/full pistol grip, semibeavertail forearm, checkered, swivels. Made before WWII.

MAGNUM RESEARCH, INC.
Minneapolis, Minnesota

Magnum Research Mountain Eagle
Bolt-Action Rifle Series
Calibers: 222 Rem., 223 Rem., 270 Win., 280 Rem., 7mm Rem. Mag., 7mm STW, 30-06, 300 Win. Mag., 338 Win. Mag., 340 Wby. Mag., 375 H&H, 416 Rem. Mag. 5-shot (Std.) or 4-shot (Mag.). 24- or 26-inch bbl. 44 to 46 inches overall. Weight: 7.75 to 9.75 lbs. Receiver drilled and tapped for scope mount; no sights. Blued finish. Fiberglass composite stock. Made 1994 to date.
Standard Model **$950**
Magnum Model **995**
Varmint Model (Intro. 1996)........................ **975**
Calibers 375 H&H, 416 Rem. Mag., **add** **295**
Left-Hand Model, **add** **95**

MAGTECH
Las Vegas, Nevada
Mfg. by CBC, Brazil

Magtech Model MT 122.2/S Bolt-Action Rifle **$75**
Calibers: 22 Short, Long, Long Rifle. 6- or 10-shot clip. Bolt action. 25-inch free-floating bbl. 43 inches overall. Weight: 6.5 lbs. Double locking bolt. Red cocking indicator. Safety lever. Brazilian hardwood finish. Double extractors. Beavertail forearm. Sling swivels. Imported 1994 to date.

Magtech Model MT 122.2/R Bolt-Action Rifle **$85**
Same as Model MT 122.2/S except, adj. rear sight and post front sight.

Magtech Model MT 122.2T Bolt-Action Rifle **$90**
Same as Model MT 122.2/S, except w/adj. micrometer-type rear sight and ramp front sight.

MANNLICHER SPORTING RIFLES
Steyr, Austria
Mfg. by Steyr-Daimler-Puch, A.-G.

Imported by Gun South, Inc. Trussville, AL
In 1967, Steyr-Daimler-Puch introduced a series of sporting rifles with a bolt action that is a departure from the Mannlicher-Schoenauer system of earlier models. In the latter, the action is locked by lugs symmetrically arranged behind the bolt head as well as by placing the bolt handle ahead of the right flank of the receiver, the rear section of which is open on top for backward movement of the bolt handle. The current action, made in four lengths to accommodate different ranges of cartridges, has a closed-top receiver; the bolt locking lugs are located toward the rear of the bolt (behind the magazine), and the Mannlicher-Schoenauer rotary magazine has been redesigned as a detachable box type of Makrolon.

> **NOTE**
>
> Certain Mannlicher-Schoenauer models were produced before WWII. Manufacture of sporting rifles and carbines was resumed at the Steyr-Daimler-Puch plant in Austria in 1950 during which time the Model 1950 rifles and carbines were introduced.

Magnum Research Mountain Eagle
Bolt-Action Rifle

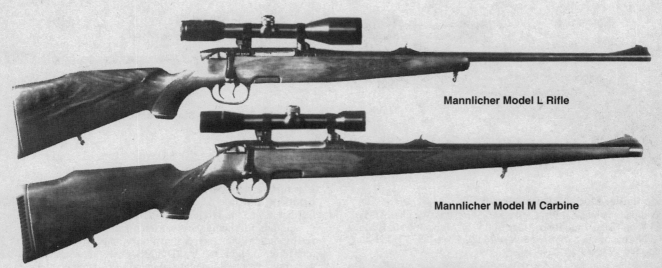

Mannlicher Model L Rifle

Mannlicher Model M Carbine

RIFLES

Mannlicher Model L Carbine $1025
Same general specifications as Model SL Carbine, except w/type "L" action, weighs 6.2 lbs. Calibers same as for Model L Rifle. Imported 1968-96.

Mannlicher Model L Rifle $1095
Same general specifications as Model SL Rifle, except w/type "L" action, weighs 6.3 lbs. Calibers: 22-250, 5.6×57 (disc. 1991), 243 Win., 6mm Rem. 308 Win. Imported 1968-96.

Mannlicher Model L Varmint Rifle $1065
Same general specifications as Model SL Varmint Rifle except w/type "L" action. Calibers: 22-250, 243 Win., 308 Win. Imported 1969-96.

Mannlicher Model Luxus Bolt-Action Rifle
Same general specifications as Models L and M, except w/3-shot detachable box magazine and single-set trigger. Full or half-stock w/low-luster oil or high-gloss lacquer finish. Disc. 1996.
Full stock $1495
Half stock 1395

Mannlicher Model M Carbine
Same general specifications as Model SL Carbine except w/type "M" action, stock w/recoil pad, weighs 6.8 lbs. Left-hand version w/additional 6.5×55 and 9.3×62 calibers intro. 1977. Imported 1969-96.
Right-Hand Carbine $1050
Left-Hand Carbine 1395

Mannlicher Model M Professional Rifle $995
Same as standard Model M Rifle, except w/synthetic (Cycolac) stock, weighs 7.5 lbs. Calibers: 6.5×55, 6.5×57, 270 Win., 7×57, 7×64, 7.5 Swiss, 30-06, 8×57JS, 9.3×62. Imported 1977-93.

Mannlicher Model M Rifle
Same general specifications as Model SL Rifle, except w/type "M" action, stock w/forend tip and recoil pad; weighs 6.9 lbs. Calibers: 6.5×57, 270 Win., 7×57, 7×64, 30-06, 8×57JS, 9.3×62. Made 1969 to date. Left-hand version also in calibers 6.5×55 and 7.5 Swiss. Imported 1977-96.
Right-Hand Rifle .... $ 995
Left-Hand Rifle 1395

Mannlicher Model S Rifle $995
Same general specifications as Model SL Rifle, except w/type "S" action, 4-round magazine, 25.63-inch bbl., stock w/forend tip and

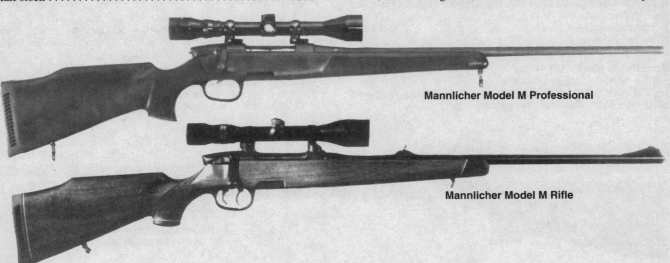

Mannlicher Model M Professional

Mannlicher Model M Rifle

**Mannlicher Model SL Rifle
w/Single-Set Trigger**

Mannlicher Model S Rifle *(Con't)*

recoil pad, weighs 8.4 lbs. Calibers: 6.5×68, 257 Weatherby Mag., 264 Win. Mag., 7mm Rem. Mag., 300 Win. Mag., 300 H&H Mag., 308 Norma Mag., 8×68S, 338 Win. Mag., 9.3×64, 375 H&H Mag. Imported 1970-96.

Mannlicher Model SL Carbine $995

Same general specifications as Model SL Rifle, except w/20-inch bbl. and full-length stock, weighs 6 lbs. Imported 1968-96.

Mannlicher Model SL Rifle $975

Steyr-Mannlicher SL bolt action. Calibers: 222 Rem., 222 Rem., 222 Rem. Mag., 223 Rem. 5-shot rotary magazine, detachable. 23.63-inch bbl. Weight: 6 lbs. Single- or double-set trigger (mechanisms interchangeable). Sights: open rear; hooded ramp front. Half stock of European walnut w/Monte Carlo comb and cheekpiece, skip-checkered forearm and pistol grip, rubber butt-pad, QD swvels. Imported 1967-96.

Mannlicher Model SL Varmint Rifle $1025

Same general specifications as Model SL Rifle, except caliber 222 Rem. only, w/25.63-inch heavy bbl., no sights, weighs 7.92 lbs. Imported 1969-96.

Mannlicher Model SSG Match Target Rifle

Type "L" action. Caliber: 308 Win. (7.62×51 NATO). 5- or 10-round magazine, single-shot plug. 25.5-inch heavy bbl. Weight: 10.25 lbs. Single trigger. Sights: micrometer peep rear; globe front. Target stock, European walnut or synthetic, w/full pistol grip, wide forearm w/swivel rail, adj. rubber buttplate. Imported 1969 to date.
W/Walnut Stock $1595
W/Synthetic Stock 1295

Mannlicher Model S/T Rifle. $1195

Same as Model S Rifle, except w/heavy 25.63-inch bbl., weighs 9 lbs. Calibers: 9.3×64, 375 H&H Mag., 458 Win. Mag. Option of 23.63-inch bbl. in latter caliber.Imported 1975-96.

Mannlicher-Schoenauer Model 1903 Bolt-Action Sporting Carbine $850

Caliber: 6.5×53mm (referred to in some European gun catalogs as 6.7×53mm, following the Austrian practice of designating calibers by bullet diameter). 5-shot rotary magazine. 450mm (17.7-inch) bbl. Weight: 6.5 lbs. Double-set trigger. Sights: two-leaf rear; ramp front. Full-length sporting stock w/cheekpiece, pistol grip, trap buttplate, swivels. Pre-WWII.

Mannlicher-Schoenauer Model 1905 Carbine $725

Same as Model 1903, except w/19.7-inch bbl.chambered 9×56mm and weighs 6.75 lbs. Pre-WWII.

Mannlicher-Schoenauer Model 1908 Carbine $775

Same as Model 1905, except calibers 7×57mm and 8×56mm Pre-WWII.

Mannlicher-Schoenauer Model 1910 Carbine $865

Same as Model 1905, except in 9.5×57mm. Pre-WWII.

Mannlicher-Schoenauer Model 1924 Carbine $795

Same as Model 1905, except caliber 30-06 (7.62×63mm). Pre-WWII.

Mannlicher-Schoenauer Model 1950 Bolt-Action Sporting Rifle $695

Calibers: 257 Roberts, 270 Win., 30-06. 5-shot rotary magazine. 24-inch bbl. Weight: 7.25 lbs. Single trigger or double-set trigger. Redesigned low bolt handle, shotgun-type safety. Sights: folding leaf open rear; hooded ramp front. Improved half-length stock w/cheekpiece, pistol grip, checkered, ebony forend tip, swivels. Made 1950-52.

Mannlicher-Schoenauer Model 1950 Carbine $725

Same general specifications as Model 1950 Rifle except w/20-inch bbl., full-length stock, weighs 7 lbs. Made 1950-52.

Mannlicher-Schoenauer Model 1950 6.5 Carbine $775

Same as other Model 1950 Carbines except caliber 6.5×53mm, w/18.25-inch bbl., weighs 6.75 lbs. Made 1950-52.

**Mannlicher Model SSG
Match Target Rifle**

Mannlicher-Schoenauer
Model 1905 Carbine

Mannlicher-Schoenauer
Model 1905 Carbine

Mannlicher-Schoenauer
Model 1905 Carbine

Mannlicher-Schoenauer
Model 1905 Carbine

Mannlicher-Schoenauer
Model 1905 Carbine

Mannlicher-Schoenauer
Model 1905 Carbine

RIFLES

Mannlicher-Schoenauer Model 1952 Improved Carbine $825
Same as Model 1950 Carbine except w/swept-back bolt handle, redesigned stock. Calibers: 257, 270, 7mm, 30-06. Made 1952-56.

Mannlicher-Schoenauer Model 1952 Improved 6.5 Carbine $795
Same as Model 1952 Carbine except caliber 6.5×53mm, w/18.25-inch bbl. Made 1952-56.

Mannlicher-Schoenauer Model 1952 Improved Sporting Rifle $695
Same as Model 1950 except w/swept-back bolt handle redesigned stock. Calibers: 257, 270, 30-06, 9.3×62mm. Made 1952-56 and imported exclusively by Stoeger Arms Corp.

Mannlicher-Schoenauer Model 1956 Custom Carbine $725
Same general specifications as Models 1950 and 1952 Carbines, except w/redesigned stock w/high comb. Drilled and tapped for scope mounts. Calibers: 243, 6.5mm, 257, 270, 7mm, 30-06, 308. Made 1956-60.

Mannlicher-Schoenauer Carbine, Model 1961-MCA $725
Same as Model 1956 Carbine, except w/universal Monte Carlo design stock. Calibers: 243 Win., 6.5mm, 270, 308, 30-06. Made 1961-71.

Mannlicher-Schoenauer Rifle, Model 1961-MCA .. $695
Same as Model 1956 Rifle, except w/universal Monte Carlo design stock. Calibers: 243, 270, 30-06. Made 1961-71.

**Mannlicher Schoenauer High Velocity
Bolt-Action Sporting Rifle**

Mannlicher-Schoenauer High Velocity
Bolt-Action Sporting Rifle **$1050**
Calibers: 7×64 Brenneke, 30-06 (7.62×63), 8×60 Magnum, 9.3×62, 10.75×68mm. 23.6-inch bbl. Weight: 7.5 lbs. Sights: British-style 3-leaf open rear; ramp front. Half-length sporting stock w/cheekpiece, pistol grip, checkered, trap buttplate, swivels. Also produced in takedown model. Pre-WWII.

Mannlicher-M72 Model L/M
Carbine **$695**
Same general specifications as M72 Model L/M Rifle except w/20-inch bbl. and full-length stock, weighs 7.2 lbs. Imported 1972 to date.

Mannlicher-M72 Model L/M Rifle **$795**
M72 bolt action, type L/M receiver front-locking bolt internal rotary magazine (5-round). Calibers: 22-250, 5.6×57, 6mm Rem., 243 Win., 6.5×57, 270 Win., 7×57, 7×64, 308 Win., 30-06. 23.63-inch bbl. Weight: 7.3 lbs. Single- or double-set trigger (mechanisms interchangeable). Sights: open rear; hooded ramp front. Half stock of European walnut, checkered forearm and pistol grip, Monte Carlo cheekpiece, rosewood forend tip, recoil pad QD swivels. Imported 1972-81.

Mannlicher-M72 Model S Rifle **$750**
Same general specifications as M72 Model L/M Rifle, except w/Magnum action, 4-round magazine, 25.63-inch bbl., weighs 8.6 lbs. Calibers: 6.5×68, 7mm Rem. Mag., 8×68S, 9.3×64, 375 H&H Mag. Imported 1972-81.

Mannlicher-M72 Model S/T Rifle **$995**
Same as M72 Model S Rifle, except w/heavy 25.63-inch bbl., weighs 9.3 lbs. Calibers: 300 Win. Mag. 9.3×64, 375 H&H Mag., 458 Win. Mag. Option of 23.63-inch bbl. in latter caliber. Imported 1975-81.

Styer/Mannlicher Model SBS Forester Rifle
Calibers: 243 Win., 25-06 Rem., 270 Win., 6.5x55mm, 6.5x57mm, 7×64mm, 7mm-08 Rem., 30-06, 308 Win. 9.3×64mm. 4-shot detachable magazine. 23.6-inch bbl. 44.5 inches overall. Weight: 7.5 lbs. No sights w/drilled and tapped for Browning A-Bolt configuration. Checkered American walnut stock w/Monte Carlo cheekpiece and Pachmayr swivels. Polished or matte blue finish. Imported 1997 to date.
SBS Forester Rifle (Standard Calibers) **$550**
SBS Forester Mountain Rifle (20-inch bbl.). 565
For Magnum Calibers, **add** 25
For Metric Calibers, **add** 100

Styer/Mannlicher Model SBS Pro-Hunter Rifle
Similar to the Forester Model, except w/ASB black synthetic stock. Matte blue finish. Imported 1997 to date.
SBS Pro-Hunter Rifle (Standard Calibers) **$545**
SBS Pro-Hunter Rifle Mountain Rifle (20-inch bbl.). 560
SBS Pro-Hunter Rifle (376 Steyr) 625
SBS Pro-Hunter Youth/Ladies Rifle. 575
For Magnum Calibers, **add** 25
For Metric Calibers, **add** 100
W/Walnut Stock **$1595**
W/Synthetic Stock 1295

MARLIN FIREARMS CO.
North Haven, Connecticut

Marlin Model 9 Semiautomatic Carbine
Calibers: 9mm Parabellum. 12-shot magazine. 16.5-inch bbl. 35.5 inches overall. Weight: 6.75 lbs. Manual bolt hold-open. Sights: hooded post front; adj. open rear. Walnut-finished hardwood stock w/rubber buttpad. Blued or nickel-Teflon finish. Made 1985 to date.
Model 9 . . . . **$185**
Model 9N, Nickel-Teflon (Disc. 1994) 195

Marlin Model 15Y/15YN
Bolt-action, single-shot "Little Buckaroo" rifle. Caliber: 22 Short, Long or LR. 16.25-inch bbl. Weight: 4.25 lbs. Thumb safety. Ramp front sight; adj. open rear. One-piece walnut Monte Carlo stock w/full pistol grip. Made 1984-88. Reintroduced in 1989 as Model 15YN.
Model 15Y . . . . **$ 85**
Model 15YN 95

Marlin Model 18 Baby Slide-Action Repeater **$220**
Exposed hammer. Solid frame. Caliber: 22 LR, Long Short. Tubular magazine holds 14 Short. 20-inch bbl., round or octagon. Weight: 3.75 lbs. Sights: Open rear; bead front. Plain straight-grip stock and slide handle. Made 1906-09.

Marlin Model 20 Slide-Action Repeating Rifle **$225**
Exposed hammer. Takedown. Caliber: 22 LR, Long, Short. Tubular magazine: half-length holds 15 Short, 12 Long, 10 LR; full-length holds 25 Short, 20 Long, 18 LR. 24-inch octagon bbl. Weight: 5 lbs. Sights: open rear; bead front. Plain straight-grip stock, grooved slide handle. Made 1907-22. *Note:* After 1920 was designated "Model 20-S."

**Mannlicher-M72
L/M Carbine**

**Marlin Model 9
9mm Carbine**

**Marlin Model 9N
Nickel-Teflon**

Marlin Model 25 Bolt-Action Rifle **$105**
Caliber: 22 Short, Long or LR.7-shot clip. 22-inch bbl. Weight: 5.5 lbs. Ramp front sight, adj. open rear. One-piece walnut Monte Carlo stock w/full pistol grip Mar-Shield® finish. Made 1984-88.

Marlin Model 25 Slide-Action Repeater **$250**
Exposed hammer. Takedown. Caliber: 22 Short (also handles 22 CB Caps). Tubular magazine holds 15 Short. 23-inch bbl. Weight: 4 lbs. Sights: open rear; beaded front. Plain straight-grip stock and slide handle. Made 1909-10.

Marlin Model 25M Bolt Action W/Scope **$115**
Caliber: 22 WMR. 7-shot clip. 22-inch bbl. Weight: 6 lbs. Ramp front sight w/brass bead, adj. open rear. Walnut-finished stock w/Monte Carlo styling and full pistol grip. Sling swivels. Made 1986-88.

Marlin Model 25MB Midget Magnum **$110**
Bolt action. Caliber: 22 WMR.7-shot capacity.16.25-inch bbl. Weight: 4.75 lbs. Walnut-finished Monte Carlo-style stock w/full pistol grip and abbreviated forend. Sights: ramp front w/brass bead, adj. open rear. Thumb safety. Made 1986-88.

Marlin Model 25MN/25N Bolt-Action Rifle
Caliber: 22 WMR (Model 25MN) or 22 LR (Model 25N). 7-shot clip magazine. 22-inch bbl. 41 inches overall. Weight: 5.5 to 6 lbs. Adj. open rear sight, ramp front; receiver grooved for scope mounts. One piece walnut-finished hardwood Monte Carlo stock w/pistol grip. Made 1989 to date.
Marlin Model 25MN **$115**
Marlin Model 25N **105**

Marlin Model 27 Slide-Action Repeating Rifle **$250**
Exposed hammer. Takedown. Calibers: 25-20, 32-20. ⅔ magazine (tubular) holds 7 shots. 24-inch octagon bbl. Weight: 5.75 lbs. Sights: open rear; bead front. Plain straight-grip stock, grooved slide handle. Made 1910-16.

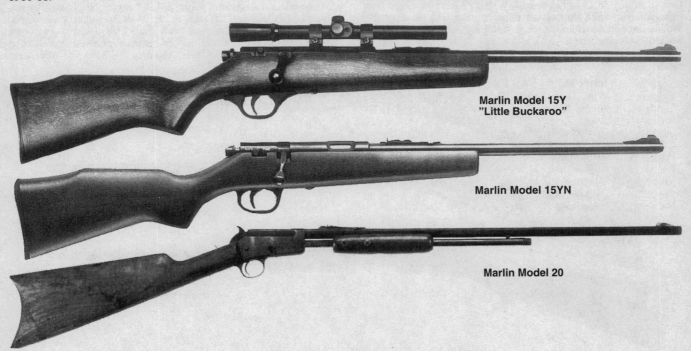

**Marlin Model 15Y
"Little Buckaroo"**

Marlin Model 15YN

Marlin Model 20

**Marlin Model 25M
Bolt-Action Rifle**

**Marlin Model 25MB
Midget Magnum**

Marlin Model 27S $265
Same as Model 27, except w/round bbl., also chambered for 25 Stevens R.F. Made 1920-32.

Marlin Model 29 Slide-Action Repeater $260
Similar to Model 20, w/23-inch round bbl., half magazine only, weighs 5.75 lbs. **Model 37** is same type except w/24-inch bbl. and full magazine. Made 1913-16.

Marlin Model 30AS Lever-Action
Caliber: 30/30 Win. 6-shot tubular magazine. 20-inch bbl. w/Micro-Groove rifling. 38.25 inches overall. Weight: 7 lbs. Brass bead front sight, adj. rear. Solid top receiver, offset hammer spur for scope use. Walnut-finished hardwood stock w/pistol grip. Mar-Shield® finish. Made 1984 to date.
Model 30AS $185
Model 30AS w/4x Scope, **add** 35

Marlin Model 32 Slide-Action Repeater $425
Hammerless. Takedown. Caliber: 22 LR, Long, Short. ⅔ tubular magazine holds 15 Short, 12 Long, 10 LR; full magazine, 25 Short, 20 Long, 18 LR. 24-inch octagon bbl. Weight: 5.5 lbs. Sights: open rear; bead front. Plain pistol-grip stock, grooved slide handle. Made 1914-15.

Marlin Model 36 Lever-Action Repeating
Carbine $395
Calibers: 30-30, 32 Special. 6-shot tubular magazine. 20-inch bbl. Weight: 6.5 lbs. Sights: Open rear; bead front. Pistol-grip stock, semibeavertail forearm w/carbine bbl. band. Made 1936-48. *Note:* In 1936, this was designated "Model 1936."

Marlin Model 36 Sporting Carbine $325
Same as Model 36A rifle except w/20-inch bbl., weighs 6.25 lbs.

Marlin Model 36A/36A-DL Lever-Action Repeating Rifle
Same as Model 36 Carbine, except has 24-inch bbl. w/hooded front sight and ⅔ magazine holding 5 cartridges. Weighs 6.75 lbs. Model 36A-DL has deluxe checkered stock, semibeavertail forearm, swivels and sling. Made 1938-48.
Model 36A (Standard) $325
Model 36A-DL (Deluxe) 395
Model 36A-DL (Blued action & bbl.) 250

Marlin Model 38 Slide-Action Repeating Rifle $245
Hammerless. Takedown. Caliber: 22 LR, Long, Short. ⅔ magazine (tubular) holds 15 Short, 12 Long, 10 LR. 24-inch octagon or round bbls. Weight: 5.5 lbs. Sights: open rear; bead front. Plain shotgun-type pistol-grip buttstock w/hard rubber buttplate, grooved slide handle. Ivory bead front sight; adj. rear. About 20,000 Model 38 rifles were made between 1920-30.

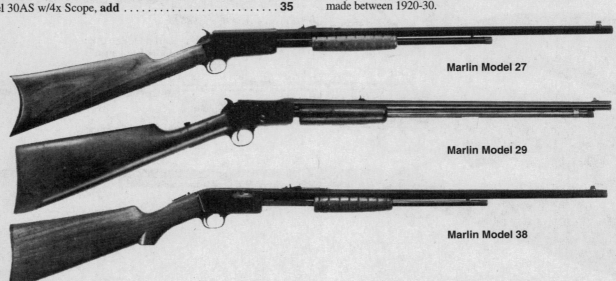

Marlin Model 27

Marlin Model 29

Marlin Model 38

Marlin Model 39 Carbine

Marlin Model 39 Century Ltd.

**Marlin Model 39 Carbine
90th Anniversary**

**Marlin Model 39 Rifle
Original First Issue**

Marlin Model 39 Carbine **$235**
Same as Model 39M, except w/lightweight bbl., ¾ magazine (capacity: 18 Short, 14 Long, 12 LR), slimmer forearm. Weight: 5.25 lbs. Made 1963-67.

Marlin Model 39 90th Anniversary Carbine **$1195**
Carbine version of 90th Anniversary Model 39A. 500 made in 1960. Value is for carbine in new, unfired condition.

Marlin Model 39 Century Ltd. **$295**
Commemorative version of Model 39A. Receiver inlaid w/brass medallion, Marlin Centennial 1870-1970. Square lever. 20-inch octagon bbl. Fancy walnut straight-grip stock and forearm; brass forend cap, buttplate, nameplate in buttstock. 35,388 made in 1970.

Marlin Model 39 Lever-Action Repeater **$1595**
Takedown. Casehardened receiver. Caliber: 22 LR, Long, Short. Tubular magazine holds 25 Short, 20 Long, 18 LR. 24-inch octagon bbl. Weight: 5.75 lbs. Sights: open rear; bead front. Plain pistol-grip stock and forearm. Made 1922-38.

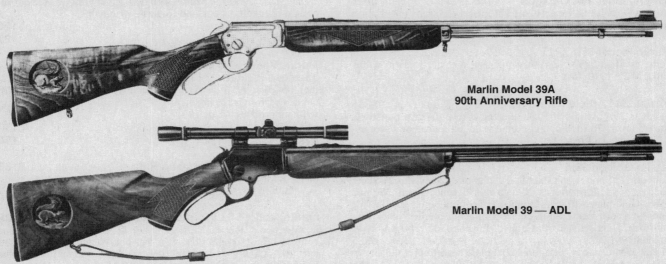

**Marlin Model 39A
90th Anniversary Rifle**

Marlin Model 39 — ADL

Marlin Model 39AS

**Marlin Model 39TDS
Takedown Carbine**

Marlin Model 39A
General specifications same as Model 39, except w/blued receiver, round bbl., heavier stock w/semibeavertail forearm, weighs 6.5 lbs. Made 1938-60.
Early Model (No prefix) **$925**
Late Model ("B" prefix) **750**

Marlin Model 39A 90th Anniversary Rifle $1150
Commemorates Marlin's 90th anniversary. Same general specifications as Golden 39A, except w/chrome-plated bbl. and action, stock and forearm of select walnut-finely checkered, carved figure of a squirrel on right side of buttstock. 500 made in 1960. Value is for rifle in new, unfired condition.

Marlin Model 39A Article II Rifle $395
Commemorates National Rifle Association Centennial 1871-1971. "The Right to Bear Arms" medallion inlaid in receiver. Similar to Model 39A. Magazine capacity: 26 Short, 21 Long, 19 LR. 24-inch octagon bbl. Fancy walnut pistol-grip stock and forearm; brass forend cap, buttplate. 6,244. Made in 1971.

Marlin Golden 39A/39AS Rifle
Same as Model 39A, except w/gold-plated trigger, hooded ramp front sight, sling swivels. Made 1960-87 (39A); Model 39AS 1988 to date.
Golden 39A **$265**
Golden 39AS (W/Hammer Block Safety) **195**

Marlin Model 39A "Mountie" Lever-Action
Repeating Rifle $295
Same as Model 39A, except w/lighter, straight-grip stock, slimmer forearm. Weight: 6.25 lbs. Made 1953-60.

Marlin Model 39A Octagon $525
Same as Golden 39A, except w/oct. bbl., plain bead front sight, slimmer stock and forearm, no pistol-grip cap or swivels. Made 1973.

Marlin Model 39D $195
Same as Model 39M, except w/pistol-grip stock, forearm w/bbl. band. Made 1970-74.

Marlin 39M Article II Carbine $395
Same as 39A Article II Rifle, except w/straight-grip buttstock, square lever, 20-inch octagon bbl., reduced magazine capacity. 3,824. Made in 1971.

Marlin Golden 39M
Calibers: 22 Short, Long and LR. Tubular magazine holds 21 Short, 16 Long or 15 LR cartridges. 20-inch bbl. 36 inches overall. Weight: 6 lbs. Gold-plated trigger. Hooded ramp front sight, adj. folding semi-buckhorn rear. Two-piece, straight-grip American black walnut stock. Sling swivels. Mar-Shield® finish. Made 1960-87.
Golden 39M **$325**
Model 39M Octagon (octagonal bbl., plain bead front sight, no swivels, made 1973) **495**

Marlin Model 39M "Mountie" Carbine
Same as Model 39A "Mountie" Rifle, except w/20-inch bbl. Weight: 6 lbs. 500 made 1960. (For values See Marlin 39 90th Anniversary Carbine)

Marlin Model 39TDS Carbine $195
Same general specifications as Model 39M, except takedown style w/16.5-inch bbl. and reduced magazine capacity. 32.63 inches overall. Weight: 5.25 lbs. Made 1988-95.

Marlin Model 45 . . . . $200
Semiautomatic action. Caliber: 45 Auto.7-shot clip.16.5-inch bbl. 35.5 inches overall. Weight: 6.75 lbs. Manual bolt hold-open. Sights: ramp front sight w/brass bead, adj. folding rear. Receiver drilled and tapped for scope mount. Walnut-finished hardwood stock. Made 1986 to date.

Marlin Model 49/49DL Autoloading Rifle
Same as Model 99C, except w/two-piece stock, checkered after 1970. Made 1968-71. **Model 49DL** w/scrollwork on sides of receiver, checkered stock and forearm; made 1971-78.
Model 49 . . . . **$120**
Model 49DL **125**

Marlin Model 50/50E Autoloading Rifle
Takedown. Caliber: 22 LR. 6-shot detachable box magazine. 22-inch bbl. Weight: 6 lbs. Sights: open rear; bead front; **Model 50E** w/peep rear sight, hooded front. Plain pistol-grip stock, forearm w/finger grooves. Made 1931-34.
Model 50 . . . . **$130**
Model 50E **150**

Marlin Model 56 Levermatic Rifle $165
Same as Model 57 except clip-loading. Magazine holds eight rounds. Weight: 5.75 lbs. Made 1955-64.

Marlin Model 57 Levermatic Rifle $185
Lever-action. Caliber: 22 LR, 22 Long, 22 Short. Tubular magazine holds 19 LR, 21 Long, 27 Short. 22-inch bbl. Weight: 6.25 lbs. Sights: open rear, adj. for windage and elevation; hooded ramp front. Monte Carlo-style stock w/pistol grip. Made 1959-65.

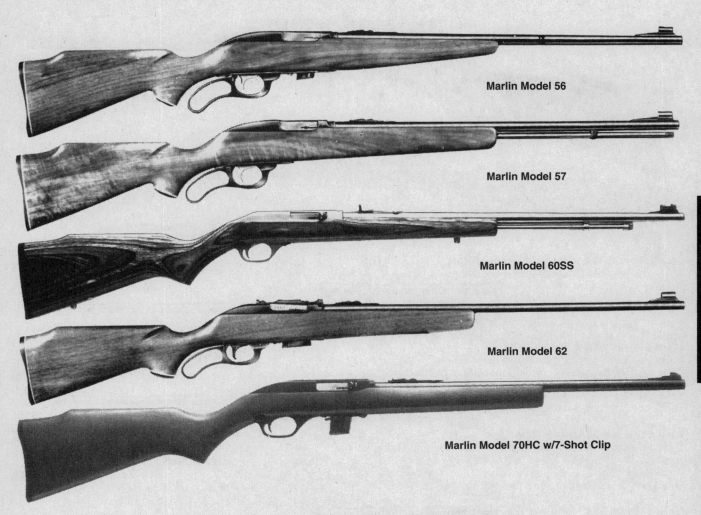

Marlin Model 56

Marlin Model 57

Marlin Model 60SS

Marlin Model 62

Marlin Model 70HC w/7-Shot Clip

Marlin Model 57M Levermatic $195
Same as Model 57, except chambered for 22 WMR cartridge, w/24-inch bbl., 15-shot magazine. Made 1960-69.

Marlin Model 60 Semiautomatic Rifle $85
Caliber: 22 LR. 14-shot tubular magazine. 22-inch bbl. 40.5 inches overall. Weight: 5.5 lbs. Grooved receiver. Ramp front sight w/removable hood; adj. open rear. Anodized receiver w/blued bbl. Monte Carlo-style walnut-finished hardwood stock w/Mar-Shield® finish. Made 1981 to date. *Note:* Marketed 1960-1980 under Glenfield promotion logo, and w/slightly different stock configuration.

Marlin Model 60SS Semiautomatic Rifle
Same general specifications as Model 60, except w/stainless bbl. and magazine tube. Synthetic, uncheckered birch or laminated black/gray birch stock w/nickel-plated swivel studs. Made 1993 to date.
Model 60SB w/uncheckered birch stock $135
Model 60SS w/laminated birch stock................... 165
Model 60SSK w/Fiberglass Stock 155

Marlin Model 62 Levermatic Rifle $325
Lever-action. Calibers: 256 Magnum, 30 Carbine. 4-shot clip magazine. 23-inch bbl. Weight: 7 lbs. Sights: open rear; hooded ramp front. Monte Carlo-style stock w/pistol grip, swivels and sling. Made in 256 Magnum 1963-66; in 30 Carbine 1966-69.

Marlin Model 65 Bolt-Action Single-Shot Rifle $75
Takedown. Caliber: 22 LR, Long, Short. 24-inch bbl. Weight: 5 lbs. Sights: open rear; bead front. Plain pistol-grip stockw/grooved forearm. Made 1932-38. **Model 65E** is same as Model 65, except w/rear peep sight and hooded front sight.

Marlin Model 70HC Semiautomatic
Caliber: 22 LR. 7- and 15-shot magazine. 18-inch bbl. Weight: 5.5 lbs. 36.75 inches overall. Ramp front sight; adj. open rear. Grooved receiver for scope mounts. Walnut-finished hardwood stock w/Monte Carlo and pistol grip. Made 1988-96.
Marlin Model $115
Glenfield Model 95

Marlin Model 70P Semiautomatic $120
"Papoose" takedown. Caliber: 22 LR. 7-shot clip. 16.25-inch bbl. 35.25 inches overall. Weight: 3.75 lbs. Sights: ramp front, adj. open rear. Side ejection, manual bolt hold-open. Cross-bolt safety. Walnut-finished hard-wood stock w/abbreviated forend, pistol grip. Made 1984-94.

Marlin Model 70SS Semiautomatic $145
Similar to Model 70P, except has stainless bbl. and breech bolt w/black synthetic stock. Made 1995 to date.

Marlin Model 70P

Marlin Model 75C

Marlin Model 80C

Marlin Model 80DL

Marlin Model 81DL

Marlin Model 88-C

Marlin Model 89-C

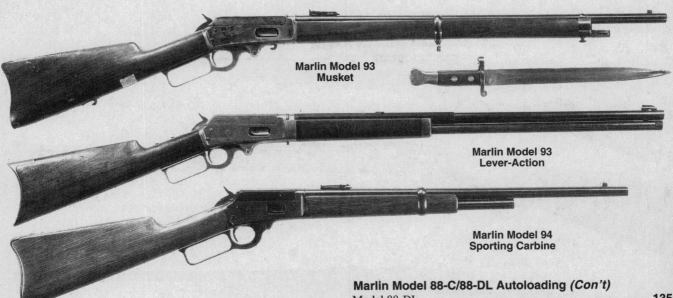

**Marlin Model 93
Musket**

**Marlin Model 93
Lever-Action**

**Marlin Model 94
Sporting Carbine**

Marlin Model 75C Semiautomatic $125

Caliber: 22 LR. 13-shot tubular magazine. 18-inch bbl. 36.5 inches overall. Weight: 5 lbs. Side ejection. Cross-bolt safety. Sights: ramp-mounted blade front; adj. open rear. Monte Carlo-style walnut-finished hardwood stock w/pistol grip. Made 1975-92.

Marlin Model 80 Bolt-Action Repeating Rifle

Takedown. Caliber: 22 LR, Long, Short. 8-shot detachable box magazine. 24-inch bbl. Weight: 6 lbs. Sights: open rear; bead front. Plain pistol-grip stock. Made 1934-39. **Model 80E,** w/peep rear sight; hooded front, made 1934-40.

Model 80 Standard $110
Model 80E 105

Marlin Model 80C/80DL Bolt-Action Repeater

Improved version of Model 80. **Model 80C** w/bead from sight, semibeavertail forearm; made 1940-70. **Model 80DL** w/peep rear sight; hooded blade front sight on ramp, swivels; made 1940-65.

Model 80C $100
Model 80DL 105

Marlin Model 81/81E Bolt-Action Repeater

Takedown. Caliber: 22 LR, Long, Short. Tubular magazine holds 24 Short, 20 Long, 18 LR. 24-inch bbl. Weight: 6.25 lbs. Sights: open rear, bead front. Plain pistol-grip stock. Made 1937-40. **Model 81E** w/peep rear sight; hooded front w/ramp.

Model 81 . . . . $115
Model 81E 135

Marlin Model 81C/81DL Bolt-Action Repeater

Improved version of Model 81 w/same general specifications. **Model 81C** w/bead front sight, semibeavertail forearm; made 1940-70. **Model 81 DL** w/peep rear sight hooded front, swivels; disc. 1965.

Model 81C $115
Model 81DL 125

Marlin Model 88-C/88-DL Autoloading

Takedown. Caliber: 22 LR. Tubular magazine in buttstock holds 14 cartridges. 24-inch bbl. Weight: 6.75 lbs. Sights: open rear; hooded front. Plain pistol-grip stock. Made 1947-56. **Model 88-DL** w/received peep sight, checkered stock and sling swivels, made 1953-56.

Model 88-C $130

Marlin Model 88-C/88-DL Autoloading *(Con't)*

Model 88-DL **135**

Sights open rear; bead front. Plain straight-grip stock and forearm (also available w/pistol-grip stock). Made 1894-1934. *Note:* Before 1906 designated "Model 1894."

Marlin Model 89-C/89-DL Autoloading Rifle

Clip magazine version of Model 88-C. 7-shot clip (12-shot in later models); other specifications same. Made 1950-61. **Model 89-DL** w/receiver peep sight, sling swivels.

Model 89-C **$125**
Model 89-DL **155**

Marlin Model 92 Lever-Action Repeating Rifle . . . $1295

Calibers: 22 Short, Long, LR; 32 Short, Long (rimfire or centerfire by changing firing pin). Tubular magazines: holding 25 Short, 20 Long, 18 LR (22); 17 Short, 14 Long (32); 16-inch bbl. model w/shorter magazine holding 15 Short, 12 Long, 10 LR. Bbl. lengths: 16 (22 cal. only) 24, 26, 28 inches. Weight: 5.5 lbs. w/24-inch bbl. Sights: open rear; blade front. Plain straight-grip stock and forearm. Made 1892-1916. *Note:* Originally designated "Model 1892."

Marlin Model 93/93SC Carbine

Same as Standard Model 93 Rifle, except in calibers 30-30 and 32 Special only. Model 93 w/7-shot magazine. 20-inch round bbl., carbine sights, weighs 6.75 lbs. Model 93SC has ⅔ magazine holding 5 shots, weighs 6.5 lbs.

Model 93 Carbine **$1450**
Model 93SC Sporting Carbine **1495**

Marlin Model 93 Lever-Action Repeating Rifle . . . $1595

Solid frame or takedown. Calibers: 25-36 Marlin, 30-30, 32 Special, 32-40, 38-55. Tubular magazine holds 10 cartridges. 26-inch round or octagon bbl. standard; also made w/28-, 30- and 32-inch bbls. Weight: 7.25 lbs. Sights: open rear; bead front. Plain straight-grip stock and forearm. Made 1893-1936. *Note:* Before 1915 designated "Model 1893."

Marlin Model 93 Musket $1650

Same as Standard Model 93, except w/30-inch bbl., angular bayonet, ramrod under bbl., musket stock, full-length military-style forearm. Weight: 8 lbs. Made 1893-1915.

Marlin Model 94 Lever-Action Repeating Rifle . . . $1475

Solid frame or takedown. Calibers: 25-20, 32-20, 38-40, 44-40. 10-shot tubular magazine. 24-inch round or octagon bbl. Weight: 7 lbs.

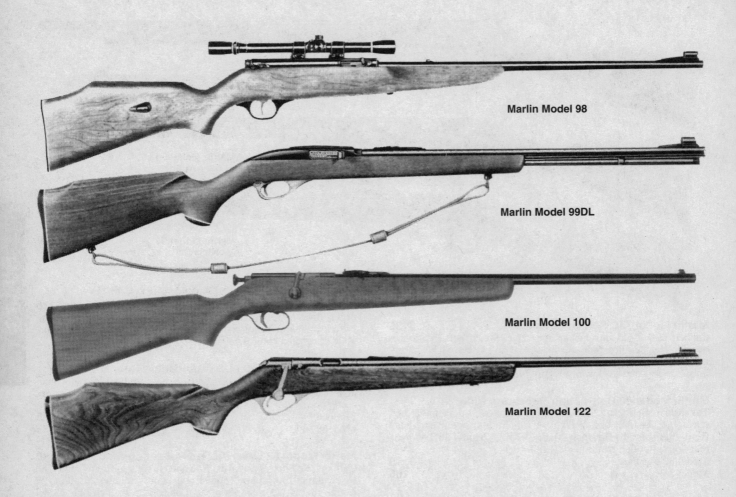

Marlin Model 98

Marlin Model 99DL

Marlin Model 100

Marlin Model 122

Marlin Model 94 Lever-Action Cowboy Series
Calibers: 357 Mag., 44-40, 44 Mag., 45 LC. 10-shot magazine. 24-inch tapered octagon bbl. Weight: 7.5 lbs. 41.5 inches overall. Marble carbine front sight, adjustable semi-buckhorn rear. Blue finish. Checkered straight-grip American black walnut stock w/hard rubber buttplate. Made 1996 to date. Cowboy II introduced in 1997.
Cowboy Model I (45 LC) $395
Cowboy Model II (357 Mag., 44-40, 44 Mag) 375

Marlin Model 97 Lever-Action Repeating Rifle . . . $1495
Takedown. Caliber: 22 LR, Long, Short. Tubular magazine; full length holds 25 Short, 20 Long, 18 LR; half length holds 16 Short, 12 Long and 10 LR. Bbl. lengths: 16, 24, 26, 28inches. Weight: 6 lbs. Sights: open rear; bead front. Plain straight-grip stock and forearm (also avail. w/pistol-grip stock). Made 1897-1922. *Note:* Before 1905 designated "Model 1897."

Marlin Model 98 Autoloading Rifle $125
Solid frame. Caliber: 22 LR. Tubular magazine holds 15 cartridges. 22-inch bbl. Weight: 6.75 lbs. Sights: open rear; hooded ramp front. Monte Carlo stock w/cheekpiece. Made 1950-61.

Marlin Model 99 Autoloading Rifle $135
Caliber: 22 LR. Tubular magazine holds 18 cartridges. 22-inch bbl. Weight: 5.5 lbs. Sights: open rear; hooded ramp front. Plain pistol-grip stock. Made 1959-61.

Marlin Model 99C $145
Same as Model 99 except w/gold-plated trigger, receiver grooved for tip-off scope mounts, Monte Carlo stock (checkered in later production). Made 1962-78.

Marlin Model 99DL $150
Same as Model 99 except w/gold-plated trigger, jeweled breech bolt, Monte Carlo stock w/pistol grip, swivels and sling. Made 1960-65.

Marlin Model 99M1 Carbine $165
Same as Model 99C except styled after U.S. 30M1 Carbine; 9-shot tubular magazine, 18-inch bbl. Sights: open rear; military-style ramp front; carbine stock w/handguard and bbl. band, sling swivels. Weight: 4.5 lbs. Made 1966-79.

Marlin Model 100 Bolt-Action Single-Shot Rifle $85
Takedown. Caliber: 22 LR, Long, Short. 24-inch bbl. Weight: 4.5 lbs. Sights: open rear; bead front. Plain pistol-grip stock. Made 1936-60.

Marlin Model 100SB $95
Same as Model 100 except smoothbore for use w/22 shot cartridges, shotgun sight. Made 1936-41.

Marlin Model 100 Tom Mix Special $300
Same as Model 100 except w/peep rear sight; hooded front; sling. Made 1936-46.

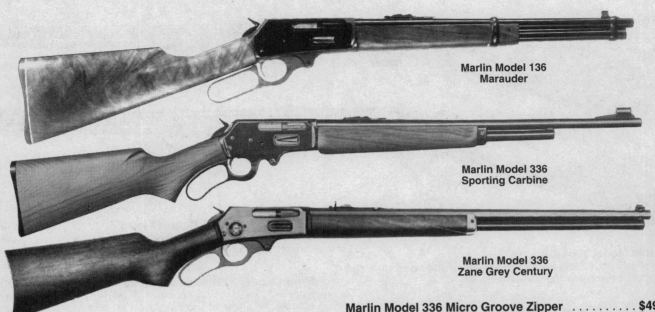

Marlin Model 136 Marauder

Marlin Model 336 Sporting Carbine

Marlin Model 336 Zane Grey Century

Marlin Model 101 . . . . **$90**
Improved version of Model 100 w/same general specifications except w/stock w/beavertail forearm, weighs 5 lbs. Intro. 1951. Disc.

Marlin Model 101 DL **$105**
Same as Model 101 except has peep rear sight; hooded front, swivels. Disc.

**Marlin Model 122 Single-Shot Junior
Target Rifle** **$75**
Bolt action. Caliber: 22 LR, 22 Long, 22 Short. 22-inch bbl. Weight: 5 lbs. Sights: open rear; hooded ramp front. Monte Carlo stock w/pistol grip, swivels, sling. Made 1961-65.

Marlin Model 322 Bolt-Action Varmint Rifle **$525**
Sako short Mauser action. Caliber: 222 Rem. 3-shot clip magazine. 24-inch medium weight bbl. Checkered stock. Sights: two-position peep rear; hooded ramp front. Weight: 7.5 lbs. Made 1954-57.

Marlin Model 336 Marauder **$350**
Same as Model 336 Texan Carbine except w/16.25-inch bbl., weighs 6.25 lbs. Made 1963-64.

Marlin Model 336 Micro Groove Zipper **$495**
General specifications same as Model 336 Sporting Carbine, except caliber 219 Zipper. Made 1955-61.

Marlin Model 336 Octagon **$350**
Same as Model 336T, except chambered for 30-30 only w/22-inch octagon bbl. Made 1973.

Marlin Model 336 Sporting Carbine **$250**
Same as Model 336A rifle, except w/20-inch bbl., weighs 6.25 lbs. Made 1948-63.

Marlin Model 336 Zane Grey Century **$325**
Similar to Model 336A, except w/22-inch octagonal bbl., caliber 30-30, Zane Grey Centennial 1872-1972 medallion inlaid in receiver; select walnut stock w/classic pistol grip and forearm; brass buttplate, forend cap. Weight: 7 lbs. 10,000 produced (numbered ZG1 through ZG10,000). Made 1972.

Marlin Model 336A Lever-Action Rifle **$245**
Improved version of Model 36A Rifle w/same general specifications, w/improved action w/round breech bolt. Calibers: 30-30, 32 Special (disc. 1963), 35 Rem. (intro. 1952). Made 1948-63; reintroduced 1973, disc. 1980.

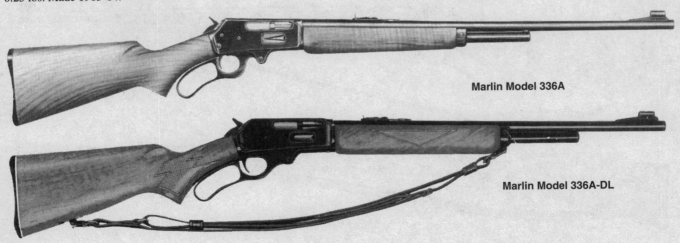

Marlin Model 336A

Marlin Model 336A-DL

RIFLES

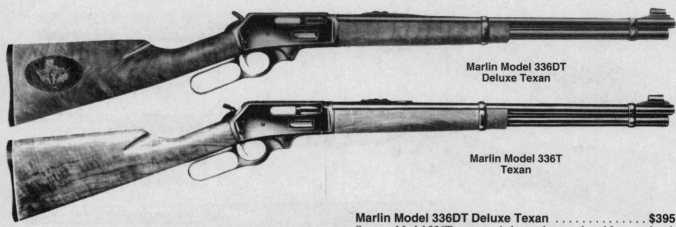

**Marlin Model 336DT
Deluxe Texan**

**Marlin Model 336T
Texan**

Marlin Model 336A-DL **$425**
Same as Model 336A Rifle except w/deluxe checkered stock and forearm, swivels and sling. Made 1948-63.

Marlin Model 336C Lever-Action Carbine **$220**
Improved version of Model 36 Carbine w/same general specifications, w/improved action w/round breech bolt. Original calibers: 30-30 and 32 Win. Spec. Made 1948-83. *Note:* Caliber 35 Rem. intro. 1953. Caliber 32 Winchester Special disc. 1963.

Marlin Model 336CS W/Scope **$265**
Lever-action w/hammer block safety. Caliber: 30/30 Win. or 35 Rem. 6-shot tubular magazine. 20-inch round bbl. w/Micro-Groove rifling. 38.5 inches overall. Weight: 7 lbs. Ramp front sight w/hood, adj. semibuckhorn folding rear. Solid top receiver drilled and tapped for scope mount or receiver sight; offset hammer spur for scope use. American black walnut stock w/pistol grip, fluted comb. Mar-Shield® finish. Made 1984 to date.

Marlin Model 336DT Deluxe Texan **$395**
Same as Model 336T, except w/select walnut stock and forearm, hand-carved longhorn steer and map of Texas on buttstock. Made 1962-64.

Marlin Model 336T Texan Carbine **$225**
Same as Model 336 Carbine, except w/straight-grip stock and is not available in caliber 32 Special. Made 1953-83. Caliber 44 Magnum made 1963-67.

Marlin Model 336TS **$195**
Lever-action w/hammer block safety. Caliber: 30-30 Win. 6-shot tubular magazine. 18.5-inch Micro-Groove bbl. 37 inches overall. Weight: 6.5 lbs. Ramp front sight, adj. semibuckhorn folding rear. Straight-grip American black walnut stock. Made 1983-87.

Marlin Model 444 Lever-Action Repeating Rifle **$275**
Action similar to Model 336. Caliber: 444 Marlin. 4-shot tubular magazine. 24-inch bbl. Weigh: 7.5 lbs. Sights: open rear; hooded ramp front. Monte Carlo stock w/straight grip, recoil pad. Carbine-style forearm w/bbl. band. Swivels, sling. Made 1965-71.

**Marlin Model 336TS
Carbine**

Marlin Model 444SS

**Marlin Model 455
Sporter**

Marlin Model 780

Marlin Model 781

Marlin Model 783

RIFLES

Marlin Model 444 Sporter $285
Same as Model 444 Rifle, except w/22-inch bbl., pistol-grip stock and forearm as on Model 336A, recoil pad, QD swivels and sling. Made 1972-83.

Marlin Model 444SS $325
Same general specifications as Model 444, except w/hammer safety. Made 1984 to date.

Marlin Model 455 Bolt-Action Sporter $335
FN Mauser action w/Sako trigger. Calibers: 270, 30-06, 308. 5-shot box magazine. 24-inch medium weight stainless-steel bbl. Monte Carlo stock w/cheekpiece, checkered pistol grip and forearm. Lyman 48 receiver sight; hooded ramp front. Weight: 8.5 lbs. Made 1957-59.

Marlin Model 780 Bolt-Action Repeater Series
Caliber: 22 LR, Long, Short. 7-shot clip magazine. 22-inch bbl. Weight: 5.5 to 6 lbs. Sights: open rear; hooded ramp front. Receiver grooved for scope mounting. Monte Carlo stock w/checkered pistol grip and forearm. Made 1971-88.
Model 780 Standard . . . . $ 90
Model 781 (w/17-shot tubular magazine) 95
Model 782 (22 WMR, w/swivels, sling) 120
Model 783 (w/12-shot tubular magazine) 125

Marlin Model 880 Bolt-Action Repeater Series
Caliber: 22 rimfire. 7-shot magazine. 22-inch bbl. 41 inches overall. Weight: 5.5 to 6 lbs. Hooded ramp front sight; adj. folding rear. Grooved receiver for scope mounts. Checkered Monte Carlo-style walnut stock w/QD studs and rubber recoil pad. Made 1989-97.
Model 880 (22 LR) $130
Model 881 (w/7-shot tubular magazine 135
Model 882 (22 WMR) 140
Model 882L (w/laminated hardwood stock) 150
Model 883 (22 WMR w/12-shot tubular magazine) 145
Model 883N (w/nickel-teflon finish) 165
Model 883SS (stainless w/laminated stock) 175

Marlin Model 980 22 Magnum $125
Bolt action. Caliber: 22 WMR. 8-shot clip magazine. 24-inch bbl. Weight: 6 lbs. Sights: open rear; hooded ramp front. Monte Carlo stock, swivels, sling. Made 1962-70.

Marlin Model 989 Autoloading Rifle $110
Caliber: 22 LR. 7-shot clip magazine. 22-inch bbl. Weight: 5.5 lbs. Sights: open rear; hooded ramp front. Monte Carlo walnut stock w/pistol grip. Made 1962-66.

Marlin Model 989M2 Carbine $115
Same as Model 99M1, except clip-loading, 7-shot magazine. Made 1966-79.

Marlin Model 990 Semiautomatic
Caliber: 22 LR. 17-shot tubular magazine. 22-inch bbl. 40.75 inches overall. Weight: 5.5 lbs. Side ejection. Cross-bolt safety. Ramp front sight w/brass bead; adj. semibuckhorn folding rear. Receiver grooved for scope mount. Monte Carlo-style American black walnut stock w/checkered pistol grip and forend. Made 1979-87.
Model 990 Semiautomatic $110
Model 990L (w/14 Rounds, Laminated hardwood
   stock, QD studs, Black recoil pad; 1992 to date) 135

Marlin Model 922 Magnum Self-Loading Rifle $255
Similar to Model 9, except chambered for 22 WMR. 7-shot magazine. 20.5-inch bbl. 39.5 inches overall. Weight: 6.5 lbs. American black walnut stock w/Monte Carlo. Blued finish. Made 1993 to date.

Marlin Model 995 Semiautomatic $125
Caliber: 22 LR. 7-shot clip magazine. 18-inch bbl. 36.75 inches overall. Weight: 5 lbs. Cross-bolt safety. Sights: ramp front w/brass bead; adj. folding semibuckhorn rear. Monte Carlo-style American black walnut stock w/checkered pistol grip and forend. Made 1979-94.

Marlin Model 1870-1970 Centennial Matched Pair, Models 336 and 39 . . . . $1495
Presentation grade rifles in luggage-style case. Matching serial numbers. Fancy walnut straight-grip buttstock and forearm brass buttplate and forend cap. Engraved receiver w/inlaid medallion; square lever. 20-inch octagon bbl. **Model 336:** 30-30, 7-shot, 7 lbs. **Model 39:** 22 Short, Long, LR, tubular magazine holds 21

Marlin Model 882L

Marlin Model 883N

Marlin Model 980

Marlin Model 989

Marlin Model 989M2

Marlin Model 990

Marlin Model 990L

Marlin Model 995

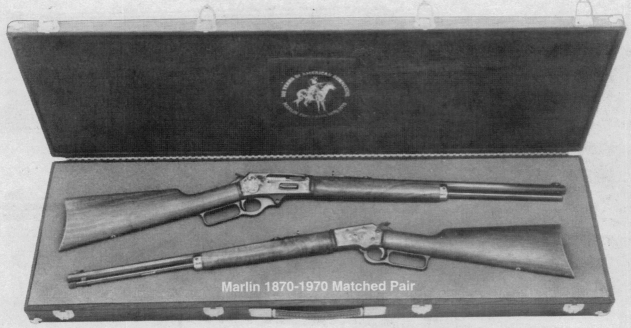

Marlin 1870-1970 Matched Pair

Marlin Model 1870-1970 Centennial Matched *(Con't)*

Short, 16 Long, 15 LR. 1,000 sets produced. Made 1970. Value is for rifles in new, unfired condition.

Marlin Model 1892 Lever-Action Rifle

See Marlin Model 92.

Marlin Model 1893 Lever-Action Rifle

See Marlin Model 93.

Marlin Model 1894 Carbine

Replica of original Model 94. Caliber: 44 Rem. 10-shot magazine. 20-inch round bbl. Weight: 6 lbs. Sight: open rear; ramp front. Straight-grip stock. Made 1969-84.

Standard Model 1894 Carbine **$285**
Model 1894 Octagon (made 1973) **275**
Model 1894 Sporter (w/22-inch bbl., made 1973) **310**

Marlin Model 1894 Lever-Action Rifle

See Marlin Model 94. Lever-Action Rifle that was listed previously under this session.

Marlin Model 1894CL Classic **$275**

Calibers: 218 Bee, 25-20 Win., 32-20 Win. 6-shot tubular magazine. 22-inch bbl. 38.75 inches overall. Weight: 6.25 lbs. Adj. semibuckhorn folding rear sight, brass bead front. Receiver tapped for scope mounts. Straight-grip American black walnut stock w/Mar-Shield® finish. Made 1988-94.

Marlin Model 1894CS Lever-Action **$250**

Caliber: 357 Magnum, 38 Special. 9-shot tubular magazine. 18.5-inch bbl. 36 inches overall. Weight: 6 lbs. Side ejection. Hammer block safety. Square finger lever. Bead front sight, adj. semibuckhorn folding rear. Offset hammer spur for scope use. Two-piece straight grip American black walnut stock w/white buttplate spacer. Mar-Shield® finish. Made 1984 to date.

Marlin Model 1894M Lever-Action **$225**

Caliber: 22 WMR. 11-shot tubular magazine. 20-inch bbl. Weight: 6.25 lbs. Sights: ramp front w/brass bead and Wide-Scan hood; adj. semibuckhorn folding rear. Offset hammer spur for scope use. Straight-grip American black walnut stock w/white buttplate spacer. Squared finger lever. Made 1986-88.

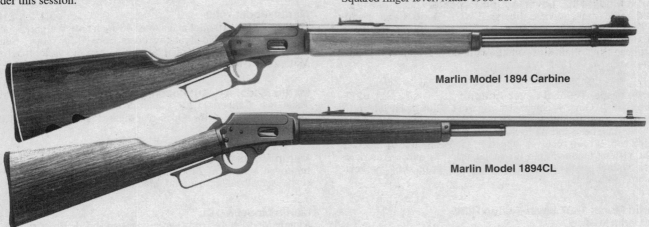

Marlin Model 1894 Carbine

Marlin Model 1894CL

RIFLES

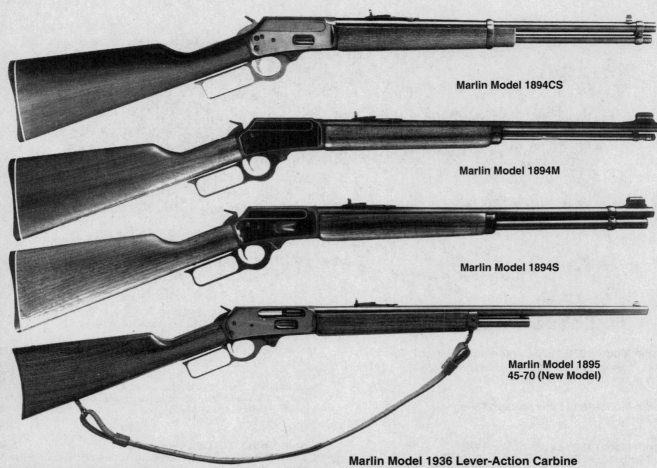

Marlin Model 1894CS

Marlin Model 1894M

Marlin Model 1894S

Marlin Model 1895
45-70 (New Model)

Marlin Model 1894S Lever-Action **$265**
Calibers: 41 Mag., 44 Rem. Mag., 44 S&W Special, 45 Colt.10-shot tubular magazine. 20-inch bbl.37.5 inches overall. Weight: 6 lbs. Sights
and stock same as Model 1894M. Made 1984 to date.

Marlin Model 1895 45-70 Repeater **$275**
Model 336-type action. Caliber: 45-70 Government. 4-shot magazine. 22-inch bbl. Weight: 7 lbs. Sights: open rear; bead front.
Straight-grip stock, forearm w/metal end cap, QD swivels, leather
sling. Made 1972-79.

Marlin Model 1895 Lever-Action Repeater **$1750**
Solid frame or takedown. Calibers: 33 WCF, 38-56, 40-65, 40-70,
40-82, 45-70. 9-shot tubular magazine. 24-inch round or octagongon
bbl. standard (other lengths available). Weight: 8 lbs. Sights: open
rear; bead front. Plain stock and forearm (also available w/pistol-
grip stock). Made 1895-1915.

Marlin Model 1895SS Lever-Action **$275**
Caliber: 45-70 Govt. 4-shot tubular magazine. 22-inch bbl. w/Micro-
Groove rifling. 40.5 inches overall. Weight: 7.5 lbs. Ramp front
sight w/brass bead and Wide-Scan hood; adj. semibuckhorn folding rear. Solid top receiver tapped for scope mount or receiver
sight. Off-set hammer spur for scope use. Two-piece American
black walnut stock w/fluted comb, pistol grip, sling swivels.
Made 1984 to date.

Marlin Model 1897 Lever-Action Rifle
See Marlin Model 97.

Marlin Model 1936 Lever-Action Carbine
See Marlin Model 36.

Marlin Model 2000 Target Rifle
Bolt-action single-shot. Caliber: 22 LR. Optional 5-shot adapter kit
available. 22-inch bbl. 41 inches overall. Weight: 8 lbs. Globe
front sight, adj. peep or aperature rear. 2-stage target trigger. Textured composite Kevlar or black/gray laminated stock . Made
1991 to date.
Model 2000 (Disc. 1995) **$375**
Model 2000A w/Adj. Comb (Made 1994 only) **395**
Model 2000L w/Laminated Stock (Intro. 1996) **415**

Marlin Model 7000 . . . . **$165**
Caliber: 22 LR. 10-shot magazine. 18-inch bbl. Weight: 5.5 lbs.
Synthetic stocks. No sights; receiver grooved for scope. Semi-
auto. Side ejection. Manual bolt hold-open. Cross bolt safety.
Matte finish. Made 1997 to date.

Marlin Model A-1 Autoloading Rifle **$150**
Takedown. Caliber: 22 LR. 6-shot detachable box magazine. 24-
inch bbl. Weight: 6 lbs. Open rear sight. Plain pistol-grip stock.
Made 1935-46.

Marlin Model A-1C Autoloading Rifle **$155**
Improved version of Model A-1 w/same general specifications,
stock w/semibeavertail forend. Made 1940-46.

Marlin Model A-1DL **$185**
Same as Model A-lC, except w/peep rear sight; hooded front, swivels.

Marlin Model 1895 Rifle
(Old Model — 1895-1915)

Marlin Model 2000

Marlin Model A-1 Autoloader

Marlin-Glenfield Model 10

Marlin-Glenfield Model 30A

Marlin-Glenfield Model 60

Marlin-Glenfield Model 70

Marlin-Glenfield Model 80G

RIFLES

Mauser Model ES340

Mauser Model ES350

Marlin Model A-1E **$175**
Same as Model A-1, except w/peep rear sight; hooded front.

Marlin Model MR-7 Bolt-Action Rifle
Calibers: 270 Win. and 30-06, 4-shot magazine. 22-inch bbl.,
43.31 inches overall. Weight: 7 lbs., 9.5 oz. Checkered American
walnut or birch stock w/recoil pad and sling-swivel studs. Jew-
eled bolt. Made from 1996 to date.
Model MR-7 . . . . **$395**
Model MR7B w/Birch Stock (Intro. 1998) 215
Open Sights, **add** 35

Marlin-Glenfield Model 10 . . . . **$80**
Same as Marlin Model 101, except w/walnut-finished hardwood
stock. Made 1966-79. *Note:* Later production w/hot-ironwood-
stamped pistol grip to simulate checkering/carving; plain forend.

Marlin-Glenfield Model 20 . . . . **$80**
Same as Marlin Model 80/780, except w/bead front sight, walnut-
finished hardwood stock. Made 1966-82. *Note:Recent production has
stamped pistol grip to simulate checkering; plain forend.*

Marlin-Glenfield Model 30 . . . . **$175**
Same as Marlin Model 336C, except chambered for 30-30 only,
w/4-shot magazine, plainer stock and forearm of walnut-finished
hardwood. Made 1966-68.

Marlin-Glenfield Model 30A **$185**
Same as Marlin Model 336C, except chambered for 30-30 only,
w/checkered stock of walnut-finished hardwood. Made 1969-83.

Marlin-Glenfield Model 36G **$225**
Same as Marlin Model 336C, except chambered for 30-30 only,
w/5-shot magazine, plainer stock. Made 1960-65.

Marlin-Glenfield Model 60 . . . . **$70**
Same as Marlin Model 99C, except w/walnut-finished hardwood
stock. Made 1960-80.

Marlin-Glenfield Model 70 . . . . **$75**
Same as Marlin Model 989M2, except w/walnut-finished hardwood stock; no handguard. Made 1966-69.

Marlin-Glenfield Model 80G **$75**
Same as Marlin Model 80C, except w/plainer stock, bead front sight.
Made 1960-65.

Marlin-Glenfield Model 81G **$80**
Same as Marlin Model 81C, except w/plainer stock, bead front sight.
Made 1960-65.

Marlin-Glenfield Model 99G **$85**
Same as Marlin Model 99C, except w/plainer stock, bead front sight.
Made 1960-65.

Marlin-Glenfield Model 101G **$70**
Same as Marlin Model 101, except w/plainer stock. Made 1960-65.

Marlin-Glenfield Model 989G Autoloading Rifle **$70**
Same as Marlin Model 989, except w/plain stock, bead front sight.
Made 1962-64.

MAUSER SPORTING RIFLES
Oberndorf am Neckar, Germany
Mfg. by Mauser-Werke GmbH
*Imported by Brolin Arms, Pomona, CA
(Previously by Gun South, Inc.; Gibbs Rifle Co.;
Precision Imports. Inc. and KDF, Inc.)*

*Before the end of WWI, the name of the Mauser firm was "Waf-
fenfabrik Mauser A.-G." Shortly after WWI, it was changed to
"Mauser-Werke A.-G." This information may be used to determine the age of genuine Original-Mauser sporting rifles made
before WWII, because all bear either of these firm names as well
as the "Mauser" banner trademark.*

*The first four rifles listed were manufactured before WWI. Those that
follow were produced between World Wars I and II. The early
Mauser models can generally be identified by the pistol grip, which is
rounded instead of capped, and the M/98 military-type magazine
floorplate and catch. The later models have hinged magazine floorplate with lever or button release.*

PRE-WORLD WAR I MODELS

Mauser Bolt-Action Sporting Carbine **$995**
Calibers: 6.5×54, 6.5×58, 7×57, 8×57, 957mm. 19.75-inch bbl.
Weight: 7 lbs. Full-stocked to muzzle. Other specifications same as
for standard rifle.

Mauser Bolt-Action Sporting Rifle **$850**
Calibers: 6.5×55, 6.5×58, 7×57, 8×57, 9×57, 9.3×62 10.75×68.5-shot
box magazine, 23.5-inch bbl. Weight: 7 to 7.5 lbs. Double-set trigger.
Sights: tangent curve rear; ramp front. Pistol-grip stock, forearm
w/Schnabel tip, swivels.

Mauser Model MS420

Mauser Standard Model

**Mauser Bolt-Action Sporting Rifle,
Military Type** **$525**
So called because of "stepped" M/98-type bbl., military front sight
and double-pull trigger. Calibers: 7×57, 8×57, 9×57mm. Other
specifications same as for standard rifle.

**Mauser Bolt-Action Sporting Rifle,
Short Model** **$850**
Calibers: 6.5×54, 8×51mm. 19.75-inch bbl. Weight: 6.25 lbs. Other
specifications same as for standard rifle.

PRE-WORLD WAR II MODELS

> **NOTE**
>
> The "B" series of Mauser 22 rifles (Model ES340B, MS350B,
> etc.) were improved versions of their corresponding models
> and were introduced about 1935.

**Mauser Model DSM34 Bolt-Action Single-Shot Sporting
Rifle** **$375**
Also called "Sport-model." Caliber: 22 LR. 26-inch bbl. Weight:
7.75 lbs. Sights: tangent curve open rear; barleycorn front. M/98
military-type stock, swivels. Intro. c. 1935.

**Mauser Model EL320 Bolt-Action Single-Shot Sporting
Rifle** **$350**
Caliber: 22 LR. 23.5-inch bbl. Weight: 4.25 lbs. Sights: adj. open
rear; bead front. Sporting stock w/checkered pistol grip, swivels.

**Mauser Model EN310 Bolt-Action Single-Shot Sporting
Rifle** **$325**
Caliber: 22 LR. ("22 Lang fur Buchsen.") 19.75-inch bbl. Weight: 4
lbs. Sights: fixed open rear, blade front. Plain pistol-grip stock.

**Mauser Model ES340 Bolt-Action Single-Shot
Target Rifle** **$375**
Caliber: 22 LR. 25.5-inch bbl. Weight: 6.5 lbs. Sights: tangent curve
rear; ramp front. Sporting stock w/checkered pistol grip and grooved
forearm, swivels.

**Mauser Model ES340B Bolt-Action Single-Shot Target
Rifle** **$375**
Caliber: 22 LR. 26.75-inch bbl. Weight: 8 lbs. Sights: tangent
curve open rear; ramp front. Plain pistol-grip stock, swivels.

**Mauser Model ES350 Bolt-Action Single-Shot
Target Rifle** **$525**
"Meistershaftsbuchse" (Championship Rifle). Caliber: 22 LR.
27.5-inch bbl. Weight: 7.75 lbs. Sights: open micrometer rear;
ramp front. Target stock w/checkered pistol-grip and forearm,
grip cap, swivels.

**Mauser Model ES350B Bolt-Action Single-Shot Target
Rifle** **$475**
Same general specifications as Model MS350B except single-shot,
weighs 8.25 lbs.

**Mauser Model KKW Bolt-Action Single-Shot
Target Rifle** **$375**
Caliber: 22 LR. 26-inch bbl. Weight: 8.75 lbs. Sights: tangent
curve open rear; barleycorn front. M/98 military-type stock,
swivels. *Note:* This rifle has an improved design Mauser 22 action
w/separate nonrotating bolt head. In addition to being pro-duced
for commercial sale, this model was used as a training rifle by the
German armed forces; it was also made by Walther and Gustloff.
Intro. just before WWII.

**Mauser Model M410 Bolt-Action Repeating
Sporting Rifle** **$675**
Caliber: 22 LR. 5-shot detachable box magazine. 23.5-inch bbl.
Weight: 5 lbs. Sights: tangent curve open rear; ramp front. Sporting
stock w/checkered pistol grip, swivels.

**Mauser Model MM410B Bolt-Action Repeating
Sporting Rifle** **$695**
Caliber: 22 LR. 5-shot detachable box magazine. 23.5-inch bbl.
Weight: 6.25 lbs. Sights: tangent curve open rear; ramp front. Light-
weight sporting stock w/checkered pistol grip, swivels.

**Mauser Model MS350B Bolt-Action Repeating
Target Rifle** **$675**
Caliber: 22 LR. 5-shot detachable box magazine. Receiver and
bbl. grooved for detachable rear sight or scope. 26.75-inch bbl.
Weight: 8.5 lbs. Sights: micrometer open rear; ramp front. Target
stock w/checkered pistol grip and forearm, grip cap, sling swivels.

**Mauser Model MS420 Bolt-Action Repeating Sporting
Rifle** **$695**
Caliber: 22 LR. 5-shot detachable box magazine. 25.5-inch bbl.
Weight: 6.5 lbs. Sights: tangent curve open rear; ramp front. Sport-
ing stock w/checkered pistol grip, grooved forearm swivels.

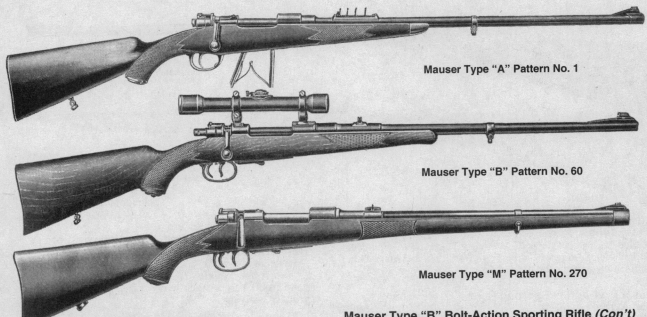

Mauser Type "A" Pattern No. 1

Mauser Type "B" Pattern No. 60

Mauser Type "M" Pattern No. 270

Mauser Model MS420B Bolt-Action Repeating Target Rifle **$625**
Caliber: 22 LR. 5-shot detachable box magazine. 26.75-inch bbl. Weight: 8 lbs. Sights: tangent curve open rear; ramp front. Target stock w/checkered pistol grip, grooved forearm, swivels.

Mauser Standard Model Rifle **$525**
Refined version of German Service Kar. 98k. Straight bolt handle. Calibers: 7mm Mauser (7×57mm), 7.9mm Mauser (8×57mm). 5-shot box magazine. 23.5-inch bbl. Weight: 8.5 lbs. Sights: blade front; adj. rear. Walnut stock of M/98 military-type. *Note:* These rifles were made for commercial sale and are of the high quality found in the Oberndorf Mauser sporters. They bear the Mauser trademark on the receiver ring.

Mauser Type "A" Bolt-Action Sporting Rifle **$1995**
Special British Model. Calibers: 7×57, 30-06 (7.62×63), 8×60, 9×57, 9.3×62mm. 5-shot box magazine. 23.5-inch round bbl. Weight: 7.25 lbs. Military-type single trigger. Sights: express rear; hooded ramp front. Circassian walnut sporting stock w/checkered pistol grip and forearm, w/ or w/o cheekpiece, buffalo horn forend tip and grip cap, detachable swivels. Variations: octagon bbl., double-set trigger, shotgun-type safety, folding peep rear sight, tangent curve rear sight, three-leaf rear sight.

Mauser Type "A" Bolt-Action Sporting Rifle, Magnum Model **$2195**
Same general specifications as standard Type "A," except w/Magnum action, weighs 7.5 to 8.5 lbs. Calibers: 280 Ross, 318 Westley Richards Express, 10.75×68mm, 404 Nitro Express.

Mauser Type "A" Bolt-Action Sporting Rifle, Short Model **$1795**
Same as standard Type "A," except w/short action, 21.5-inch round bbl., weighs 6 lbs. Calibers: 250-3000, 6.5×54, 8×51mm.

Mauser Type "B" Bolt-Action Sporting Rifle **$1350**
Normal Model. Calibers: 7×57, 30-06 (7.62×63), 8×57, 8×60, 9×57, 9.3×62, 10.7568mm. 5-shot box magazine. 23.5-inch round bbl. Weight: 7.25 lbs. Double-set trigger. Sights: three-leaf rear, ramp front. Fine walnut stock w/checkered pistol grip,

Mauser Type "B" Bolt-Action Sporting Rifle *(Con't)*
Schnabel forend tip, cheekpiece, grip cap, swivels. Variations: octagon or half-octagon bbl., military-type single trigger, shotgun-type safety, folding peep rear sight, tangent curve rear sight, telescopic sight.

Mauser Type "K" Bolt-Action Sporting Rifle **$2850**
Light Short Model. Same specifications as Normal Type "B" model except w/short action, 21.5-inch round bbl., weighs 6 lbs. Calibers: 250-3000, 6.5×54, 8×51mm.

Mauser Type "M" Bolt-Action Sporting Carbine **$1795**
Calibers: 6.5×54, 7×57, 30-06 (7.62×63), 8×51, 8×60, 9×57mm. 5-shot box magazine. 19.75-inch round bbl. Weight: 6 to 6.75 lbs. Double-set trigger, flat bolt handle. Sights: three-leaf rear; ramp front. Stocked to muzzle, cheekpiece, checkered pistol grip and forearm, grip cap, steel forend cap, swivels. Variations: military-type single trigger, shotgun-type trigger, shotgun-type safety, tangent curve rear sight, telescopic sight.

Mauser Type "S" Bolt-Action Sporting Carbine . . **$1850**
Calibers: 6.5×54 7×57, 8×51, 8×60, 9×57mm. 5-shot box magazine. 19.75-inch round bbl. Weight: 6 to 6.75 lbs. Double-set trigger. Sights: three-leaf rear; ramp front. Stocked to muzzle, Schnabel forend tip, cheekpiece, checkered pistol grip w/cap, swivels. Variations: same as listed for Normal Model Type "B."

POST-WORLD WAR II MODELS

Mauser Model 66S Bolt-Action Standard Sporting Rifle
Telescopic short action. Bbls. interchangeable within caliber group. Single- or double-set trigger (interchangeable). Calibers: 243 Win., 6.5×57, 270 Win., 7×64, 308 Win., 30-06. 3-round magazine. 23.6-inch bbl. (25.6-inch in 7×64). Weight: 7.3 lbs. (7.5 lbs. in 7×64). Sights: adj. open rear, hooded ramp front. Select European walnut stock, Monte Carlo w/cheekpiece, rosewood forend tip and pistol-grip cap, skip checkering, recoil pad, sling swivels. Made 1965 to date, export to U.S. discontinued 1974. *Note:* U.S. designation, 1971-73, was "Model 660."
Model 66S **$1195**
W/Extra Bbl. Assembly, **add** 500

**Mauser Model 66S
Standard**

**Mauser Model 66S
Deluxe**

**Mauser Model 66SP
Super Match Target Rifle**

**Mauser Model 66ST
Carbine**

Mauser Model 66S Deluxe Sporter
Limited production special order. Model 66S rifles and carbines are available w/elaborate engraving, gold and silver inlays and carved select walnut stocks. Added value is upward of **$1000**.

Mauser Model 66S Ultra
Same general specifications as Model 66S Standard, except w/20.9-inch bbl., weighs 6.8 lbs.
Model 66S Ultra **$1295**
W/Extra Bbl. Assembly, **add** 500

Mauser Model 66SG Big Game
Same general specifications as Model 66S Standard, except w/25.6-inch bbl., weighs 9.3 lbs. Calibers: 375 H&H Mag., 458 Win. Mag. *Note:* U.S. designation, 1971-73, was "Model 660 Safari."
Model 66SG **$1495**
W/ Extra Bbl. Assembly, **add** 550

Mauser Model 66SH High Performance **$1095**
Same general specifications as Model 66S Standard, except w/25.6-inch bbl., weighs 7.5 lbs. (9.3 lbs. in 9.3×64). Calibers: 6.5×68, 7mm Rem. Mag., 7mm S.E.v. Hofe, 300 Win. Mag., 8×68S, 9.3×64.

Mauser Model 66SP Super Match Bolt-Action
Target Rifle **$3195**
Telescopic short action. Adj. single-stage trigger. Caliber: 308 Win. (chambering for other cartridges available on special order). 3-shot magazine. 27.6-inch heavy bbl. w/muzzle brake, dovetail rib for special scope mount. Weight: 12 lbs. Target stock w/wide and deep forearm, full pistol grip, thumbhole adj. cheekpiece, adj. rubber buttplate.

Mauser Model 66ST Carbine
Same general specifications as Model 66S Standard, except w/20.9-inch bbl., full-length stock, weighs 7 lbs.
Model 66ST **$1095**
W/Extra Bbl. Assembly, **add** 500

Mauser Model 83 Bolt-Action Rifle **$1850**
Centerfire single-shot, bolt-action rifle for 300-meter competition. Caliber: 308 Win. 25.5-inch fluted bbl. Weight: 10.5 lbs. Adj. micrometer rear sight globe front. Fully adj. competition stock. Disc. 1988.

Mauser Model 96 . . . . **$525**
Calibers: 25-06, 270 Win., 7×64, 308 Win., 30-06, 7mm Rem. Mag., 300 Win. Mag. 22-inch bbl.; magnums 24-inch. Weight: 6.25 lbs. No sights; drilled and tapped for scope. Walnut stock. 5-shot top-loading magazine. 3-position safety.

RIFLES

Mauser Model 99

Mauser Model 201

Mauser Model 3000

Mauser Model 4000

Mauser Model 99 Classic Bolt-Action Rifle
Calibers: 243 Win., 25-06, 270 Win., 30-06, 308 Win., 257 Wby., 270 Wby., 7mm Rem. Mag., 300 Win., 300 Wby. 375 H&H. 4-shot magazine (standard), 3-shot (Magnum). Bbl.: 24-inch (standard) or 26-inch (Magnum). 44 inches overall (standard). Weight: 8 lbs. No sights. Checkered European walnut stock w/rosewood grip cap available in Classic and Monte Carlo styles w/High-Luster or oil finish. Disc. importing 1994.

Standard Classic or Monte Carlo (Oil Finish)	**$895**
Magnum Classic or Monte Carlo (Oil Finish)	950
Standard Classic or Monte Carlo (H-L Finish)	925
Magnum Classic or Monte Carlo (H-L Finish)	995

Mauser Model 107 Bolt-Action Rifle **$265**
Caliber: 22 LR. Mag. 5-shot magazine. 21.5-inch bbl. 40 inches overall. Weight: 5 lbs. Receiver drilled and tapped for rail scope mounts. Hooded front sight, adj. rear. Disc. importing 1994.

Mauser Model 201/201 Luxus Bolt-Action Rifle
Calibers: 22 LR, 22 Win. Mag. 5-shot magazine. 21-inch bbl. 40 inches overall. Weight: 6.5 lbs. Receiver drilled and tapped for scope mounts. Sights optional. Checkered walnut-stained beech stock w/Monte Carlo. Model 201 Luxus w/checkered European walnut stock QD swivels, rosewood forend and rubber recoil pad. Made 1989 to date. Disc. importing 1994.

Model 201 Standard	**$470**
Model 201 Magnum	495
Model 201 Luxus Standard	575
Model 201 Luxus Magnum	625

Mauser Model 2000 Bolt-Action Sporting Rifle **$425**
Modified Mauser-type action. Calibers: 270 Win., 308 Win., 30-06. 5-shot magazine. 24-inch bbl. Weight: 7.5 lbs. Sights: folding leaf rear; hooded ramp front. Checkered walnut stock w/Monte Carlo comb and cheekpiece, forend tip, sling swivels. Made 1969-71. *Note:* Model 2000 is similar in appearance to Model 3000.

Mauser Model 3000 Bolt-Action Sporting Rifle **$475**
Modified Mauser-type action. Calibers: 243 Win., 270 Win., 308 Win., 30-06. 5-shot magazine. 22-inch bbl. Weight: 7 lbs. No sights. Select European walnut stock, Monte Carlo style w/cheekpiece, rosewood forend tip and pistol-grip cap, skip checkering, recoil pad, sling swivels. Made 1971-74.

Mauser Model 3000 Magnum **$495**
Same general specifications as standard Model 3000, except w/3-shot magazine, 26-inch bbl., weighs 8 lbs. Calibers: 7mm Rem. Mag., 300 Win. Mag., 375 H&H Mag.

Mauser Model 4000 Varmint Rifle **$395**
Same general specifications as standard Model 3000, except w/smaller action, folding leaf rear sight; hooded ramp front, rubber buttplate instead of recoil pad, weighs 6.75 lbs. Calibers: 222 Rem., 223 Rem. 22-inch bbl. Select European walnut stock w/rosewood forend tip and pistol-grip cap. French checkering and sling swivels.

McMILLAN GUN WORKS
Phoenix, Arizona
Harris Gunworks

See Harris Gunworks.

**Midland Model 2700
Bolt-Action Rifle**

GEBRÜDER MERKEL
Suhl, Germany

For Merkel combination guns and drillings, see listings under Merkel shotguns.

Merkel Model 220

Merkel Over/Under Rifles ("Bock-Doppelbüchsen")

Calibers: 5.6×35 Vierling, 6.5×58r5, 7×57r5, 8×57JR, 8×60R Magnum, 9.3×53r5, 9.3×72r5, 9.3×74r5, 10.3×60R as well as most of the British calibers for African and Indian big game. Various bbl. lengths, weights. In general, specifications correspond to those of Merkel over/under shotguns. Values of these over/under rifles (in calibers for which ammunition is obtainable) are about the same as those of comparable shotgun models. Currently manufactured. For more specific data, *see* Merkel shotgun models indicated below.

Model 220	$ 6,250
Model 220E	7,350
Model 221	5,950
Model 221E	7,850
Model 320	6,500
Model 320E	12,000
Model 321	13,250
Model 321E	14,000
Model 322	14,500
Model 323E	18,500
Model 324	21,500

MEXICAN MILITARY RIFLE
Mfd. by Government Arsenal, Mexico, D.F.

Mexican Model 1936 Mauser Military Rifle **$175**
Same as German Kar.98k w/minor variations and U.S. M/1903 Springfield-type knurled cocking piece.

MIDLAND RIFLES
Mfg. by Gibbs Rifle Company, Inc.
Martinsburg, WV

Midland Model 2100 Bolt-Action Rifle **$265**
Calibers: 22-250, 243 Win., 6mm Rem., 270 Win., 6.5×55, 7×57, 7×64, 308 Win., and 30-06. Springfield 1903 action. 4-shot magazine. 22-inch bbl. 43 inches overall. Weight: 7 lbs. Flip-up rear sight; hooded ramp front. Finely finished and checkered walnut stock w/pistol-grip cap and sling swivels. Steel recoil bar. Action drilled and tapped for scope mounts. Production Disc.1997.

Midland Model 2600 Bolt-Action Rifle **$250**
Same general specifications as Model 2100 except no pistol-grip cap, and stock is walnut-finished hardwood. Made 1992-97.

Midland Model 2700 Bolt-Action Rifle **$275**
Same general specifications as Model 2100, except the weight of this rifle as been reduced by utilizing a tapered bbl., anodized aluminum trigger housing and lightened stock. Weight: 6.5 lbs. Disc.

Midland Model 2800 Lightweight Rifle **$285**
Same general specifications as Model 2100, except W/laminated birch stock. Made 1992-94 and 1996-97.

MITCHELL ARMS. INC.
Fountain Valley, California
(Formerly Santa Ana, CA)

Mitchell Model 15/22 Semiautomatic
High Standard-style action. Caliber: 22 LR. 15-shot magazine (10-shot after 10/13/94). 20.5-inch bbl. 37.5 inches overall. Weight: 6.25 lbs. Ramp front sight; adj. open rear. Blued finish. Mahogany stock; Monte Carlo-style American walnut stock on Deluxe Model. Made 1994-96.
Model 15/22 SP (Special) w/plastic buttplate . . . . **$ 75**
Model 15/22 Carbine **105**
Model 15/22D Deluxe **125**

**Mitchell CAR-15 22
Carbine**

RIFLES

Mossberg Model 25

Mossberg Model 35A

Mossberg Model 40

Mossberg Model L42A

Mossberg Model 42B

Mitchell Model 9300 Series Bolt-Action Rifle
Calibers: 22 LR, 22 Mag. 5- or 10-shot magazine. 22.5-inch bbl. 40.75 inches overall. Weight: 6.5 lbs. Beaded ramp front sight; adj. open rear. Blued finish. American walnut stock. Made 1994-95.
Model 9302 (22 LR, Checkered, Rosewood Caps) **$185**
Model 9302 (22 Mag., Checkered, Rosewood Caps) **195**
Model 9303 (22 LR, Plain Stock) **145**
Model 9304 (22 Mag., Checkered, No Rosewood Caps) **155**
Model 9305 (22 LR, Special Stock) **125**

Mitchell AK-22 Semiautomatic Rifle **$185**
Replica of AK-47 rifle. Calibers: 22 LR, 22 WMR. 20-shot magazine (22 LR), 10-shot (22 WMR). 18-inch bbl. 36 inches overall. Weight: 6.5 lbs. Sights: post front; open adj. rear. European walnut stock and forend. Matte black finish. Made 1985-94.

Mitchell CAR-15 22 Semiautomatic Rifle **$185**
Replica of AR-15 CAR rifle. Caliber: 22 LR. 15-shot magazine. 16.25-inch bbl. 32 inches overall. Sights: adj. post front; adj. aperture rear. Telescoping buttstock and ventilated forend. Matte black finish. Made 1990-94.

Mitchell Galil 22 Semiautomatic Rifle **$215**
Replica of Israeli Galil rifle. Calibers: 22 LR, 22 WMR. 20-shot magazine (22 LR), 10-shot (22 WMR). 18-inch bbl. 36 inches overall. Weight: 6.5 lbs. Sights: adj. post front; rear adj. for windage. Folding metal stock w/European walnut grip and forend. Matte black finish. Made 1987-93.

Mitchell M-16A 22 Semiautomatic Rifle **$185**
Replica of AR-15 rifle. Caliber: 22 LR. 15-shot magazine. 20.5-inch bbl. 38.5 inches overall. Weight: 7 lbs. Sights: adj. post front; adj. aperture rear. Black composition stock and forend. Matte black finish. Made 1990-94.

Mitchell MAS 22 Semiautomatic Rifle **$220**
Replica of French MAS bullpup rifle. Caliber: 22 LR. 20-shot magazine. 18-inch bbl. 28 inches overall. Weight: 7.5 lbs. Sights: adj. post front, folding aperture rear. European walnut buttstock and forend. Matte black finish. Made 1987-93.

Mitchell PPS Semiautomatic Rifle
Caliber: 22 LR. 20-shot magazine, 50-shot drum. 16.5-inch bbl. 33.5 inches overall. Weight: 5.5 lbs. Sights: blade front; adj. rear. European walnut stock w/ventilated bbl. shroud. Matte black finish. Made 1989-94.
Model PPS (20-shot) **$200**
Model PPS/50 (50-shot drum) **325**

O.F. MOSSBERG & SONS, INC.
North Haven, Connecticut
(Formerly New Haven, CT)

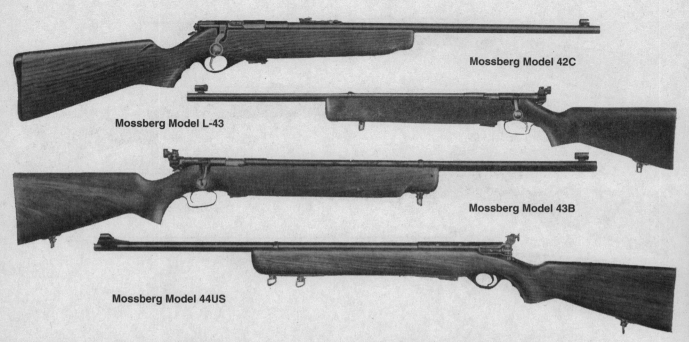

Mossberg Model 42C

Mossberg Model L-43

Mossberg Model 43B

Mossberg Model 44US

Mossberg Model 10 Bolt-Action Single-Shot
Rifle **$105**
Takedown. Caliber: 22 LR, Long, Short. 22-inch bbl. Weight: 4 lbs. Sights: open rear; bead front. Plain pistol-grip stock w/swivels, sling. Made 1933-35.

Mossberg Model 14 Bolt-Action Single-Shot
Rifle **$110**
Takedown. Caliber: 22 LR, Long, Short. 24-inch bbl. Weight: 5.25 lbs. Sights: peep rear; hooded ramp front. Plain pistol-grip stock w/semibeavertail forearm, 1.25-inch swivels. Made 1934-35.

Mossberg Model 20 Bolt-Action Single-Shot
Rifle **$115**
Takedown. Caliber: 22 LR, Long, Short. 24-inch bbl. Weight: 4.5 lbs. Sights: open rear; bead front. Plain pistol-grip stock and forearm w/finger grooves, sling and swivels. Made 1933-35.

Mossberg Model 25/25A Bolt-Action Single-Shot Rifle
Takedown. Caliber: 22 LR, Long, Short. 24-inch bbl. Weight: 5 lbs. Sights: peep rear; hooded ramp front. Plain pistol-grip stock w/semibeavertail forearm. 1.25-inch swivels. Made 1935-36.
Model 25 . . . . **$100**
Model 25A (Improved Model 25, 1936-38) **125**

Mossberg Model 26B/26C Bolt-Action Single-Shot
Takedown. Caliber: 22 LR, Long, Short. 26-inch bbl. Weight: 5.5 lbs. Sights; Rear, micrometer click peep or open; hooded ramp front. Plain pistol-grip stock swivels. Made 1938-41.
Model 26B **$125**
Model 26C (No rear sight/swivels) **100**

Mossberg Model 30 Bolt-Action Single-Shot
Rifle **$100**
Takedown. Caliber: 22 LR, Long, Short. 24-inch bbl. Weight: 4.5 lbs. Sights: peep rear; bead front, on hooded ramp. Plain pistol-grip stock, forearm w/finger grooves. Made 1933-35.

Mossberg Model 34 Bolt-Action Single-Shot
Rifle **$110**
Takedown. Caliber: 22 LR, Long, Short. 24-inch bbl. Weight: 5.5 lbs. Sights: peep rear; hooded ramp front. Plain pistol-grip stock w/semibeavertail forearm, 1.25-inch swivels. Made 1934-35.

Mossberg Model 35 Target Grade Bolt-Action
Single-Shot Rifle **$220**
Caliber: 22 LR. 26-inch heavy bbl. Weight: 8.25 lbs. Sights: micrometer click rear peep; hooded ramp front. Large target stock w/full pistol grip, cheekpiece, full beavertail forearm, 1.25-inch swivels. Made 1935-37.

Mossberg Model 35A Bolt-Action Single-Shot
Rifle **$225**
Caliber: 22 LR. 26-inch heavy bbl. Weight: 8.25 lbs. Sights: micrometer click peep rear; hooded front. Target stock w/cheekpiece full pistol grip and forearm, 1.25-inch sling swivels. Made 1937-38.

Mossberg Model 35A-LS **$235**
Caliber 22 LR.Same as Model 35A but w/Lyman 57 rear sight, 17A front. Target stock w/checkpiece, full pistol grip and forearm.

Mossberg Model 35B **$200**
Same specifications as Model 44B, except single-shot. Made 1938-40.

Mossberg Model 40 Bolt-Action Repeater **$95**
Takedown. Caliber: 22 LR, Long, Short, 16-round tubular magazine. 24-inch bbl. Weight: 5 lbs. Sights: peep rear; bead front, on hooded ramp. Plain pistol-grip stock, forearm w/finger grooves. Made 1933-35.

Mossberg Model 42 Bolt-Action Repeater **$100**
Takedown. Caliber: 22 LR, Long, Short.7-shot detachable box magazine. 24-inch bbl. Weight: 5 lbs. Sights: receiver peep, open rear; hooded ramp front. Pistol-grip stock. 1.25-inch swivels. Made 1935-37.

Mossberg Model L45A
Left-Hand Model

Mossberg Model 45B

Mossberg Model L46A-LS

Mossberg Model 46B

Mossberg Model 46M

Mossberg Model 50

Mossberg Model 51

Mossberg Model 51M

Mossberg Model 42A/L42A Bolt-Action Repeaters
Takedown. Caliber: 22 LR, Long, Short. 7-shot detachable box magazine. 24-inch bbl. Weight: 5 lbs. Sights: receiver peep, open rear; ramp front. Plain pistol-grip stock. Made 1937-38. **Model L42A** (left-hand action) made 1937-1941.

Model 42A $120
Model L42A 185

Mossberg Model 42B/42C Bolt-Action Repeaters
Takedown. Caliber: 22 LR, Long, Short. 5-shot detachable box magazine. 24-inch bbl. Weight: 6 lbs. Sights: micrometer click receiver peep, open rear hooded ramp front. Plain pistol-grip stock, swivels. Made 1938-41.

Model 42B $100
Model 42C (No rear peep sight) 95

Mossberg Model 42M Bolt-Action Repeater $145
Caliber: 22 LR, Long, Short. 7-shot detachable box magazine. 23-inch bbl. Weight: 6.75 lbs. Sights: microclick receiver peep, open rear; hooded ramp front. Two-piece Mannlicher-type stock w/cheekpiece and pistol grip, swivels. Made 1940-50.

Mossberg Model 43/L43 Bolt-Action Repeaters $325
Speedlock, adj. trigger pull. Caliber: 22 LR. 7-shot detachable box magazine. 26-inch heavy bbl. Weight: 8.25 lbs. Sights: Lyman 57 rear; selective aperture front. Target stock w/cheekpiece, full pistol grip, beavertail forearm, adj. front swivel. Made 1937-38. Model L43 is same as Model 43 except w/left-hand action.

Mossberg Model 43B $225
Same as Model 44B, except w/Lyman 57 receiver sight and 17A front sight. Made 1938-39.

Mossberg Model 44 Bolt-Action Repeater $150
Takedown. Caliber: 22 LR, Long, Short. Tubular magazine holds 16 LR. 24-inch bbl. Weight: 6 lbs. Sights: peep rear; hooded ramp front. Plain pistol-grip stock w/semibeavertail forearm, 1.25-inch swivels. Made 1934-35. *Note:* Do not confuse this rifle w/later Models 44B and 44US, which are clip repeaters.

Mossberg Model 44B Bolt-Action Target Rifle $215
Caliber: 22 LR. 7-shot detachable box magazine. Made 1938-41.

Mossberg Model 44US Bolt-Action Repeater
Caliber: 22 LR. 7-shot detachable box magazine. 26-inch heavy bbl. Weight: 8.5 lbs. Sights: micrometer click receiver peep, hooded front. Target stock, swivels. Made 1943-48. *Note:* This model was used as a training rifle by the U.S. Armed Forces during WWII.

Model 44US $185
Model 44US (Marked U.S. Property) 295

Mossberg Model 45 Bolt-Action Repeater $145
Takedown. Caliber: 22 LR, Long, Short. Tubular magazine holds 15 LR, 18 Long, 22 Short. 24-inch bbl. Weight: 6.75 lbs. Sights: rear peep; hooded ramp front. Plain pistol-grip stock, 1.25-inch swivels. Made 1935-37.

Mossberg Model 45A, L45A, 45AC Bolt-Action Repeaters
Takedown. Caliber: 22 LR, Long, Short. Tubular magazine holds 15 LR, 18 Long, 22 Short. 24-inch bbl. Weight: 6.75 lbs. Sights: receiver peep, open rear; hooded blade front sight mounted on ramp. Plain pistol-grip stock, 1.25-inch sling swivels. Made 1937-38.

Model 45A $140
Model L45A (Left-Hand Action) 180
Model 45AC (No receiver peep sight) 120

Mossberg Model 45B/45C Bolt-Action Repeaters
Takedown. Caliber: 22 LR, Long, Short. Tubular magazine holds 15 LR, 18 Long, 22 Short. 24-inch bbl. Weight: 6.25 lbs. Open rear sight; hooded blade front sight, mounted on ramp. Plain pistol-grip stock w/sling swivels. Made 1938-40.

Model 45B $125
Model 45C (No sights, made 1935-37) 110

Mossberg Model 46 Bolt-Action Repeater $145
Takedown. Caliber: 22 LR, Long, Short. Tubular magazine holds 15 LR, 18 Long, 22 Short. 26-inch bbl. Weight: 7.5 lbs. Sights: a micrometer click rear peep; hooded ramp front. Pistol-grip stock w/cheekpiece, full beavertail forearm, 1.25-inch swivels. Made 1935-37.

Mossberg Model 46A, 46A-LS, L46A-LS Bolt-Action Repeaters
Takedown. Caliber: 22 LR, Long, Short. Tubular magazine holds 15 LR, 18 Long, 22 Short. 26-inch bbl. Weight: 7.25 lbs. Sights: micrometer click receiver peep, open rear; hooded ramp front. Pistol-grip stock w/cheekpiece and beavertail forearm, quick-detachable swivels. Made 1937-38.

Model 46A $145
Model 46A-LS (w/Lyman 57 Receiver Sight) 195
Model L46A-LS (Left-Hand Action) 285

Mossberg Model 46B Bolt-Action Repeater $115
Takedown. Caliber: 22 LR, Long, Short. Tubular magazine holds 15 LR, 18 Long, 22 Short. 26-inch bbl. Weight: 7 lbs. Sights: micrometer click receiver peep open rear, hooded front. Plain pistol-grip stock w/cheekpiece, swivels. *Note:* Postwar version of this model has full magazine holding 20 LR, 23 Long, 30 Short. Made 1938-50.

Mossberg Model 46BT $165
Same as Model 46B, except w/heavier bbl. and stock weighs 7.75 lbs. Made 1938-39.

Mossberg Model 46C $125
Same as Model 46 except w/a heavier bbl. and stock than that model, weighs 8.5 lbs. Made 1936-37.

Mossberg Model 46M Bolt-Action Repeater $145
Caliber: 22 LR, Long, Short. Tubular magazine holds 22 Short, 18 Long, 15 LR. 23-inch bbl. Weight: 7 lbs. Sights: microclick receiver peep, open rear; hooded ramp front. Two-piece Mannlicher-type stock w/cheekpiece and pistol grip, swivels. Made 1940-52.

Mossberg Model 50 Autoloading Rifle $140
Same as Model 51, except w/plain stock w/o beavertail cheekpiece, swivels or receiver peep sight. Made 1939-42.

Mossberg Model 51 Autoloading Rifle $150
Takedown. Caliber: 22 LR. 15-shot tubular magazine in buttstock. 24-inch bbl. Weight: 7.25 lbs. Sights: micrometer click receiver peep, open rear; hooded ramp front. Cheekpiece stock w/full pistol grip and beavertail forearm, swivels. Made 1939 only.

Mossberg Model 51M Autoloading Rifle $155
Caliber: 22 LR. 15-shot tubular magazine. 20-inch bbl. Weight: 7 lbs. Sights: microclick receiver peep, open rear; hooded ramp front. Two-piece Mannlicher-type stock w/pistol grip and cheekpiece, hard-rubber buttplate and sling swivels. Made 1939-46.

Mossberg Model 140B Sporter-Target Rifle $145
Same as Model 140K, except w/peep rear sight, hooded ramp front sight. Made 1957-58.

RIFLES

Mossberg Model 140B

Mossberg Model 140K

Mossberg Model 144LS

Mossberg Model 146B

Mossberg Model 151K

Mossberg Model 151M

Mossberg Model 140K Bolt-Action Repeater **$120**
Caliber: 22 LR, 22 Long, 22 Short. 7-shot clip magazine. 24.5-inch bbl. Weight: 5.75 lbs. Sights: open rear; bead front. Monte Carlo stock w/cheekpiece and pistol grip, sling swivels. Made 1955-58.

**Mossberg Model 142-A Bolt-Action
Repeating Carbine** **$150**
Caliber: 22 Short Long, LR. 7-shot detachable box magazine. 18-inch bbl. Weight: 6 lbs. Sights: peep rear, military-type front. Monte Carlo stock w/pistol grip, hinged forearm pulls down to form hand grip, sling swivels mounted on left side of stock. Made 1949-57.

Mossberg Model 142K **$105**
Same as Model 142, except w/open rear sight. Made 1953-57.

Mossberg Model 144 Bolt-Action Target Rifle **$190**
Caliber: 22 LR. 7-shot detachable box magazine. 26-inch heavy bbl. Weight: 8 lbs. Sights: microclick receiver peep; hooded front. Pistol-grip target stock w/beavertail forearm, adj. hand stop, swivels. Made 1949-54. *Note:* This model designation was resumed c.1973 to replace Model 144LS, and then disc. again in 1985.

Mossberg Model 144LS **$225**
Same as Model 144 except w/Lyman 57MS or Mossberg S331 receiver sight and Lyman 17A front sight. Made 1954 to date. *Note:* This model since c.1973 w/been marketed as Model 144.

Mossberg Model 146B Bolt-Action Repeater **$145**
Takedown. Caliber: 22 LR, Long, Short. Tubular magazine holds 30 Short, 23 Long, 20 LR. 26-inch bbl. Weight: 7 lbs. Sights: micrometer click rear peep, open rear; hooded front. Plain stock w/pistol grip, Monte Carlo comb and cheekpiece, knob forend tip, swivels. Made 1949-54.

Mossberg Model 152

Mossberg Model 151K **$130**
Same as Model 151M except w/24-inch bbl., weighs 6 lbs., w/o peep sight, plain stock w/Monte Carlo comb and cheekpiece, pistol-grip knob, forend tip, w/o swivels. Made 1950-51.

Mossberg Model 151M Autoloading Rifle **$155**
Improved version of Model 51M w/same general specifications, complete action is instantly removable w/o use of tools. Made 1946-58.

Mossberg Model 152 Autoloading Carbine **$150**
Caliber: 22 LR. 7-shot detachable box magazine. 18-inch bbl. Weight: 5 lbs. Sights: peep rear; military-type front. Monte Carlo stock w/pistol grip, hinged forearm pulls down to form hand grip, sling mounted on swivels on left side of stock. Made 1948-57.

Mossberg Model 152K **$120**
Same as Model 152, except w/open instead of peep rear sight. Made 1950-57.

Mossberg Model 320B Boy Scout Target Rifle **$130**
Same as Model 340K, except single-shot w/auto. safety. Made 1960-71.

Mossberg Model 320K Hammerless Bolt-Action Single-Shot **$100**
Same as Model 346K except single-shot, w/drop-in loading platform, automatic safety. Weight: 5.75 lbs. Made 1958-60.

Mossberg Model 321B **$115**
Same as Model 321K, except w/receiver peep sight. Made 1972-75.

Mossberg Model 321K Bolt-Action Single-Shot **$120**
Same as Model 341, except single-shot. Made 1972-80.

Mossberg Model 333 Autoloading Carbine **$130**
Caliber: 22 LR. 15-shot tubular magazine. 20-inch bbl. Weight: 6.25 lbs. Sights: open rear; ramp front. Monte Carlo stock w/checkered pistol grip and forearm, bbl. band, swivels. Made 1972-73.

Mossberg Model 340B Target Sporter **$135**
Same as Model 340K, except w/peep rear sight, hooded ramp front sight. Made 1958-81.

Mossberg Model 340K Hammerless Bolt-Action Repeater **$120**
Same as Model 346K, except clip type, 7-shot magazine. Made 1958-71.

Mossberg Model 340M **$225**
Same as Model 340K, except w/18.5-inch bbl., Mannlicher-style stock w/swivels and sling. Weight: 5.25 lbs. Made 1970-71.

Mossberg Model 341 Bolt-Action Repeater **$105**
Caliber: 22 Short, Long, LR. 7-shot clip magazine. 24-inch bbl. Weight: 6.5 lbs. Sights: open rear, ramp front. Monte Carlo stock w/checkered pistol grip and forearm, sling swivels. Made 1972-85.

Mossberg Model 342K Hammerless Bolt-Action Carbine **$100**
Same as Model 340K, except w/18-inch bbl., stock w/no cheekpiece, extension forend is hinged, pulls down to form hand grip; sling swivels and web strap on left side of stock. Weight: 5 lbs. Made 1958-74.

Mossberg Model 346B **$135**
Same as Model 346K, except w/peep rear sight, hooded ramp front sight. Made 1958-67.

Mossberg Model 346K Hammerless Bolt-Action Repeater **$115**
Caliber: 22 Short, Long, LR. Tubular magazine holds 25 Short, 20 Long, 18 LR. 24-inch bbl. Weight: 6.5 lbs. Sights: open rear; bead front. Walnut stock w/Monte Carlo comb, cheekpiece, pistol grip, sling swivels. Made 1958-71.

Mossberg Model 350K Autoloading Rifle — Clip Type **$105**
Caliber: 22 Short (high speed), Long, LR. 7-shot clip magazine. 23.5-inch bbl. Weight: 6 lbs. Sights: open rear; bead front. Monte Carlo stock w/pistol grip. Made 1958-71.

Mossberg Model 351C Autoloading Carbine **$135**
Same as Model 351K, except w/18.5-inch bbl., Western carbine-style stock w/barrel band and sling swivels. Weight: 5.5 lbs. Made 1965-71.

Mossberg Model 351K Autoloading Sporter **$125**
Caliber: 22 LR. 15-shot tubular magazine in buttstock. 24-inch bbl. Weight: 6 lbs. Sights: open rear; bead front. Monte Carlo stock w/pistol grip. Made 1960-71.

Mossberg Model 352K Autoloading Carbine **$115**
Caliber: 22 Short, Long, LR. 7-shot clip magazine. 18-inch bbl. Weight: 5 lbs. Sights: open rear; bead front. Monte Carlo stock w/pistol grip; extension forend of Tenite is hinged, pulls down to form hand grip; sling swivels, web strap. Made 1958-71.

Mossberg Model 353 Autoloading Carbine **$125**
Caliber: 22 LR. 7-shot clip magazine. 18-inch bbl. Weight: 5 lbs. Sights: open rear; ramp front. Monte Carlo stock w/checkered pistol grip and forearm; black Tenite extension forend pulls down to form hand grip. Made 1972-85.

Mossberg Model 377 Plinkster Autoloader **$145**
Caliber: 22 LR. 15-shot tubular magazine. 20-inch bbl. Weight: 6.25 lbs. 4× scope sight. Thumbhole stock w/rollover cheekpiece, Monte Carlo comb, checkered forearm; molded of modified polystyrene foam in walnut-finish; sling swivel studs. Made 1977-79.

RIFLES

Mossberg Model 320B

Mossberg Model 320K

Mossberg Model 333

Mossberg Model 340B

Mossberg Model 340K

Mossberg Model 341

Mossberg Model 342K

Mossberg Model 346B

Mossberg Model 346K

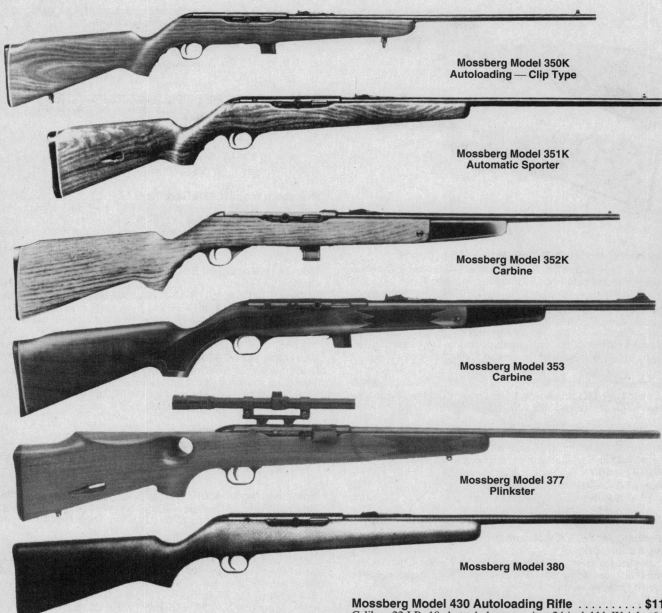

Mossberg Model 350K
Autoloading — Clip Type

Mossberg Model 351K
Automatic Sporter

Mossberg Model 352K
Carbine

Mossberg Model 353
Carbine

Mossberg Model 377
Plinkster

Mossberg Model 380

Mossberg Model 380 Semiautomatic Rifle **$120**
Caliber: 22 LR. 15-shot buttstock magazine. 20-inch bbl. Weight: 5.5 lbs. Sights: open rear; bead front. Made 1980-85.

Mossberg Model 400 Palomino Lever-Action **$185**
Hammerless. Caliber: 22 Short, Long, LR. Tubular magazine holds 20 Short, 17 Long, 15 LR. 24-inch bbl. Weight: 5.5 lbs. Sights; open rear; bead front. Monte Carlo stock w/checkered pistol grip; beavertail forearm. Made 1959-64.

Mossberg Model 402 Palomino Carbine **$190**
Same as Model 400, except w/18.5-inch (1961-64) or 20-inch bbl. (1964-71), forearm w/bbl. band, swivels; magazine holds two less rounds. Weight: 4.75 lbs. Made 1961-71.

Mossberg Model 430 Autoloading Rifle **$115**
Caliber: 22 LR. 18-shot tubular magazine. 24-inch bbl. Weight: 6.25 lbs. Sights: open rear; bead front. Monte Carlo stock w/checkered pistol grip; checkered forearm. Made 1970-71.

Mossberg Model 432 Western-Style Auto **$110**
Same as Model 430 except w/plain straight-grip carbine-type stock and forearm, bbl. band, sling swivels. Magazine capacity: 15 cartridges. Weight: 6 lbs. Made 1970-71.

Mossberg Model 472 Brush Gun **$175**
Same as Model 472 Carbine w/straight-grip stock, except w/18-inch bbl., weighs 6.5 lbs. Caliber: 30-30. Magazine capacity: 5 rounds. Made 1974-76.

Mossberg Model 472 Lever-Action Carbine **$165**
Calibers: 30-30, 35 Rem. centerfire. 6-shot tubular magazine. 20-inch bbl. Weight: 6.75 to 7 lbs. Sights: open rear; ramp front. Pistol-

Mossberg Model 400

Mossberg Model 402

Winchester Model 52-B Bolt-Action Rifle *(Con't)*
grip or straight-grip stock, forearm w/bbl. band; sling swivels on pistol-grip model saddle ring on straight-grip model. Made 1972-79.

Mossberg Model 472 One in Five Thousand $325
Same as Model 472 Brush Gun, except w/Indian scenes etched on receiver; brass buttplate, saddle ring and bbl. bands, gold-plated trigger-bright blued finish-select walnut stock and forearm. Limited edition of 5,000; serial numbered 1 to 5,000. Made in 1974.

Mossberg Model 472 Rifle $165
Same as Model 472 Carbine w/pistol-grip stock, except w/24-inch bbl., 5-shot magazine, weighs 7 lbs. Made 1974-76.

Mossberg Model 479
Calibers: 30-30 centerfire. 6-shot tubular magazine. 20-inch bbl. Weight: 6.75 to 7 lbs. Sights: open rear; ramp front. Made 1983-85.
Model 479 Rifle **$175**
Model 479PCA (Carbine w/20-inch Bbl.) 165
Model 479RR (Roy Roger, 5000 Ltd. Edition) 295

Mossberg Model 620K Hammerless Single Shot Bolt-Action Rifle $145
Single shot. Caliber: 22 WMR. 24-inch bbl. Weight: 6 lbs. Sights: open rear; bead front. Monte Carlo stock w/cheekpiece, pistol grip, sling swivels. Made 1959-60.

Mossberg Model 620K-A $125
Same as Model 640K, except w/sight modification. Made 1960-68.

Mossberg Model 640K Chuckster Hammerless Bolt-Action Rifle $145
Caliber: 22 WMR. 5-shot detachable clip magazine. 24-inch bbl. Weight: 6 lbs. Sights: open rear; bead front. Monte Carlo stock w/cheekpiece, pistol grip, sling swivels. Made 1959-84.

Mossberg Model 640KS $175
Deluxe version of Model 640K, w/select walnut stock hand checkering; gold-plated front sight, rear sight elevator, and trigger. Made 1960-64.

Mossberg Model 640M $225
Similar to Model 640K, except w/heavy receiver and jeweled bolt. 20-inch bbl., full length Mannlicher-style stock w/Monte Carlo comb and cheekpiece, swivels and leather sling. 40.75 inches overall. Weight: 6 lbs. Made 1971-73.

Mossberg Model 642K Carbine $195
Caliber: 22 WMR. 5-shot detachable clip magazine. 18-inch bbl. Weight: 5 lbs. 38.25 inches overall. Sights: open rear; bead front. Monte Carlo walnut stock w/black Tenite forearm extension that pulls down to form hand grip. Made 1961-68.

Mossberg Model 800 Bolt-Action Centerfire Rifle $175
Calibers: 222 Rem., 22-250, 243 Win., 308 Win. 4-shot magazine, 3-shot in 222. 22-inch bbl. Weight: 7.5 lbs. Sights: folding leaf rear; ramp front. Monte Carlo stock w/cheekpiece, checkered pistol grip and forearm, sling swivels. Made 1967-79.

Mossberg Model 800D Super Grade $275
Deluxe version of Model 800, except w/stock w/rollover comb and cheekpiece, rosewood forend tip and pistol-grip cap. Weight: 6.75 lbs. Chambered for all calibers listed for the Model 800, except for 222 Rem. Sling swivels. Made 1970-73.

Mossberg Model 800M $250
Same as Model 800, except w/flat bolt handle, 20-inch bbl., Mannlicher-style stock. Weight: 6.5 lbs. Calibers: 22-250, 243 Win., 308 Win. Made 1969-72.

Mossberg Model 800VT Varmint/Target $185
Similar to Model 800, except w/24-inch heavy bbl., no sights. Weight: 9.5 lbs. Calibers: 222 Rem., 22-250, 243 Win. Made 1968-79.

Mossberg Model 810 Bolt-Action Centerfire Rifle
Calibers: 270 Win., 30-06, 7mm Rem. Mag., 338 Win. Mag. Detachable box magazine (1970-75) or internal magazine w/hinged floorplate (1972 to date). Capacity: 4-shot in 270 and 30-06, 3-shot in Magnums. 22-inch bbl. in 270 and 30-06, 24-inch in Magnums. Weight: 7.5 to 8 lbs. Sights: leaf rear; ramp front. Stock w/Monte Carlo comb and cheekpiece, checkered pistol grip and forearm, grip cap, sling swivels. Made 1970-79.
Standard Calibers $260
Magnum Calibers 295

Mossberg Model 1500 Mountaineer Grade I Centerfire Rifle $255
Calibers: 223, 243, 270, 30-06, 7mm Mag. 22-inch or 24-inch (7mm Mag.) bbl. Weight: 7 lbs. 10 oz. Hardwood walnut-finished checkered stock. Sights: hooded ramp front w/gold bead; fully adj. rear. Drilled and tapped for scope mounts. Sling swivel studs. Imported 1986-87.

Mossberg Model 1500 Varmint Bolt-Action Rifle
Same as Model 1500 Grade I, except w/22-inch heavy bbl. Chambered in 222, 22-250, 223 only. High-luster blued finish or Parkerized satin finished stock. Imported from Japan 1986-87.
High-Luster Blue $295
Parkerized Satin Finish 315

**Mossberg Model 472
Brush Gun**

**Mossberg Model 472 Carbine
(Pistol Grip)**

**Mossberg Model 472 Carbine
(Straight Grip)**

**Mossberg Model 472
"One in Five Thousand"**

Mossberg Model 640K

Mossberg Model 640KS

Mossberg Model 640M

Mossberg Model 642K

RIFLES

Mossberg Model 800

Mossberg Model 800D

Mossberg Model 800M

Mossberg Model 810

Mossberg Model L

Mossberg Model R

**Mossberg Model 1700LS Classic Hunter
Bolt-Action Rifle** **$345**
Same as Model 1500 Grade I, except w/checkered classic-style stock
and Schnabel forend. Chambered in 243, 270, 30-06 only. Imported
from Japan 1986-87.

Mossberg Model B Bolt-Action Rifle **$125**
Takedown. Caliber: 22 LR, Long, Short. Single-shot. 22-inch bbl.
Sights: open rear; bead front. Plain pistol-grip stock. Made 1930-32.

Mossberg Model K Slide-Action Repeater **$225**
Hammerless. Takedown. Caliber: 22 LR, Long, Short. Tubular
magazine holds 20 Short, 16 Long, 14 LR. 22-inch bbl. Weight: 5
lbs. Sights: open rear; bead front. Plain straight-grip stock. Grooved
slide handle. Made 1922-31.

Mossberg Models L42A, L43, L45A, L46A-LS
See Models 42A, 43, 45A and 46A-LS respectively; "L" refers to a
left-hand version of those rifles.

Mossberg Model L Single-Shot Rifle **$345**
Martini-type falling-block lever-action. Takedown. Caliber: 22
LR, Long, Short. 24-inch bbl. Weight: 5 lbs. Sights: open rear;
bead front. Plain pistol-grip stock and forearm. Made 1929-32.

Mossberg Model M Slide-Action Repeater **$225**
Specifications same as for Model K except w/24-inch octagon bbl.,
pistol-grip stock, weighs 5.5 lbs. Made 1928-31.

Mossberg Model R Bolt-Action Repeater **$215**
Takedown. Caliber: 22 LR, Long, Short. Tubular magazine. 24-
inch bbl. Sights: open rear; bead front. Plain pistol-grip stock.
Made 1930-32.

MUSGRAVE RIFLES
MUSGRAVE MFRS. & DIST. (PTY) LTD.
Bloemfontein, South Africa

*The following Musgrave bolt-action rifles were manufactured
1971-76.*

**Musgrave Premier NR5 Bolt-Action
Hunting Rifle** **$350**
Calibers: 243 Win., 270 Win., 30-06, 308 Win., 7mm Rem. Mag.
5-shot magazine. 25.5-inch bbl. Weight: 8.25 lbs. Furnished w/o

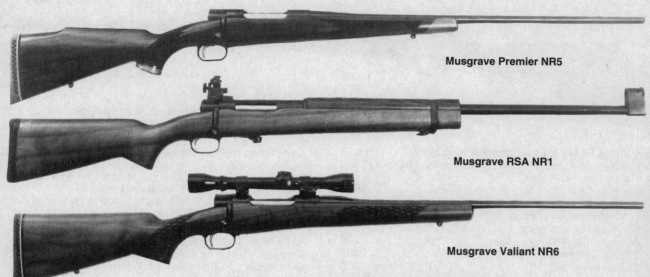

Musgrave Premier NR5

Musgrave RSA NR1

Musgrave Valiant NR6

Musgrave Premier NR5 Bolt-Action *(Con't)*
sights, but drilled and tapped for scope mount. Select walnut Monte Carlo stock w/cheekpiece, checkered pistol grip and forearm, contrasting pistol-grip cap and forend tip, recoil pad, swivel studs.

Musgrave RSA NR1 Bolt-Action Single-Shot
Target Rifle **$345**
Caliber: 308 Win. (7.62mm NATO). 26.4-inch heavy bbl. Weight: 10 lbs. Sights: aperture receiver; tunnel front. Walnut target stock w/beavertail forearm, handguard, bbl. band, rubber buttplate, sling swivels.

Musgrave Valiant NR6 Hunting Rifle **$295**
Similar to Premier, except w/24-inch bbl.; stock w/straight comb, skip French-style checkering, no grip cap or forend tip. Sights: leaf rear; hooded ramp front bead sight. Weight: 7.75lbs. Unloaded.

MUSKETEER RIFLES
Washington, D.C.
Mfd. by Firearms International Corp.

Musketeer Mauser Sporter
FN Mauser bolt action. Calibers: 243, 25-06, 270, 264 Mag., 308, 30-06, 7mm Mag., 300 Win. Mag. Magazine holds 5 standard, 3 Magnum cartridges. 24-inch bbl. Weight: 7.25 lbs. No sights. Monte Carlo stock w/checkered pistol grip and forearm, swivels. Made 1963-72.
Standard Sporter	**$295**
Deluxe Sporter	**325**
Standard Carbine.	**285**

NAVY ARMS
Ridgefield, New Jersey

Navy Arms 45-70 Mauser Carbine **$175**
Same as 45-70 Mauser Rifle, except w/18-inch bbl., straight-grip stock w/low comb, weighs 7.5 lbs. Disc.

Navy Arms 45-70 Mauser Rifle **$160**
Siamese Mauser bolt action. Caliber: 45-70 Gov't 3-round magazine. 24- 0r 26-inch bbl. Weight: 8.5 lbs. w/26-inch bbl. Sights: open rear; ramp front: Checkered stock w/Monte Carlo comb. Intoduced 1973. Disc.

Navy Arms Model 1873 Carbine. **$555**
Similar to Model 1873 Rifle, except w/blued receiver, 10-shot magazine, 19-inch round bbl. carbine-style forearm w/bbl. band, weighs 6.75 lbs. Disc. Reissued in 1991 in 44-40 or Colt.

Navy Arms Model 1873 Lever-Action Rifle **$585**
Replica of Winchester Model 1873. Casehardened receiver. Calibers: 22 LR, 357 Magnum, 44-40. 15-shot magazine. 24-inch octagon bbl. Weight: 8 lbs. Sights: open rear; blade front. Straight-grip stock, forearm w/end cap. Disc. Reissued in 1991 in 44-40 or 45 Colt w/12-shot magazine. Disc. 1994.

Navy Arms Model 1873 Trapper's **$525**
Same as Model 1873 Carbine, except w/16.5-inch bbl., 8-shot magazine, weighs 6.25 lbs. Disc.

Navy Arms Model 1874 Sharps Cavalry Carbine . . . **$455**
Replica of Sharps 1874 Cavalry Carbine. Similar to Sniper Model, except w/22-inch bbl. and carbine stock. Caliber: 45-70. Imported 1994 to date.

Navy Arms Model 1874 Sharps Sniper Rifle
Replica of Sharps 1874 Sharpshooter's Rifle. Caliber: 45-70. Falling breech, single-shot. 30-inch bbl. 46.75 inches overall. Weight: 8.5 lbs. Double-set triggers. Color casehardened receiver. Blade front sight; rear sight w/elevation leaf. Polished blued bbl. Military three-band stock w/patch box. Imported 1994 to date.
Infantry Model (Single Trigger)	**$665**
Sniper Model (DST)	**725**

Navy Arms Engraved Models
Yellowboy and Model 1873 rifles are available in deluxe models w/select walnut stocks and forearms and engraving in three grades. Grade "A" has delicate scrollwork in limited areas. Grade "B" is more elaborate with 40 percent coverage. Grade "C" has highest grade engraving. Add to value:
Grade "A".	**$175**
Grade "B"	**195**
Grade "C".	**495**

Navy Arms Henry Lever-Action Rifle
Replica of the Winchester Model 1860 Henry Rifle. Caliber: 44-40. 12-shot magazine. 16.5-, 22- or 24.25-inch octagon bbl. Weight: 7.5 to 9 lbs. 35.4 to 43.25 inches overall. Sights: blade front; adjustable ladder rear. European walnut straight-grip buttstock w/bbl. and side

Musketeer Mauser Sporter

Navy Arms Henry Lever-Action Rifle *(Con't)*

stock swivels. Imported 1985 to date. Brass or steel receiver. Blued or color casehardened metal.

Carbine Model w/22-inch bbl., introduced 1992)	$475
Military Rifle Model (w/brass frame)	460
Trapper Model (w/brass frame)	450
Trapper Model (w/iron frame)	575
W/"A" Engraving, **add**	300
W/"B" Engraving, **add**	500
W/"C" Engraving, **add**	900

Navy Arms Martini Target Rifle $375

Martini single-shot action. Calibers: 444 Marlin, 45-70. 26- or 30-inch half-octagon or full-octagon bbl. Weight: 9 lbs. w/26-inch bbl. Sights: Creedmore tang peep, open middle, blade front. Stock w/cheekpiece and pistol grip, forearm w/Schnabel tip, both checkered. Intro. 1972. Disc.

Navy Arms Revolving Carbine $450

Action resembles that of Remington Model 1875 Revolver. Casehardened frame. Calibers: 357 Magnum, 44-40, 45 Colt. 6-shot cylinder. 20-inch bbl. Weight: 5 lbs. Sights: open rear; blade front. Straight-grip stock brass trigger guard and buttplate. Intro. 1968. Disc.

Navy Arms Rolling Block Baby Carbine $175

Replica of small Remington Rolling Block single-shot action. Casehardened frame, brass trigger guard. Calibers: 22 LR, 22 Hornet, 357 Magnum, 44-40. 20-inch octagon or 22-inch round bbl. Weight: 5 lbs. Sights: open rear; blade front. Straight-grip stock, plain forearm, brass buttplate. Imported 1968-81.

Navy Arms Rolling Block Buffalo Carbine $295

Same as Buffalo Rifle, except w/18-inch bbl., weighs 10 lbs.

Navy Arms Rolling Block Buffalo Rifle $345

Replica Remington Rolling Block single-shot action. Casehardened frame, brass trigger guard. Calibers: 444 Marlin, 45-70, 50-70. 26- or 30-inch heavy half-octagon or full-octagon bbl. Weight: 11 to 12 lbs. Sights: open rear; blade front. Straight-grip stock w/brass buttplate, forearm w/brass bbl. band. Made 1971 to date.

Navy Arms Rolling Block Creedmoor Rifle

Same as Buffalo Rifle, except calibers 45-70 and 50-70 only, 28- or 30-inch heavy half-octagon or full-octagon bbl., Creedmoor tang peep sight.

Target Model	$495
Deluxe Target Model (Disc. 1998)	1195

Navy Arms Yellowboy Carbine $425

Similar to Yellowboy Rifle, except w/19-inch bbl., 10-shot magazine (14-shot in 22 Long Rifle), carbine-style forearm. Weight: 6.75 lbs. Disc. Reissued 1991 in 44-40 only.

Navy Arms Yellowboy Lever-Action Repeater $495

Replica of Winchester Model 1866. Calibers: 38 Special, 44-40. 15-shot magazine. 24-inch octagon bbl. Weight: 8 lbs. Sights; folding leaf rear; blade front. Straight-grip stock, forearm w/end cap. Intro. 1966. Disc. Reissued 1991 in 44-40 only w/12-shot magazine and adj. ladder-style rear sight.

Navy Arms Yellowboy Trapper's Model $445

Same as Yellowboy Carbine, except w/16.5-inch bbl., magazine holds two fewer rounds, weighs 6.25 lbs. Disc.

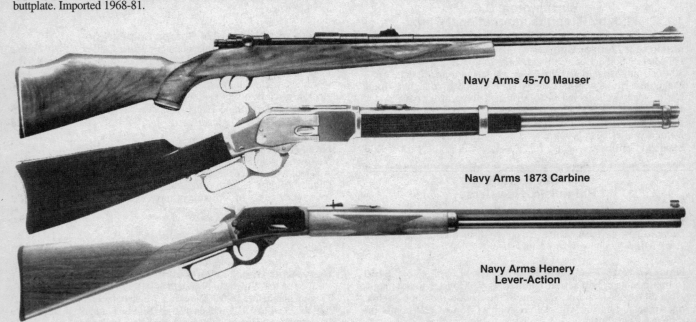

Navy Arms 45-70 Mauser

Navy Arms 1873 Carbine

**Navy Arms Henery
Lever-Action**

Navy Arms Model 1873 Carbine

Navy Arms 1874 Sharps Cavalry Carbine

Navy Arms 1874 Sharps Sniper Rifle

Navy Arms Revolving Carbine

Navy Arms Rolling Block Baby Carbine

Navy Arms Rolling Block Buffalo Rifle

Navy Arms Yellowboy Carbine

RIFLES

NEW ENGLAND FIREARMS
Gardner, Massachusetts

New England Firearms Handi-Rifle

Single-shot, break-open action w/side-lever release. Calibers: 22 Hornet, 22-250, 223, 243, 270, 30-30, 30-06, 45-70. 22-inch bbl. Weight: 7 lbs. Sights: ramp front; folding rear. Drilled and tapped for scope mounts. Walnut-finished hardwood or synthetic stock. Blued finish. Made 1989 to date.

Calibers 22-250, 243, 270 and 30-06 . . . . **$150**

Calibers 22 Hornet, 223, 30-30 and 45-70 . . . . **140**

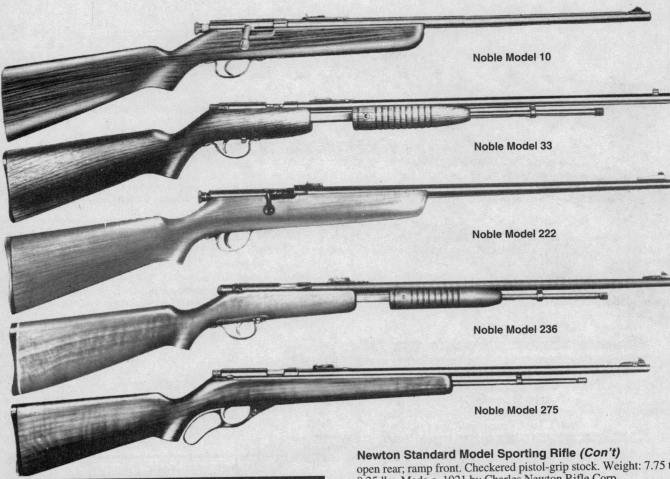

Noble Model 10

Noble Model 33

Noble Model 222

Noble Model 236

Noble Model 275

NEWTON SPORTING RIFLES
Buffalo, New York

Mfd. by Newton Arms Co., Charles Newton Rifles Corp. and Buffalo Newton Rifle Co.

Newton Buffalo Sporting Rifle **$725**
Same general specifications as Standard Model — Second Type. Made c. 1922-32 by Buffalo Newton Rifle Co.

Newton-Mauser Sporting Rifle **$595**
Mauser (Oberndorf) action. Caliber: 256 Newton. 5-shot box magazine, hinged floorplate. Double-set triggers. 24-inch bbl. Open rear sight, ramp front sight. Sporting stock w/checkered pistol grip. Weight: 7 lbs. Made c. 1914 by Newton Arms Co.

**Newton Standard Model Sporting Rifle —
First Type** **$825**
Newton bolt action, interrupted screw-type breech-locking mechanism, double-set triggers. Calibers: 22, 256, 280, 30, 33, 35 Newton; 30-06. 24-inch bbl. Sights: open rear or cocking-piece peep; ramp front. Checkered pistol-grip stock. Weight: 7 to 8 lbs., depending on caliber. Made c. 1916-18 by Newton Arms Co.

**Newton Standard Model Sporting Rifle —
Second Type** **$735**
Newton bolt action, improved design; distinguished by reversed-set trigger and 1917-Enfield-type bolt handle. Calibers:256, 30, 35 Newton and 30-06. 5-shot box magazine. 24-inch bbl. Sights:

Newton Standard Model Sporting Rifle *(Con't)*
open rear; ramp front. Checkered pistol-grip stock. Weight: 7.75 to 8.25 lbs. Made c. 1921 by Charles Newton Rifle Corp.

NIKKO FIREARMS LTD.
Tochiga, Japan

See listings under Golden Eagle Rifles.

NOBLE MFG. CO.
Haydenville, Massachusetts

Noble Model 10 Bolt-Action Single-Shot Rifle **$65**
Caliber: 22 LR, Long, Short. 24-inch bbl. Plain pistol-grip stock. Sights: open rear, bead front. Weight: 4 lbs. Made 1955-58.

Noble Model 20 Bolt-Action Single-Shot Rifle **$65**
Manually cocked. Caliber: 22 LR, Long, Short. 22-inch bbl. Weight: 5 lbs. Sights: open rear; bead front. Walnut stock w/pistol grip. Made 1958-63.

Noble Model 33 Slide-Action Repeater **$75**
Hammerless. Caliber: 22 LR, Long, Short. Tubular magazine holds 21 Short, 17 Long, 15 LR. 24-inch bbl. Weight: 6 lbs. Sights: open rear; bead front. Tenite stock and slide handle. Made 1949-53.

Noble Model 33A **$65**
Same general specifications as Model 33 except w/wood stock and slide handle. Made 1953-55.

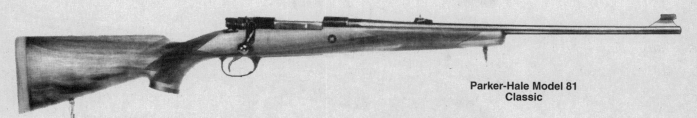

**Parker-Hale Model 81
Classic**

Noble Model 222 Bolt-Action Single-Shot Rifle **$75**
Manually cocked. Caliber: 22 LR, Long, Short. Barrel integral w/receiver. Overall length: 38 inches. Weight: 5 lbs. Sights: interchangeable V-notch and peep rear; ramp front. Scope mounting base. Pistol-grip stock. Made 1958-71.

Noble Model 236 Slide-Action Repeating Rifle **$85**
Hammerless. Caliber: 22 Short, Long, LR. Tubular magazine holds 21 Short, 17 Long, 15 LR. 24-inch bbl. Weight: 5.5 lbs. Sights: open rear; ramp front. Pistol-grip stock, grooved slide handle. Made 1951 to date.

Noble Model 275 Lever-Action Rifle **$100**
Hammerless. Caliber: 22 Short, Long, LR. Tubular magazine holds 21 Short, 17 Long, 15 LR. 24-inch bbl. Weight: 5.5 lbs. Sights: open rear; ramp front. Stock w/semipistol grip. Made 1958-71.

NORINCO
Mfd. by Northern China Industries Corp., Beijing, China
Imported by Century International Arms; Interarms; KBI and China Sports, Inc.

Norinco Model 81S/AK Semiautomatic Rifle
Semiautomatic Kalashnikov style AK-47 action. Caliber: 7.62×39mm. 5-, 30- or 40-shot magazine. 17.5-inch bbl. 36.75 inches overall. Weight: 8.5 lbs. Hooded post front sight, 500 meters leaf rear sight. Oil-finished hardwood (military style) buttstock, pistol grip, forearm and handguard or folding metal stock. Black oxide finish. Imported 1988-89.
Model 81S (w/wood stock)........................... **$475**
Model 81S-1 (w/under-folding metal stock)............. 525
Model 81S-5/56S-2 (w/side-folding metal stock)......... 550

Norinco Model 84S/AK Semiautomatic Rifle
Semiautomatic Kalashnikov style AK-47 action. Caliber: 223 (5.56mm). 30-shot magazine. 16.25-inch bbl. 35.5 inches overall. Weight: 8.75 lbs. Hooded post front sight, 800 meters leaf rear sight. Oil-finished hardwood (military style) buttstock, pistol grip, forearm and handguard; sporterized composite fiberglass stock or folding metal stock. Black oxide finish. Imported 1988-89.
Model 84S (w/wood stock)........................... **$495**
Model 84S-1 (w/under-folding metal stock)............. 550
Model 84S-3 (w/fiberglass stock)..................... 425
Model 84S-5 (w/side-folding metal stock)............. 595

Norinco Model AK-47 Thumbhole Sporter
Semiautomatic AK-47 sporterized variant. Calibers: 223 (5.56mm) or 7.62×39mm. 5-shot magazine. 16.25-inch bbl. or 23.25-inch bbl. 35.5 or 42.5 inches overall. Weight: 8.5 to 10.3 lbs. Adj. post front sight, open adj. rear. Forged receiver w/black oxide finish. Walnut-finished thumbhole stock w/recoil pad. Imported 1991-93
Model AK-47 Sporter (5.56mm) **$395**
Model AK-47 Sporter (7.62x39mm) 375

Norinco Model Mak 90/91 Sport
Similar to Model AK-47 Thumbhole Sporter, except w/minor modifications implemented to meet importation requirements. Imported 1994-95.
Model Mak 90 (w/16.25-inch bbl.).................... **$425**
Model Mar 91 (w/23.25-inch bbl.) 395

OLYMPIC ARMS
Olympia, Washington

Olympic Arms PCR Series
Gas-operated semi-auto action. Calibers: 17 Rem., 223, 7.62×39, 6×45, 6PPC or 9mm, 40S&W, 45ACP (in carbine version only). 10-shot magazine. 16-, 20- or 24-inch bbl. Weight: 7 to 10.2 lbs. Black composite stocks. Post front, rear adj. sights; scope ready flattop. Barrel fluting. William set trigger. Made 1994 to date.
PCR-1/Ultra Match **$750**
PCR-2/MultiMatch ML-1.......................... 795
PCR-3/MultiMatch ML-2.......................... 825
PCR-4/AR-15 Match 595
PCR-5/CAR-15 (223 Rem.) 575
PCR-5/CAR-15 (9mm, 40S&W, 45ACP) 815
PCR-5/CAR-15 (223 Rem.) 775
PCR-6/A-2 (7.62x39mm)......................... 585

PARKER-HALE LIMITED
Birmingham, England

Parker-Hale Model 81 African **$675**
Same general specifications as Model 81 Classic, except in caliber 375 H&H only. Sights: African express rear; hooded blade front. Barrel-band swivel. All-steel trigger guard. Checkered European walnut stock w/pistol grip and recoil pad. Engraved receiver. Imported 1986-91.

Parker-Hale Model 81 Classic Bolt-Action Rifle **$495**
Calibers: 22-250, 243 Win., 270 Win., 6mm Rem., 6.5×55, 7×57, 7×64, 308 Win., 30-06, 300 Win. Mag., 7mm Rem. Mag. 4-shot magazine. 24-inch bbl. Weight: 7.75 lbs. Sights: adj. open rear, hooded ramp front. Checkered pistol-grip stock of European walnut. Imported 1984-91.

Parker-Hale Model 85 Sniper Rifle **$1350**
Caliber: 308 Win. 10- or 20-shot M-14-type magazine. 24.25-inch bbl. 45 inches overall. Weight: 12.5 lbs. Blade front sight, folding aperture rear. McMillan fiberglass stock w/detachable bipod. Imported 1989-91.

Parker-Hale Model 87 Bolt-Action Repeating Target Rifle **$995**
Calibers: 243 Win., 6.5×55, 308 Win., 30-06 Springfield, 300 Win. Mag. 5-shot detachable box magazine. 26-inch bbl. 45 inches overall. Weight: 10 lbs. No sights; grooved for target-style scope mounts. Stippled walnut stock w/adj. buttplate. Sling swivel studs. Parkerized finish. Folding bipod. Imported 1988-91. *See* Illustration next page.

RIFLES

Parker-Hale Model 87

Parker-Hale Model 1000 Standard Rifle **$295**
Calibers: 22-250, 243 Win., 270 Win., 6mm Rem., 308 Win., 30-06. 4-shot magazine. Bolt action. 22-inch or 24-inch (22-250) bbl. 43 inches overall. 7.25 lbs. Checkered walnut Monte Carlo-style stock w/satin finish. Imported 1984-88.

Parker-Hale Model 1100 Lightweight
Bolt-Action Rifle **$375**
Same general specifications as Model 1000 Standard except w/22-inch lightweight profile bbl., hollow bolt handle, alloy trigger guard and floorplate, 6.5 lbs., Schnabel forend. Imported 1984-91.

Parker-Hale Model 1100M African Magnum Rifle **$625**
Same as Model 1000 Standard, except w/24-inch bbl. in calibers 404 Jeffery, 458 Win. Mag. Weight: 9.5 lbs. Sights: adj. rear; hooded post front. Imported 1984-91.

Parker-Hale Model 1200 Super Clip
Bolt-Action Rifle **$425**
Same as Model 1200 Super, except w/detachable box magazine in calibers 243 Win., 6mm Rem., 270 Win. 30-06 and 308 Win., 300 Win. Mag., 7mm Rem. Mag. Imported 1984-91.

Parker-Hale Model 1200 Super Bolt-Action
Sporting Rifle **$395**
Mauser-type bolt action. Calibers: 22-250, 243 Win., 6mm Rem., 25-06, 270 Win., 30-06, 308 Win. 4-shot magazine. 24-inch bbl. Weight: 7.25 lbs. Sights: folding open rear, hooded ramp front. European walnut stock w/rollover Monte Carlo cheekpiece, rosewood forend tip and pistol-grip cap, skip checkering, recoil pad, sling swivels. Imported 1968-91.

Parker-Hale Model 1200 Super Magnum **$495**
Same general specifications as 1200 Super, except calibers 7mm Rem. Mag. and 300 Win. Mag., 3-shot magazine. Mported 1988-91.

Parker-Hale Model 1200P Presentation **$395**
Same general specifications as 1200 Super, except w/scroll-engraved action, trigger guard and floorplate, no sights. QD swivels. Calibers: 243 Win. and 30-06. Imported 1969-75.

Parker-Hale Model 1200V Varmint **$385**
Same general specifications as 1200 Super, except w/24-inch heavy bbl., no sights, weighs 9.5 lbs. Calibers: 22-250, 6mm Rem., 25-06, 243 Win. Imported 1969-89.

Parker-Hale Model 1300C Scout **$475**
Calibers: 243, 308 Win. 10-round magazine. 20-inch bbl. w/muzzle brake. 41 inches overall. Weight: 8.5 lbs. No sights, drilled and tapped for scope. Checkered laminated birch stock w/QD swivels. Imported 1991.

Parker-Hale Model 2100 Midland Bolt
Action Rifle **$265**
Calibers: 22-250, 243 Win., 6mm Rem., 270 Win., 6.5×55, 7×57, 7×64, 308 Win, 30-06. 4-shot box magazine. 22-inch or 24-inch (22-250) bbl. 43-inches overall. Weight: 7 lbs. Sights: adj. folding rear; hooded ramp front. Checkered European walnut Monte Carlo stock w/pistol grip. Imported 1984-91.

Parker-Hale Model 2700 Lightweight **$275**
Same general specifications as Model 2100 Midland, except w/tapered lightweight bbl. and aluminum trigger guard. Weight: 6.5 lbs. Imported 1991.

Parker-Hale Model 1100 Lightweight

Parker-Hale Model 1200 Super Clip

**Pedersen Model 3000
Grade I Bolt-Action Rifle**

Parker-Hale Model 2800 Midland **$290**
Same general specifications as model 2100, except w/laminated birch stock. Imported 1991.

PEDERSEN CUSTOM GUNS
North Haven, Connecticut
Division of O.F. Mossberg & Sons, Inc.

Pedersen Model 3000 Grade I Bolt-Action Rifle **$795**
Richly engraved w/silver inlays, full-fancy American black walnut stock. Mossberg Model 810 action. Calibers: 270 Win., 30-06, 7mm Rem. Mag., 338 Win. Mag. 3-shot magazine, hinged floorplate. 22-inch bbl. in 270 and 30-06, 24-inch in Magnums. Weight: 7 to 8 lbs. Sights: open rear; hooded ramp front. Monte Carlo stock w/roll-over cheekpiece, wraparound hand checkering on pistol grip and forearm, rosewood pistol-grip cap and forend tip, recoil pad or steel buttplate w/trap, detachable swivels. Imported 1973-75.

Pedersen Model 3000 Grade II **$595**
Same as Model 3000 Grade I, except less elaborate engraving, no inlays, fancy grade walnut stock w/recoil pad. Imported 1973-75.

Pedersen Model 3000 Grade III **$495**
Same as Model 3000 Grade I, except no engraving or inlays, select grade walnut stock w/recoil pad. Imported 1973-74.

**Pedersen Model 4700 Custom
Deluxe Lever-Action Rifle** **$195**
Mossberg Model 472 action. Calibers: 30-30, 35 Rem. 5-shot tubular magazine. 24-inch bbl. Weight: 7.5 lbs. Sights: open rear, hooded ramp front. Hand-finished black walnut stock and beavertail forearm, barrel band swivels. Imported 1975.

J.C. PENNEY CO., INC.
Dallas, Texas

Firearms sold under the J.C. Penney label were mfd. by Marlin, High Standard, Stevens, Savage and Springfield.

J.C. Penney Model 2025 Bolt-Action Repeater **$50**
Takedown. Caliber: 22 RF. 8-shot detachable box magazine. 24-inch bbl. Weight: 6 lbs. Sights: open rear; bead front. Plain pistol-grip stock. Mfd. by Marlin.

J.C. Penney Model 2035 Bolt-Action Repeater **$50**
Takedown. Caliber: 22 RF. 8-shot detachable box magazine. 24-inch bbl. Weight: 6 lbs. Sights: open rear; bead front. Plain pistol-grip stock. Mfd. by Marlin.

J.C. Penney Model 2935 Lever-Action Rifle **$145**
Same general specifications as Marlin Model 336.

**J.C. Penney Model 6400 Bolt-Action
Centerfire Rifle** **$120**
Same general specifications as Savage Model 340.

J.C. Penney Model 6660 Autoloading Rifle **$75**
Caliber: 22 RF. Tubular magazine. 22-inch bbl. Weight: 5.5 lbs. Sights: open rear; hooded ramp front. Plain pistol-grip stock. Mfd. by Marlin.

PLAINFIELD MACHINE COMPANY
Dunellen, New Jersey

Plainfield M-1 Carbine **$175**
Same as U.S. Carbine, Cal. 30, M-1, except also available in caliber 5.7mm (22 w/necked-down 30 Carbine cartridge case). Current production w/ventilated metal handguard and barrel band w/o bayonet lug; earlier models have standard military-type fittings. Made 1960-77.

Plainfield M-1 Carbine, Commando Model **$195**
Same as M-1 Carbine, except w/paratrooper-type stock w/telescoping wire shoulderpiece. Made 1960-77.

Plainfield M-1 Carbine, Military Sporter **$185**
Same as M-1 Carbine, except w/unslotted buttstock and wood handguard. Made 1960-77.

Plainfield M-1 Deluxe Sporter **$195**
Same as M-1 Carbine, except w/Monte Carlo sporting stock Made 1960-73.

RIFLES

**Pedersen Model 3000
Grade III**

Plainfield M-1 Carbine

POLISH MILITARY RIFLES
Manufactured by Government Arsenals at Radom and Warsaw, Poland

Polish Model 1898 (Karabin 98, K98) Mauser Military Carbine **$195**
Same as German Kar. 98a, except for minor details. First manufactured during early 1920s.

Polish Model 1898 (Karabin 98, WZ98A) Mauser Military Rifle **$165**
Same as German Kar. 98a, used in WWI, except for minor details. Manufacture began c. 1921.

Polish Model 1929 (Karabin 29, WZ29) Mauser Military Rifle **$195**
Same as Czech Model 24, Mfd. 1929 thru WWII, except for minor details. A similar model produced during German occupation was designated Gew. 29/40.

WILLIAM POWELL & SON LTD.
Birmingham, England

Powell Double-Barrel Rifle **$25,000**
Boxlock. Made to order in any caliber during the time that rifle was manufactured. Bbls.: Made to order in any legal length, but 26 inches recommended. Highest grade French walnut buttstock and forearm w/fine checkering. Metal is elaborately engraved. Imported by Stoeger 1938-51.

Powell Bolt-Action Rifle **$2100**
Mauser-type bolt action. Calibers: 6×54 through 375 H&H Magnum. 3- and 4-shot shot magazine, depending upon chambering. 24-inch bbl. Weight: 7.5 to 8.75 lbs. Sights: folding leaf rear; hooded ramp front. Cheekpiece stock, checkered forearm and pistol grip, swivels. Imported by Stoeger 1938-51.

JAMES PURDEY & SONS LTD.
London, England

Purdey Double Rifle
Sidelock action, hammerless, ejectors. Almost any caliber is available but the following are the most popular: 375 Flanged Magnum Nitro Express, 500/465 Nitro Express 470 Nitro Express, 577 Nitro Express. 25.5-inch bbls. (25-inch in 375). Weight: 9.5 to 12.75 lbs. Sights: folding leaf rear; ramp front. Cheekpiece stock, checkered forearm and pistol grip, recoil pad, swivels. Currently manufactured to individual measurements and specifications; same general specifications apply to pre-WWII model.
H & H Calibers **$45,500**
NE Calibers . . . . **35,500 to 67,000**

Purdey Bolt-Action Rifle **$12,950**
Mauser-type bolt action. Calibers: 7×57, 300 H&H Magnum, 375 H&H Magnum, 10.75×73. 3-shot magazine. 24-inch bbl. Weight: 7.5 to 8.75 lbs. Sights: folding leaf rear; hooded ramp front. Cheekpiece stock, checkered forearm and pistol grip, swivels. Currently manufactured; same general specifications apply to pre-WWII model.

RAPTOR ARMS COMPANY, INC.
Newport, New Hampshire

Raptor Bolt-Action Rifle
Calibers: 243 Win., 270 Win., 30-06 or 308 Win. 4-shot magazine. 22-inch sporter or heavy bbl. Weight: 7.3 to 8 lbs. 42.5 inches overall. No sights w/drilled and tapped receiver or optional blade front, adjustable rear. Blue, stainless or "Taloncote" rust-resistant finish. Checkered black synthetic stock w/Monte Carlo cheepiece and vented recoil pad. Imported 1997 to date..
Rartor Sporter Model **$195**
Rartor Deluxe Peregrine Model (Disc. 1998). **235**
Rartor Heavy Barrel Model **225**
Rartor Stainless Barrel Model **245**
W/Optional Sights, **add** **30**

Polish Model 1929 Mauser

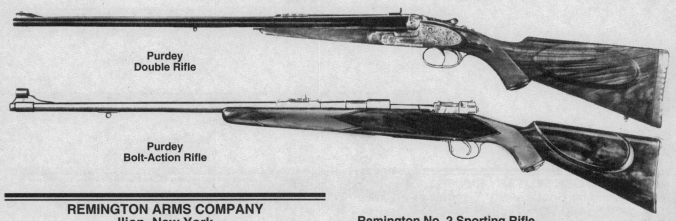

**Purdey
Double Rifle**

**Purdey
Bolt-Action Rifle**

REMINGTON ARMS COMPANY
Ilion, New York

To facilitate locating Remington firearms, models are grouped into four categories: Single-shot rifles, bolt-action repeating rifles, slide-action (pump) rifles, and semiautomatic rifles. For a complete listing, please refer to the index.

Remington Single-Shot Rifles

Remington No. 1 Sporting Rifle **$1295**
Single-Shot, rolling-block action. Calibers: 40-50, 40-70, 44-77, 50-45, 50-70 Gov't. centerfire and 44 Long, 44 Extra Long, 45-70, 46 Long, 46 Extra Long, 50-70 rimfire. Bbl. lengths: 28- or 30-inch part octagon. Weight: 5 to 7.5 lbs. Sights: folding leaf rear sight; sporting front, dovetail bases. Plain walnut straight stock; flanged-top, semicarbine buttplate. Plain walnut forend with thin rounded front end. Made 1868 to 1902.

Remington No. 1½ Sporting Rifle **$895**
Single-Shot, rolling-block action. Calibers: 22 Short, Long, or Extra Long. 25 Stevens, 32, and 38 rimfire cartridges. 32-20, 38-40 and 44-40 centerfire. Bbl. lengths: 24-, 26-, 28- or 30-inch part octagon. Remaining features similar to Remington No. 1. Made from 1869 to 1902.

Remington No. 2 Sporting Rifle
Single-shot, rolling-block action. Calibers: 22, 25, 32, 38, 44 rimfire or centerfire. Bbl. lengths: 24, 26, 28 or 30 inches. Weight: 5 to 6 lbs. Sights: open rear; bead front. Straight-grip sporting stock and knob-tip forearm of walnut. Made 1873-1910.
Calibers: 22, 25, 32 . . . . $625
Calibers: 38, 44 695

Remington No. 3 Creedmoor and Schuetzen Rifles . . . . **$5,500 to 25,000**
Produced in a variety of styles and calibers, these are collector's items and bring far higher prices than the sporting types. The Schuetzen Special, which has an under-lever action, is especially rare — perhaps less than 100 having been made.

Remington No. 3 High Power Rifle
Single shot. Hepburn falling-block action w/side lever. Calibers: 30-30, 30-40, 32 Special, 32-40, 38-55, 38-72 (high-power cartridges). Bbl. lengths: 26-, 28-, 30-inch. Weight: about 8 lbs. Open sporting sights. Checkered pistol-grip stock and forearm. Made 1893-1907.
Calibers: 30-30, 30-40, 32 Special, 32-40 . . . . $1495
Calibers: 38-55, 38-72 1750

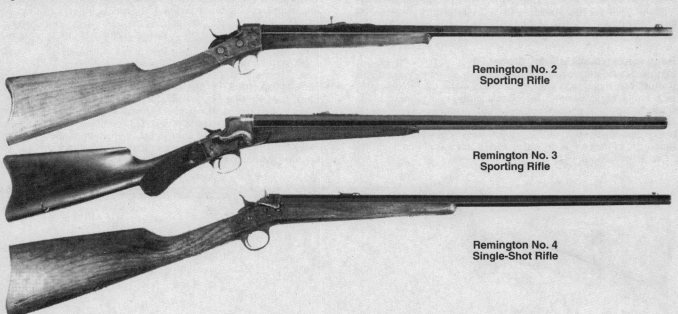

**Remington No. 2
Sporting Rifle**

**Remington No. 3
Sporting Rifle**

**Remington No. 4
Single-Shot Rifle**

RIFLES

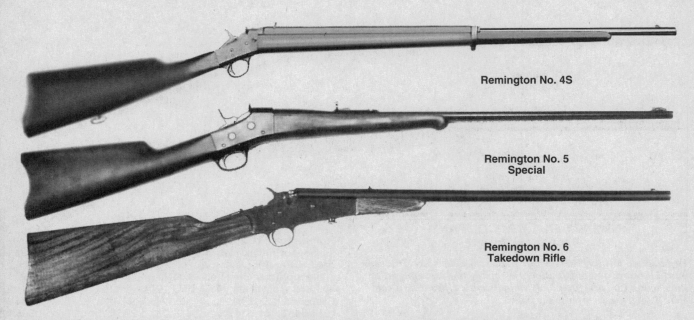

Remington No. 4S

Remington No. 5
Special

Remington No. 6
Takedown Rifle

Remington No. 3 Sporting Rifle $1295
Single shot. Hepburn falling-block action w/side lever. Calibers: 22 WCF, 22 Extra Long, 25-20 Stevens, 25-21 Stevens, 25-25 Stevens, 32 WCF, 32-40 Ballard & Marlin, 32-40 Rem., 38 WCF, 38-40 Rem., 38-50 Rem., 38-55 Ballard & Marlin, 40-60 Ballard & Marlin, 40-60 WCF, 40-65 Rem. Straight, 40-82 WCF, 45-70 Gov., 45-90 WCF, also was supplied on special order in bottle-necked 40-50, 40-70, 40-90, 44-77, 44-90, 44-105, 50-70 Gov., 50-90 Sharps Straight. Bbl. lengths: 26-inch (22, 25, 32 cal. only), 28-inch, 30-inch; half-octagon or full-octagon. Weight: 8 to 10 lbs. Sights: open rear; blade front. Checkered pistol-grip stock and forearm. Made 1880 to c. 1911.

Remington No. 4 Single-Shot Rifle $450
Rolling-block action. Solid frame or takedown. Calibers: 22 Short and Long, 22 LR, 25 Stevens R.F., 32 Short and Long R.F. 22.5-inch octagon bbl., 24-inch available in 32 caliber only. Weight: about 4.5 lbs. Sights: open rear; blade front. Plain walnut stock and forearm. Made 1890-1933.

Remington No. 4S Military Model 22
Single-Shot Rifle $725
Rolling-block action. Calibers: 22 Short only, 22 LR only. 28-inch bbl. Weight: about 5 lbs. Sights: military-type rear; blade front. Military-type stock w/handguard, stacking swivel, sling. Has a bayonet stud on the barrel; bayonet and scabbard were regularly supplied. *Note:* At one time the "Military Model" was the official rifle of the Boy Scouts of America and was called the "Boy Scout Rifle." Made 1913-33.

Remington No. 5 Special Single-Shot Rifle
Single-shot, rolling-block action. Calibers: 7mm Mauser, 30-30, 30-40 Krag, 303 British, 32-40, 32 Special, 38-55 (high-power cartridges). Bbl. lengths: 24, 26 and 28 inches. Weight: about 7 lbs. Open sporting sights. Plain straight-grip stock and forearm. Made 1902-18. *Note:* Models 1897 and 1902 Military Rifles, intended for the export market, are almost identical with the No. 5 except for 30-inch bbl. full military stock and weight (about 8.5 lbs.); a carbine was also supplied. The military rifles were produced in caliber 8mm Lebel for France, 7.62mm Russian for Russia and 7mm Mauser for the Central and South American government trade. At one time, Remington also offered these military models to retail buyers.
Sporting Model $650
Military Model 475

Remington No. 6 Takedown Rifle $325
Single-shot, rolling-block action. Calibers: 22 Short, 22 Long, 22 LR, 32 Short/Long RF. 20-inch bbl. Weight: avg. 4 lbs. Sights: open front and rear; tang peep. Plain straight-grip stock, forearm. Made 1901-33.

Remington No. 7 Target and Sporting Rifle $1950
Single shot. Rolling-block Army Pistol frame. Calibers: 22 Short, 22 LR, 25-10 Stevens R.F. (other calibers as available in No. 2 Rifle were supplied on special order). Half-octagon bbls.: 24-, 26-, 28-inch. Weight: about 6 lbs. Sights: Lyman combination rear; Beach combination front. Fancy walnut stock and forearm, Swiss buttplate available as an extra. Made 1903-11.

Remington No. 7
Target and Sporting Rifle

**Remington Model 33
Single-Shot Rifle**

Remington Model 33 Bolt-Action Single-Shot Rifle $145

Takedown. Caliber: 22 Short, Long, LR. 24-inch bbl. Weight: about 4.5 lbs. Sights: open rear, bead front. Plain, pistol-grip stock, forearm with grasping grooves. Made 1931-36.

Remington Model 33 NRA Junior Target Rifle $225

Same as Model 33 Standard, except has Lyman peep rear sight, Patridge-type front sight, 0.88-inch sling and swivels, weighs about 5 lbs.

Remington Model 40X Centerfire Rifle $425

Specifications same as for Model 40X Rimfire (heavy weight). Calibers: 222 Rem., 222 Rem. Mag., 7.62mm NATO, 30-06 (others were available on special order). Made 1961-64. Value shown is for rifle w/o sights.

Remington Model 40X Heavyweight Bolt-Action Target Rifle (Rimfire)

Caliber: 22 LR. Single shot. Action similar to Model 722. Click adj. trigger. 28-inch heavy bbl. Redfield Olympic sights. Scope bases. High-comb target stock bedding device, adj. swivel, rubber buttplate. Weight: 12.75 lbs. Made 1955-1964.
With sights $625
Without sights 565

Remington Model 40-X Sporter $1250

Same general specifications as Model 700 C Custom (*see* that listing in this section of the Rifle Section), except in caliber 22 LR. Made 1972-77.

Remington Model 40X Standard Barrel

Same as Model 40X Heavyweight except has lighter bbl. Weight: 10.75 lbs.
With sights $395
Without sights 350

Remington Model 40-XB Centerfire Match Rifle $725

Bolt action, single shot. Calibers: 222 Rem., 222 Rem. Mag., 223 Rem., 22-250, 6×47mm, 6mm Rem., 243 Win., 25-06, 7mm Rem. Mag., 30-06, 308 Win. (7.62mm NATO), 30-338, 300 Win. Mag. 27.25-inch standard or heavy bbl. Target stock w/adj. front swivel block on guide rail, rubber buttplate. Weight w/o sights: standard bbl., 9.25 lbs.; heavy bbl., 11.25 lbs. Value shown is for rifle without sights. Made 1964 to date.

Remington Model 40-XB Rangemaster Centerfire

Single-shot target rifle with same basic specifications as Model 40-XB Centerfire Match. Additional calibers in 220 Swift, 6mm BR Rem. and 7mm BR Rem., and stainless bbl. only. American walnut or Kevlar (weighs 1 lb. less) target stock with forend stop. Discontinued 1994.
Model 40-XB Right-hand Model $795
Model 40-XB Left-hand Model 850
For 2-oz. trigger, **add** 100
Model 40-XB KS (Kevlar Stock, R.H.) 895
Model 40-XB KS (Kevlar Stock, L.H.) 850
For 2-oz. trigger, **add** 100
For Repeater Model, **add** 80

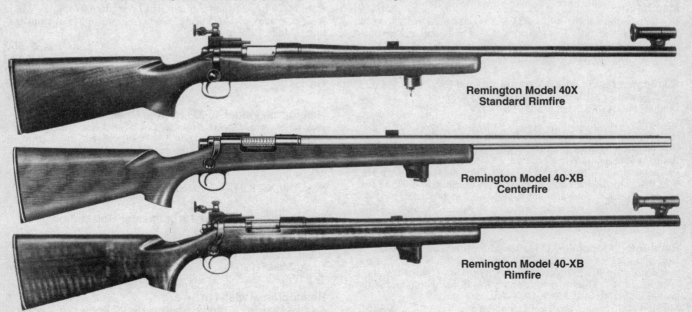

**Remington Model 40X
Standard Rimfire**

**Remington Model 40-XB
Centerfire**

**Remington Model 40-XB
Rimfire**

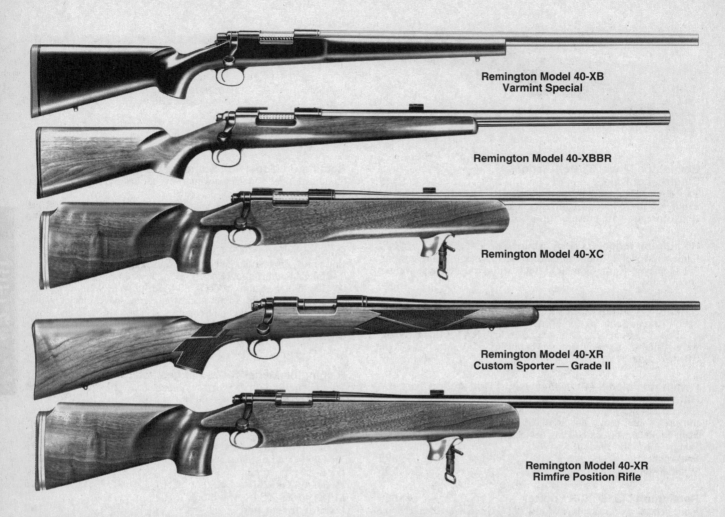

Remington Model 40-XB Varmint Special

Remington Model 40-XBBR

Remington Model 40-XC

Remington Model 40-XR Custom Sporter — Grade II

Remington Model 40-XR Rimfire Position Rifle

Remington Model 40-XB Rangemaster Rimfire Match Rifle $450
Bolt action, single shot. Caliber: 22 LR. 28-inch standard or heavy bbl. Target stock with adj. front swivel block on guide rail, rubber buttplate. Weight w/o sights: standard bbl., 10 lbs.; heavy bbl., 11.25 lbs. Value shown is for rifle without sights. Made 1964-74.

Remington Model 40-XB Varmint Special Rifle $825
Same general specifications as Model 40-XB Repeater, except has synthetic stock of Kevlar. Made 1987-94.

Remington Model 40-XBBR Bench Rest Rifle
Bolt action, single shot. Calibers: 222 Rem., 222 Rem. Mag., 223 Rem., 6×47mm, 308 Win. (7.62mm NATO). 20- or 26-inch unblued stainless-steel bbl. Supplied w/o sights. Weight: with 20-inch bbl. 9.25 lbs., with 26-inch bbl.,12 lbs. (heavy Varmint class; 7.25 lbs. w/Kevlar stock (light Varmint class). Made 1969 to date.
Model 40-XBBR (Discontinued) $695
Model 40-XBBR KS (Kevlar Stock) 895

Remington Model 40-XC National Match Course Rifle
Bolt-action repeater. Caliber: 308 Win. (7.62mm NATO). 5-shot magazine, clip slot in receiver. 24-inch bbl. Supplied w/o sights. Weight: 11 lbs. Thumb groove stock w/adj. hand stop and sling swivel, adj. buttplate. Made 1974 to date.
Model 40-XC (Wood Stock) (Discontinued) $650
Model 40-XC KS (Kevlar Stock) (disc. 1994) 825

Remington Model 40-XR Custom Sporter Rifle
Caliber: 22 RF. 24-inch contoured bbl. Supplied w/o sights. Made in four grades of checkering, engraving and other custom features. Made 1987 to date.
Grade I .... $ 850
Grade II 1495
Grade III 1950
Grade IV 3595

Remington Model 40-XR Rimfire Position Rifle
Bolt action, single shot. Caliber: 22 LR. 24-inch heavy bbl. Supplied w/o sights. Weight: about 10 lbs. Position-style stock w/thumb groove, adj. hand stop and sling swivel on guide rail, adj. buttplate. Made 1974 to date.
Model 40-XR $850
Model 40-XR KS (Kevlar Stock) 895

Remington Model 41A Targetmaster Bolt-Action Single-Shot Rifle $150
Takedown. Caliber: 22 Short, Long, LR. 27-inch bbl. Weight: about 5.5 lbs. Sights: open rear; bead front. Plain pistol-grip stock. Made 1936-40.

Remington Model 41AS $175
Same as Model 41A, except chambered for 22 Remington Special (22 W.R.F.).

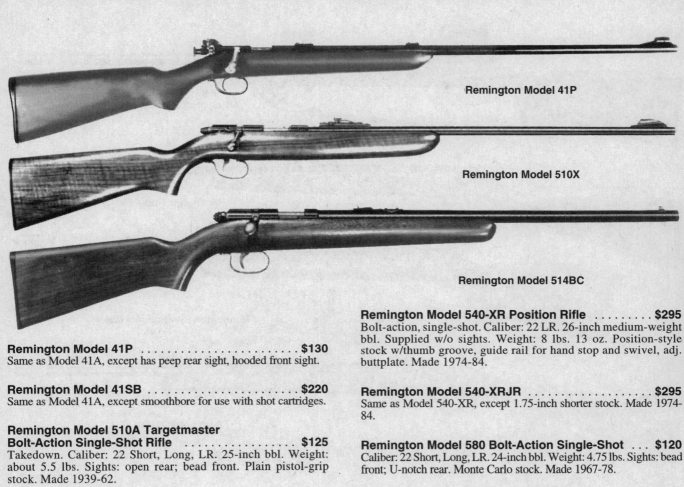

Remington Model 41P

Remington Model 510X

Remington Model 514BC

Remington Model 41P $130
Same as Model 41A, except has peep rear sight, hooded front sight.

Remington Model 41SB $220
Same as Model 41A, except smoothbore for use with shot cartridges.

Remington Model 510A Targetmaster
Bolt-Action Single-Shot Rifle $125
Takedown. Caliber: 22 Short, Long, LR. 25-inch bbl. Weight: about 5.5 lbs. Sights: open rear; bead front. Plain pistol-grip stock. Made 1939-62.

Remington Model 510P $135
Same as Model 510A, except has peep rear sight, Patridge front, on ramp.

Remington Model 510SB $250
Same as Model 510A, except (Routledge) smoothbore for use with shot cartridges, shotgun bead front sight, no rear sight.

Remington Model 510X Bolt-Action
Single-Shot Rifle $145
Same as Model 510A, except improved sights. Mfd.1964-66.

Remington Model 514 Bolt-Action Single-Shot $135
Takedown. Caliber: 22 Short, Long, LR. 24-inch bbl. Weight: 4.75 lbs. Sights: open rear; bead front. Plain pistol-grip stock. Made 1948-71.

Remington Model 514BC Boy's Carbine $145
Same as Model 514 except has 21-inch bbl., 1-inch shorter stock. Made 1961-71.

Remington Model 514P $185
Same as Model 514, except has receiver peep sight.

Remington Model 540-X Rimfire Target Rifle $265
Bolt-action, single-shot. Caliber: 22 R. 26-inch heavy bbl. Supplied w/o sights. Weight: about 8 lbs. Target stock w/Monte Carlo cheekpiece and thumb groove, guide rail for hand stop and swivel, adj. buttplate. Made 1969-74.

Remington Model 540-XR Position Rifle $295
Bolt-action, single-shot. Caliber: 22 LR. 26-inch medium-weight bbl. Supplied w/o sights. Weight: 8 lbs. 13 oz. Position-style stock w/thumb groove, guide rail for hand stop and swivel, adj. buttplate. Made 1974-84.

Remington Model 540-XRJR $295
Same as Model 540-XR, except 1.75-inch shorter stock. Made 1974-84.

Remington Model 580 Bolt-Action Single-Shot . . . $120
Caliber: 22 Short, Long, LR. 24-inch bbl. Weight: 4.75 lbs. Sights: bead front; U-notch rear. Monte Carlo stock. Made 1967-78.

Remington Model 580BR Boy's Rifle $130
Same as Model 580, except w/1-inch shorter stock. Made 1971-78.

Remington Model 580SB Smooth Bore $175
Same as Model 580, except smooth bore for 22 Long Rifle shot cartridges. Made 1967-78.

Remington International Free Rifle $695
Same as Model 40-XB rimfire and centerfire, except has "free rifle" -type stock with adj. buttplate and hook, adj. palmrest, movable front sling swivel, 2-oz. trigger. Weight: about 15 lbs. Made 1964-74. Value shown is for rifle with professionally finished stock, no sights.

Remington International Match Free Rifle $795
Calibers: 22 LR, 222 Rem., 222 Rem. Mag., 7.62mm NATO, 30-06 (others were available on special order). Model 40X-type bolt action, single shot. 2-oz. adj. trigger. 28-inch heavy bbl. Weight: about 15.5 lbs. "Free rifle" -style stock with thumbhole (furnished semifinished by mfr.); interchangeable and adj. rubber buttplate and hook buttplate, adj. palmrest, sling swivel. Made 1961-64. Value shown is for rifle with professionally finished stock, no sights.

Remington Nylon 10 Bolt-Action
Single-Shot Rifle $150
Caliber: 22 Short, Long, LR. 19.13-inch bbl. Weight: 4.25 lbs. Open rear sight; ramped blade front. Receiver grooved for scope mount. Brown nylon stock. Made 1962-1966.

Remington Model 540X

Remington Model 580

Remington International Match

Remington Nylon 10

Remington Model 7

Remington Model Seven FS

Remington Model Seven KS
Custom Rifle

Bolt-Action Repeating Rifles

Remington Model Seven (7) CF Bolt-Action Rifle
Calibers: 17 Rem., 222 Rem., 223 Rem., 243 Win., 6mm Rem., 7mm-08 Rem., 308 Win. Magazine capacity: 5-shot in 17 Rem., 222 Rem., 223 Rem., 4-shot in other calibers. 18.5-inch bbl. Weight: 6.5 lbs. Walnut stock checkering, and recoil pad. Made 1983 to date. 223 Rem. added in 1984.

Standard calibers except 17 Rem. **$355**
Caliber 17 Rem. **395**

Remington Model Seven (7) FS Rifle **$425**
Calibers: 243, 7mm-08 Rem., 308 Win. 18.5-inch bbl. 37.5 inches overall. Weight: 5.25 lbs. Hand layup fiberglass stock, reinforced with DuPont Kevlar at points of bedding and stress. Made 1987-90.

Remington Model Seven (7) KS Rifle **$645**
Calibers: 223 Rem., 7mm-08, 308, 35 Rem. and 350 Rem. Mag. 20-inch bbl. Custom made in Remington's Custom shop with Kevlar stock. Made 1987 to date.

Remington Model Seven (7) MS Custom Rifle **$665**
Similar to the standard Model 7, except fitted with a laminated full Mannlicher-style stock. Weight: 6.75 lbs. Calibers: 222 Rem., 22-250, 243, 6mm Rem.,7mm-08, 308, 350 Rem. Additional calibers available on special order. Made 1993 to date.

Remington Model Seven (7) SS Rifle **$375**
Same as Model 7, except 20-inch stainless bbl., receiver and bolt; black synthetic stock. Calibers: 243, 7mm-08 or 308. Made 1994 to date.

Remington Model Seven (7) Youth Rifle **$315**
Similar to the standard Model 7, except fitted with hardwood stock with a 12.19-inch pull. Calibers: 243, 6mm, 7mm-08 only. Made 1993 to date.

Remington Model 30A Bolt-Action
Express Rifle **$445**
Standard Grade. Modified M/1917 Enfield Action. Calibers: 25, 30, 32 and 35 Rem., 7mm Mauser, 30-06. 5-shot box magazine. 22-inch bbl. Weight: about 7.25 lbs. Sights: open rear; bead front. Walnut stock w/checkered pistol grip and forearm. Made 1921-40. *Note:*

RIFLES

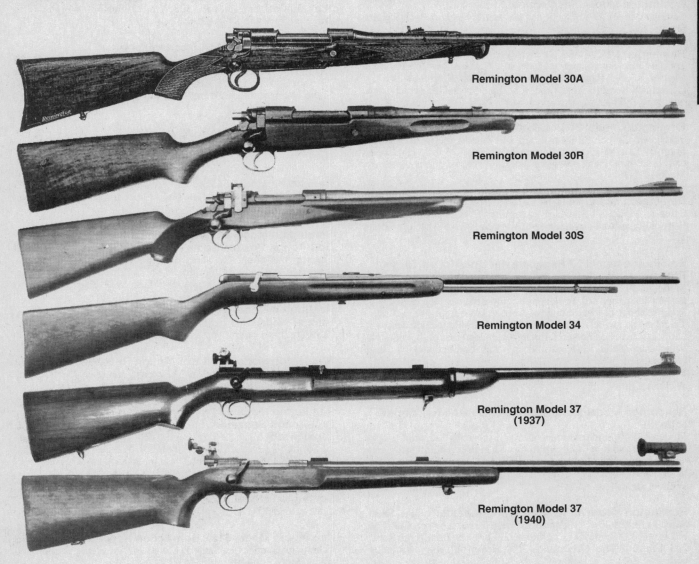

Remington Model 30A

Remington Model 30R

Remington Model 30S

Remington Model 34

Remington Model 37
(1937)

Remington Model 37
(1940)

**Remington Model 78
Sportsman**

Remington Model 341A

Remington Model 30A Bolt-Action *(Con't)*
Early Model 30s had a slender forend with schnabel tip, military-type double-pull trigger.

Remington Model 30R Carbine **$475**
Same as Model 30A, except has 20-inch bbl., plain stock weighs about 7 lbs.

Remington Model 30S Sporting Rifle **$545**
Special Grade. Same action as Model 30A. Calibers: 257 Roberts, 7mm Mauser, 30-06. 5-shot box magazine. 24-inch bbl. Weight: about 8 lbs. Lyman #48 Receiver sight, bead front sight. Special high comb stock with long, full forearm, checkered. Made 1930-40.

Remington Model 34 Bolt-Action Repeater **$140**
Takedown. Caliber: 22 Short, Long, LR. Tubular magazine holds 22 Short, 17 Long or 15 LR. 24-inch bbl. Weight: 5.25 lbs. Sights: open rear; bead front. Plain, pistol-grip stock, forearm w/grasping grooves. Made 1932-36.

Remington Model 34 NRA Target Rifle **$365**
Same as Model 34 Standard, except has Lyman peep rear sight, Patridge-type front sight, .88-inch sling and swivels, weighs about 5.75 lbs.

Remington Model 37 Rangemaster Bolt-Action Target Rifle (I)
Model of 1937. Caliber: 22 LR. 5-shot box magazine, single shot adapter also supplied as standard equipment. 28-inch heavy bbl. Weight: about 12 lbs. Remington front and rear sights, scope bases. Target stock, swivels, sling. *Note:* Original 1937 model had a stock with outside bbl. hand similar in appearance to that of the old-style Winchester Model 52, forearm design was modified and bbl. band eliminated in 1938. Made 1937-40.
With factory sights **$585**
Without sights **425**

Remington Model 37 Rangemaster Bolt-Action Target Rifle (II)
Model of 1940. Same as Model of 1937, except has "Miracle" trigger mechanism and Randle design stock with high comb, full pistol grip and wide beavertail forend. Made 1940-54.
With factory sights **$550**
Without sights **395**

Remington Model 40-XB Centerfire Repeater **$835**
Same as Model 40-XB Centerfire except 5-shot repeater. Calibers: 222 Rem., 222 Rem. Mag., 223 Rem., 22-250, 6×47mm, 6mm Rem., 243 Win., 308 Win. (7.62mm NATO). Heavy bbl. only. Discontinued.

**Remington Model 78 Sportsman
Bolt-Action Rifle** **$225**
Similar to Model 700 ADL, except with straight-comb walnut-finished hardwood stock in calibers 223 Rem., 243 Win, 270 Win., 30-06 Springfield and 308 Win. 22-inch bbl. Weight: 7 lbs. Adj. sights. Made 1984-91.

**Remington Model 341A Sportsmaster
Bolt-Action Repeater** **$135**
Takedown. Caliber: 22 Short, Long, LR. Tubular magazine holds 22 Short, 17 Long, 15 LR. 27-inch bbl. Weight: about 6 lbs. Sights; open rear; bead front. Plain pistol-grip stock. Made 1936-40.

Remington Model 341P **$155**
Same as Model 341A, except has peep rear sight, hooded front sight.

Remington Model 341SB **$275**
Same as Model 341A, except smoothbore for use with shot cartridges.

**Remington Model 511A Scoremaster
Bolt-Action Box Magazine Repeater** **$140**
Takedown. Caliber: 22 Short, Long, LR. 6-shot detachable box magazine. 25-inch bbl. Weight: about 5.5 lbs. Sights: open rear; bead front. Plain pistol-grip stock. Made 1939-62.

Remington Model 511P **$150**
Same as Model 511A, except has peep rear sight, Patridge-type blade front, on ramp.

Remington Model 511X Bolt-Action Repeater **$150**
Clip type. Same as Model 511A, except improved sights. Made 1964-66.

**Remington Model 512A Sportsmaster
Bolt-Action Repeater** **$135**
Takedown. Caliber: 22 Short, Long, LR. Tubular magazine holds 22 Short, 17 Long, 15 LR. 25-inch bbl. Weight: about 5.75 lbs. Sights: open rear; bead front. Plain pistol-grip stock w/semibeavertail forend. Made 1940-62.

Remington Model 512P **$175**
Same as Model 512A, except has peep rear sight, blade front, on ramp.

Remington Model 512X Bolt-Action Repeater **$185**
Tubular magazine type. Same as Model 512A, except has improved sights. Made 1964-66.

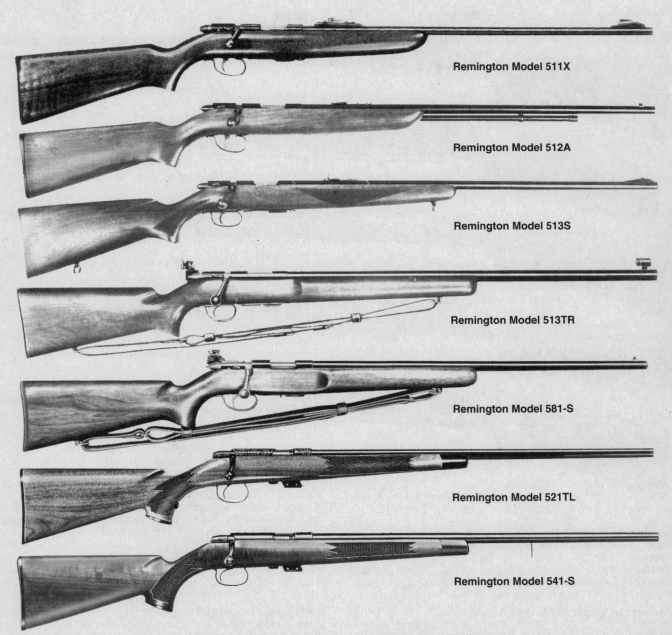

Remington Model 511X

Remington Model 512A

Remington Model 513S

Remington Model 513TR

Remington Model 581-S

Remington Model 521TL

Remington Model 541-S

RIFLES

Remington Model 513S Bolt-Action Rifle **$495**
Caliber: 22 LR. 6-shot detachable box magazine. 27-inch bbl.
Weight: about 6.75 lbs. Marble open rear sight, Patridge-type
front. Checkered sporter stock. Made 1941-56.

Remington Model 513TR Matchmaster
Bolt-Action Target Rifle **$265**
Caliber: 22 LR. 6-shot detachable box magazine. 27-inch bbl. Weight:
about 9 lbs. Sights: Redfield No. 75 rear; globe front. Target stock. Sling
and swivels. Made 1941-69.

Reminaton Model 521TL Junior Target
Bolt-Action Repeater **$235**
Takedown. Caliber: 22 LR. 6-shot detachable box magazine. 25-inch
bbl. Weight: about 7 lbs. Sights: Lyman No. 57RS rear; dovetailed
blade front. Target stock. Sling and swivels. Made 1947-69.

Remington Model 522 Viper **$120**
Calibers: 22 LR. 10-shot magazine. 20-inch bbl. 40 inches overall.
Weight: 4.63 lbs. Checkered black PET resin stock with beavertail
forend. Dupont high-tech synthetic lightweight receiver. Matte black
finish on all exposed metal. Made 1993 to date.

Remington Model 541-S Custom Sporter **$525**
Bolt-action repeater. Scroll engraving on receiver and trigger guard.
Caliber: 22 Short, Long, LR. 5-shot clip magazine. 24-inch bbl. Weight:
5.5 lbs. Supplied w/o sights. Checkered walnut stock w/rosewood-
finished forend tip, pistol-grip cap and buttplate. Made 1972-84.

Remington Model 541-T Bolt-Action Rifle
Caliber: 22 RF. Clip-fed, 5-shot. 24-inch bbl. Weight: 5.88 lbs. Check-
ered walnut stock. Made 1986 to date; heavy bbl. model intro. 1993.
Model 541-T Standard **$250**
Model 541-T-HB Heavy bbl. **295**

Remington Model 581

Remington Model 581-S

Remington Model 582

Remington Model 591

Remington Model 592

Remington Model 581 Clip Repeater
Same general specifications as Model 580, except has 5-shot clip magazine. Made 1967-84.
Model 581 .... **$125**
Model 581 Left Hand (made 1969-1984) **150**

Remington Model 581-S Bolt-Action Rifle $155
Caliber: 22 RF. Clip-fed, 5-shot. 24-inch bbl. Weight: about 4.75 lbs. Plain walnut-colored stock. Made 1987-92.

Remington Model 582 Tubular Repeater $130
Same general specifications as Model 580, except has tubular magazine holding 20 Short,15 Long,14 LR. Weight: about 5 lbs. Made 1967-84.

Remington Model 591 Bolt-Action
Clip Repeater $195
Caliber: 5mm Rimfire Magnum. 4-shot clip magazine. 24-inch bbl. Weight: 5 lbs. Sights: bead front; U-notch rear. Monte Carlo stock. Made 1970-73.

Remington Model 592 Tubular Repeater $145
Same as Model 591, except has tubular magazine holding 10 rounds, weighs 5.5 lbs. Made 1970-73.

Remington Model 600 Bolt-Action Carbine
Calibers: 222 Rem., 223 Rem., 243 Win., 6mm Rem., 308 Win., 35 Rem., 5-shot magazine. (6-shot in 222 Rem.) 18.5-inch bbl. with ventilated rib. Weight: 6 lbs. Sights: open rear; blade ramp front. Monte Carlo stock w/pistol grip. Made 1964-67.
222 Rem... **$550**
223 Rem **1050**
35 Rem.. **625**
Standard Calibers **495**

Remington Model 600 Magnum $795
Same as Model 600, except calibers 6.5mm Mag. and 350 Rem. Mag., 4-shot magazine, special Magnum-type bbl. with racket for scope back-up, laminated walnut and beech stock w/recoil pad. QD swivels and sling; weight: about 6.5 lbs. Made 1965-67.

Remington Model 600 Montana Territorial
Centennial $750
Same as Model 600, except has commemorative medallion embedded in buttstock. Made 1964. Value is for rifle in new, unfired condition.

Remington Model 660 STP
Calibers: 222 Rem., 6mm Rem., 243 Win., 308 Win., 5-shot magazine. (6-shot in 222 Rem.) 20-inch bbl. Weight: 6.5 lbs. Sights: open rear; bead front on ramp. Monte Carlo stock, checkered, black pistol-grip cap and forend tip. Made 1968-71.
222 Rem **$485**
Other Calibers **450**

Remington Model 600

Remington Model 600
Montana Territorial Centennial

Remington Model 660

Remington Model 700 ADL

Remington Model 700 ADL
w/Laminated Stock

Remington Model 700 ADL
Left Hand

Remington Model 660 Magnum **$675**
Same as Model 660, except calibers 6.5mm Rem. Mag., and 350 Rem. Mag., 4-shot magazine, laminated walnut-and-beech stock with recoil pad. QD swivels and sling. Made 1968-71.

Remington Model 700 ADL Centerfire Rifle **$285**
Calibers: 22-250, 222 Rem., 25-06, 6mm Rem., 243 Win., 270 Win., 30-06, 308 Win., 7mm Rem. Mag. Magazine capacity: 6-shot in 222 Rem.; 4-shot in 7mm Rem. Mag. 5-shot in other calibers. Bbl. lengths: 24-inch in 22-250, 222 Rem., 25-06, 7mm Rem. Mag.; 22-inch in other calibers. Weight: 7 lbs. standard; 7.5 lbs. in 7mm Rem. Mag. Sights: ramp front; sliding ramp open rear. Monte Carlo stock w/cheekpiece, skip checkering, recoil pad on Magnum. Laminated stock also avail. Made 1962-93.

Remington Model 700 APR Bolt-Action Rifle **$950**
Acronym for African Plains Rifle. Calibers: 7mm Rem. Mag., 7mm STW, 300 Win. Mag., 300 Wby. Mag., 300 Rem. Ultra Mag., 338 Win. Mag., 375 H&H. 3-shot magazine. 26-inch bbl. on a magnum action. 46.5 inches overall. Weight: 7.75 lbs. Matte blue finish. Checkered classic-style laminated wood stock w/black magnum recoil pad. Made 1994 to date.

Remington Model 700 AS Bolt-Action Rifle
Similar to the Model 700 BDL, except with nonreflective matte black metal finish, including the bolt body. Weight: 6.5 lbs. Straight comb synthetic stock made of Arylon, a fiberglass-reinforced thermoplastic resin with nonreflective matte finish. Made 1988-92.
Standard Caliber **$365**
Magnum Caliber **395**

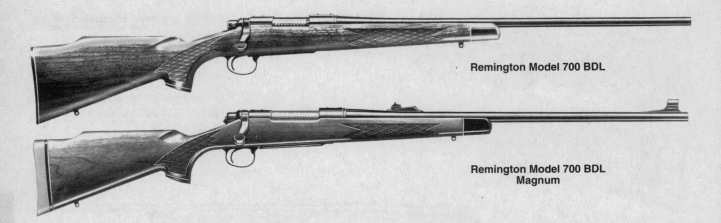

Remington Model 700 BDL

Remington Model 700 BDL Magnum

Remington Model 700 AWR Bolt-Action Rifle...... $850
Acronym for Alaskan Wilderness Rifle similar to Model 700 APR, except w/24-inch stainless bbl. and black chromed action. Matte gray or black Kevlar stock w/straight comb and raised cheekpiece fitted w/black magnum recoil pad. Made 1994 to date.

Remington Model 700 BDL Centerfire Rifle
Same as Model 700 ADL, except has hinged floorplate hooded ramp front sight, stock w/black forend tip and pistol-grip cap, cut checkering, QD swivels and sling. Additional calibers: 17 Rem., 223 Rem., 264 Win. Mag., 7mm-08, 280, 300 Sav., 300 Win. Mag., 8mm Rem. Mag., 338 Win. Mag., 35 Whelen. All have 24-inch bbls. Magnums have 4-shot magazine, recoil pad, weighs 7.5 lbs; 17 Rem. has 6-shot magazine, weighs 7 lbs. Made 1962 to date. Made 1973 to date.

Standard calibers except 17 Rem.	$385
Magnum calibers and 17 Rem.	445
Left-hand, 270 Win. and 30-06	395
Left-hand, 7mm Rem. Mag and 222 Rem.	475

Remington Model 700 BDL European Rifle
Same general specifications as Model 700 BDL, except has oil-finished walnut stock. Calibers: 243, 270, 7mm-08, 7mm Mag., 280 Rem., 30-06. Made 1993-95.

Standard Calibers	$395
Magnum Calibers	450

Remington Model 700 BDL SS Bolt-Action Rifle
Same as Model 700 BDL, except w/24-inch stainless bbl., receiver and bolt plus black synthetic stock. Calibers: 223 Rem., 243 Win., 6mm Rem., 25-06 Rem., 270 Win. 280 Rem., 7mm-08, 7mm Rem. Mag., 7mm Wby. Mag., 30-06, 300 Win. Mag., 308 Win., 338 Win. Mag. 375 H&H. Made 1992 to date.

Standard Calibers	$425
Magnum Calibers, **add**	30
DM, **add**	40
DM-B (Muzzle Break), **add**	90

Remington Model 700 BDL Varmint Special $395
Same as Model 700 BDL, except has 24-inch heavy bbl., no sights, weighs 9 lbs. (8.75 lbs. in 308 Win.). Calibers: 22-250, 222 Rem., 223 Rem., 25-06, 6mm Rem., 243 Win., 308 Win. Made 1967-94.

Remington Model 700 CS Bolt-Action Rifle
Similar to Model 700 BDL, except with nonreflective matte black metal finish, including the bolt body. Straight comb synthetic stock camouflaged in mossy Oak Mottomland pattern. Made 1992-94.

Standard Caliber	$395
magnum Caliber	425

Remington Model 700 Classic
Same general specifications as Model 700 BDL, except has "Classic" stock of high-quality walnut with full-pattern cut-checkering, special satin wood finish; schnabel forend. Brown rubber buttpad. Hinged floorplate. No sights. Weight: 7 lbs. Also chambered for "Classic" cartridges such as 257 Roberts and 250-3000. Intro. 1981.

Standard Calibers	$355
Magnum Calibers	395

Remington Model 700 Custom Bolt-Action Rifle
Same general specifications as Model 700 BDL, except custom built, and available in choice of grades — each with higher quality wood, different checkering patterns, engraving, high-gloss blued finish. Introduced in 1965.

Model 700 C Grade I	$ 950
Model 700 C Grade II	1650
Model 700 C Grade III	2150
Model 700 C Grade IV	3950
Model 700 D Peerless	1395
Model 700 F Premier	2655

Remington Model 700 FS Bolt-Action Rifle
Similar to Model 700 ADL, except with straight comb fiberglass stock reinforced with DuPont Kevlar finished in gray or gray como with Old English style recoil pad. Made 1987-89.

Standard Calibers	$445
Magnum Calibers	475

Remington Model 700 KS Custom MTN Rifle
Similar to standard Model 700 MTN Rifle, except with custom Kevlar reinforced resin synthetic stock with standard or wood-grain finish. Calibers: 270 Win., 280 Rem., 7mm Rem Mag., 30-06, 300 Win. Mag., 300 Wby. Mag., 8mm Rem. Mag., 338 Win. Mag., 35 Whelen, 375 H&H. 4-shot magazine. 24-inch bbl. Weight: 6.75 lbs. Made 1986 to date.

Standard KS Stock (Disc. 1993)	$695
Wood-grained KS Stock	725
SS Model Stainless Synthetic (1995-97)	765
Left-hand, Model, add	70

Remington Model 700 LS Bolt-Action Rifle
Similar to Model 700 ADL, except with Checkered Monte Carlo style laminated wood stock with alternating grain and wood color impregnated with phenolic resin and finished with a low satin luster. Made 1988-93

Standard Caliber	$295
Magnum Caliber	325

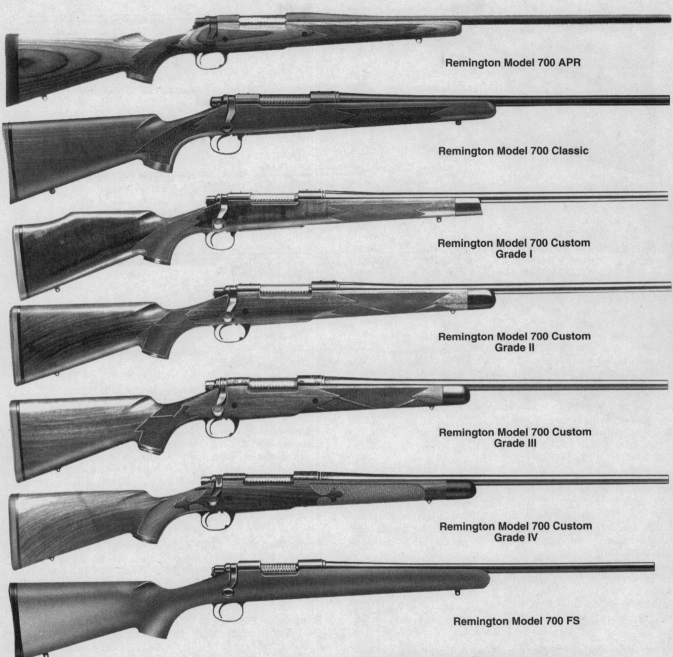

Remington Model 700 APR

Remington Model 700 Classic

Remington Model 700 Custom Grade I

Remington Model 700 Custom Grade II

Remington Model 700 Custom Grade III

Remington Model 700 Custom Grade IV

Remington Model 700 FS

Remington Model 700 LSS Bolt-Action Rifle

Similar to Model 700 BDL, except with stainless steel barrel and action. Checkered Monte Carlo style laminated wood stock with alternating grain and gray tinted color impregnated with phenolic resin and finished with a low satin luster. Made 1996 to date.

Standard Caliber **$395**
Magnum Caliber **425**

Remington Model 700 MTN Rifle

Lightweight version of Model 700. Calibers: 243 Win., 25.06, 257 Roberts, 270 Win., 7×57, 7mm-08 Rem., 280 Rem., 30-06 and 308 Win. 4-shot magazine. 22-inch bbl. Weight: 6.75 lbs. Satin blue or stainless finish. Checkered walnut stock and redesigned pistol grip, straight comb, contoured cheekpiece, Old Eng-

Remington Model 700 MTN Rifle *(Con't)*

lish style recoil pad and satin oil finish or black synthetic stock with pressed checkering and blind magazine. Made 1986 to date.

Standard w/Hinged Floorplate (Disc. 1994). **$350**
DM Model (New 1995) **425**
SS Model Stainless Synthetic (Disc. 1993) **365**

Remington Model 700 RS Bolt-Action Rifle

Similar to the Model 700 BDL, except with straight comb DUPont Rynite stock finished in gray or gray como with Old English style recoil pad. made 1987-90.

Standard Caliber **$425**
Magnum Caliber **445**
280 Rem. Caliber (Limited production) **550**

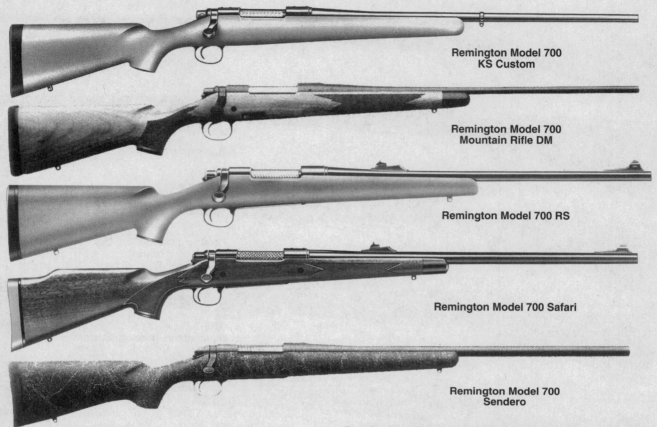

Remington Model 700
KS Custom

Remington Model 700
Mountain Rifle DM

Remington Model 700 RS

Remington Model 700 Safari

Remington Model 700
Sendero

Remington Model 700 Safari Grade
Big game heavy magnum version of the Model 700 BDL. Cailbers: 8mm Rem. Mag., 375 H&H Mag., 416 Rem. Mag. and 458 Win. mag. 24-inch heavy bbl. Weight: 9 lbs. Blued or stainless finish. Checkered walnut stock in synthetic/Kevlar stock with standard matte or wood-grain finish with old English style recoil pad. made 1962 to date.

Safari Classic/Monte Carlo	$695
Safari KS (Kevlar Stock) intro. 1989	795
Safari KS (Wood-grain) intro. 1992	850
Safari KS SS (Stainless) intro. 1993	950
Safari Left-hand, **add**	95

Remington Model 700 Sendero Bolt-Action Rifle
Same as Model 700 VS, except chambered in long action and magnum calibers: 25-06 Rem., 270 Win., 280 Rem., 7mm Rem. Mag., 300 Win. Made 1994 to date.

Standard Calibers	$450
Magnum Calibers, **add**	30
SF Model, **add**	565

Remington Model 700 VLS (Varmit Laminated Stock)
Bolt-Action Rifle **$395**
Same as Modle 700 BDL Varmint Special, except with 26-inch polished blue barrel. Laminated wood stock with alternating grain and wood color impregnated with phenolic resin and finished with a satin luster. Calibers: 222 Rem., 223Rem., 22-250 Rem., 243 Win., 7mm-08 Rem., 308 Win. Weight: 9.4 lbs. Made 1995 to date.

Remington Model 700 VS Bolt-Action Rifle
Same as Model 700 BDL Varmint Special, except w/26-inch matte blue or fluted stainless barrel. Textured black or gray synthetic stock reinforced with Kevlar, fiberglass and graphite with full length aluminum bedding block. Calibers: 22-250 Rem., 220 Swift, 223 Rem., 308 Win. Made 1992 to date.

Model 700 VS	$425
Model 700 VS SF (Fluted Barrel)	525

Remington Model 720A Bolt-Action High Power $925
Modified M/1917 Enfield action. Calibers: 257 Roberts, 270 Win., 30-06. 5-shot box magazine. 22-inch bbl. Weight: about 8 lbs. Sights: open rear; bead front, on ramp. Pistol-grip stock, checkered. **Model 720R** has 20-inch bbl.; **Model 720S** has 24-inch bbl. Made 1941.

Remington Model 721A Standard Grade
Bolt-Action High-Power Rifle **$295**
Calibers: 264 Win., 270 Win., 30-06. 4-shot box magazine. 24-inch bbl. Weight: about 7.25 lbs. Sights: open rear; bead front, on ramp. Plain sporting stock. Made 1948-62.

Remington Model 721A Magnum
Standard Grade **$495**
Caliber: 264 Win. Mag. or 300 H&H mag. Same as standard model, except has 26-inch bbl. 3-shot magazine and recoil pad. Weight: 8.25 lbs.

Remington Model 721ADL/BDL Deluxe
Same as Model 721A Standard or Magnum, except has deluxe checkered stock and/or select wood.

Model 721ADL Deluxe Grade	$360
Model 721ADL 300 Magnum Deluxe	475
Model 721BDL Deluxe Special Grade	425
Model 721BDL 300 Magnum Deluxe	450

Remington Model 721A Deluxe

Remington Model 722A

Remington Model 722A Standard Grade Sporter
Same as Model 721A Bolt-Action, except shorter action. Calibers: 222 Rem. mag., 243 Win., 257 Roberts, 308 Win., 300 Savage. 4- or 5-shot magazine. Weight: 7-8 lbs. 222 Rem. introduced 1950; 244 Rem. introduced 1955. Made 1948-62.

222 Rem	**$325**
244 Rem	295
222 Rem. Mag. & 243 Win.	375
Other Calibers	265

Remington Model 722ADL Deluxe Grade
Same as Model 722A, except has deluxe checkered stock.

Standard Calibers	**$365**
222 Rem. Deluxe Grade	425
244 Rem. Deluxe Grade	550

Remington Model 722BDL Deluxe Special Grade
Same as Model 722ADL, except select wood.

Standard Calibers	**$425**
222 Rem. Deluxe Special Grade	450
224 Rem. Deluxe Special Grade	465

Remington Model 725 Kodiak Magnum Rifle $2950
Similar to Model 725ADL. Calibers: 375 H&H Mag., 458 Win. Mag. 3-shot magazine. 26-inch bbl. with recoil reducer built into muzzle. Weight: about 9 lbs. Deluxe, reinforced Monte Carlo stock with recoil pad, black forend tip swivels, sling. Less than 100 made in 1961.

Remington Model 725ADL Bolt-Action
Repeating Rifle **$575**
Calibers: 222, 243, 244, 270, 280, 30-06. 4-shot box magazine (5-shot in 222). 22-inch bbl. (24-inch in 222). Weight: about 7 lbs. Sights: open rear, hooded ramp front. Monte Carlo comb stock w/pistol grip, checkered, swivels. Made 1958-61.

Remington Model 788 Centerfire Bolt-Action
Calibers: 222 Rem., 22-250, 223 Rem., 6mm Rem., 243 Win., 308 Win., 30-30, 44 Rem. Mag. 3-shot clip magazine (4-shot in 222 and 223 Rem.). 24-inch bbl. in 22s, 22-inch in other calibers. Weight: 7.5 lbs. with 24-inch bbl.; 7.25 lbs. with 22-inch bbl. Sights: blade front, on ramp; U-notch rear. Plain Monte Carlo stock. Made 1967-84.

Standard R.H. Model	**$275**
Left Hand (6mm Rem. and 308 Win. only made 1972-79)	295

Remington Nylon 11 Bolt-Action Repeater $165
Clip type. Caliber: 22 Short, Long, LR. 6- or 10-shot clip mag. 19.63-inch bbl. Weight: 4.5 lbs. Sights: open rear; blade front. Nylon stock. Made 1962-66.

Remington Nylon 12 Bolt-Action Repeater $175
Same as Nylon 11 except has tubular magazine holding 22 Short, 17 Long, 15 LR. Made 1962-66.

Slide- and Lever-Action Rifles

Remington Model Six (6) Slide-Action Repeater $325
Hammerless. Calibers: 6mm Rem., 243 Win., 270 Win. 7mm Express Rem., 30-06, 308 Win. 22-inch bbl. Weight: 7.5 lbs. Checkered Monte Carlo stock and forearm. Made 1981-88.

Remington Nylon 11

Remington Nylon 12

RIFLES

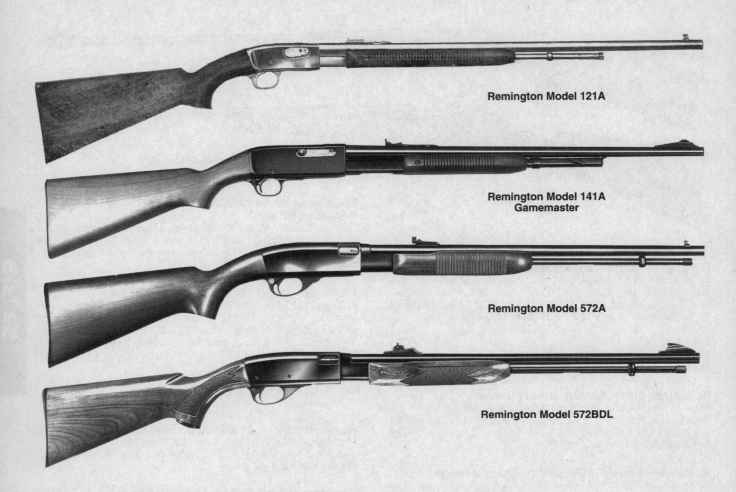

Remington Model 121A

Remington Model 141A
Gamemaster

Remington Model 572A

Remington Model 572BDL

Remington Model Six (6) Slide-Action Repeater, Peerless Grade **$1350**
Same as Model Six Standard, except has engraved receiver. Made 1981-88.

Remington Model Six (6) Slide-Action Repeater, Premier Grade **$3250**
Same as Model Six Standard, except has engraved receiver with gold inlay. Made 1981-88.

Remington Model 12A, 12B, 12C, 12CS Slide-Action Repeaters
Standard Grade. Hammerless. Takedown. Caliber: 22 Short, Long or LR. Tubular magazine holds 15 Short, 12 Long or 10 LR cartridges. 22- or 24-inch round or octagonal bbl. Open rear sight, bead front. Plain, half-pistol-grip stock and grooved slide handle of walnut. Made 1909-36.
Model 12A **$295**
Model 12B (22 Short only w/octagon bbl.) 450
Model 12C (w/24-inch octagon bbl.) 395
Model 12CS (22 WRF w/24-inch octagon bbl.) 325

Remington Model 14A High Power Slided-Action Repeating Rifle **$325**
Standard Grade. Hammerless. Takedown. Calibers: 25, 30, 32 and 35 Rem. 5-shot tubular magazine. 22-inch bbl. Weight: about 6.75 lbs. Sights: open rear; bead front. Plain half-pistol-grip stock and grooved slide handle of walnut. Made 1912-35.

Remington Model 14R Carbine **$570**
Same as Model 14R, except has 18.5-inch bbl., straight-grip stock, weighs about 6 lbs.

Remington Model 14½ Carbine **$575**
Same as Model 14.5 Rifle, except has 9-shot magazine, 18.5-inch bbl.

Remington Model 14½ Rifle **$675**
Similar to Model 14A, except calibers: 38-40 and 44-40, 11-shot full magazine, 22.5-inch bbl. Made 1912 to early 1920's.

Remington Model 25A Slide-Action Repeater **$350**
Standard Grade. Hammerless. Takedown. Calibers: 25-20, 32-20. 10-shot tubular magazine. 24-inch bbl. Weight: about 5.5 lbs. Sights: open rear; bead front. Plain, pistol-grip stock, grooved slide handle. Made 1923-36.

Remington Model 25R Carbine **$425**
Same as Model 25A, except has 18-inch bbl. 6-shot magazine, straight-grip stock, weighs about 4.5 lbs.

Remington Model 121A Fieldmaster Slide-Action Repeater **$295**
Standard Grade. Hammerless. Takedown. Caliber: 22 Short, Long, LR. Tubular magazine holds 20 Short, 15 Long or 14 LR cartridges. 24-inch round bbl. Weight: 6 lbs. Plain, pistol-grip stock and grooved semibea-vertail slide handle. Made 1936-54.

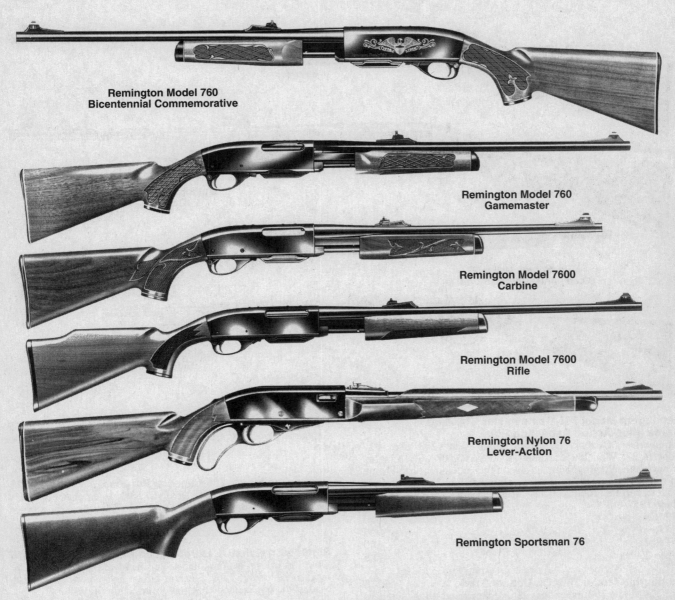

Remington Model 760
Bicentennial Commemorative

Remington Model 760
Gamemaster

Remington Model 7600
Carbine

Remington Model 7600
Rifle

Remington Nylon 76
Lever-Action

Remington Sportsman 76

RIFLES

Remington Model 121S **$425**
Same as Model 121A, except chambered for 22 Remington Special
(22 W.R.F.). Magazine holds 12 rounds. Disc.

Remington Model 121SB **$495**
Same as Model 121A, except smoothbore. Discontinued.

Remington Model 141A Gamemaster
Slide-Action Repeater **$295**
Standard Grade. Hammerless. Takedown. Calibers: 30, 32 and 35 Rem.
5-shot tubular magazine. 24-inch bbl. Weight: about 7.75 lbs. Sights:
open rear; bead front, on ramp. Plain, pistol-grip stock, semibeavertail
forend (slide-handle). Made 1936-50.

Remington Model 572A Fieldmaster
Slide-Action Repeater **$150**
Hammerless. Caliber: 22 Short, Long, LR. Tubular magazine holds
20 Short, 17 Long, 15 LR. 23-inch bbl. Weight: about 5.5 lbs.
Sights: open rear; ramp front. Pistol-grip stock, grooved forearm.
Made 1955-88.

Remington Model 572BDL Deluxe **$185**
Same as Model 572A, except has blade ramp front sight, sliding ramp
rear; checkered stock and forearm. Made 1966 to date.

Remington Model 572SB Smooth Bore **$250**
Same as Model 572A except smooth bore for 22 LR shot cartridges.
Made 1961 to date.

Remington Model 760 Bicentennial
Commemorative **$495**
Same as Model 760, except has commemorative inscription on re-
ceiver. Made 1976.

Remington Model 760 Carbine **$385**
Same as Model 760 Rifle, except made in calibers 270 Win., 280 Rem.,
30-06 and 308 Win. only, has 18.5-inch bbl., weighs 7.25 lbs. Made
1961-80.

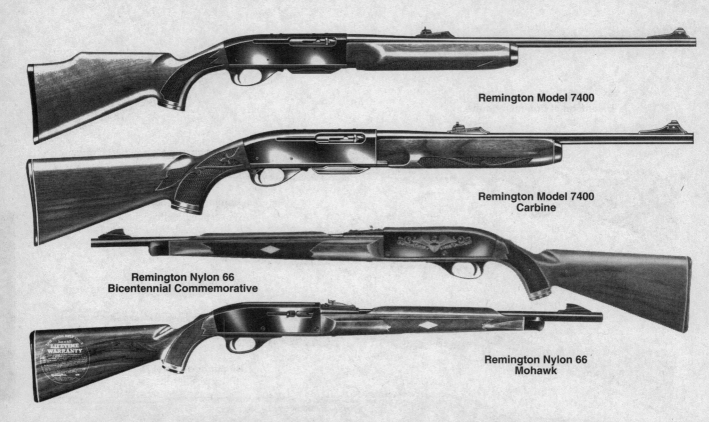

Remington Model 7400

Remington Model 7400 Carbine

Remington Nylon 66 Bicentennial Commemorative

Remington Nylon 66 Mohawk

Remington Model 760 Gamemaster Standard Grade Slide-Action Repeating Rifle

Hammerless. Calibers: 223 Rem., 6mm Rem., 243 Win., 257 Roberts, 270 Win. 280 Rem., 30-06, 300 Sav., 308 Win., 35 Rem. 22-inch bbl. Weight: about 7.5 lbs. Sights: open rear; bead front, on ramp. Plain pistol-grip stock, grooved slide handle on early models, current production has checkered stock and slide handle. Made 1952-80.

222 Rem	**$850**
223 Rem	895
257 Roberts	650
Other Calibers	295

Remington Model 760ADL Deluxe Grade $375
Same as Model 760, except has deluxe checkered stock, standard or high comb, grip cap, sling swivels. Made 1953-63.

Remington Model 760BDL Custom Deluxe $365
Same as Model 760 Rifle, except made in calibers 270, 30-06 and 308 only, has Monte Carlo cheekpiece stock forearm with black tip, basket-weave checkering. Available also in left-hand model. Made 1953-80.

Remington Model 760D Peerless Grade $950
Same as Model 760, except scroll engraved, fancy wood. Made 1953-80.

Remington Model 760F Premier Grade $1950
Same as Model 760, except extensively engraved with game scenes and scroll, finest grade wood. Also available with receiver inlaid with gold; adds 50 percent to value. Made 1953-80.

Remington Model 7600 Slide-Action Carbine $335
Same general specifications as Model 7600 Rifle, except has 18.5-inch bbl. and weighs 7.25 lbs. Made 1987 to date.

Remington Model 7600 Slide-Action Rifle $345
Similar to Model Six, except has lower grade finishes. Made 1981 to date.

Remington Model 7600 Special Purpose $325
Same general specification as the Model 7600, except chambered only in 270 or 30-06. Special Purpose matte black finish on all exposed metal. American walnut stock with SP nonglare finish.

Remington Nylon 76 Lever-Action Repeater $235
Short-throw lever action. Caliber: 22 LR, 14-shot buttstock tubular magazine. Weight: 4 lbs. Black or brown nylon stock and forend. Made 1962-64. Remington's only lever-action rifle.

Remington Sportsman 76 Slide-Action Rifle $225
Caliber: 30-06, 4-shot magazine. 22-inch bbl. Weight: 7.5 lbs. Open rear sight; front blade mounted on ramp. Uncheckered hardwood stock and forend. Made 1985-87.

Semiautomatic Rifles

Remington Model Four (4) Autoloading Rifle
Hammerless. Calibers: 6mm Rem., 243 Win., 270 Win. 7mm Express Rem., 30-06, 308 Win. 22-inch bbl. Weight: 7.5 lbs. Sights: open rear; bead front, on ramp. Monte Carlo checkered stock and forearm. Made 1981-88.

Standard	**$ 345**
Peerless Grade (Engr. receiver)	1395
Premier Grade (Engr. receiver)	2950
Premier Grade (Engr. receiver, Gold Inlay)	4850

Remington Model Four Diamond Anniversary Ltd. Edition. $995
Same as Model Four Standard, except has engraved receiver w/inscription, checkered high-grade walnut stock and forend. Only 1,500 produced. Made 1981 only. (Values for new condition)

Remington Model 8A Autoloading Rifle $395
Standard Grade. Takedown. Calibers: 25, 30, 32 and 35 Rem. Five-shot, clip-loaded magazine. 22-inch bbl. Weight: 7.75 lbs. Sights: adj. and dovetailed open rear; dovetailed bead front. Half-moon metal butt-plate on plain straight-grip walnut stock; plain walnut forearm with thin curved end. Made 1906-1936.

Remington Model 16 Autoloading Rifle $300
Takedown. Closely resembles the Winchester Model 03 semiautomatic rifle. Calibers: 22 Short, 22 LR, 22 Rem. Auto. 15-shot tubular magazine in buttstock. 22-inch bbl. Weight: 5.75 lbs. Sights: open rear; dovetailed bead front. Plain straight-grip stock and forearm. Made 1914-1928. *Note:* In 1918 this model was discontinued in all calibers except 22 Rem. Auto; specifications given are for that.

Remington Model 24A Autoloading Rifle $295
Standard Grade. Takedown. Calibers: 22 Short only, 22 LR only. Tubular magazine in buttstock, holds 15 Short or 10 LR. 21-inch bbl. Weight: about 5 lbs. Sights: dovetailed adj. open rear; dovetailed bead front. Plain walnut straight-grip buttstock; plain walnut forearm. Made 1922-35.

Remington Model 81A Woodsmaster Autoloader $350
Standard Grade. Takedown. Calibers: 30, 32 and 35 Rem., 300 Sav. 5-shot box magazine (not detachable). 22-inch bbl. Weight: 8.25 lbs. Sights: open rear; bead front. Plain walnut pistol-grip stock, forearm. Made 1936-50.

Remington Model 241A Speedmaster Autoloader $275
Standard Grade. Takedown. Calibers: 22 Short only, 22 LR only. Tubular magazine in buttstock, holds 15 Short or 10 LR. 24-inch bbl. Weight: about 6 lbs. Sights; open rear, bead front. Plain walnut stock and forearm. Made 1935-51.

Remington Model 550A Autoloader $145
Has "Power Piston" or floating chamber, which permits interchangeable use of 22 Short, Long or LR cartridges. Tubular magazine holds 22 Short, 17 Long, 15 LR. 24-inch bbl. Weight: about 6.25 lbs. Sights: open rear; bead front. Plain, one-piece pistol-grip stock. Made 1941-71.

Remington Model 550P $155
Same as Model 550A, except has peep rear sight, blade front, on ramp.

Remington Model 550-2G $185
"Gallery Special." Same as Model 550A, except has 22-inch bbl., screweye for counter chain and fired shell deflector.

Remington Model 552A Speedmaster Autoloader $145
Caliber: 22 Short, Long, LR. Tubular magazine holds 20 Short, 17 Long, 15 LR. 25-inch bbl. Weight: about 5.5 lbs. Sights: open rear; bead front. Pistol-grip stock, semibeavertail forearm. Made 1957-88.

Remington Model 552BDL Deluxe $185
Same as Model 552A, except has checkered walnut stock and forearm. Made 1966 to date.

Remington Model 552C Carbine $160
Same as Model 552A, except has 21-inch bbl. Made 1961-77.

Remington Model 552GS Gallery Special $175
Same as Model 552A, except chambered for 22 Short only. Made 1957-77.

Remington Model 740A Woodsmaster Autoloader Rifle $275
Standard Grade. Gas-operated. Calibers: 30-06 or 308. 4-shot detachable box magazine. 22-inch bbl. Weight: about 7.5 lbs. Plain pistol-grip stock, semibeavertail forend with finger grooves. Sights: open rear; ramp front. Made 1955-60. *See* Illustration page 336.

Remington Model 740ADL/BDL Deluxe
Same as Model 740A, except has deluxe checkered stock, standard or high comb, grip cap, sling swivels. Model 740 BDL also has select wood. Made 1955-60.
Model 740 ADL Deluxe Grade $325
Model 740 BDL Deluxe Special Grade 350

Remington Model 742 Bicentennial Commemorative $495
Same as Model 742 Woodsmaster rifle, except has commemorative inscription on receiver. Made 1876. (Value for new condition)

Remington Sportsman 742 Canadian Centennial

Remington Model 742 Canadian Centennial $495
Same as Model 742 rifle, except has commemorative inscription on receiver. Made 1967. Value is for rifle in new, unfired condition.

Remington Model 742 Carbine $285
Same as Model 742 Woodsmaster Rifle, except made in calibers 30-06 and 308 only, has 18.5-inch bbl., weighs 6.75 lbs. Made 1961-80.

Remington Model 742 Woodsmaster Automatic Big Game Rifle $300
Gas-operated semiautomatic. Calibers: 6mm Rem., 243 Win., 280 Rem., 30-06, 308 Win. 4-shot clip magazine. 22-inch bbl. Weight: 7.5 lbs. Sights: open rear; bead front, on ramp. Checkered pistol-grip stock and forearm. Made 1960-80.

Remington Model 742BDL Custom Deluxe $325
Same as Model 742 Rifle, except made in calibers 30-06 and 308 only, has Monte Carlo cheekpiece stock, forearm with black tip, basket-weave checkering. Available in left-hand model. Made 1966-80.

Remington Model 742D Peerless Grade $1350
Same as Model 742 except scroll engraved, fancy wood. Made 1961-80.

RIFLES

Remington Sportsman 74

**Rigby 275
Magazine Rifle**

Rigby 350 Magnum

Remington Model 742F Premier Grade **$2950**
Same as Model 742 except extensively engraved with game scenes and scroll, finest grade wood. Also available with receiver inlaid with gold; adds 50 percent to value. Made 1961-1980.

Remington Model 7400 Autoloader
Similar to Model Four w/lower grade finishes. Made 1981 to date.
Model 7400 Standard **$300**
Model 7400 HG (High Gloss finish) **345**

Remington Model 7400 Carbine............... **$305**
Caliber: 30-06 only. Similar to the Model 7400 rifle except has 18.5-inch bbl. and weighs 7.25 lbs. Made 1988 to date.

Remington Model 7400 Special Purpose **$325**
Same general specification as the Model 7400, except chambered only in 270 or 30-06. Special Purpose matte black finish on metal. American walnut stock with SP nonglare finish. Made 1993-95.

Remington Nylon 66 Apache Black **$135**
Same as Nylon 66 Mohawk Brown, except bbl. and receiver cover chrome-plated, black stock. Made 1962-84.

**Remington Nylon 66 Bicentennial
Commemorative**............................... **$175**
Same as Nylon 66 except has commemorative inscription on receiver. Made 1976.

Remington Nylon 66MB Autoloading Rifle **$125**
Similar to the early production Nylon 66 Black Apache, except with blued bbl. and receiver cover. Made 1978-87.

Remington Nylon 66 Mohawk Brown Autoloader **$95**
Caliber: 22 LR. Tubular magazine in buttstock holds 14 rounds. 19.5-inch bbl. Weight: about 4 lbs. Sights: open rear; blade front. Brown nylon stock and forearm. Made 1959-87.

Remington Nylon 77 Clip Repeater **$150**
Same as Nylon 66, except has 5-shot clip magazine. Made 1970-71.

Remington Sportsman 74 Autoloading Rifle **$245**
Caliber: 30-06, 4-shot magazine. 22-inch bbl. Uncheckered buttstock and forend. Open rear sight; ramped blade front sight. Made 1985-88.

JOHN RIGBY & CO.
London, England

Rigby Model 275 Lightweight Magazine Rifle ... **$4000+**
Same as standard 275 rifle, except has 21-inch bbl. and weighs only 6.75 lbs.

Rigby 275 Magazine Sporting Rifle .... **$5000+**
Mauser action. Caliber: 275 High Velocity or 7×57mm 5-shot box magazine. 25-inch bbl. Weight: about 7.5 lbs. Sights: folding leaf rear; bead front. Sporting stock w/half-pistol grip, checkered. Specifications given are those of current model; however, in general, they apply also to prewar model.

**Rigby Model 350 Magnum Magazine
Sporting Rifle** .... **$3500+**
Mauser action. Caliber: 350 Magnum. 5-shot box magazine. 24-inch bbl. Weight: about 7.75 lbs. Sights: folding leaf rear; bead front. Sporting stock with full pistol grip, checkered. Currently mfd.

**Rigby Model 416 Big Game Magazine
Sporting Rifle** .... **$6000+**
Mauser action. Caliber: 416 Big Game. 4-shot box magazine. 24-inch bbl. Weight: 9 to 9.25 lbs. Sights: folding leaf rear; bead front. Sporting stock with full pistol grip, checkered. Currently mfd.

**Rigby Best Quality Hammerless Ejector
Double Rifle** .... **$39,000+**
Sidelocks. Calibers: 275 Magnum, 350 Magnum, 470 Nitro Express. 24- to 28-inch bbls. Weight: 7.5 to 10.5 lbs. Sights: folding leaf rear; bead front. Checkered pistol-grip stock and forearm.

**Rigby Second Quality Hammerless Ejector
Double Rifle** **$17,500**
Same general specifications as Best Quality double rifle, except boxlock.

**Rigby Third Quality Hammerless Ejector
Double Rifle** **$9900**
Same as Second Quality double rifle, except plainer finsh and not of as high quality.

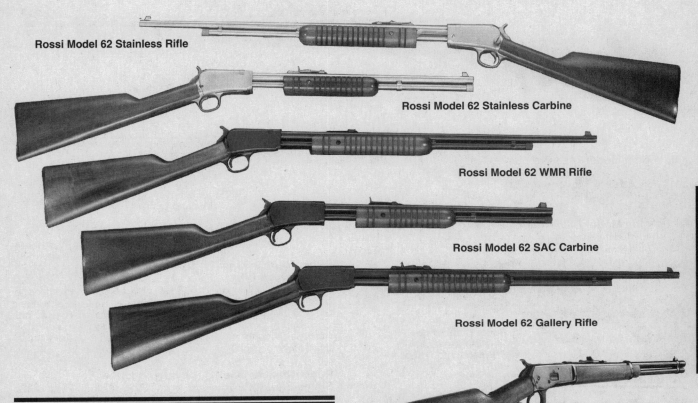

Rossi Model 62 Stainless Rifle

Rossi Model 62 Stainless Carbine

Rossi Model 62 WMR Rifle

Rossi Model 62 SAC Carbine

Rossi Model 62 Gallery Rifle

Rossi Lever-Action Carbine

ROSS RIFLE CO.
Quebec, Canada

Ross Model 1910 Bolt-Action Sporting Rifle **$325**
Straight-pull bolt action with interrupted-screw-type lugs. Calibers: 280 Ross, 303 British. 4-shot or 5-shot magazine. Bbl. lengths: 22, 24, 26 inches. Sights: two-leaf open rear; bead front. Checkered sporting stock. Weight: about 7 lbs. Made c. 1910 to end of World War I. *Note:* Most firearm authorities agree that this and other Ross models with interrupted-screw-type lugs are unsafe to fire.

ROSSI RIFLES
Sao Leopoldo, Brazil
Manufactured by Amadeo Rossi, S.A.

Rossi 62 Gallery Model SAC Carbine
Same as standard Gallery Model, except in 22 LR only with 16.25-inch bbl.; weighs 5.5 lbs. Imported 1975-98.
Blued Finish **$145**
Nickel Finish **165**
Stainless **185**

Rossi 62 Gallery Model Magnum **$145**
Same as standard Gallery Model, except chambered for 22 WMR, 10-shot magazine. Imported 1975-98.

Rossi 62 Gallery Model Slide-Action Repeater
Similar to Winchester Model 62. Calibers: 22 LR, Long, Short or 22 WMR. Tubular magazine holds 13 LR, 16 Long, 20 Short. 23-inch bbl. 39.25 inches overall. Weight: 5.75 lbs. Sights: open rear; bead front. Straight-grip stock, grooved slide handle. Blued, nickel or stainless finish. Imported 1970-98. Values same as SAC Model.

Rossi Lever-Action 65/92 Carbine **$225**
Similar to Winchester Model 92. Caliber: 38 Special/357 Mag., 44 Mag., 44-40, 45 LC. 8- or 10-shot magazine. 16-, 20- or 24-inch round or half-octagonal bbl. Weight: 5.5 to 6 lbs. 33.5- to 41.5-inches overall. Satin blue, chrome or stainless finish. Brazilian hardwood buttstock and forearm. Made 1978-98.
Model M92 SRC 45LC **$275**
Model M92 SRC 38/357, 44 Mag. **250**
Model M92 w/Octagon Bbl. **295**
Model M92 LL Lever **235**
Engraved, **add** **70**
Chrome, **add** **50**
Stainless, **add** **65**

RUGER RIFLES
Southport, Connecticut
Manufactured by Sturm, Ruger & Co.

Ruger Number One (1) Light Sporter **$415**
Same as No.1 Standard, except has 22-inch bbl., folding leaf rear sight on quarter-rib and ramp front sight, Henry pattern forearm. Made 1966 to date.

Ruger Number One (1) Medium Sporter **$425**
Same as No. 1 Light Sporter, except has 26-inch bbl. 22-inch in 45-70); weight is 8 lbs (7.25 lbs. in 45-70). Calibers: 7mm Rem. Mag., 300 Win. Mag., 45-70. Made 1966 to date.

**Ruger Number One (1)
Light Sporter**

**Ruger Number One (1)
Medium Sporter**

**Ruger Number One (1)
"North Americans"
Presentation Rifle**

**Ruger Number One (1) "North Americans"
Presentation Rifle** **$52,000**
Same general specifications as the Ruger No. 1 Standard, except highly customized with elaborate engravings, carvings, fine-line checkering and gold inlays. A series of 21 is planned, each rifle depicting a North American big-game animal, chambered in the caliber appropriate to the game. Stock is of Northern California English walnut. Comes in trunk-style Huey case with Leupold scope and other accessories.

**Ruger Number One (1) RSI International
Single-Shot Rifle** **$425**
Similar to the No. 1 Light Sporter, except with lightweight 20-inch bbl. and full Mannlicher-style forend, in calibers 243 Win., 270 Win., 7×57mm, 30-06. Weight: 7.25 lbs.

Ruger Number One "North Americans" Presentation Set

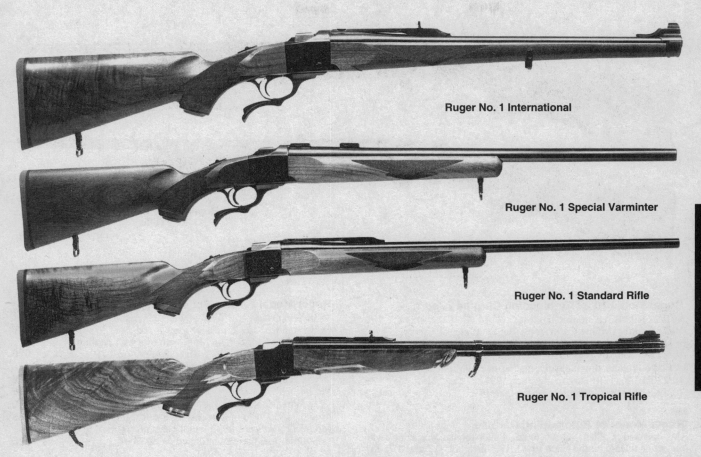

Ruger No. 1 International

Ruger No. 1 Special Varminter

Ruger No. 1 Standard Rifle

Ruger No. 1 Tropical Rifle

RIFLES

Ruger Number One (1) Special Varminter $405
Same as No. 1 Standard, except has heavy 24-inch bbl. with target scope bases, no quarter-rib. Weight: 9 lbs. Calibers: 22-250, 25-06, 7mm Rem. Mag., 300 Win. Mag. Made 1966 to date.

Ruger Number One (1) Standard Rifle $395
Falling-block single-shot action with Farquharson-type lever. Calibers: 22-250, 243 Win., 6mm Rem., 25-06, 270 Win., 30-06, 7mm Rem. Mag., 300 Win. Mag. 26-inch bbl. Weight: 8 lbs. No sights, has quarter-rib for scope mounting. Checkered pistol-grip buttstock and semibeavertail forearm, QD swivels, rubber buttplate. Made 1966 to date.

Ruger Number One (1) Tropical Rifle $475
Same as No. 1 Light Sporter, except has heavy 24-inch bbl.; calibers are 375 H&H 404 Jeffery, 416 Rigby, and 458 Win. Mag. Weight: 8.25 lbs. to 9 lbs. Made 1966 to date.

Ruger Number Three (3) Single-Shot Carbine $365
Falling-block action with American-style lever. Calibers: 22 Hornet 233 Rem., 30-40 Krag, 357 Win., 44 Mag., 45-70. 22-inch bbl. Weight: 6 lbs. Sights: folding leaf rear; gold bead front. Carbine-style stock w/curved buttplate, forearm with bbl. band. Made 1972-87.

Ruger Model 10/22 Autoloading Carbine
Caliber: 22 LR. Detachable 10-shot rotary magazine. 18.5-inch bbl. Weight: 5 lbs. Sights: folding leaf rear; bead front. Carbine-style stock with bbl. band and curved buttplate (walnut stock discontinued 1980). Made 1964 to date. International and Sporter versions discontinued 1971.
10/22 Standard Carbine (Walnut stock) $175
10/22 Int'l (w/Mannlicher-style stock, swivels)
   Discontinued 1971 425
10/22 RB (Birch stock, Blued) 125

Ruger No. 3 Single-Shot Carbine

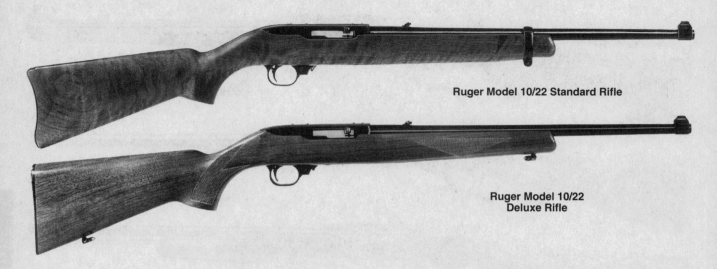

Ruger Model 10/22 Standard Rifle

Ruger Model 10/22 Deluxe Rifle

Ruger Model 10/22 Autoloading Carbine *(Con't)*
K10/22 RB (Birch stock, Stainless) **150**
10/22 Sporter (Monte Carlo stock, flat buttplate
 sling swivels). Discontinued 1971 . . . . **165**
10/22 SP Deluxe Sporter (w/Checkered stock
 Flat buttplate, sling swivels; made since 1966) **185**
10/22 RBI Int'l (Blued); made since 1994 . . . . **175**
10/22 RBI Int'l. (Stainless); made since 1995 . . . . **195**

Ruger Model 44 Autoloading Carbine
Gas-operated. Caliber: 44 Magnum. 4-shot tubular magazine (with magazine release button since 1967). 18.5-inch bbl. Weight: 5.75 lbs. Sights: folding leaf rear; gold bead front. Carbine-style stock w/bbl. band and curved buttplate. Made 1961-86. International and Sporter versions discontinued 1971.
Model 44 Standard Autoloading Carbine **$365**
Model 44 Int'l (w/Mannlicher-style stock, swivels) **625**
Model 44 Sporter (w/Monte Carlo stock, flat buttplate,
 fingergroove, sling swivels) **525**
Model 44RS Carbine (w/Rear peep sight,
 sling swivels; disc. 1978) **425**

Ruger Model 77 Bolt-Action Rifle
Receiver with integral scope mount base or with round top. Short stroke or magnum length action (depending on caliber) in the former type receiver, magnum only in the latter. Calibers: 22-250, 220 Swift, 6mm Rem., 243 Win., 250-3000, 25-06, 257 Roberts, 6.5 Rem. Mag., 270 Win., 7×57mm, 7mm-08 7mm Rem. Mag., 280 Rem., 284 Win., 308 Win., 30-06, 300 Win. Mag. 338 Win. Mag., 350 Rem. Mag., 458 Win. Mag. 5-shot magazine standard, 4-shot in 220 Swift, 3-shot in magnum calibers. 22-, 24- or 26-inch bbl. (depending on caliber). Weight: about 7 lbs.; 458 Mag. model, 8.75 lbs. Round-top model furnished w/folding leaf rear sight and ramp front; integral base model furnished w/scope rings, with or w/o open sights. Stock w/checkered pistol grip and forearm, pistol-grip cap, rubber recoil pad, QD swivel studs. Made 1968-92.
Model 77, integral base, no sights **$335**
6.5 Rem. Mag., **add** **175**
 284 Win., **add** **45**
 338 Win. Mag., **add** **50**
 350 Rem. Mag., 6.5 Rem. Mag., **add** **95**
Model 77RL Ultra Light, no sights, ... **350**
Model 77RL Ultra Light, open sights, ... **375**
Model 77RS, integral base, open sights **365**

Ruger Model 77 Bolt-Action Rifle *(Con't)*
 338 Win. Mag., 458 Win. Mag.,
 with standard stock, **add** **75**
Model 77RSC, 458 Win. Mag. with fancy Circassian
 walnut stock, **add** **525**
Model 77RSI International, Mannlicher stock,
 short action, 18.5-inch bbl., 7 lbs. **350**
Model 77ST, round top, open sights **355**
 338 Win. Mag., add **50**
Model 77V, Varmint, integral base, not sights **335**
Model 77NV, Varmint, integral base, no sights,
 stainless steel barrel, laminated wood stock **385**

Ruger Model 77 Mark II All-Weather Rifle $335
Revised Model 77 action. Same general specifications as Model M-77 Mark II, except with stainless bbl. and action. Zytel injection-molded stock. Calibers: 223, 243, 270, 308, 30-06, 7mm Mag., 300 Win. Mag., 338 Win. Mag. Made 1990 to date.

Ruger Model 77 Mark II Bolt-Action Rifle
Revised Model 77 action. Same general specifications as Model M-77, except with new 3-position safety and fixed blade ejector system. Calibers 22 PPC, 223 Rem., 6mm PPC, 6.5×55 Swedish, 375 H&H, 404 Jeffery nad 416 Rigby also available. Weight 6 to 10.25 lbs. Made 1989 to date.
Model 77 MKIIR, integral base, no sights **$335**
 Left-hand Model 77LR MKII, **add** **25**
Model 77RL MKII, Ultra Light, no sights **350**
Model 77RLS MKII, Ultra Light, open sights **375**
Model 77RS MKII, integral base, open sights. **365**
Model 77RS MKII, Express, with fancy French
 walnut stock, integral base, open sights. **950**
Model 77RSI MKII, International, Mannlicher. **395**
Model 77RSM MKII magnum, with fancy Circassian
 walnut stock, integral base, open sights. **995**
Model 77VT (VBZ or VTM) MKII Varmint/Target
 stainless steel action, laminated wood stock **385**

Ruger Model 77/22 Hornet Bolt-Action Rifle
Mini-Sporter built on the 77/22 action in caliber 22 Hornet. 6-shot rotary magazine. 20- inch bbl. 40 inches overall. Weight: 6 lbs. Receiver machined for Ruger rings (included). Beaded front sight and open adj. rear, or no sights. Blued or stainless finish. Checkered American walnut stock. Made 1994 to date.

Ruger Model 44 Carbine

**Ruger Model 77
Round Top Receiver**

**Ruger Model 77
Ultra-Light Carbine**

**Ruger Model 77
International Carbine**

**Ruger Model 77
Varmint Rifle**

**Ruger Model 77
Mark II — All-Weather Rifle**

**Ruger Model 77
Mark II**

**Ruger Model 77/22
Rimfire**

RIFLES

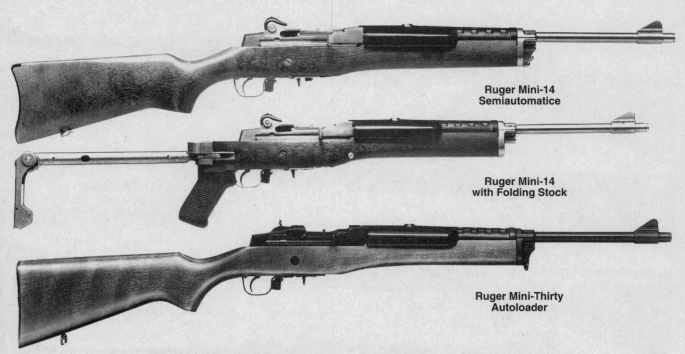

**Ruger Mini-14
Semiautomatice**

**Ruger Mini-14
with Folding Stock**

**Ruger Mini-Thirty
Autoloader**

Ruger Model 77/22 Hornet Bolt-Action Rifle *(Con't)*
Model 77/22RH (rings, no sights) $315
Model 77/22RSH (rings & sights) 325
Model 77/22VHZ (S/S w/Laminated wood stock) 350

Ruger Model 77/22 Rimfire Bolt-Action Rifle
Calibers: 22 LR or 22 WMR. 10-shot (22 LR) or 9-shot (22 WMR) rotary magazine. 20-inch bbl. 39.75 inches overall. Weight: 5.75 lbs. Integral scope bases; with or w/o sights. Checkered American walnut or Zytel injection-molded stock. Stainless or blued finish. Made 1983 to date. (Blued); stainless. Introduced 1989.
77/22 R, rings, no sights, walnut stock $275
77/22 RS, rings, sights, walnut. 285
77/22 RP, rings, no sights, synthetic stock 225
77/22 RSP, rings, sights, synthetic stock 245
K77/22 RP, S/S rings, no sights, synthetic 285
K77/22 RSP, S/S, rings, sights, synthetic 295
77/22 RM, 22 WMR, rings, no sights, walnut 270
77/22 RSM, 22 WMR, rings, sights, walnut 285
K77/22 RSMP, 22 WMR, S/S, rings, sights, synthetic. 295
K77/22 RMP, 22 WMR, S/S, no sights, synthetic 275
K77/22 VBZ, 22 WMR, no sights, Laminated. (1993) 290

Ruger Model 77/44 Bolt-Action
Short-action carbine-style 77 similar to the 77/22RH. Chambered 44 Rem. Mag. 4-shot rotary magazine. 18.5-inch bbl. 38.25 inches overall. Weight: 6 lbs. Gold bead front sight, folding adjustable rear w/integral scope base and Ruger rings. Blue or stainless finish. Synthetic or checkered American walnut stock w/rubber buttpad and swivels. Made 1997 to date.

Ruger Model 77/44 Bolt-Action *(Con't)*
Model 77/44 Blue $295
Model 77/44 Stainless............................. 315

Ruger Model 96 Lever Action Carbine
Caliber: .22 LR, .22 Mag., .44 Mag. Detachable 10-, 9- or 4-shot magazine. 18.5-inch bbl. Weight: 5.25 lbs. Front gold bead sights. Drilled and tapped for scope. American hardwood stock. Made 1996 to date.
.22 Long Rifle................................... $195
.22 Magnum 215
.44 Magnum 245

Ruger Mini-14 Semiautomatic Rifle
Gas-operated. Caliber: 223 Rem. (5.56mm). 5-, 10- or 20-shot box magazine. 18.5-inch bbl. Weight: about 6.5 lbs. Sights: peep rear; blade front mounted on removable barrel band. Pistol-grip stock w/curved buttplate, handguard. Made 1976 to date.
Mini-14/5 Blued $350
K-Mini-14/5 Stainless Steel 375
Mini-14/5F Blued, Folding Stock 550
K-Mini-14/5F Stainless, Folding Stock 600
Mini-14 Ranch Rifle, Scope Model, 6.25 lbs. 385
K-Mini-1H Ranch Rifle, Scope Model, Stainless 395

Ruger Mini-Thirty (30) Autoloader
Caliber: 7.62 × 39mm. 5-shot detachable magazine. 18.5-inch bbl. 37.25 inches overall. Weight: 7 lbs. 3 oz. Designed for use with telescopic sights. Walnut stained stock. Sights: Peep rear;

**Ruger Mini-Thirty
Autoloader**

Ruger Mini-Thirty (30) Autoloader *(Con't)*
blade front mounted on bbl. band. Blued or stainless finish. Made 1986 to date.
Blued Finish **$385**
Stainless Finish **415**

Ruger PC Series Semiautomatic Carbines
Calibers: 9mm Parabellum or 40 S&W. 10-shot magazine. 15.25-inch bbl. Weight: 6.25 lbs. Integral Ruger scope mounts with or without sights. Optional blade front sight, adjustable open rear. Matte black oxide finish. Matte black Zytel stock w/checkered pistol-grip and forearm. Made 1997 to date.
Model PC9 (w/o sights) **$350**
Model PC40 (w/o sights) **365**
W/Adjustable Sights, **add** **30**

RUSSIAN MILITARY RIFLES
Principal U.S.S.R. Arms Plant is located at Tula

Russian Model 1891 Mosin Military Rifle **$95**
Nagant system bolt action. Caliber: 7.62mm Russian. 5-shot box magazine. 31.5-inch bbl. Weight: about 9 lbs. Sights: open rear; blade front. Full stock w/straight grip. Specifications given are for WWII version; earlier types differ slightly. *Note:* In 1916, Remington Arms Co. and New England Westinghouse Co. produced 250,000 of these rifles on a contract from the Imperial Russian Government. Few were delivered to Russia and the balance bought by the U.S. Government for training in 1918. Eventually, many of these rifles were sold to N.R.A. members for about $3 each by the director of Civilian Marksmanship.

Russian Tokarev Model 40 Semiautomatic Military Rifle **$495**
Gas-operated. Caliber: 7.62mm Russian. 10-shot detachable box magazine. 24.5-inch bbl. Muzzle brake. Weight: about 9 lbs. Sights: leaf rear, hooded post front. Full stock w/pistol grip. Differences among Models 1938, 1940 and 1941 are minor.

SAKO RIFLES
Riihimaki, Finland
Manufactured by Sako L.T.D.

Imported by Stoeger Industries, Wayne NJ (formerly by Garcia Corp.)

Sako Model 72 . . . . **$735**
Single model designation replacing Vixen Sporter, Vixen Carbine, Vixen Heavy Barrel, Forester Sporter, Forester Carbine, Forester Heavy Barrel, Finnbear Sporter, and Finnbear Carbine, with same specifications except all but heavy barrel models fitted with open rear sight. Values same as for corresponding earlier models. Imported 1972-74.

Sako Model 73 Lever-Action Rifle **$755**
Same as Finnwolf, except has 3-shot clip magazine, flush floorplate; stock has no cheekpiece. Imported 1973-75.

Sako Model 74 Carbine **$550**
Long Mauser-type bolt action. Caliber: 30-06. 5-shot magazine. 20-inch bbl. Weight: 7.5 lbs. No sights. Checkered Mannlicher-type full stock of European walnut, Monte Carlo cheekpiece. Imported 1974-78.

Sako Model 74 Heavy Barrel Rifle, Long Action **$525**
Same specifications as short action, except w/24-inch heavy bbl., weighs 8.75 lbs.; magnum w/4-shot magazine. Calibers: 25-06, 7mm Rem. Mag. Imported 1974-78.

Sako Model 74 Heavy Barrel Rifle, Medium Action **$545**
Same specifications as short action, except w/23-inch heavy bbl., weighs 8.5 lbs. Calibers: 220 Swift, 22-250, 243 Win., 308 Win. Imported 1974-78.

Sako Model 74 Heavy Barrel Rifle, Short Action **$560**
Mauser-type bolt action. Calibers: 222 Rem., 223 Rem. 5-shot magazine. 23.5-inch heavy bbl. Weight: 8.25 lbs. No sights. Target-style checkered European walnut stock w/beavertail forearm. Imported 1974-78.

Sako Model 74 Super Sporter, Long Action **$535**
Same specifications as short action, except w/24-inch bbl., weighs 8 lbs.; magnums have 4-shot magazine, recoil pad. Calibers: 25-06, 270 Win. 7mm Rem. Mag., 30-06, 300 Win. Mag., 338 Win. Mag., 375 H&H Mag. Imported 1974-78.

Sako Model 74 Super Sporter, Medium Action **$595**
Same specifications as short action, except weighs 7.25 lbs. Calibers: 220 Swift, 22-250, 243 Win. Imported 1974-78.

Sako Model 74 Super Sporter, Short Action **$615**
Mauser-type bolt action. Calibers: 222 Rem., 223 Rem. 5-shot magazine. 23.5-inch bbl. Weight: 6.5 lbs. No sights. Checkered European walnut stock w/Monte Carlo cheekpiece, QD swivel studs. Imported 1974, now disc.

Sako Model 75 Deluxe **$950**
Same specifications as Sako 75 Hunter Model, except w/hinged floor plate, deluxe high gloss checkered walnut stock w/rosewood forend cap and grip cap w/silver inlay. Imported 1998 to date.

Sako Model 75 Hunter **$795**
New bolt action design available in four action lengths fitted with a new bolt featuring three front locking lugs with an external extractor positioned under the bolt. Calibers: 17 Rem., 222 Rem., 223 Rem., (I); 22-250 Rem., 243 Win., 7mm-08 Rem., 308 Win., (III); 25-06 Rem., 270 Win., 280 Rem., 30-06, (IV); 7mm Rem Mag., 300 Win. Mag., 300 Wby. Mag., 338 Win. Mag. 7mm STW, 300 Wby. Mag., 340 Wby. Mag., 375 H&H Mag. and 416 Rem. Mag.,(V). 4-, 5- or 6-shot magazine w/detachable magazine. 22-, 24-, and 26-inch bbls. 41.75 to 45.6 inches over all. Weight: 6.3 to 9 lbs. Sako dovetail scope base integral with receiver with no sights. Checkered high-grade walnut stock w/recoil pad and sling swivels. Made 1997 to date.

Sako Model 75 Stainless Synthetic **$850**
Similar to Model 75 Hunter, except chambered 22-250 Rem., 243 Win., 25-06 Rem., 270 Win., 7mm-08 Rem., 7mm STW, 30-06, 308 Win., 7mm Rem Mag., 300 Win. Mag., 338 Win. Mag. or 375 H&H Mag. 22-, 24-, and 26-inch bbls. Black composite stock w/soft rubber grips inserts. Matte stainless steel finish. Made 1997 to date.

Sako Model 75 Varmint Rifle **$895**
Similar to Model 75 Hunter, except chambered 17 Rem., 222 Rem., 223 Rem. and 22-250 Rem. 24-inch bbl. Matte lacquered walnut stock w/beavertail forearm. Made 1998 to date.

Sako Model 78 Super Hornet Sporter **$425**
Same specifications as Model 78 Rimfire, except caliber 22 Hornet, 4-shot magazine. Imported 1977-87.

Sako Model 78 Super Rimfire Sporter **$365**
Bolt action. Caliber: 22 LR. 5-shot magazine. 22.5-inch bbl. Weight, 6.75 lbs. No sights. Checkered European walnut stock, Monte Carlo cheekpiece. Imported 1977-86.

RIFLES

Sako Model 73

Sako Model 74
Carbine

Sako Model 74
Super Sporter

Sako Model 74
Super Rimfire

Sako Model 75
Hunter

Sako 75
Stainless Synthetic

Sako Model 75
Varmint Rifle

Sako Classic Bolt-Action Rifle
Medium Action (243. Win.) or Long Action (270 Win. 30-06, 7mm Rem. Mag.). American walnut stock. Made 1980-86. Reintroduced in 1993 w/matte lacquer finish stock, 22- or 24-inch bbl., overall length of 42 to 44 inches, weight 6.88 to 7.25lbs. Imported 1992-97.

Standard Calibers $675
Magnum Caliber   725
Left-Hand Models   735

Sako Deluxe Grade AI $625
Same specifications as Standard Grade, except w/22 lines to the inch French checkering, rosewood grip cap and forend tip, semibeavertail forend. Disc.

Sako Deluxe Grade AII $650
Same specifications as Standard Grade, except w/22 lines per inch French checkering, rosewood grip cap and forend tip, semi beavertail forend. Disc.

Sako Deluxe Grade AIII $895
Same specifications as w/standard, except w/French checkering, rosewood grip cap and forend tip, semibeavertail forend. Disc.

Sako Deluxe Lightweight Bolt-Action Rifle $845
Same general specifications as Hunter Lightweight, except w/beautifully grained French walnut stock; superb high-gloss finish, fine hand-cut checkering, rosewood forend tip and grip cap. Imported 1985-97.

Sako Classic

Sako Deluxe Lightweight

Sako Fiberglass

Sako Finsport 2700

Sako Fiberclass Bolt-Action Rifle $825
All-weather fiberglass stock version of Sako barreled long action. Calibers: 25-06, 270, 30-06, 7mm Rem. Mag., 300 Win. Mag., 338 Win. Mag., 375 H&H Mag. Bbl. length: 22.5 inches. Overall length: 44.25 inches. Weight: 7.25 lbs. Imported 1984-96.

Sako Finnbear Carbine $795
Same as Finnbear Sporter, except w/20-inch bbl., Mannlicher-type full stock. Imported 1971. Disc.

Sako Finnbear Sporter $750
Long Mauser-type bolt action. Calibers: 25-06, 264 Mag. 270, 30-06, 300 Win. Mag., 338 Mag., 7mm Mag., 375 H&H Mag. Magazine holds 5 standard or 4 magnum cartridges. 24-inch bbl. Weight: 7 lbs. Hooded ramp front sight. Sporter stock w/Monte Carlo cheekpiece, checkered pistol grip and forearm, recoil pad, swivels. Imported 1961-71.

Sako Finnfire Bolt-Action Rifle
Mini-Sporter built for rimfires on a scaled-down Sako design. Caliber: 22 LR. 5- or 10-shot magazine. 22-inch bbl. 39.5 inches overall. Weight: 5.25 lbs. Receiver machined for 11mm dovetail scope rings. Beaded blade front sight, open adj. rear. Blued finish. Checkered European walnut stock. Imported 1994 to date.
Hunter Model $655
Varmint Model $695
Sporter Model $740

Sako Finnwolf Lever-Action Rifle $775
Hammerless. Calibers: 243 Win., 308 Win. 4-shot clip magazine. 23-inch bbl. Weight: 6.75 lbs. Hooded ramp front sight. Sporter stock w/Monte Carlo cheekpiece, checkered pistol grip and forearm, swivels (available w/right- or left-hand stock). Imported 1963-72.

Sako Finsport 2700 . . . . $675
Bolt-action centerfire rifle. Calibers: 270, 30-06, 7mm Rem. Mag., 300 Win. Mag. Bbl. length: 24 inches. Weight: 8 lbs. Imported 1984-86.

Sako Forester Carbine $725
Same as Forester Sporter, except w/20-inch bbl., Mannlicher-type full stock. Imported 1958-71.

Sako Forester Heavy Barrel $695
Same as Forester Sporter, except w/24-inch heavy bbl. weighs 7.5 lbs. Imported 1958-71.

Sako Forester Sporter $700
Medium-length Mauser-type bolt action. Calibers: 22-250, 243 Win., 308 Win. 5-shot magazine. 23-inch bbl. Weight: 6.5 lbs. Hooded ramp front sight. Sporter stock w/Monte Carlo cheekpiece, checkered pistol grip and forearm, swivels. Imported 1957-71.

Sako Golden Anniversary Model $1850
Special presentation-grade rifle issued in 1973 to commemorate Sako's 50th anniversary. 1,000 (numbered 1 to 1,000) made. Same specifications as Deluxe Sporter, long action, 7mm Rem. Mag. Receiver, trigger guard and floorplate decorated w/gold oak leaf and acorn motif. Stock of select European walnut, checkering bordered w/hand-carved oak leaf pattern.

Sako High-Power Mauser Sporting Rifle $575
FN Mauser action. Calibers: 270, 30-06. 5-shot magazine. 24-inch bbl. Sights: open rear leaf; Patridge front; hooded ramp. Checkered stock w/Monte Carlo comb and cheekpiece. Weight: 7.5 lbs. Imported 1950-57.

Sako Hunter Lightweight Bolt-Action Rifle
5- or 6-shot magazine. Bbl. length: 21.5 inches, AI; 22 inches, AII; 22.5 inches, AIII. Overall length: 42.25-44.5 inches. Weight: 5.75 lbs., AI; 6.75 lbs. AII; 7.25 lbs., AIII. Monte Carlo-style European walnut stock, oil finished. Hand-checkered pistol grip and forend. Imported 1985-97. Left-hand version intro. 1987.
AI (Short Action) 17 Rem. . . . . $ 695
   222 Rem., 223 Rem. **750**
AII (Medium Action)
   22-250 Rem., 243 Win., 308 Win. **675**
AIII (Long Action) 25-06 Rem., 270 Win., 30-06 . . . . **725**
   338 Win. Mag. **775**
   375 H&H Mag. **850**

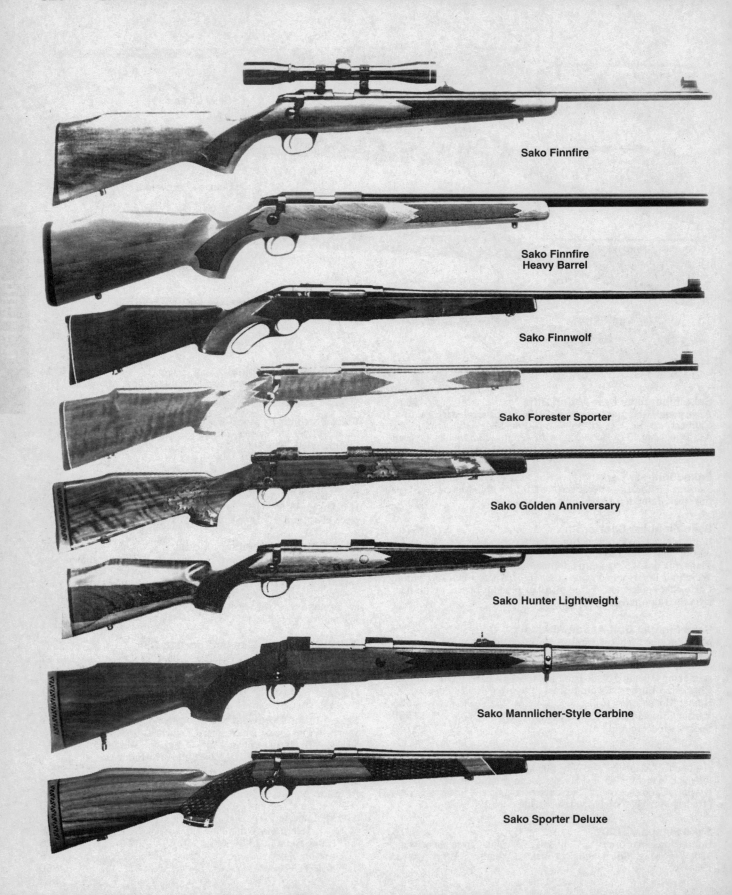

Sako Finnfire

Sako Finnfire Heavy Barrel

Sako Finnwolf

Sako Forester Sporter

Sako Golden Anniversary

Sako Hunter Lightweight

Sako Mannlicher-Style Carbine

Sako Sporter Deluxe

**Sako TRG-21
Target Rifle**

Sako Hunter Lightweight Bolt-Action Rifle *(Con't)*
Left-Hand Model (Standard Cal.) **930**
Magnum Calibers **1070**

Sako Laminated Stock Bolt-Action Rifles
Similar in style and specifications to Hunter Grade, except w/stock of resin-bonded, hardwood veneers. Available 18 calibers in AI (Short), AII (Medium) or AV action, left-hand version in 10 calibers, AV only. Imported 1987-95.
Short or Medium Action **$750**
Long Action/Magnum **775**

Sako Magnum Mauser **$1095**
Similar specifications as Standard Model, except w/recoil pad and redesigned longer AIII action to handle longer magnum cartridges. Calibers: 300 H&H Magnum, 375 H&H Magnum, standard at time of introduction. Disc.

Sako Mannlicher-Style Carbine
Similar to Hunter Model, except w/full Mannlicher-style stock and 18.5-inch bbl. Weighs 7.5 lbs. Chambered in 243, 25-06, 270, 308, 30-06, 7mm Rem. Mag., 300 Win. Mag.,338 Win. Mag., 375 H&H. Intro. in 1977. Disc.
Standard Calibers **$860**
Magnum Calibers (except 375) **895**
375 H&H **900**

Sako Safari Grade **$1695**
Classic bolt action. Calibers: 300 Win. Mag., 338 Win. Mag., 375 H&H. Oil-finished European walnut stock w/hand-checkering. Barrel band swivel, express-type sight rib; satin or matte blue finish. Imported 1980-96.

Sako Sporter Deluxe **$875**
Same as Vixen, Forester, Finnbear and Model 74, except w/fancy French walnut stock w/skip checkering, rosewood forend tip and pistol-grip cap, recoil pad, inlaid trigger guard and floorplate. Disc.

Sako Standard Grade AI **$575**
Short bolt action. Calibers: 17 Rem.,222 Rem.,223 Rem. 5-shot magazine. 23.5-inch bbl. Weight: 6.5 lbs. No sights. Checkered European walnut stock w/Monte Carlo cheekpiece, QD swivel studs. Imported 1978-85.

Sako Standard Grade AII **$595**
Medium bolt action. Calibers: 22-250 Rem., 243 Win., 308 Win. 23.5-inch bbl. in 22-250; 23-inch bbl. in other calibers. 5-shot magazine. Weight: 7.25 lbs. Checkered European walnut stock w/Monte Carlo cheekpiece, QD swivel studs. Imported 1978-85.

Sako Standard Grade AIII **$625**
Long bolt action. Calibers: 25-06 Rem., 270 Win., 30-06, 7mm Rem. Mag., 300 Win. Mag., 338 Win. Mag., 375 H&H. 24-inch bbl. 4-shot magazine. Weight: 8 lbs. Imported 1978-84.

Sako Super Deluxe Rifle..................... **$1725**
Available in AI, AII, AIII action calibers. Select European walnut stock, hand-checkered, deep oak leaf hand-engraved design. Disc.

Sako TRG-Bolt-Action Target Rifle
Caliber: 308 Win or 338 Lapua Mag. Detachable 10-shot magazine. 25.75- or 27.2-inch bbl. Weight: 10.5 to 11 lbs. Blued action w/stainless barrel. Adjustable two-stage trigger. modular reinforced polyurethane target stock w/adj. cheekpiece and buttplate. Options: Muzzle break; detachable bipod; Q/D sling swivels and scope mounts w/1-inch or 30mm rings. Imported 1993 to date.
TGR-21 308 Win **$2650**
TGR-41 338 Lapua.............................. **3150**

Sako TRG-S Bolt-Action Rifle
Calibers: 243, 7mm-08, 270, 30-06, 7mm Rem. Mag., 300 Win. Mag., 338 Win. Mag. 5-shot magazine (standard calibers), 4-shot (magnum), 22- or 24-inch bbl. 45.5 inches overall. Weight: 7.75 lbs. No sights. Reinforced polyurethane stock w/Monte Carlo. Intro. in 1993.
Standard Calibers **$525**
Magnum Calibers **575**

Sako Vixen Carbine **$775**
Same as Vixen Sporter, except w/20-inch bbl., Mannlicher-type full stock. Imported 1947-71.

**Sako TRG-S
Bolt-Action Rifle**

Sako Vixen Heavy Barrel

Sako Vixen Sporter

Sako Vixen Heavy Barrel **$750**
Samee as Vixen Sporter, except calibers 222 Rem., 222 Rem. Mag., 223 Rem., heavy bbl., target-style stock w/beavertail forearm. Weight: 7.5 lbs. Imported 1947-71.

Sako Vixen Sporter **$795**
Short Mauser-type bolt action. Calibers: 218 Bee, 22 Hornet, 222 Rem., 222 Rem. Mag., 223 Rem. 5-shot magazine. 23.5-inch bbl. Weight: 6.5 lbs. Hooded ramp front sight. Sporter stock w/Monte Carlo cheekpiece, checkered pistol grip and forearm, swivels. Imported 1946-71.

J. P. SAUER & SOHN
Eckernforde, Germany
(Formerly Suhl, Germany)
Imported by Sigarms Exeter, NH
(Previously by Paul Company Inc. and G.U. Inc.)

Sauer Mauser Bolt-Action Sporting Rifle **$995**
Calibers: 7×57 and 8×57mm most common, but these rifles were produced in a variety of calibers, including most of the popular Continental numbers as well as our 30-06. 5-shot box magazine. 22- or 24-inch Krupp steel bbl., half-octagon w/raised matted rib. Double-set trigger. Weight: 7.5 lbs. Sights: three-leaf open rear; ramp front. Sporting stock w/cheekpiece, checkered pistol grip, raised side-panels, Schnabel tip, swivels. Also made w/20-inch bbl. and full-length stock. Mfd before WWII.

Sauer Model S-90 Bolt-Action Rifles
Calibers: 243 Win., 308 Win. (Short action); 25-06, 270 Win.,30-06 (Medium action); 7mm Rem. Mag.,300 Win. Mag., 300 Wby., 338 Win., 375 H&H (Magnum action). 4-shot (standard) or 3-shot magazine (magnum). Bbl. length: 20-inch (Stutzen), 24-inch. Weight: 7.6 to 10.75 lbs. Adjustable (Supreme) checkered Monte Carlo style stock. contrasting forend and pistol grip cap w/high-gloss finish or European (Lux) checkered Classic style European walnut stock w/satin oil finish. Imported 1983-89.

S-90 Standard **$675**
S-90 Lux 950
S-90 Safari 875
S-90 Stutzen 695
S-90 Supreme 965
Grade I Engraving, **add** 600
Grade II Engraving, **add** 800
Grade III Engraving **add** 1000
Grade IV Engraving **add** 1500

Sauer Model 200 Bolt-Action Rifles
Calibers: 243 Win., 25-06, 270 Win., 7mm Rem Mag.,30-06, 308 Win., 300 Win. Mag., Detachable box magazine. 24-inch (American) or 26-inch (European) interchangeable bbl. Standard (steel) or lightweight (alloy) action. Weight: 6.6 to 7.75 lbs. 44 inches overall. Stock options: American Model w/checkered Monte Carlo style 2-piece stock contrasting forend and pistol grip cap w/high gloss finish and no sights.European walnut stock w/Schnabel forend, satin oil finish and iron sights. Contemporary Model w/synthetic carbon fiber stock. Imported 1986-93.

Standard Model **$825**
Lightweight Model 795
Contemporary Model 875
American Model 950
European Model 975
Left-Hand Model, **add** 125
Magnum Calibers, **add** 115
Interchangeable Barrel Assembly, **add** 295

Sauer Model 202 Bolt-Action Rifles
Calibers: 243 Win., 6.5×55, 6.5×57, 6.6×68, 25-06, 270 Win., 280 7×64,308, 30-06, Springfield, 7mm Rem. Mag., 300 Win. Mag., 300 Wby. mag., 8×68S, 338 Win. Mag., 375 H&H Mag. Removalbe 3-shot box magazine. 23.6- and 26-inch interchangable bbl. 44.3 and 46 inches overall. Modular receiver drilled and tapped for scope bases. Adjustable 2-stage trigger w/dual release safety. Weight: 7.7 to 8.4 lbs. Stock options: Checkered Monte carlo style select American walnut 2-piece stock; Euro-classic French walnut 2-piece stock w/semi Schnabel forend and satin oil finish; Super Grade Claro walnut 2-piece stock fitted w/rosewood forend and grip cap w/high-gloss epoxy finish. Imported 1994 to date.

Standard Model **$695**
Euro-Classic Model 725
Super Grade model 775
Left-Hand Model, **add** 150
Magnum Calibers, **add** 125
Interchangeable Barrel Assembly, **add** 295

SAVAGE INDUSTRIES
Westfield, Massachusetts
(Formerly Chicopee Falls, MA and Utica, NY)

Savage Model 3 Bolt-Action Single-Shot Rifle **$135**
Takedown. Caliber: 22 Short, Long, LR. 26-inch bbl. on prewar rifles, postwar production w/24-inch bbl. Weight: 5 lbs. Sights: open rear; bead front. Plain pistol-grip stock. Made 1933-52.

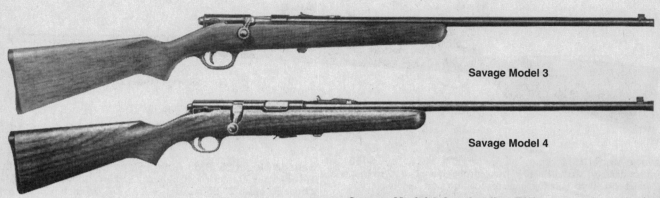

Savage Model 3

Savage Model 4

Savage Model 3S **$175**
Same as Model 3, except w/peep rear sight, hooded front. Made 1933-42.

Savage Model 3ST **$180**
Same as Model 3S, except fitted w/swivels and sling. Made 1933-42.

Savage Model 4 Bolt-Action Repeater **$135**
Takedown. Caliber: 22 Short, Long, LR. 5-shot detachable box magazine. 24-inch bbl. Weight: 5.5 lbs. Sights: open rear; bead front. Checkered pistol-grip stock on prewar models, early production had grooved forearm; postwar rifles have plain stocks. Made 1933-65.

Savage Model 4M **$125**
Same as Model 4, except chambered for 22 Rimfire Magnum. Made 1961-65.

Savage Model 4S **$145**
Same as Model 4, except w/peep rear sight, hooded front. Made 1933-42.

Savage Model 5 Bolt-Action Repeater **$135**
Same as Model 4 except w/tubular magazine (holds 21 Short, 17 Long, 15 LR), weighs 6 lbs. Made 1936-61.

Savage Model 5S **$140**
Same as Model 5, except w/peep rear sight, hooded front. Made 1936-42.

Savage Model 6 Autoloading Rifle **$165**
Takedown. Caliber: 22 Short, Long, LR. Tubular magazine holds 21 Short, 17 Long, 15 LR. 24-inch bbl. Weight: 6 lbs. Sights: open rear; bead front. Checkered pistol-grip stock on prewar models, postwar rifles have plain stocks. Made 1938-68.

Savage Model 6S **$180**
Same as Model 6, except w/peep rear sight, bead front. Made 1938-42.

Savage Model 7 Autoloading Rifle **$150**
Same general specifications as Model 6, except w/5-shot detachable box magazine. Made 1939-51.

Savage Model 7S **$175**
Same as Model 7, except w/peep rear sight, hooded front. Made 1938-42.

Savage Model 19 Bolt-Action Target Rifle **$225**
Model of 1933. Speed lock. Caliber: 22-LR. 5-shot detachable box magazine. 25-inch bbl. Weight: 8 lbs. Adj. rear peep sight, blade front on early models, later production equipped w/extension rear sight, hooded front. Target stock w/full pistol grip and beavertail forearm. Made 1933-46.

Savage Model 19 NRA Bolt-Action Match Rifle **$295**
Model of 1919. Caliber: 22 LR. 5-shot detachable box magazine. 25-inch bbl. Weight: 7lbs. Sights: adj. rear peep; blade front. Full military stock w/pistol grip. Made 1919-33.

Savage Model 5

Savage Model 6

**Savage Model 19
NRA (1933)**

RIFLES

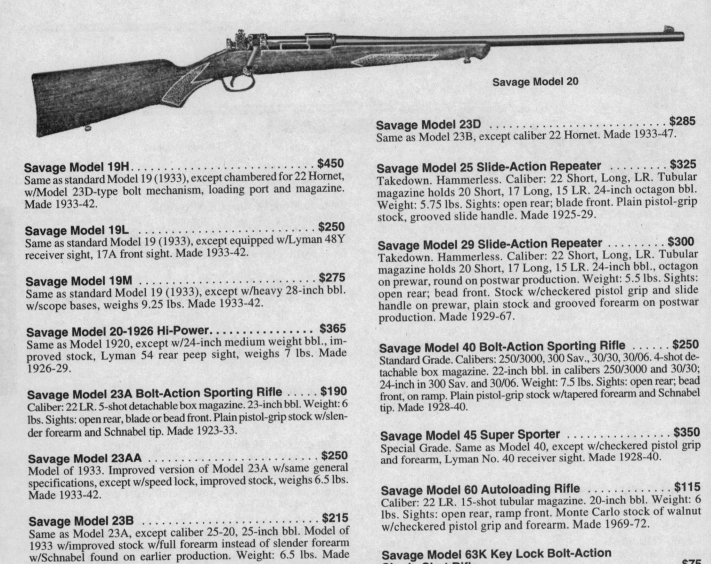

Savage Model 20

Savage Model 19H **$450**
Same as standard Model 19 (1933), except chambered for 22 Hornet, w/Model 23D-type bolt mechanism, loading port and magazine. Made 1933-42.

Savage Model 19L **$250**
Same as standard Model 19 (1933), except equipped w/Lyman 48Y receiver sight, 17A front sight. Made 1933-42.

Savage Model 19M **$275**
Same as standard Model 19 (1933), except w/heavy 28-inch bbl. w/scope bases, weighs 9.25 lbs. Made 1933-42.

Savage Model 20-1926 Hi-Power **$365**
Same as Model 1920, except w/24-inch medium weight bbl., improved stock, Lyman 54 rear peep sight, weighs 7 lbs. Made 1926-29.

Savage Model 23A Bolt-Action Sporting Rifle **$190**
Caliber: 22 LR. 5-shot detachable box magazine. 23-inch bbl. Weight: 6 lbs. Sights: open rear, blade or bead front. Plain pistol-grip stock w/slender forearm and Schnabel tip. Made 1923-33.

Savage Model 23AA **$250**
Model of 1933. Improved version of Model 23A w/same general specifications, except w/speed lock, improved stock, weighs 6.5 lbs. Made 1933-42.

Savage Model 23B **$215**
Same as Model 23A, except caliber 25-20, 25-inch bbl. Model of 1933 w/improved stock w/full forearm instead of slender forearm w/Schnabel found on earlier production. Weight: 6.5 lbs. Made 1923-42.

Savage Model 23C **$225**
Same as Model 23B, except caliber 32/20. Made 1923-42.

Savage Model 23D **$285**
Same as Model 23B, except caliber 22 Hornet. Made 1933-47.

Savage Model 25 Slide-Action Repeater **$325**
Takedown. Hammerless. Caliber: 22 Short, Long, LR. Tubular magazine holds 20 Short, 17 Long, 15 LR. 24-inch octagon bbl. Weight: 5.75 lbs. Sights: open rear; blade front. Plain pistol-grip stock, grooved slide handle. Made 1925-29.

Savage Model 29 Slide-Action Repeater **$300**
Takedown. Hammerless. Caliber: 22 Short, Long, LR. Tubular magazine holds 20 Short, 17 Long, 15 LR. 24-inch bbl., octagon on prewar, round on postwar production. Weight: 5.5 lbs. Sights: open rear; bead front. Stock w/checkered pistol grip and slide handle on prewar, plain stock and grooved forearm on postwar production. Made 1929-67.

Savage Model 40 Bolt-Action Sporting Rifle **$250**
Standard Grade. Calibers: 250/3000, 300 Sav., 30/30, 30/06. 4-shot detachable box magazine. 22-inch bbl. in calibers 250/3000 and 30/30; 24-inch in 300 Sav. and 30/06. Weight: 7.5 lbs. Sights: open rear; bead front, on ramp. Plain pistol-grip stock w/tapered forearm and Schnabel tip. Made 1928-40.

Savage Model 45 Super Sporter **$350**
Special Grade. Same as Model 40, except w/checkered pistol grip and forearm, Lyman No. 40 receiver sight. Made 1928-40.

Savage Model 60 Autoloading Rifle **$115**
Caliber: 22 LR. 15-shot tubular magazine. 20-inch bbl. Weight: 6 lbs. Sights: open rear, ramp front. Monte Carlo stock of walnut w/checkered pistol grip and forearm. Made 1969-72.

**Savage Model 63K Key Lock Bolt-Action
Single-Shot Rifle** **$75**
Trigger locked w/key. Caliber: 22 Short, Long, LR. 18-inch bbl. Weight: 4 lbs. Sights: open rear; hooded ramp front. Full-length stock w/pistol grip, swivels. Made 1970-72.

Savage Model 23AA

Savage Model 29

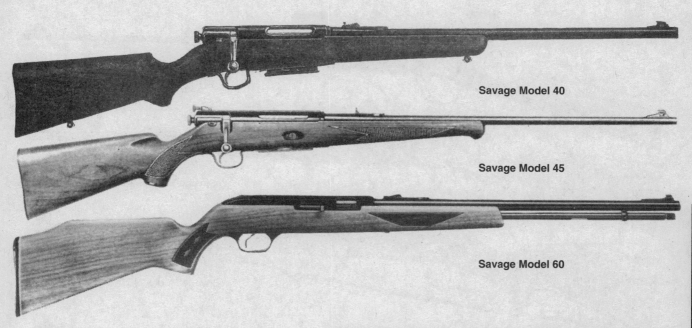

Savage Model 40

Savage Model 45

Savage Model 60

Savage Model 63KM **$90**
Same as Model 63K, except chambered for 22 WMR. Made 1970-72.

Savage Model 64F Autoloading Rifle **$95**
Same general specifications as Model 64, except w/black graphite/polymer stock. Weight: 5 lbs. Made 1997 to date.

Savage Model 64G Autoloading Rifle **$100**
Caliber: 22 LR. 10-shot magazine. 20-inch bbl. Weight: 5.5 lbs. 40 inches overall. Sights: adj. open rear; bead front. Grooved receiver for scope mounts. Stamped checkering on walnut-finished hardwood stock w/Monte Carlo cheekpiece. Made 1996 to date.

Savage Model 71 "Stevens Favorite" Single-Shot Lever-Action Rifle **$150**
Replica of original Stevens Favorite issued as a tribute to Joshua Stevens, "Father of 22 Hunting." Caliber: 22 LR. 22-inch full-octagon bbl. Brass-plated hammer and lever. Sights: open rear; brass blade front. Weight: 4.5 lbs. Plain straight-grip buttstock and Schnabel forend; brass commemorative medallion inlaid in buttstock, brass crescent-shaped buttplate. 10,000 produced. Made in 1971 only. Value is for new, unfired specimen.

Savage Model 90 Autoloading Carbine **$120**
Similar to Model 60, except w/16.5-inch bbl. w/folding leaf rear sight, bead front. 10-shot tubular magazine. Unchecked, carbine-style walnut stock w/bbl. Band and sling swivels. Weight: 5.75 lbs. Made 1969-72.

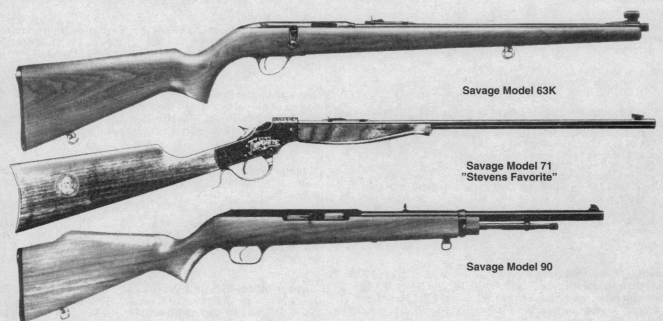

Savage Model 63K

Savage Model 71 "Stevens Favorite"

Savage Model 90

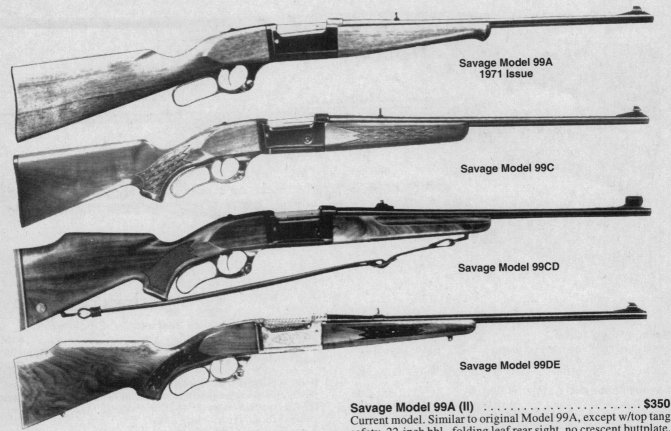

Savage Model 99A
1971 Issue

Savage Model 99C

Savage Model 99CD

Savage Model 99DE

Savage Model 93G Bolt-Action Rifle **$125**
Caliber: 22 Win Mag. 5-shot magazine. 20.75-inch bbl. 39.5
inches overall. Weight: 5.75 lbs. Sights: adj. open rear; bead
front. Grooved receiver for scope mounts. Stamped checkering
on walnut-finished hardwood stock w/Monte Carlo cheekpiece.
Made 1996 to date.

Savage Model 93F Bolt-Action Rifle **$105**
Same general specifications as Model 93G, except w/black graph-
ite/polymer stock. Weight: 5.2 lbs. Made 1997 to date.

NOTE

Savage Model 99 Lever-Action Repeater

Introduced in 1899, this model has been produced in a variety of styles and calibers. Original designation "Model 1899"
was changed to "Model 99" c.1920. Earlier rifles and carbines — similar to Models 99A, 99B and 99H — were supplied in calibers 25-35, 30-30, 303 Sav., 32-40 and 38-55.
Post-WWII Models 99A, 99C, 99CD, 99DE, 99DL, 99F and
99PE have top tang safety other 99s have slide safety on
right side of trigger guard. Models 99C and 99CD have detachable box magazine instead of traditional Model 99 rotary
magazine.

Savage Model 99A (I) **$525**
Hammerless. Solid frame. Calibers: 30/30, 300 Sav., 303 Sav.
Five-shot rotary magazine. 24-inch bbl. Weight: 7.25 lbs. Sights:
open rear; bead front, on ramp. Plain straight-grip stock, tapered
forearm. Made 1920-36.

Savage Model 99A (II) **$350**
Current model. Similar to original Model 99A, except w/top tang
safety, 22-inch bbl., folding leaf rear sight, no crescent buttplate.
Calibers: 243 Win., 250 Sav., 300 Sav., 308 Win. Made 1971-82.

Savage Model 99B **$650**
Takedown. Otherwise same as Model 99A, except weight 7.5 lbs.
Made 1920-36.

Savage Model 99C **$425**
Current model. Same as Model 99F, except w/clip magazine instead of
rotary. Calibers: 243 Win., 284 Win., 308 Win. (4-shot detachable
magazine holds one round less in 284). Weight: 6.75 lbs. Made 1965 to
date.

Savage Model 99CD **$395**
Deluxe version of Model 99C. Calibers: 243 Win.,250 Sav., 308 Win.
Hooded ramp front sight. Weight: 8.25 lbs. Stock w/Monte Carlo comb
and cheekpiece, checkered pistol grip, grooved forearm, swivels and
sling. Made 1975 to 81.

Savage Model 99DE Citation Grade **$525**
Same as Model 99PE, except w/less elaborate engraving. Made
1968-70.

Savage Model 99DL Deluxe **$315**
Postwar model. Calibers: 243 Win., 308 Win. Same as Model 99F,
except w/high comb Monte Carlo stock, sling swivels. Weight: 6.75
lbs. Made 1960-73.

Savage Model 99E Carbine (I) **$525**
Pre-WWII type. Solid frame. Calibers: 22 Hi-Power, 250/3000,
30/30, 300 Sav., 303 Sav. w/22-inch bbl.; 300 Sav. 24-inch. Weight:
7 lbs. Other specifications same as Model 99A. Made 1920-36.

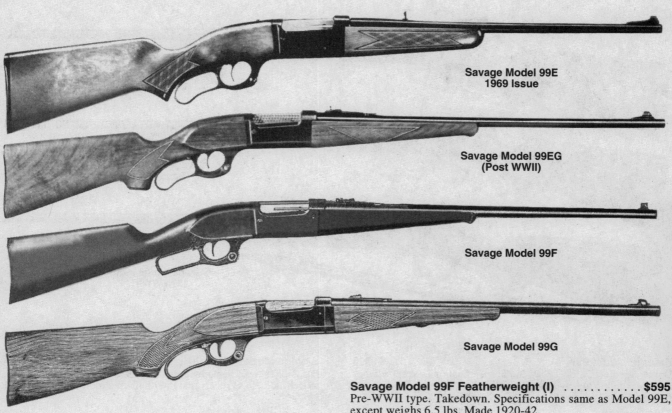

Savage Model 99E
1969 Issue

Savage Model 99EG
(Post WWII)

Savage Model 99F

Savage Model 99G

Savage Model 99E Carbine (II) **$275**
Current model. Solid frame. Calibers: 250 Sav., 243 Win., 300 Sav., 308 Win. 20- or 22-inch bbl. Checkered pistol-grip stock and forearm. Made 1960-89.

Savage Model 99EG (I) **$535**
Pre-WWII type. Solid frame. Plain pistol-grip stock and forearm. Otherwise same as Model G. Made 1936-41.

Savage Model 99EG (II) **$425**
Post-WWII type. Same as prewar model, except w/checkered stock and forearm. Calibers: 250 Sav., 300 Sav., 308 Win. (intro. 1955), 243 Win., and 358 Win. Made 1946-60.

Savage Model 99F Featherweight (I) **$595**
Pre-WWII type. Takedown. Specifications same as Model 99E, except weighs 6.5 lbs. Made 1920-42.

Savage Model 99F Featherweight (II) **$325**
Postwar model. Solid frame. Calibers: 243 Win., 300 Sav., 308 Win. 22-inch bbl. Checkered pistol-grip stock and forearm. Weight: 6.5 lbs. Made 1955-73.

Savage Model 99G **$625**
Takedown. Checkered pistol-grip stock and forearm. Weight: 7.25 lbs. Other specifications same as Model 99E. Made 1920-42.

Savage Model 99H Carbine **$475**
Solid frame. Calibers: 250/3000, 30/30, 303 Sav. 20-inch special weight bbl. Walnut carbine stock w/metal buttplate; walnut fore-arm w/bbl. band. Weight: 6.5 lbs. Open rear sights; ramped blade front sight. Other specifications same as Model 99A. Made 1931-42.

Savage Model 99PE
Early Issue

Savage Model 99PE
Late Issue

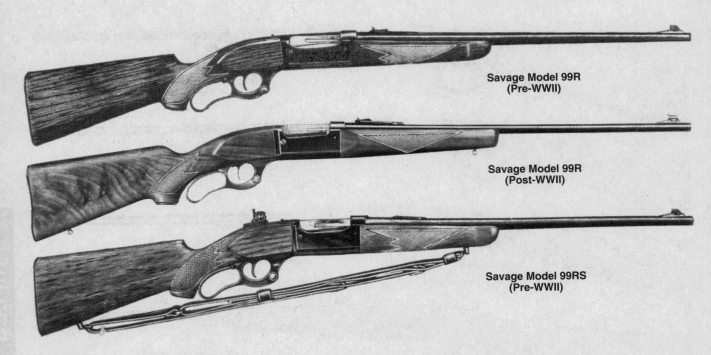

Savage Model 99R
(Pre-WWII)

Savage Model 99R
(Post-WWII)

Savage Model 99RS
(Pre-WWII)

Savage Model 99K **$1395**
Deluxe version of Model G w/similar specifications, except w/fancy stock and engraving on receiver and bbl. Lyman peep rear sight and folding middle. Made 1931-42.

Savage Model 99PE Presentation Grade **$1425**
Same as Model 99DL, except w/engraved receiver (game scenes on sides), tang and lever, fancy walnut Monte Carlo stock and forearm w/hand checkering, QD swivels. Calibers: 243, 284, 308. Made 1968-70.

Savage Model 99R (I) **$575**
Pre-WWII type. Solid frame. Calibers: 250/3000 (22-inch bbl.), 300 Sav. (24-inch bbl.). Weight: 7.5 lbs. Special large pistol-grip stock and forearm, checkered. General specifications same as other Model 99 rifles. Made 1936-42.

Savage Model 99R (II) **$395**
Post-WWII type. Same as prewar model, except w/24-inch bbl. only, w/screw eyes for sling swivels. Calibers: 250 Sav., 300 Sav., 308 Win., 243 Win. and 358 Win. Made 1946-60.

Savage Model 99RS (I) **$595**
Pre-WWII type. Same as prewar Model 99R, except equipped w/Lyman rear peep sight and folding middle sight, quick detachable swivels and sling. Made 1936-42.

Savage Model 99RS (II) **$350**
Post-WWII type. Same as postwar Model 99RS, except equipped w/Redfield 70LH receiver sight, blank in middle sight slot. Made 1946-58.

Savage Model 99T **$435**
Featherweight. Solid frame. Calibers: 22 Hi-Power, 30/30, 303 Sav. w/20-inch bbl.; 300 Sav. w/22-inch bbl. Checkered pistol-grip stock and beavertail forearm. Weight: 7 lbs. General specifications same as other Model 99 rifles. Made 1936-42.

Savage Model 99-358 . . . . **$395**
Similar to current Model 99A, except caliber 358 Win. has grooved forearm, recoil pad, swivel studs. Made 1977-80.

Savage Model 110 Sporter Bolt-Action Rifle **$160**
Calibers: 243, 270, 308, 30-06. 4-shot box magazine. 22-inch bbl. Weight: about 6.75 lbs. Sights: open rear; ramp front. Standard sporter stock with pistol grip checkered. Made 1958-63.

Savage Model 110B Bolt-Action Rifle
Same as Model 110E, except with checkered select walnut Monte Carlo-style stock (early models) or brown laminated stock (late models). Calibers: 243 Win., 270 Win. 30-06, 7mm Rem. Mag., 338 Win. Mag. Made 1976 to date.
Early Model **$250**
Laminated Stock Model **265**

Savage Model 99T

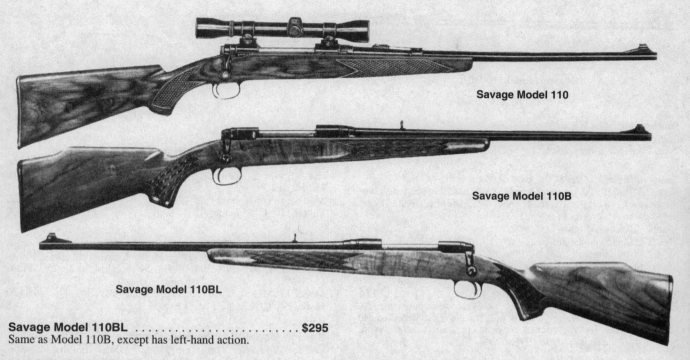

Savage Model 110

Savage Model 110B

Savage Model 110BL

RIFLES

Savage Model 110BL $295
Same as Model 110B, except has left-hand action.

Savage Model 110C
Calibers: 22-250, 243, 25-06, 270, 308, 30-06, 7mm Rem. Mag., 300 Win. Mag. 4-shot detachable clip magazine (3-shot in Magnum calibers). 22-inch bbl. (24-inch in 22-250 Magnum calibers). Weight: 6.75 lbs., Magnum, 7.75 to 8 lbs. Sights: open rear; ramp front. Checkered Monte Carlo-style walnut stock (Magnum has recoil pad). Made 1966-88.
Standard Calibers $285
Magnum Calibers 295

Savage Model 110CL
Same as Model 110C, except has left-hand action. (Available only in 243 Win., 30-06, 270 and 7mm Mag.) Made 1963-66.
Standard Calibers $275
Magnum Calibers 295

Savage Model 110CY Youth/Ladies Rifle $265
Same as Model 110 G except with walnut-finished hardwood stock with 12.5-inch pull. Calibers: 243 and 300 Savage. Made 1991 to date.

Savage Model 110D
Similar to Model 110C, except has internal magazine with hinged floorplate. Calibers: 243 Win., 270 Win., 30-06, 7mm Rem. Mag., 300 Win. Mag. Made 1972-88.
Standard Calibers $275
Magnum Calibers 295

Savage Model 110DL
Same as Model 110D, except has left-hand action. Discontinued.
Standard Calibers $325
Magnum Calibers $350

Savage Model 110E $225
Calibers: 22-250, 223 Rem., 243, 270 Win., 308, 7mm Rem. Mag., 30-06. 4-shot box magazine (3-shot in Magnum). 20- or 22-inch bbl. (24-inch stainless steel in Magnum). Weight: 6.75 lbs.; Magnum, 7.75 lbs. Sights: open rear; ramp front. Plain Monte Carlo stock on early production; current models have checkered stocks of walnut-finished hardwood (Magnum has recoil pad). Made 1963-89.

Savage Model 110C

Savage Model 110E

Savage Model 110 MCL

Savage Model 110EL $245
Same as Model 110E, except has left-hand action, made in 30-06 and
7mm Rem. Mag. only. Made 1969-73.

Savage Model 110F/110K Bolt-Action Rifle
Same as Model 110E, except **Model** 110F has black Rynite® syn-
thetic stock, swivel studs; made 1988-93. **Model** 110K has lami-
nated camouflage stock; made 1986-88.
Model 110F, Adj. Sights $275
Model 110FNS, No Sights 285
Model 110K Standard Calibers 295
Model 110K Magnum Calibers 335

Savage Model 110FM Sierra Ultra Light $275
Calibers: 243 Win., 270 Win., 30-06, 308 Win. 5-shot magazine. 20-
inch bbl. 41.5 inches overall. Weight: 6.25 lbs. No sights w/drilled and
tapped receiver. Black graphite/fiberglass composition stock. Nonglare matte blue finish. Made 1996 to date.

Savage Model 110FP Police Rifle $295
Calibers: 223, 308 Win. 4-shot magazine. 24-inch bbl. 45.5 inches
overall. Weight: 9 lbs. Black Rynite composition stock. Matte blue
finish. Made 1990 to date.

Savage Model 110G Bolt-Action Rifle
Calibers: 223, 22-250, 243, 270, 7mm Rem. Mag., 308 Win., 30-06,
300 Win. Mag. 5-shot (standard) or 4-shot magazine (magnum). 22-
or 24-inch bbl. 42.38 overall (standard). Weight: 6.75 to 7.5 lbs.
Ramp front sight, adj. rear. Checkered walnut-finished hardwood
stock with rubber recoil pad. Made 1989-93.
Model 110G Standard Calibers $260
Model 110G Magnum Calibers 285
Model 110GLNS, Left-hand, No Sights 305

Savage Model 110GV Varmint Rifle $275
Similar to the Model 110 G, except fitted with medium-weight
varmint bbl. with no sights. Receiver drilled and tapped for scope
mount. Calibers 22-250 and 223 only. Made 1989-93.

Savage Model 110M Magnum
Same as Model 110MC, except calibers: 7mm Rem. Mag. 264, 300
and 338 Win. 24-inch bbl. Stock with recoil pad. Weight: 7.75 to 8
lbs. Made 1963-69.
Model 110M Magnum $235
Model 110ML Magnum, Left-Hand Action 260

Savage Model 110MC
Same as Model 110, except has Monte Carlo-style stock. Calib-
ers: 22-250, 243, 270, 308, 30-06. 24-inch bbl. in 22-250. Made
1959-69.
Model 110MC $170
Model 110MCL w/Left-hand Action 185

Savage Model 110P Premier Grade
Calibers: 243 Win., 7mm Rem. Mag., 30-06. 4-shot magazine (3-
shot in Magnum). 22-inch bbl. (24-inch stainless steel in Magnum). Weight: 7 lbs.; Magnum, 7.75 lbs. Sights: open rear folding
leaf; ramp front. French walnut stock w/Monte Carlo comb and
cheekpiece, rosewood forend tip and pistol-grip cap, skip checkering, sling swivels (Magnum has recoil pad). Made 1964-70.
Calibers 243 Win. and 30-06 . . . . $335
Caliber 7mm Rem. Mag. 355

Savage Model 110PE Presentation Grade
Same as Model 110P, except has engraved receiver, floorplate
and trigger guard, stock of choice grade French walnut. Made
1968-70.
Calibers 243 and 30-06 . . . . $550
Caliber 7mm Rem. Mag. 595

Savage Model 110PEL Presentation Grade
Same as Model 110PE, except has left-hand action.
Calibers 243 and 30-06 . . . . $595
Caliber 7mm Rem. Mag. 650

Savage Model 110P

Savage Model 110PE

**Savage Model 111
Chieftain**

Savage Model 110PL Premier Grade
Same as Model 110P, except has left-hand action.
Calibers 243 Win. and 30-06 . . . . **$355**
Caliber 7mm Rem. Mag. 375

Savage Model 110S/110V
Same as Model 110E, except **Model 110S** in 308 Win. only; discontinued 1985. **Model 110V** in 22-250 and 223 Rem. w/heavy 2-inch barrel, 47 inches overall, 9 lbs. Discontinued. 1989.
Model 110S **$275**
Model 110V 295

Savage Model 111 Chieftain Bolt-Action Rifle
Calibers: 243 Win., 270 Win., 7×57mm, 7mm Rem. Mag. 30-06. 4-shot clip magazine (3-shot in Magnum). 22-inch bbl. (24-inch in Magnum). Weight: 7.5 lbs., 8.25 lbs. Magnum. Sights: leaf rear; hooded ramp front. Select walnut stock w/Monte Carlo comb and cheekpiece, checkered, pistol-grip cap, QD swivels and sling. Made 1974-79.
Standard Calibers **$325**
Magnum Calibers 345

Savage Models 111 F, 111 FC, 111 FNS Classic Hunters
Similar to the Model 111G, except with graphite/fiberglass composition stock. Weight: 6.25 lbs. Made 1994 to date.
Model 111F (Box Mag., Right/Left Hand) **$265**
Model 111FC (Detachable Magazine) 295
Model 111FNS (Box Mag., No Sights, R/L Hand) 260

Savage Models 111 G, 111 GC, 111 GNS Classic Hunters
Calibers: 22-250 Rem., 223 Rem., 243 Win., 25-06 Rem. 250 Sav., 270 Win., 7mm-08 Rem., 7mm Rem. Mag., 30-06, 300 Sav., 300 Win. Mag., 308 Win., 338 Win. 22- or 24-inch bbl. Weight: 7 lbs. Ramp front sight, adj. open rear. Walnut-finished hardwood stock. Blued finish. Made 1994 to date.
Model 111G (Box Mag., Right/Left Hand) **$270**
Model 111GC (Detachable Mag., R/L Hand) 290
Model 111GNS (Box Mag., No Sights) 260

Savage Model 112BV, 112BVSS Heavy Varmint Rifles
Similar to the Model 110G, except fitted with 26-inch heavy bbl. Laminated wood stock with high comb. Calibers 22-250 and 223 only.
Model 112BV (Made 1993-94) **$385**
Model 112BVSS (Fluted stainless bbl.; made
  since 1994) 395

Savage Model 112FV, 112FVS, 112FVSS Varmint Rifles
Similar to the Model 110G, except fitted with 26-inch heavy bbl. and Dupont Rynite stock. Calibers: 22-250, 223 and 220 Swift (112FVS only). Blued or stainless finish. Made 1991 to date..
Model 112FV (blued) **$275**
Model 112FV-S (blued, single shot) Disc. 1993 315
Model 112FVSS (stainless) 365
Model 112 FVSS-S (Stainless, Single Shot). 375

Savage Model 112V Varmint Rifle **$305**
Bolt action, single shot. Caliber: 220 Swift, 222 Rem., 223 Rem., 22-250, 243 Win., 25-06. 26-inch heavy bbl. with scope bases. Supplied w/o sights. Weight: 9.25 lbs. Select walnut stock in varmint style w/checkered pistol grip, high comb, QD sling swivels. Made 1975-79.

Savage Model 111F

Savage Model 112V

Savage Model 116FCSAK

RIFLES

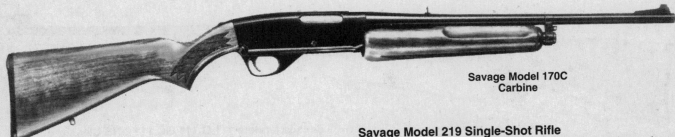

**Savage Model 170C
Carbine**

Savage Model 114C, 114CE, 114CU Rifles
Calibers: 270 Win., 7mm Rem. Mag., 30-06, 300 Win. Mag. 22- or 24-inch bbl. Weight: 7 lbs. Detachable 3- or 4-shot magazine. Ramp front sight; adjustable. open rear; (114CU) no sights. Checkered select walnut stock w/oil finish, red butt pad. Schnabel forend and skip-line checkering (114CE). High-luster blued finish. Made 1991 to date.
Model 114C (Classic)..................................$360
Model 114CE (Classic European) 360
Model 114CU (Classic Ultra)........................375

Savage Models 116FSAK, 116FCSAK Bolt-Action Rifles
Similar to the Model 116FSK, except in calibers 270 Win., 30-06, 7mm Mag., 300 Win. Mag., 338 Win. Mag. Fluted 22-inch stainless bbl. w/adj. muzzle brake. Weight: 6.5 lbs. Made 1994 to date.
Model 116FSAK $395
Model 116FCSAK (Detachable Mag.) 450

Savage Models 116FSC, 116FSS Bolt-Action Rifles
Improved Model 110 with satin stainless action and bbl. Calibers: 223, 243, 270,30-06, 7mm Rem. Mag., 300 Win. Mag., 338 Win. Mag. 22- or 24-inch bbl. 4- or 5-shot capacity. Weight: about 7.5 lbs. Black Rynite® stock w/recoil pad and swivel studs. Receiver drilled and tapped for scope mounts, no sights. Made 1991 to date.
Model 116FSS $375
Model 116FSC, Detachable Magazine 395

Savage Model 116FSK Kodiak Rifle $395
Similar to the Model 116FSS, except with 22-inch bbl. chambered for 338 Win. Mag. only. "Shock Suppressor" recoil reducer. Made 1993 to date.

Savage Model 116SE, 116US Rifles
Calibers: 270 Win., 7mm Rem Mag., 30-06, 300 Win. Mag. (116US); 300 Win. Mag., 338 Win. mag., 425 Express, 458 Win. Mag. (116SE). 24-inch stainless barrel (with muzzle brake 116SE only). 45.5 inches overall. Weight: 7.2 to 8.5 lbs. 3-shot magazine. 3-leaf Express sights 116SE only. Checkered Classic style select walnut stock with ebony forend tip. Stainless finish. Made 1994 to date.
Model 116SE (Safari Express)........................695
Model 116US (Ultra Stainless).......................525

Savage Model 170 Pump-Action Centerfire Rifle . . . $190
Calibers: 30-30, 35 Rem. 3-shot tubular magazine. 22-inch bbl. Weight: 6.75 lbs. Sights: folding leaf rear; ramp front. Select walnut stock w/checkered pistol grip Monte Carlo comb, grooved slide handle. Made 1970-81.

Savage Model 170C Carbine $195
Same as Model 170 Rifle, except has 18.5-inch bbl., straight comb stock, weighs 6 lbs.; caliber 30-30 only. Made 1974-81.

Savage Model 219 Single-Shot Rifle
Hammerless. Takedown. Shotgun-type action with top lever. Calibers: 22 Hornet, 25-20, 32-20, 30-30. 26-inch bbl. Weight: about 6 lbs. Sights: open rear; bead front. Plain pistol-grip stock and forearm. Made 1938-65.
Model 219 ....$175
Model 219L (w/side lever, made 1965-67) 125

Savage Model 221-229 Utility Guns
Same as Model 219, except in various calibers, supplied in combination with an interchangeable shotgun bbl. All versions discontinued.
Model 221 (30-30,12-ga. 30-inch bbl.) $140
Model 222 (30-30,16-ga. 28-inch bbl.) 125
Model 223 (30-30, 20-ga. 28-inch bbl.) 110
Model 227 (22 Hornet, 12-ga. 30-inch bbl.) 150
Model 228 (22 Hornet, 16-ga. 28-inch bbl.) 145
Model 229 (22 Hornet, 20-ga. 28-inch bbl.) 140

Savage Model 340 Bolt-Action Repeater
Calibers: 22 Hornet, 222 Rem., 223 Rem., 225 Win., 30-30. Clip magazine; 4-shot capacity (3-shot in 30-30). Bbl. lengths: originally 20-inch in 30-30, 22-inch in 22 Hornet; later 22-inch in 30-30, 24-inch in other calibers. Weight: 6.5 to 7.5 lbs. depending on caliber and vintage. Sights: open rear (folding leaf on recent production); ramp front. Early models had plain pistol-grip stock checkered since 1965. Made 1950-85. (*Note:* Those rifles produced between 1947-1950 were 22 Hornet Stevens Model 322 and 30-30 Model 325. The Savage Model, however, was designated Model 340 for all calibers.)
Pre-1965 with plain stock $175
Current model 160
Savage Model 340C Carbine 185

Savage Model 340S Deluxe $225
Same as Model 340, except has checkered stock, screw eyes for sling, peep rear sight, hooded front. Made 1955-60.

Savage Model 342 .... $250
Designation, 1950 to 1955, of Model 340 22 Hornet.

Savage Model 342S Deluxe $275
Designation, 1950 to 1955, of Model 340S 22 Hornet.

Savage Anniversary Model 1895 Lever-Action..... $475
Replica of Savage Model 1895 Hammerless Lever-Action Rifle issued to commemorate the 75th anniversary (1895-1970) of Savage Arms. Caliber: 308 Win. 5-shot rotary magazine. 24-inch full-octagon bbl. Engraved receiver. Brass-plated lever. Sights: open rear; brass blade front. Plain straight-grip buttstock, schnabel-type forend; brass medallion inlaid in buttstock, brass crescent-shaped buttplate. 9,999 produced. Made in 1970 only. Value is for new, unfired specimen.

Savage Model 1903 Slide-Action Repeater $225
Hammerless. Takedown. Caliber: 22 Short, Long, LR. Detachable box magazine. 24-inch octagon bbl. Weight: about 5 lbs. Sights: open rear; bead front. Pistol-grip stock, grooved slide handle. Made 1903-21.

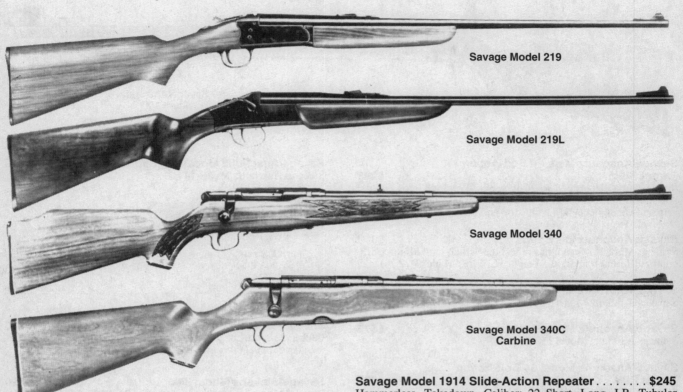

Savage Model 219

Savage Model 219L

Savage Model 340

Savage Model 340C Carbine

RIFLES

Savage Model 1904 Bolt-Action Single-Shot Rifle $100
Takedown. Caliber: 22 Short, Long, LR. 18-inch bbl. Weight: about 3 lbs. Sights: open rear; bead front. Plain, straight-grip, one-piece stock. Made 1904-17.

Savage Model 1905 Bolt-Action Single-Shot Rifle $100
Takedown. Caliber: 22 Short, Long, LR. 22-inch bbl. Weight: about 5 lbs. Sights: open rear; bead front. Plain, straight-grip one-piece stock. Made 1905-19.

Savage Model 1909 Slide-Action Repeater $195
Hammerless. Takedown. Similar to Model 1903, except has 20-inch round bbl., plain stock and forearm, weighs about 4.75 lbs. Made 1909-15.

Savage Model 1912 Autoloading Rifle $295
Takedown. Caliber: 22 LR. only. 7-shot detachable box magazine. 20-inch bbl., plain stock and forearm. Made 1912-16.

Savage Model 1914 Slide-Action Repeater $245
Hammerless. Takedown. Caliber: 22 Short, Long, LR, Tubular magazine holds 20 Short, 17 Long, 15 LR. 24-inch octagon bbl. Weight: about 5.75 lbs. Sights: open rear; bead front. Plain pistol-grip stock, grooved slide handle. Made 1914-24.

Savage Model 1920 Hi-Power Bolt-Action Rifle $375
Short Mauser-type action. Calibers: 250/3000, 300 Sav. 5-shot box magazine. 22-inch bbl. in 250 cal.; 24-inch in 300 cal. Weight: about 6 lbs. Sights: open rear; bead front. Checkered pistol-grip stock w/slender forearm and schnabel tip. Made 1920-26.

> **NOTE**
>
> In 1965, Savage began the importation of rifles manufactured by J. G. Anschutz GmbH, Ulm, West Germany. Models designated "Savage/Anschutz" are listed in this section, those marketed in the U.S. under the "Anschutz" name are included in that firm's listings. Anschutz rifles are now distributed in the U.S. by Precision Sales Int'l., Westfield, Mass. *See* "Anschutz" for detailed specifications.

Savage Model 1895 Replica

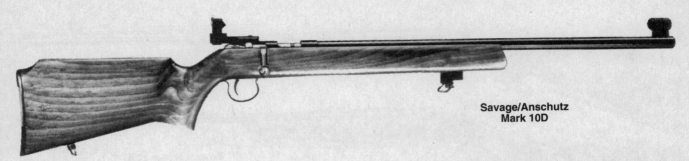

**Savage/Anschutz
Mark 10D**

**Savage/Anschutz Mark 10 Bolt-Action
Target Rifle** **$295**
Single shot. Caliber: 22 LR. 26-inch bbl. Weight: 8.5 lbs. Sights: Anschutz micrometer rear; globe front. Target stock w/full pistol grip and cheekpiece, adj. hand stop and swivel. Made 1967-72.

Savage/Anschutz Mark 10D **$300**
Same as Mark 10, except has redesigned stock with Monte Carlo comb, different rear sight. Weight: 7.75 lbs. Made 1972.

Savage/Anschutz Model 54 Custom Sporter **$535**
Same as Anschutz Model 1422D.

Savage/Anschutz Model 54M **$575**
Same as Anschutz Model 1522D.

**Savage/Anschutz Model 64 Bolt-Action
Target Rifle** **$395**
Same as Anschutz Model 1403.

**Savage/Anschutz Model 153 Bolt-Action
Sporter** **$425**
Caliber: 222 Rem. 3-shot clip magazine. 24-inch bbl. Sights: folding leaf open rear; hooded ramp front. Weight: 6.75 lbs. French walnut stock w/cheekpiece, skip checkering, rosewood forend tip and grip cap, swivels. Made 1964-67.

Savage/Anschutz Model 153S **$495**
Same as Model 153, except has double-set trigger. Made 1965-67.

Savage/Anschutz Model 164 Custom Sporter **$275**
Same as Anschutz Model 1416.

Savage/Anschultz Model 164M **$375**
Same as Anchultz Model 1516.

Savage/Anschultz Model 184 Sporter **$315**
Same as Anschultz Model 1441.

> **NOTE**
>
> Since J. Stevens Arms (see *also* separate listing) is a division of Savage Industries, certain Savage models carry the "Stevens" name.

Savage-Stevens Model 34 Bolt-Action Repeater **$85**
Caliber: 22 Short, Long, LR. 20-inch bbl. Weight: 4.75 lbs. Sights: open rear; bead front. Plain pistol-grip stock. Made 1965-80.

Savage-Stevens Model 34M **$90**
Same as Model 34, except chambered for 22 WMR. Made 1969-73.

Savage-Stevens Model 35 .... **$80**
Bolt-action repeater. Caliber: 22 LR. 6-shot clip magazine. 22-inch bbl. Weight: about 5 lbs. Sights: open rear; ramp front. Monte Carlo stock w/checkered pistol grip and forearm. Made 1982 to date.

Savage-Stevens Model 35M **$90**
Same as Model 35, except chambered for 22 WMR. Made 1982 to date.

Savage-Stevens Model 46 Bolt-Action Rifle **$95**
Caliber: 22 Short, Long, LR. Tubular magazine holds 22 Short, 17 Long, 15 LR. 20-inch bbl. Weight: 5 lbs. Plain pistol-grip stock on early production; later models have Monte Carlo stock w/checkering. Made 1969-73.

**Savage-Stevens
Model 34**

**Savage-Stevens
Model 35**

RIFLES

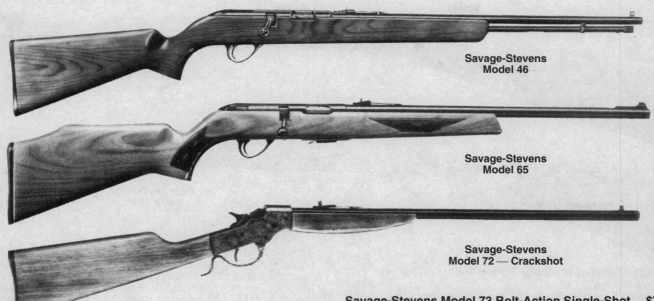

**Savage-Stevens
Model 46**

**Savage-Stevens
Model 65**

**Savage-Stevens
Model 72 — Crackshot**

Savage-Stevens Model 65 Bolt-Action Rifle **$90**
Caliber: 22 Short, Long, LR. 5-shot clip magazine. 20-inch bbl. Weight: 5 lbs. Sights: open rear; ramp front. Monte Carlo stock w/checkered pistol grip and forearm. Made 1969-73.

Savage-Stevens Model 65M **$105**
Same as Model 65, except chambered for 22 WMR, has 22-inch bbl., weighs 5.25 lbs. Made 1969-81.

> ### NOTE
>
> The Model 72 is a "Favorite"-type single-shot, unlike the smaller original "Crackshot" made by Stevens 1913-39.)

**Savage-Stevens Model 72 Crackshot
Single-Shot Lever-Action Rifle**. **$125**
Falling-block action. Casehardened frame. Calibeer: 22 Short, Long, LR. 22-inch octagon bbl. Weight: 4.5lbs. Sights: open rear; bead front. Plain straight-grip stock and forend of walnut. Made 1772 to date.

Savage-Stevens Model 73 Bolt-Action Single-Shot . . **$75**
Caliber: 22 Short, Long, LR. 20-inch bbl. Weight: 4.75 lbs. Sights: open rear; bead front. Plain pistol-grip stock. Made 1965-80.

Savage-Stevens Model 73Y Youth Model **$85**
Same as Model 73, except has 18-inch bbl., 1.5-inch shorter butt-stock, weighs 4.5 lbs. Made 1965-80.

Savage-Stevens Model 74 Little Favorite **$130**
Same as Model 72 Crackshot, except has black-finished frame, 22-inch round bbl., walnut-finished hardwood stock. Weight: 4.75 lbs. Made 1972-74.

Savage-Stevens Model 80 Autoloading Rifle **$120**
Caliber: 22 LR. 15-shot tubular magazine. 20-inch bbl. Weight: 6 lbs. Sights: open rear, bead front. Monte Carlo stock of walnut w/checkered pistol grip and forearm. Made 1976 to date. *(Note: This rifle is essentially the same as Model 60 of 1969-72, except for a different style of checkering, side instead of top safety and plain bead instead of ramp front sight.)*

Savage-Stevens Model 88 Autoloading Rifle **$125**
Similar to Model 60, except has walnut-finished hardwood stock, plain bead front sight; weight, 5.75 lbs. Made 1969-72.

**Savage-Stevens
Model 73**

**Savage-Stevens
Model 80**

**Savage-Stevens
Model 89**

**Savage-Stevens
Model 987-T**

**Savage-Stevens Model 89 Single-Shot
Lever-Action Carbine** **$100**
Martini-type action. Caliber: 22 Short, Long, LR. 18.5-inch bbl.
Weight: 5 lbs. Sights: open rear; bead front. Western-style carbine
stock w/straight grip, forearm with bbl. band. Made 1976-89.

**Savage-Stevens Model 987-T Autoloading
Rifle** **$120**
Caliber: 22 LR. 15-shot tubular magazine. 20-inch bbl. Weight: 6
lbs. Sights: open rear; ramp front. Monte Carlo stock w/checkered
pistol grip and forearm. Made 1981-89.

Savage "Stevens Favorite"
See **Savage Model 71.**

**Savage-Stevens Model 987-T Autoloading
Rifle.** **$120**
Caliber: 22 LR. 15-shot tubular magazine. 20-inch bbl. Weight: 6
lbs. Sights: open rear, ramp front. Monte Carlo stock w/checkered
pistol grip and forearm. made 1981-89.

Savage "Stevens Favorite"
See **Savage Model 71.**

V.C. SCHILLING
Suhl, Germany

**Schilling Mauser-Mannlicher Bolt Action
Sporting Rifle.** **$650**
Same general specifications as given for the Haenel Mauser-
Mannlicher Sporter. *See* separate listing.

Schilling '88 Mauser Sporter **$525**
Same general specifications as Haenel '88 Mauser Sporter. *See*
separate listing.

SCHULTZ & LARSEN GEVAERFABRIK
Otterup, Denmark

Schultz & Larsen Match Rifle No. 47 . . . . **$550**
Caliber: 22 LR. Bolt action, single-shot, set trigger. 28.5-inch
heavy bbl. Weight: 14 lbs. Sights: micrometer receiver, globe
front. Free-rifle stock w/cheekpiece, thumbhole, adj. Schuetzen-type
buttplate, swivels, palmrest.

Schultz & Larsen Free Rifle Model 54 . . . . **$725**
Calibers: 6.5×55mm or any standard American centerfire caliber.
Schultz & Larsen M54 bolt-action, single-shot, set trigger. 27.5-inch
heavy bbl. Weight: 15.5 lbs. Sights: micrometer receiver; globe front.
Free-rifle stock w/cheekpiece, thumbhole, adj. Schuetzen-type
buttplate, swivels, palmrest.

Schultz & Larsen Model 54J Sporting Rifle **$545**
Calibers: 270 Win., 30-06, 7×61 Sharpe & Hart. Schultz & Larsen bolt
action. 3-shot magazine. 24-inch bbl. in 270 and 30-06, 26-inch in 7×61
S&H. Checkered stock w/ Monte Carlo comb and cheekpiece. Value
shown is for rifle less sights.

SEARS, ROEBUCK & COMPANY
Chicago, Illinois

The most encountered brands or model designations used by Sears
are J. C. Higgins and Ted Williams. Firearms sold under these designations have been mfd. by various firms including Winchester, Marlin, Savage, Mossberg, etc.

Sears Model 2C Bolt-Action Rifle **$85**
Caliber: 22RF. 7-shot clip mag. 21-inch bbl. Weight: 5 lbs.
Sights: open rear; ramp front. Plain Monte Carlo stock. Mfd. by
Win.

Sears Model 42 Bolt-Action Repeater **$80**
Takedown. Caliber: 22RF. 8-shot detachable box magazine. 24-inch
bbl. Weight: 6 lbs. Sights: open rear; bead front. Plain pistol-grip
stock. Mfd. by Marlin.

Sears Model 42DL Bolt-Action Repeater **$85**
Same general specifications as Model 42 except fancier grade w/peep
sight, hooded front sight and swivels.

Sears Model 44DL Lever-Action Rifle **$145**
Caliber: 22RF. Tubular magazine holds 19 LR cartridges. 22-inch
bbl. Weight: 6.25 lbs. Sights: open rear; hooded ramp front. Monte
Carlo-style stock w/pistol grip. Mfd. by Marlin.

Sears Model 53 Bolt-Action Rifle **$195**
Calibers: 243, 270, 308, 30-06. 4-shot magazine. 22-inch bbl.
Weight: 6.75 lbs. Sights: open rear; ramp front. Standard sporter
stock w/pistol grip, checkered. Mfd. by Savage.

Sears Model 54 Lever-Action Rifle **$155**
Similar general specifications as Winchester Model 94 carbine.
Made in 30-30 caliber only. Mfd. by Winchester.

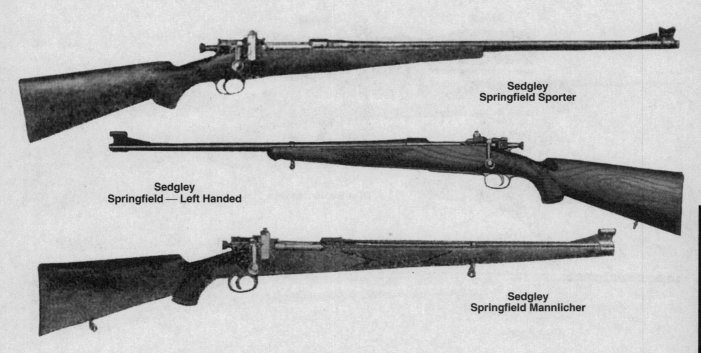

**Sedgley
Springfield Sporter**

**Sedgley
Springfield — Left Handed**

**Sedgley
Springfield Mannlicher**

Sears Model 103 Series Bolt-Action Repeater $85
Same general specifications as Model 103.2 w/minor changes. Mfd. by Marlin.

Sears Model 103.2 Bolt-Action Repeater $85
Takedown. Caliber: 22RF. 8-shot detachable box magazine. 24-inch bbl. Weight: 6 lbs. Sights: open rear; bead front. Plain pistol-grip stock. Mfd. by Marlin.

R. F. SEDGLEY, INC.
Philadelphia, Pennsylvania

Sedgley Springfield Sporter $1025
Springfield '03 bolt action. Calibers: 220 Swift, 218 Bee, 22-3000, R2, 22-4000, 22 Hornet, 25-35, 250-3000, 257 Roberts, 270 Win., 7mm, 30-06. 24-inch bbl. Weight: 7.5 lbs. Sights: Lyman No. 48 receiver; bead front on matted ramp. Checkered walnut stock, grip cap, sling swivels. Disc. 1941.

Sedgley Springfield Left-Hand Sporter $1150
Bolt action reversed for left-handed shooter; otherwise the same as standard Sedgley Springfield Sporter. Disc. 1941.

Sedgley Springfield Mannlicher-Type Sporter $1225
Same as standard Sedgley Springfield Sporter, except w/20-inch bbl., Mannlicher-type full stock w/cheekpiece, weighs 7.75 lbs. Disc. 1941.

SHILEN RIFLES, INC.
Ennis, Texas

Shilen DGA Benchrest Rifle $1025
DGA single-shot bolt action. Calibers as listed for Sporter. 26-inch medium-heavy or heavy bbl. Weight: from 10.5 lbs. No sights. Fiberglass or walnut stock, classic or thumbhole pattern. Currently manufactured.

Shilen DGA Sporter $995
DGA bolt action. Calibers: 17 Rem.,222 Rem.,223 Rem. 22-250, 220 Swift, 6mm Rem., 243 Win., 250 Sav., 257 Roberts, 284 Win., 308 Win., 358 Win. 3-shot blind magazine. 24-inch bbl. Average weight: 7.5 lbs. No sights. Select Claro walnut stock w/cheekpiece, pistol grip, sling swivel studs. Currently manufactured.

Shilen DGA Varminter $950
Same as Sporter, except w/25-inch medium-heavy bbl. Weight: 9 lbs.

SHILOH RIFLE MFG. Co.
Big Timber, Montana
(Formerly Shiloh Products)

Shiloh Sharps Model 1874 Business Rifle $725
Replica of 1874 Sharps similar to No. 3 Sporting Rifle. Calibers: 32-40, 38-55, 40-50 BN, 40-70 BN, 40-90 BN, 45-70 ST, 45-90 ST, 50-70 ST, 50-100 ST. 28-inch round heavy bbl. Blade front sight, buckhorn rear. Double-set triggers. Straight-grip walnut stock w/steel cresent buttplate. Made 1986 to date.

**Shiloh Sharps Model 1874 Long Range
Express Rifle** $795
Replica of 1874 Sharps w/single-shot falling breech action. Calibers: 32-40, 38-55, 40-50 BN, 40-70 BN, 40-90 BN, 45-70 ST, 45-90 ST, 45-110 ST, 50-70 ST, 50-90 ST, 50-110 ST. 34-inch tapered octagon bbl. 51 inches overall. Weight: 10.75 lbs. Globe front sight, sporting tang peep rear. Walnut buttstock w/pistol grip and Schnabel style forend. Color casehardened action w/double-set triggers Made 1986 to date.

Shiloh Sharps Model 1874 Saddle Rifle $750
Similar to 1874 Express Rifle, except w/30-inch bbl., blade front sight and buckhorn rear. Made 1986 to date.

Shiloh Sharps Model 1874 Sporting Rifle No. 1 . . . $785
Similar to 1874 Express Rifle, except w/30-inch bbl., blade front sight and buckhorn rear. Made 1986 to date.

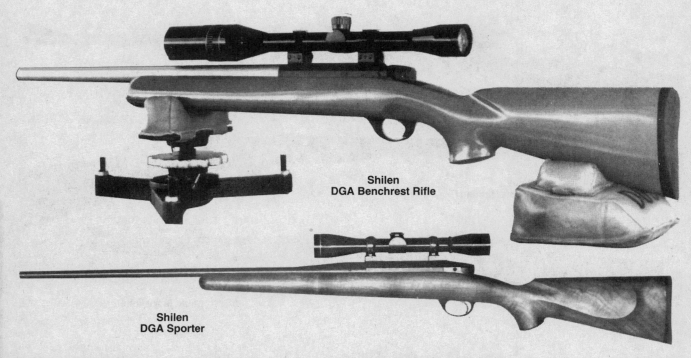

**Shilen
DGA Benchrest Rifle**

**Shilen
DGA Sporter**

Shiloh Sharps Model 1874 Sporting Rifle No. 3 . . . **$725**
Similar to 1874 Sporting Rifle No. 1, except w/straight-grip stock
w/steel cresent buttplate. Made 1986 to date.

Shiloh Sharps Model 1874 Montana Roughrider
Similar to 1874 Sporting Rifle No. 1, except w/24- to 34-inch
half-octogon or full-octogon bbl. Standard or deluxe walnut stock
w/pistol grip or military style buttstock. Made 1989 to date.
Standard Model. **$695**
Deluxe Model 775

SIG SWISS INDUSTRIAL COMPANY
Neuhausen-Rhine Falls, Switzerland

SIG-AMT Semiautomatic Rifle **$3950**
Caliber: 308 Win.(7.62 NATO). 5, 10, or 20-shot magazine. 18.5-
inch bbl. w/flash suppressor. Weight: 9.5 lbs. Sights: adj. aperture
rear, post front. Walnut buttstock and forend w/synthetic pistol grip.
Imported 1980-88.

SIG-AMT Sporting Rifle **$2550**
Semiautomatic version of SG510-4 automatic assault rifle based on
Swiss Army SIGW57. Roller-delayed blowback action. Caliber:
7.62×51mm NATO (308 Win.). 5-, 10- and 20-round magazines.
19-inch bbl. Weight: 10 lbs. Sights, aperture rear, post front.
Wood buttstock and forearm, folding bipod. Made 1960-74.

SIG-PE57 Semiautomatic Rifle **$2895**
Caliber: 7.5 Swiss. 24-shot magazine. 23.75-inch bbl. Weight: 12.5
lbs. Sights: adj. aperture rear; post front. High-impact synthetic
stock. Imported from Switzerland during the 1980s.

SMITH & WESSON
Springfield, Massachusetts
Mfd. by Husqvarna Vapenfabrik A.B.,
Huskvarna, Sweden & Howa Machonery LTD.,
Shinkawa-Chonear, Nagota 452, Japan

Smith & Wesson Model 1500 . . . . **$275**
Bolt action. Calibers: 243 Win., 270 Win., 30-06, 7mm Rem.
Mag. 22-inch bbl. (24-inch in 7mm Rem. Mag.). Weight: 7.5 lbs.
American walnut stock w/Monte Carlo comb and cheekpiece, cut
checkering. Sights: open rear, hooded ramp, gold beadfront. This
model was also imported by Mossberg *(see* separate listings); Imported 1979-84.

Smith & Wesson Model 1500DL Deluxe **$295**
Same as standard model, except w/o sights; w/engine-turned bolt,
decorative scroll on floorplate, French checkering. Imported
1983-84.

**Shilen
DGA Varminter**

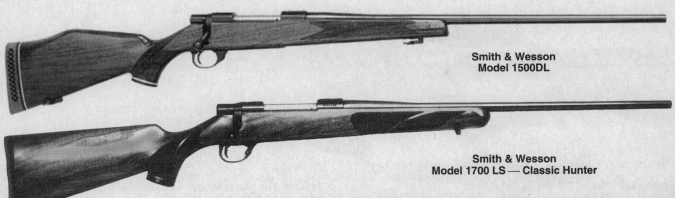

**Smith & Wesson
Model 1500DL**

**Smith & Wesson
Model 1700 LS — Classic Hunter**

Smith & Wesson Model 1700 LS "Classic Hunter" $325
Bolt action. Calibers: 243 Win., 270 Win., 30-06, 5-shot magazine. 22-inch bbl. Weight: 7.5 lbs. Solid recoil pad, no sights, Schnabel forend, checkered walnut stock. Imported 1983-84.

Smith & Wesson Model A Bolt-Action Rifle $335
Similar to Husqvarna Model 9000 Crown Grade. Mauser-type bolt action. Calibers: 22-250, 243 Win., 270 Win., 308 Win., 30-06, 7mm Rem. Mag., 300 Win. Mag. 5-shot magazine, except 3-round capacity in latter two calibers. 23.75-inch bbl. Weight: 7 lbs. Sights: folding leaf rear; hooded ramp front. Checkered walnut stock w/Monte Carlo cheekpiece, rosewood forend tip and pistol-grip cap, swivels. Made 1969-72.

Smith & Wesson Model B $325
Same as Model A, except w/20.25-inch extra-light bbl., Monte Carlo cheekpiece w/Schnabel-style forearm, weighs 6 lbs. 10 oz. Calibers: 243 Win., 30-06.

Smith & Wesson Model C $325
Same as Model B, except w/cheekpiece stock w/straight comb.

Smith & Wesson Model D $395
Same as Model C, except w/full-length Mannlicher-style forearm.

Smith & Wesson Model E $425
Same as Model B, except w/full-length Mannlicher-style forearm.

SPRINGFIELD, INC.
Colona, Illinois
(Formerly Springfield Armory of Geneseo, Ill.)

This is a private firm, not to be confused with the former U.S. Government facility in Springfield, Mass.

Springfield Armory BM-59

Springfield Armory BM-59 Semiautomatic Rifle
Gas-operated. Caliber: 308 Win. (7.62mm NATO). 20-shot detachable box magazine. 19.3-inch bbl. w/flash suppressor. About 43 inches overall. Weight: 9.25 lbs. Adj. military aperture rear sight, square post front; direct and indirect grenade launcher sights. European walnut stock w/handguard or folding buttstock (Alpine Paratrooper). Made 1981-90.
Standard Model $1595
Paratrooper Model .... — 1950

Springfield Armory M-1 Garand Semiautomatic Rifle
Gas-operated. Calibers: 308 Win. (7.62 NATO), 30-06. 8-shot stripper clip. 24-inch bbl. 43.5 inches overall. Weight: 9.5 lbs. Adj. aperture rear sight, military square blade front. Standard "Issue-grade" walnut stock or folding buttstock. Made 1979-90.
Standard Model .... $ 725
National Match 875
Ultra Match 950
Sniper Model 1025
Paratrooper w/folding stock 925

Springfield Armory Match M1A
Same as Standard M1A, except w/National Match grade bbl. w/modified flash suppressor, National Match sights, turned trigger pull, gas system assembly in one unit, modified mainspring guide glass-bedded walnut stock. Super Match M1A w/premium-grade heavy bbl. (weighs 10 lbs).
Match M1A $1050
Super Match M1A 1195

Springfield Armory Standard M1A Semiautomatic
Gas-operated. Similar to U.S. M14 service rifle, except w/o provision for automatic firing. Caliber: 7.65mm NATO (308 Win.). 5-, 10- or 20-round detachable box magazine. 25.13-inch bbl w/flash suppressor. Weight: 9 lbs. Sights: adj. aperture rear; blade front. Fiberglass, birch or walnut stock, fiberglass handguard, sling swivels. Made 1974 to date.
W/fiberglass or birch stock $695
W/walnut stock................................. 765

Springfield Model M-6 Scout Rifle/Shotgun Combo
Similar to (14-inch) short-barrel Survival Gun provided as backup weapon to U.S. combat pilots. Calibers: 22 LR/.410 and 22 Hornet/.410. 18.5-inch bbl. 32 inches overall. Weight: 4 lbs. Parkerized or stainless steel finish. Folding detachable stock w/storgae for fifteen 22 LR cartridges and four 410 shells. Drilled and tapped for scope mounts. Intro. 1982 and imported from Czech Republic 1995 to date.
First Issue (no trigger guard) $150
Second Issue (w/trigger guard)..................... 125

RIFLES

**Springfield Armory SAR-8
Sporter Rifle**

Springfield Model SAR-8 Sporter Rifle
Similar to H&K 911 semiautomatic rifle. Calibers: 308 Win., (7.62×51) NATO. Detachable 5- 10- or 20-shot magazine. 18- or 20-inch bbl. 38.25 or 45.3 inches overall. Weight: 8.7 to 9.5 lbs. Protected front post and rotary adj. rear sight. Delayed roller-locked blow-back action w/fluted chamber. Kevlar reinforced fiberglass thumb-hole style wood stock. Imported 1990 to date.

SAR-8 w/wood stock (Disc. 1994). $950
SAR-8 w/thumb-hole stock. 875

Springfield Model SAR-48, SAR-4800
Similar to Browning FN FAL/LAR semiautomatic rifle. Calibers: 233 Rem. (5.56×45) and 308 Win. (7.62×51) NATO. Detachable 5- 10- or 20-shot magazine. 18- or 21-inch chrome-lined bbl. 38.25 or 45.3 inches overall. Weight: 9.5 to 13.25 lbs. Protected post front and adj. rear sight. Forged receiver and bolt w/adj. gas system. Pistol-grip ir thumb-hole style synthetic or wood stock. Imported 1985; reintroduced 1995.

SAR-48 w/pistol-grip stock (Disc. 1989). $1595
SAR-48 w/wood stock (Disc. 1989). 1750
SAR-48 w/folding stock (Disc. 1989) 1995
SAR-48 w/thumb-hole stock 925

**SQUIRES BINGHAM CO., INC.
Makati, Rizal, Philippines**

Squires Bingham Model 14D Deluxe Bolt-Action Repeating Rifle **$125**
Caliber: 22 LR. 5-shot box magazine. 24-inch bbl. Sights: V-notch rear; hooded ramp front. Receiver grooved for scope mounting. Pu-long Dalaga stock w/contrasting forend tip and grip cap, checkered forearm and pistol grip. Weight: 6 lbs. Disc.

Squires Bingham Model 15 . . . . **$150**
Same as Model 14D, except chambered for 22 WMR. Importation. Disc.

Squires Bingham Model M16 Semiautomatic Rifle **$175**
Styled after U.S. M16 military rifle. Caliber: 22 LR. 15-shot box magazine. 19.5-inch bbl. w/muzzle brake/flash hider. Rear sight in carrying handle, post front on high ramp. Black-painted mahogany buttstock and forearm. Weight: 6.5 lbs. Importation. Disc.

Squires Bingham Model M20D Deluxe. **$180**
Caliber: 22 LR. 15-shot box magazine. 19.5-inch bbl. w/muzzle brake/flash hider. Sights: V-notch rear; blade front. Receiver grooved for scope mounting. Pulong Dalaga stock w/contrasting forend tip and grip cap, checkered forearm/pistol grip. Weight: 6 lbs. Importation. Disc.

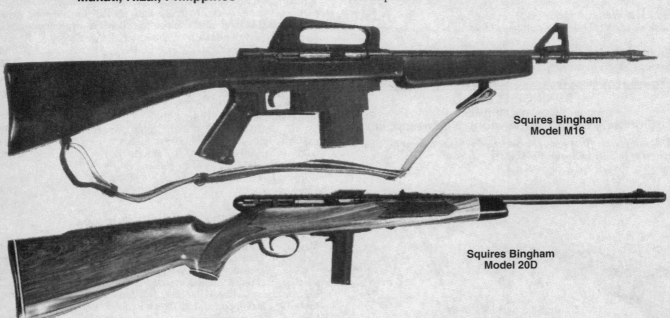

**Squires Bingham
Model M16**

**Squires Bingham
Model 20D**

Star
Rolling Block Carbine

STANDARD ARMS COMPANY
Wilmington, Delaware

Standard Model G Automatic Rifle **$425**
Gas-operated. Autoloading. Hammerless. Takedown. Calibers:
25-35, 30-30, 25 Rem., 30 Rem., 35 Rem. Magazine capacity: 4 rounds
in 35 Rem., 5 rounds in other calibers. 22.38-inch bbl. Weight: 7.75 lbs.
Sights: open sporting rear; ivory bead front. Shotgun-type stock. Made
c. 1910. *Note:* This was the first gas-operated rifle manufactured in the
U.S. While essentially an autoloader, the gas port can be closed and the
rifle operated as a slide-action repeater.

Standard Model M Hand-Operated Rifle **$295**
Slide-action repeater w/same general specifications as Model G, except lacks autoloading feature. Weight: 7 lbs.

STAR
Eibar, Spain
Mfd. by Bonifacio Echeverria, S.A.

Star Rolling Block Carbine **$160**
Single-shot action similar to Remington Rolling Block. Calibers: 30-
30, 357 Mag. ,44 Mag. 20-inch bbl. Weight: 6 lbs. Sights: folding leaf
rear; ramp front. Walnut straight-grip stock w/crescent buttplate, forearm w/bbl. band. Imported 1973-75.

STERLING
Imported by Lanchester U.S.A., Inc., Dallas, Texas

Sterling Mark 6 Semiautomatic Carbine **$625**
Caliber: 9mm Para. 34-shot magazine. Bbl.: 16.1 inches. Weight: 7.5
lbs. Flip-type rear peep sight, ramp front. Folding metal skeleton
stock. Made 1983-94.

J. STEVENS ARMS CO.
Chicopee Falls, Massachusetts
Div. of Savage Industries, Westfield, Mass.

*J. Stevens Arms eventually became a division of Savage Industries.
Consequently, the "Stevens" brand name is used for some rifles by
Savage; see separate Savage-Stevens listings under Savage.*

Stevens No. 12 Marksman Single-Shot Rifle **$145**
Lever-action, tip-up. Takedown. Calibers: 22 LR, 25 R.F., 32
R.F. 22-inch bbl. Plain straight-grip stock, small tapered forearm.

Stevens No. 14 Little Scout Single-Shot Rifle **$125**
Caliber: 22 RF. 18-inch bbl. One-piece slab stock readily distin-
guishes it from the No. 14½ that follows. Made 1906-1910.

**Stevens No. 14½ Little Scout Single-Shot
Rifle** **$150**
Rolling block. Takedown. Caliber: 22 LR. 18- or 20-inch bbl. Weight:
2.75 lbs. Sights: open rear; blade front. Plain straight-grip stock, small
tapered forearm.

Stevens Model 15 . . . . **$135**
Same as Stevens-Springfield Model 15, except w/24-inch bbl.,
weighs 5 lbs., w/redesigned stock. Made 1948-65.

Stevens Model 15Y Youth's Rifle **$130**
Same as Model 15, except w/21-inch bbl., short buttstock, weighs
4.75 lbs. Made 1958-65.

Stevens No. 44 Ideal Single-Shot Rifle **$625**
Rolling block. Lever-action. Takedown. Calibers: 22 LR 25 R.F.,
32 R.F., 25-20 S.S., 32-20, 32-40, 38-40, 38-55, 44-40. Bbl.
lengths: 24-inch, 26-inch (round, half-octagon, full-octagon). Weight:
7 lbs w/26-inch round bbl. Sights: open rear; Rocky Mountain
front. Plain straight-grip stock and forearm. Made 1894-32.

Stevens No. 44½ Ideal Single-Shot Rifle **$750**
Falling-block. Lever-action rifle. Aside from the new design action intro. 1903, specifications of this model are the same as those of Model
44. Model 44½ disc. 1916.

Stevens Nos. 45 to 54 Ideal Single-Shot Rifles
These are higher-grade models, differing from the standard No. 44 and
44½ chiefly in finish, engraving, set triggers, levers, bbls., stock, etc.
The Schuetzen types (including the "Stevens-Pope" models) are in this
series. Model Nos. 45 to 54 were intro. 1896 and originally had the No.
44-type rolling-block action, which was superseded in 1903 by the No.
44½-type falling-block action. These models were all disc. about 1916.
Generally speaking, the 45-54 series rifles, particularly the "Stevens
Pope" and higher grade Schuetzen models, are collector's items, bring-
ing much higher prices than the ordinary No. 44 and 44½.

Stevens Model 66 Bolt-Action Repeating Rifle **$115**
Takedown. Caliber: 22 Short, Long, LR. Tubular magazine holds 13
LR, 15 Long, 19 Short. 24-inch bbl. Weight: 5 lbs. Sights: open rear,
bead front. Plain pistol-grip stock w/grooved forearm. Made 1931-35.

Stevens No. 70 Visible Loading Slide-Action **$225**
Exposed hammer. Caliber: 22 LR., Long, Short. Tubular magazine
holds 11 LR., 13 Long, 15 Short. 22-inch bbl. Weight: 4.5 lbs. Sights:
open rear; bead front. Plain straight-grip stock, grooved slide handle.
Made 1907-34. *Note:* Nos. 70½, 71, 71½, 72, 722 essentially the same
as No. 70, differing chiefly in bbl. length or sight equipment.

Stevens Model 87 Autoloading Rifle **$150**
Takedown. Caliber: 22 LR. 15-shot tubular magazine. 24-inch bbl.
(20-inch on current model). Weight: 6 lbs. Sights: open rear, bead front.
Pistol-grip stock. Made 1938 to date. *Note:* This model originally bore
the "Springfield" brand name, disc. in 1948.

Stevens No. 14
Little Scout

Stevens No. 14½
Little Scout

Stevens No. 44
Ideal

Stevens Ideal
Schuetzen Rifle

**Stevens Model 322 Hi-Power Bolt-Action
Carbine** **$225**
Caliber: 22 Hornet. 4-shot detachable magazine. 21-inch bbl. Weight: 6.75 lbs. Sights: open rear; ramp front. Pistol-grip stock. Made 1947-50 (See Savage Models 340, 342.)

Stevens Model 322-S **$275**
Same as Model 325, except w/peep rear sight. (*See* Savage Models 340S, 342S.)

**Stevens Model 325 Hi-Power Bolt-Action
Carbine** **$225**
Caliber: 30-30. 3-shot detachable box magazine. 21-inch bbl. Weight: 6.75 lbs. Sights: open rear; bead front. Plain pistol-grip stock. Made 1947-50. (*See* Savage Model 340.)

Stevens Model 325-S **$265**
Same as Model 325, except w/peep rear sight. (*See* Savage Model 340S.)

**Stevens No. 414 Armory Model
Single-Shot Rifle** **$425**
No. 44-type lever-action. Calibers: 22 LR only, 22 Short only. 26-inch bbl. Weight: 8 lbs. Sights: Lyman receiver peep; blade front. Plain straight-grip stock, military-type forearm, swivels. Made 1912-32.

Stevens Model 416 Bolt-Action Target Rifle **$195**
Caliber: 22 LR. 5-shot detachable box magazine. 26-inch heavy bbl. Weight: 9.5 lbs. Sights: receiver peep; hooded front. Target stock, swivels, sling. Made 1937-49.

**Stevens No. 419 Junior Target Model Bolt-Action
Single-Shot Rifle** **$250**
Takedown. Caliber: 22 LR. 26-inch bbl. Weight: 5.5 lbs. Sights: Lyman No. 55 rear peep; blade front. Plain junior target stock w/pistol grip and grooved forearm, swivels, sling. Made 1932-36.

**Stevens Buckhorn Model 053 Bolt-Action
Single-Shot Rifle** **$110**
Takedown. Calibers: 22 Short, Long, LR, 22 WMR. 25 Stevens R.F. 24-inch bbl. Weight: 5.5 lbs. Sights: receiver peep; open middle; hooded front. Sporting stock w/pistol grip and black forend tip. Made 1935-48.

Stevens Buckhorn Model 53 . . . . **$125**
Same as Buckhorn Model 053, except w/open rear sight and plain bead front sight.

Stevens Model 87

Stevens No. 414
Armory

Stevens Model 416

Stevens Buckhorn
Model 53

Stevens Buckhorn
Model 055

Stevens Buckhorn
Model 56

RIFLES

Stevens Buckhorn 055 . . . . **$140**
Takedown. Same as Model 056, except in single-shot configuration. Weighs:5.5 lbs. Caliber: 22 LR., Long, Short. 24-inch bbl. Weight: 6 lbs. Sights: receiver peep, open middle, hooded front. Made 1935-48.

Stevens Buckhorn Model 056 Bolt-Action **$150**
Takedown. Caliber: 22 LR., Long, Short. 5-shot detachable box magazine. 24-inch bbl. Weight: 6 lbs. Sights: receiver peep, open middle, hooded front. Sporting stock w/pistol grip and black forend tip. Made 1935-48.

Stevens Buckhorn Model 56 . . . . **$115**
Same as Buckhorn Model 056, except w/open rear sight and plain bead front sight.

Stevens Buckhorn No. 057 . . . . **$110**
Same as Buckhorn Model 076, except w/5-shot detachable box magazine. Made 1939-48.

Stevens Buckhorn No. 57 . . . . **$125**
Same as Buckhorn Model 76, except w/5-shot detachable box magazine. Made 1939-48.

**Stevens Buckhorn Model 066 Bolt-Action
Repeating Rifle** **$150**
Takedown. Caliber: 22 LR, Long, Short. Tubular magazine holds 21 Short, 17 Long, 15 LR. 24-inch bbl. Weight: 6 lbs. Sights: receiver peep; open middle; hooded front. Sporting stock w/pistol grip and black forend tip. Made 1935-48.

Stevens Buckhorn Model 66 . . . . **$125**
Same as Buckhorn Model 066, except w/open rear sight, plain bead front sight.

Stevens Buckhorn No. 076 Autoloading Rifle **$150**
Takedown. Caliber: 22 LR. 15-shot tubular magazine. 24-inch bbl. Weight: 6 lbs. Sights: receiver peep; open middle; hooded front. Sporting stock w/pistol grip, black forend tip. Made 1938-48.

Stevens Buckhorn No. 76 . . . . **$145**
Same as Buckhorn No. 076, except w/open rear sight, plain bead front sight.

Stevens Crack Shot No. 26 Single-Shot Rifle **$155**
Lever-action. Takedown. Calibers: 22 LR, 32 R.F.18-inch or 22-inch bbl. Weight: 3.25 lbs. Sights: open rear; blade front. Plain straight-grip stock, small tapered forearm. Made 1913-39.

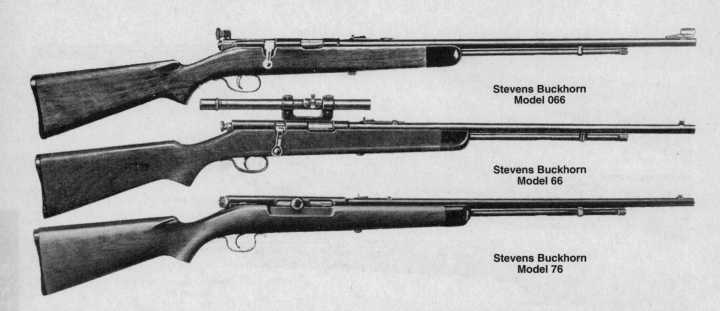

Stevens Buckhorn Model 066

Stevens Buckhorn Model 66

Stevens Buckhorn Model 76

Stevens Crack Shot No. 26½ $180
Same as Crack Shot No.26, except w/smoothbore bbl. for shot cartridges.

Stevens Favorite No. 17 Single-Shot Rifle $165
Lever-action. Takedown. Calibers: 22 LR, 25 R.F., 32 R.F. 24-inch round bbl., other lengths were available. Weight: 4.5 lbs. Sights: open rear; Rocky Mountain front. Plain straight-grip stock, small tapered forearm. Made 1894-1935.

Stevens Favorite No. 18 . . . . $245
Same as Favorite No. 17 except w/Vernier peep rear sight, leaf middle sight, Beach combination front sight.

Stevens Favorite No. 19 . . . . $275
Same as Favorite No. 17 except w/Lyman combination rear sight, leaf sight, Lyman front sight.

Stevens Favorite No. 20 . . . . $175
Same as Favorite No. 17, except w/smoothbore barrel.

Stevens Favorite No. 27 . . . . $195
Same as Favorite No. 17, except w/octagon bbl.

Stevens Favorite No. 28 . . . . $240
Same as Favorite No. 18, except w/octagon bbl.

Stevens Favorite No. 29 . . . . $250
Same as Favorite No. 19, except w/octagon bbl.

Stevens Walnut Hill No. 417-0 Single-Shot Target Rifle $625
Lever-action. Calibers: 22 LR only, 22 Short only, 22 Hornet. 28-inch heavy bbl. (extra heavy 29-inch bbl. also available). Weight: 10.5 lbs. Sights: Lyman 52L extension rear; 17A front, scope bases mounted on bbl. Target stock w/full pistol grip, beavertail forearm, bbl. band, swivels, sling. Made 1932-47.

Stevens Walnut Hill No. 417-1 . . . . $675
Same as No. 417-0, except w/Lyman 48L receiver sight.

Stevens Walnut Hill No. 417-2 . . . . $775
Same as No. 417-0, except w/Lyman No. 144 tang sight.

Stevens Walnut Hill No. 417-3 . . . . $625
Same as No. 417-0, except w/o sights.

Stevens Walnut Hill No. 417½ Single-Shot Rifle $695
Lever-action. Calibers: 22 LR, 22 WMR, 25 R.F., 22 Hornet. 28-inch bbl. Weight: 8.5 lbs. Sights: Lyman No. 144 tang peep, folding middle; bead front. Sporting stock w/pistol grip, semibeavertail forearm, swivels, sling. Made 1932-40.

Stevens Walnut Hill No. 418 Single-Shot Rifle $395
Lever-action. Takedown. Calibers: 22 LR only, 22 Short only. 26-inch bbl. Weight: 6.5 lbs. Sights: Lyman No. 144 tang peep; blade front. Pistol-grip stock, semibeavertail forearm, swivels, sling. Made 1932-40.

Stevens Crack Shot No. 26

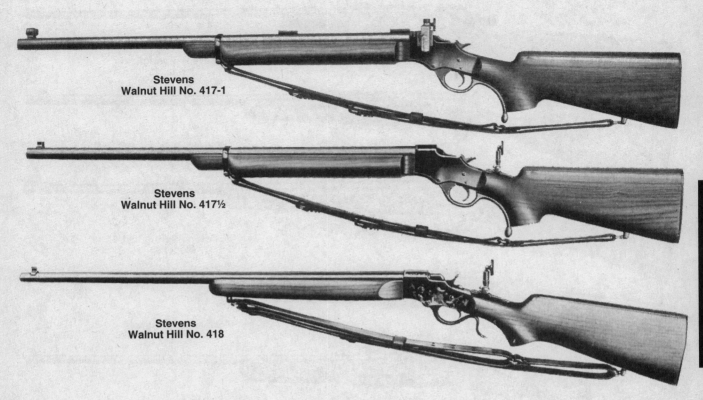

Stevens
Walnut Hill No. 417-1

Stevens
Walnut Hill No. 417½

Stevens
Walnut Hill No. 418

Stevens Walnut Hill No. 418½ **$485**
Same as No. 418, except also available in calibers 22 WMR and 25 Stevens R.F., w/Lyman No. 2A tang peep sight, bead front sight.

Stevens-Springfield Model 15 Single-Shot Bolt-Action Rifle **$125**
Takedown. Caliber: 22 LR, Long, Short. 22-inch bbl. Weight: 4 lbs. Sights: open rear, bead front. Plain pistol-grip stock. Made 1937-48.

Stevens-Springfield Model 82 Bolt-Action Single-Shot Rifle **$110**
Takedown. Caliber: 22 LR, Long, Short. 22-inch bbl. Weight: 4 lbs. Sights: open rear; gold bead front. Plain pistol-grip stock w/grooved forearm. Made 1935-39.

Stevens-Springfield Model 83 Bolt-Action Single-Shot Rifle **$115**
Takedown. Calibers: 22 LR, Long, Short; 22 WMR, 25 Stevens R.F. 24-inch bbl. Weight: 4.5 lbs. Sights: peep rear; open middle; hooded front. Plain pistol-grip stock w/grooved forearm. Made 1935-39.

Stevens-Springfield Model 84 . . . . **$130**
Same as Model 86, except w/5-shot detachable box magazine. Pre-1948 rifles of this model were designated Springfield Model 84, later known as Stevens Model 84. Made 1940-65.

Stevens-Springfield Model 84-S (084) **$135**
Same as Model 84, except w/peep rear sight and hooded front sight. Pre-1948 rifles of this model were designated Springfield Model 084, later known as Stevens Model 84-S. Disc.

Stevens-Springfield Model 85 . . . . **$165**
Same as Stevens Model 87, except w/5-shot detachable box magazine. Made 1939 to date. Pre-1948 rifles of this model were designated Springfield Model 85, currently known as Stevens Model 85. Earlier models command slight premiums.

Stevens-Springfield Model 85-S (085) **$150**
Same as Model 85, except w/peep rear sight and hooded front sight. Pre-1948 models were designated Springfield Model 085; also known as Stevens Model 85-S.

Stevens-Springfield Model 86 Bolt-Action **$115**
Takedown. Caliber: 22 LR, Long, Short. Tubular magazine holds 15 LR, 17 Long, 21 Short. 24-inch bbl. Weight: 6 lbs. Sights: open rear, gold bead front. Pistol-grip stock, black forend tip on later production. Made 1935-65. *Note:* The "Springfield" brand name was disc. in 1948.

Stevens-Springfield Model 86-S (086) **$125**
Same as Model 86, except w/peep rear sight and hooded front sight. Pre-1948 rifles of this model were designated as Springfield Model 086, later known as Stevens Model 86-S. Disc.

Stevens-Springfield Model 87-S (087) **$145**
Same as Stevens Model 87, except w/peep rear sight and hooded front sight. Pre-1948 rifles of this model were designated as Springfield Model 087, later known as Stevens Model 87-S. Disc.

STEYR-DAIMLER-PUCH A.-G.
Steyr, Austria

See also **listings under Mannlicher.**

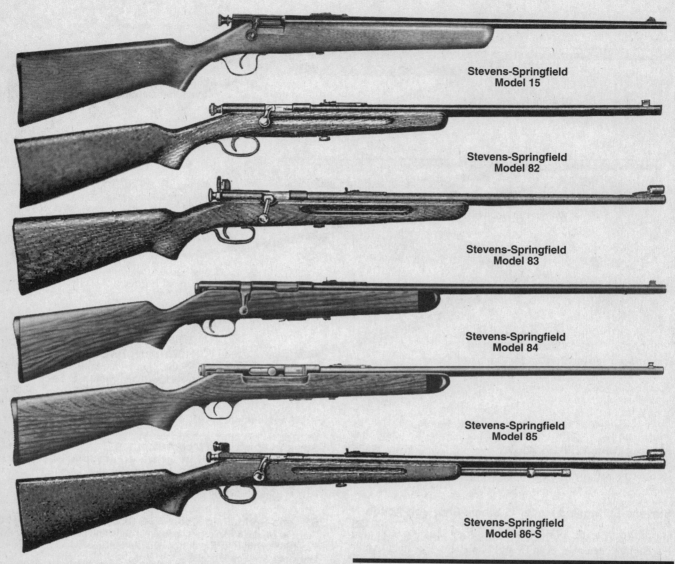

Stevens-Springfield
Model 15

Stevens-Springfield
Model 82

Stevens-Springfield
Model 83

Stevens-Springfield
Model 84

Stevens-Springfield
Model 85

Stevens-Springfield
Model 86-S

Steyr AUG-SA Semiautomatic Rifle $2995
Gas-operated. Caliber: 223 Rem. (5.56mm). 30- or 40-shot magazine. 20-inch bbl. standard; optional 16-inch or 24-inch heavy bbl.
w/folding bipod. 31 inches overall. Weight: 8.5 lbs. Sights: Integral
1.5X scope and mount. Green high-impact synthetic stock w/folding
vertical grip.

Steyr Small Bore Carbine $425
Bolt-action repeater. Caliber: 22 LR. 5-shot detachable box maga-
zine. 19.5-inch bbl. Sights: leaf rear; hooded bead front.
Mannlicher-type stock, checkered, swivels. Made 1953-67.

STOEGER RIFLE
Mfd. by Franz Jaeger & Co., Suhl, Germany; dist.
in the U.S. by A. F. Stoeger, Inc., New York, N.Y.

Stoeger Hornet Rifle $950
Same specifications as Herold Rifle design and built on a Miniature
Mauser-type action. *See* listing under Herold Bolt-Action Repeating
Sporting Rifle for additional specifications. Imported during the
1930s.

SURVIVAL ARMS
Orange, CT

Survival Arms AR-7 Explorer $110
Caliber: 22 LR. 8-shot magazine. Weight: 3 lbs. Polymer stocks.
Drift adj. sights. Disassembles into five separate elements allowing barrel, action and magazine to fit in buttstock; assembles
quickly w/o tools. Choice of camo, silvertone or black matte finishes. Made 1992-95.

THOMPSON/CENTER ARMS
Rochester, New Hampshire

Thompson/Center Contender Carbine
Calibers: 22 LR, 22 Hornet, 222 Rem., 223 Rem., 7mm T.C.U.,
7×30 Waters, 30-30 Win., 35 Rem., 44 Mag., 357 Rem. Max. and
410 bore. 21-inch interchangeable bbls. 35 inches overall. Adj.
iron sights. Checkered American walnut or Rynite stock and
forend. Made 1985 to date.
Standard Model (rifle calibers) $335
Standard Model (410 bore) 350

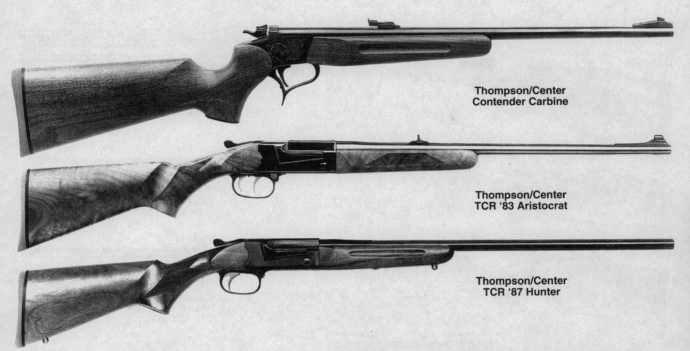

Thompson/Center
Contender Carbine

Thompson/Center
TCR '83 Aristocrat

Thompson/Center
TCR '87 Hunter

Thompson/Center Contender Carbine *(Con't)*

Rynite Stock Model (rifle calibers) 295
Rynite Stock Model (410 bore) 335
Extra Bbls. (rifle calibers). 175
Extra Bbls. (410 bore). 195
Youth Model (all calibers and 410 bore) 285

Thompson/Center Contender Carbine Survival System $415
Similar to standard Contender Carbine w/Rynite stock and forend. Comes w/two 16.25-inch bbls., chambered in 223 and 45/.410 bore. Camo Cordura case.

Thompson/Center Stainless Contender Carbine
Same as standard Contender Carbine Model, except stainless steel w/blued sights. Calibers: 22 LR, 22 Hornet, 223 Rem., 7-30 Waters, 30-30 Win., .410 Ga. Walnut or Rynite stock and forend. Made 1993 to date.
Walnut Stock Model $395
Rynite Stock Model 375
Youth Stock Model 350
Extra Bbls. (rifle calibers). 175

Thompson/Center TCR '83 Aristocrat Model
Break frame, overlever action. Calibers: 223 Rem., 22/250 Rem., 243 Win., 7mm Rem. Mag., 30-06 Springfield. Interchangeable bbls.: 23 inches in length. Weight: 6 lbs., 14 oz. American walnut stock and forearm, checkered, black rubber recoil pad, cheekpiece. Made 1983-87.
TCR '83 Standard Model $295
TCR '83 Aristocrat 365
Extra Bbls. (rifle calibers). 175

Thompson/Center TCR '87 Hunter Rifle
Similar to TCR '83, except in calibers 22 Hornet, 222 Rem., 223 Rem., 22-250 Rem., 243 Win., 270 Win., 7mm-08, 308 Win., 30-06, 32-40 Win. Also 12-ga. slug and 10- and 12-ga. field bbls. 23-inch standard or 25.88-inch heavy bbl. interchangeable. 39.5 to 43.38 inches overall.

Thompson/Center TCR '87 Hunter Riflee *(Con't)*
Weight: 6 lbs. 14 oz. to 7.5 lbs. Iron sights optional. Checkered American black walnut buttstock w/fluted end. Disc. 1993.
Standard Model $375
Extra Bbl. (rifle calibers and 10- or 12-ga. Field) 190
Extra Bbl. (12-ga. slug) 225

TIKKA RIFLES
Mfg. by Sako, Ltd. of Riihimaki, Finland & Armi Marocchi in Italy
Imported by Stoeger Industries, Inc., Wayne, NJ

NOTE

Tikka New Generation, Battue and Continental series bolt action rifles are being manufactured by Sako, Ltd., in Finland. Tikka O/U rifles (previously Valmet) are being manufactured in Italy by Armi Marocchi. For earlier importation *see* additional listings under Ithaca LSA and Valmet 412S models.

Tikka Model 412S Double Rifle
Calibers: 308 Win., 30-06, 9.3×74R. 24-inch bbl. w/quarter rib machined for scope mounts; automatic ejectors (9.3×74R only). 40 inches overall. Weight: 8.5 lbs. Ramp front and folding adj. rear sight. Barrel selector on trigger and cocking indicators in tang. European walnut buttstock and forearm. Model 412S was replaced by the 512S version in 1994. Imported 1989-93.
Model 412S (Disc. 1993) $1050
Extra Barrel Assembly (O/U Shotgun), **add**. 650
Extra Barrel Assembly (O/U Combo), **add** 775
Extra Barrel Assembly (O/U Rifle), **add** 995

Tikka Model 512S Double Rifle
Formerly Valmet. 412S. In 1994 following the joint-venture of 1989, the model designation was changed to 512s. Imported 1994 to date.
Model 512S. 1095
Extra Barrel Assembly (O/U Rifle), **add** 995

Tikka Model 412S
Double Rifle

Tikka Model LSA55
Deluxe

Tikka Model LSA65
Deluxe

Tikka Model M55
Deluxe

Tikka Model M55
Standard

Tikka LSA55 Deluxe **$365**
Same as LSA55 Standard, except w/rollover cheekpiece, rosewood
grip cap and forend tip, skip checkering, high-luster blue. Imported
1965-88.

Tikka LSA55 Sporter **$395**
Same as LSA55, except has 22.8-inch heavy bbl. w/o sights, spe-
cial stock w/beavertail forearm, not available in 6mm Rem.
Weighs 9 lbs. Imported 1965-88.

Tikka LSA55 Standard Bolt-Action Repeater **$350**
Mauser-type action. Calibers: 222 Rem., 22-250, 6mm Rem.
Mag., 243 Win., 308 Win. 3-shot clip magazine. 22.8-inch bbl.
Weight: 6.8 lbs. Sights: folding leaf rear; hooded ramp front.
Checkered walnut stock w/Monte Carlo cheekpiece, swivels.
Made 1965-88.

Tikka LSA65 Deluxe **$385**
Same as LSA65 Standard, except w/special features of LSA55 De-
luxe. Imported 1970-88.

Tikka LSA65 Standard **$335**
Same as LSA55 Standard, except calibers: 25-06, 6.5×55 270 Win.,
30-06. 5-shot magazine, 22-inch bbl., weighs 7.5 lbs. Imported 1970-
88.

Tikka Model M 55
Bolt action. Calibers: 222 Rem., 22-250 Rem., 223 Rem. 243 Win., 308
Win. (6mm Rem. and 17 Rem. available in Standard and Deluxe models only). 23.2-inch bbl. (24.8-inch in Sporter and Heavy Barrel models). 42.8 inches overall (44 inches in Sporter and Heavy Barrel
models). Weight: 7.25 to 9 lbs. Monte Carlo-style stock w/pistol grip.
Sling swivels. Imported 1965-88.

Continental	**$495**
Deluxe Model	**500**
Sporter	**475**
Sporter w/sights	**495**
Standard	**450**
Super Sporter	**540**
Super Sporter w/sights	**575**
Trapper	**465**

Tikka Model M65 Sporter

Tikka Model M65 Wildboar

Tikka Model M65
Bolt action. Calibers: 25-06, 270 Win., 308 Win., 30-06, 7mm Rem. Mag., 300 Win. Mag. (Sporter and Heavy Bbl. models in 270 Win., 308 Win. and 30-06 only). 22.4-inch bbl. (24.8-inch in Sporter and Heavy Bbl. models). 43.2 inches overall (44 inches in Sporter, 44.8 inches in Heavy Bbl.). Weight: 7.5 to 9.9 lbs. Monte Carlo-style stock w/pistol grip. Disc. 1989.

Continental	**$530**
Deluxe Magnum	540
Deluxe Model	495
Magnum	485
Sporter	470
Sporter w/sights	525
Standard	450
Super Sporter	575
Super Sporter w/sights	595
Super Sporter Master	730

Tikka Model M65 Wildboar $525
Same general specifications as Model M 65, except 20.8-inch bbl., overall length of 41.6 inches and weight of 7.5 lbs. Disc. 1989.

Tikka New Generation Rifles
Short throw bolt available in three action lengths. Calibers: 22-250 Rem., 223 Rem., 243 Win., 308 Win., (medium action) 25-06 Rem., 270 Win., 30.06, (Long Action) 7mm Rem. Mag., 300 Win. Mag., 338 Win. Mag., (Magnum Action). 22- to 26-inch bbl. 42.25 to 46 inches overall. Weight: 7.2 to 8.5 lbs. Available w/o sights or w/hooded front and open rear sight on quarter rib. Quick release 3- or 5-shot detachable magazine w/recessed side release. Barrel selector on trigger and cocking indicators in tang. European walnut buttstock and forearm matte lacquer finish. Imported 1989-94.

Standard Calibers	**$595**
Magnum Calibers	640

Tikka Premium Grade Rifle
Similar to New Generation rifles, except w/hand checkered deluxe wood stock w/roll-over check-piece and rosewood grip cap and forend tip. High polished blued finish. Imported 1989-94.

Standard Calibers	**$625**
Magnum Calibers	675

Tikka Whitetail Bolt-Action Rifle Series
New Generation design in multiple model configurations and three action lengths chambered 22-250 to 338 Win. Mag. See specifications of individual models next page.

Tikka Battue Model
Similar to Hunter Model, except designed for snapshooting w/hooded front and open rear sights on quarter rib. Blued finish. Checkered select walnut stock w/matt lacquered finish. Imported 1991-96.

Battue Model (standard w/sights)	**$450**
Magnum Calibers, **add**	40

Tikka Continental Model
Similar to Hunter Model, except w/prone-style stock w/wider forearm and 26-inch heavy bbl. chambered 22-250 Rem., 223 Rem., 308 Win. (Varmint); 25-06 Rem., 270 Win., 7mm Rem. Mag., 300 Win. Mag. (Long Range). Imported 1991 to date.

Continental Long-Range Model	**$495**
Continental Varmint Model	485
Magnum Calibers, add	40

Tikka Whitetail Hunter Model
Calibers: 22-250 Rem., 223 Rem., 243 Win., 25-06 Rem., 270 Win., 7mm Rem. Mag., 308 Win 30.06, 300 Win. Mag., 338 Win. Mag. 3- or 5-shot detachable box magazine. 20.5- to 24.5-inch bbl. with no sights. 42 to 44.5 inches overall. Weight: 7 to 7.5 lbs. Adj. single-stage or single-set trigger. Blued or stainless finish. All-Weather synthetic or checkered select walnut stock w/matt lacquered finish. Imported 1991 to date.

Standard Model	**$450**
Synthetic Model	495
Stainless Model	520
Magnum Calibers, **add**	40

UBERTI RIFLES
Lakeville, Connecticut
Mfd. By Aldo Uberti, Ponte Zanano, Italy
Imported by Uberti USA, Inc.

Uberti Model 1866 Sporting Rifle
Replica of Winchester Model 1866 lever-action repeater. Calibers: 22 LR, 22 WMR, 38 Spec., 44-40, 45 LC. 24.25-inch octagonal bbl. 43.25 inches overall. Weight: 8.25 lbs. Blade front sight, rear elevation leaf. Brass frame and buttplate. Bbl., magazine tube, other metal parts blued. Walnut buttstock and forearm.

Model 1866 Rifle	**$495**
Model 1866 Carbine (19-inch Round Bbl.)	485
Model 1866 Trapper (16-inch Bbl.) Disc. 1989	470
Model 1866 Rimfire, **deduct**	85

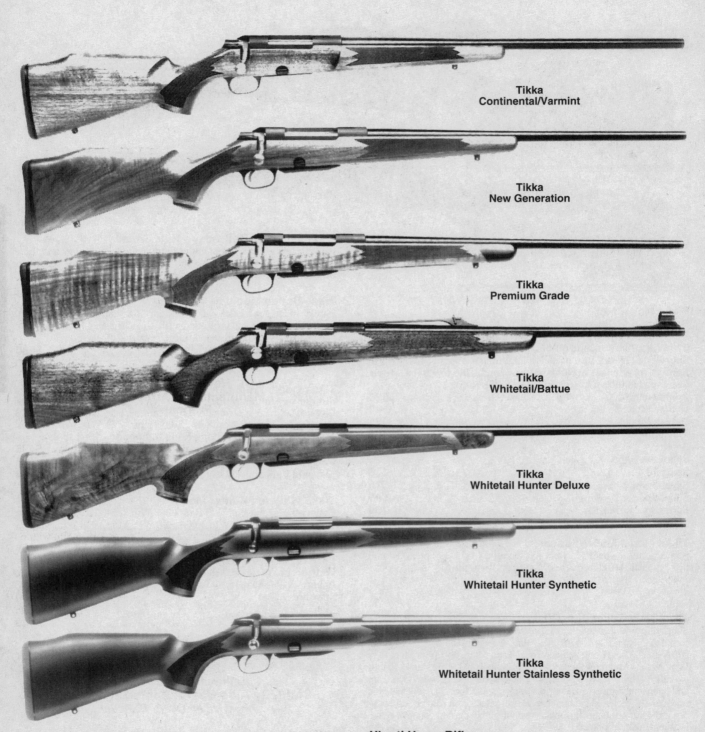

Tikka
Continental/Varmint

Tikka
New Generation

Tikka
Premium Grade

Tikka
Whitetail/Battue

Tikka
Whitetail Hunter Deluxe

Tikka
Whitetail Hunter Synthetic

Tikka
Whitetail Hunter Stainless Synthetic

Uberti Model 1873 Sporting Rifle

Replica of Winchester Model 1873 lever-action repeater. Calibers: 22 LR, 22 WMR, 38 Spec., 357 Mag., 44-40, 45 LC. 24.25- or 30-inch octagonal bbl. 43.25 inches overall. Weight: 8 lbs. Blade front sight; adj. open rear. Color casehardened frame. Bbl., magazine tube, hammer, lever and buttplate blued. Walnut buttstock and forearm.

Model 1873 Rifle **$575**
Model 1873 Carbine (19-inch Round Bbl.) **550**
Model 1873 Trapper (16-inch Bbl.) Disc. 1989 . . . . **525**

Uberti Henry Rifle

Replica of Henry lever-action repeating rifle. Calibers: 44-40, 45 LC. 24.5-inch half-octagon bbl. 43.75 inches overall. Weight: 9.25 lbs. Blade front sight; rear sight adj. for elevation. Brass frame, buttplate and magazine follower. Bbl., magazine tube and remaining parts blued. Walnut buttstock.

Henry Rifle **$550**
Henry Carbine (22.5-inch Bbl.) **525**
Henry Trapper (16- or 18-inch Bbl.) **560**
Steel Frame, **add** **85**

**Uberti Model 1873
Carbine**

ULTRA-HI PRODUCTS COMPANY
Hawthorne, New Jersey

**Ultra-Hi Model 2200 Single-Shot
Bolt-Action Rifle** **$125**
Caliber: 22 LR, Long, Short. 23-inch bbl. Weight: 5 lbs. Sights: open rear; blade front. Monte Carlo stock w/pistol grip. Made in Japan. Intro. 1977; Disc.

ULTRA LIGHT ARMS COMPANY
Granville, West Virginia

Ultra Light Arms Model 20 Bolt-Action Rifle
Calibers: 22-250 Rem., 243 Win., 6mm Rem., 250-3000 Savage, 257 Roberts, 257 Ack., 7mm Mauser, 7mm Ack., 7mm-08 Rem., 284 Win., 300 Savage, 308 Win., 358 Win. Box magazine. 22-inch ultra light bbl. Weight: 4.75 lbs. No sights. Synthetic stock of Kevlar or graphite finished seven different colors. Nonglare matte or bright metal finish. Medium-length action available L.H. models. Made 1985 to date.
Standard Model **$1395**
Left-Hand Model 1525

Ultra Light Arms Model 20S Bolt-Action Rifle
Same general specifications as Model 20, except w/short action in calibers 17 Rem., 222 Rem., 223 Rem., 22 Hornet only.
Standard Model **$1495**
Left-Hand Model 1595

Ultra Light Arms Model 24 Bolt-Action Rifle
Same general specifications as Model 20, except w/long action in calibers 25-06, 270 Win., 30-06 and 7mm Express only.
Standard Model **$1425**
Left-Hand Model 1550

Ultra Light Arms Model 28 Bolt-Action Rifle **$1650**
Same general specifications as Model 20, except w/long Magnum action in calibers 264 Win. Mag., 7mm Rem. Mag., 300 Win. Mag., 338 Win. Mag. only. Offered w/recoil arrestor. Left-hand model available.

Ultra Light Arms Model 40 Bolt-Action Rifle
Similar to Model 28, except in calibers 300 Wby. and 416 Rigby. Weight: 5.5 lbs. Made 1994 to date.
Standard Model **$1765**
Left Hand, Model 1795

UNIQUE RIFLE
Hendaye, France
Mfd. by Manufacture d'Armes des Pyrénées Francaises

Unique T66 Match Rifle **$395**
Single-shot bolt-action rifle. Caliber: 22 LR. 25.5-inch bbl. Weight: 10.5 lbs. Sights: micrometer aperture rear; globe front. French walnut target stock w/Monte Carlo comb, bull pistol grip, wide and deep forearm, stippled grip surfaces, adj. swivel on accessory track, adj. rubber buttplate. Made 1966. Disc.

U.S. MILITARY RIFLES
Mfd. by Springfield Armory, Remington Arms Co., Winchester Repeating Arms Co., Inland Mfg. Div. of G.M.C., and other contractors. *See* notes.

Unless otherwise indicated, the following U.S. Military rifles were mfg. at Springfield Armory, Springfield, Mass.

RIFLES

Ultra Model 2200

Ultra Light Model 28

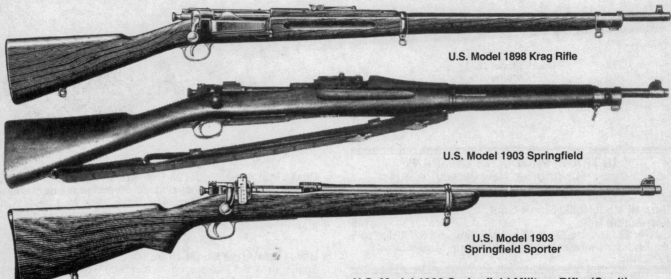

U.S. Model 1898 Krag Rifle

U.S. Model 1903 Springfield

U.S. Model 1903
Springfield Sporter

U.S. Model 1898 Krag-Jorgensen Carbine $1395
Same general specifications as Model 1898 Rifle, except w/22-inch bbl., weighs 8 lbs., carbine-type stock. *Note:* The foregoing specifications apply, in general, to Carbine Models 1896 and 1899, which differed from Model 1898 only in minor details.

U.S. Model 1898 Krag-Jorgensen Military Rifle $895
Bolt action. Caliber: 30-40 Krag. 5-shot hinged box magazine. 30-inch bbl. Weight: 9 lbs. Sights: adj. rear; blade front. Military-type stock, straight grip. *Note:* The foregoing specifications apply, in general, to Rifle Models 1892 and 1896, which differed from Model 1898.

U.S. Model 1903 Mark I Springfield $995
Same as Standard Model 1903, except altered to permit use of the Pedersen Device. This device, officially designated "U.S. Automatic Pistol Model 1918," converted the M/1903 to a semiautomatic weapon firing a 30 caliber cartridge similar to 32 automatic pistol ammunition. Mark I rifles have a slot milled in the left side of the receiver to serve as an ejection port when the Pedersen Device was in use; these rifles were also fitted w/a special sear and cutoff. Some 65,000 of these devices were manufactured and presumably a like number of M/1903 rifles converted to handle them. During the early 1930s all Pedersen Devices were ordered destroyed and the Mark I rifles were reconverted by replacement of the special sear and cut-off w/standard components. Some 20-odd specimens are known to have escaped destruction and are in government museums and private collections. Probably more are extant. Rarely is a Pedersen Device offered for sale, so a current value cannot be assigned. However, many of the altered rifles were bought by members of the National Rifle Association through the Director of Civilian Marksmanship. Value shown is for the Mark I rifle w/o the Pedersen Device.

U.S. Model 1903 National Match Springfield $1150
Same general specifications as Standard Model 1903, except specially selected w/star-gauged bbl., Type C pistol-grip U.S. Model 1903 National Match Springfield *(Con't)* stock, polished bolt assembly; early types have headless firing pin assembly and reversed safety lock. Produced especially for target shooting.

U.S. Model 1903 Springfield Military Rifle
Modified Mauser-type bolt action. Caliber: 30-06. 5-shot box magazine. 23.79-inch bbl. Weight: 8.75 lbs. Sights: addition. rear; blade front. Military-type stock straight grip. *Note:* M/1903 rifles of Spring-

U.S. Model 1903 Springfield Military Rifle *(Con't)*
field manufacture w/serial numbers under 800,000 (1903-1918) have casehardened receivers; those between 800,000 and 1,275,767 (1918-1927) were double-heat-treated; rifles numbered over 1,275,767 have nickel-steel bolts and receivers. Rock Island production from No. 1 to 285,507 have case-hardened receivers. Improved heat treatment was adopted in May 1918 with No. 285,207; about three months later with No. 319,921 the use of nickel steel was begun, but the production of some double-heat-treated carbon-steel receivers and bolts continued. Made 1903-30 at Springfield Armory; during WWI, M/1903 rifles were also made at Rock Island Arsenal, Rock Island, Ill.
W/casehardened receiver **$185**
W/double-heat-treated receiver **225**
W/nickel steel receiver **350**

U.S. Model 1903 Springfield Sporter $995
Same general specifications as National Match, except w/sporting design stock, Lyman No. 48 receiver sight.

U.S. Model 1903 Style T Springfield Match Rifle $1150
Same specifications as Springfield Sporter, except w/heavy bbl. (26-, 28- or 30-inch), scope bases, globe front sight, weighs 12.5 lbs. w/26-inch bbl.

U.S. Model 1903 Type A Springfield Free Rifle $1375
Same as Style T, except made w/28-inch bbl. only, w/Swiss buttplate, weighs 13.25 lbs.

U.S. Model 1903 Type B Springfield Free Rifle ... $1825
Same as Type A, except w/cheekpiece stock, palmrest, Woodie double-set triggers, Garand fast firing pin, weighs 14.75 lbs.

U.S. Model 1903-A1 Springfield
Same general specifications as Model 1903, except may have Type C pistol-grip stock adopted in 1930. The last springfields produced at the Springfield Armory were of this type, final serial number was 1,532,878 made in 1939. *Note:* Late in 1941, Remington Arms Co., Ilion, N.Y., began production under government contract of Springfield rifles of this type w/a few minor modifications. These rifles are numbered 3,000,001-3,348,085 and were manufactured before the adoption of Mode 1903-A3.
Springfield manufacture **$750**
Remington manufacture **650**

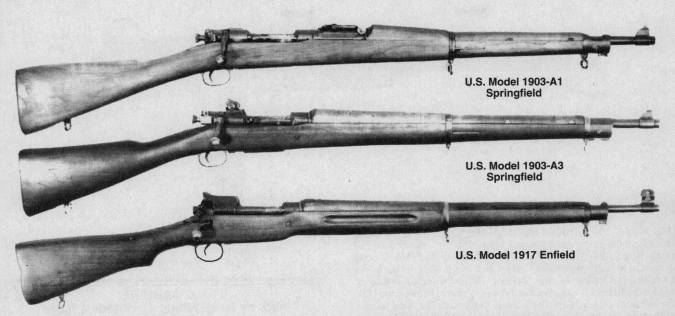

U.S. Model 1903-A1 Springfield

U.S. Model 1903-A3 Springfield

U.S. Model 1917 Enfield

U.S. Model 1903-A3 Springfield $425

Same general specifications as Model 1903-A1, except modified to permit increased production and lower cost; may have either straight-grip or pistol-grip stock, bolt is not interchangeable w/earlier types, w/receiver peep sight, many parts are stamped sheet steel, including the trigger guard and magazine assembly. Quality of these rifles, lower than that of other 1903 Springfields, reflects the emergency conditions under which they were produced. Mfd. during WWII by Remington Arms Co. and L. C. Smith Corona Typewriters, Inc.

U.S. Model 1922-M1 22 Springfield Target Rifle $875

Modified Model 1903. Caliber: 22 LR. 5-shot detachable box magazine. 24.5-inch bbl. Weight: 9 lbs. Sights: Lyman No. 48C receiver, blade front. Sporting-type stock similar to that of Model 1903 Springfield Sporter. Issued 1927. *Note:* The earlier Model 1922, which is seldom encountered, differs from the foregoing chiefly in the bolt mechanism and magazine.

U.S. M2 22 Springfield Target Rifle $895

Same general specifications as Model 1922-M1, except w/speedlock, improved bolt assembly adj. for headspace. *Note:* These improvements were later incorporated in many rifles of the preceding models (M1922, M1922MI) and arms so converted were marked "M1922M2" or "M1922MII."

U.S. Rifle, Caliber 30, M1 (Garand) Mil. Rifle $750

Clip-fed, gas-operated, air-cooled semiautomatic. Uses a clip containing 8 rounds. 24-inch bbl. Weight: w/o bayonet, 9.5 lbs. Sights: adj. peep rear; blade front w/guards. Pistol-grip stock, handguards. Made 1937-57. *Note:* Garand rifles have also been produced by Winchester Repeating Arms Co., Harrington & Richardson Arms Co., and International Harvester Co. Deduct 25% for arsenal-assembled mismatches.

U.S. Rifle, Caliber 30, M1, National Match $1250

Accurized target version of the Garand. Glass-bedded stock; match grade bbl., sights, gas cylinder. "NM" stamped on bbl. forward of handguard.

NOTE
The U.S. Model 1917 Enfield was mfd. 1917-18 by Remington Arms Co. of Delaware (later Midvale Steel & Ordnance Co.), Eddystone, PA; Remington Arms Co., Ilion, NY; Winchester Repeating Arms Co., New Haven, CT.

U.S. Carbine Caliber 30, M1 (Grand)

U.S. Carbine Caliber 30, M1

Uzi Semiautomatic
Model B Carbine

U.S. Model 1917 Enfield Military Rifle **$395**
Modified Mauser-type bolt action. Caliber: 30-06. 5-shot box magazine. 26-inch bbl. Weight: 9.25 lbs. Sights: adj. rear; blade front
w/guards. Military-type stock w/semi-pistol grip. This design origi-
nated in Great Britain as their "Pattern '14" and was mfd. in caliber
303 for the British Government in three U.S. plants. In 1917, the U.S.
Government contracted w/these firms to produce the same rifle in
caliber 30-06; over two million of these Model 1917 Enfields were
mfd. While no more were produced after WWI, the U.S. supplied
over a million of them to Great Britain during WWII.

> **NOTE**
>
> The WWII-vintaqe 30-caliber U.S. Carbine was mfd. by Inland
> Mfg. Div. of G.M.C., Dayton, OH; Winchester Repeating Arms
> Co., New Haven, CT, and other contractors: International Busi-
> ness Machines Corp., Poughkeepsie, NY; National Postal Meter
> Co., Rochester, NY; Quality Hardware & Machine Co., and
> Rock-Ola Co., Chicago, IL; Saginaw Steering Gear Div. of
> G.M.C., Saginaw, M1; Standard Products Co., Port Clinton, OH;
> Underwood-Elliott-Fisher Co., Hartford, CT.

U.S. Carbine, Caliber 30, M1 . . . . **$575**
Gas-operated (short-stroke piston), semiautomatic. 15- or 30-round
detachable box magazine. 18-inch bbl. Weight: 5.5 lbs. Sights: adj.
rear; blade front sight w/guards. Pistol-grip stock w/handguard,
side-mounted web sling. Made 1942-45. In 1963, 150,000 surplus
M1 Carbines were sold at $20 each to members of the National Rifle
Assn. by the Dept. of the Army. *Note:* For Winchester and Rock-Ola,
add 30%; for Irwin Pedersen, add 80%. Quality Hardware did not
complete its production run. Guns produced by other manufacturers
were marked "Unquality" & command premium prices.

U.S. REPEATING ARMS CO.

See **Winchester Rifle listings.**

UNIVERSAL SPORTING GOODS, INC.
Miami, Florida

Universal Deluxe Carbine **$250**
Same as standard model, except also available in caliber 256, w/deluxe
walnut Monte Carlo stock and handguard. Made 1965 to date.

Universal Standard M-1 Carbine **$195**
Same as U.S. Carbine, Cal.30, M1, except may have either wood or
metal handguard, bbl. band w/ or w/o bayonet lug; 5-shot magazine
standard. Made 1964 to date.

UZI CARBINE
Mfd. by Israel Military Industries, Israel

Uzi Semiautomatic Model B Carbine **$995**
Calibers: 9mm Parabellum, 41 Action Express, 45 ACP. 20- to
50-round magazine. 16.1-inch bbl. Weight: 8.4 lbs. Metal folding
stock. Front post-type sight, open rear, both adj. Imported by Action Arms 1983-89. NFA (Selective Fire) models imported by
UZI America, INC., 1994 to date.

VALMET OY
Jyväskylä, Finland

Valmet M-62S Semiautomatic Rifle **$1195**
Semiautomatic version of Finnish M-62 automatic assault rifle
based on Russian AK-47. Gas-operated rotating bolt action. Caliber: 7.62mm×39 Russian. 15- and 30-round magazines. 16.63-
inch bbl. Weight: 8 lbs. w/metal stock. Sights: tangent aperture
rear; hooded blade front w/luminous flip-up post for low-light
use. Tubular steel or wood stock. Intro. 1962. Disc.

Valmet M-71S **$1050**
Same specifications as M-62S, except caliber 5.56mm×45 (223
Rem.), w/open rear sight, reinforced resin or wood stock, weighs
7.75 lbs. w/former. Made 1971-89.

Valmet M-76 Semiautomatic Rifle
Semiautomatic assault rifle. Gas-operated, rotating bolt action. Caliber:
223 Rem. 15- and 30-shot magazines. Made 1984-89.
Wooden stock . . . . **$ 1095**
Folding stock 1395

Valmet M-78 Semiautomatic Rifle **$1295**
Caliber: 7.62×51 (NATO). 24.13-inch bbl. Overall length: 43.25
inches. Weight: 10.5 lbs.

Valmet M-82 Semiautomatic Carbine **$1750**
Caliber: 223 Rem. 15- or 30-shot magazine. 17-inch bbl. 27 inches
overall. Weight: 7.75 lbs.

Valmet Model 412 S Double Rifle **$1050**
Boxlock. Manual or automatic extraction. Calibers: 243, 308, 30-06,
375 Win., 9.3×74R. Bbls.: 24-inch over/under. Weight: 8.63 lbs.
American walnut checkered stock and forend.

Valmet M-62S

Valmet Hunter

Valmet Hunter Semiautomatic Rifle **$750**
Similar to M-78, except in calibers 223 Rem. (5.56mm), 243 Win., 308 Win. (7.62 NATO) and 30-06. 5-, 9- or 15-shot magazine. 20.5-inch plain bbl. 42 inches overall. Weight: 8 lbs. Sights: adj. 412 combination scope mount/rear; blade front, mounted on gas tube. Checkered European walnut buttstock and extended checkered forend and handguard. Imported 1986-89.

VICKERS LTD.
Crayford, Kent, England

Vickers Empire Model **$325**
Similar to Jubilee Model, except w/27- or 30-inch bbls., straight-grip stock, weighs 9.25 lbs. w/30-inch bbl. Made before WWII.

Vickers Jubilee Model Single-Shot Target Rifle **$350**
Round-receiver Martini-type action. Caliber: 22 LR. 28-inch heavy bbl. Weight: 9.5 lbs. Sights: Parker-Hale No.2 front; Perfection rear peep. One-piece target stock w/full forearm and pistol grip. Made before WWII.

VOERE
Manufactured in Kufstein, Austria

Imported by JagerSport, Cranston, Rhode Island

Voere VEC-91 Lightning Bolt-Action Rifle **$1950**
Features unique electronic ignition system to activate or fire caseless ammunition. Calibers: 5.56 UCC (222 Cal.), 6mm UCC caseless. 5-shot magazine. 20-inch bbl. 39 inches overall. Weight: 6 lbs. Open adj. rear sight. Drilled and tapped for scope mounts. European walnut stock w/cheekpiece. Twin forward locking lugs. Imported 1992 to date.

VOERE, VOELTER & COMPANY
Vaehrenbach, Germany

Mauser-Werke acquired Voere in 1987 and all models are now marketed under new designations.

Voere Model 1007 Biathlon Repeater **$255**
Caliber: 22 LR. 5-shot magazine. 19.5-inch bbl. 39 inches overall. Weight: 5.5 lb. Sights: adj. rear, blade front. Plain beechwood stock. Imported 1984-86.

Voere Model 1013 Bolt-Action Repeater **$445**
Same as Model 1007, except w/military-style stock in 22 WMR caliber. Double-set triggers optional. Imported 1984-86 by KDF, Inc.

Voere Model 2107 Bolt-Action Repeater
Caliber: 22 LR. 5 or 8-shot magazine. 19.5-inch bbl. 41 inches overall. Weight: 6 lbs. Sights: adj. rear sight, hooded front. European hardwood Monte Carlo-style stock. Imported 1986 by KDF, Inc.
Standard Model **$195**
Deluxe Model 225

WALTHER RIFLES
Mfd. by the German firms of Waffenfabrik Walther and Carl Walther Sportwaffenfabrik

The following Walther rifles were mfd. before WWII by Waffenfabrik Walther, Zella-Mehlis (Thür.), Germany.

Walther Model 1 Autoloading Rifle (Light) **$350**
Similar to Standard Model 2, but w/20-inch bbl., lighter stock, weighs 4.5 lbs.

RIFLES

**Vickers Jubilee
Single-Shot Target Rifle**

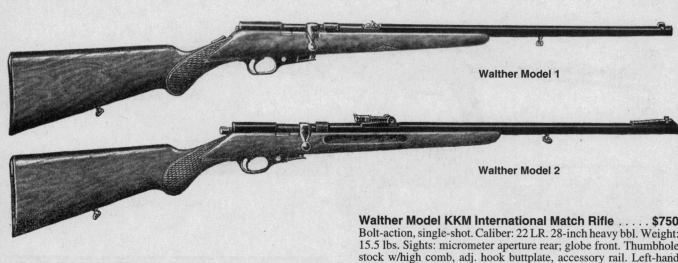

Walther Model 1

Walther Model 2

Walther Model 2 Autoloading Rifle **$445**
Bolt action, may be used as autoloader, manually operated repeater or single-shot. Caliber: 22 LR. 5- or 9-shot detachable box magazine. 24.5-inch bbl. Weight: 7 lbs. Sights: tangent-curve rear; ramp front. Sporting stock w/checkered pistol grip, grooved forearm, swivels. Disc.

Walther Olympic Bolt-Action Match Rifle **$925**
Single-shot. Caliber: 22 LR. 26-inch heavy bbl. Weight: 13 lbs. Sights: micrometer extension rear; interchangeable front. Target stock w/checkered pistol grip, thumbhole, full beavertail forearm covered w/corrugated rubber, palmrest, adj. Swiss-type buttplate, swivels. Disc.

Walther Model V Bolt-Action Single-Shot Rifle **$395**
Caliber: 22 LR. 26-inch bbl. Weight: 7 lbs. Sights: open rear; ramp front. Plain pistol-grip stock w/grooved forearm. Disc.

Walther Model V Meisterbüchse (Champion) **$435**
Same as standard Model V, except w/micrometer open rear sight and checkered pistol grip. Disc.

Post WWII Models

The Walther rifles listed below have been manufactured since WWII by Carl Walther Sportwaffenfabrik, Ulm (Donau), Germany.

Walther Model GX-1 Free Rifle **$1325**
Bolt-action, single-shot. Caliber: 22 LR. 25.5-inch heavy bbl. Weight: 15.9 lbs. Sights: micrometer aperture rear; globe front. Thumbhole stock w/adj. cheekpiece and buttplate w/removable hook, accessory rail. Left-hand stock available. Accessories furnished include hand stop and sling swivel, palmrest, counterweight assembly.

Walther Model KKJ Sporter **$725**
Bolt action. Caliber: 22 LR. 5-shot box magazine. 22.5-inch bbl. Weight: 5.5 lbs. Sights: open rear; hooded ramp front. Stock w/cheekpiece, checkered pistol grip and forearm, sling swivels. Disc.

Walther Model KKJ-Ho **$795**
Same as Model KKJ, except chambered for 22 Hornet. Disc.

Walther Model KKJ-Ma **$695**
Same as Model KKJ, except chambered for 22 WMR. Disc.

Walther Model KKM International Match Rifle **$750**
Bolt-action, single-shot. Caliber: 22 LR. 28-inch heavy bbl. Weight: 15.5 lbs. Sights: micrometer aperture rear; globe front. Thumbhole stock w/high comb, adj. hook buttplate, accessory rail. Left-hand stock available. Disc.

Walther Model KKM-S **$795**
Same specifications as Model KKM, except w/adj. cheekpiece. Disc.

Walther Moving Target Match Rifle **$650**
Bolt-action, single-shot. Caliber: 22 LR. 23.6-inch bbl. w/weight. Weight: 8.6 lbs. Supplied w/o sights. Thumbhole stock w/adj. cheekpiece and buttplate. Left-hand stock available.

Walther Prone 400 Target Rifle **$625**
Bolt-action, single-shot. Caliber: 22 LR. 25.5-inch heavy bbl. Weight: 10.25 lbs. Supplied w/o sights. Prone stock w/adj. cheekpiece and buttplate, accessory rail. Left-hand stock available. Disc.

Walther Model SSV Varmint Rifle **$550**
Bolt-action, single-shot. Calibers: 22 LR, 22 Hornet. 25.5-inch bbl. Weight: 6.75 lbs. Supplied w/o sights. Monte Carlo stock w/high cheekpiece, full pistol grip and forearm. Disc.

Walther Model U.I.T. Special Match Rifle **$895**
Bolt-action, single-shot. Caliber: 22 LR. 25.5-inch bbl. Weight: 10.2 lbs. Sights: Micrometer aperture rear; globe front. Target stock w/high comb, adj. buttplate, accessory rail. Left-hand stock avail. Disc. 1993.

Walther Model U.I.T. Super Match Rifle **$925**
Bolt-action, single-shot. Caliber: 22 LR. 25.5-inch heavy bbl. Weight: 10.2 lbs. Micrometer aperture rear; globe front. Target stock w/support for off-hand shooting, high comb, adj. buttplate and swivel. Left-hand stock available. Disc. 1993.

MONTGOMERY WARD
Chicago, Illinois
Western Field and Hercules Models

Firearms under the "private label" names of Western Field and Hercules are manufactured by such firms as Mossberg, Stevens, Marlin, and Savage for distribution and sale by Montgomery Ward.

Montgomery Ward Model 14M-497B Western Field Bolt-Action Rifle **$80**
Caliber: 22 RF. 7-shot detachable box magazine. 24-inch bbl. Weight: 5 lbs. Sights: receiver peep; open rear; hooded ramp front. Pistol-grip stock. Mfg. by Mossberg.

Walther Model GX-1

Walther Model KKM-S

**Walther Model U.I.T.
Super Match**

**Montgomery Ward Model M771 Western Field
Lever-Action Rifle** **$155**
Calibers: 30-30, 35 Rem. 6-shot tubular magazine. 20-inch bbl.
Weight: 6.75 lbs. Sights: open rear; ramp front. Pistol-grip or
straight stock, forearm w/barrel band. Mfg. by Mossberg.

**Montgomery Ward Model M772 Western Field
Lever-Action Rifle** **$175**
Calibers: 30-30, 35 Rem. 6-shot tubular magazine. 20-inch bbl. Weight:
6.75 lbs. Sights: open rear; ramp front. Pistol-grip or straight stock, forearm w/bbl. band. Mfg. by Mossberg.

**Montgomery Ward Model M775 Bolt-Action
Rifle** **$190**
Calibers: 222 Rem., 22-250, 243 Win., 308 Win. 4-shot magazine.
Weight: 7.5 lbs. Sights: folding leaf rear; ramp front. Monte Carlo stock
w/cheekpiece, pistol grip. Mfg by Mossberg.

**Montgomery Ward Model M776 Bolt-Action
Rifle** **$195**
Calibers: 222 Rem., 22-250, 243 Win., 308 Win. 4-shot magazine.
Weight: 7.5 lbs. Sights: folding leaf rear; ramp front. Monte Carlo
stock w/cheekpiece, pistol grip. Mfg. by Mossberg.

Montgomery Ward Model M778 Lever-Action **$165**
Calibers: 30-30, 35 Rem. 6-shot tubular magazine. 20-inch bbl.
Weight: 6.75 lbs. Sights: open rear; ramp front. Pistol-grip or
straight stock, forearm w/bbl. band. Mfg. by Mossberg.

**Montgomery Ward Model M780 Bolt-Action
Rifle** **$195**
Calibers: 222 Rem., 22-250, 243 Win., 308 Win. 4-shot magazine.
Weight: 7.5 lbs. Sights: folding leaf rear; ramp front. Monte Carlo
stock w/cheekpiece, pistol grip. Mfg. by Mossberg.

**Montgomery Ward Model M782 Bolt-Action
Rifle** **$195**
Same general specifications as Model M780.

Montgomery Ward Model M808 . . . . **$85**
Takedown. Caliber: 22RF. 15-shot tubular magazine. Bbls.: 20- and
24-inch. Weight: 6 lbs. Sights: open rear; bead front. Pistol-grip
stock. Mfg. by Stevens.

Montgomery Ward Model M832 Bolt-Action Rifle **$90**
Caliber: 22 RF. 7-shot clip magazine. 24-inch bbl. Weight: 6.5
lbs. Sights: open rear; ramp front. Mfg. by Mossberg.

Montgomery Ward Model M836 . . . . **$95**
Takedown. Caliber: 22RF. 15-shot tubular magazine. Bbls.: 20- and
24-inch. Weight: 6 lbs. Sights: open rear; bead front. Pistol-grip
stock. Mfg. by Stevens.

**Montgomery Ward Model M865 Lever-Action
Carbine** **$125**
Hammerless. Caliber: 22RF. Tubular magazine. Made w/both
18.5-inch and 20-inch bbls., forearm w/bbl. band, swivels.
Weight: 5 lbs. Mfg. by Mossberg.

**Montgomery Ward Model M894 Autoloading
Carbine** **$115**
Caliber: 22 RF. 15-shot tubular magazine. 20-inch bbl. Weight: 6
lbs. Sights: open rear; ramp front. Monte Carlo stock w/pistol grip.
Mfg. by Mossberg.

Montgomery Ward Model M-SD57 . . . . **$90**
Takedown. Caliber: 22RF. 15-shot tubular magazine. Bbls.: 20- and
24-inch. Weight: 6 lbs. Sights: open rear; bead front. Pistol-grip
stock. Mfg. by Stevens.

Weatherby Mark V Classicmark I

Weatherby Crown Custom

Weatherby Fiberguard™

Weatherby Fibermark™

WEATHERBY, INC.
Atascadero, CA
(Formerly South Gate, CA)

Weatherby Crown Custom Rifle................. **$3250**
Calibers: 240, 30-06, 257, 270, 7mm, 300, and 340. Bbl.: made to order. Super fancy walnut stock. Also available w/engraved barreled action including gold animal overlay.

Weatherby Deluxe 378 Magnum Rifle **$1750**
Same general specifications as Deluxe Magnum in other calibers, except caliber 378 W. M. Schultz & Larsen action; 26-inch bbl. Disc. 1958.

Weatherby Deluxe Magnum Rifle **$1290**
Calibers: 220 Rocket, 257 Weatherby Mag., 270 W.M. 7mm W.M., 300 W.M., 375 W.M. Specially processed FN Mauser action. 24-inch bbl. (26-inch in 375 cal.). Monte Carlo-style stock w/cheekpiece, black forend tip, grip cap, checkered pistol grip and forearm, quick-detachable sling swivels. Value shown is for rifle w/o sights. Disc. 1958.

Weatherby Deluxe Rifle **$995**
Same general specifications as Deluxe Magnum, except chambered for standard calibers such as 270, 30-06, etc. Disc. 1958.

Weatherby Fiberguard™ Rifle **$425**
Same general specifications as Vanguard except for fiberglass stock and matte metal finish. Disc, 1988.

Weatherby Fibermark™ Rifle **$850**
Same general specifications as Mark V except w/molded fiberglass stock, finished in a nonglare black wrinkle finish. The metal is finished in a non-glare matte finish. Disc. 1993.

Weatherby Mark V Accumark Bolt Action Repeating
Weatherby Mark V magnum action. Calibers: 257 Wby., 270 Wby., 7mm Rem. Mag., 7mm Wby., 7mm STW, 300 Win. Mag., 300 Wby. Mag., 30-338 Wby., 30-378 Wby. and 340 Wby. 26- or 28-inch stainless bbl. w/black oxide flutes. 46.5 or 48.5 inches overall. Weight: 8 to 8.5 lbs. No sights, drilled and tapped for scope. Stainless finish w/blued receiver. H-S Precision black synthetic stock w/aluminum bedding plate, recoil pad and sling swivels. Imported 1996 to date.
Mark V Accumark (30-338 & 30-378 Wby. Mag.) **$1050**
Mark V Accumark Lightweight (All other calibers) **850**
Mark V Left-handed Action, **add**..................... **75**

Weatherby Mark V Accumark
Light Weight Rifle **$550**
Similar to the Mark V Accuramark, except w/Light Weight Mark V action designed for standard calibers w/six-locking lugs (rather than nine). 24-inch stainless bbl. Weight: 5.75 lbs. Gray or black Monte Carlo-style composite Kevlar/fiberglass stock w/ Pachmayr "Decelerator" pad. No sights. Imported 1997 to date.

Weatherby Mark V Classicmark I Rifle
Same general specifications as Mark V, except w/checkered select American Claro walnut stock w/oil finish and presentation recoil pad. Satin metal finish. Imported 1992-93.
Calibers 240 to 300 Wby. **$750**
Caliber 340 Wby. **795**

Weatherby Mark V Deluxe

Weatherby Mark V Euromark

Weatherby Mark V Lazermark

Weatherby Mark V Safari Grade

Weatherby Mark V Classicmark I Rifle *(Con't)*
Caliber 378 Wby.   825
Caliber 416 Wby.   850
Caliber 460 Wby.   895

Weatherby Mark V Classicmark II Rifle
Same general specifications as Classicmark I, except w/checkered se-
lect American walnut stock w/oil finish steel grip cap and Old English
recoil pad. Satin metal finish. Right-hand only. Imported 1992-93.
Calibers 240 to 340 Wby. (26-inch bbl.) **$1095**
Caliber 378 Wby. 1150
Caliber 416 Wby. 1295
Caliber 460 Wby. 1350

Weatherby Mark V Deluxe Rifle $595
Similar to Mark V Sporter, except w/Light Weight Mark V action
designed for standard calibers w/six-locking lugs (rather than
nine). 4- or 5-shot magazine. 24-inch bbl. 44 inches overall.
Weight: 6.75 lbs. Checkered Monte Carlo American walnut stock
w/rosewood forend and pistol grip and diamond inlay. Imported
1997 to date.

Weatherby Mark V Deluxe Bolt-Action Sporting Rifle
Mark V action, right or left hand. Calibers: 22-250, 30-06- 224
Weatherby Varmintmaster; 240, 257, 270, 7mm, 300, 340, 375,
378, 416, 460 Weatherby Magnums. Box magazine holds 2 to 5
cartridges depending on caliber. 24- or 26-inch bbl. Weight: 6.5 to
10.5 lbs. Monte Carlo-style stock w/cheekpiece, skip checkering,
forend tip, pistol-grip cap, recoil pad, QD swivels. Values shown
are for rifles w/o sights. Made in Germany 1958-69; in Japan
1970-94. Values shown for Japanese production.

Weatherby Mark V Deluxe Bolt-Action *(Con't)*
Calibers 22-250, 224. . . . . **$795**
Caliber 378 Weatherby Magnum   895

Weatherby Mark V Euromark Bolt-Action Rifle
Same general specifications as other Mark V rifles, except w/hand-
rubbed, satin oil finish Claro walnut stock and nonglare special pro-
cess blue matte barreled action. Left-hand models available. Im-
ported 1986-93. Reintroduced 1995.
Caliber 378 Wby. Mag. **$825**
Caliber 416 Wby. Mag.   850
Caliber 460 Wby Mag.   995
Other Calibers   695

Weatherby Mark V Lazermark Rifle
Same general specifications as Mark V except w/lazer-carved stock.
Caliber 378 Wby. Mag. **$925**
Caliber 416 Wby. Mag.   995
Caliber 460 Wby. Mag. 1150
Other Calibers   825

Weatherby Mark V Safari Grade Rifle $1695
Same general specifications as Mark V except extra capacity
magazine, bbl. sling swivel, and express rear sight typical "Safari" style.

Weatherby Mark V Sporter Rifle
Sporter version of Mark V Magnum, w/low-luster metal finish.
Checkered Carlo walnut stock w/o grip cap or forend tip. No
sights. Imported 1993 to date.
Calibers 257 to 300 Wby **$595**
340 Weatherby   650
375 H&H   695

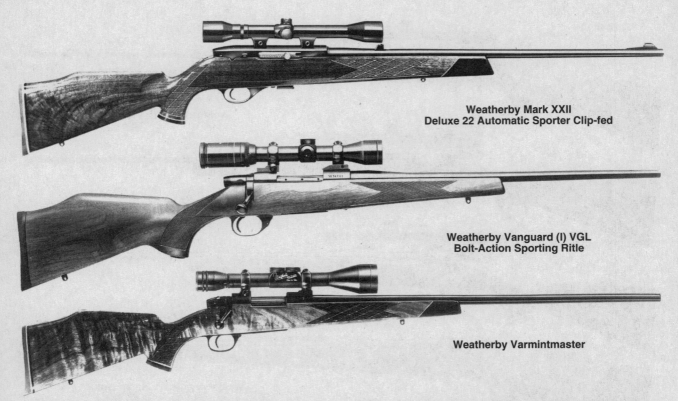

Weatherby Mark XXII
Deluxe 22 Automatic Sporter Clip-fed

Weatherby Vanguard (I) VGL
Bolt-Action Sporting Ritle

Weatherby Varmintmaster

Weatherby Mark V Light Weight Sporter Rifle $575
Similar to Mark V Sporter, except w/Light Weight Mark V action designed for standard calibers w/six-locking lugs (rather than nine). Imported 1997 to date.

Weatherby Mark V Stainless Rifle
Similar to the Mark V Magnum, except in 400-series stainless steel w/bead-blasted matte finish. Weight: 8 lbs. Monte Carlo synthetic stock w/aluminum bedding block. Imported 1995 to date.

Mark V Stainless (30-378 Wby.) $625
Mark V Stainless (375 H&H.) 650
All Other Calibers 595
W/Flutted Bbl., **add** 100

Weatherby Mark V (LW) Stainless Rifle. $495
Similar to the Mark V (standard calibers), except in 400-series stainless steel w/bead-blasted matte finish. 5-shot magazine. 24-inch bbl. 44 inches overall. Weight: 6.5 lbs. Monte Carlo synthetic stock w/aluminum bedding block. Imported 1995 to date.

Weatherby Mark V SLS Rifle
Acronym for Stainless Laminated Sporter. Similar to the Mark V Magnum Sporter, except w/stainless 400-series action and 24- or 26-inch stainless bbl. Laminated wood stock. Weight: 8.5 lbs. Black oxide bead-blasted matte blue finish. Imported 1997 to date.

Mark V SLS (340 Wby.) $725
All Other Calibers 695

Weatherby Mark V Synthetic Rifle
Similar to the Mark V Magnum, except w/Monte Carlo synthetic stock w/aluminum bedding block. 24- or 26-inch standard tapper or fluted bbl. Weight: 7.75 to 8 lbs. Matte blue finish. Imported 1995 to date.

Mark V Synthetic (340 Wby.) $625
Mark V Synthetic (30-378 Wby.) 650
All Other Calibers 595
W/Flutted Bbl., **add** 100

Weatherby Mark V Ultra Light Weight Rifle
Similar to the Mark V Magnum, except w/skeletonized bolt handle. 24- or 26-inch fluted stainless bbl. chambered 257 Wby., 270 Wby., 7mm Rem. Mag., 7mm Wby., 300 Win. Mag., 300 Wby. Monte Carlo synthetic stock w/aluminum bedding block. Weight: 6.75 lbs. Imported 1998 to date.

Mark V Ultra Light Weight (Standard Calibers) $850
Mark V Ultra Light Weight (Weatherby Calibers). 950

Weatherby Mark XXII Deluxe 22 Automatic
Sporter Clip-fed Model $295
Semiautomatic w/single-shot selector. Caliber: 22 LR. 5- and 10-shot clip magazines. 24-inch bbl. Weight: 6 lbs. Sights: folding leaf open rear; ramp front. Monte Carlo-type stock w/cheekpiece, pistol grip, forend tip, grip cap, skip checkering, QD swivels. Intro. 1964. Made in Italy 1964-69; in Japan, 1970-1981; in the U.S., 1982-90.

Weatherby Mark XXII, Tubular Magazine Model $275
Same as Mark XXII, Clip-fed Model, except w/15-shot tubular magazine. Made in Japan 1973-81; in the U.S.,1882-90.

Weatherby Vanguard (I) Bolt-Action Sporting Rifle
Mauser-type action. Calibers: 243 Win., 25-06, 270 Win., 7mm Rem. Mag., 30-06, 300 Win. Mag. 5-shot magazine; (3-shot in Magnum calibers).24-inch bbl. Weight: 7 lbs. 14 oz. No sights. Monte Carlo-type stock w/cheekpiece, rosewood forend tip and pistol-grip cap, checkering, rubber buttpad, QD swivels. Imported 1970-84.

Vanguard Standard $325
Vanguard VGL (w/shorter 20-inch bbl., plain checkered
   stock matte finish, 6.5 lbs 335
Vanguard VGS (w/24-inch bbl., plain checkered stock,
   matte finish 355
Vanguard VGX (w/higher grade finish) 395

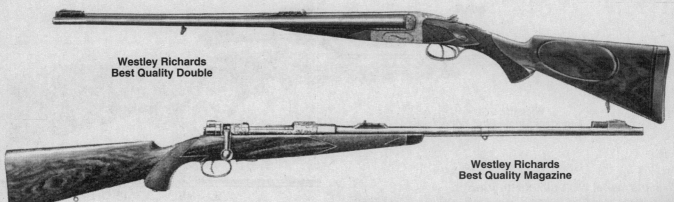

**Westley Richards
Best Quality Double**

**Westley Richards
Best Quality Magazine**

Weatherby Vanguard Classic I Rifle **$365**
Same general specifications as Vanguard VGX Deluxe, except w/hand-checkered classic-style stock, black buttpad and satin finish. Calibers 223 Rem., 243 Win. 270 Win., 7mm-08, 7mm Rem. Mag., 30-06 and 308 Win. Imported 1989-94.

Weatherby Vanguard Classic II Rifle **$525**
Same general specifications as Vanguard VGX Deluxe, except custom checkered classic-style American walnut stock w/black forend tip, grip cap and solid black recoil pad, satin finish. Imported 1989-94.

Weatherby Vanguard VGX Deluxe **$475**
Calibers: 22-250 Rem., 243 Rem., 270 Wby. Mag., 270 Win., 7mm Rem. Mag., 30-06, 300 Win. Mag., 300 Wby. Mag., 338 Win. Mag. 3- or 5-round capacity. 24-inch bbl. About 44 inches overall. Weight: 7 to 8.5 lbs. Custom checkered American walnut stock w/Monte Carlo and recoil pad. Rosewood forend tip and pistol-grip cap. High-luster finish. Disc. 1994.

Weatherby Varmintmaster Bolt-Action Rifle **$895**
Calibers: 224 Wby., 22-250, 4-shot magazine. 26-inch bbl. 45 inches overall. Weight: 7.75 lbs. Checkered walnut stock. No sights. Disc.

Weatherby Weathermark Rifle
Same general specifications as Classicmark, except w/checkered black Weathermark™ composite stock. Mark V bolt action. Calibers: 240, 257, 270, 300, 340, 378, 416 and 460 Weatherby Magnums; plus 270 Win., 7mm Rem. Mag., 30-06 and 375 H&H Mag. Weight: 8 to 10 lbs. Right-hand only. Imported 1992-94.
Calibers 257 to 300 Wby **$495**
Caliber 340 Weatherby **525**
375 H&H **625**
Other non-Wby. calibers **475**

Weatherby Weathermark Alaskan Rifle **$635**
Same general specifications as Weathermark, except w/nonglare electroless nickel finish. Right-hand only. Imported 1992-94.

WEIHRAUCH
Melrichstadt, West Germany

Imported by European American Armory, Sharpes, FL.

Weihrauch Model HW 60 Target Rifle **$485**
Single-shot. Caliber: 22 LR. 26.75-inch bbl. Walnut stock. Adj. buttplate and trigger. Hooded ramp front sight. Push button safety. Imported 1995-97.

Weihrauch Model HW 66 Bolt-Action Rifle **$425**
Caliber: 22 Hornet. 22.75-inch bbl. 41.75 inches overall. Weight: 6.5 lbs. Walnut stock w/cheekpiece. Hooded blade ramp front sight. Checkered pistol grip and forend. Imported 1989-90.

**Weihraugh Model HW 660 Match
Bolt-Action Rifle** **$625**
Caliber: 22 LR. 26-inch bbl. 45.33 inches overall. Weight: 10.75 lbs. Walnut or laminated stock w/adj. cheekpiece and buttplate. Checkered pistol grip and forend. Adj. trigger. Imported 1991 to date.

WESTERN FIELD RIFLES

See listings under "W" for Montgomery Ward.

WESTLEY RICHARDS & CO., LTD.
London, England

Westley Richards Best Quality Double Rifle **$27,500**
Boxlock, hammerless, ejector. Hand-detachable locks. Calibers: 30-06, 318 Accelerated Express, 375 Mag., 425 Mag. Express, 465 Nitro Express, 470 Nitro Express. 25-inch bbls. Weight: 8.5 to 11 lbs. Sights: leaf rear; hooded front. French walnut stock w/cheekpiece, checkered pistol grip and forend.

Westley Richards Best Quality Magazine Rifle
Mauser or Magnum Mauser action. Calibers: 7mm High Velocity, 30-06, 318 Accelerated Express, 375 Mag., 404 Nitro Express, 425 Mag. Bbl. lengths: 24-inch; 7mm, 22-inch; 425 caliber, 25-inch. Weight 7.25 to 9.25 lbs. Sights: leaf rear; hooded front.
Standard Action **$6,950**
Magnum Action **10,250**

WICHITA ARMS
Wichita, Kansas

Wichita Model WCR Classic Bolt-Action Rifle
Single-shot. Calibers: 17 Rem through 308 Win. 21-inch octagon bbl. Hand-checkered walnut stock. Drilled and tapped for scope w/no sights. Right or left-hand action w/Canjar trigger. Non-glare blued finish. Made 1978 to date.
Right-Hand Model **$1650**
Left-Hand Model **1795**

Wichita Model WSR Silhouette Bolt-Action Rifle
Single-shot, bolt action, chambered in most standard calibers. Right or left-hand action w/fluted bolt. Drilled and tapped for scope mount with no sights. 24-inch bbl. Canjar trigger. Metallic gray fiberthane stock w/vented rubber recoil pad. Made 1983-95.

**Wickliffe '76
Standard Single-Shot Rifle**

Wichita Model WSR Silhouette *(Con't)*
Right-Hand Model $1550
Left-Hand Model 1775

**Wichita Model WMR Stainless Magnum
Bolt-Action Rifle** $1295
Single-shot or w/blind magazine action chambered 270 Win.
through 458 Win. Mag. Drilled and tapped for scope with no sights.
Fully adj. trigger. 22- or 24-inch bbl. Hand-checkered select walnut
stock. Made 1980-84.

Wichita Model WVR Varmint Rifle
Calibers: 17 Rem through 308 Win. 3-shot magazine. Right or left-
hand action w/jeweled bolt. 21-inch bbl. w/o sights. Drilled and
tapped for scope. Hand-checkered American walnut pistol-grip
stock. Made 1978 to date.
Right-Hand Model $1595
Left-Hand Model 1825

WICKLIFFE RIFLES
Wickliffe, Ohio
Mfd. by Triple S Development Co., Inc.

Wickliffe '76 Commemorative Model $650
Limited edition of 100. Same as Deluxe Model, except w/filled
etching on receiver sidewalls, U.S. silver dollar inlaid in stock,
26-inch bbl. only, comes in presentation case. Made in 1976 only.

Wickliffe '76 Deluxe Model $395
Same as Standard Model, except w/22-inch bbl. in 30-06 only;
high-luster blued finish, fancy-grade figured American walnut stock
w/nickel silver grip cap.

Wickliffe '76 Standard Model Single-Shot Rifle $335
Falling-block action. Calibers: 22 Hornet, 223 Rem., 22-250, 243
Win., 25-06, 308 Win., 30-06, 45-70. 22-inch lightweight bbl.
(243 and 308 only) or 26-inch heavy sporter bbl. Weight: 6.75 or
8.5 lbs., depending on bbl. No sights. Select American walnut
Monte Carlo stock w/right or left cheekpiece and pistol grip,
semi-beavertail forearm. Intro. 1976. Disc.

Wickliffe Stinger Model. $350
Falling block, single-shot. Calibers: 22 Hornet and 223 Rem. 22-
inch bbl. w/no sights. American walnut Monte Carlo stock
w/continental-type forend. Made 1979-80.

Wickliffe Traditionalist Model. $345
Falling block single-shot. Calibers: 30-06, 45-70. 24-inch bbl.
w/open sights. Hand checkered. American walnut classic-style buttstock and forearm. Made 1979-80.

WILKINSON ARMS CO.
Covina, California

Wilkinson Terry Carbine. $370
Caliber: 9mm Para. Semiautomatic. 30-shot magazine. 16-inch
bbl. 30 inches overall. Weight: 6 lbs. Dovetailed receiver for
scope mounting. Bolt-type safety. Ejection port w/automatic trap
door. Blowback action. Fires from closed bolt. Made 1975 to date.

TED WILLIAMS RIFLES

See **Sears, Roebuck and Company.**

WINCHESTER RIFLES
Winchester Repeating Arms Company
New Haven, Connecticut

> **NOTE**
>
> Most Winchester rifles manufactured prior to 1918 used the
> date of approximate manufacture as the Model number. For
> example, the Model 1894 repeating rifle was manufactured
> from 1894 to 1937. When Winchester started using two-digit
> model numbers after 1918, the "18" was dropped and the rifle
> was then called the Model 94. The Model 1892 was called the
> Model 92, etc.

Early Models 1873 – 1918

Winchester Model 1873 Lever-Action Carbine $6950
Same as Standard Model 1873 Rifle, except w/20-inch bbl., 12-shot
magazine, weighs 7.25 lbs.

Winchester Model 1873 Lever-Action Rifle $6250
Calibers: 32-20, 38-40, 44-40; a few were chambered for 22 rimfire.
15-shot magazine, also made w/6-shot half magazine. 24-inch bbl.
(round, half-octagon, octagon). Weight: 8.5 lbs. Sights: open rear; bead
or blade front. Plain straight-grip stock and forearm. Made 1873-1924.
720,610 rifles of this model were mfd.

**Winchester Model 1873 — One of One
Thousand** . . . . **$50,000+**
During the late 1870s Winchester offered Model 1873 rifles of supe-
rior accuracy and extra finish, designated "One of One Thousand"
grade, at $100. These rifles are marked "1 of 1000" or "One of One
Thousand." Only 136 of this model are known to have been manufactured. This is one of the rarest of shoulder arms and, because so
very few have been sold in recent years, it is extremely difficult to as-
sign a value; however, in the author's opinion, an "excellent" specimen would probably bring a price upward of $100,000.

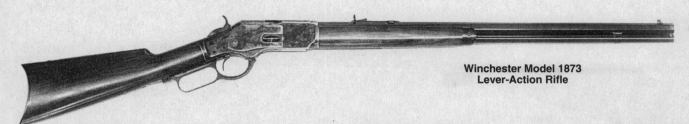

Winchester Model 1873
Lever-Action Rifle

Close-up barrel engraving on Winchesster Model 1873 One of One Thousand.

Alternate barrel inscription designating a Winchester Model 1873 "1 of 1000."

Winchester Model 1873 Special Sporting Rifle. $6850
Same as Standard Model 1873 Rifle, except this type has receiver casehardened in colors, pistol-grip stock of select walnut, octagon bbl. only.

Winchester Model 1885 Single-Shot Rifle

Designed by John M. Browning, this falling-block, lever-action rifle was manufactured from 1885 to 1920 in a variety of models and chambered for most of the popular cartridges of the period — both rimfire and centerfire — from 22 to 50 caliber. There are two basic styles of frames, low-wall and high-wall. The low-wall was chambered only for the lower-powered cartridges, while the high-wall was supplied in all calibers and made in three basic types: the standard model for No. 3 and heavier barrels is the type commonly encountered; the thin-walled version was supplied with No. 1 and No. 2 light barrels and the thick-walled action in the heavier calibers. Made in both solid frame and takedown versions. Barrels were available in five weights ranging from the lightweight No. 1 to the extra heavy No. 5 in round, half-octagon and full-octagon styles. Many other variations were also offered.

Winchester Model 1885 High-Wall
Sporting Rifle $1895
Solid frame or takedown. No. 3, 30-inch bbl., standard. Weight: 9.5 lbs. Standard trigger and lever. Open rear sights; blade front sight. Plain stock and forend.

Winchester Model 1885 Low-Wall
Sporting Rifle $1295
Solid frame. No. 1, 28-inch round or octagon bbl. Weight: 7 lbs. Open rear sight; blade front sight. Plain stock and forend.

RIFLES

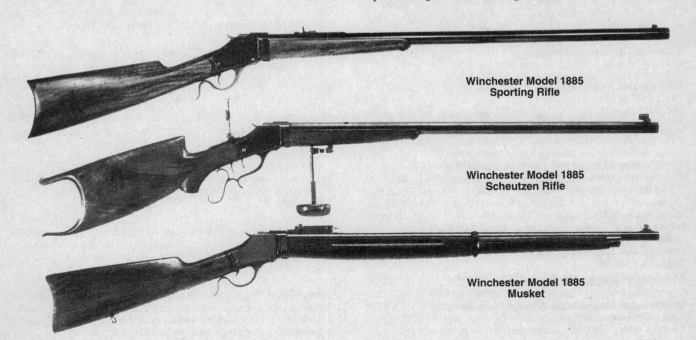

Winchester Model 1885
Sporting Rifle

Winchester Model 1885
Scheutzen Rifle

Winchester Model 1885
Musket

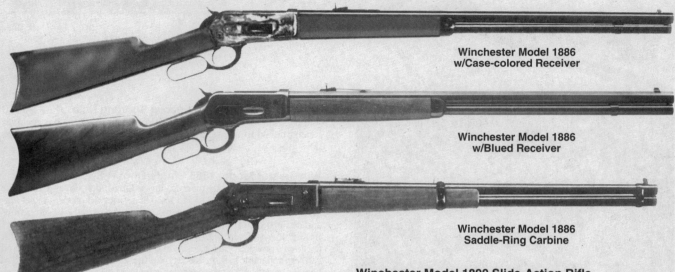

**Winchester Model 1886
w/Case-colored Receiver**

**Winchester Model 1886
w/Blued Receiver**

**Winchester Model 1886
Saddle-Ring Carbine**

Winchester Model 1885 Scheutzen Rifle $4395
Solid frame or takedown. High-wall action. Scheutzen double-set trigger. Spur finger lever. No. 3, 30-inch octagon bbl. Weight: 12 lbs. Vernier rear peep sight; wind-gauge front sight. Fancy walnut Scheutzen stock with checkered pistol-grip and forend. Scheutzen buttplate; adj. palmrest.

Winchester Model 1885 Special Sporting Rifle $1950
Same general specifications as the standard high-wall model except with checkered fancy walnut stock and forend.

Winchester Model 1885 Single-Shot Musket $995
Solid frame. Low-wall. Calibers: 22 Short and Long Rifle. 28-inch round bbl. Weight: 8.6 lbs. Lyman rear peep sight; blade front sight. Military-type stock and forend. *Note:* The U.S. Government purchased a large quantity of these muskets during World War I for training purposes.

Winchester Model 1885 Single-Shot "Winder" Musket $750
Solid frame or takedown. High-wall. Plain trigger. 28-inch round bbl. Weight: 8.5 lbs. Musket rear sight; blade front sight. Military-type stock and forend w/bbl. band and sling stud/rings.

Winchester Model 1886 Lever-Action Rifle
Solid frame or takedown. Calibers: 33 Win., 38-56, 38-70, 40-65, 40-70, 40-82, 45-70, 45-90, 50-100, 50-110. The 33 Win. and 45-70 were the last calibers in which this model was supplied. 8-shot tubular magaine; also 4-shot half-magazine. 26-inch bbl. (round, half-octagon, octagon). Weight: 7.5 lbs. Sights: open rear; bead or blade front. Plain straight-grip stock and forend or standard models. Made 1886-1935.
Standard Model $3895
Takedown Model 4395
Deluxe Model (pistol grip and high-quality walnut) 6995

Winchester Model 1886 Saddle-Ring Carbine $8795
Same as standard rifle, except with 22-inch bbl., carbine buttstock and forend. Carbine rear sight. Saddle ring on left side of receiver.

Winchester Model 1890 Slide-Action Rifle
Visible hammer. Calibers: 22 Short, Long, LR; 22 WRF (not interchangeable). Tubular magazine holds 15 Short, 12 Long, 11 LR; 12 WRF. 24-inch octagon bbl. Weight: 5.75 lbs. Sights: open rear; bead front. Plain straight-grip stock, grooved slide handle. Originally solid frame; after No. 15,499, all rifles of this model were takedown type. Fancy checkered pistol-grip stock, nickel-steel bbl. supplied at extra cost, which can also increase the value by 100% or more. Made 1890-1932.
Blue . . . . $ 695
Blue (22 WRF) 1150
Case-Colored Receiver 4395

Winchester Model 1892 Lever-Action Rifle $1895
Solid frame or takedown. Calibers: 25-20, 32-20, 38-40, 44-40. 13-shot tubular magazine; also 7-shot half-magazine. 24-inch bbl. (round, octagon, half-octagon). Weight: from 6.75 lbs. up. Sights: open rear; bead front. Plain straight-grip stock and forend. Pistol-grip fancy walnut stocks were available at extra cost and also doubles the value of the current value for standard models. *See* Illustration next page.

Winchester Model 1892 Saddle-Ring Carbine $2350
Same general specifications as the Model 1892 rifle except carbine buttstock, forend and sights. 20-inch bbl. Saddle ring on left side of receiver.

Winchester Model 1894 Lever-Action Rifle $1695
Solid frame or takedown. Calibers: 25-35, 30-30, 32-40, 32 Special, 38-55. 7-shot tubular magazine or 4-shot half-magazine. 26-inch bbl. (round, octagon, half-octagon). Weight: about 7.35 lbs. Sights: open rear; bead front. Plain straight-grip stock and forearm on standard model; crescent-shaped or shotgun-style buttplate. Made 1894-1937. *See* also Winchester Model 94 for later variations of this Model.

Winchester Model 1894 Lever-Action Deluxe $2850
Same general specifications as the standard rifle except checkered pistol-grip buttstock and forend using high-grade walnut. Engraved versions are considerably higher in value.

Winchester Model 1894 Saddle-Ring Carbine $1395
Same general specifications as the Model 1894 standard rifle except 20-inch bbl., carbine buttstock, forend, and sights. Saddle ring on left side of receiver. Weighs about 6.5 lbs.

Winchester Model 1890

Winchester Model 1892

Winchester Model 1894 Standard Carbine **$995**
Same general specifications as Saddle-Ringle Carbine except shotgun type buttstock and plate, no saddle ring, standard open rear sight. Sometimes called "Eastern Carbine." *See* also Winchester Model 94 carbine.

Winchester Model 1895 Lever-Action Carbine **$2150**
Same as Model 95 Standard Rifle (below), except has 22-inch bbl., carbine-style buttstock and forend, weighs about 8 lbs., calibers 30-40 Krag, 30-03, 30-06 and 303, solid frame only.

Winchester Model 1895 Lever-Action Rifle **$1550**
Calibers: 30-40 Krag, 30-03, 30-06, 303 British, 7.62mm Russian, 35 Win., 38-72, 40-72, 405 Win. 4-shot box magazine, except 30-40 and 303, which have 5-shot magazines. Bbl. lengths: 24-, 26-, 28-inches (round, half-octagon, octagon). Weight: about 8.5 lbs. Sights: open rear; bead or blade front. Plain straight-grip stock and forend (standard). Both solid frame and takedowns avail. Made 1897-1931.

Winchester Model (1897) Lee Bolt-Action Rifle
Straight-pull bolt action. Caliber: 236 U.S. Navy, 5-shot box magazine, clip loaded. 24- and 28-inch bbl. Weight: 7.5 to 8.5 lbs. Sights: folding leaf rear sight on musket; open sporting sight on sporting rifle.
Musket Model **$1095**
Sporting Rifle 1150

Winchester Model 1900 Bolt-Action Single-Shot Rifle **$425**
Takedown. Caliber: 22 Short and Long. 18-inch bbl. Weight: 2.75 lbs. Open rear sight; blade front sight. One-piece, straight-grip stock. Made from 1899 to 1902.

Winchester Model 1902 Bolt-Action Single-Shot Rifle **$265**
Takedown. Basically the same as Model 1900 with minor improvements. Calibers: 22 Short and Long, 22 Extra Long, 22 LR. Weight: 3 lbs. Made 1902-1931.

Winchester Model 1903 Self-Loading Rifle **$795**
Takedown. Caliber: 22 WRA. 10-shot tubular magazine in buttstock. 20-inch bbl. Weight: 5.75 lbs. Sights: open rear; bead front. Plain straight-grip stock and forearm (fancy grade illustrated). Made 1903-36.

Winchester Model (1904) 99 Thumb-Trigger Bolt-Action Single-Shot Rifle **$595**
Takedown. Same as Model 1902 except fired by pressing a button behind the cocking piece. Made 1904-23.

Winchester Model 1904 Bolt-Action Single-Shot Rifle **$225**
Similar to Model 1902. Takedown. Caliber: 22 Short, Long Extra Long, LR. 21-inch bbl. Weight: 4 lbs. Made 1904-31.

RIFLES

Winchester Model 1894

Winchester Model 1894 Fancy-Grade Takedown

Winchester Model 1894 Saddle-Ring Carbine

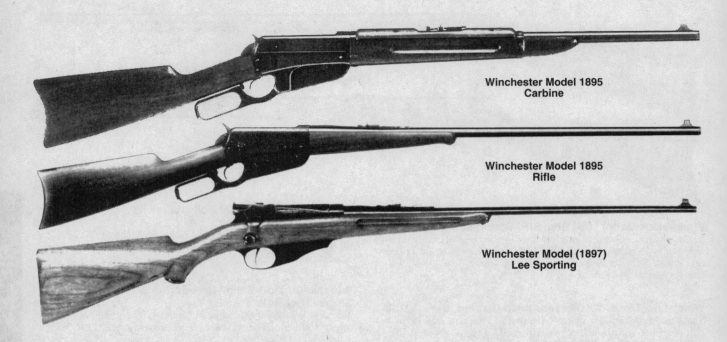

Winchester Model 1895 Carbine

Winchester Model 1895 Rifle

Winchester Model (1897) Lee Sporting

Winchester Model 1905 Self-Loading Rifle **$550**
Takedown. Calibers: 32 Win. S.L., 35 Win. S.L. 5- or 10-shot detachable box magazine. 22-inch bbl. Weight: 7.5 lbs. Sights: open rear; bead front. Plain pistol-grip stock and forearm. Made 1905-20.

Winchester Model 1906 Slide-Action Repeater **$595**
Takedown. Visible hammer. Caliber: 22 Short, Long, LR. Tubular magazine holds 20 Short, 16 Long or 14 LR. 20-inch bbl. Weight: 5 lbs. Sights: open rear; bead front. Straight-grip stock and grooved forearm. Made 1906-32.

Winchester Model 1907 Self-Loading Rifle **$495**
Takedown. Caliber: 351 Win. S.L. 5- or 10-shot detachable box magazine. 20-inch bbl. Weight: 7.75 lbs. Sights: open rear; bead front. Plain pistol-grip stock and forearm. Made 1907-57.

Winchester Model 1910 Self-Loading Rifle **$595**
Takedown. Caliber: 401 Win. S.L. 4-shot detachable box magazine. 20-inch bbl. Weight: 8.5 lbs. Sights: open rear; bead front. Plain pistol-grip stock and forearm. Made 1910-36.

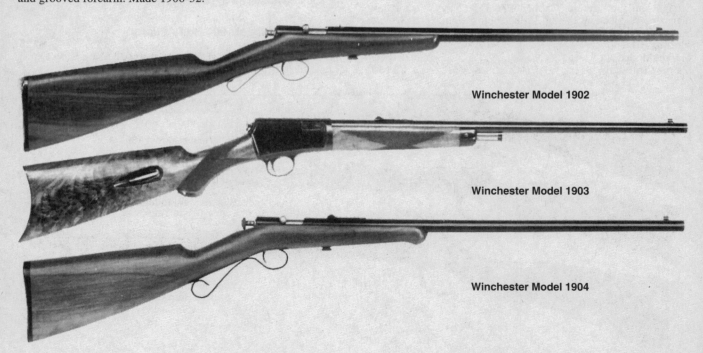

Winchester Model 1902

Winchester Model 1903

Winchester Model 1904

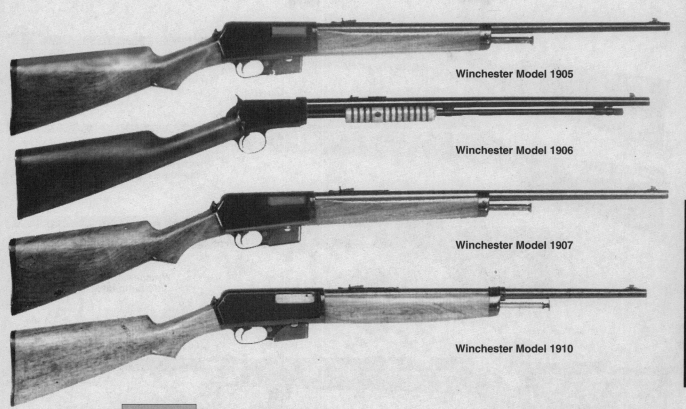

Winchester Model 1905

Winchester Model 1906

Winchester Model 1907

Winchester Model 1910

NOTE

Following WWI, Winchester had financial difficulties and like many other firearm firms of the day, failed. However, Winchester continued to operate in the hands of receivers. Then, in 1931, The Western Cartridge Co. — under the leadership of John Olin — purchased all assets of the firm. After that, Winchester leaped ahead of all other firms of the day in firearm and ammunition development.

The first sporting firearm to come out of the Winchester plant after WWI was the Model 20 shotgun, but this was quickly followed by the famous Model 52 bolt-action rifle. This was also a time when Winchester dropped the four-digit model numbers and began using two-digit numbers instead. This model-numbering procedure, with one exception (Model 677), continued for the next several years.

Winchester Model 43 Bolt-Action Sporting Rifle $595
Standard Grade. Calibers: 218 Bee, 22 Hornet, 25-20, 32-20 (latter two discontinued 1950). 3-shot detachable box magazine. 24-inch bbl. Weight: 6 lbs. Sights: open rear, bead front on hooded ramp. Plain pistol-grip stock with swivels. Made 1949-57.

Winchester Model 43 Special Grade $650
Same as Standard Model 43, except has checkered pistol grip and forearm, grip cap.

Winchester Model 47 Bolt-Action Single-Shot Rifle $275
Caliber: 22 Short, Long, LR. 25-inch bbl. Weight: 5.5 lbs. Sights: peep or open rear; bead front. Plain pistol-grip stock. Made 1949-54.

Winchester Model 52 Bolt-Action Target Rifle
Standard bbl. First type. Caliber: 22 LR. 5-shot box magazine. 28-inch bbl. Weight: 8.75 lbs. Sights: folding leaf peep rear; blade front sight; standard sights various other combinations available. Scope bases. Semi-military-type target stock w/pistol grip; original model has grasping grooves in forearm; higher comb and semibeavertail forearm on later models. Numerous changes were made in this model, the most important was the adoption of the speed lock in 1929; Model 52 rifles produced before this change are generally referred to as "slow lock" models. Last arms of this type bore serial numbers followed by the letter "A." Made 1919-37.
Slow Lock Model $425
Speed Lock Model 575

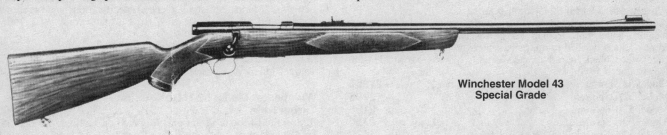

**Winchester Model 43
Special Grade**

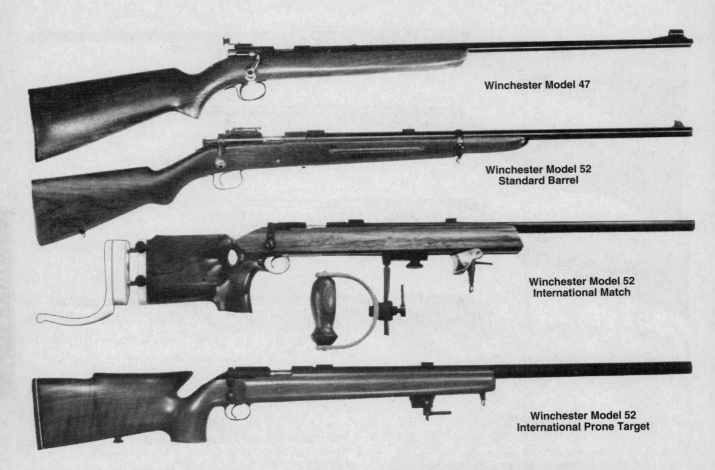

Winchester Model 47

Winchester Model 52 Standard Barrel

Winchester Model 52 International Match

Winchester Model 52 International Prone Target

Winchester Model 52 Heavy Barrel $625
First type. Speed lock. Same general specifications as Standard Model 52 of this type, except has heavier bbl., Lyman 17G front sight, weighs 10 lbs.

Winchester Model 52 International Match Rifle
Similar to Model 52-D Heavy Barrel, except has special lead-lapped bbl., laminated "free rifle" style stock with high comb, thumbhole, hook buttplate, accessory rail, handstop/swivel assembly, palmrest. Weight: 13.5 lbs. Made 1969-78.
With standard trigger $495
With Kenyon or I.S.U. trigger 550

Winchester Model 52 International Prone $450
Similar to Model 52-D Heavy Barrel, except has special lead-lapped bbl., prone stock with fuller pistol grip rollover cheekpiece removable for bore-cleaning. Weight 11.5 lbs. Made 1975-80.

Winchester Model 52 Sporting Rifle
First type. Same as Standard Model 52 of this type, except has light-weight 24-inch bbl., Lyman No. 48 receiver sight and gold bead front sight on hooded ramp, deluxe checkered sporting stock with cheekpiece, black forend tip, etc. Weight: 7.75 lbs. Made 1934-58. Reintroduced 1993.
Model 52 Sporter $1950
Model 52A Sporter 2795
Model 52B Sporter 2195
Model 52C Sporter 2750
Model 52 C Sporter (1993 BAC reissue) 475

Winchester Model 52-B Bolt-Action Rifle
Standard bbl. Extensively redesigned action. Supplied with choice of "Target" stock, an improved version of the previous Model 52 stock, or "Marksman" stock with high comb, full pistol grip and beavertail forearm. Weight: 9 lbs. Offered with a wide choice of target sight combinations (Lyman, Marble-Goss, Redfield, Vaver, Winchester), value shown is for rifle less sight equipment. Other specifications as shown for first type. Made 1935-47. Reintroduced 1997. *See* Illustration next page.
Target Model $595
BAC Model (1997 BAC reissue) 495
USRAC Sporting Model 450

Winchester Model 52-B Bull Gun/Heavy Barrel $675
Same specifications as Standard Model 52-B, except Bull Gun has extra heavy bbl., Marksman stock only, weighs 12 lbs. Heavy Bbl. model weighs 11 lbs. Made 1940-47.

Winchester Model 52-C Bolt-Action Rifle
Improved action with "Micro-Motion" trigger mechanism and new-type "Marksman" stock. General specifications same as shown for previous models. Made 1947-61, Bull Gun from 1952. Value shown is for rifle less sights.
Bull Gun (Extra Heavy Barrel, wt. 12 lbs.) $650
Standard Barrel (Wt. 9.75 lbs.) 575
Target Model (Heavy Barrel) 595

Winchester Model 52-D Bolt-Action
Target Rifle $495
Redesigned Model 52 action, Single-Shot. Caliber: 22 LR. 28-inch standard or heavy bbl., free-floating, with blocks for stan-

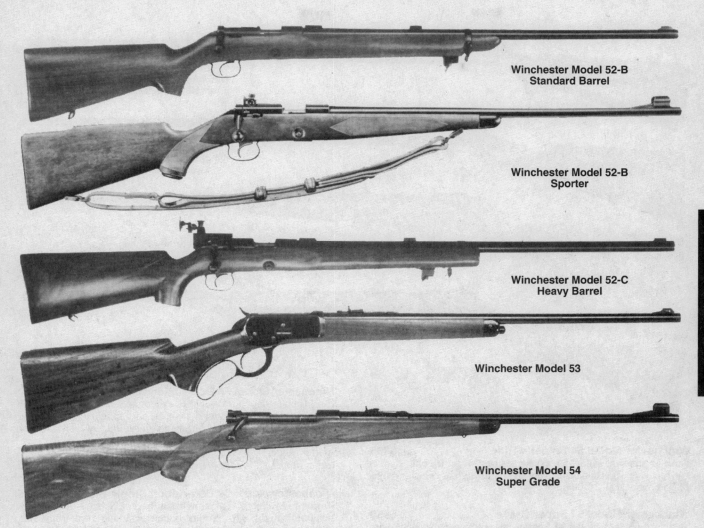

Winchester Model 52-B
Standard Barrel

Winchester Model 52-B
Sporter

Winchester Model 52-C
Heavy Barrel

Winchester Model 53

Winchester Model 54
Super Grade

Model 52-D Bolt-Action *(Con't)*

dard target scopes. Weight: with standard bbl., 9.75 lbs., with heavy barrel, 11 lbs. Restyled Marksman stock with accessory channel and forend stop, rubber buttplate. Made 1961-78. Value shown is for rifle without sights.

Winchester Model 53 Lever-Action Repeater $1695

Modification of Model 92. Solid frame or takedown. Calibers: 25-20, 32-20, 44-40. 6-shot tubular half-magazine in solid frame model. 7-shot in takedown. 22-inch nickel steel bbl. Weight: 5.5 to 6.5 lbs. Sights: open rear; bead front. Redesigned straight-grip stock and forearm. Made 1924-32.

Winchester Model 54 Bolt-Action
High Power Sporting Rifle (I) $625

First type. Calibers: 270 Win., 7×57mm, 30-30, 30-06, 7.65×53mm, 9×57mm. 5-shot box magazine. 24-inch bbl. Weight: 7.75 lbs. Sights: open rear; bead front. Checkered stock w/pistol grip, tapered forearm w/schnabel tip. This type has two-piece firing pin. Made 1925-30.

Winchester Model 54 Bolt-Action
High Power Sporting Rifle (II) $675

Standard Grade. Improved type with speed lock and one-piece firing pin. Calibers: 22 Hornet, 220 Swift, 250/3000, 257 Roberts, 270 Win., 7×57mm, 30-06. 5-shot box magazine. 24-inch bbl., 26-inch in cal. 220

Winchester Model 54 Bolt-Action *(Con't)*

Swift. Weight: about 8 lbs. Sights: open rear, bead front on ramp. NRA-type stock w/checkered pistol grip and forearm. Made 1930-36. **Add** $200 for 22 Hornet caliber.

Winchester Model 54 Carbine (I) $695

First type. Same as Model 54 rifle, except has 20-inch bbl., plain lightweight stock with grasping grooves in forearm. Weight: 7.25 lbs.

Winchester Model 54 Carbine (II). $795

Improved type. Same as Model 54 Standard Grade Sporting Rifle of this type, except has 20-inch bbl. Weight: about 7.5 lbs. This model may have either NRA-type stock or the lightweight stock found on the first-type Model 54 Carbine.

Winchester Model 54 National Match Rifle $825

Same as Standard Model 54, except has Lyman sights, scope bases, Marksman-type target stock, weighs 9.5 lbs. Same calibers as Standard Model.

Winchester Model 54 Sniper's Match Rifle $995

Similar to the earlier Model 54 Sniper's Rifle, except has Marksman-type target stock, scope bases, weighs 12.5 lbs. Available in same calibers as Model 54 Standard Grade.

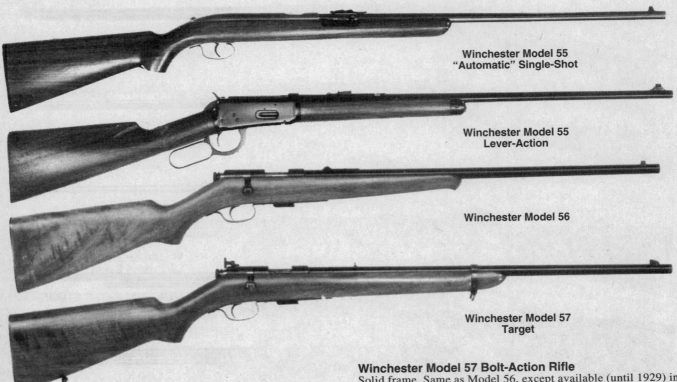

Winchester Model 55 "Automatic" Single-Shot

Winchester Model 55 Lever-Action

Winchester Model 56

Winchester Model 57 Target

Winchester Model 54 Sniper's Rifle **$895**
Same as Standard Model 54, except has heavy 26-inch bbl., Lyman #48 rear peep sight and blade front sight semi-military stock, weighs 11.75 pounds, cal. 30-06 only.

Winchester Model 54 Super Grade **$950**
Same as Standard Model 54 Sporter, except has deluxe stock with cheekpiece, black forend tip, pistol-grip cap, quick detachable swivels, 1-inch sling strap. *See* Illustration previous page.

Winchester Model 54 Target Rifle **$825**
Same as Standard Model 54, except has 24-inch medium-weight bbl. (26-inch in cal. 220 Swift), Lyman sights, scope bases, Marksman-type target stock, weighs 10.5 lbs., same calibers as Standard Model.

Winchester Model 55 "Automatic" Single-Shot . . . **$225**
Caliber: 22 Short, Long, LR. 22-inch bbl. Sights: open rear, bead front. One-piece walnut stock. Weight: about 5.5 lbs. Made 1958-60.

Winchester Model 55 Lever-Action Repeater
Modification of Model 94. Solid frame or takedown. Calibers: 25-35, 30-30, 32 Win. Special. 3-shot tubular half magazine. 24-inch nickel steel bbl. Weight: about 7 lbs. Sights: open rear; bead front. Made 1924-32.
Standard Model (Straight Grip) . . . . **$ 950**
DeluxeModel (Pistol Grip) **2495**

Winchester Model 56 Bolt-Action Sporting Rifle **$795**
Solid frame. Caliber: 22 LR, 22 Short. 5- or 10-shot detachable box magazine. 22-inch bbl. Weight: 4.75 lbs. Sights: open rear; bead front. Plain pistol-grip with schnabel forend. Made 1926-29.

Winchester Model 57 Bolt-Action Rifle
Solid frame. Same as Model 56, except available (until 1929) in 22 Short as well as LR with 5- or 10-shot magazine. Has semi-military style target stock, bbl. band on forend, swivels and web sling, Lyman peep rear sight, weighs 5 lbs. Mfd. 1926-36.
Sporter Model **$550**
Target Model **495**

Winchester Model 58 Bolt-Action Single-Shot **$325**
Similar to Model 02. Takedown. Caliber. 22 Short, Long LR. 18-inch bbl. Weight: 3 lbs. Sights, open rear; blade front. Plain, flat, straight-grip hardwood stock. Not serial numbered. Made 1928-31.

Winchester Model 59 Bolt-Action Single-Shot **$425**
Improved version of Model 58, has 23-inch bbl., redesigned stock w/pistol grip, weighs 4.5 lbs. Made 1930.

Winchester Model 60, 60A Bolt-Action Single-Shot
Redesign of Model 59. Caliber: 22 Short, Long, LR. 23-inch bbl. (27-inch after 1933). Weight: 4.25 lbs. Sights: open rear, blade front. Plain pistol grip stock. Made 1930-34 (60), 1932-39 (60A). *See* Illustration next page.
Model 60 . . . . **$225**
Model 60A **265**

Winchester Model 60A Target Rifle **$375**
Essentially the same as Model 60, except has Lyman peep rear sight and square top front sight, semi-military target stock and web sling, weighs 5.5 lbs. Made 1932-39.

Winchester Model 61 Hammerless Slide-Action Repeater
Takedown. Caliber: 22 Short, Long, LR. Tubular magazine holds 20 Short, 16 Long, 14 LR. 24-inch round bbl. Weight: 5.5 lbs. Sights: open rear; bead front. Plain pistol-grip stock, grooved semibeavertail slide handle. Also available with 24-inch full-

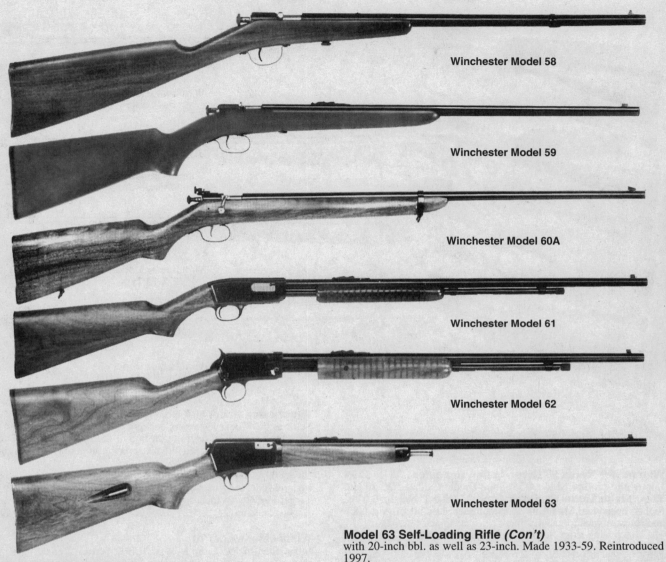

Winchester Model 58

Winchester Model 59

Winchester Model 60A

Winchester Model 61

Winchester Model 62

Winchester Model 63

Model 61 Hammerless *(Con't)*
octagon bbl. and only calibers 22 Short, 22 LR or 22 WRF. *Note*: Octagon barrel model discontinued 1943-44; assembled 1948.

Model 61 (round barrel) **$595**
Model 61 (grooved receiver) **750**
Model 61 (octagon barrel) **1095**

Winchester Model 61 Magnum **$625**
Same as Standard Model 61, except chambered for 22 WMR; magazine holds 12 rounds. Made 1960-63.

Winchester Model 62 Visible Hammer **$495**
Modernized version of Model 1890. Caliber: 22 Short, Long, LR. 23-inch bbl. Weight: 5.5 lbs. Plain straight-grip stock, grooved semibeavertail slide handle. Also available in Gallery Model chambered for 22 Short only. Made 1932-1959. *Note:* Pre-WWII model (small forearm) commands 25% higher price.

Winchester Model 63 Self-Loading Rifle
Takedown. Caliber: 22 LR High Speed only. 10-shot tubular magazine in buttstock. 23-inch bbl. Weight: 5.5 lbs. Sights: open rear; bead front. Plain pistol-grip stock and forearm. Originally available

Model 63 Self-Loading Rifle *(Con't)*
with 20-inch bbl. as well as 23-inch. Made 1933-59. Reintroduced 1997.

Model 66 . . . . **$695**
Model 66 Grade I (1997 BAC reissue) **565**
Model 66 High Grade (1997 BAC reissue) **850**

Winchester Model 64 Deluxe Deer Rifle **$1095**
Same as Standard Model 64, calibers 30-30 and 32 Win. Special, except has checkered pistol grip and semibeavertail forearm, swivels and sling, weighs 7.75 lbs. Made 1933-56.

Winchester Model 64 Lever-Action Repeater
Standard Grade. Improved version of Models 94 and 55. Solid frame. Calibers: 25-35, 30-30, 32 Win. Special. 5-shot tubular two-thirds magazine. 20- or 24-inch bbl. Weight: about 7 lbs. Sights: open rear; bead front on ramp w/sight cover. Plain pistol-grip stock and forearm. Made 1933-56. Production resumed in 1972 (caliber 30-30, 24-inch bbl.). Discontinued 1974.

Original model **$595**
1972-74 model **450**

Winchester Model 64 — 219 Zipper **$1695**
Same as Standard Grade Model 64, except has 26-inch bbl., peep rear sight. Made 1937-47.

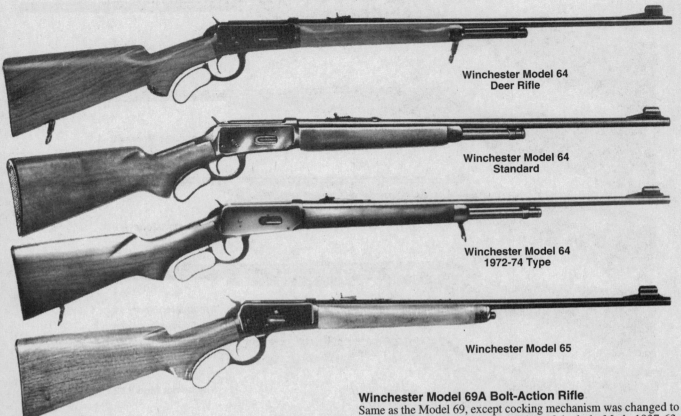

Winchester Model 64 Deer Rifle

Winchester Model 64 Standard

Winchester Model 64 1972-74 Type

Winchester Model 65

Winchester Model 65 Lever-Action Repeater $195
Improved version of Model 53. Solid frame. Calibers: 25-20 and 32-20. Six-shot tubular half-magazine. 22-inch bbl. Weight: 6.5 lbs. Sights: open rear, bead front on ramp base. Plain pistol-grip stock and forearm. made 1933-47.

Winchester Model 65 — 218 Bee $2150
Same as Standard Model 65, except has 24-inch bbl., peep rear sight. Made 1938-47.

Winchester Model 67 Bolt-Action Single-Shot Rifle $150
Takedown. Calibers: 22 Short, Long, LR, 22 LR shot (smoothbore), 22 WRF. 27-inch bbl. Weight: 5 lbs. Sights: open rear, bead front. Plain pistol-grip stock (original model had grasping grooves in forearm). Made 1934-63.

Winchester Model 67 Boy's Rifle $175
Same as Standard Model 67, except has shorter stock, 20-inch bbl., weighs 4.25 lbs.

Winchester Model 68 Bolt-Action Single-Shot $195
Same as Model 67, except has rear peep sight. Made 1934-1946.

Winchester Model 69 Bolt-Action Rifle $275
Takedown. Caliber: 22 S, L, LR. 5- or 10-shot box magazine. 25-inch bbl. Weight: 5.5 lbs. Peep or open rear sight. Plain pistol-grip stock. Rifle cocks on closing motion of the bolt. Made 1935-37.

Winchester Model 69A Bolt-Action Rifle
Same as the Model 69, except cocking mechanism was changed to cock the rifle by the opening motion of the bolt. Made 1937-63. *Note:* Models with grooved receivers command 20% higher prices.
Model 69A Standard **$325**
Match Model w/Lyman #57EW receiver sight **395**
Target Model w/Winchester peep rear sight,
    swivels, sling **475**

Winchester Model 70
Introduced in 1937, the Model 70 Bolt-Action Repeating Rifle was offered in several styles and calibers. Only minor design changes were made over a period of 27 years, and more than one-half million of these rifles were sold. The original model was dubbed "The Rifleman's Rifle." In 1964, the original Model 70 was superseded by a revised version with redesigned action, improved bolt, swaged (free-floating) barrel, restyled stock. This model again underwent major changes in 1972 — most visible: new stock with contrasting forend tip and grip cap, cut checkering (instead of impressed as in predecessor) knurled bolt handle. The action was machined from a solid block of steel with barrels made from chrome molybdenum steel. Other changes in the design and style of the Model 70 continued. The XTR models were added in 1978 along with the Model 70A, the latter omitting the white liners, forend caps and floor plates. In 1981, an XTR Featherweight Model was added to the line, beginning with serial number G1,440,000. This version featured lighter barrels, fancy-checkered stocks with schnabel forend. After U.S. Repeating Arms took over the Winchester plant, the Model 70 went through even more changes as described under that section of Winchester rifles.

PRE-1964 MODEL 70

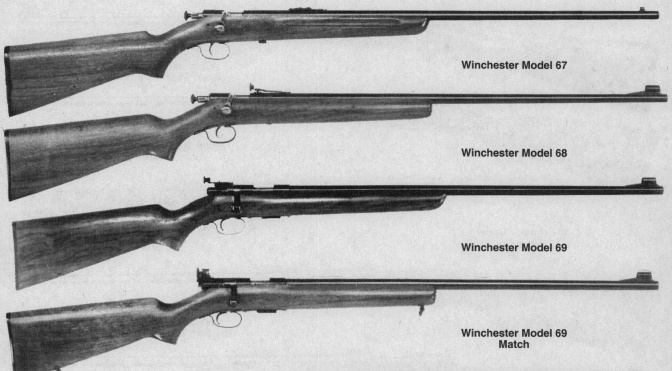

Winchester Model 67

Winchester Model 68

Winchester Model 69

Winchester Model 69 Match

RIFLES

Winchester Model 70 African Rifle $3595
Same general specifications as Super Grade Model 70, except w/25-inch bbl., 3-shot magazine, Monte Carlo stock w/recoil pad. Weight: 9.5 lbs. Caliber: 458 Winchester Magnum. Made 1956–63.

Winchester Model 70 Alaskan
Same as Standard Model 70, except calibers 338 Win. Mag., 375 H&H Mag.; 3-shot magazine in 338, 4-shot in 375 caliber; 25-inch bbl.; stock w/recoil pad. Weight: 8 lbs. in 338; 8.75 lbs. in 375 caliber. Made 1960–63.

338 Win Mag......................................	**$1495**
375 H&H	**1795**

Winchester Model 70 Bull Gun................. $1995
Same as Standard Model 70, except w/heavy 28-inch bbl., scope bases, Marksman stock, weighs 13.25 lbs., caliber 300 H&H Magnum and 30-06 only. Disc. 1963.

Winchester Model 70 Featherweight Sporter
Same as Standard Model 70, except w/redesigned stock and 22-inch bbl., aluminum trigger guard, floorplate and buttplate. Calibers: 243 Win., 264 Win. Mag., 270 Win., 308 Win., 30-06, 358 Win. Weight: 6.5 lbs. Made 1952–63.

243 Win...	**$795**
264 Win...	**1195**
270 Win...	**995**
30.06 Springfield	**750**
308 Win...	**695**
358 Win...	**1695**

Winchester Model 70 National Match Rifle $1250
Same as Standard Model 70, except w/scope bases, Marksman-type target stock, weighs 9.5 lbs. caliber 30-06 only. Disc. 1960.

Winchester Model 70 Standard Grade
Calibers: 22 Hornet, 220 Swift, 243 Win., 250-3000, 257 Roberts, 270 Win., 7×57mm, 30-06, 308 Win., 300 H&H Mag., 375 H&H Mag. 5-shot box magazine (4-shot in Magnum calibers). 24-inch bbl. standard; 26-inch in 220 Swift and 300 Mag.; 25-inch in 375 Mag.; at one time a 20-inch bbl. was available. Sights: open rear; hooded ramp front. Checkered walnut stock; Monte Carlo comb standard on later production. Weight: from 7.75 lbs. depending on caliber and bbl. length. Made 1937–63.

22 Hornet (1937-58)...............................	**$1550**
220 Swift (1937-63)	**1095**
243 Win. (1955-63)	**895**
250-3000 Sav. (1937-49)	**Rare**
257 Roberts (1937-59)	**1295**
264 Win. Mag. (1959-63) limited....................	**950**
270 Win. (1937-63)	**795**
7×57mm Mauser (1937-49)	**Rare**
7.65 Argentine (1937 only) limited	**Very Rare**
30.06 Springfield (1937-63).........................	**650**
308 Win. (1952-63) special order....................	**Very Rare**
300 H&H (1937-63).................................	**1150**
300 Sav. (1944-50) limited	**Rare**
300 Win. Mag. (1962-63)	**1495**
338 Win. Mag. (1959-63) special order only	**1425**
35 Rem. (1941-47) limited	**Very Rare**
358 Win. (1955-58).................................	**Rare**
375 H&H (1937-63)	**1795**
458 Win. Mag. (1956-63) Super Grade only............	**3595**
9×57mm Mauser (1937 only) limited	**Very Rare**

Winchester Model 70 Super Grade
Same as Standard Grade Model 70, except w/deluxe stock w/cheekpiece, black forend tip, pistol-grip cap, quick detachable swivels, sling. Disc. 1960. Prices for Super Grade models also reflect rarity in both production and caliber. Values are generally twice that of standard models of similar configuration.

**Winchester Model 70
Basic Post WWII Model**

**Winchester Model 70
Standard Model**

**Winchester Model 70
Super Grade**

**Winchester Model 70
Standard Weight Target Rifle**

**Winchester Model 70
Heavy Weight Target Rifle**

**Winchester Model 70
Bull Gun**

**Winchester Model 70
(Pre-1964) Standard Model**

**Winchester Model 70
African (1964)**

**Winchester Model 70
Deluxe (1964)**

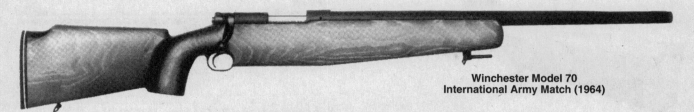

**Winchester Model 70
International Army Match (1964)**

Winchester Model 70 Super Grade Featherweight
Same as Standard Grade Featherweight, except w/deluxe stock w/cheekpiece, black forend tip, pistol-grip cap, quick detachable swivels, sling. Disc. 1960. *Note:* SG-FWs are very rare, but unless properly documented, will not command premium prices. Prices for authenticated Super Grades Featherweight Models are generally 4 to 5 times that of a standard production Featherweight Model w/similar chambering.

Winchester Model 70 Target Rifle
Same as Standard Model 70, except w/24-inch medium-weight bbl., scope bases, Marksman stock, weight 10.5 lbs. Originally offered in all of the Model 70 calibers, this rifle was available later in calibers 243 Win. and 30-06. Disc. 1963. Values are generally twice that of standard models of similar configuration.

Winchester Model 70 Target Heavy Weight $1695
Same general specifications as Standard Model 70, except w/either 24- or 26-inch heavy weight bbl. and weighs 10.5 lbs. No checkering. 243 and 30-06 calibers.

Winchester Model 70 Target Bull Barrel $2150
Same general specifications as Standard Model 70 except 28-inch heavy weight bbl. and chambered for either 30-06 or 300 H&H Mag. Drilled and tapped for front sight base. Receiver slotted for clip loading. Weight: 13.25 lbs.

Winchester Model 70 Varmint Rifle $925
Same general specifications as Standard Model 70, except w/26-inch heavy bbl., scope bases, special varminter stock. Calibers: 220 Swift, 243 Win. Made 1956–63.

Winchester Model 70 Westerner $975
Same as Standard Model 70, except calibers 264 Win. Mag., 300 Win. Mag.; 3-shot magazine; 26-inch bbl. in former caliber, 24-inch in latter. Weight: 8.25 lbs. Made 1960–63.

1964-TYPE MODEL 70

Winchester Model 70 African $675
Caliber: 458 Win. Mag. 3-shot magazine. 22-inch bbl. Weight: 8.5 lbs. Special "African" sights. Monte Carlo stock w/ebony forend tip, hand-checkering, twin stock-reinforcing bolts, recoil pad, QD swivels. Made 1964–71.

Winchester Model 70 Deluxe $595
Calibers: 243, 270 Win., 30-06, 300 Win. Mag. 5-shot box magazine (3-shot in Magnum). 22-inch bbl. (24-inch in Magnum). Weight: 7.5 lbs. Sights: open rear; hooded ramp front. Monte Carlo stock w/ebony forend tip, hand-checkering, QD swivels, recoil pad on Magnum. Made 1964–71.

RIFLES

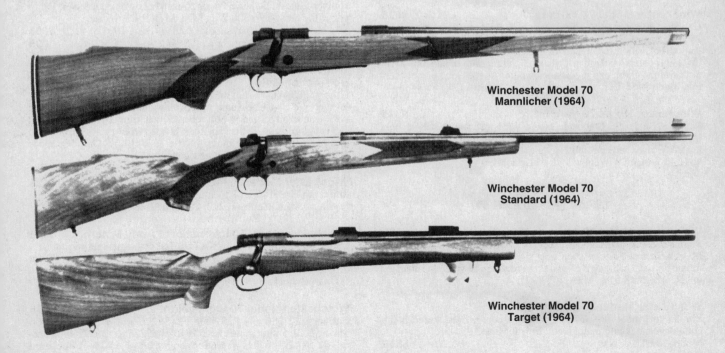

**Winchester Model 70
Mannlicher (1964)**

**Winchester Model 70
Standard (1964)**

**Winchester Model 70
Target (1964)**

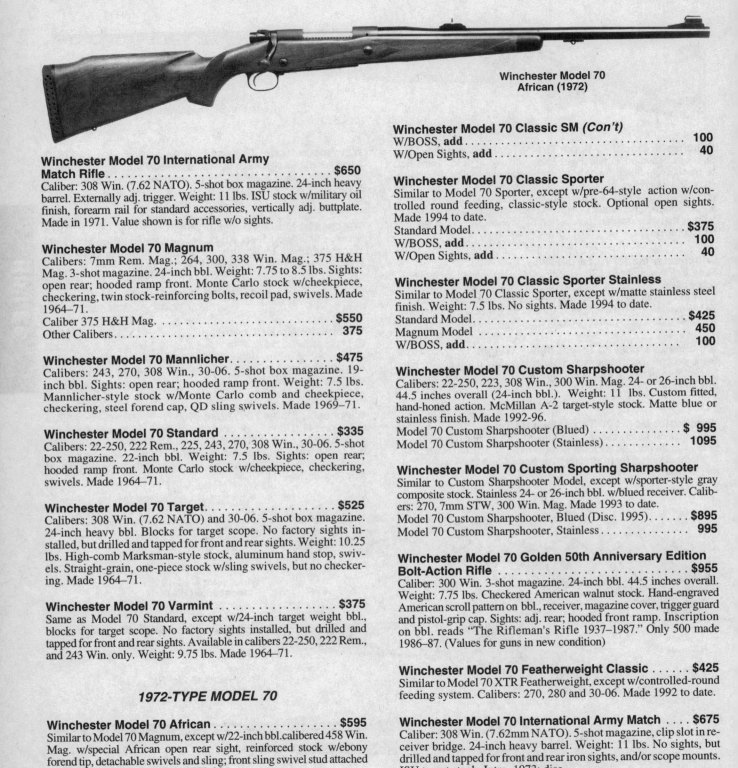

**Winchester Model 70
African (1972)**

Winchester Model 70 International Army
Match Rifle **$650**
Caliber: 308 Win. (7.62 NATO). 5-shot box magazine. 24-inch heavy barrel. Externally adj. trigger. Weight: 11 lbs. ISU stock w/military oil finish, forearm rail for standard accessories, vertically adj. buttplate. Made in 1971. Value shown is for rifle w/o sights.

Winchester Model 70 Magnum
Calibers: 7mm Rem. Mag.; 264, 300, 338 Win. Mag.; 375 H&H Mag. 3-shot magazine. 24-inch bbl. Weight: 7.75 to 8.5 lbs. Sights: open rear; hooded ramp front. Monte Carlo stock w/cheekpiece, checkering, twin stock-reinforcing bolts, recoil pad, swivels. Made 1964–71.
Caliber 375 H&H Mag. **$550**
Other Calibers **375**

Winchester Model 70 Mannlicher **$475**
Calibers: 243, 270, 308 Win., 30-06. 5-shot box magazine. 19-inch bbl. Sights: open rear; hooded ramp front. Weight: 7.5 lbs. Mannlicher-style stock w/Monte Carlo comb and cheekpiece, checkering, steel forend cap, QD sling swivels. Made 1969–71.

Winchester Model 70 Standard **$335**
Calibers: 22-250, 222 Rem., 225, 243, 270, 308 Win., 30-06. 5-shot box magazine. 22-inch bbl. Weight: 7.5 lbs. Sights: open rear; hooded ramp front. Monte Carlo stock w/cheekpiece, checkering, swivels. Made 1964–71.

Winchester Model 70 Target **$525**
Calibers: 308 Win. (7.62 NATO) and 30-06. 5-shot box magazine. 24-inch heavy bbl. Blocks for target scope. No factory sights installed, but drilled and tapped for front and rear sights. Weight: 10.25 lbs. High-comb Marksman-style stock, aluminum hand stop, swivels. Straight-grain, one-piece stock w/sling swivels, but no checkering. Made 1964–71.

Winchester Model 70 Varmint **$375**
Same as Model 70 Standard, except w/24-inch target weight bbl., blocks for target scope. No factory sights installed, but drilled and tapped for front and rear sights. Available in calibers 22-250, 222 Rem., and 243 Win. only. Weight: 9.75 lbs. Made 1964–71.

1972-TYPE MODEL 70

Winchester Model 70 African **$595**
Similar to Model 70 Magnum, except w/22-inch bbl. calibered 458 Win. Mag. w/special African open rear sight, reinforced stock w/ebony forend tip, detachable swivels and sling; front sling swivel stud attached to bbl. Weight: 8.5 lbs. Made 1972-92.

Winchester Model 70 Classic SM
Similar to Model 70 Classic Sporter, except w/ checkered black composite stock and matte metal finish. Made 1994-96.
Model 70 Classic SM **$350**
Caliber 375 H&H **425**

Winchester Model 70 Classic SM *(Con't)*
W/BOSS, **add** **100**
W/Open Sights, **add** **40**

Winchester Model 70 Classic Sporter
Similar to Model 70 Sporter, except w/pre-64-style action w/controlled round feeding, classic-style stock. Optional open sights. Made 1994 to date.
Standard Model. **$375**
W/BOSS, **add** **100**
W/Open Sights, **add** **40**

Winchester Model 70 Classic Sporter Stainless
Similar to Model 70 Classic Sporter, except w/matte stainless steel finish. Weight: 7.5 lbs. No sights. Made 1994 to date.
Standard Model. **$425**
Magnum Model **450**
W/BOSS, **add** **100**

Winchester Model 70 Custom Sharpshooter
Calibers: 22-250, 223, 308 Win., 300 Win. Mag. 24- or 26-inch bbl. 44.5 inches overall (24-inch bbl.). Weight: 11 lbs. Custom fitted, hand-honed action. McMillan A-2 target-style stock. Matte blue or stainless finish. Made 1992-96.
Model 70 Custom Sharpshooter (Blued) . . . . **$ 995**
Model 70 Custom Sharpshooter (Stainless) **1095**

Winchester Model 70 Custom Sporting Sharpshooter
Similar to Custom Sharpshooter Model, except w/sporter-style gray composite stock. Stainless 24- or 26-inch bbl. w/blued receiver. Calibers: 270, 7mm STW, 300 Win. Mag. Made 1993 to date.
Model 70 Custom Sharpshooter, Blued (Disc. 1995). **$895**
Model 70 Custom Sharpshooter, Stainless **995**

Winchester Model 70 Golden 50th Anniversary Edition
Bolt-Action Rifle **$955**
Caliber: 300 Win. 3-shot magazine. 24-inch bbl. 44.5 inches overall. Weight: 7.75 lbs. Checkered American walnut stock. Hand-engraved American scroll pattern on bbl., receiver, magazine cover, trigger guard and pistol-grip cap. Sights: adj. rear; hooded front ramp. Inscription on bbl. reads "The Rifleman's Rifle 1937–1987." Only 500 made 1986–87. (Values for guns in new condition)

Winchester Model 70 Featherweight Classic **$425**
Similar to Model 70 XTR Featherweight, except w/controlled-round feeding system. Calibers: 270, 280 and 30-06. Made 1992 to date.

Winchester Model 70 International Army Match **$675**
Caliber: 308 Win. (7.62mm NATO). 5-shot magazine, clip slot in receiver bridge. 24-inch heavy barrel. Weight: 11 lbs. No sights, but drilled and tapped for front and rear iron sights, and/or scope mounts. ISU target stock. Intro. 1973; disc.

Winchester Model 70 Lightweight **$345**
Calibers: 22-250 and 223 Rem.; 243, 270 and 308 Win.; 30-06 Springfield. 5-shot mag. capacity (6-shot 223 Rem.). 22-inch barrel. 42 to 42.5 inches overall. Weight: 6 to 6.25 lbs. Checkered classic straight stock. Sling swivel studs. Made 1986-95.

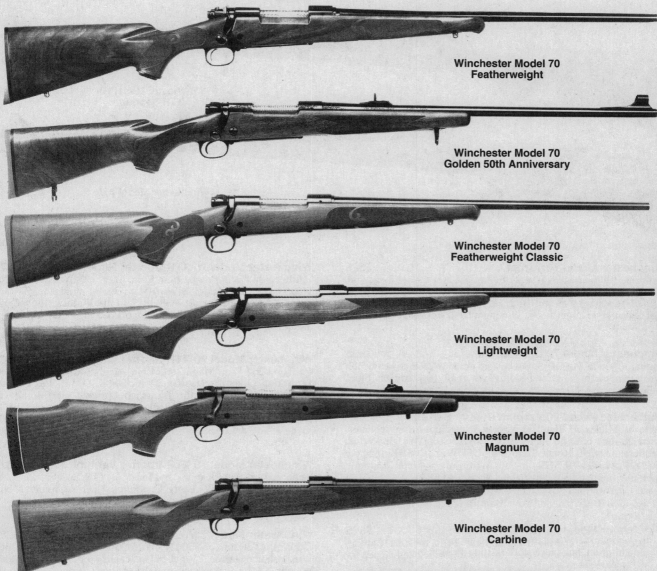

Winchester Model 70
Featherweight

Winchester Model 70
Golden 50th Anniversary

Winchester Model 70
Featherweight Classic

Winchester Model 70
Lightweight

Winchester Model 70
Magnum

Winchester Model 70
Carbine

RIFLES

Winchester Model 70 Magnum

Same as Model 70, except w/3-shot magazine, 24-inch bbl., reinforced stock w/recoil pad. Weight: 7.75 lbs. (except 8.5 lbs. in 375 H&H Mag.). Calibers: 264 Win. Mag., 7mm Rem. Mag., 300 Win. Mag., 338 Win. Mag., 375 H&H Mag. Made 1972-80.
375 H&H Magnum...................................**$445**
Other Magnum Calibers **395**

Winchester Model 70 Standard $325

Same as Model 70A, except w/5-shot magazine, Monte Carlo stock w/cheekpiece, black forend tip and pistol-grip cap w/white spacers, checkered pistol grip and forearm, detachable sling swivels. Same calibers plus 225 Win. Made 1972-80.

Winchester Model 70 Standard Carbine $330

Same general specifications as Standard Model 70 except 19-inch bbl. and weighs 7.25 lbs. Shallow recoil pad. Walnut stock and forend w/traditional Model 70 checkering. Swivel studs. No sights, but drilled and tapped for scope mount.

Winchester Model 70 Sporter DBM

Same general specifications as Model 70 Sporter SSM, except w/detachable box magazine. Calibers: 22-250 (disc. 1994), 223 (disc. 1994), 243 (disc. 1994), 270, 7mm Rem. Mag., 308 (disc. 1994), 30-06, 300 Win. Mag. Made 1992-94.
Model 70 DBM.......................................**$365**
Model 70 DBM-S (w/iron sights) **395**

Winchester Model 70 Stainless Sporter SSM $395

Same general specifications as Model 70 XTR Sporter, except w/checkered black composite stock and matte finished receiver, bbl. and other metal parts. Calibers: 270, 7mm Rem. Mag., 30-06, 300 Win. Mag., 338 Win. Mag. Weight: 7.75 lbs. Made 1992-94.

Winchester Model 70 Classic Super Grade $495

Calibers: 270, 7mm Rem. Mag., 30-06, 300 Win. Mag., 338 Win. Mag. 5-shot magazine (standard), 3-shot (magnum). 24-inch bbl. 44.5 inches overall. Weight: 7.75 lbs. Checkered walnut stock w/sculptured cheekpiece and tapered forend. Scope bases and rings, no sights. Controlled-round feeding system. Made 1990 to date.

**Winchester Model 70
XTR Sporter**

Winchester Model 70A

Winchester Model 70 Target. **$525**
Calibers: 30-06 and 308 Win. (7.62mm NATO). 5-shot magazine. 26-inch heavy bbl. Weight: 10.5 lbs. No sights, but drilled and tapped for scope mount and also open sights. High-comb Marksman-style target stock, aluminum hand stop and swivels. Intro. 1972. Disc.

Winchester Model 70 Ultra Match **$595**
Similar to Model 70 Target, but custom grade w/26-inch heavy bbl. w/deep counterbore, glass bedding, externally adj. trigger. Intro. 1972. Disc.

Winchester Model 70 Varmint (Heavy Barrel)
Same as Model 70 Standard, except w/medium-heavy, counterbored 26-inch bbl., no sights, stock w/less drop. Weight: 9 lbs. Calibers: 22-250 Rem., 223 Rem., 243 Win., 308 Win. Made 1972-93. **Model 70 SHB**, in 308 Win. only w/black synthetic stock and matte blue receiver/bbl. Made 1992-93.
Model 70 Varmint. **$435**
Model 70 SHB (Synthetic Heavy Barrel). **375**

Winchester Model 70 Win-Cam Rifle. **$345**
Caliber: 270 Win. and 30-06 Springfield. 24-inch barrel. Camouflage one-piece laminated stock. Recoil pad. Drilled and tapped for scope. Made 1986 to date.

Winchester Model 70 Winlite Bolt-Action Rifle. **$450**
Calibers: 270 Win., 280 Rem., 30-06 Springfield, 7mm Rem., 300 Win. Mag., and 338 Win. Mag. 5-shot magazine; 3-shot for Magnum calibers. 22-inch bbl.; 24-inch for Magnum calibers. 42.5 inches overall; 44.5, Magnum calibers. Weight: 6.25 to 7 lbs. Fiberglass stock w/rubber recoil pad, sling swivel studs. Made 1986-90.

Winchester Model 70 Win-Tuff Bolt-Action Rifle
Calibers: 22-250, 223, 243, 270, 308 and 30-06 Springfield. 22-inch bbl. Weight: 6.25–7 lbs. Laminated dye-shaded brown wood stock w/recoil pad. Barrel drilled and tapped for scope. Swivel studs. FWT Model made 1986–94. LW Model intro. 1992.
Featherweight Model **$395**
Lightweight Model (Made 1992–93) **345**

Winchester Model 70 XTR Featherweight. **$365**
Similar to Standard Win. Model 70, except lightweight American walnut stock w/classic Schnabel forend, checkered. 22-inch bbl., hooded blade front sight, folding leaf rear sight. Stainless-steel magazine follower. Weight: 6.75 lbs. Made 1984-94.

Winchester Model 70 XTR Sporter Rifle **$375**
Calibers: 264 Win. Mag., 7mm Rem. Mag., 300 Win. Mag., 200 Weatherby Mag., and 338 Win. Mag. 3-shot magazine. 24-inch barrel. 44.5 inches overall. Weight: 7.75 lbs. Walnut Monte Carlo stock. Rubber buttpad. Receiver tapped and drilled for scope mounting. Made 1986-94.

Winchester Model 70 XTR Sporter Magnum. **$395**
Calibers: 264 Win. Mag., 7mm Rem. Mag., 300 Win. Mag., 338 Win. Mag. 3-shot magazine. 24-inch bbl. 44.5 inches overall. Weight: 7.75 lbs. No sights furnished, optional adj. folding leaf rear; hooded ramp. Receiver drilled and tapped for scope. Checkered American walnut Monte Carlo-style stock w/satin finish. Made 1986-94.

Winchester Model 70 XTR Sporter Varmint **$375**
Same general specifications as Model 70 XTR Sporter, except in calibers 223, 22-250, 243 only. Checkered American walnut Monte Carlo-style stock w/cheekpiece. Made 1986-94.

Winchester Model 70A **$285**
Calibers: 222 Rem., 22-250, 243 Win., 25-06, 270 Win., 30-06, 308 Win. 4-shot magazine. 22-inch bbl. (except 24- or 26-inch in 25-06). Weight: 7.5 lbs. Sights: open rear; hooded ramp front. Monte Carlo stock w/checkered pistol grip and forearm, sling swivels. Made 1972-78.

Winchester Model 70A Magnum **$295**
Same as Model 70A, except w/3-shot magazine, 24-inch bbl., recoil pad. Weight: 7.75 lbs. Calibers: 264 Win. Mag., 7mm Rem. Mag., 300 Win. Mag. Made 1972-78

**Winchester Model 70 Ultimate Classic
Bolt-Action Rifle**
Calibers: 25-06 Rem., 264 Win., 270 Win., 270 Wby. Mag., 280 Rem., 7mm Rem. Mag., 7mm STW, 30-06, Mag., 300 Win. Mag., 300 Wby. Mag., 300 H&H Mag., 338 Win. Mag., 340 Wby. Mag., 35 Whelen, 375 H&H Mag., 416 Rem. Mag. and 458 Win. Mag. 3-, 4- or 5-shot magazine. 22- 24- 26-inch stainless bbl. in various configurations including: full-fluted tapered round, half round and half octagonal or tapered full octagonal. Weight: 7.75 to 9.25 lbs. Checkered fancy walnut stock. Made 1995 to date.
Model 70 Ultimate Classic **$1395**
For Mag. Calibers (375 H&H, 416 and 458), **add** **250**

Winchester Model 70 Laminated Stainless
Bolt-Action Rifle **$495**
Calibers: 270 Win., 30-06 Spfld., 7mm Rem. Mag., 300 Win. Mag., and 338 Win. Mag. 5-shot magazine. 24-inch bbl. 44.75 inches overall. Weight: 8 to 8.525 lbs. Gray/Black laminated stock. Made 1998 to date.

Winchester Model 70 Characteristics

Winchester Model 70 First Model
(Serial Numbers 1 – 80,000)
First manufactured in 1936; first sold in 1937. Receiver drilled for Lyman No. 57W or No. 48WJS receiver peep sights. Also drilled and tapped for Lyman or Fecker scope-sight block. Weight w/24-inch bbl. in all calibers except 375 H&H Mag.: 8.25 lbs. 9 lbs. in H&H Mag. Early type safety located on bolt top. Production of this model ended in 1942 near serial number 80,000 due to World War II.

Winchester Model 70 Second Model
(Serial Numbers 80,000 – 350,000)
All civilian production of Winchester Model 70 rifles halted during World War II. Production resumed in 1947 w/improved safety and integral front-sight ramp. Serial numbers started at around 80,000. This model type was produced until 1954, ending around serial number 350,000.

Winchester Model 70 Third Model
(Serial Numbers 350,000 – 400,000)
This variety was manufactured from 1954 to 1960 and retained many features of the Second Model, except that a folding rear sight replaced the earlier type and front-sight ramps were brazed onto the bbl. rather than being an integral part of the bbl. The Model 70 Featherweight Rifle was intro. in 1954 in 308 WCF caliber. It was fitted w/light 22-inch bbl. and was also available w/either a Monte Carlo or Standard stock. The 243 Win. cartridge was added in 1955 in all grades of the Winchester Model 70 except the National Match and Bull Gun models. The 358 Win. cartridge was also intro. in 1955, along w/new Varmint model chambered for the 243 caliber only.

Winchester Model 70 Fourth Model
(Serial Numbers 400,000 – 500,000)
Different markings were inscribed on the barrels of these models and new magnum calibers were added; that is, 264 Win Mag., 338 Win. Mag, and 458 Win. Mag. All bbls. of this variation were about 0.13 inch shorter than previous ones. The 22 Hornet and 257 Roberts were disc. in 1962; the 358 Win. caliber in 1963.

Winchester Model 70 Fifth Model
(Serial Numbers 500,000 to about 570,000)
These rifles may be recognized by slightly smaller checkering patterns and slightly smaller lightweight stocks. Featherweight bbls. were marked "Featherweight." Webbed recoil pads were furnished on magnum calibers.

Post-1964 Winchester Model 70 Rifles

In 1964, the Winchester-Western Division of Olin Industries claimed that they were losing money on every Model 70 they produced. Both labor and material costs had increased to a level that could no longer be ignored. Other models followed suit. Consequently, sweeping changes were made to the entire Winchester line. Many of the older, less popular, models were discontinued. Models that were to remain in production were modified for lower production costs.

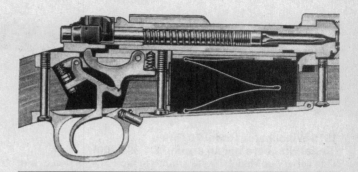

Cross-sectional view of the pre-1964 Winchester Model 70's speed lock action. This action cocks on the opening movement of the bolt with polished, smooth-functioning cams and guide lug, insuring fast and smooth operation.

1964 Winchester Model 70 Rifles
(Serial Numbers 570,000 to about 700,000)
The first version of the "New Model 70s" utilized a free-floating barrel, swaged rifle bore, new stock and sights, new type of bolt and receiver, and a different finish throughout on both the wood and metal parts. The featherweight grade was dropped, but six other grades were available in this new line:

- Standard
- Deluxe (Replaced Previous Super Grade)
- Magnum
- Varmint
- Target
- African

1966 Winchester Model 70 Rifles
(Serial Numbers 700,000-G to about 1,005,000)
In general, this group of Model 70s had fancier wood checkering, cross-bolt stock reinforcement, improved wood finish, and improved action. One cross-bolt reinforcement was used on standard guns. Magnum calibers, however, used an additional forward cross-bolt and red recoil pad. The free-floating barrel clearance forward of the breech taper was reduced in thickness. Impressed checkering was used on the deluxe models until 1968. Hand checkering was once again used on deluxe and carbine models in 1969; the big, red "W" was removed from all grip caps. A new red safety-indicator and undercut cheekpiece was introduced in 1971.

1972 Winchester Model 70 Rifles
(Serial Numbers G1,005,000 to about G1,360,000)
Both the barrels and receivers for this variety of Model 70s were made from chrome molybdenum (C-M) steel. The barrels were tapered w/spiral rifling ranging in length from 22 to 24 inches. Calibers 222 Rem., 225 Win. 22-250, 243 Win., 25-06, 270, 308 Win., 30-06 and 458 WM used the 22-inch length, while the following calibers used the 24-inch length: 222 Rem., 22-250, 243 Win., 264 Win. Mag., 7mm Mag., 300 and 375 H&H Mag. The 225 Win caliber was dropped in 1973; Mannlicher stocks were also disc. in 1973. The receiver for this variety of Model 70s was machined from a block of C-M steel. A new improved anti-bind bolt was introduced along with a new type of ejector. Other improvements included hand-cut checkering, pistol-grip stocks with pistol-grip and dark forend caps. An improved satin wood finished was also utilized.

RIFLES

**Winchester Model 70
Black Shadow**

1978 Winchester Model 70 Rifles
(Serial Numbers began around G1,360,000)

This variety of Model 70 was similar to the 1972 version except that a new XTR style was added which featured high-luster wood and metal finishes, fine-cut checkering, and similar embellishments. All Model 70 rifles made during this period used the XTR style; no standard models were available. In 1981, beginning with serial number G1,440,000 (approximately), a Featherweight version of the Model 70 XTR was introduced. The receiver was identical to the 1978 XTR, but lighter barrels were fitted. Stocks were changed to a lighter design with larger scroll checkering patterns and a Schnabel forend with no Monte Carlo comb. A satin sheen stock finish on the featherweight version replaced the high-luster finish used on the other XTR models. A new style red buttplate with thick, black rubber liner was used on the featherweight models. The grip cap was also redesigned for this model.

U.S. Repeating Arms
Model 70s — 1982 to date

In the early 1980s, negotiations began between Olin Industries and an employee-based corporation. The result of these negotiations ended with Olin selling all tools, machinery, supplies, etc. at the New Haven plant to the newly-formed corporation which was eventually named *U.S. Repeating Arms Company.* Furthermore, U.S. Repeating Arms Company purchased the right to use the Winchester name and logo. Winchester Model 70s went through very few changes the first two years after the transistion. However, in 1984, the Featherweight Model 70 XTR rifles were offered in a new short action for 22-250 Rem., 223 Rem., 243 Win. and 308 Win. calibers, in addition to their standard action which was used for the longer cartridges. A new Model 70 lightweight carbine was also introduced this same year. Two additional models were introduced in 1985 — the Model 70 Lightweight Mini-Carbine Short Action and the Model 70 XTR Sporter Varmint; but this was just the beginning. The Model 70 Winlite appeared in the 1986 "Winchester" catalog, along with two economy versions of the Model 70 — the Winchester Ranger and the Ranger Youth Carbine. Five or six different versions of the Winchester Model 70 had been sufficient for 28 years (1937 - 1964). Now, changes in design and the addition of new models each year seemed to be necessary to keep the rifle alive. New models were added, old models dropped, changed in design, etc. on a regular basis. The trend continues. Still, the Winchester Model 70 Bolt-Action Repeating Rifle — in any of its variations — is the most popular bolt-action rifle ever built.

Winchester Model 70 Black Shadow $295

Calibers: 243 Win., 270 Win., 300 Win. Mag., 308 Win., 338 Win. Mag., 30-06 Spfld., 7mm STW., 7mm Rem. Mag. and 7mm-08 Rem. 3- 4- or 5-shot magazine. 20- 24- 25- or 26-inch bbls. 39.5 to 46.75 inches overall. Weight: 6.5 to 8.25 lbs. Composite, Walnut or Gray/Black laminated Stocks. Made 1998 to date.

Winchester Model 70 Classic Camo
Bolt-Action Rifle $525

Calibers: 270 Win., 30-06 Spfld., 7mm Rem. Mag., 300 Win. Mag. 3- or 5-shot magazine. 24- or 26-inch bbl. 44.75 to 46.75 inches overall. Weight; 7.25 to 7.5 lbs. Mossy Oak® finish and composite stock. Made 1998 to date.

Winchester Model 70 Classic Compact
Bolt-Action Rifle $395

Calibers: 243 Win., 308 Win., and 7mm-08 Rem. 3-shot magazine. 20-inch bbl., 39.5 inches overall. Weight: 6.5 lbs. Walnut stock. Made 1998 to date.

Winchester Model 70 Classic Laredo™
Range Hunter Bolt-Action Rifle

Calibers: 7mm STW, 7mm Rem. mag., 300 Win. Mag. 3-shot magazine. 26-inch bbl. 46.75 inches overall. Weight: 9.5 lbs. Composite stock. Made 1996 to date.

Classic Laredo	**$495**
Classic Laredo Fluted (Made 1998 to date)	550
Boss® Classic Laredo.	525

Winchester Model 71 Lever-Action Repeater

Solid frame. Caliber: 348 Win. 4-shot tubular magazine. 20- or 24-inch bbl. Weight: 8 lbs. Sights: open or peep rear; bead front on ramp w/hood. Walnut stock. Made 1935–57.

Special Grade (checkered pistol grip and forearm, grip cap, quick-detachable swivels and sling	**$1200**
Special Grade Carbine (20-inch bbl.; Disc. 1940)	1895
Standard Grade (lacks checkering, grip cap, sling and swivels)	850
Standard Grade Carbine (20-inch bbl.; Disc. 1940)	1400

Winchester Model 72 Bolt-Action Repeater $335

Tubular magazine. Takedown. Caliber: 22 Short, Long, LR. Magazine holds 20 Short, 16 Long or 15 LR. 25-inch bbl. Weight: 5.75 lbs. Sights: peep or open rear; bead front. Plain pistol-grip stock. Made 1938–59.

**Winchester Model 70
Classic Camo**

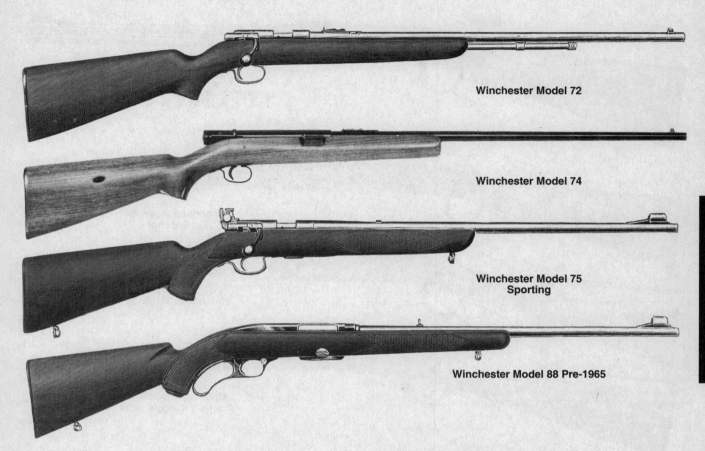

Winchester Model 72

Winchester Model 74

Winchester Model 75 Sporting

Winchester Model 88 Pre-1965

Winchester Model 73 Lever-Action Repeater
See Model 1873 rifles, carbines, "One of One Thousand" and other variations of this model at the beginning of Winchester Rifle Section. *Note:* The Winchester Model 1873 was the first lever-action repeating rifle bearing the Winchester name.

Winchester Model 74 Self-Loading Rifle $215
Takedown. Calibers: 22 Short only, 22 LR only. Tubular magazine in buttstock holds 20 Short, 14 LR. 24-inch bbl. Weight: 6.25 lbs. Sights: open rear; bead front. Plain pistol-grip stock, one-piece. Made 1939–55.

Winchester Model 75 Sporting Rifle $625
Same as Model 75 Target, except has 24-inch bbl., checkered sporter stock, open rear sight; bead front on hooded ramp, weighs 5.5 lbs.

Winchester Model 75 Target Rifle $395
Caliber: 22 LR. 5- or 10-shot box magazine. 28-inch bbl. Weight: 8.75 lbs. Target sights (Lyman, Redfield or Winchester). Target stock w/pistol grip and semibeavertail forearm, swivels and sling. Made 1938–59.

Winchester Model 77 Semiautomatic Rifle, Clip Type $215
Solid frame. Caliber: 22 LR. 8-shot clip magazine. 22-inch bbl. Weight: about 5.5 lbs. Sights: open rear; bead front. Plain, one-piece pistol-grip stock. Made 1955–63.

Winchester Model 77, Tubular Magazine Type $275
Same as Model 77. Clip type, except has tubular magazine holding 15 rounds. Made 1955–63.

Winchester Model 86 Carbine and Rifle
See Model 1886 at beginning of Winchester Rifle Section.

Winchester Model 88 Carbine
Same as Model 88 Rifle, except has 19-inch bbl., plain carbine-style stock and forearm with bbl. band. Weight: 7 lbs. Made 1968-73.
88 Carbine. **$495**
284 Win. Caliber. **895**

Winchester Model 88 Lever-Action Rifle
Hammerless. Calibers: 243 Win., 284 Win., 308 Win., 358 Win. 4-shot box magazine. 3-shot in pre-1963 models and in 284. 22-inch bbl. Weight: about 7.25 lbs. One-piece walnut stock with pistol grip, swivels (1965 and later models have basket-weave ornamentation instead of checkering). Made 1955–1973. *Note:* 243 and 358 introduced 1956, latter discontinued 1964; 284 introduced 1963.
Model 88 (Checkered Stock) **$495**
Model 88 (Basketweave Stock) **450**
284 Win. Caliber. **850**
358 Win. Caliber. **995**

Winchester Model 92 Lever-Action Rifle
Similar to the original Model 1892. Calibers: 357 Mag., 44-40, 44 Mag., 45 LC. 10-shot magazine. 24-inch round bbl. Weight: 6.25 lbs. 41.25 inches overall. Bead front sight, adjustable buckhorn rear. Etched receiver and gold trigger. Blue finish. Smooth straight-grip walnut stock and forewarn w/ metal grip cap. Made 1997 to date.
Standard Rifle. **$450**
Short Rifle w/20-inch bbl. (44 Mag. only) **435**

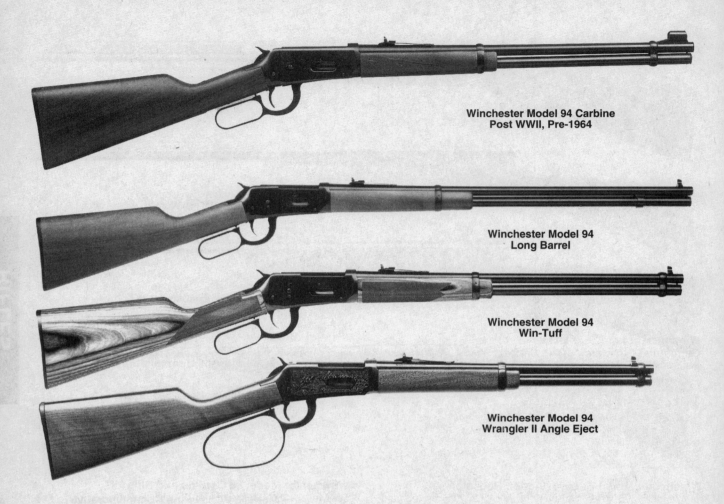

**Winchester Model 94 Carbine
Post WWII, Pre-1964**

**Winchester Model 94
Long Barrel**

**Winchester Model 94
Win-Tuff**

**Winchester Model 94
Wrangler II Angle Eject**

Winchester Model 94 Antique Carbine $225
Same as standard Post-64 Model 94 Carbine, except has decorative scrollwork and casehardened receiver, brass-plated loading gate, saddle ring; caliber 30-30 only. Made 1964–84.

Winchester Model 94 Carbine
Same as Model 1894 Rifle, except 20-inch round bbl., 6-shot full-length magazine. Weight: about 6.5 lbs. Originally made in calibers 25-35, 30-30, 32 Special and 38-55. Original version discontinued 1964. *See* 1894 Models at beginning of Winchester Rifle Section.
Pre-World War II (under No. 1,300,000) **$995**
Postwar, pre-1964 (under No. 2,700,000) **550**

Winchester Model 94 Classic Carbine. $295
Same as Canadian Centennial '67 Commemorative Carbine, except without commemorative details; has scroll-engraved receiver, gold-plated loading gate. Made 1968–70.

Winchester Model 94 Classic Rifle $295
Same as Model 67 Rifle, except without commemorative details; has scroll-engraved receiver, gold-plated loading gate. Made 1968–70.

Winchester Model 94 Deluxe Carbine $285
Caliber: 30-30. 6-shot magazine. 20-inch bbl. 37.75 inches overall. Weight: 6.5 lbs. Semi-fancy American walnut stock with rubber buttpad, long forearm and specially cut checkering. Engraved with "Deluxe" script. Made 1987 to date.

Winchester Model 94 Long Barrel Rifle. $250
Caliber: 30-30. 7-round magazine. 24-inch bbl. 41.75 inches overall. Weight: 7 lbs. American walnut stock. Blade front sight. Made 1987 to date.

Winchester Model 94 Trapper $225
Same as Winchester Model 94 Carbine, except 16-inch bbl. and weighs 6 lbs. 2 oz. Made 1980 to date.

Winchester Model 94 Win-Tuff Rifle $235
Caliber: 30-30. 6-round magazine. 20-inch bbl. 37.75 inches overall. Weight: 6.5 lbs. Brown laminated wood stock. Made 1987 to date.

Winchester Model 94 Wrangler Carbine
Same as standard Model 94 Carbine, except has 16-inch bbl., engraved receiver and chambered for 32 Special & 38-55 Win.
Wrangler, Top Eject (Disc.1984) **$285**
Wrangler II, Angle Eject (Disc. 1985) **215**

Winchester Model 94 XTR Big Bore $225
Modified Model 94 action for added strength. Caliber: 375 Win. 20-inch bbl. Rubber buttpad. Checkered stock and forearm. Weight: 6.5 lbs. Made 1978 to date. *See* Illustration next page.

Winchester Model 94 XTR Lever-Action Rifle. $215
Same general specifications as standard Angle Eject M94 except chambered 30-30 and 7-30 Waters and has 20- or 24-inch bbl. Weight: 7 lbs. Made 1985 to date by U.S. Repeating Arms.

**Winchester Model 94
XTR Big Bore**

**Winchester Model 94
XTR 7-30 Waters**

MODEL 94 COMMEMORATIVES

Values indicated are for commemorative Winchesters in new condition.

Winchester Model 94 Alaskan Purchase
Centennial Commemorative Carbine **$1595**
Same as Wyoming issue, except different medallion and inscription. 1,501 made in 1967.

Winchester Model 94 Antlered Game **$525**
Standard Model 94 action. Gold-colored medallion inlaid in stock. Antique gold-plated receiver, lever tang and bbl. bands. Medallion and receiver engraved with elk, moose, deer and caribou. 20.5-inch bbl. Curved steel buttplate. In 30-30 caliber. 19,999 made in 1978.

Winchester Model 94 Bicentennial '76 Carbine **$625**
Same as Standard Model 94 Carbine, except caliber 30-30 only; antique silver-finished, engraved receiver; stock and forearm of fancy walnut, checkered, Bicentennial medallion embedded in buttstock, curved buttplate. 20,000 made in 1976.

Winchester Model 94 Buffalo Bill Commemorative
Same as Centennial '66 Rifle, except receiver is black-chromed, scroll-engraved and bears name "Buffalo Bill"; hammer, trigger, loading gate, saddle ring, forearm cap, and buttplate are nickel-plated; Buffalo Bill Memorial Assn. commemorative medallion embedded in buttstock; "Buffalo Bill Commemorative" inscribed on bbl., facsimile signature "W.F. Cody, Chief of Scouts" on tang. Carbine has 20-inch bbl., 6-shot magazine, 7-lb. weight. 112,923 made in 1968.
Carbine **$450**
Rifle 475
Matched Carbine/Rifle Set 1095

Winchester Canadian Centennial '67 Commemorative
Same as Centennial '66 Rifle, except receiver engraved — with maple leaves — and forearm cap are black-chromed, buttplate is blued, commemorative inscription in gold on barrel and top tang: "Canadian Centennial 1867–1967." Carbine has 20-inch bbl., 6-shot magazine, 7-lb. weight. 90,398 made in 1967.
Carbine **$450**
Rifle 475
Matched Carbine/Rifle Set 1050

Winchester Centennial '66 Commemorative
Commemorates Winchester's 100th anniversary. Standard Model 94 action. Caliber: 30-30. Full-length magazine holds 8 rounds. 26-inch octagon bbl. Weight: 8 lbs. Gold-plated receiver and forearm cap. Sights: open rear; post front. Saddle ring. Walnut buttstock and forearm with high-gloss finish, solid brass buttplate. Commemorative inscription on bbl. and top tang of receiver. 100,478 made in 1966.
Carbine **$450**
Rifle 475
Matched Carbine/Rifle Set 1095

Winchester Model 94 Cheyenne Commemorative . . **$825**
Available in Canada only. Same as Standard Model 94 Carbine, except chambered for 44-40. 11,225 made in 1977.

Winchester Model 94 Chief Crazy Horse
Commemorative **$550**
Cailber: 38-55, 7-shot tubular magazine. 24-inch bbl., 41.75 inches overall. Walnut stock with medallion of the United Sioux Tribes; buttstock and forend also decorated with brass tacks. Engraved receiver. Open rear sights; bead front sight. 19,999 made in 1983.

Winchester Model 94 Colt Commemorative
Carbine Set **$2995**
Standard Model 94 action. Caliber: 44-40 Win. 20-inch bbl. Weight: 6.25 lbs. Features a horse-and-rider trademark and distinctive WC monogram in gold etching on left side of receiver. Sold in set with Colt Single Action Revolver chambered for same caliber.

**Winchester Model '66
Commemorative**

RIFLES

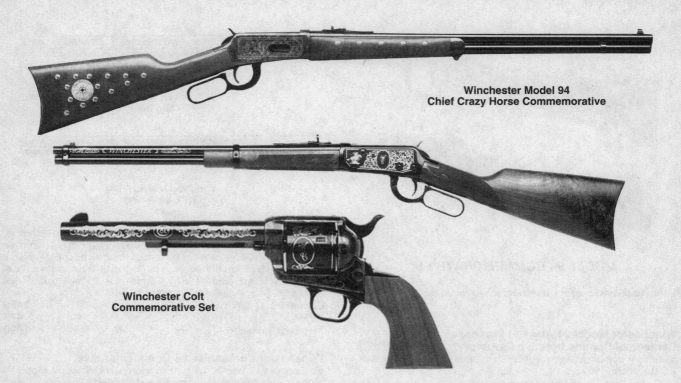

**Winchester Model 94
Chief Crazy Horse Commemorative**

**Winchester Colt
Commemorative Set**

Winchester Model 94 Cowboy Commemorative Carbine

Same as Standard Model 94 Carbine, except caliber 30-30 only; nickel-plated receiver, tangs, lever, bbl. bands; engraved receiver, "Cowboy Commemorative" on bbl., commemorative medallion embedded in buttstock; curved buttplate. 20,915 made in 1970. Nickel-silver medallion inlaid in stock. Antique silver-plated receiver engraved with scenes of the old frontier. Checkered walnut stock and forearm. 19,999 made in 1970.

Cowboy Carbine. **$450**
Cowboy Carbine (1 of 300) **2895**

Winchester Model 94 Golden Spike Commemorative Carbine **$395**

Same as Standard Model 94 Carbine, except caliber 30-30 only; gold-plated receiver, tangs and bbl. bands; engraved receiver, commemorative medallion embedded in stock. 64,758 made in 1969.

Winchester Model 94 Illinois Sesquicentennial Commemorative Carbine **$395**

Same as Standard Model 94 Carbine, except caliber 30-30 only; gold-plated buttplate, trigger, loading gate, and saddle ring; receiver engraved with profile of Lincoln, commemorative inscription on receiver, bbl.; souvenir medallion embedded in stock. 31,124 made in 1968.

Winchester Model 94 Legendary Frontiersmen Commemorative **$525**

Standard Model 94 action. Caliber: 39-55. 24-inch round bbl. Nickel-silver medallion inlaid in stock. Antique silver-plated receiver engraved with scenes of the old frontier. Checkered walnut stock and forearm. 19,999 made in 1979.

Winchester Model 94 Legendary Lawmen Commemorative **$525**

Same as Standard Model 94 Carbine, except 30-30 only; antique silver-plated receiver engraved with action law-enforcement scenes. 16-inch Trapper bbl., antique silver-plated bbl. bands. 19,999 made in 1978.

Winchester Model 94 Lone Star Commemorative

Same as Theodore Roosevelt Rifle, except yellow-gold plating; "Lone Star" engraving on receiver and bbl., commemorative medallion embedded in buttstock. 30,669 made in 1970.

Rifle or Carbine **$450**
Matched Carbine/Rifle Set **1095**

Winchester Model 94 NRA Centennial Musket **$425**

Commemorates 100th anniversary of National Rifle Association of America. Standard Model 94 action. Caliber: 30-30. 7-shot magazine. 26-inch bbl. Sights: military folding rear; blade front. Black chrome-finished receiver engraved "NRA 1871–1971"

**Winchester Model 94
NRA Centennial**

Winchester Model 100

Winchester Model 94 NRA Centennial Musket *(Con't)*
plus scrollwork. Barrel inscribed "NRA Centennial Musket." Musket-style buttstock and full-length forearm; commemorative medallion embedded in buttstock. Weight: 7.13 lbs. Made in 1971.

Winchester Model 94 NRA Centennial Rifle $450
Same as Model 94 Rifle, except has commemorative details as in NRA Centennial Musket (barrel inscribed "NRA Centennial Rifle"); caliber 30-30, 24-inch bbl., QD sling swivels. Made in 1971.

Winchester Model 94 NRA Centennial
Matched Set $950
Rifle and musket were offered in sets with consecutive serial numbers. *Note:* Production figures not available. These rifles offered in Winchester's 1972 catalog.

Winchester Model 94 Nebraska Centennial
Commemorative Carbine $1325
Same as Standard Model 94 Carbine, except caliber 30-30 only; gold-plated hammer, loading gate, bbl. band, and buttplate; souvenir medallion embedded in stock, commemorative inscription on bbl. 2,500 made in 1966.

Winchester Model 94 Theodore Roosevelt
Commemorative Rifle/Carbine
Standard Model 94 action. Caliber: 30-30. Rifle has 6-shot half-magazine, 26-inch octagon bbl., 7.5-lb. weight. Carbine has 6-shot full magazine, 20-inch bbl., 7-lb. weight. White gold-plated receiver, upper tang, and forend cap; receiver engraved with American Eagle, "26th President 1901–1909," and Roosevelt's signature. Commemorative medallion embedded in buttstock. Saddle ring. Half pistol grip, contoured lever. 49,505 made in 1969.
Carbine $425
Rifle. 450
Matched Set 1050

Winchester Model 94 Texas Ranger
Association Carbine $2595
Same as Texas Ranger Commemorative Model 94, except special edition of 150 carbines, numbered 1 through 150, with hand-checkered full-fancy walnut stock and forearm. Sold only through Texas Ranger Association. Made in 1973.

Winchester Model 94 Texas Ranger
Commemorative Carbine $745
Same as Standard Model 94 Carbine, except caliber 30-30 only, stock and forearm of semi-fancy walnut, replica of Texas Ranger star embedded in buttstock, curved buttplate. 5,000 made in 1973.

Winchester Model 94 Trapper
Same as Winchester Model 94 Carbine, except w/16-inch bbl. and weighs 6 lbs. 2 oz. Angle Eject introduced in 1985 also chambered 357 Mag., 44 Mag. and 45 LC. Made 1980 to date.
94 Trapper, Top Eject (Disc. 1984) $225
94 Trapper, Angle Eject (30-30). 195
Chambered 357 Mag., 44 Mag. or 45 LC., **add** 25

Winchester Model 94 John Wayne
Commemorative Carbine. $950
Standard Model 94 action. Caliber: 32-40. 18.5-inch bbl. Receiver is pewter-plated with engraving of Indian attack and cattle drive scenes. Oversized bow on lever. Nickel-silver medallion in buttstock bears a bas-relief portrait of Wayne. Selected American walnut stock with deep-cut checkering. Introduced by U.S. Repeating Arms in 1981.

Winchester Model 94 Wells Fargo & Co.
Commemorative Carbine $525
Same as Standard Model 94 Carbine, except 30-30 only; antique silver-finished, engraved receiver; stock and forearm of fancy walnut, checkered, curved buttplate. Nickel-silver stagecoach medallion (inscribed "Wells Fargo & Co. —1852–1977—125 Years") embedded in buttstock. 20,000 made in 1977.

Winchester Model 94 Oliver F. Winchester Com . . . $725
Standard Model 94 action. Caliber: 38-55. 24-inch octagonal bbl. Receiver is satin gold-plated with distinctive engravings. Stock and forearm semi-fancy American walnut with high grade checkering. 19,999 made in 1980.

Winchester Model 94 Wrangler Carbine
Same as standard Model 94 Carbine, except w/16-inch bbl., engraved receiver and chambered for 32 Special & 38-55 Win. Angle Eject introduced in 1985 as Wrangler II also chambered 30-30 Win., 44 Mag. and 45 LC. Made 1980-86. Reintroduced in 1992.
94 Wrangler, Top Eject (Disc. 1984) $235
94 Wrangler II, Angle Eject (30-30). 200
Chambered 44 Mag. or 45 LC., **add**. 25

Winchester Model 94 Wyoming Diamond
Jubilee Commemorative Carbine $1495
Same as Standard Model 94 Carbine, except caliber 30-30 only, receiver engraved and casehardened in colors, brass saddle ring and loading gate, souvenir medallion embedded in buttstock, commemorative inscription on bbl. 1,500 made in 1964.

Winchester Model 94 XTR Big Bore
Modified Model 94 action for added strength. Calibers: 307 Win., 356 Win., 375 Win. or 444 Marlin. 20-inch bbl. 6-shot magazine. Rubber buttpad. Checkered stock and forearm. Weight: 6.5 lbs. Made 1978 to date.
94 XTR BB, Top Eject (Disc. 1984) $225
94 XTR BB, Angle Eject (Intro. 1985). 195
Chambered 356 Win. or 375 Win., **add** 50

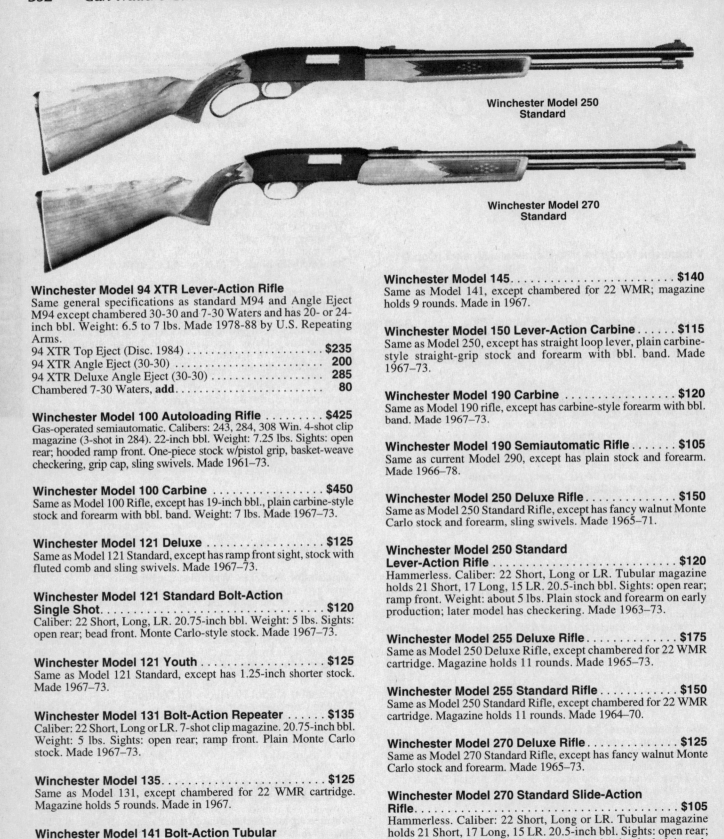

Winchester Model 250
Standard

Winchester Model 270
Standard

Winchester Model 94 XTR Lever-Action Rifle
Same general specifications as standard M94 and Angle Eject M94 except chambered 30-30 and 7-30 Waters and has 20- or 24-inch bbl. Weight: 6.5 to 7 lbs. Made 1978-88 by U.S. Repeating Arms.

94 XTR Top Eject (Disc. 1984) **$235**
94 XTR Angle Eject (30-30) **200**
94 XTR Deluxe Angle Eject (30-30) **285**
Chambered 7-30 Waters, **add** **80**

Winchester Model 100 Autoloading Rifle $425
Gas-operated semiautomatic. Calibers: 243, 284, 308 Win. 4-shot clip magazine (3-shot in 284). 22-inch bbl. Weight: 7.25 lbs. Sights: open rear; hooded ramp front. One-piece stock w/pistol grip, basket-weave checkering, grip cap, sling swivels. Made 1961–73.

Winchester Model 100 Carbine $450
Same as Model 100 Rifle, except has 19-inch bbl., plain carbine-style stock and forearm with bbl. band. Weight: 7 lbs. Made 1967–73.

Winchester Model 121 Deluxe $125
Same as Model 121 Standard, except has ramp front sight, stock with fluted comb and sling swivels. Made 1967–73.

Winchester Model 121 Standard Bolt-Action Single Shot. $120
Caliber: 22 Short, Long, LR. 20.75-inch bbl. Weight: 5 lbs. Sights: open rear; bead front. Monte Carlo-style stock. Made 1967–73.

Winchester Model 121 Youth $125
Same as Model 121 Standard, except has 1.25-inch shorter stock. Made 1967–73.

Winchester Model 131 Bolt-Action Repeater $135
Caliber: 22 Short, Long or LR. 7-shot clip magazine. 20.75-inch bbl. Weight: 5 lbs. Sights: open rear; ramp front. Plain Monte Carlo stock. Made 1967–73.

Winchester Model 135. . . . . $125
Same as Model 131, except chambered for 22 WMR cartridge. Magazine holds 5 rounds. Made in 1967.

Winchester Model 141 Bolt-Action Tubular Repeater $140
Same as Model 131, except has tubular magazine in buttstock; holds 19 Short, 15 Long, 13 LR. Made 1967–73.

Winchester Model 145. . . . . $140
Same as Model 141, except chambered for 22 WMR; magazine holds 9 rounds. Made in 1967.

Winchester Model 150 Lever-Action Carbine $115
Same as Model 250, except has straight loop lever, plain carbine-style straight-grip stock and forearm with bbl. band. Made 1967–73.

Winchester Model 190 Carbine $120
Same as Model 190 rifle, except has carbine-style forearm with bbl. band. Made 1967–73.

Winchester Model 190 Semiautomatic Rifle $105
Same as current Model 290, except has plain stock and forearm. Made 1966–78.

Winchester Model 250 Deluxe Rifle $150
Same as Model 250 Standard Rifle, except has fancy walnut Monte Carlo stock and forearm, sling swivels. Made 1965–71.

Winchester Model 250 Standard Lever-Action Rifle $120
Hammerless. Caliber: 22 Short, Long or LR. Tubular magazine holds 21 Short, 17 Long, 15 LR. 20.5-inch bbl. Sights: open rear; ramp front. Weight: about 5 lbs. Plain stock and forearm on early production; later model has checkering. Made 1963–73.

Winchester Model 255 Deluxe Rifle $175
Same as Model 250 Deluxe Rifle, except chambered for 22 WMR cartridge. Magazine holds 11 rounds. Made 1965–73.

Winchester Model 255 Standard Rifle $150
Same as Model 250 Standard Rifle, except chambered for 22 WMR cartridge. Magazine holds 11 rounds. Made 1964–70.

Winchester Model 270 Deluxe Rifle $125
Same as Model 270 Standard Rifle, except has fancy walnut Monte Carlo stock and forearm. Made 1965–73.

Winchester Model 270 Standard Slide-Action Rifle. $105
Hammerless. Caliber: 22 Short, Long or LR. Tubular magazine holds 21 Short, 17 Long, 15 LR. 20.5-inch bbl. Sights: open rear; ramp front. Weight: about 5 lbs. Early production had plain walnut stock and forearm (slide handle); latter also furnished in plastic (Cycolac); last model has checkering. Made 1963–73.

Winchester Model 310

Winchester Model 320

Winchester Model 490 Rifle

Winchester Model 670
Bolt Action Rifle

Winchester Model 670
Magnum

RIFLES

Winchester Model 275 Deluxe Rifle **$165**
Same as Model 270 Deluxe Rifle, except chambered for 22 WMR cartridge. Tubular magazine holds 11 rounds. Made 1965–70.

Winchester Model 275 Standard Rifle **$125**
Same as Model 270 Standard Rifle, except chambered for 22 WMR cartridge. Magazine holds 11 rounds. Made 1964–70.

Winchester Model 290 Deluxe Rifle **$175**
Same as Model 290 Standard Rifle, except has fancy walnut Monte Carlo stock and forearm. Made 1965–73.

Winchester Model 290 Standard Semiautomatic Rifle
Caliber: 22 Long or LR. Tubular magazine holds 17 Long, 15 LR. 20.5-inch bbl. Sights: open rear; ramp front. Weight: about 5 lbs. Plain stock and forearm on early production; current model has checkering. Made 1963–77.
W/plain stock/forearm **$175**
W/checkered stock/forearm **210**

Winchester Model 310 Bolt-Action Single Shot **$195**
Caliber: 22 Short, Long, LR. 22-inch bbl. Weight: 5.63 lbs. Sights: open rear; ramp front. Monte Carlo stock w/checkered pistol grip and forearm, sling swivels. Made 1972–75.

Winchester Model 320 Bolt-Action Repeater **$295**
Same as Model 310, except has 5-shot clip magazine. Made 1972–74.

Winchester Model 490 Semiautomatic Rifle **$250**
Caliber: 22 LR. 5-shot clip magazine. 22-inch bbl. Weight: 6 lbs. Sights: folding leaf rear; hooded ramp front. One-piece walnut stock w/checkered pistol grip and forearm. Made 1975–77.

Winchester Model 670 Bolt-Action Sporting Rifle . . **$265**
Calibers: 225 Win., 243 Win., 270 Win., 30-06, 308 Win. 4-shot magazine. 22-inch bbl. Weight: 7 lbs. Sights: open rear; ramp front. Monte Carlo stock w/checkered pistol grip and forearm. Made 1967–73.

Winchester Model 670 Carbine **$275**
Same as Model 670 Rifle, except has 19-inch bbl. Weight: 6.75 lbs. Calibers: 243 Win., 270 Win., 30-06. Made 1967–70.

Winchester Model 670 Magnum **$295**
Same as Model 670 Rifle, except has 24-inch bbl., reinforced stock with recoil pad with slightly different checkering pattern. Weight: 7.25 lbs. Calibers: 264 Win. Mag., 7mm Rem. Mag., 300 Win. Mag. Open rear sight; ramp front sight with hood. Made 1967–70.

Winchester Model 770

Winchester Model 770 Bolt-Action Sporting Rifle $295

Model 70-type action. Calibers: 22-250, 222 Rem., 243, 270 Win., 30-06. 4-shot box magazine. 22-inch bbl. Sights: open rear; hooded ramp front. Weight: 7.13 lbs. Monte Carlo stock, checkered pistol grip and forend; sling swivels. Made 1969–71.

Winchester Model 770 Magnum $315

Same as Standard Model 770, except 24-inch bbl., weight 7.25 lbs., recoil pad. Calibers: 7mm Rem. Mag., 264 and 300 Win. Mag. Made 1969–71.

Winchester Model 9422 Lever-Action Rimfire Rifles

Similar to the standard Model 94 except, chambered 22 Rimfire. Calibers: 22 Short, Long, LR. (9422) or 22 WMR (9422M). Tublar magazine holds 21 or 15 Short.17 or 12 Long, 15 or 11 LR (9422 or Trapper) or 11 or 8 WRM (9422M or Trapper M). 16.5- or 20.5 inch bbl. 33.125- to 37.125 inches overall. Weight: 5.75 to 6.25 lbs. Open

Winchester Model 9422 *(Con't)*

rear sight; hooded ramp front. Carbine-style stock and barrel-band forearm. Stock options: Walnut (Standard), laminated brown (Win-Tuff), laminated green (Win-Cam). Made 1972 to date.

Walnut (Standard)	$225
WinCam™	245
WinTuff™	235
Legacy	250
Trapper (16.5-inch bbl.)	230
XTR Classic	425
High Grade Series I	325
High Grade Series II	295
25th Anniversary Edition Grade I(1of 2500)	495
25th Anniversary Edition High Grade (1of 250)	995
Boy Scout Commerative (1of 15,000)	550
Eagle Scout Commerative (1of 1000)	2695
22 WRM, **add**	10%

Winchester Double Xpress Rifle $2195

Over/under double rifle. Caliber: 30-06. 23.5-inch bbl. Weight: 8.5 lbs. Made for Olin Corp. by Olin-Kodensha in Japan. Introduced 1982.

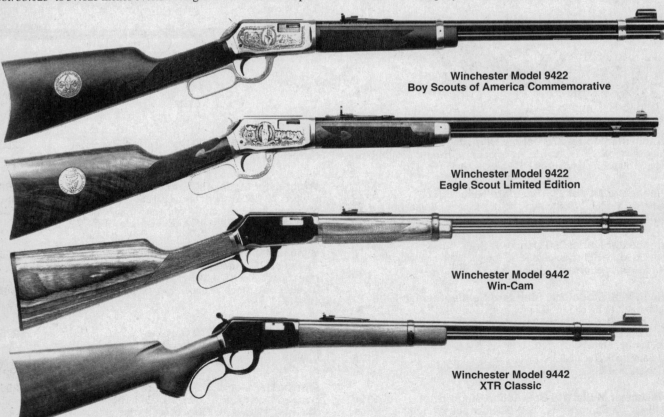

Winchester Model 9422
Boy Scouts of America Commemorative

Winchester Model 9422
Eagle Scout Limited Edition

Winchester Model 9442
Win-Cam

Winchester Model 9442
XTR Classic

**Winchester Ranger
Bolt-Action Rifle**

**Winchester Ranger
Lever-Action Rifle**

Winchester Ranger Youth Bolt-Action Carbine $265
Calibers: 223 (discontinued 1989), 243, and 308 Win. 4- and 5-shot magazine. Bbl.: 20-inch. Weight: 5.75 lbs. American hardwood stock. Open rear sight. Made 1985 to date by U.S. Repeating Arms.

Winchester Ranger Lever-Action
Carbine $195
Caliber: 30-30. 5-shot tubular magazine. Bbl.: 20-inch round. Weight: 6.5 lbs. American hardwood stock. Economy version of Model 94. Made 1985 to date by U.S. Repeating Arms.

Winchester Ranger Bolt-Action Carbine $260
Calibers: 223 Rem., 243 Win., 270, 30-06, 7mm Rem. (discontinued 1985), Mag. 3- and 4-shot magazine. Bbl.: 24-inch in 7mm; 22-inch in 270 and 30-06. Open sights. American hardwood stock. Made 1985 to date by U.S. Repeating Arms.

Winchester Model 1892 Grade I Lever-Action Rifle
Similar to the original Model 1892. Calibers: 357 Mag., 44-40, 44 Mag., 45 LC. 10-shot magazine. 24-inch round bbl. Weight: 6.25 lbs. 41.25 inches overall. Bead front sight, adjustable buckhorn rear. Etched receiver and gold trigger. Blue finish. Smooth straight-grip walnut stock and forearm w/ metal grip cap. Made 1997 to date.
Standard Rifle $450
Short Rifle w/20-inch bbl. (44 Mag. only) 435

Winchester Model 1892 Grade II
Lever-Action Rifle $795
Similar to the Grade I Model 1892, except w/gold appointments and receiver game scene. Chambered 45 LC only. Limited production of 1,00 in 1997.

WINSLOW ARMS COMPANY
Camden, South Carolina

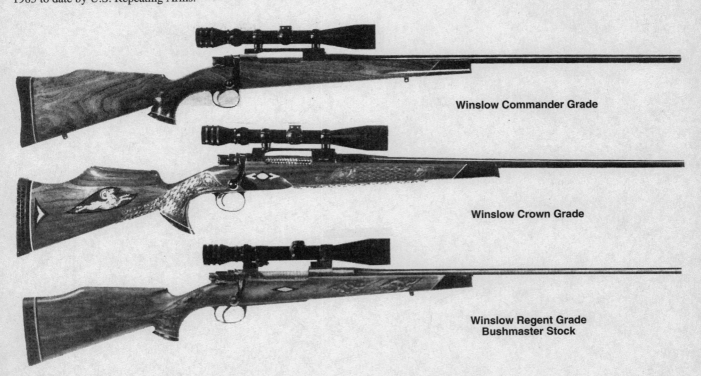

Winslow Commander Grade

Winslow Crown Grade

**Winslow Regent Grade
Bushmaster Stock**

Winslow Bolt-Action Sporting Rifle

Action: FN Supreme Mauser, Mark X Mauser, Remington 700 and 788, Sako, Winchester 70. *Standard calibers:* 17-222, 17-223, 222 Rem., 22-250, 243 Win., 6mm Rem., 25-06, 257 Roberts, 270 Win., 7×57, 280 Rem., 284 Win., 308 Win., 30-06, 358 Win. *Magnum calibers:* 17-222 Mag., 257 Weatherby, 264 Win., 270 Weath., 7mm Rem., 7mm Weath., 300 H&H, 300 Weath., 300 Win., 308 Norma, 8mm Rem., 338 Win., 358 Norma, 375 H&H, 375 Weath., 458 Win. 3-shot magazine in standard calibers, 2-shot in magnum. 24-inch barrel in standard calibers, 26-inch in magnum. Weight: with 24-inch bbl., 7 to 7.5 lbs.; with 26-inch bbl., 8 to 9 lbs. No sights. Stocks: "Bushmaster" with slender pistol grip and beavertail forearm, "Plainsmaster" with full curl pistol grip and flat forearm; both styles have Monte Carlo cheekpiece; Values shown are for basic rifle in each grade; extras such as special fancy wood, more elaborate carving, inlays and engraving can increase these figures considerably. Made 1962–89. *See Illustrations next page.*

Commander Grade . . . . $ 495
Regal Grade 595

Winslow Bolt-Action Sporting Rifle *(Con't)*

Regent Grade 695
Regimental Grade. 895
Crown Grade 1250
Royal Grade 1495
Imperial Grade 2950
Emperor Grade 5595

ZEPHYR DOUBLE RIFLES
Manufactured by Victor Sarasqueta Company
Eibar, Spain

Zephyr Double Rifle $16,500
Boxlock. Calibers: Available in practically every caliber from .22 Hornet to .505 Gibbs. Bbls.: 22 to 28 inches standard, but any lengths were available on special order. Weight: 7 lbs. for the smaller caliber up to 12 or more lbs. for the larger calibers. Imported by Stoeger from about 1938 to 51.

Section III
SHOTGUNS

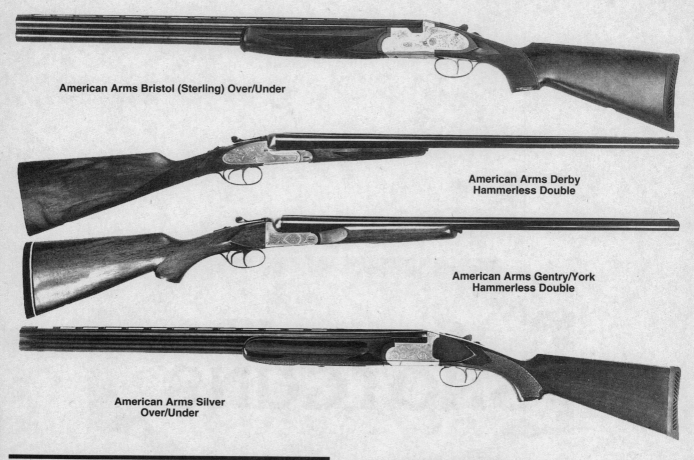

American Arms Bristol (Sterling) Over/Under

American Arms Derby Hammerless Double

American Arms Gentry/York Hammerless Double

American Arms Silver Over/Under

ALDENS SHOTGUN
Chicago, Illinois

Aldens Model 670 Chieftain Slide Action **$165**
Hammerless. Gauges: 12, 20 and others. 3-shot tubular magazine. Bbl.: 26- to 30-inch; various chokes. Weight: 6.25 to 7.5 lbs. depending on bbl. length and ga. Walnut-finished hardwood stock.

AMERICAN ARMS
N. Kansas City, Missouri

See also Franchi Shotguns.

American Arms Bristol (Sterling) Over/Under **$550**
Boxlock w/Greener crossbolt and engraved sideplates. Single selective trigger. Selective automatic ejectors. Gauges: 12, 20; 3-inch chambers. 26-, 28-, 30-, or 32-inch vent-rib bbls. w/screw in choke tubes (IC/M/F). Weight: 7 lbs. Antique-silver receiver w/game scene or scroll engraving. Checkered full pistol-grip style buttstock and forearm w/high-gloss finish. Imported 1986-88 with the model designation Bristol, 1989-90 redesignated Sterling.

American Arms Brittany Hammerless Double **$565**
Boxlock w/engraved case-colored receiver. Single selective trigger. Selective automatic ejectors. Gauges: 12, 20. 3-inch chambers. Bbls.: 25- or 27-inch w/screw in choke tubes (IC/M/F). Weight: 6.5 lbs (20 ga.). Checkered English-style walnut stock w/semibeavertail forearm or pistol-grip stock w/high-gloss finish. Imported 1989 to date.

American Arms Camper Special **$85**
Similar to the Single Barrel, except a takedown model w/21-inch bbl., M choke and pistol-grip stock. Made in 1989 only.

American Arms Camper Special

American Arms Combo **$155**
Similar to the Single-Barrel model, except available w/interchangeable rifle and shotgun bbls. 22 LR/20-ga. shotgun or 22 Hornet/12-ga. shotgun. Rifle bbl. has adj. rear sights; blade-type front sight. Made in 1989.

American Arms Derby Hammerless Double
Sidelock w/engraved sideplates. Single non-selective or double triggers . Selective automatic ejectors. Gauges: 12, 20, 28 and 410. 3 inch chambers. Bbls.: 26-inch (IC/M) or 28-inch (M/ F). Weight: 6 lbs. (20 ga.). Checkered English-style walnut stock and splinter forearm w/hand-rubbed oil finish. Engraved frame/sideplates w/antique silver finish. Imported 1986-94.
12 or 20 Gauge **$675**
28 or .410 (Disc. 1991) **750**

American Arms WS/SS
Hammerless Double

American Arms F.S. Series Over/Under

Greener crossbolt in Trap and Skeet configuration. Single selective trigger. Selective automatic ejectors. 12 gauge only. 26-, 28-, 30-, or 32-inch separated bbls. Weight: 6.5 to 7.25 lbs. Black or chrome receiver. Checkered walnut buttstock and forearm. Imported 1986-87.

Model F.S. 200 Boxlock	**$525**
Model F.S. 300 Boxlock	625
Model F.S. 400 Sidelock	895
Model F.S. 500 Sidelock	925

American Arms Gentry/York Hammerless Double

Chrome, coin-silver or color casehardened boxlock receiver w/scroll engraving. Double triggers. Extractors. Gauges: 12, 16, 20, 28, .410. 3-inch chambers (16 and 28 have 2.75-inch). Bbls.: 26-inch (IC/M) or 28-inch (M/F, 12, 16 and 20). Weight: 6.75 lbs. (12 ga.). Checkered walnut buttstock w/pistol grip and beavertail forearm; both w/semi-gloss oil finish. Imported as York 1986-88, redesignated Gentry 1989 to date.

Gentry 12, 16 or 20 Gauge	**$450**
Gentry 28 or .410 Gauge	500
York 12, 16 or 20 Gauge (Disc. 1988)	395
York 28 or .410 Gauge (Disc. 1988)	475

American Arms Grulla #2 Hammerless Double

True sidelock w/engraved detachable sideplates. Double triggers. Extractors and cocking indicators. Gauges: 12, 20, .410 w/3-inch chambers; 28 w/2.75-inch. 26-inch bbl. Imported 1989 to date.

Standard Model	**$2450**
Two-bbl. Set (Disc. 1995)	3195

American Arms Silver I Over/Under

Boxlock. Single selective trigger. Extractors. Gauges: 12, 20 and .410 w/3-inch chambers; 28 w/2.75 inch. Bbls.: 26-inch (IC/M), 28-inch (M/F, 12 and 20 ga. only). Weight: 6.75 lbs. (12 ga.). Checkered walnut stock and forearm. Antique-silver receiver w/scroll engraving. Imported 1987 to date.

12 or 20 Gauge	**$425**
28 or .410 Gauge	450

American Arms Silver II Over/Under

Similar to Model Silver I, except w/selective automatic ejectors and 26-inch bbls. w/screw-in tubes (12 and 20 ga.). Fixed chokes (28 and .410). Made 1987 to date.

12 or 20 Gauge	**$550**
28 or .410	595
Upland Lite II	650
Two-bbl. Set	850

American Arms Silver Lite Over/Under

Similar to Model Silver II, except w/blued, engraved alloy receiver. Available in 12 and 20 ga. only. Imported from 1990-92.

Standard Model	**$575**
Two-bbl. Set	850

American Arms Silver Skeet/Trap $675

Similar to the Silver II Model, except has 28-inch (Skeet) or 30-inch (Trap) ported bbls. w/target-style rib and mid-bead sight. Imported 1992-94.

American Arms Silver Sporting Over/Under...... $645

Boxlock. Single selective trigger. Selective automatic ejectors. Gauges: 12, 2.75-inch chambers. 28-inch bbls.w/Franchoke tubes (SK, IC, M and F). Weight: 7.5 lbs. Checkered walnut stock and forearm. Special broadway rib and vented side ribs. Engraved receiver w/chrome-nickel finish. Imported from 1990 to date.

American Arms Single-Shot Shotgun

Break-open action. Gauges: 10 (3.5), 12, 20, .410, 3-inch chamber. Weight: about 6.5 lbs. Bead front sight. Walnut-finished hardwood stock w/checkered grip and forend. Made from 1988 to 1990.

10 ga. (3.5-inch)l.	**$100**
12, 20 or .410	85
Multi-choke bbl., **add**	30

American Arms Slugger Single-Shot Shotgun..... $95

Similar to the Single-Shot model, except in 12 and 20 ga. only w/24-inch slug bbl. Rifle-type sights and recoil pad. Made from 1989 to 1990.

American Arms TS/OU 12 Shotgun $595

Turkey Special. Boxlock. Single selective trigger. Selective automatic ejectors. Gauge: 12, 3.5-inch chambers. Bbls.: 24-inch O/U w/screw-in choke tubes (IC, M, F). Weight: 6 lbs. 15 oz. Checkered European walnut stock and beavertail forearm. Matte blue metal finish. Imported 1987 to date.

American Arms TS/SS 10 Hammerless Double $525

Turkey Special. Same general specifications as Model WS/SS 10, except w/26-inch side-by-side bbls., screw-in choke tubes (F/F) and chambered for 10-ga. 3.5-inch shells. Weight: 10 lbs. 13 oz. Imported 1987 to 1993.

American Arms TS/SS 12 Hammerless Double $475

Same general specifications as Model WS/SS 10, except in 12 ga. w/26-inch side-by-side bbls. and 3 screw-in choke tubes (IC/M/F). Weight: 7 lbs. 6 oz. Imported 1987 to date.

American Arms WS/OU 12 Shotgun $495

Waterfowl Special. Boxlock. Single selective trigger. Selective automatic ejectors. Gauge: 12; 3.5-inch chambers. Bbls.: 28-inch O/U w/screw-in tubes (IC/M/F). Weight: 7 lbs. Checkered European walnut stock and beavertail forearm. Matte blue metal finish. Imported 1987 to date.

American Arms WS/SS 10 Hammerless Double $595

Waterfowl Special. Boxlock. Double triggers. Extractors. Gauge: 10; 3.5-inch chambers. Bbls.: 32-inch side/side choked F/F. Weight: about 11 lbs. Checkered walnut stock and beavertail forearm w/satin finish. Parkerized metal finish. Imported from 1987 to 1995.

SHOTGUNS

**Armalite AR-17
Golden Gun**

American Arms WT/OU 10 Shotgun **$695**
Same general specifications as Model WS/OU 12, except chambered for 10-ga. 3.5-inch shells. Extractors. Satin wood finish and matte blue metal. Imported 1987 to date.

ARMALITE, INC.
Costa Mesa, California

Armalite AR-17 Golden Gun **$550**
Recoil-operated semiautomatic. High-test aluminum bbl. and receiver housing. 12 ga. only. 2-shot. 24-inch bbl. w/interchangeable choke tubes: IC/M/F. Weight: 5.6 lbs. Polycarbonate stock and forearm recoil pad. Gold anodized finish standard, also made w/black finish. Made 1964-65. Less than 2,000 produced.

ARMSCOR (Arms Corp.)
Manila, Philippines
Imported until 1991 by Armscor Precision, San Mateo, CA; 1991-95 by Ruko Products, Inc., Buffalo NY: Currently improted by K.B.I., Harrisburg, PA

Armscor Model M-30 Field Pump Shotgun
Double slide-action bars w/damascened bolt. Gauge: 12 only w/3-inch chamber. Bbl.: 28-inch w/fixed chokes or choke tubes. Weight: 7.6 lbs. Walnut or walnut finished hardwood stock.
Model M30-F (w/hardwood stock and fixed chokes). **$125**
Model M-30F (w/hardwood stock and choke tubes) **145**
Model M-30F/IC (w/walnut stock and choke tubes) **165**

Armscor Model M-30 Riot Pump
Double-action slide bar w/damascened bolt. Gauge: 12 only w/3-inch chamber. Bbls: 18.5 and 20-inch w/IC bore. 5- or 7-shot magazine. Weight: 7 lbs, 2 ozs. Walnut finished hardwood stock.
Model M-30R6 (5-shot magazine) **$145**
Model M-30R8 (7-shot magazine) **150**

Armscor Model M-30 Special Combo
Simlar to Special Purpose Model, except has detachable synthetic stock that removes to convert to pistol-grip configuration.
Model M-30C (Disc. 1995) **$165**
Model M-30RP (Disc. 1995) **175**

Armscor Model M-30 Special Purpose
Double-action slide bar w/damascened bolt. 7-shot magazine. Gauge: 12 only w/3-inch chamber. 20-inch bbl. w/cylinder choke. Iron sights (DG Model) or venter handguard (SAS Model). Weight: 7.5 lbs. Walnut finished hardwood stock.
Model M-30DG (Deer Gun). **$165**
Model M-30SAS (Special Air Services). **185**

ARMSPORT, INC.
Miami, Florida

Armsport 1000 Series Hammerless Doubles
Side-by-side w/engraved receiver, double triggers and extractors. Gauges: 10 (3.5), 12, 20, .410- 3-inch chambers. **Model 1033**: 10 ga., 32-inch bbl. **Model 1050/51**: 12 ga., 28-inch bbl., M/F choke. **Model 1052/53**: 20 ga., 26-inch bbl., I/M choke. **Model 1054/57**: .410 ga., 26-inch bbl., I/M. **Model 1055**: 28 ga., Weight: 5.75 to 7.25 lbs. European walnut buttstock and forend. Made in Italy. Importation disc. 1993.
Model 1033 (10 ga. Disc. 1989). **$550**
Model 1050 (12 ga. Disc. 1993). **495**
Model 1051 (12 ga. Disc. 1985). **295**
Model 1052 (20 ga. Disc. 1985). **275**
Model 1053 (20 ga. Disc. 1993). **495**
Model 1054 (.410 Disc. 1992) **550**
Model 1055 (28 ga. Disc. 1992). **325**
Model 1057 (410 Disc. 1985). **350**

Armsport 1050

Armsport 1125

Armsport 2741

Armsport Model 1125 Single-Shot Shotgun $75
Bottom-opening lever. Gauges: 12, 20. 3-inch chambers. Bead front sight. Plain stock and forend. Imported 1987-89.

Armsport Model 2700 Goose Gun
Similar to the 2700 Standard Model, except 10 ga. w/3.5-inch chambers. Double triggers w/28-inch bbl. choked IC/M or 32-inch bbl., F/F. 12mm wide vent rib. Weight: 9.5 lbs. Canada geese engraved on receiver. Antiqued silver-finished action. Checkered European walnut stock w/rubber recoil pad. Imported from Italy 1986 to 1993.

W/Fixed Chokes **$795**
W/Choke Tubes 875

Armsport Model 2700 Over/Under Series
Hammerless, takedown shotgun w/engraved receiver. Selective single or double triggers. Gauges: 10, 12, 20, 28 and .410. Bbl.: 26-or 28-inch w/fixed chokes or choke tubes. Weight: 8 lbs. Checkered European walnut buttstock and forend. Made in Italy. Importation disc. 1993.

Model 2701 12 ga. (Disc. 1985) **$395**
Model 2702 12 ga. 425
Model 2703 20 ga. (Disc. 1985) 415
Model 2704 20 ga. 450
Model 2705 (.410, DT, fixed chokes) 535
Model 2730/31 (Boss-style action, SST Choke tubes) 595
Model 2733/35 (Boss-style action, extractors) 550
Model 2741 (Boss-style action, ejectors) 495
Model 2742 Sporting Clays (12 ga./choke tubes) 575
Model 2744 Sporting Clays (20 ga./choke tubes) 585
Model 2750 Sporting Clays (12 ga./sideplates) 625
Model 2751 Sporting Clays (20 ga./sideplates) 650

Armsport Model 2755 Slide-Action Shotgun
Gauge: 12 w/3-inch chamber. Tubular magazine. Bbls.: 28- or 30-inch w/fixed choke or choke tubes. Weight: 7 lbs. European walnut stock. Made in Italy 1986-87.

Standard Model, Fixed choke **$250**
Standard Model, Choke Tubes 395
Police Model, 20-inch Bbl. 225

Armsport Model 2900 Tri-Barrel (Trilling)Shotgun
Boxlock. Double triggers w/top-tang bbl. selector. Extractors. Gauge: 12; 3-inch chambers. Bbls.: 28-inch (IC, M and F). Weight: 7.75 lbs. Checkered European walnut stock and forearm. Engraved silver receiver. Imported 1986-87 and 1990-93.

Model 2900 (Fixed Chokes) 1595
Model 2900 (Choke Tubes) 2150
Deluxe Grades, add 500

ASTRA SHOTGUNS
Guernica, Spain
Manufactured by Unceta y Compania

Astra Model 650 Over/Under Shotgun
Hammerless, takedown w/double triggers. 12 ga. w/.75-inch chambers. Bbls.: 28-inch (M/F or SK/SK); 30-inch (M/F). Weight: 6.75 lbs. Checkered European walnut buttstock and forend.

W/Extractors **$450**
W/Ejectors 545

Astra Model 750 Over/Under Shotgun
Similar to the Model 650, except w/selective single trigger and ejectors. Made in field, Skeet and Trap configurations since 1980.

Field Model w/Extractors **$500**
Field Model w/Ejectors 595
Trap or Skeet 695

Armsport Model 2900 Tri-Barrel

AYA (Aguirre Y Aranzabal)
Eibar, Spain
(Previously Mfd. by Diarm)
Imported by Armes De Chasse, Hertford, NC

AyA Model 1 Hammerless Double
A Holland & Holland sidelock similar to the Model 2 except in 12 and 20 ga. only, w/special engraving and exhibition-grade wood. Weight: 5-8 lbs., depending on ga. Imported by Diarm until 1987, since 1992 by Armes de Chasse.

Model 1 Standard **$2395**
Model 1 Deluxe 3250
Extra Set of Bbls., **add** 2550

AyA Model 2 Hammerless Double
Sidelock action w/selective single or double triggers automatic ejectors and safety. Gauges: 12, 20, 28, (2.75-inch chambers); .410 (3-inch chambers). Bbls.: 26- or 28-inch w/various fixed choke combinations. Weight: 7 lbs. (12 ga.). English-style straight walnut buttstock and splinter forend. Imported by Diarm until 1987, since 1992 by Armes de Chasse.

12 or 20 ga. w/Double Triggers **$1450**
12 or 20 ga. w/Single Trigger 1595
28 or .410 ga. w/Double Triggers 1650
28 or .410 ga. w/Single Trigger 1895
Extra Set of Bbls., **add** 1250

AyA Model 4 Hammerless Double
Lightweight Anson & Deely boxlock w/a scalloped frame. Gauges: 12, 16, 20, 28, and .410. Bbls.: 25- to 28-inch w/concave rib. Importation disc. 1987 and resumed in 1992 by Armes de Chasse.

12 Gauge **$650**
16 Gauge (early importation) 475
20 Gauge 635
28 Gauge 725
.410 Bore 795
Deluxe Grades, **add** 500

AyA Model 37 Super Over/Under Shotgun $2595
Sidelock. automatic ejectors. Selective single trigger. Made in all gauges, bbl. lengths and chokes. Vent-rib bbls. Elaborately engraved. Checkered stock (w/straight or pistol grip) and forend. Discontinued 1895.

AyA Model 37 Super A Over/Under Shotgun $4595
Similar to the Standard Model 37 Super except has nickel steel frame and fitted w/detachable sidelocks engraved w/game scenes. Importation disc. 1987 and resumed 1992 by Armes de Chasse.

SHOTGUNS

AyA Model 53E $1750
Same general specifications as Model 117, except more elaborate engraving and select figured wood. Importation disc. 1987 and resumed in 1992 by Armes de Chasse.

AyA Model 56 Hammerless Double
Pigeon weight Holland & Holland sidelock w/Purdey-style third lug and sideclips. Gauges: 12, 16, 20. Receiver has fine-line scroll and rosette engraving; gold-plated locks. Importation disc. 1987 and resumed 1992 by Armes de Chasse.

12 Gauge.	$2850
16 Gauge (early importation)	2250
20 Gauge (early importation)	2995

AyA Model 76 Hammerless Double $550
Anson & Deeley boxlock. Auto ejectors. Selective single trigger. Gauges: 12, 20 (3-inch). Bbls.: 26-, 28-, 30-inch (latter in 12 ga. only), any standard choke combination. Checkered pistol-grip stock/beavertail forend. Discontinued.

AyA Model 76—.410 Gauge $650
Same general specifications as 12 and 20 ga. Model 76, except chambered for 3-inch shells in .410, has extractors, double triggers, 26-inch bbls. only, English-style stock w/straight grip and small forend. Discontinued.

AyA Model 117 Hammerless Double $725
Holland & Holland-type sidelocks, hand-detachable. Engraved action. Automatic ejectors. Selective single trigger. Gauges: 12, 20 (3-inch). Bbls.: 26-, 27-, 28-, 30-inch; 27- and 30-inch in 12 ga. only; any standard choke combination. Checkered pistol-grip stock and beavertail forend of select walnut. Manufactured in 1985.

AyA Bolero $375
Same general specifications as Matador except non-selective single trigger and extractors. Gauges: 12 16, 20, 20 Magnum (3-inch), .410 (3-inch). *Note:* This model, prior to 1956, was designated F. I. Model 400 by the importer. Made 1955-63.

AYA Model XXV Boxlock

AyA Contento Over/Under Shotgun
Boxlock w/Woodward side lugs and double internal bolts. Gauge: 12 (2.75-inch chambers). Bbls.: 26-, 28-inch field; 30-, 32-inch trap; either fixed chokes as required or screw-in choke tubes. Hand-checkered European Walnut stock and forend. Single-selective trigger and automatic ejectors.

M.K.1 Field	$ 575
M.K.2	725
M.K.3	1260
W/Interchangeable Single Bbl., **add**	400

AyA Matador Hammerless Double $425
Anson & Deeley boxlock. Selective automatic ejectors. Selective single trigger. Gauges: 12, 16, 20, 20 Magnum (3-inch). Bbls.: 26-, 28-, 30-inch; any standard choke combination. Weight: 6.5 to 7.5 lbs., depending on ga. and bbl. length. Checkered pistol-grip stock and beavertail forend. *Note:* This model, prior to 1956, was designated F. I. Model 400E by the U.S. importer, Firearms Int'l. Corp. of Washington, D.C. Made 1955-63.

AyA Matador II $495
Improved version of Matador w/same general specifications, except has vent-rib bbls.. Made 1964-69.

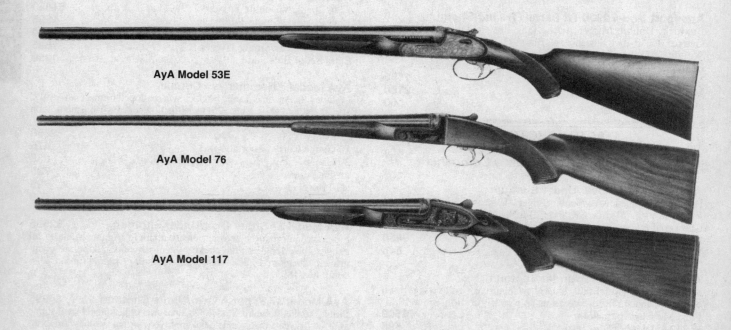

AyA Model 53E

AyA Model 76

AyA Model 117

AyA Matador II

AyA Matador III $695
Same general specifications as AyA Matador II. Made 1970-85.

AYA Model XXV Boxlock
Anson & Deeley boxlock w/double locking lugs. Gauges: 12 and 20. 25-inch chopper lump, satin blued bbls. w/Churchill rib. Weight: 5 to 7 lbs. Double triggers. Automatic safety and ejectors. Color-casehardened receiver w/continental style scroll and floral engraving. European walnut stock. Imported 1979-86 and 1991 to date.
12 or 20 Gauge **$1095**
Extra Set of Bbls., add **1050**

AYA Model XXV Sidelock
Holland & Holland-type sidelock. Gauges: 12, 20, 28 and 410. 25-, 26-, 27- 28-, 29-, and 32-inch bbls. chopper lump, satin blued bbls. w/Churchill rib. Weight: 5 to 7 lbs. Double triggers standard or selective or non-selective single trigger optional. Automatic safety and ejectors. Cocking indicators. Color-casehardened or coin-silver finished receiver w/continental style scroll and floral engraving. Select European walnut stock w/hand-cut checkering and oil finish. Imported 1979-86 and 1991 to date.
12 or 20 Gauge **$1450**
28 Gauge (disc. 1997) **1595**
.410 Bore (disc. 1997) **1750**
Single Trigger, add **75**
Single Selective Trigger, add **120**
Extra Set of Bbls., add **1250**

BAIKAL SHOTGUNS
Izhevsk and Tula, Russia

Baikal Model IJ-18M Single Shot $50
Hammerless w/cocking indicator. Automatic ejector. Manual safety. Gauges: 12 , 20, 16 w/2.75-inch chamber or .410 w/3-inch chamber. Bbls.: 26-, 28-inch w/fixed chokes (IC, M, F). Weight: 5.5 to 6 lbs. Made in Russia.

Baikal Model IJ-27 Field Over/Under $265
Boxlock. Double triggers w/extractors. 12 ga.; 2.75-inch chambers. Bbls.: 26-inch, IC/M; 28-inch, M/F w/fixed chokes. Weight: 6.75 lbs. Made in Russia.

Baikal Model IJ-43 Field Side-by-side
Side-by-side; boxlock. Double triggers; extractors. 12 or 20 ga. w/2.75-inch chambers. Bbls: 20-inch cylinder bbl and 26-or 28-inch modified full bbl. Weight: 6.75 to 7 lbs. Checkered walnut stock, forend. Blued, engraved receiver. Imported 1994-96.
Model IJ-43 Field w/20-inch bbls. **$185**
Model IJ-43 Field w/26- or 28-inch bbls **165**

Baikal IZH-43 Series Side-by-side
Boxlock. Gauges: 12, 16, 20 or .410 w/2.75- or 3-inch chambers. Bbls.: 20-, 24-, 26- or 28-inch w/fixed chokes or choke tubes. Single selective or double triggers. Weight: 6.75 lbs. Checkered

Baikal IZH-43 Series Side-by-Side *(Con't)*
hardwood (standard on Hunter II Model) or walnut stock and forend (standard on Hunter Model). Blued, engraved receiver. Imported 1994 to date.
Model IZH-43 Hunter (12ga. w/walnut stock) **$195**
Model IZH-43 Hunter (20, 16 or .410ga.) **265**
Model IZH-43 Hunter II (12ga. w/external hammers) **275**
Model IZH-43 Hunter II (12ga. hammerless) **175**
Model IZH-43 Hunter II (20, 16 or .410ga.) **185**
Hunter II w/walnut stock, add **35**
Hunter II w/single selective trigger, add **45**

Baikal Model IJ-27 Over/Under
Boxlock. Single selective trigger w/automatic ejectors or double triggers w/extractors. Gauges: 12 or 20 w/2.75-inch chambers. Bbls.: 26-inch or 28-inch w/fixed chokes. Weight: 7 lbs. Checkered European hardwood stock and forearm. Made in Russia.
Model IJ-27 (w/double triggers and extractors.) **$255**
Model IJ-27 (single selective trigger and ejectors) **295**

Baikal Model IJ-18M Single Shot $50
Hammerless w/cocking indicator. Automatic ejector. Manual safety. Gauges: 12 , 20, 16 w/2.75-inch chamber or .410 w/3-inch chamber. Bbls.: 26-, 28-inch w/fixed chokes (IC, M, F). Weight: 5.5 to 6 lbs. Made in Russia.

BAKER SHOTGUNS
Batavia, New York
Made 1903-1933 by Baker Gun Company

Baker Batavia Ejector $795
Same general specifications as the Batavia Leader, except higher quality and finer finish throughout; has Damascus or homotensile steel bbls., checkered pistol-grip stock and forearm of select walnut; automatic ejectors standard; 12 and 16 ga. only. **Deduct 60%** for Damascus bbls.

Baker Batavia Leader Hammerless Double
Sidelock. Plain extractors or automatic ejectors. Double triggers. Gauges: 12, 16, 20. Bbls.: 26- to 32-inch; any standard boring. Weight: about 7.75 lbs. (12 ga. w/30-inch bbls.). Checkered pistol-grip stock and forearm.
W/Plain Extractors **$395**
W/Automatic Ejectors **450**

Baker Batavia Special $325
Same general specifications as the Batavia Leader, except 12 and 16 ga. only; extractors, homotensile steel bbls.

Baker Black Beauty
Same general specifications as the Batavia Leader, except higher quality and finer finish throughout; has line engraving, special steel bbls., select walnut stock w/straight, full or half-pistol grip.
Black Beauty w/Plain Extractors **$395**
Black Beauty Special w/Plain Extractors **695**
Black Beauty Special w/Automatic Ejectors **795**

SHOTGUNS

Baker Black Beauty Special

Baker Grade R

High-grade gun w/same general specifications as the Batavia Leader, except has fine Damascus or Krupp fluid steel bbls., engraving in line, scroll and game scene designs, checkered stock and forearm of fancy European walnut; 12 and 16 ga. only. **Deduct 60%** for Damascus bbls.

Non-ejector **$995**
W/Automatic Ejectors **1250**

Baker Grade S

Same general specifications as the Batavia Leader, except higher quality and finer finish throughout; has Flui-tempered steel bbls., line and scroll engraving, checkered stock w/half-pistol grip and forearm of semifancy imported walnut; 10, 12 and 16 ga.

Non-ejector **$795**
W/Automatic Ejectors **1025**

Baker Paragon, Expert and Deluxe Grades

Made to order only, these are the higher grades of Baker hammerless sidelock double-bbl. shotguns. After 1909, the Paragon Grade, as well as the Expert and Deluxe Intro. that year, had a cross bolt in addition to the regular Baker system taper wedge fastening. There are early Paragon guns w/Damascus bbls. and some are non-ejector, but this grade was also produced w/automatic ejectors and w/the finest fluid steel bbls., in lengths to 34 inches, standard on Expert and Deluxe guns.

Differences among the three models are in overall quality, finish, elaborateness of engraving and grade of fancy figured walnut in the stock and forearm; Expert and Deluxe wood may be carved as well as checkered. Choice of straight, full or half-pistol grip was offered. A single trigger was available in the two higher grades. The Paragon was available in 10 ga (Damascus bbls. only), this and the other two models were regularly produced in 12, 16 and 20 ga.

Paragon Grade, Non-ejector **$1395**
Paragon Grade, Automatic Ejectors **1595**
Expert Grade **2495**
Deluxe Grade **3725**
For Single Trigger, **add** **250**

BELKNAP SHOTGUNS
Louisville, Kentucky

Belknap Model B-63 Single-Shot Shotgun **$85**
Takedown. Visible hammer. Automatic ejector. Gauges: 12, 20 and .410. Bbls.: 26- to 36-inch, F choke. Weight: average 6 lbs. Plain pistol-grip stock and forearm.

Belknap Model B-63E Single-Shot Shotgun **$85**
Same general specifications as Model B-68, except has side lever opening instead of top lever.

Belknap Model B-64 Slide-Action Shotgun **$175**
Hammerless. Gauges: 12, 16, 20 and .410. 3-shot tubular magazine. Various bbl. lengths and chokes from 26-inch to 30-inch. Weight: 6.25 to 7.5 lbs. Walnut-finished hardwood stock.

Belknap Model B-65C Autoloading Shotgun **$285**
Browning-type lightweight alloy receiver. 12 ga. only. 4-shot tubular magazine. Bbl.: plain, 28-inch. Weight: about 8.25 lbs. Discontinued 1949.

Belknap Model B-68 Single-Shot Shotgun **$95**
Takedown. Visible hammer. Automatic ejector. Gauges: 12, 16, 20 and .410. Bbls.: 26-inch to 36-inch; F choke. Weight: 6 lbs. Plain pistol-grip stock and forearm.

BENELLI SHOTGUNS
Urbino, Italy
Imported by Benelli USA, Accokeek, MD
(Previously by Heckler & Koch)

Benelli Model 121 M1 Military/Police Autoloading Shotgun **$395**
Gauge: 12. 7-shot magazine. 19.75-inch bbl. 39.75 inches overall. Cylinder choke, 2.75-inch chamber. Weight: 7.4 lbs. Matte black finish and European hardwood stock. Post front sight, fixed buckhorn rear sight. Imported in 1985.

Baker Paragon
Expert and Deluxe Grades

Benelli Black Eagle Autoloading Shotgun

Two-piece aluminum and steel receiver. Gauge: 12; 3-inch chamber. 4-shot magazine. Screw-in choke tubes (SK, IC, M, IM, F). Bbls.: ventilated rib; 21, 24, 26 or 28 inches w/bead front sight; 24-inch rifled slug. 42.5 to 49.5 inches overall. Weight: 7.25 lbs. (28-inch bbl.). Matte black lower receiver w/blued upper receiver and bbl. Checkered walnut stock w/high-gloss finish and drop adjustment. Imported from 1989-90 and 1997-98.

Limited Edition **$1350**
Competition Model **725**
Slug Model (Discontinued 1992) **525**
Standard Model (Discontinued 1992) **565**

Benelli Black Eagle Executive

Custom Black Eagle Series. Montefeltro-style rotating bolt w/three locking lugs. All steel lower receiver engraved and gold inlayed by Bottega Incisione di Cesare Giovanelli. 12 ga. only. 21-, 24-, 26-, or 28-inch vent-rib bbl. w/5 screw-in choke tubes (Type I) or fixed chokes. Custom deluxe walnut stock and forend. Built to customer specifications on special order.

Executive Type I **$2795**
Executive Type II **3150**
Executive Type III **3795**

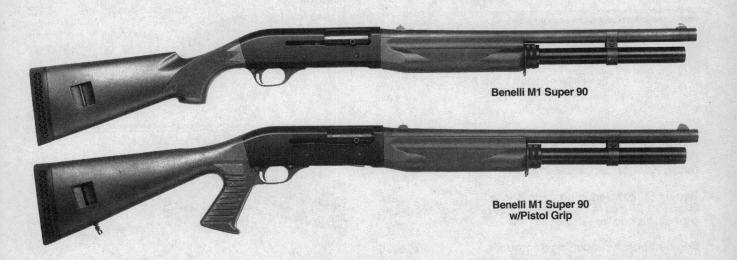

Benelli M1 Super 90

Benelli M1 Super 90 w/Pistol Grip

Benelli Lagecy Autoloading Shotgun $750
Same general specifications as Black Eagle Competition Model, except available in both 12 and 20 ga. w/3-inch chambers. 26- or 28-inch bbl. w/screw-in choke tubes. Imported 1998 to date.

Benelli M1 Super 90 Autoloading Shotgun $495
Gauge: 12. 7-shot magazine. Cylinder choke. 19.75-inch bbl. 39.75 inches overall. Weight: 7 lbs. 4 oz. to 7 lbs. 10 oz. Matte black finish. Stock and forend made of fiberglass reinforced polymer. Sights: post front, fixed buckhorn rear, drift adj. Introduced 1985; when the model line expanded in 1989 this configuration was discontinued..

Benelli M1 Super 90 Defense Autoloader $525
Same general specifications as Model Super 90, except w/pistol-grip stock. Available w/Ghost-Ring sight option. Imported 1986-98.

Benelli M1 Super 90 Entry Autoloader $535
Same gen. specifications as Model Super 90, except w/5-shot magazine. 14-inch bbl. 35.5 inches overall. Weight: 6.5 lbs. Standard or pistol-grip stock. Imported 1992 to date. *Special permit required for under 18 inch bbl.*

Benelli M1 Super 90 Field
Inertia-recoil semiautomatic shotgun. Gauge: 12; 3-inch chamber. 3-shot magazine. Bbl.: 21, 24, 26 or 28 inches. 42.5 to 49.5 inches overall. Choke: SK, IC, M, IM, F. Matte receiver. Standard polymer stock or satin walnut (26- or 28-inch bbl. only). Bead front sight. Imported from 1990 to date.
W/Real Tree Camo $550
W/Polymer Stock 450
W/Walnut Stock 475

Benelli M1 Super 90 Slug Autoloader
Same general specifications as M1 Super 90 Field, except w/5-shot magazine. 18.5-inch bbl. Cylinder bore. 39.75 inches overall. Weight: 6.5 lbs. Polymer standard stock. Rifle or Ghost-Ring sights. Imported 1986-98.
W/Rifle Sights $550
W/Ghost-Ring Sights 595

Benelli M1 Super 90 Sporting Special Autoloader . $545
Same general specifications as M1 Super 90 Field, except w/18.5-inch bbl. 39.75 inches overall. Weight: 6.5 lbs. Ghost-ring sights. Polymer stock. Imported 1994 to date.

Benelli M1 Super 90 Tactical Autoloader $565
Same general specifications as M1 Super 90 Field, except w/18.5-inch bbl. 5-shot magazine. IC, M, or F choke. 39.75 inches overall. Weight: 6.5 lbs. Rifle or Ghost-Ring sights. Polymer pistol-grip or standard stock. Imported 1994 to date.

Benelli M3 Super 90 Pump/Autoloader
Inertia-recoil semiautomatic and/or pump action. Gauge: 12. 7-shot magazine. Cylinder choke. 19.75-inch bbl. 41 inches overall (31 inches folded). Weight: 7 to 7.5 lbs. Matte black finish. Stock: standard synthetic, pistol-grip or folding tubular steel. Standard rifle or Ghost-Ring sights. Imported 1989 to date. *Caution: Increasing the magazine capacity to more than 5 rounds in M3 shotguns w/pistol-grip stocks violates provisions of the 1994 Crime Bill. This model may be used legally only by the military and law-enforcement agencies.*
Standard Model. $650
Pistol-grip Model 750
W/Folding Stock 795
W/Laser Sight 995
For Ghost Ring Sights, **add** 50

Benelli Montefeltro Super 90 Semiautomatic
Same general specifications as Model M1 Super 90, except available in both 12 and 20 ga. w/3-inch chamber and 4-shot magazine. Bbls.: 21, 24, 26 or 28 inches, w/screw-in choke tubes (IC, M, IM, F). Weight: 7.5 lbs. Blued metal finish. Imported 1987 to date.
Standard Hunter Model $550
Slug Model (Discontinued 1992) 465
Turkey Model 450
Uplander Model 495
Left-hand Model 575
Limited Edition 1425

Benelli SL 121V Semiauto Shotgun $325
Gauge: 12. 5-shot capacity. Bbls.: 26-, 28- and 30-inch. 26-inch choked M, IM, IC, 28-inch, F, M, IM; 30-inch, F choke (Mag.). Straight walnut stock w/hand-checkered pistol grip and forend. Ventilated rib. Imported in 1985.

Benelli SL 121V Slug Shotgun $350
Same general specifications as Benelli SL 121V, except designed for rifled slugs and equipped w/rifle sights. Imported in 1985.

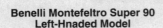

**Benelli Montefeltro Super 90
Left-Hnaded Model**

Benelli SL 123V Semiauto Shotgun **$375**
Gauge: 12. 26- and 28-inch bbls. 26-inch choked IM, M, IC; 28-inch choked F, IM, M. Imported in 1985.

Benelli SL 201 Semiautomatic Shotgun **$320**
Gauge: 20. 26-inch bbl. Mod. choke. Weight: 5 lbs., 10 oz. Ventilated rib. Imported in 1985.

Benelli Sport Autoloading Shotgun **$735**
Similar to the Black Eagle Competition Model, except has one-piece matte finished alloy receiver w/inscribed red Benelli logo. 26 or 28 inches bbl. w/2 inchangable carbon fiber vent ribs. Oil-finished checkered walnut stock w/adjustable buttpad and buttstock. Imported 1997 to date.

Benelli Super Black Eagle Autoloading Shotgun
Same general specifications as Model Black Eagle, except w/3.5-inch chamber that accepts 2.75-, 3- and 3.5-inch shells. 2-shot magazine (3.5-inch), 3-shot magazine (2.75- or 3-inch). High-gloss or satin finish stock. Matte black or blued metal finish. Imported 1991 to date.
Standard Model **795**
Realtree Camo **885**
Custom Slug Model **775**
Limited Edition **$1425**

BERETTA USA CORP.
Accokeek, Maryland
Mfd. by Fabbrica D'Armi Pietro Beretta S.p.A. in
the Gardone Val Trompia (Brescia), Italy
Imported by Beretta USA
(Previously by Garcia Corp.)

Beretta Model 57E

Beretta Model 57E Over-and-Under
Same general specifications as Golden Snipe, but higher quality throughout. Imported 1955-67.
W/Non-selective Single Trigger **$695**
W/Selective Single Trigger **795**

Beretta Model 409PB Hammerless Double **$650**
Boxlock. Double triggers. Plain extractors. Gauges: 12, 16, 20, 28. Bbls.: 27.5-, 28.5- and 30-inch, IC/M choke or M/F choke. Weight: from 5.5 to 7.75 lbs., depending on ga. and bbl. length. Straight or pistol-grip stock and beavertail forearm, checkered. Imported 1934-64.

Beretta Model 409PB

Beretta Model 410E

Beretta Model 410, 10-Gauge Magnum **$950**
Same as Model 410E, except heavier construction. Plain extractors. Double triggers. 10-ga. Magnum, 3.5-inch chambers. 32-inch bbls., both F choke. Weight: about 10 lbs. Checkered pistol-grip stock and forearm, recoil pad. Imported 1934-84.

Beretta Model 410E
Same general specifications as Model 409PB, except has automatic ejectors and is of higher quality throughout. Imported 1934-64.
Model 410E (12 ga.) . . . . **$ 950**
Model 410E (20 ga.) **1350**
Model 410E (28 ga.) **3500**

Beretta Model 411E
Same general specifications as Model 409PB except has sideplates, automatic ejectors and is of higher quality throughout. Imported 1934-64.
Model 411E (12 ga.) **$1495**
Model 411E (20 ga.) **1950**
Model 411E (28 ga.) **3595**

Beretta Model 424 Hammerless Double **$1095**
Boxlock. Light border engraving. Plain extractors. Gauges: 12, 20; chambers 2.75-inch in former, 3-inch in latter. Bbls.: 28-inch M/F choke, 26-inch IC/M choke. Weight: 5 lbs. 14 oz. to 6 lbs. 10 oz., depending on ga. and bbl. length. English-style straight-grip stock and forearm, checkered. Imported 1977-84.

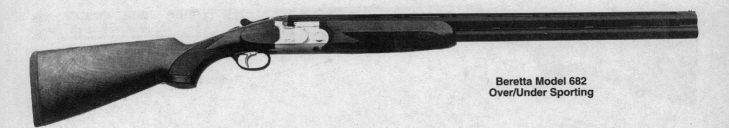

**Beretta Model 682
Over/Under Sporting**

Beretta Model 426E $1395
Same as Model 424, except action body is finely engraved, silver pigeon inlaid in top lever; has selective automatic ejectors and selective single trigger, stock and forearm of select European walnut. Imported 1977-84.

Beretta Model 450 Series Hammerless Doubles
Custom English-style sidelock. Single, non-selective trigger or double triggers. Manual safety. Selective automatic ejectors. Gauge: 12; 2.75- or 3-inch chambers. Bbls.: 26, 28 or 30 inches choked to customers specifications. Weight: 6.75 lbs. Checkered high-grade walnut stock. Receiver w/coin-silver finish. Imported 1948 to date.

Model 450 EL (Disc. 1982)	**$5,895**
Model 450 EELL (Disc. 1982).......................	**6,250**
Model 451 (Disc. 1987)	**5,095**
Model 451 E (Disc. 1989)	**9,595**
Model 451 EL (Disc. 1985)	**4,950**
Model 451 EELL (Disc. 1990).......................	**10,925**
Model 452 (Intro. 1990)............................	**20,000**
Model 452 EELL (Intro. 1992)......................	**28,250**
Model 455	**30,000**
Model 455 EELL	**44,000**
Extra Set of Bbls., add	**6,750**

Beretta Model 625 S/S Hammerless Double
Boxlock. Gauges: 12 or 20. Bbls.: 26-, 28- or 30-inch w/fixed choke combinations. Single selective or double triggers w/extractors. Checkered English-style buttstock and forend. Imported 1984-87.

W/Double Triggers.................................	**$725**
W/Single Selective Trigger..........................	**850**
20 gauge, **add**	**150**

Beretta Model 626 S/S Hammerless Double
Field Grade side-by-side. Boxlock action w/single selective trigger, extractors and automatic safety. Gauges: 12 (2.75-inch chambers); 20 (3-inch chambers). Bbls.: 26- or 28-inch w/Mobilchoke® or various fixed-choke combinations. Weight: 6.75 lbs. (12 ga.). Bright chrome finish. Checkered European walnut buttstock and forend in straight English style. Imported 1985-94.

Model 626 Field (Discontinued 1988)	**$ 825**
Model 626 Onyx	**950**
Model 626 Onyx (3.5-inch Magnum, Disc. 1993)........	**1095**
20 Ga., **add**	**200**

Beretta Model 627 S/S Hammerless Double
Same as Model 626 S/S, except w/engraved sideplates and pistol-grip or straight English-style stock. Imported 1985-94.

Model 627 EL Field................................	**$1925**
Model 627 EL Sport	**1950**
Model 627 EELL...................................	**3750**

Beretta Model 682 Over/Under Shotgun
Hammerless takedown w/single selective trigger. Gauges: 12, 20, 28, .410. Bbls.: 26- to 34-inch w/fixed chokes or Mobilchoke® tubes. Checkered European walnut buttstock/forend in various grades and configurations. Imported 1984 to date.

Model 682 Comp Skeet	**$1495**
Model 682 Comp Skeet Deluxe	**2250**
Model 682 Comp Super Skeet	**1795**
Model 682 Comp Skeet 2-Bbl. Set (Disc. 1989)	**3850**
Model 682 Comp Skeet 4-Bbl. Set (Disc. 1996)	**4595**
Model 682 Sporting Continental	**1795**
Model 682 Sporting Combo	**2350**
Model 682 Sporting Gold	**1395**
Model 682 Sporting Super Sport	**1525**
Model 682 Comp Trap Gold X	**1495**
Model 682 Comp Trap Top Single (1986-95)	**1450**
Model 682 Comp Trap Live Pigeon (1990-98)	**1925**
Model 682 Comp Mono/ComboTrap Gold X	**2150**
Model 682 Comp Mono Trap (1985-88)	**1295**
Model 682 Super Trap Gold X (1991-95)	**1550**
Model 682 Super Trap Combo Gold X (1991-97)........	**2195**
Model 682 Super Trap Top Single Gold X (1991-95).....	**1595**
Model 682 Super Trap Unsingle (1992-94)	**1495**

Beretta Model 686 Over/Under Shotgun
Low-profile improved boxlock action. Single selective trigger. Selective automatic ejectors. Gauges: 12, 20, 28 w/3.5- 3- or 2.75-inch chambers, depending upon ga. Bbls.: 26-, 28-, 30-inch w/fixed chokes or Mobilchoke® tubes. Weight: 5.75 to 7.5 lbs. Checkered American walnut stock and forearm of various qualities, depending upon model. Receiver finishes also vary, but all have blued bbls. Sideplates to simulate sidelock action on EL models. Imported 1988 to date.

Model 686 Field Onyx	**$ 825**
Model 686 (3.5-inch Mag., Disc. 1993 & Reintro.1996) ...	**950**
Model 686 EL Gold Perdiz (1992-97)..................	**1295**

SHOTGUNS

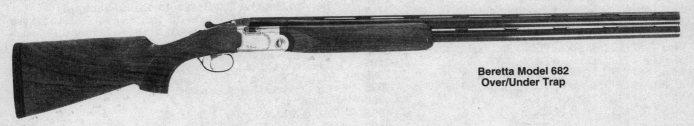

**Beretta Model 682
Over/Under Trap**

Beretta Model 687EL

Beretta Model 687EEL

Beretta Model 686 Over/Under Shotgun *(Con't)*

Model 686 Essential (1994-96)	695
Model 686 Silver Essential (1997-98)	750
Model 686 Silver Pigeon Onyx (Intro. 1996)	950
Model 686 Silver Perdiz Onyx (Disc. 1996)	925
Model 686 Silver Pigeon/Perdiz Onyx Combo	1325
Model 686 L Silver Perdiz (Disc. 1994)	775
Model 686 Skeet Silver Pigeon (1996-98)	975
Model 686 Skeet Silver Perdiz (1994-96)	950
Model 686 Skeet Silver Pigeon/Perdiz Combo	1325
Model 686 Sporting Special (1987-93)	1095
Model 686 Sporting English (1991-92)	1150
Model 686 Sporting Onyx w/fixed chokes (1991-92)	1095
Model 686 Sporting Onyx w/tubes (Intro. 1992)	825
Model 686 Sporting Onyx Gold (Disc. 1993)	1195
Model 686 Sporting Silver Pigeon (Intro. 1996)	975
Model 686 Sporting Silver Perdiz (1993-96)	950
Model 686 Sporting Collection Sport (1996-97)	725
Model 686 Sporting Combo	1695
Model 686 Trap International (1994-95)	695
Model 686 Trap Silver Pigeon (Intro. 1997)	750
Model 686 Trap Top Mono (Intro. 1998)	775
Model 686 Ultralight Onyx (Intro. 1992)	995
Model 686 Ultralight Deluxe Onyx (Intro. 1998)	1250

Beretta Model 687 Over/Under Shotgun

Same as Model 686, except w/decorative sideplates and varying grades of engraving and game-scene motifs.

Model 687 L Onyx (Disc. 1991)	$ 895
Model 687 L Onyx Gold Field (1988-89)	950
Model 687 L Onyx Silver Pigeon	1350
Model 687 EL Onyx (Disc. 1990)	1895
Model 687 EL Gold Pigeon	2095
Model 687 EL Gold Pigeon Small Frame	2350
Model 687 EL Gold Pigeon Sporting (Intro. 1993)	2795
Model 687 EELL Diamond Pigeon	3650
Model 687 EELL Diamond Pigeon Skeet	2695
Model 687 EELL Diamond Pigeon Sporting	3850
Model 687 EELL Diamond Pigeon X Trap	2595
Model 687 EELL Diamond Pigeon Mono Trap	2695
Model 687 EELL Diamond Pigeon Trap Combo	3595
Model 687 EELL Field Combo	4500
Model 687 EELL Skeet 4-bbl. Set	4995
Model 687 EELL Gallery Specials	4500

Beretta Model 687 Over/Under Shotgun *(Con't)*

Model 687 EELL Gallary Special Combo	5500
Model 687 EELL Gallary Special Pairs	11,000
Model 687 Sporting English (1991-92)	1450
Model 687 Sporting Silver Pigeon (Intro. 1996)	1395
Model 687 Sporting Silver Perdiz (1993-96)	1350

Beretta Model 1200 Semiautoloading Shotgun

Short recoil action. Gauge: 12; 2.75- or 3-inch chamber. 6-shot magazine. 24-, 26- or 28-inch vent-rib bbl. w/fixed chokes or Mobilchoke® tubes. Weight: 7.25 lbs. Matte black finish. Adj. technopolymer stock and forend. Imported 1988 to date.

Model 1200 w/Fixed Choke (Disc. 1989)	$375
Model 1200 Riot (Disc. 1989)	425
Model 1201 w/Mobilchoke® (Disc. 1994)	395
Model 1201 Riot	450
W/Tritium Sights, **add**	50

Beretta Model A-301 Autoloading Shotgun $345

Field Gun. Gas-operated. Scroll-decorated receiver. Gauge: 12 or 20; 2.75-inch chamber in former, 3-inch in latter. 3-shot magazine. Bbl.: ventilated rib; 28-inch F or M choke, 26-inch IC. Weight: 6 lbs. 5 oz. – 6 lbs. 14 oz., depending on gauge and bbl. length. Checkered pistol grip stock/forearm. Imported 1977-82.

Beretta Model A-301 Magnum $355

Same as Model A-301 Field Gun, except chambered for 12 ga. 3-inch Magnum shells, 30-inch F choke bbl. only, stock w/recoil pad. Weight: 7.25 lbs.

Beretta Model A-301 Skeet Gun $375

Same as Model A-301 Field Gun, except 26-inch bbl. SK choke only, skeet-style stock, gold-plated trigger.

Beretta Model A-301 Slug Gun $350

Same as Model A-301 Field Gun, except has plain 22-inch bbl., slug choke, w/rifle sights. Weight: 6 lbs. 14 oz.

Beretta Model A-301 Trap Gun $375

Same as Model A-301 Field Gun, except has 30-inch bbl. in F choke only, checkered Monte Carlo stock w/recoil pad, gold-plated trigger. Blued bbl. and receiver. Weight: 7 lbs. 10 oz. Imported 1978-82.

Beretta Model A-302 Semiautoloading Shotgun

Similar to gas-operated Model 301. Hammerless, takedown shotgun w/tubular magazine and Mag-Action that handles both 2.75- and 3-inch Magnum shells. Gauge: 12 or 20; 2.75- or 3-inch Mag. chambers. Bbl.: vent or plain; 22-inch/Slug (12 ga.); 26-inch/IC (12 or 20) 28-inch/M (20 ga.), 28-inch/Multi-choke (12 or 20 ga.) 30-inch/F (12 ga.). Weight: 6.5 lbs., 20 ga.; 7.25.lbs., 12 ga. Blued/ black finish. Checkered European walnut, pistol-grip stock and forend. Imported 1983 to c. 1987.

Standard Model w/Fixed Choke	$355
Standard Model w/Multi-choke	395

Beretta Model A-303

Beretta Model A-302 Super Lusso $1895
A custom A-302 in presentation grade w/hand-engraved receiver and custom select walnut stock.

Beretta Model A-303 Semiautoloader
Similar to Model 302, except w/target specifications in Trap, Skeet and Youth configurations, and weighs 6.5 to 8 lbs. Imported 1983-96.
Field and Upland Models **$395**
Skeet and Trap (Discontinued 1994) **415**
Slug Model (Discontinued 1992) **455**
Sporting Clays.................................... **475**
Super Skeet..................................... **695**
Super Trap **660**
Waterfowl/Turkey (Discontinued 1992)............... **435**
For Mobilchoke®, **add** **50**

Beretta Model A-303 Youth Gun $395
Locked-breech, gas-operated action. Gauges: 12 and 20; 2-shot magazine. Bbls.: 24, 26, 28, 30 or 32 inches, vent rib. Weight: 7 lbs. (12 ga.), 6 lbs. (20 ga.). Crossbolt safety. Length of pull shortened to 12.5 inches. Imported 1988-96.

Beretta Model A-390 Semiautomatic Shotgun
Gas-operated, self-regulating action designed to handle any size load. Gauge: 12; 3-inch chamber. 3-shot magazine. Bbl.: 24, 26, 28 or 30 inches w/vent rib and Mobilchoke® tubes. Weight: 7.5 lbs. Select walnut stock w/adj. comb. Blued or matte black finish. Imported 1992-97. Superseded by AL-390 series.
Standard Model/Slug Model **$435**
Field Model/Silver Mallard......................... **445**
Deluxe Model/Gold Mallard........................ **485**
Turkey/Waterfowl Model (Matte finish) **455**

Beretta Model A-390 Target
Similar to the Model 390 Field, except in 12 ga. only w/2.75-inch chamber. Skeet: 28-inch ported bbl. w/wide vent rib and fixed choke (SK). Trap: 30- or 32-inch w/Mobilchoke® tubes. Weight: 7.5 lbs. Fully adj. buttstock. Imported from 1993-97.
Sport Trap Model................................. **$445**
Sport Skeet Model................................ **435**
Sporting Clays Model (Unported) **450**

Beretta Model A-390 Target *(Con't)*
Super Trap Model (Ported)......................... **595**
Super Skeet Model (Ported) **585**
W/Ported Bbl., **add**............................... **85**

Beretta Model AL-1 Field Gun $355
Same as Model AL-2 gas-operated Field Gun, except has bbl. w/orib, no engraving on receiver. Imported 1971-73.

Beretta Model AL-2 Autoloading Shotgun
Field Gun. Gas-operated. Engraved receiver (1968 version, 12 ga. only, had no engraving). Gauge: 12 or 20. 2.75-inch chamber. 3-shot magazine. Bbls.: vent rib; 30-inch F choke, 28-inch F or M choke, 26-inch IC. Weight: 6.5 to 7.25 lbs, depending on ga. and bbl. length. Checkered pistol-grip stock and forearm. Imported 1968-75.
W/Plain Receiver **$285**
W/Engraved Receiver.............................. **395**

Beretta Model AL-2 Magnum $355
Same as Model AL-2 Field Gun, except chambered for 12 ga. 3-inch Magnum shells; 30-inch F or 28-inch M choke bbl. only. Weight: about 8 lbs. Imported 1973-75.

Beretta Model AL-2 Skeet Gun $335
Same as Model AL-2 Field Gun, except has wide rib, 26-inch bbl. in SK choke only, checkered pistol-grip stock and beavertail forearm. Imported 1973-75.

Beretta Model AL-2 Trap Gun $325
Same as Model AL-2 Field Gun, except has wide rib, 30 inch bbl. in F choke only, beavertail forearm. Monte Carlo stock w/recoil pad. Weight: about 7.75 lbs. Imported 1973-75.

Beretta Model AL-3
Similar to corresponding AL-2 models in design and general specifications. Imported 1975-76.
Field Model....................................... **$335**
Magnum Model **345**
Skeet Model **350**
Trap Model **315**

SHOTGUNS

Beretta Model A390 Field

Beretta Model AL-2

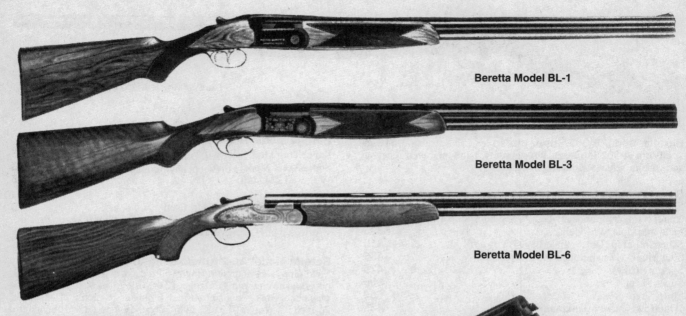

Beretta Model BL-1

Beretta Model BL-3

Beretta Model BL-6

Beretta Model ASEL

Beretta Model AL-3 Deluxe Trap Gun $685
Same as standard Model AL-3 Trap Gun, except has fully engraved receiver, gold-plated trigger and safety, stock and forearm of premium grade European walnut, gold monogram escutcheon inlaid in buttstock. Imported 1975-76.

Beretta AL390 Field Shotgun
Lightweight version of A-390 series. Gauges: 12 or 20 ga. 22- 24-, 26-, 28-, or 30-inch bbl. 41.7 to 47.6 inches overall. Weight: 6.4 to 7.5 lbs. Imported 1997 to date.
Model AL390 Field Lioness........................	**$2450**
Model AL390 Filed 470th Anniversary	**$2450**
Model AL390 Field/Silver Mallard (12 or 20 ga.)	**435**
Model AL390 Field/Silver Mallard Youth (20 ga.)	**440**
Model AL390 Field/Slug (12 ga. Only)	**425**
Model AL390 Silver Mallard Synthetic	**435**
Model AL390 Gold Mallard (12 or 20 ga.)	**495**
Model AL390 Camouflage (12 ga. Only)	**455**

Beretta AL390 Sport Sporting Shotgun
Similar to Model AL-390 Sport Skeet. Gauges: 12 or 20 ga. 28- or 30-inch bbls. Weight: 6.8 to 8 lbs. Imported 1997 to date.
Model AL390 Sport Diamond Sporting	**$2250**
Model AL390 Sport Sporting Collection	**675**
Model AL390 Sport Gold Sporting	**795**
Model AL390 Sport Sporting Youth (20 ga. Only)	**575**

Beretta AL390 Sport Skeet Shotgun
Gauges: 12 ga. only. 26- or 28-inch bbl. w/3-shot chamber. Weight: 7.6 to 8 lbs. Matte finish wood and metal. Imported 1997 to date.
Model AL390 Sport Skeet	**$525**
Model AL390 Sport Super Skeet (Semi-Auto)...........	**775**

Beretta AL390 Sport Trap Shotgun
Gauges: 12 ga. only. 30- or 32-inch bbl. w/3-shot chamber. Weight: 7.8 to 8.25 lbs. Matte finish wood and metal. Black recoil rubber pad. Imported 1997 to date.
Model AL390 Sport Trap	**$535**
Model AL390 Sport Super Trap	**765**
Multi-choke bbl. (30-inch only), **add**	**15**
Ported Bbl., **add**	**100**

Beretta Model ASE 90 Over-and-Under Shotgun
Competition-style receiver w/coin-silver finish and gold inlay featuring drop-out trigger group. Gauge: 12; 2.75-inch chamber. Bbls.: 28- or 30-inch w/fixed or Mobilchoke® tubes; vent rib. Weight: 8.5 lbs. (30-inch bbl.). Checkered high-grade walnut stock. Imported 1992 to date.
Pigeon, Skeet, Trap Models	**$5850**
Sporting Clays Model............................	**5995**
Trap Combo Model..............................	**9250**
Deluxe (Introduced 1996)	**14,750**

Beretta Model ASE Series Over-and-Under Shotgun
Boxlock. Single non-selective trigger. Selective automatic ejectors. gauges: 12 and 20. Bbls. 26-, 28-, 30-inch; IC and M choke or M and F choke. Weight: about 5.75-7 lbs. Checkered pistol-grip stock and forearm. Receiver w/various grades of engraving. Imported 1947-64.
Model ASE (Light Scroll Engraving)...................	**$1495**
Model ASEL (Half Coverage Engraving)	**2150**
Model ASEELL (Full Coverage Engraving)	**2895**
For 20 ga. Models, **add**	**95%**

Beretta Model BL-1/BL-2 Over/Under
Boxlock. Plain extractors. Double triggers. 12 gauge, 2.75-inch chambers only. Bbls.: 30-and 28-inch M/F choke, 26-inch IC/M choke. Weight: 6.75-7 lbs., depending on bbl. length. Checkered pistol-grip stock and forearm. Imported 1968-73.
Model BL-1	**$325**
Model BL-2 (Single Selective Trigger)	**375**

Beretta Model GR-2

Beretta Model BL-2/S **$385**
Similar to Model BL-1, except has selective "Speed-Trigger," vent-rib bbls., 2.75- or 3-inch chambers. Weight: 7-7.5 lbs. Imported 1974-76.

Beretta Model BL-3 . . . . **$555**
Same as Model BL-1, except has deluxe engraved receiver, selective single trigger, vent-rib bbls., 12 or 20 ga., 2.75-inch or 3-inch chambers in former, 3-inch in latter. Weight: 6-7.5 lbs. depending on ga. and bbl. length. Imported 1968-76.

Beretta Models BL-4, BL-5 and BL-6
Higher grade versions of Model BL-3 w/more elaborate engraving and fancier wood; Model BL-6 has sideplates. Selective automatic ejectors standard. Imported 1968-76.
Model BL-4 . . . . $ 675
Model BL-5 895
Model BL-6 (1973-76) 1125

Beretta Series BL Skeet Guns
Models BL-3, BL-4, BL-5 and BL-6 w/standard features of their respective grades plus wider rib and skeet-style stock, 26-inch bbls. SK choked. Weight: 6 – 7.25 lbs. depending on Ga.
Model BL-3 Skeet Gun . . . . $ 600
Model BL-4 Skeet Gun 695
Model BL-5 Skeet Gun 925
Model BL-6 Skeet Gun 1255

Beretta Series BL Trap Guns
Models BL-3, BL-4, BL-5 and BL-6 w/standard features of their respective grades plus wider rib and Monte Carlo stock w/recoil pad; 30-inch bbls., improved M/F or both F choke. Weight: about 7.5 lbs.
Model BL-3 Trap Gun . . . . $ 535
Model BL-4 Trap Gun 595
Model BL-5 Trap Gun 925
Model BL-6 Trap Gun 1195

Beretta FS-1 Folding

Beretta Model FS-1 Folding Single **$155**
Formerly "Companion." Folds to length of bbl. Hammerless. Underlever. Gauge: 12, 16, 20, 28 or .410. Bbl.: 30-inch in 12 ga., 28-inch in 16 and 20 ga.; 26-inch in 28 and .410 ga.; all F choke. Checkered semipistol-grip stock/forearm. Weight: 4.5-5.5 lbs. depending on ga. Discontinued 1971.

Beretta Golden Snipe Over-and-Under
Same as Silver Snipe, except has automatic ejectors, vent rib is standard feature. Imported 1959-67.
W/Non-Selective Single Trigger **$595**
Extra for Selective Single Trigger **100**

Beretta Model GR-2 Hammerless Double **$595**
Boxlock. Plain extractors. Double triggers. Gauges: 12, 20; 2.75-inch chambers in former, 3-inch in latter. Bbls.: vent rib; 30-inch M/F choke (12 ga. only); 28-inch M/F choke, 26-inch IC/M choke. Weight: 6.5 to 7.5 lbs. depending on ga. and bbl. length. Checkered pistol-grip stock and forearm. Imported 1968-76.

Beretta Model GR-3 . . . . **$675**
Same as Model GR-2, except has selective single trigger chambered for 12-ga. 3-inch or 2.75-inch shells. Magnum model has 30-inch M/F choke bbl., recoil pad. Weight: about 8 lbs. Imported 1968-76.

Beretta Model GR-4 . . . . **$925**
Same as Model GR-2, except has automatic ejectors and selective single trigger, higher grade engraving and wood. 12 ga., 2.75-inch chambers only. Imported 1968-76.

Beretta Grade 100 Over-and-Under Shotgun **$1495**
Sidelock. Double triggers. Automatic ejectors. 12 ga. only. Bbls.: 26-, 28-, 30-inch, any standard boring. Weight: about 7.5 lbs. Checkered stock and forend, straight or pistol grip. Discontinued.

Beretta Grade 200 . . . . **$1995**
Same general specifications as Grade 100 except higher quality, bores and action parts hard chrome-plated. Discontinued.

Beretta Mark II Single-Barrel Trap Gun **$450**
Boxlock action similar to that of Series "BL" Over-and-Unders. Engraved receiver. Automatic ejector. 12 ga. only. 32- or 34-inch bbl. w/wide vent rib. Weight: about 8.5 lbs. Monte Carlo stock w/pistol grip and recoil pad, beavertail forearm. Imported 1972-76.

Beretta Model S55B Over-and-Under Shotgun **$475**
Boxlock. Plain extractors. Selective single trigger. Gauges: 12, 20; 2.75- or 3-inch chambers in former, 3-inch in latter. Bbls. vent rib; 30-inch M/F choke or both F choke in 12-ga. 3-inch Magnum only; 28-inch M/F choke; 26 inch IC/M choke. Weight: 6.5 to 7.5 lbs. depending on ga. and bbl. length. Checkered pistol-grip stock and forearm. Introduced in 1977. Discontinued.

Beretta Model S56E **$550**
Same as Model S55B, except has scroll-engraved receiver selective automatic ejectors. Introduced in 1977. Discontinued.

Beretta Model S58 Skeet Gun **$675**
Same as Model S56E, except has 26-inch bbls. of Boehler Antinit Anticorro steel, SK choked, w/wide vent rib; skeet-style stock and forearm. Weight: 7.5 lbs. Introduced in 1977.

SHOTGUNS

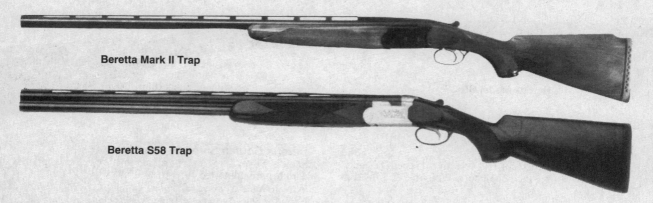

Beretta Mark II Trap

Beretta S58 Trap

Beretta Model S58 Trap Gun $595
Same as Model S58 Skeet Gun, except has 30-inch bbls. bored IM/F Trap, Monte Carlo stock w/recoil pad. Weight: 7 lbs. 10 oz. Introduced in 1977. Discontinued.

Beretta Silver Hawk Featherweight Hammerless Double-Barrel Shotgun
Boxlock. Double triggers or non-selective single trigger. Plain extractor. Gauges: 12, 16, 20, 28, 12 Mag. Bbls.: 26- to 32-inch w/high matted rib, all standard choke combinations. Weight: 7 lbs. (12 ga. w/26-inch bbls.). Checkered walnut stock w/beavertail forearm. Discontinued 1967.
W/Double Triggers $425
For Non-selective Single Trigger, **add** 65

Beretta Silver Snipe Over-and-Under Shotgun
Boxlock. Non-selective or selective single trigger. Plain extractor. Gauges: 12, 20, 12 Mag., 20 Mag. Bbls.: 26-, 28-, 30-inch; plain or vent rib; chokes IC/M, M/F, SK #1 and #2, F/F. Weight: from about 6 lbs. in 20 ga. to 8.5 lbs. in 12 ga. (Trap gun). Checkered walnut pistol-grip stock and forearm. Imported 1955-67.
W/Plain Bbl., Non-selective Trigger $395
W/Vent-rib Bbl., Non-selective. Single Trigger 455
For Selective Single Trigger, **add** 65

Beretta Model SL-2 Pigeon Series Pump Gun
Hammerless. Takedown. 12 ga. only. 3-shot magazine. Bbls.: vent rib; 30-inch F choke, 28-inch M, 26-inch IC. Weight: 7-7.25 lbs., depending on bbl. length. Receiver w/various grades of en-

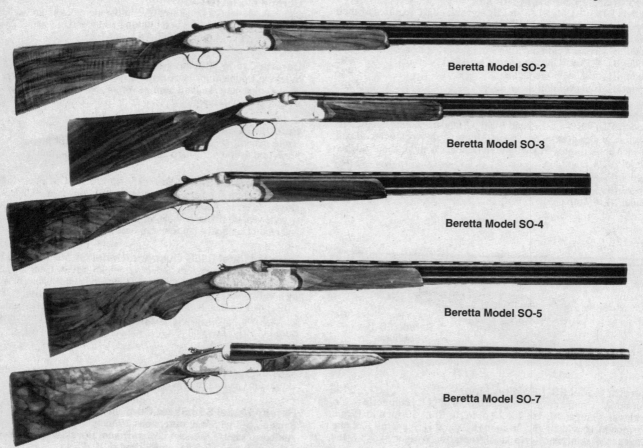

Beretta Model SO-2

Beretta Model SO-3

Beretta Model SO-4

Beretta Model SO-5

Beretta Model SO-7

Beretta Model SL-2 Pigeon Series Pump Gun *(Con't)*

graving. Checkered pistol-grip stock and forearm. Imported 1968-71.

Model SL-2	$275
Silver Pigeon	225
Gold Pigeon	350
Ruby Pigeon	450

Beretta Series "SO" Over-and-Under Shotguns

Jubilee Series Introduced in 1998. The finest Beretta Boxlock is made with mechanical works from a single block of hot forged high resistance steel. The gun is richly engraved in Botta scroll and game scenes. All engraving is signed by master engravers. High quality finishing on the inside with high polishing of all internal points. This is so quality wood has a true oil finish. Sidelock. Selective automatic ejectors. Selective single trigger or double triggers. 12 ga. only, 2.75- or 3-inch chambers. Bbls.: vent rib (wide type on skeet and trap guns); 26-, 27-, 29-, 30-inch; any combination of standard chokes. Weight: 7 to 7.75 lbs., depending on bbl. length, style of stock and density of wood. Stock and forearm of select walnut, finely checkered; straight or pistol grip, field, skeet and trap guns have appropriate styles of stock and forearm. Models differ chiefly in quality of wood and grade of engraving. Models SO-3EL, SO-3EELL, SO4 and SO-5 have hand-detachable locks. "SO-4" is used to designate skeet and trap models derived from Model SO-3EL but with less elaborate engraving. Models SO3EL and SO-3EELL are similar to the earlier SO-4 and SO-5, respectively. Imported 1933 to date.

Jubilee .410-28-20-12	$10,000
Jubilee Pairs	23,000
Jubilee II Side-by-side Model 20 or 12 gauge	10,000
Jubilee Pairs	23,000
Model SO-2 (Discontinued 1986)	4000
Model SO-3 (Discontinued 1986)	6000
Model SO-3EL (Discontinued 1985)	7500
Model SO-3EELL (Discontinued 1987)	10,950
Model SO-4 Skeet or Trap Gun (Disc. 1986)	6250
Model SO-5 Sporting, Skeet or Trap	12,500

Beretta Models SO-6 and SO-9 Premium Grade Shotguns

High-grade over/unders in the SO series. Gauges: 12 ga. only (SO-6); 12, 20, 28 and .410 (SO-9). Fixed or Mobilchoke® (12 ga. only). Sidelock action. Silver or casehardened receiver (SO-6); English custom hand-engraved scroll or game scenes (SO-9). Supplied w/leather case and accessories. Imported from about 1990 to date.

SO-6 Over/Under	$19,000
SO-6 EELL Over/Under	26,000
SO-9 Over/Under	29,000
SO-9 EELL Special Engraver	50,000

Beretta Model SO6/SO-7 S/S Doubles

Side-by-side shotgun w/same general specifications as SO Series over/unders, except higher grade w/more elaborate engraving, fancier wood.

SO-6 SxS (Imported 1948-82)	$5,395
SO-7 SxS (Imported 1948-90)	7,500

Beretta Model TR-1 Single-Shot Trap Gun $245

Hammerless. Underlever action. Engraved frame.12 ga. only. 32-inch bbl. w/vent rib. Weight: about 8.25 lbs. Monte Carlo stock w/pistol grip and recoil pad, beavertail forearm. Imported 1968-71.

Beretta Model TR-2 .... $275

Same as Model TR-1, except has extended ventilated rib. Imported 1969-73.

Beretta Victoria Pintail (ES100) Semiautoloader

Short Montefeltro-type recoil action. Gauge: 12 w/3-inch chamber. Bbl.: 24-inch slug, 24-, 26- or 28-inch vent rib w/Mobilchoke® tubes. Weight: 7 lbs. to 7 lbs. 5 oz. Checkered synthetic or walnut buttstock and forend. Matte finish on both metal and stock. Imported 1993 to date.

Field Model w/synthetic stock (Intro. 1998)	$495
Field Model w/walnut stock (Disc. 1998)	425
Rifled Slug Model w/synthetic stock (Intro. 1998)	525
Standard Slug Model w/walnut stock (Disc. 1998)	450

VINCENZO BERNARDELLI
Gardone V.T. (Brescia), Italy

Previously imported by Armsport, Inc., Miami, FL (formerly byMagnum Research, Inc., Quality Arms, Stoeger Industries, Inc. & Action Arms, LTD).

Bernardelli 115 Series O/U Shotguns

Boxlock w/single trigger and ejectors. Gauges: 12 ga. only. 25.5-, 26.75-, and 29.5-inch bbls. Concave top and vented middle rib. Anatomical grip stock. Blued or coin-silver finish w/various gredes of engraving. Imported 1985-97.

Hunting Model 115 (Disc. 1990)	$1330
Hunting Model 115E (Disc. 1990)	3895
Hunting Model 115L (Disc. 1990)	2150
Hunting Model 115S (Disc. 1990)	1825
Target Model 115 (Disc. 1992)	1450
Target Model 115E (Disc. 1992)	4465
Target Model 115L (Disc. 1992)	2775
Target Model 115S (Disc. 1992)	1950
Trap/SkeetModel 115S (Imported 1996-97)	1995
Sporting Clays Model 115S (Imported 1995-97)	2950

Bernardelli Brescia Hammer Double $795

Back-action sidelock. Plain extractors. Double triggers. Gauges: 12, 20. Bbls.: 27.5 or 29.5-inch M/F choke in 12 ga. 25.5-inch IC/M choke in 20 ga.. Weight: from 5.75 to 7 lbs., depending on ga. and bbl. length. English-style stock and forearm, checkered. No longer imported.

Bernardelli Elio

Bernardelli Elio $925

Lightweight game gun, 12 ga. only, w/same general specifications as Standard Gamecock (S. Uberto 1), except weighs about 6 to 6.25 lbs., has automatic ejectors, fine English-pattern scroll engraving. No longer imported.

Bernardelli Gamecock, Premier (Rome 3) $905

Same general specifications as Standard Gamecock (S. Uberto 1), except has sideplates, auto ejectors, single trigger. Currently imported.

Bernardelli Gamecock, Standard (S. Uberto 1)
Hammerless Double-Barrel Shotgun $750

Boxlock. Plain extractors. Double triggers. Gauges: 12, 16, 20; 2.75-inch chambers in 12 and 16, 3-inch in 20 ga. Bbls. 25.5-inch IC/M choke; 27.5-inch M/F choke. Weight: 5.75-6.5 lbs., depending on ga. and bbl. length. English-style straight-grip stock and forearm, checkered. No longer imported.

SHOTGUNS

Bernardelli Gamecock

Bernardelli Standard Gamecock

Bernardelli Gardone

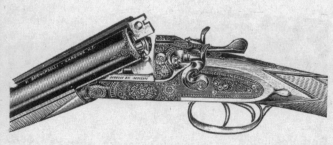

Bernardelli Italia

Bernardelli Roma 6

Bernardelli Gardone Hammer Double $1850
Same general specifications as Brescia, except for higher grade engraving and wood, but not as high as the Italia. Half-cock safety. Discontinued 1956.

Bernardelli Hemingway Hammerless Double
Boxlock. Single or double triggers w/hinged front. Selective automatic ejectors. Gauges: 12 and 20 w/2.75- or 3-inch chambers, 16 and 28 w/2.75-inch. Bbls.: 23.5- to 28-inch w/fixed chokes. Weight: 6.25 lbs. Checkered English-style European walnut stock. Silvered and engraved receiver.
Standard Model $1295
Deluxe Model w/Sideplates (Disc. 1993) 1495
For Single Trigger, **add** 100

Bernardelli Italia $2350
Same general specifications as Brescia, except higher grade engraving and wood. Discontinued 1986.

Bernardelli Roma 4 and Roma 6
Same as Premier Gamecock (Rome 3), except higher grade engraving and wood, double triggers. Currently manufactured; Roma 6 disc. 1993.
Roma 4 . . . . $1095
Roma 6 1150

Bernardelli Roma 7, 8, and 9
Side-by-side. Anson & Deeley boxlock; hammerless. Ejectors; double triggers. 12 ga. Barrels: 27.5- or 29.5- inch. M/F chokes. Fancy hand-checkered European walnut straight or pistol grip stock, forearm. Elaborately engraved, sivler finished sideplates. Imported 1994 to date.
Roma 7 . . . . $1950
Roma 8 2250
Roma 9 2795

Bernardelli Uberto 2

Bernadelli S. Uberto 2 . . . . $995
Same as Standard Gamecock (S. Uberto 1), except higher grade engraving and wood. Currently imported.

Bernardelli S. Uberto F.S. $1295
Same as Standard Gamecock, except w/higher grade engraving, wood and has auto ejectors. Currently imported.

Bernardelli Holland V.B. Series Shotguns
Holland & Holland-type sidelock action. Auto ejectors. Double triggers. Gauge: 12 ga. only. bbl. length or choke to custom specification. Silver finish receiver (Liscio) or engraved coin finish receiver (Incisio). Extra select wood and game scene engraving (Lusso). Checkered stock (straight or pistol grip). Imported 1992-97.
Model V.B. Liscio . . . . $ 4,950
Model V.B. Incisio 6,250
Model V.B. Lusso 7,550
Model V.B. Extra 9,250

Bernardelli Uberto F.S.

Bernardelli V.B. Holland Liscio

Bernardelli Holland V.B. Series Shotguns *(Con't)*
Model V.B. Gold 28,950
Engraving Pattern No. 4, **add** 950
Engraving Pattern No. 12, **add** 4,450
Engraving Pattern No. 20, **add** 8,550
Single Trigger, **add** 465

BOSS & COMPANY
London, England

Boss Hammerless Double-Barrel

Boss Hammerless Double-Barrel Shotgun $25,000
Sidelock. Automatic ejectors. Double triggers, non-selective or selective single trigger. Made in all gauges, bbl. lengths and chokes. Checkered stock and forend, straight or pistol grip.

Boss Hammerless Over/Under Shotgun $38,500
Sidelock. Automatic ejectors. Selective single trigger. Made in all gauges, bbl. lengths and chokes. Checkered stock and forend, straight or pistol grip. Discontinued.

BREDA MECCANICA BRESCIANA
Brescia, Italy
formerly ERNESTO BREDA
Milan, Italy

Breda Autoloading Shotgun
Recoil-operated.12 ga, 2.75-inch chamber. 4-shell tubular magazine. Bbls.: 25.5- and 27.5-inch, plain, matted or vent rib, IC, M or F choke; current model has 26-inch vent-rib bbl. w/interchangeable choke tubes. Weight: about 7.25 lbs. Checkered straight or pistol-grip stock and forearm. Discontinued 1988.
W/Plain Bbl. **$355**
W/Raised Matted-rib Bbl. 395
W/Ventilated-rib Bbl. 415
W/Vent Rib, Interchangeable Choke Tubes. 430

Breda Magnum
Same general specifications as standard model, except chambered for 12-ga. 3-inch Magnum, 3-shot magazine; latest model has 29-inch vent-rib bbl. Discontinued 1988.
W/Plain Bbl. **$465**
W/Ventilated-rib Bbl. 475

BRETTON SHOTGUNS
St. Etienne (Cedex1), France

Bretton Baby Standard Sprint Over/Under $635
Inline sliding breech action. 12 or 20 gauge w/2.75-inch chambers. 27.5-inch separated bbls. w/vent rib and choke tubes. Weight: 4.8 to 5 lbs. Engraved alloy receiver. Checkered walnut buttstock and forearm w/satin oil finish. Imported 1992 to date.

Bretton Sprint Deluxe Over/Under $675
Similar to the Standard Model, except w/engraved coin finished receiver and chambered 12, 16 and 20 ga. Discontinued 1994.

Bretton Fair Play Over/Under $600
Lightweight action similar to the Sprint Model, except w/hinged action that pivots open and is chambered 12 or 20 gauge only.

BRNO SHOTGUNS
Brno and Uherski Brod, Czech Republic
(formerly Czechoslovakia)

Brno 500 Over/Under Shotgun $625
Hammerless boxlock w/double triggers and ejectors.12 ga. w/2.75-inch chambers. 27.5-inch bbls.. choked M/F. 44 inches overall. Weight: 7 lbs. Etched receiver. Checkered walnut stock w/classic style cheekpiece. Imported 1987-91.

Brno 500 Series O/U Combination Guns
Similar to the 500 Series over/under shotgun, except w/lower bbl. chambered in rifle calibers and set trigger option. Imported 1987-95.
Model 502 12/222, 12/243 (Disc. 1991) . . . . $ 695
Model 502 12/308, 12/30.06 (Disc. 1991). 695
Model 571 12/6×65R (Disc. 1993) 695
Model 572 12/7×65R (Imported since 1992) 725
Model 584 12/7×57R (Imported since 1992) 695
Sport Series 4-Bbl. Set (Disc. 1991) 1850

Brno CZ 581 Solo Over/Under Shotgun $575
Hammerless boxlock w/double triggers, ejectors and automatic safety. 12 ga. w/2.75- or 3-inch chambers. 28-inch bbls. choked M/F. Weight: 7.5 lbs. Checkered walnut stock. Discontinued 1996.

SHOTGUNS

Brno Super Series Over/Under Shotgun

Hammerless sidelock w/selective single or double triggers and ejectors. 12 ga. w/2.75- or 3-inch chambers. 27.5-inch bbls. choked M/F. 44.5 inches overall. Weight: 7.25 lbs. Etched or engraved sideplates. Checkered European walnut stock w/classic-style cheekpiece. Imported 1987-91.

Super Series Shotgun (Disc. 1992) . . . . $ 725
Super Series Combo (Disc. 1992) 850
Super Series 3-Bbl. Set (Disc. 1990) 1695
Super Series Engraving, **add** 1250

Brno ZH 300 Series Over/Under Shotguns

Hammerless boxlock w/double triggers. Gauge: 12 or 16 w/2.75- or 3-inch chambers. Bbls.: 26, 27.5 or 30 inches; choked M/F. Weight: 7 lbs. Skip-line checkered walnut stock w/classic-style cheekpiece. Imported 1986-93.

Model 300 (Discontinued 1993). $415
Model 301 Field (Discontinued 1991) 425
Model 302 Skeet (Discontinued 1992) 475
Model 303 Trap (Discontinued 1992) 495

Brno ZH 300 Series O/U Combination Guns

Similar to the 300 Series over/under shotgun, except lower bbl. chambered in rifle calibers.

Model 300 Combo 8-Bbl. Set (Disc. 1991) $2595
Model 304 12 Ga./7×57R (Disc. 1995) 525
Model 305 12 Ga./5.6×52R (Disc. 1993) 595
Model 306 12 Ga./5.6×50R (Disc. 1993) 625
Model 307 12 Ga./22 Hornet (Imported since 1995) 550
Model 324 16 Ga./7×57R (Disc. 1987) 575

Brno ZP 149 Hammerless Double

Sidelock action w/double triggers, automatic ejectors and automatic safety.12 ga. w/2.75- or 3-inch chambers. 28.5-inch bbls. choked M/F. Weight: 7.25 lbs. Checkered walnut buttstock w/cheekpiece.

Standard Model $475
Engraved Model 520

BROLIN ARMS, INC.
Pomona, California

Brolin Arms Hawk Pump Shotgun— Field Series

Slide-action. Gauge: 12 ga. w/3-inch chamber. 24-, 26-, 28- or 30-inch bbl. 44 and 50 inches overall. Weight: 7.3 to 7.6 lbs. Cross-bolt safety. Vent rib bbl. w/screw-in choke tube and bead sights. Non reflective metal finish. Synthetic or oil-finished wood stock w/swivel studs. Made 1997 to date.

Synthetic Stock Model $175
Wood Stock Model 165

Brolin Arms Hawk Pump Shotgun— Combo Model

Similar to the Field Model, except w/extra 18.5- or 22-inch bbl. w/bead or rifle sight. Made 1997 to date.

Synthetic Stock Model $195
Wood Stock Model 185

Brolin Arms Hawk Pump Shotgun— Lawman Model

Similar to the Field Model, except has 18.5-inch bbl. w/cylinder bore fixed choke. Weight: 7 lbs. Dual operating bars. Bead, rifle or ghost ring sights. Black synthetic or wood stock. Matte chrome or satin nickel finish. Made 1997 to date.

Synthetic Stock Model $165
Wood Stock Model 155
Rifle Sights, **add** 20
Ghost Ring Sights, **add** 35
Satin Nickel Finish (disc. 1997), **add** 25

Brolin Arms Hawk — Slug Model

Similar to the Field Model, except has 18.5- or 22-inch bbl. w/IC fixed choke or 4-inch extended rifled choke. Rifle or ghost ring sights or optional cantilevered scope mount. Black synthetic or wood stock. Matte blued finish. Made 1998 to date.

Synthetic Stock Model $165
Wood Stock Model 155
W/Rifled Bbl., **add** 20
W/Cantilevered Scope Mount, **add** 25

Brolin Arms Hawk — Turkey Special

Similar to the Field Model, except has 22-inch vent-rib bbl. w/extended extra-full choke. Rifle or ghost ring sights or optional cantilevered scope mount. Black synthetic or wood stock. Matte blued finish. Made 1998 to date.

Synthetic Stock Model $160
Wood Stock Model 150
W/Cantilevered Scope Mount, **add** 25

BROWNING SHOTGUNS
Morgan (formerly Ogden), Utah

NOTE

Fabrique Nationale Herstal (formerly Fabrique Nationale d'Armes de Guerre) of Herstal, Belgium, is the longtime manufacturer of Browning shotguns, dating back to 1900. Miroku Firearms Mfg. Co. of Tokyo, Japan, bought into the Browning company and has, since the early 1970s, undertaken some of the production. The following shotguns were manufactured for Browning by these two firms.

AMERICAN BROWNING SHOTGUNS

Designated "American" Browning because they were produced in Ilion, New York, the following Remington-made Brownings are almost identical to the Remington Model 11 A and Sportsman and the Browning Auto-5. They are the only Browning shotguns manufactured in the U.S. during the 20th century and were made for Browning Arms when production was suspended in Belgium because of WW II.

American Browning Grade I Autoloader $350

Recoil-operated autoloader. Similar to the Remington Model 11A, except w/different style engraving and identified w/the Browning logo. Gauges: 12, 16 or 20. Plain barrel, 26- to 32-inch, any standard boring. 2 or 4 shell tubular magazine w/magazine cut-off. Weight: about 6.88 lbs. (20 ga.) to 8 lbs. (12 ga.). Checkered pistol-grip stock and forearm. Made 1940-49.

American Browning Special

Same general specifications as Grade I, except supplied w/raised matted rib or vent rib. Discont. 1949.

W/raised matted rib $395
W/ventilated rib 425

American Browning Special Skeet Model $525

Same general specifications as Grade I, except has 26-inch bbl. w/vent rib and Cutts Compensator. Discontinued1949.

American Browning Utility Field Gun $345

Same general specifications as Grade I, except has 28-inch plain bbl. w/Poly Choke. Discontinued 1949.

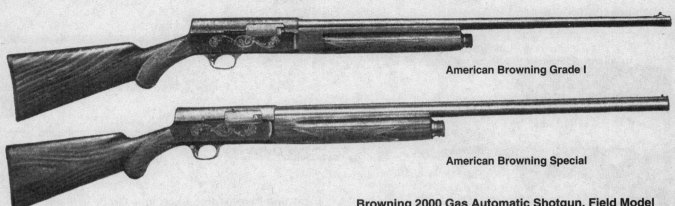

American Browning Grade I

American Browning Special

Browning Model 12 Pump Shotgun
Special limited edition Winchester Model 12. Gauge: 20 or 28. Five-shot tubular magazine. 26-inch bbl., M choke. 45 inches overall. Weight: about 7 lbs. Grade I has blued receiver, checkered walnut stock w/matte finish. Grade V has engraved receiver, checkered deluxe walnut stock w/high-gloss finish. Made 1988-92.

Grade I, 20 Gauge (8600)	**$425**
Grade I, 28 Gauge	**445**
Grade V, 20 Gauge (4000)	**755**
Grade V, 28 Gauge	**790**

Browning Model 42 Limited Edition Shotgun
Special limited edition Winchester Model 42 pump shotgun. Same general specifications as Model 12, except w/smaller frame in .410 ga. and 3-inch chamber. Made 1991-93.

Grade I (6000 produced)	**$475**
Grade V (6000 produced)	**750**

Browning 2000 Buck Special $315
Same as Field Model, except has 24-inch plain bbl. Bored for rifled slug and buck shot, fitted w/rifle sights (open rear, ramp front). 12 ga., 2.75-inch or 3-inch chamber; 20 ga., 2.75-inch chamber. Weight: 12 ga., 7 lbs. 8 oz.; 20 ga., 6 lbs. 10 oz. Made 1974-81 by FN.

Browning 2000 Gas Automatic Shotgun, Field Model
Gas-operated. Gauge: 12 or 20. 2.75-inch chamber. 4-shot magazine. Bbl.: 26-, 28-, 30-inch, any standard choke plain matted bbl. (12 ga. only) or vent rib. Weight: 6 lbs. 11 oz.-7 lbs. 12 oz. depending on ga. and bbl. length. Checkered pistol-grip stock/forearm. Made 1974-81 by FN; Assembled in Portigal.

W/Plain Matted Bbl.	**$335**
W/Vent-rib Bbl.	**350**

Browning 2000 Magnum Model $350
Same as Field Model, except chambered for 3-inch shells, 3-shot magazine. Bbl.: 26- (20 ga. only), 28-, 30- or 32-inch (latter two 12 ga. only); any standard choke; vent rib. Weight: 6 lbs. 11 oz.-7 lbs. 13 oz. depending on ga. and bbl. Made 1974-81 by FN.

Browning 2000 Skeet Model $345
Same as Field Model, except has skeet-style stock w/recoil pad, 26-inch vent-rib bbl., SK choke. 12 or 20 ga., 2.75-inch chamber. Weight: 8 lbs. 1 oz. (12 ga.); 6 lbs. 12 oz.(20 ga.) Made 1974-81 by FN.

Browning 2000 Trap Model $345
Same as Field Model except has Monte Carlo stock w/recoil pad, 30- or 32-inch bbl. w/high-post vent rib and receiver extension, M/I/F chokes. 12 ga., 2.75-inch chamber. Weight: about 8 lbs. 5 oz. Made 1974-81 by FN.

SHOTGUNS

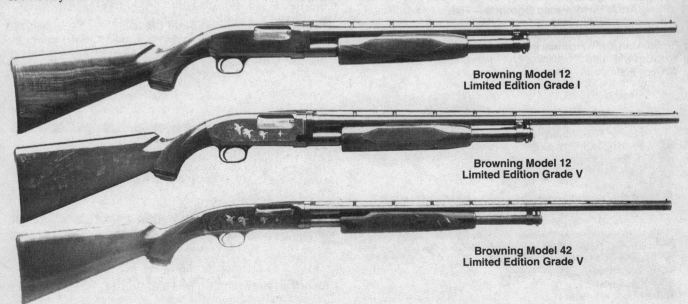

**Browning Model 12
Limited Edition Grade I**

**Browning Model 12
Limited Edition Grade V**

**Browning Model 42
Limited Edition Grade V**

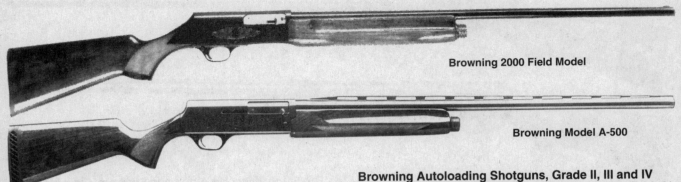

Browning 2000 Field Model

Browning Model A-500

Browning A-500G Gas-Operated Semiautomatic
Same general specifications as Browning Model A-500R except gas-operated. Made 1990-93.
Buck Special **$425**
Hunting Model **435**

Browning A-500G Sporting Clays **$445**
Same general specifications as Model A-500G except has matte blued receiver w/"Sporting Clays" logo. 28- or 30-inch bbl. w/Invector choke tubes. Made 1992-93.

Browning A-500R Semiautomatic
Recoil-operated. Gauge: 12. 26- to 30-inch vent-rib bbls. 24-inch Buck Special. Invector choke tube system. 2.75- or 3-inch Magnum cartridges. Weight: 7 lbs. 3 oz.-8 lbs. 2 oz. Cross-bolt safety. Gold-plated trigger. Scroll-engraved receiver. Gloss-finished walnut stock and forend. Made by FN from 1987-93.
Hunting Model **$455**
Buck Special **495**

Browning A-Bolt Series Shotgun
Bolt-action repeating single-barrel shotgun. 12 ga. only w/3-inch chambers, 2-shot magazine. 22- or 23-inch rifled bbl., w/or w/o a rifled invector tube. Receiver drilled and tapped for scpoe mounts. Bbl. w/ or w/o open sights. Checkered walnut or graphite/fiberglass composite stock. Matte black metal finish. Imported 1995-98.
Stalker Model w/Composite Stock **$315**
Hunter Model w/Walnut stock **300**
W/Rifled Bbl., **add** **95**
W/Open Sights, **add** **25**

Browning Autoloading Shotguns, Grade II, III and IV
These higher grade models differ from the Standard or Grade I in general quality, grade of wood, checkering, engraving, etc., otherwise specifications are the same. Grade IV guns, sometimes called Midas Grade, are inlaid w/yellow and green gold. Discontinued in 1940.
Grade II, Plan Bbl. **$1150**
Grade III, Plain Bbl. **2300**
Grade IV, Plain Bbl. **3700**
For Raised Matte-rib Bbl., **add** **225**
For Vent-rib Bbl., **add** **450**

Browning Automatic-5, Buck Special Models
Same as Light 12, Magnum 12, Light 20, Magnum 20, in respective gauges, except 24-inch plain bbl. bored for rifled slug and buckshot, fitted w/rifle sights (open rear, ramp front). Weight: 6.13-8.25 lbs. depending on ga. Made 1964-76 by FN, since then by Miroku.
FN Manufacture, w/Plain Bbl. **$525**
Miroku Manufacture **465**

Browning Automatic-5 Classic **$750**
Gauge: 12. 5-shot capacity. 28-inch vent-rib bbl./M choke. 2.75-inch chamber. Engraved silver gray receiver. Gold-plated trigger. Cross-bolt safety. High-grade, hand checkered select American walnut stock w/rounded pistol grip. 5,000 issued; made in Japan in 1984, engraved in Belgium.

Browning Automatic-5 Gold Classic **$2750**
Same general specifications as Automatic-5 Classic, except engraved receiver inlaid w/gold. Pearl border on stock and forend plus fine-line hand-checkering. Each gun numbered "1 of Five Hundred," etc. 500 issued in 1984; made in Belgium.

Browning
Model A-Bolt Hunter

Browning
Model A-Bolt Stalker

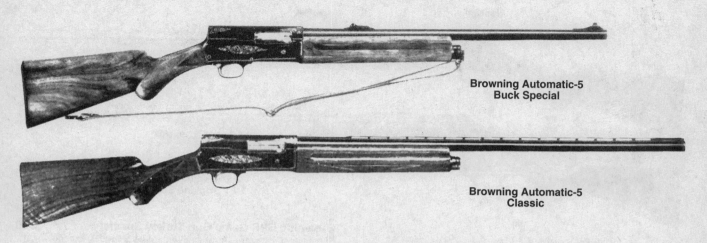

**Browning Automatic-5
Buck Special**

**Browning Automatic-5
Classic**

**Browning Automatic-5
Gold Classic**

Browning Automatic-5, Light 12

12 ga. only. Same general specifications as Standard Model, except lightweight (about 7.25 lbs.), has gold-plated trigger. Guns w/orib have striped matting on top of bbl. Fixed chokes or Invector tubes. Made 1948-76 by FN, since then by Miroku.

FN Manufacture, Plain Bbl. $450
FN Manufacture, Raised Matte Rib 525
FN Manufacture, Ventilated Rib 575
Miroku Manufacture, Vent Rib, Fixed Choke 450
Miroku manufacture, Vent Rib, Invectors 495

Browning Automatic-5, Light 20

Same general specifications as Standard Model, except lightweight and 20 ga. Bbl.: 26- or 28-inch; plain or vent rib. Weight: about 6.25-6.5 lbs. depending on bbl. Made 1958-76 by FN, since then by Miroku.

FN Manufacture, Plain Bbl. $515
FN Manufacture, Vent-rib Bbl. 555
Miroku Manufacture, Vent Rib, Fixed Choke 415
Miroku Manufacture, Vent Rib, Invectors 475

Browning Automatic-5, Magnum 12 Gauge

Same general specifications as Standard Model. Chambered for 3-inch Magnum 12-ga. shells. Bbl.: 28-inch M/F, 30- or 32-inch F/F, plain or vent rib. Weight: 8.5-9 lbs. depending on bbl. Buttstock has recoil pad. Made 1958-76 by FN, since then by Miroku. Fixed chokes or Invector tubes.

FN Manufacture, Plain Bbl. $535
FN Manufacture, Vent-rib Bbl. 580
Miroku Manufacture, Vent Rib, Fixed Chokes 435
Miroku Manufacture, Vent Rib, Invectors 495

Browning Automatic-5, Magnum 20 Gauge

Same general specifications as Standard Model, except chambered for 3-inch Magnum 20-ga. shell. Bbl.: 26- or 28-inch, plain or vent rib. Weight: 7 lbs. 5 oz.-7 lbs. 7 oz. depending on bbl. Made 1967-76 by FN, since then by Miroku.

FN Manufacture, Plain Bbl. $545
FN Manufacture, Vent-rib Bbl. 575
Miroku Manufacture, Vent Rib, Invectors 450

Browning Automatic-5, Skeet Model

12 ga. only. Same general specifications as Light 12. Bbl.: 26-or 28-inch, plain or vent rib, SK choke. Weight: 7 lbs. 5 oz.-7 lbs. 10 oz. depending on bbl. Made by FN prior to 1976, since then by Miroku.

FN Manufacture, Plain Bbl. $495
FN Manufacture, Vent-rib Bbl. 525
Miroku Manufacture, Vent-rib Bbl. 395

Browning Automatic-5 Stalker

Same general specifications as Automatic-5 Light and Magnum models, except w/matte blue finish and black graphite fiberglass stock and forearm. Made 1992 to date.

Light Model $495
Magnum Model 525

Browning Automatic-5, Standard (Grade I)

Recoil-operated. Gauge: 12 or 16 (16-gauge guns made prior to WW II were chambered for 2⁹⁄₁₆-inch shells; standard 16 disc. 1964). 4-shell magazine in 5-shot model, prewar guns were also available in 3-shot model. Bbls.: 26- to 32-inch; plain, raised matted or vent rib; choice of standard chokes. Weight: about 8 lbs., 12 ga.,7.5 lbs., 16 ga. Checkered pistol-grip stock and forearm. (*Note:* Browning Special, disc. about 1940.) Made 1900-73 by FN.

Grade I, Plain Bbl. $395
Grade I or Browning Special, Raised Matted Rib 505
Grade I or Browning Special, Vent Rib 520

Browning Automatic-5, Sweet 16

16 ga. Same general specifications as Standard Model, except lightweight (about 6.75 lbs.), has goldplated trigger. Guns w/orib have striped matting on top of bbl. Made 1937-76 by FN.

W/Plain Bbl. $450
W/Raised Matted or Ventilated Rib 495

Browning Auto-5, Sweet Sixteen New Model $495

Reissue of popular 16-gauge Hunting Model w/5-shot capacity, 2.75-inch chamber, scroll-engraved blued receiver, high-gloss French walnut stock w/rounded pistol grip. 26- or 28-inch vent-rib bbl. F choke tube. Weight: 7 lbs. 5 oz. ReIntro. 1987-93.

SHOTGUNS

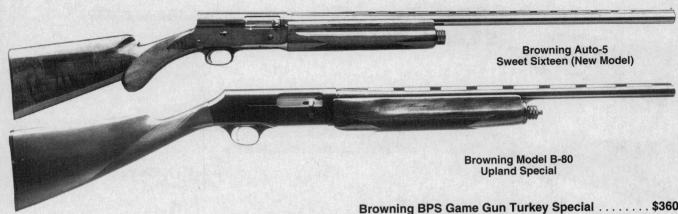

**Browning Auto-5
Sweet Sixteen (New Model)**

**Browning Model B-80
Upland Special**

Browning Automatic-5, Trap Model **$550**
12 gauge only. Same general specifications as Standard Model, except has trap-style stock, 30-inch vent-rib bbl. F choke. Weight: 8.5 lbs. Discontinued 1971.

Browning Model B-80 Gas-Operated Automatic . . . **$395**
Gauge: 12 or 20; 2.75-inch chamber. 4-shot magazine. Bbl.: 26-, 28- or 30-inch, any standard choke, vent-rib bbl. w/fixed chokes or vector tubes. Weight: 6 lbs. 12 oz.-8 lbs. 1 oz. depending on ga. and bbl. Checkered pistol-grip stock and forearm. Made 1981-87.

Browning Model B-80 Plus **$435**
Same general specifications as Browning Model B-80, except chambered for 3-inch shotshells. Made in 1988 only.

Browning Model B-80 Superlight **$395**
Same as Standard Model, except weighs 1 lb. less.

Browning Model B-80 Upland Special **$405**
Gauge: 12 or 20. 22-inch vent-rib bbl. Invector choke tube system. 2.75-inch chambers. 42 inches overall. Weight: 5 lbs. 7 oz. (20 ga.); 6 lbs. 10 oz. (12 ga.). German nickel silver sight bead. Cross-bolt safety. Checkered walnut straight-grip stock and forend. Discontinued 1988.

Browning BPS Game Gun Deer Special **$350**
Same general specifications as Standard BPS Model, except has 20.5-inch bbl. w/adj. rifle-style sights. Solid scope mounting system. Checkered walnut stock w/sling swivel studs. Made from 1992 to date.

Browning BPS Game Gun Turkey Special **$360**
Same general specifications as Standard BPS Model, except w/matte blue metal finish and satin-finished stock. Chambered 12 ga. 3-inch only. 20.5-inch bbl. w/extra full invector choke system. Receiver drilled and tapped for scope. Made 1992 to date.

Browning BPS Pigeon Grade **$475**
Same general specifications as Standard BPS Model, except w/select grade walnut stock and gold-trimmed receiver. Available in 12 ga. only w/26- or 28-inch vent-rib bbl. Made 1992 to date.

Browning BPS Pump Invector Stalker
Same general specifications as BPS Pump Shotgun, except in 10 and 12 ga. w/Invector choke system, 22-, 26-, 28- or 30-inch bbls.; matte blue metal finish w/matte black stock. Made 1987 to date.
12-Gauge Model (3-inch) **$335**
10- & 12-Gauge Model (3.5-inch) **465**

Browning BPS Pump Shotgun
Takedown. Gauges: 10, 12 (3.5-inch chamber); 12 or 20 (3-inch), 2.75-inch in 28 ga. and target models. Bbls.: 22-, 24-, 26-, 28-, 30-, or 32-inch; fixed choke or Invector tubes. Weight: 7.5 lbs. (w/28-inch bbl.). Checkered select walnut pistol-grip stock and semibeavertail forearm, recoil pad. Introduced in 1977. Made by Miroku.
Hunting Model (10 or 12 ga., 3.5") **$475**
Hunting, Upland (12, 20 or 28 ga.). **325**
Buck Special (10 or 12 ga., 3.5") **450**
Buck Special (12 or 20 ga.) **315**
Waterfowl (10 or 10 Ga., 3.5-inch). **495**
W/Fixed Choke, **deduct** **50**

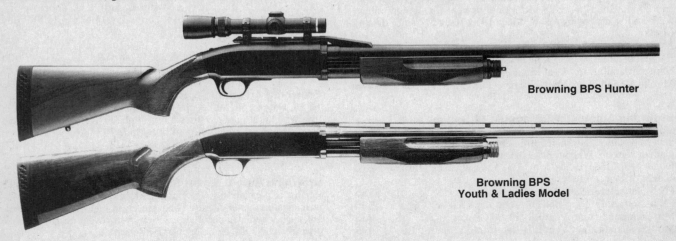

Browning BPS Hunter

**Browning BPS
Youth & Ladies Model**

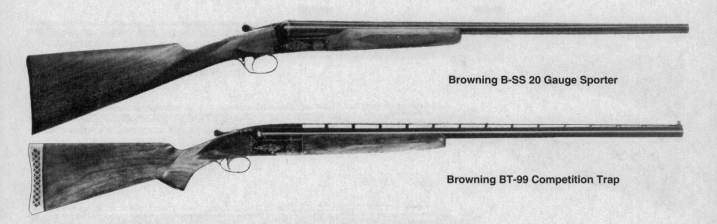

Browning B-SS 20 Gauge Sporter

Browning BT-99 Competition Trap

Browning BPS Youth and Ladies' Model
Lightweight (6 lbs. 11 oz.) version of BPS Pump Shotgun in 20 ga. w/22-inch bbl. and floating vent rib, F choke (invector) tube. Made 1986 to date.

Standard Invector Model (Disc. 1994)	**$325**
Invector Plus Model	**325**

Browning BSA 10 Semiautomatic Shotgun
Gas-operated short-stroke action. 10 ga.; 3.5-inch chamber. 5-shot magazine. Bbls.: 26-, 28-or 30-inch w/Invector tubes and vent rib. Weight: 10.5 lbs. Checkered select walnut buttstock and forend. Blued finish. Made 1993 to date. *Note:* Although Intro. as the BSA 10, this model is now marketed as the Gold Series. *See* separate listing for pricing.

Browning B-SS Side-by-Side
Boxlock. Automatic ejectors. Non-selective single trigger (early production) or selective-single trigger (late production). Gauges: 12 or 20 w/3-inch chambers. Bbls. 26-, 28-, or 30-inch; IC/M, M/F, or F/F chokes; matte solid rib. Weight: 7 to 7.5 lbs. Checkered straight-grip stock and beavertail forearm. Made 1972-88 by Miroku.

Standard Model (early/NSST)	**$425**
Standard Model (late/SST)	**495**
Grade II (Antique Silver Receiver)	**850**
20 Ga. Models, add	**100**

Browning B-SS Side-by-Side Sidelock
Same general specifications as B-SS boxlock models, except side-lock version available in 26- or 28-inch bbl. lengths. 26-inch choked IC/M; 28-inch, M/F. Double triggers. Satin grayed receiver engraved w/rosettes and scrolls. German nickel-silver sight bead. Weight: 6.25 lbs. to 6 lbs. 11 oz. 12 ga. made in 1983; 20 ga. Made in 1984. Discontinued 1988.

12 Ga. Model	**$1395**
20 Ga. Model	**1595**

Browning B-SS S/S 20 Gauge Sporter **$525**
Same as standard B-SS 20 gauge, except has selective single trigger, straight-grip stock. Introduced 1977. Discontinued 1987.

Browning BT-99 Competition Trap Special
Same as BT-99, except has super-high wide rib and standard Monte Carlo or fully adj. stock. Available w/adj. choke or Invector Plus tubes w/optional porting. Made 1976-94.

Grade I w/Fixed Choke (Disc. 1992)	**$685**
Grade I w/Invectors	**850**
Grade I Stainless (Disc. 1994)	**895**
Grade I Pegeon Grade (Disc. 1994)	**825**

Browning BT-99 Grade I Single Bbl. Trap **$525**
Boxlock. Automatic ejector. 12 ga. only. 32- or 34-inch vent-rib bbl., M, IM or F choke. Weight: about 8 lbs. Checkered pistol-grip stock and beavertail forearm, recoil pad. Made 1971-76 by Miroku.

Browning BT-99 Max
Boxlock. 12 ga. only w/ejector selector and no safety. 32- or 34-inch ported bbl. w/high post vent-rib. Checkered select walnut buttstock and finger-grooved forend w/high luster finish. Engraved receiver w/blued or stainless metal finish. Made 1995-96.

Blued	**$ 795**
Stainless	**1195**

Browning BT-99 Plus
Similar to the BT-99 Competition, except w/Browning Recoil Reduction System. Made 1989-95.

Grade I	**$ 895**
Pigeon Grade	**1025**
Signature Grade	**995**
Stainless Model	**1095**
Golden Clays	**2195**

Browning BT-99 Plus Micro **$925**
Same general specifications as BT-99 Plus, except scaled down for smaller shooters. 30-inch bbl. w/adj. rib and Browning's recoil reducer system. Made 1991 to date.

Browning BT-100 Single-Shot Trap
Similar to the BT-99 Max except w/additional stock options and removable trigger group. Made 1995 to 94.

Blued	**$ 995**
Stainless	**1295**
Thumbhole Stock, **add**	**295**
Removable Trigger Group, **add**	**495**
Fixed Choke, **deduct**	**50**

Browning Citori Hunting Over/Under Models
Boxlock. Gauges: 12, 16 (disc. 1989), 20, 28 (disc. 1992) and .410 bore (disc. 1989). Bbl. lengths: 24-, 26-, 28-, or 30-inch w/vent rib. Chambered 2.75-, 3- or 3.5-inch mag. Chokes: IC/M, M/F (Fixed Chokes); Standard Invector, or Invector plus choke systems. Overall length ranges from 41-47 inches. 2.75-, 3- or 3-inch Mag. loads, depending on ga. Weight: 5.75 lbs. to 7 lbs. 13 oz. Single selective, gold-plated trigger. Medium raised German nickel-silver sight bead. Checkered, rounded pistol-grip walnut stock w/beavertail forend. Invector Chokes and Invector Plus became standard in 1988 and 1995 respectively. Made from 1973 to date by Miroku.

Grade I (Disc. 1994)	**$ 625**

SHOTGUNS

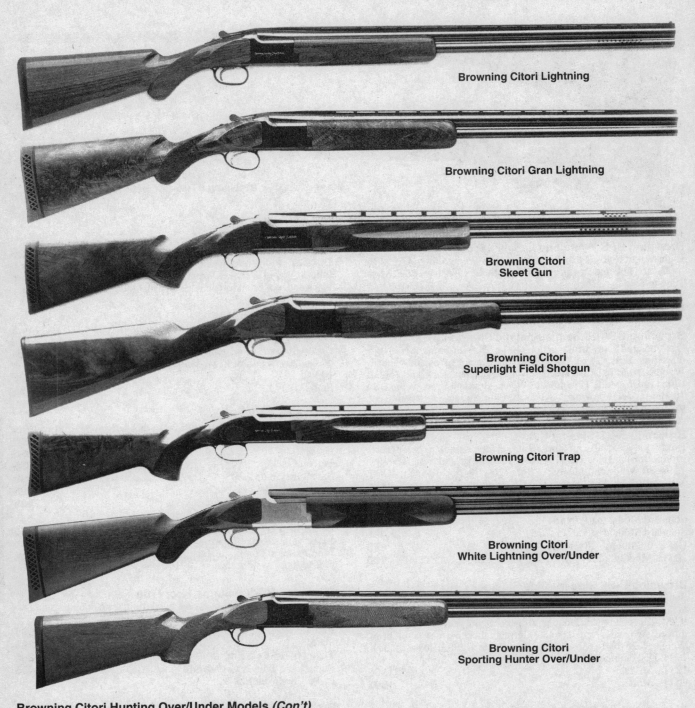

Browning Citori Lightning

Browning Citori Gran Lightning

Browning Citori
Skeet Gun

Browning Citori
Superlight Field Shotgun

Browning Citori Trap

Browning Citori
White Lightning Over/Under

Browning Citori
Sporting Hunter Over/Under

Browning Citori Hunting Over/Under Models *(Con't)*

Grade I - 3.5 Mag.(1989 to date). 825
Grade II (Disc. 1983) 775
Grade III (1985-95) 895
Grade V (Disc. 1984) 1295
Grade VI (1985-95). 1450
Model Sporting Hunter (12 and 20 ga.; 1998 to date). . . . 1075
Model Satin Hunter (12 ga. only; 1998 to date). 950
Model Upland Special (12, 16 or 20 ga.; 1984 to date) 1050
Models w/o Invector Choke System, **deduct** 120
For 3.5 Mag., **add** 90
For Disc. Gauges (16, 28 and .410), **add** 15%

Browning Citori Lightning O/U Models

Same general specifications as the Citori Hunting Models except
w/classic Browning rounded pistol-grip stock. Made from 1988 to
date, by Miroku.

Grade I. . . . . **$ 640**
Grade III. 925
Grade VI 1350
Gran Lightning Model 995
Micro Model, **add** 10%
Models w/o Invector Choke System, **deduct** 120
28 and .410 Ga., **add** 15%

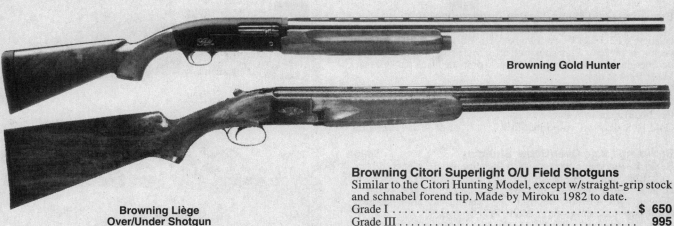

Browning Gold Hunter

**Browning Liège
Over/Under Shotgun**

Browning Citori Skeet Gun
Same as Hunting Model, except has skeet-style stock and forearm, 26- or 28-inch bbls., both bored SK choke. Available w/either standard vent rib or special target-type, high-post, wide vent rib. Weight (w/26-inch bbls.): 12 ga., 8 lbs., 20 ga., 7 lbs. Made 1974 to date by Miroku.

Grade I	$ 795
Grade II	850
Grade III.	975
Grade VI (Disc. 1995)	1495
Golden Clays	1650
3-Bbl. Set Grade I (Disc. 1996)	1750
3-Bbl. Set Grade III (Disc. 1996)	1895
3-Bbl. Set Grade VI (Disc. 1994)	2150
3-Bbl. Set, Golden Clays (Disc. 1995)	2795
4-Bbl. Set, Grade I	2559
4-Bbl. Set, Grade III	2850
4-Bbl. Set, Grade VI (Disc. 1994)	2995
4-Bbl. Set, Golden Clays (Disc. 1995)	3550

Browning Citori Sporting Clays
Similar to the standard Citori Lightning Model, except Classic-style stock with rounded pistol grip. 30-inch back-bored bbls. with Invector Plus tubes. Receiver with "Lightning Sporting Clays Edition" logo. Made 1989 to date.

GTI Model (Disc. 1995)	$ 750
GTI Golden Clays Model (1993-94)	750
Lightning Model (Intro. 1989)	865
Lightning Golden Clays (1993-98)	1550
Lightning Pigeon Grade (1993-94)	925
Micro Citori Lightning Model (w/low rib)	850
Special Sporting Model (Intro. 1989)	875
Special Sporting Golden Clays (1993-98)	1775
Special Sporting Pigeon Grade (1993-94)	950
Ultra Model (Intro. 1995-Previously GTI)	975
Ultra Golden Clays Model (Intro. 1995)	1795
Model 325 (1993-94)	925
Model 325 Golden Clays (1993-94)	1525
Model 425 Grade I (Intro. 1995)	995
Model 425 Golden Clays (Intro. 1995)	1795
Model 425 WSSF (Intro. 1995)	950
Model 802 Sporter (ES) Extended Swing (Intro. 1996)	1050
For 2 bbl. Set add	895
For Adjustable Stock, add	200
For High Rib, add	85
For Ported Barrels, add	65

Browning Citori Superlight O/U Field Shotguns
Similar to the Citori Hunting Model, except w/straight-grip stock and schnabel forend tip. Made by Miroku 1982 to date.

Grade I	$ 650
Grade III	995
Grade V (Disc. 1985)	1295
Grade VI	1495
Models w/o Invector Choke System, **deduct**	120
28 and 410 Ga., **add**	15%

Browning Citori Trap Gun
Same as Hunting Model, except 12 ga. only, has Monte Carlo or fully adjustable stock and beavertail forend, trap-style recoil pad; 20- or 32-inch bbls.; M/F, IM/F, or F/F. Available with either standard vent rib or special target-type, high-post, wide vent rib. Weight: 8 lbs. Made from 1974 to date by Miroku.

Grade I Trap	$ 950
Grade I Trap Pigeon Grade (Disc. 1994)	995
Grade I Trap Signature Grade (Disc. 1994)	975
Grade I Plus Trap (Disc. 1994)	1025
Grade I Plus trap w/ported bbls. (Disc. 1994)	1095
Grade I Plus Trap Combo (Disc. 1994)	2375
Grade I Plus Trap Golden Clays (Disc. 1994)	1025
Grade II w/HP Rib (Disc. 1984)	895
Grade III Trap	1095
Grade V Trap (Disc. 1984)	1195
Grade VI Trap (Disc. 1994)	1395
Grade VI Trap Golden Clays (Disc. 1994)	1695

Browning Citori Upland Special O/U Shotgun
A shortened version of the Hunting Model, fitted with 24-inch bbls. and straight-grip stock.

Upland Special 12, 20 Ga. Models	$675
Upland Speical 16, 410 Ga. Models (Disc. 1989)	795
Models w/o invector Choke Syste, **deduct**	120

Browning Citori White Lightning Shotgun $865
Gauges: 12 or 20 w/3-inch chambers. 26- or 28-inch bbls. Weight: 6 lbs. 9 oz. to 8 lbs. 1 oz. Silver nitride receiver w/scroll and rosette engraving. Satin wood finish w/round pistol grip. Made 1998 to date.

Browning Double Automatic (Steel Receiver)
Short recoil system. Takedown. 12 ga. only. Two shots. Bbls.: 26-, 28-, 30-inch; any standard choke. Checkered pistol-grip stock and forend. Weight: about 7.75 lbs. Made 1955-1961.

With plain bbl.	$465
With recessed-rib bbl.	590

Browning Gold Deer Hunter Autoloading Shotgun
Similar to the Standard Gold Hunter Model, except chambered 12 ga. only. 22-inch bbl. W/rifled bore or smooth bore w/5-inch rifled invector tube. Cantilevered scope mount. Made 1997 to date.

Gold Deer Hunter (w/standard finish)	$425
Field Model (w/Mossy Oak finish)	495

SHOTGUNS

Browning Gold Hunter Series Autoloading Shotgun
Self-cleaning, gas-operated, short-stroke action. Gauges: 10 (3.5-inch chamber); 12 or 20 (3-inch chamber). 26-, 28-, or 30-inch bbl. w/Invector or Invector Plus choke tubes. Checkered walnut or graphite/fiberglass composite stock. Polished or matte black metal finish. Made 1994 to date

Gold Hunter Model w/Walnut Stock	**$455**
Gold Stalker Model w/Composite Stock	**480**
Gold 10 Hunter w/Walnut Stock	**695**
Gold 10 Stalker w/Composite Stock	**725**

Browning Liège Over/Under Shotgun $625
Boxlock. Automatic ejectors. Non-selective single trigger. 12 ga. only. Bbls.: 26.5-, 28-, or 30-inch; 2.75-inch chambers in 26.5- and 28-inch, 3-inch in 30-inch, IC/M, M/F, or F/F chokes; vent rib. Weight: 7 lbs. 4 oz.to 7 lbs. 14 oz., depending on bbls. Checkered pistol-grip stock and forearm. Made 1973-75 by FN.

Browning Light Sporting 802ES $995
Over-under. Invector-plus choke tubes. 12 ga. only with 28-inch bbl. Weight: 7 lbs, 5 oz.

Browning Lightning Sporting Clays
Similar to the standard Citori Lightning Model, except Classic-style stock with rounded pistol grip. 30-inch back-bored bbls. with Invector Plus tubes. Receiver with "Lightning Sporting Clays Edition" logo. Made 1989 to date.

Standard Model	**$850**
Pigeon Grade	**925**

Browning Over/Under Classic $1650
Gauge: 20, 2.75-inch chambers. 26-inch blued bbls. choked IC/M. Gold-plated, single selective trigger. Manual, top-tang-mounted safety. Engraved receiver. High grade, select American walnut straight-grip stock with schnabel forend. Fine-line checkering with pearl borders. High-gloss finish. 5,000 issued in 1986; made in Japan, engraved in Belgium.

**Browning Over/Under
Gold Classic**

Browning Over/Under Gold Classic $3995
Same general specifications as Over/Under Classic, except more elaborate engravings, enhanced in gold, including profile of John M. Browning. Fine oil finish. 500 issued; made in 1986 in Belgium.

Browning Recoilless Trap Shotgun
The action and bbl. are driven forward when firing to achieve 70 percent less recoil. 12 ga, 2.75-inch chamber. 30-inch bbl. with Invector Plus tubes; adjustable vent rib. 51.63 inches overall. Weight: 9 lbs. Adj. checkered walnut buttstock and forend. Blued finish. Made from 1993 to date.

Standard Model	**$1095**
Micro Model (27-inch bbl.)	**1125**

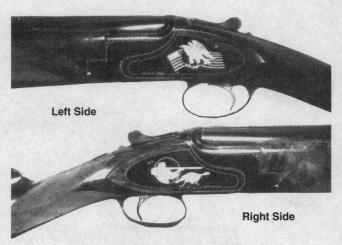

Left Side

Right Side

Browning Superposed Bicentennial

Browning Superposed Bicentennial Commemorative $12,950
Special limited edition issued to commemorate U.S. Bicentennial. 51 guns, one for each state in the Union plus one for Washington, D.C. Receiver with sideplates has engraved and gold-inlaid hunter and wild turkey on right side, U.S. flag and bald eagle on left side, together with state markings inlaid in gold, on blued background. Checkered straight-grip stock and schnabel-style forearm of highly figured American walnut. Velvet-lined wooden presentation case. Made in 1976 by FN. Value shown is for gun in new, unfired condition.

Browning Superposed BROADway 12 Trap $1525
Same as standard Trap Gun, except has 30- or 32-inch bbls. with wider BROADway rib. Discontinued 1976.

Browning Superposed Shotguns, Hunting Models
Over/under boxlock. Selective automatic ejectors. Selective single trigger; earlier models (worth 25 % less) supplied w/double triggers, twin selective triggers or non-selective single trigger. Gauges: 12, 20 (intro. 1949, 3-inch chambers in later production), 28, .410 (latter two ga. intro. 1960). Bbls.: 26.5-,28-,30-,32-inch, raised matted or vent rib, prewar Lightning Model made w/oribbed bbl., postwar version supplied only w/vent rib; any combination of standard chokes. Weight (w/26.5-inch vent-rib bbls.): *Standard 12,* 7 lbs. 11 oz., *Lightning 12,* 7 lbs. 6 oz.; *Standard 20,* 6 lbs. 8 oz.; *Lightning 20,* 6 lbs. 4 oz.; *Lightning 28,* 6 lbs. 7 oz.; *Lightning .410,* 6 lbs. 10 oz. Checkered pistol-grip stock/forearm.

Higher grades — Pigeon, Pointer, Diana, Midas, Grade VI — differ from standard Grade I models in overall quality, engraving, wood and checkering, otherwise, specifications are the same. Midas Grade and Grade VI guns are richly gold-inlaid. Made by FN 1928-1976. Prewar models may be considered as disc. in 1940 when Belgium was occupied by Germany. Grade VI offered 1955-1960. Pointer Grade disc. in 1966, Grade I Standard in 1973, Pigeon Grade in 1974. Lightning Grade I, Diana and Midas Grades were not offered after 1976.

Grade I Standard	**$1195**
Grade I Lightning	**1450**
Grade I Lightning, prewar, matted bbl., no rib	**2095**
Grade II—Pigeon	**2495**
Grade III—Pointer	**2950**
Grade IV—Diana	**3795**
Grade V—Midas	**4650**
Grade VI	**6350**
Add for 28 or .410 gauge	**995**
Values shown are for models w/ventilated rib, if gun has raised matted rib, **deduct**	**200**

**Browning Superposed
Grade IV Diana (Postwar)**

**Browning Superposed
Grade V Midas (Postwar)**

Browning Superposed Lightning and Superlight Models (Reissue)

Reissue of popular 12-and 20-ga. superposed shotguns. Lightning models available in 26.5- and 28-inch bbl. lengths w/2.75- or 3-inch chambering, full pistol grip. Superlight models available in 26.5-inch bbl. lengths w/2.75-inch chambering only, and straight-grip stock w/schnabel forend. Both have hand-engraved receivers, fine-line checkering, gold-plated single selective trigger, automatic selective ejectors, manual safety. Weight: 6 to 7.5 lbs. ReIntro. 1985-86.

Grade II, Pigeon	$2595
Grade III, Pointer	2995
Grade IV, Diana	3195
Grade V, Midas	4195

Browning Superposed Magnum $1350
Same as Grade I, except chambered for 12-ga. 3-inch shells, 30-inch vent-rib bbls., stock w/recoil pad. Weight: about 8.25 lbs. Discontinued 1976.

Browning Superposed Ltd. Black Duck Issue $4600
Gauge: 12. Superposed Lightning action. 28-inch vent-rib bbls. Choked M/F. 2.75-inch chambers. Weight: 7 lbs. 6 oz. Gold inlaid receiver and trigger guard engraved w/Black Duck scenes. Gold-plated, single selective trigger. Top-tang mounted manual safety. Automatic, selective ejectors. Front and center ivory sights. High-grade, hand-checkered, hand-oiled select walnut stock and forend. 500 issued in 1983.

Browning Superposed Ltd. Mallard Duck Issue $4650
Same general specifications as Ltd. Black Duck Issue, except Mallard Duck scenes engraved on receiver and trigger guard, dark French walnut stock w/rounded pistol grip. 500 issued in 1981.

Browning Superposed Ltd. Pintail Duck Issue $4695
Same general specifications as Ltd. Black Duck Issue, except Pintail Duck scenes engraved on receiver and trigger guard, stock is of dark French walnut w/rounded pistol grip. 500 issued in 1982.

Browning Superposed, Presentation Grades
Custom versions of Super-Light, Lightning Hunting, Trap and Skeet Models, w/same general specifications as those of standard guns, but of higher overall quality. The four Presentation Grades differ in receiver finish (grayed or blued), engraving gold inlays, wood and checkering Presentation 4 has sideplates. Made by FN, these models were Intro. in 1977.

Presentation 1	$2595
Presentation 1, gold-inlaid	2950
Presentation 2	3095
Presentation 2, gold-inlaid	3650
Presentation 3, gold-inlaid	5125
Presentation 4	5600
Presentation 4, gold-inlaid	6750

Browning Superposed Skeet Guns, Grade I
Same as standard Lightning 12, 20, 28 and .410 Hunting Models, except has skeet-style stock and forearm, 26.5- or 28-inch vent-rib bbls. w/SK choke. Available also in All Gauge Skeet Set: Lightning 12 w/one removable forearm and three extra sets of bbls. in 20, 28 and .410 ga. in fitted luggage case. Discontinued 1976.

12 or 20 gauge	$1450
28 or .410 gauge	1795
All Gauge Skeet Set	4595

SHOTGUNS

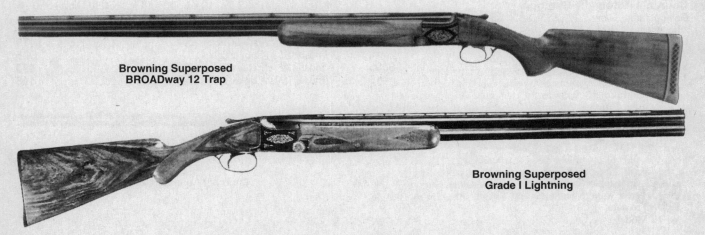

**Browning Superposed
BROADway 12 Trap**

**Browning Superposed
Grade I Lightning**

**Browning Superposed Ltd.
Pintail Duck Issue**

Browning Superposed Super-Light Model $1425
Ultralight field gun version of Standard Lightning Model has classic straight-grip stock and slimmer forearm. Available only in 12 and 20 gauges (2.75-inch chambers), w/26.5-inch vent-rib bbls. Weight: 6.5 lbs., 12 ga.; 6 lbs., 20 ga. Made 1967-76.

Browning Superposed Trap Gun $1475
Same as Grade I, except has trap-style stock, beavertail forearm, 30-inch vent-rib bbls., 12 ga. only. Discontinued 1976.

Browning Twelvette Double Automatic
Lightweight version of Double Automatic w/same general specifications except aluminum receiver. Bbl. w/plain matted top or vent rib. Weight: 6.75 to 7 lbs., depending on bbl. Receiver is finished in black w/gold engraving; 1956-1961 receivers were also anodized in gray, brown and green w/silver engraving. Made 1955-71.
W/Plain Bbl. $425
W/Vent-rib Bbl. 525

CENTURY
INTERNATIONAL ARMS INC.
St. Albans, VT

Century Centurion $265
Over/Under boxlock action with double triggers and extractors. 12 ga. w/2.75-inch chamber. Bbls: 28-inch choked modified/full with ventilated rib . Weight: 7-1/4 lbs. Checkered European walnut buttstock and forend. Polished blue finish. Imported 1993 to date.

CHURCHILL SHOTGUNS
Italy and Spain
Imported by Ellett Brothers, Inc., Chapin, SC;
previously by Kassnar Imports, Inc., Harrisburg, PA

Churchill Automatic Shotgun
Gas-operated. Gauge: 12, 2.75- or 3-inch chambers. 5-shot magazine w/cutoff. Bbl.: 24-, 25-, 26-, 28-inch w/ICT Choke tubes. Checkered walnut stock w/satin finish. Imported from Italy 1990-94.
Standard Model $395
Turkey Model 415

Churchill Monarch Over/Under Shotgun
Hammerless, takedown w/engraved receiver. Selective single or double triggers. Gauges: 12, 20, 28, .410; 3-inch chambers. Bbls.: 25- or 26-inch (IC/M); 28-inch (M/F). Weight: 6.5-7.5 lbs. Checkered European walnut buttstock and forend. Made in Italy 1986-93.
W/Double Triggers $335
W/Single Trigger 375

Churchill Regent Over/Under Shotguns
Gauges: 12 or 20; 2.75-inch chambers. 27-inch bbls. w/interchangeable choke tubes and wide vent rib. Single selective trigger, selective automatic ejectors. Checkered pistol-grip stock in fancy walnut. Imported from Italy 1984-88 and 1990-94.
Regent V(Disc. 1988) $750
Regent VII w/Sideplates (Disc. 1994) 850

Churchill Regent Skeet $625
12 or 20 ga. w/2.75-inch chambers. Selective automatic ejectors, single-selective trigger. 26-inch over/under bbls. w/vent rib. Weight: 7 lbs. Made in Italy 1984-88 and.

Churchill Regent Trap $635
12-ga. competition shotgun w/2.75-inch chambers. 30 inch over/under bbls. choked IM/F, vent side ribs. Weight: 8 lbs. Selective automatic ejectors, single selective trigger. Checkered Monte Carlo stock w/Supercushion recoil pad. Made in Italy 1984-88.

Churchill Sporting Clays Over/Under $665
Same general specifications as Windsor IV, except in 12 ga. only w/28-inch ported bbls. and choke tubes. Selective automatic ejectors. Weight: 7.5 lbs. Made from 1992-94.

Churchill Windsor Over/Under Shotguns
Hammerless, boxlock w/engraved receiver, selective single trigger. Extractors or ejectors. Gauges: 12, 20, 28 or .410; 3-inch chambers. Bbls.: 24 to 30 inches w/fixed chokes or choke tubes. Weight: 6 lbs. 3 oz. (Flyweight) to 7 lbs. 10 oz. (12 ga.). Checkered straight (Flyweight) or pistol-grip stock and forend of European walnut. Imported from Italy 1984-93.
Windsor III w/Fixed Chokes $475
Windsor III w/Choke Tubes 595
Windsor IV w/Fixed Chokes (Disc. 1993) 555
Windsor IV w/Choke Tubes 495

Churchill Automatic Shotgun

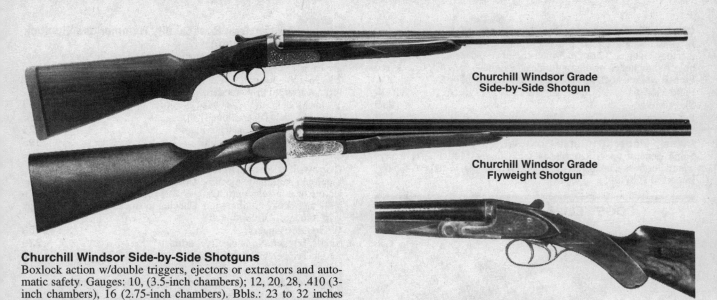

**Churchill Windsor Grade
Side-by-Side Shotgun**

**Churchill Windsor Grade
Flyweight Shotgun**

E.J. Churchill Premier

Churchill Windsor Side-by-Side Shotguns

Boxlock action w/double triggers, ejectors or extractors and automatic safety. Gauges: 10, (3.5-inch chambers); 12, 20, 28, .410 (3-inch chambers), 16 (2.75-inch chambers). Bbls.: 23 to 32 inches w/various fixed choke or choke tube combinations. Weight: 5 lbs. 12 oz. (Flyweight) to 11.5 lbs. (10 ga.). European walnut buttstock and forend. Imported from Spain 1984-90.

Windsor I 10 ga.	$625
Windsor I 12 thru .410 ga.	475
Windsor II 12 or 20 ga.	445
Windsor VI 12 or 20 ga.	650

E.J. CHURCHILL, LTD.
Surrey (previously London), England

All of the E.J. Churchill shotguns listed below are no longer imported.

E.J. Churchill Field Model Hammerless Double

Sidelock Hammerless ejector gun w/same general specifications as Premiere Model but of lower quality.

W/Double Triggers	$7590
Selective Single Trigger, **add**	400

E.J. Churchill Premiere Quality Hammerless Double

Sidelock. Automatic ejectors. Double triggers or selective single trigger. Gauges: 12, 16, 20, 28. Bbls.: 25-, 28- 30-, 32-inch; any degree of boring. Weight: 5-8 lbs. depending on ga. and bbl. length. Checkered stock and forend, straight or pistol grip.

W/Double Triggers	$13,750
Selective Single Trigger, **extra**	900

E.J. Churchill Premiere Quality Under-and-Over Shotgun

Sidelock. Automatic ejectors. Double triggers or selective single trigger. Gauges: 12, 16, 20, 28. Bbls.: 25-, 28-, 30-, 32-inch, any degree of boring. Weight: 5-8 lbs. depending on ga. and bbl. length. Checkered stock and forend, straight or pistol grip.

W/Double Triggers	$14,995
Selective Single Trigger, **add**	700
Raised Vent Rib, **add**	400

E.J. Churchill Utility Model Hammerless Double Barrel

Anson & Deeley boxlock action. Double triggers or single trigger. Gauges: 12, 16, 20, 28, .410. Bbls.: 25-, 28-, 30-, 32-inch, any degree of boring. Weight: 4.5-8 lbs. depending on ga. and bbl. length. Checkered stock and forend, straight or pistol grip.

W/Double Triggers	$3950
Selective Single Trigger, **add**	450

E.J. Churchill XXV Premiere Hammerless Double

$13,250

Sidelock. Assisted opening. Automatic ejectors. Double triggers. Gauges: 12, 20. 25-inch bbls. w/narrow, quick-sighting rib; any standard choke combination. English-style straight-grip stock and forearm, checkered.

E.J. Churchill XXV Imperial $10,500

Similar to XXV Premiere, but no assisted opening feature.

E.J. Churchill XXV Hercules $7695

Boxlock, otherwise specs same as for XXV Premiere.

E.J. Churchill XXV Regal $4250

Similar to XXV Hercules, but w/oassisted opening feature. Gauges: 12, 20, 28, .410.

CLASSIC DOUBLES
Tochigi, Japan

Imported by Classic Doubles International, St. Louis, MO, and previously by Olin as Winchester Models 101 and 23.

Classic Model 101 Over/Under Shotgun

Boxlock. Engraved receiver w/single selective trigger, auto ejectors and combination bbl. selector and safety. Gauges: 12, 20, 28 or.410, 2.75-, 3-inch chambers. 25.5- 28- or 30-inch vent-rib bbls. Weight: 6.25 – 7.75 lbs. Checkered French walnut stock. Imported 1987-90.

Classic I Field	$1295
Classic II Field	1450
Classic Sporter	1550
Classic Sporter Combo	2450
Classic Trap	1200
Classic Trap Single	1295
Classic Trap Combo	1950
Classic Skeet	1450
Classic Skeet 2-Bbl. Set	1895
Classic Skeet 4-Bbl. Set	3750
Classic Waterfowler	1095
For Grade II (28 Ga.), **add**	750
For Grade II (.410 Ga.), **add**	250

Classic Model 201 Side-by-Side Shotgun

Boxlock. Single selective trigger, automatic safety and selective ejectors. Gauges: 12 or 20; 3-inch chambers.26- or 28-inch vent-rib bbl., fixed chokes or internal tubes. Weight: 6 to 7 lbs. Checkered French walnut stock and forearm. Imported 1987-90.

Field Model	$1095
Skeet Model	1295
For Internal Choke Tubes, **add**	100

Classic Model 201 Small Bore Set $3595

Same general specifications as the Classic Model 201, except w/smaller frame, in 28 ga. (IC/M) and .410 (F/M). Weight: 6-6.5 lbs. Imported 1987-90.

COGSWELL & HARRISON, LTD.
London, England

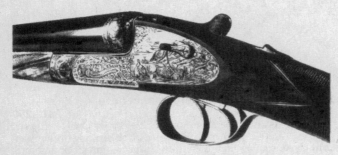

Cogswell & Harrison Best Quality Hammerless Sidelock
Double-Barrel Shotgun

Cogswell & Harrison Ambassador
Hammerless Double-Barrel Shotgun $3195

Boxlock. Sideplates w/game scene or rose scroll engraving. Automatic ejectors. Double triggers. Gauges: 12, 16, 20. Bbls.: 26-, 28-, 30-inch; any choke combination. Checkered straight-grip stock and forearm. Discontinued.

Cogswell & Harrison Avant Tout Series Hammerless
Double-Barrel Shotguns

Boxlock. Sideplates (except Avant Tout III Grade). Automatic ejectors. Double triggers or single trigger (selective or non-selective). Gauges: 12, 16, 20. Bbls.: 25-, 27.5-, 30-inch, any choke combination. Checkered stock and forend, straight grip standard. Made in three models— Avant Tout I or Konor, Avant Tout II or Sandhurst, Avant Tout III or Rex—which differ chiefly in overall quality engraving, grade of wood, checkering, etc.; general specifications are the same. Discontinued.

Avant Tout I	$2695
Avant Tout II	2350
Avant Tout III	1795
Single Trigger, Non-selective, **add**	225
Single Trigger, Selective, **add**	395

Cogswell & Harrison Best Quality Hammerless Sidelock Double-Barrel Shotgun

Hand-detachable locks. Automatic ejectors. Double triggers or single trigger (selective or non-selective). Gauges: 12, 16, 20. Bbls.: 25-, 26-, 28-, 30-inch, any choke combination. Checkered stock and forend, straight grip standard.

Victor Model	$6355
Primic Model (Discontinued)	4395
Single Trigger, Non-selective, **add**	225
Single Trigger, Selective, **add**	395

Cogswell & Harrison Huntic Model Hammerless Double

Sidelock. Automatic ejectors. Double triggers or single trigger (selective or non-selective). Gauges: 12, 16, 20. Bbls.: 25-, 27.5-, 30-inch; any choke combination. Checkered stock and forend, straight grip standard. Discontinued.

W/Double Triggers	$3250
Single Trigger, Non-selective, **add**	225
Single Trigger, Selective, **add**	350

Cogswell & Harrison Markor Hammerless Double

Boxlock. Non-ejector or ejector. Double triggers. Gauges: 12, 16, 20. Bbls.: 27.5 or 30-inch; any choke combination. Checkered stock and forend, straight grip standard. Discontinued.

Non-ejector	$1325
Ejector Model	1650

Cogswell & Harrison Regency
Hammerless Double $2550

Anson & Deeley boxlock action. Automatic ejectors. Double triggers. Gauges: 12, 16, 20. Bbls.: 26-, 28-, 30-inch, any choke combination. Checkered straight-grip stock and forearm. Introduced in 1970 to commemorate the firm's bicentenary, this model has deep scroll engraving and the name "Regency" inlaid in gold on the rib. Discontinued.

COLT INDUSTRIES
Hartford, Connecticut

Colt Auto Shotguns were made by Luigi Franchi S.p.A. and are similar to corresponding models of that manufacturer.

Colt Auto Shotgun — Magnum

Same as Standard Auto, except steel receiver, handles 3-inch Magnum shells, 30- and 32-inch bbls. in 12 ga., 28-inch in 20 ga. Weight: 12 ga., about 8.25 lbs. Made 1964-66.

W/Plain Bbl.	$345
W/Solid-rib Bbl.	375
W/Ventilated-rib Bbl.	395

Colt Auto Shotgun — Magnum Custom

Same as Magnum, except has engraved receiver, select walnut stock and forearm. Made 1964-66.

W/Solid-rib Bbl.	$425
W/Ventilated-rib Bbl.	475

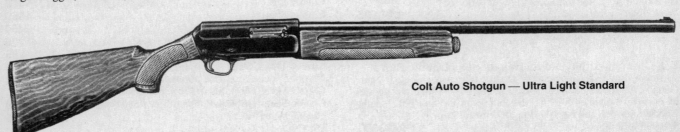

Colt Auto Shotgun — Ultra Light Standard

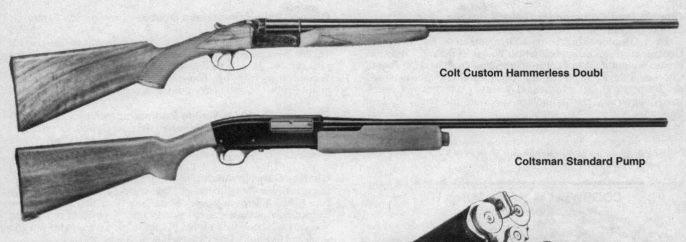

Colt Custom Hammerless Doubl

Coltsman Standard Pump

Colt-Sauer Drilling

Colt Auto Shotgun — Ultra Light Custom
Same as Standard Auto, except has engraved receiver, select walnut stock and forearm. Made 1964-66.
W/Solid-rib Bbl. **$350**
W/Ventilated-rib Bbl. **375**

Colt Auto Shotgun — Ultra Light Standard
Recoil-operated. Takedown. Alloy receiver. Gauges: 12, 20. Magazine holds 4 shells. Bbls.: plain, solid or vent rib, chrome-lined; 26-inch IC or M choke, 28-inch M or F choke, 30-inch F choke, 32-inch F choke. Weight: 12 ga., about 6.25 lbs. Checkered pistol-grip stock and forearm. Made 1964-66.
W/Plain Bbl. **$245**
W/Solid-rib Bbl. **275**
W/Vent-rib Bbl. **295**

Colt Custom Hammerless Double **$595**
Boxlock. Double triggers. Auto ejectors. Gauges: 12 Mag., 16. Bbls.: 26-inch IC/M; 28-inch M/F; 30-inch F/F. Weight: 12 ga., about 7.5 lbs. Checkered pistol-grip stock and beavertail forearm. Made in 1961.

Coltsman Custom Pump **$320**
Same as Standard Pump shotgun except has checkered stock, vent-rib bbl. Weight: about 6.5 lbs. Made 1961-63 by Manufrance.

Coltsman Standard Pump Shotgun **$295**
Takedown. Gauges: 12, 16, 20. Magazine holds 4 shells. Bbls.: 26-inch IC; 28-inch M or F choke; 30-inch F choke. Weight: about 6 lbs. Plain pistol-grip stock and forearm. Made 1961-65 by Manufrance.

Colt-Sauer Drilling **$2750**
Three-bbl. combination gun. Boxlock. Set rifle trigger. Tang bbl. selector, automatic rear sight positioner. 12 ga. over 30-06 or 243 rifle bbl. 25-inch bbls., F and M choke. Weight: about 8 lbs. Folding leaf rear sight, blade front w/brass bead. Checkered pistol-grip stock and beavertail forearm, recoil pad. Made 1974 to date by J. P. Sauer & Sohn, Eckernförde, Germany.

CONNECTICUT VALLEY CLASSICS
Westport, Connecticut

CVC Classic Field Over/Under **$1495**
Similar to the standard Classic Sporter Over/Under Model, except w/30-inch bbls. only and non-reflective matte blued finish on both bbls. and receiver for Waterfowler; other Grades w/different degrees of embellishment; Grade I the lowest and Grade III the highest. Made 1993 to date.

CVC Classic Field Over/Under *(Con't)*
Grade I **$1755**
Grade II **1895**
Grade III **2065**
Waterfowler **1695**

CVC Classic Sporter Over/Under
Gauge: 12; 3-inch chamber. Bbls.: 28-, 30- or 32-inch w/ screw-in tubes. Weight: 7.75 lbs. Engraved stainless or nitrided receiver; blued bbls. Checkered American black walnut buttstock and forend w/low-luster satin finish. Made from 1993 to date.
Classic Sporter **$1595**
Stainless Classic Sporter **1895**

CHARLES DALY, INC.
New York, New York

The pre-WWII Charles Daly shotguns, w/the exception of the Commander, were manufactured by various firms in Suhl, Germany. The postwar guns, except for the Novamatic series were produced by Miroku Firearms Mfg. Co., Tokyo. Miroku production ceased production in 1976 and the Daly trademark was acquired by Outdoor Sports Headquarters located in Dayton, Ohio. OSHI continued to market O/U shotguns from both Italy and Spain under the Daly logo. Automatic models were produced in Japan for distribution in the USA. In 1996, KBI, Inc. Located in Harrisburg, PA acquired the Daly trademark and currently imports firearms under that logo.

SHOTGUNS

Charles Daly Commander Over/Under Shotgun

Daly pattern Anson & Deeley system boxlock action. Automatic ejectors. Double triggers or Miller selective single trigger. Gauges: 12, 16, 20, 28, .410. Bbls.: 26- to 30-inch, IC/M or M/F choke. Weight: 5.25 to 7.25 lbs. depending on ga. and bbl. length. Checkered stock and forend, straight or pistol grip. The two models, 100 and 200, differ in general quality, grade of wood, checkering, engraving, etc.; otherwise specs are the same. Made in Belgium c. 1939.

Model 100	**$395**
Model 200	**525**
Miller Single Trigger, **add**	**100**

Charles Daly Hammerless Regent Diamond Quality

Charles Daly Hammerless Double-Barrel Shotgun

Daly pattern Anson & Deeley system boxlock action. Automatic ejectors—except "Superior Quality" is non-ejector. Double triggers. Gauges: 10, 12, 16, 20, 28, .410. Bbls.: 26- to 32-inch, any combination of chokes. Weight: from 4 to 8.5 lbs. depending on ga. and bbl. length. Checkered pistol-grip stock and forend. The four grades—Regent Diamond, Diamond, Empire, Superior—differ in general quality, grade of wood, checkering, engraving, etc.; otherwise specifications are the same. Discontinued about 1933.

Diamond Quality	**$ 8,750**
Empire Quality	**4,250**
Regent Diamond Quality	**10,500**
Superior Quality	**995**

Charles Daly Hammerless Drilling

Daly pattern Anson & Deeley system boxlock action. Plain extractors. Double triggers, front single set for rifle bbl. Gauges: 12, 16, 20, 25-20, 25-35, 30-30 rifle bbl. Supplied in various bbl. lengths and weights. Checkered pistol-grip stock and forend. Auto rear sight operated by rifle bbl. selector. The three grades — Regent Diamond, Diamond, Superior—differ in general quality, grade of wood, checkering, engraving, etc.; otherwise, specs. are the same. Discont. about 1933.

Diamond Quality	**$ 4750**
Regent Diamond Quality	**10,250**
Superior Quality	**2595**

Charles Daly Hammerless Double— Empire Grade $495

Boxlock. Plain extractors. Non-selective single trigger. Gauges: 12, 16, 20; 3-inch chambers in 12 and 20, 2.75-inch in 16 ga. Bbls.: vent rib; 26-, 28-, 30-inch (latter in 12 ga. only); IC/M, M/F, F/F. Weight: 6 to 7.75 lbs., depending on ga. and bbls. Checkered pistol-grip stock and beavertail forearm. Made 1968-71.

Charles Daly 1974 Wildlife Commemorative $1695

Limited issue of 500 guns. Similar to Diamond Grade over/under. 12-ga. trap and skeet models only. Duck scene engraved on right side of receiver, fine scroll on left side. Made in 1974.

Charles Daly Novamatic Lightweight Autoloader

Same as Breda. Recoil-operated. Takedown.12 ga., 2.75-inch chamber. 4-shell tubular magazine. Bbls.: plain vent rib; 26-inch IC or Quick-Choke w/three interchangeable tubes, 28-inch M or F choke. Weight (w/26-inch vent-rib bbl.): 7 lbs. 6 oz. Checkered pistol-grip stock and forearm. Made 1968 by Ernesto Breda, Milan, Italy.

W/Plain Bbl.	**$325**
W/Vent-rib Bbl.	**350**
For Quick-Choke, **add**	**20**

Charles Daly Novamatic Super Lightweight

Lighter version of Novamatic Lightweight. Gauges: 12, 20. Weight (w/26-inch vent-rib bbl.): 12 ga., 6 lbs. 10 oz., 20 ga., 6 lbs. SK choke available in 26-inch vent-rib bbl. 28-inch bbls. in 12 ga. only. Quick-Choke in 20 ga. w/plain bbl. Made 1968 by Ernesto Breda, Milan, Italy.

12 Ga., Plain Bbl.	**$295**
12 Ga., Vent-rib Bbl.	**325**
20 Ga., Plain Bbl.	**310**
20 Ga., Plain Bbl. w/Quick-Choke	**325**
20 Ga., Vent-rib Bbl.	**345**

Charles Daly Novamatic Super Lightweight 20-Gauge Magnum $345

Same as Novamatic Super Lightweight 20, except 3-inch chamber, has 3-shell magazine, 28-inch vent-rib bbl., F choke.

Charles Daly Novamatic 12-Gauge Magnum $350

Same as Novamatic Lightweight, except chambered for 12-ga. Magnum 3-inch shell, has 3-shell magazine, 30-inch vent-rib bbl., F choke, and stock w/recoil pad. Weight: 7.75 lbs.

Charles Daly Novamatic Trap Gun $375

Same as Novamatic Lightweight, except has 30-inch vent-rib bbl., F choke and Monte Carlo stock w/recoil pad. Weight: 7.75 lbs.

Charles Daly Over/Under Shotguns (Prewar)

Daly-pattern Anson & Deeley-system boxlock action. Sideplates. Auto ejectors. Double triggers. Gauges: 12, 16, 20. Supplied in various bbl. lengths and weights. Checkered pistol-grip stock and forend. The two grades — Diamond and Empire — differ in gen-

Charles Daly Over/Under Field Grade (Postwar)

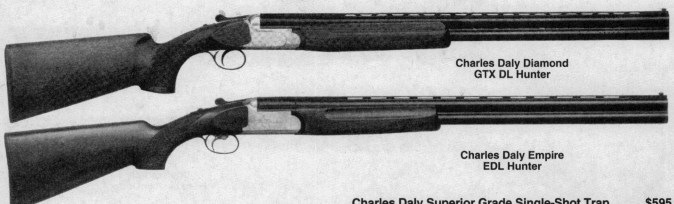

Charles Daly Diamond GTX DL Hunter

Charles Daly Empire EDL Hunter

Charles Daly Over/Under Shotguns *(Con't)*
eral quality, grade of wood, checkering, engraving, etc.; otherwise specifications are the same. Discontinued about 1933.
Diamond Quality $4250
Empire Quality 3295

NOTE

Post War Charles Daly shotguns were imported by Sloan's Sporting Goods trading as Charles Daly in New York. In 1976, Outdoor Sports Headquarters acquired the Daly trademark and continued to import Europeam made shotguns under that logo. In 1996, KBI, Inc. located in Harrisburg, PA acquired the Daly trademark and currently imports firearms under that logo.

Charles Daly Over/Under Shotguns (Postwar)
Boxlock. Auto ejectors or selective auto/manual ejection. Selective single trigger. Gauges: 12, 12 Magnum (3- inch chambers), 20 (3-inch chambers), 28, .410. Bbls.: vent rib; 26-, 28-, 30-inch; standard choke combinations. Weight: 6 to 8 lbs. depending on ga. and bbls. Select walnut stock w/pistol grip, fluted forearm checkered; Monte Carlo comb on trap guns; recoil pad on 12-ga. Mag. and trap models. The various grades differ in quality of engraving and wood. Made 1963-76.
Diamond Grade $995
Field Grade 550
Superior Grade 725
Venture Grade 525

Charles Daly Sextuple Model Single-Barrel Trap Gun
Daly pattern Anson & Deeley system boxlock action. Six locking bolts. Auto ejector. 12 ga. only. Bbls.: 30-, 32-, 34-inch, vent rib. Weight: 7.5 to 8.25 lbs. Checkered pistol-grip stock and forend. The two models made — Empire and Regent Diamond — differ in general quality, grade of wood, checkering, engraving, etc., otherwise specs are the same. Discontinued about 1933.
Regent Diamond Quality (Linder) $4250
Empire Quality (Linder) 3795
Regent Dimond Quality (Sauer) 2695
Empire Quality (Sauer) 1995

Charles Daly Single-Shot Trap Gun
Daly pattern Anson & Deeley system boxlock action. Auto ejector. 12 ga. only. Bbls.: 30-, 32-, 34-inch, vent rib. Weight: 7.5 to 8.25 lbs. Checkered pistol-grip stock and forend. This model was made in Empire Quality only. Discontinued about 1933.
Empire Grade (Linder) $3350
Empire Grade (Sauer) 1650

Charles Daly Superior Grade Single-Shot Trap $595
Boxlock. Automatic ejector. 12 ga. only. 32- or 34-inch vent-rib bbl., F choke. Weight: about 8 lbs. Monte Carlo stock w/pistol grip and recoil pad, beavertail forearm, checkered. Made 1968-76.

Charles Daly Diamond Grade Over/Under
Boxlock. Single selective trigger. Selective automatic ejectors. Gauges: 12 and 20, 3-inch chambers (2.75 target grade). Bbls.: 26-, 27- or 30-inch w/fixed chokes or screw-in tubes. Weight: 7 lbs. Checkered European walnut stock and forearm w/oil finish. Engraved antique silver receiver and blued bbls. Made 1984-90.
Standard Model $595
Skeet Model 650
Trap Model 695

Charles Daly Diamond GTX DL Hunter O/U Series
Sidelock. Single selective trigger and selective auto ejectors. Gauges: 12, 20, 28 ga. or .410 bore. 26-, 28- and 30-inch bbls w/3-inch chambers (2 ¾-inch 28 ga.). Choke tubes (12 and 20 ga.), Fixed chokes (28 and 410). Weight: 5-8 lbs. Checkered European walnut stock w/hand-rubbed oil finish and recoil pad. Made 1997 to date.
Model Diamond GTX DL Hunter $7,550
Model Diamond GTX EDL Hunter 9,325
Model Diamond GTX Sporting (12 or 20 ga.) 4,195
Model Diamond GTX Skeet 12 or 20 ga.) 3,895
Model Diamond GTX Trap (12 ga. only) 4,050

Charles Daly Empire DL Hunter O/U $850
Boxlock. Ejectors. Single selective trigger. Gauges:12, 20, 28 ga. and .410 bore. 26- or 28- inch bbls. w/3-inch chambers (2 ¾ -inch 28 ga.). Choke tubes (12 and 20 ga.), Fixed chokes (28 and 410).Engraved coin-silver receiver w/game scene. Imported 1997 to date.

Charles Daly Empire EDL Hunter Series
Similar to Model Empire DL Hunter, except engraved sideplates. Made 1998 to date.
Model Empire EDL Hunter $925
Model Empire Sporting 900
Model Empire Skeet 875
Model Empire Trap 925
28 Ga., **add** 95
.410 Bore, **add** 120
Multi-Chokes w/Monte Carlo Stock, **add** 125

Charles Daly DSS Hammerless Double $525
Boxlock. Single selective trigger. Selective automatic ejectors. Gauges: 12 and 20; 3-inch chambers. 26-inch bbls. w/screw-in choke tubes. Weight: 6.75 lbs. Checkered walnut pistol-grip stock and semibeavertail forearm w/recoil pad. Engraved antique silver receiver and blued bbls. Made from 1990 to date.

SHOTGUNS

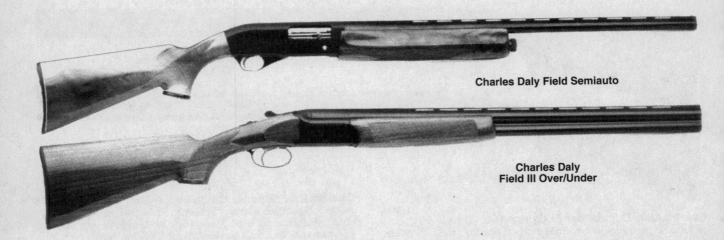

Charles Daly Field Semiauto

Charles Daly Field III Over/Under

Charles Daly Field Grade Over/Under **$385**
Boxlock. Single selective trigger. Extractors. Gauges: 12 and 20; 3-inch chambers. Bbls.: 26-inch, IC/M; 28-inch, M/F. Weight: 6.75 lbs. (12 ga.). Checkered walnut stock and forearm w/semi-gloss finish and recoil pad. Engraved color-casehardened receiver and blued bbls. Made from 1989 to date.

Charles Daly Field Semiauto Shotgun **$295**
Recoil-operated. Takedown. 12-ga. and 12-ga. Magnum. Bbls.: 27- and 30-inch; vent rib. Made 1982-88.

Charles Daly Field III Over/Under Shotgun **$395**
Boxlock. Plain extractors. Non-selective single trigger. Gauges: 12 or 20. Bbls.: vent rib; 26- and 28-inch; IC/M, M/F. Weight: 6 to 7.75 lbs. depending on ga. and bbls. Chrome-molybdenum steel bbls. Checkered pistol-grip stock and forearm. Made from 1982 to date.

Charles Daly Lux Over/Under **$565**
Similar to the Field Grade, except w/selective automatic ejectors and choke tubes. Gauges: 12, 20, 28 and .410. Receiver w/antique silver finish and blued bbls. Made from 1989 to date.

Charles Daly Multi-XII Self-loading Shotgun **$395**
Similar to the gas-operated field semiauto, except w/new Multi-Action gas system designed to shoot all loads w/o adjustment. 12 ga. w/3-inch chamber. 27-inch bbl. w/Invector choke tubes, vent rib. Made in Japan from 1987 to date.

Charles Daly Over/Under Presentation Grade **$825**
Purdey boxlock w/double cross-bolt. Gauges: 12 or 20. Engraved receiver w/single selective trigger and auto ejectors. 27-inch chrome-molybdenum steel, rectified, honed and internally chromed, vent-rib bbls. Hand-checkered deluxe European walnut stock. Made 1982-86.

Charles Daly Over/Under Superior II Shotgun **$575**
Boxlock. Plain extractors. Non-selective single trigger. Gauges: 12 or 20. Bbls.: chrome-molybdenum vent rib 26-,28-, 30-inch, latter in Magnum only, assorted chokes. Silver engraved receiver. Checkered pistol-grip stock and forearm. Made from 1982 to date.

Charles Daly Sporting Clays Over/Under **$695**
Similar to the Field Grade, except in 12 ga. only w/ported bbls. and internal choke tubes. Made from 1990 to date.

DAKOTA ARMS, INC.
Sturgis, South Dakota

**Dakota Arms Classic
Field Grade S/S Shotgun** **$6250**
Boxlock. Gauge: 20 ga. 27-inch bbl. w/fixed chokes. Double triggers. Selective ejectors. Color-casehardened receiver. Weight: 6 lbs. Checkered English walnut stock and splinter forearm w/hand-rubbed oil finish.. Made 1996 to date.

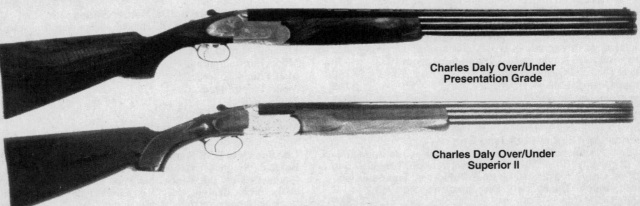

Charles Daly Over/Under Presentation Grade

Charles Daly Over/Under Superior II

Dakota Arms Premium Grade S/S Shotgun $7850
Similar to Classic Field Grade model, except w/50% engraving coverage. Exhibition grade English walnut stock. Made 1996 to date.

**Dakota Arms American Legend
S/S Shotgun** $12,500
Limited Edition built to customer's specifications. Gauge: 20 ga. 27-inch bbl. Double triggers. Selective ejectors. Fully engraved, coin-silver finished receiver w/gold inlays. Weight: 6 lbs. Hand checkered special selection English walnut stock and forearm. Made 1996 to date.

DARNE S.A.
Saint-Etienne, France

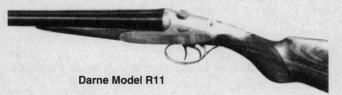

Darne Model R11

Darne Hammerless Double-Barrel Shotguns
Sliding-breech action w/fixed bbls. Auto ejectors. Double triggers. Gauges: 12, 16, 20, 28; also 12 and 20 Magnum w/3-inch chambers. Bbls.: 27.5-inch standard, 25.5- to 31.5-inch lengths available; any standard choke combination. Weight: 5.5 to 7 lbs., depending on ga. and bbl. length. Checkered straight-grip or pistol-grip stock and forearm. The various models differ in grade of engraving and wood. Manufactured 1881-1979.
Model R11 (Bird Hunter) . . . . $ 795
Model R15 (Pheasant Hunter) 1950
Model R16 (Magnum) 1425
Model V19 (Quail Hunter) 2850
Model V22 3250
Model V Hors Série No. 1 3500

DAVIDSON GUNS
Mfd. by Fabrica de Armas ILJA, Eibar, Spain;
distributed by Davidson Firearms Co.,
Greensboro, North Carolina

Davidson Model 63B Double-Barrel Shotgun $250
Anson & Deeley boxlock action. Frame-engraved and nickel plated. Plain extractors. Auto safety. Double triggers. Gauges: 12, 16, 20, 28, .410. Bbl. lengths: 25 (.410 only), 26, 28, 30 inches (latter 12 ga. only). Chokes: IC/ M, M/F, F/F. Weight: 5 lbs. 11 oz. (.410) to 7 lbs. (12 ga.). Checkered pistol-grip stock and forearm of European walnut. Made 1963. Discontinued.

Davidson Model 63B Magnum
Similar to standard Model 63B, except chambered for 10 ga. 3.5-inch, 12 and 20 ga. 3-inch Magnum shells; 10 ga. has 32-inch bbls., choked F/F. Weight: 10 lb. 10 oz. Made from 1963. Discontinued.
12-and 20-gauge Magnum $335
10-gauge Magnum 395

Davidson Model 69SL Double-Barrel Shotgun $415
Sidelock action w/detachable sideplates, engraved and nickel-plated. Plain extractors. Auto safety. Double triggers. 12 and 20 ga. Bbls.: 26-inch IC/M, 28-inch M/F. Weight: 12 ga., 7 lbs., 20 ga., 6.5 lbs. Pistol-grip stock and forearm of European walnut, checkered. Made 1963-76.

**Davidson Model 73 Stagecoach
Hammer Double** $265
Sidelock action w/detachable sideplates and exposed hammers. Plain extractors. Double triggers. Gauges: 12, 20, 3-inch chambers. 20-inch bbls, M/F chokes. Weight: 7 lbs., 12 ga.; 6.5 lbs., 20 ga. Checkered pistol-grip stock and forearm. Made from 1976. Discontinued.

EXEL ARMS OF AMERICA
Gardner, Massachusetts

Exel Series 100 Over/Under Shotgun
Gauge: 12. Single selective trigger. Selective auto ejectors. Hand-checkered European walnut stock w/full pistol grip, tulip forend. Black metal finish. Chambered for 2.75-inch shells (Model 103 for 3-inch). Weight: 6. 88 to 7.88 lbs. Discontinued 1988.
Model 101, 26-inch bbl., IC/M $375
Model 102, 28-inch bbl., IC/IM 380
Model 103, 30-inch bbl., M/F 395
Model 104, 28-inch bbl., IC/IM 435
Model 105, 28-inch bbl., 5 choke tubes 550
Model 106, 28-inch bbl., 5 choke tubes 660
Model 107 Trap, 30-inch bbl., Full + 5 tubes 695

Exel Series 200 Side-by-Side Shotgun $395
Gauges: 12, 20, 28 and .410. Bbls.: 26-, 27- and 28-inch; various choke combinations. Weight: 7 lbs. average. American or European-style stock and forend. Made 1985 to 87.

Exel Series 300 Over/Under Shotgun $450
Gauge: 12. Bbls.: 26-, 28- and 29-inch. Non-glare black-chrome matte finish. Weight: 7 lbs. average. Selective auto ejectors, engraved receiver. Hand-checkered European walnut stock and forend. Made 1985-86.

FIAS
Fabrica Italiana Armi Sabatti
Gardone Val Trompia, Italy

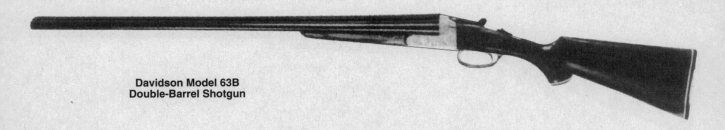

Davidson Model 63B
Double-Barrel Shotgun

SHOTGUNS

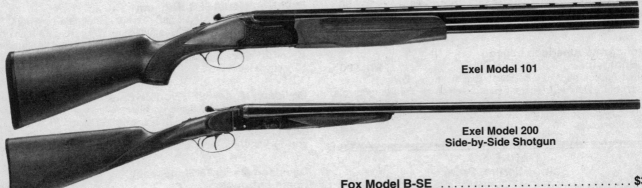

Exel Model 101

**Exel Model 200
Side-by-Side Shotgun**

Fias Grade I Over/Under

Boxlock. Single selective trigger. Gauges: 12, 20, 28, .410; 3-inch chambers. Bbls.: 26-inch IC/M; 28-inch M/F; screw-in choke tubes. Weight: 6.5 to 7.5 lbs. Checkered European walnut stock and forearm. Engraved receiver and blued finish.

12 Gauge Model	$375
20 Gauge Model	425
28 and .410 Models	550

FOX SHOTGUNS
Made by A. H. Fox Gun Co., Philadelphia, PA, 1903 to 1930, and since then by Savage Arms, originally of Utica, NY, now of Westfield, MA. In 1993 Connecticut Manufacturing Co. of New Britain, CT, reintroduced selected models.

Values shown are for 12 and 16 ga. doubles made by A. H. Fox. Twenty gauge guns often are valued up to 75% higher. Savage-made Fox models generally bring prices 25% lower. With the exception of Model B, production of Fox shotguns was discontinued about 1942.

Fox Model B Hammerless Double $235
Boxlock. Double triggers. Plain extractor. Gauges: 12, 16, 20, .410. 24- to 30-inch bbls., vent rib on current production; chokes: M/F, C/M, F/F (.410 only). Weight: about 7.5 lbs., 12 ga. Checkered pistol-grip stock and forend. Made about 1940-85.

Fox Model B-DE $265
Same as Model B-ST except frame finished in satin chrome, select walnut buttstock w/checkered pistol grip and beavertail forearm. Made 1965-66.

Fox Model B-DL $295
Same as Model B-ST except frame finished in satin chrome, select walnut buttstock w/checkered pistol grip side panels, beavertail forearm. Made 1962-66.

Fox Model B-SE $375
Same as Model B except has selective ejectors and single trigger. Made 1966-89.

Fox Model B-ST $255
Same as Model B except has non-selective single trigger. Made 1955-66.

Fox Hammerless Double-Barrel Shotguns
The higher grades have the same general specifications as the standard Sterlingworth model w/differences chiefly in workmanship and materials. Higher grade models are stocked in fine select walnut; quantity and quality of engraving increases w/grade and price. Except for Grade A, all other grades have auto ejectors.

Grade A	$ 1,425
Grade AE	1,750
Grade BE	3,595
Grade CE	4,150
Grade DE	9,295
Grade FE	16,750
Grade XE	5,825
Fox-Kautzky selective single trigger, **extra**	300
Ventilated rib, **extra**	400
Beavertail forearm, **extra**	200

Fox Single-Barrel Trap Guns
Boxlock. Auto ejector.12 ga. only. 30- or 32-inch vent-rib bbl. Weight: 7.5 to 8 lbs. Trap-style stock and forearm of select walnut, checkered, recoil pad optional. The four grades differ chiefly in quality of wood and engraving; Grade M guns, built to order, have finest Circassian walnut. Stock and receiver are elaborately engraved and inlaid w/gold. Discontinued 1942. *Note:* In 1932 the Fox Trap Gun was redesigned and those manufactured after that date have a stock w/full pistol grip and Monte Carlo comb; at the same time frame was changed to permit the rib line to extend across it to the rear.

Grade JE	$2995
Grade KE	3750
Grade LE	5195
Grade ME	9550

Fox Model B

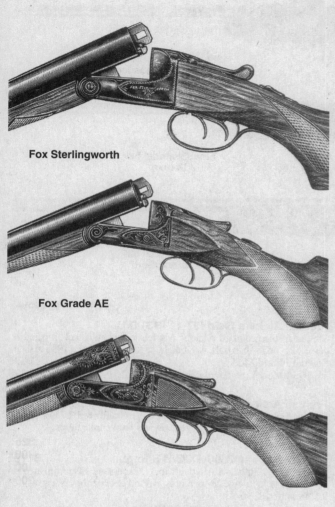

Fox Sterlingworth

Fox Grade AE

Fox Grade CE

Fox "Skeeter" Double-Barrel Shotgun $3595
Boxlock. Gauge: 12 or 20. Bbls.: 28 inches w/full-length vent rib. Weight: approx. 7 lbs. Buttstock and beavertail forend of select American walnut, finely checkered. Soft rubber recoil pad and ivory bead sights. Made in early 1930s.

Fox Sterlingworth Deluxe
Same general specifications as Sterlingworth, except 32-inch bbl. also available, recoil pad, ivory bead sights.
W/plain extractors $1295
W/automatic ejectors 1595

Fox Sterlingworth Hammerless Double
Boxlock. Double triggers (Fox-Kautzky selective single trigger extra). Plain extractors (auto ejectors extra). Gauges: 12,16, 20. Bbl. lengths: 26-, 28-, 30-inch; chokes F/F, M/F, C/M (any combination of C to F choke borings was available at no extra cost). Weight: 12 ga., 6.88 to 8.25 lbs.; 16 ga., 6 to 7 lbs.; 20 ga., 5.75 to 6.75 lbs. Checkered pistol-grip stock and forearm.
W/plain extractors $1195
W/automatic ejectors 1525
Selective single trigger, **extra** 300

Fox Sterlingworth Skeet and Upland Game Gun
Same general specifications as the standard Sterlingworth except has 26- or 28-inch bbls. w/skeet boring only, straight-grip stock. Weight: 7 lbs., 12 ga.
W/plain extractors $1625
W/automatic ejectors 1950

Fox Super HE Grade $3795
Long-range gun made in 12 ga. only (chambered for 3-inch shells on order), 30- or 32-inch full choke bbls., auto ejectors standard. Weight: 8.75 to 9.75 lbs. General specifications same as standard Sterlingworth.

Fox CMC Hammerless Double-Barrel Shotguns
High-grade doubles similar to the original Fox models. 20 ga. only. 26-, 28- or 30-inch bbls. Double triggers automatic safety and ejectors. Weight: 5.5 to 7 lbs. Custom Circassian walnut stock w/hand-rubbed oil finish. Custom stock configuration: straight, semi- or full pistol-grip stock w/traditional pad, hard rubber plate checkered or skeleton butt; schnabel, splinter or beavertail forend. Made 1993 to date.
CE Grade.. $ 5,995
XE Grade 7,650
DE Grade 9,275
FE Grade 12,250
Exhibition Grade 18,995

LUIGI FRANCHI S.P.A.
Brescia, Italy

Franchi Model 48/AL Ultra Light Shotgun
Recoil-operated, takedown, hammerless shotgun w/tubular magazine. Gauges: 12 or 20 (2.75-inch); 12-ga. Magnum (3-inch chamber). Bbls.: 24- to 32-inch w/various choke combinations. Weight: 5 lbs. 2 oz. (20 ga.) to 6.25 lbs. (12 ga.). Checkered pistol-grip walnut stock and forend w/high-gloss finish.
Standard Model.................................... $365
Hunter or Magnum Models 395

Franchi Model 500 Standard Autoloader $315
Gas-operated. 12 gauge. 4-shot magazine. Bbls.: 26-, 28-inch; vent rib; IC, M, IM, F chokes. Weight: about 7 lbs. Checkered pistol-grip stock and forearm. Made 1976-80.

Franchi Model 520 Deluxe $350
Same as Model 500, except higher grade w/engraved receiver. Made 1975-79.

Franchi Model 520 Eldorado Gold $745
Same as Model 520, except custom grade w/engraved and gold-inlaid receiver, finer quality wood. Intro. 1977.

Franchi Model 610VS Semiautomatic Shotgun
Gas-operated Variopress system adjustable to function w/2.75- or 3-inch shells. 12 gauge. 4-shot magazine. 26- or 28-inch vent rib bbls. w/Franchoke tubes. Weight: 7 lbs. 2 oz. 47.5 inches overall. Alloy receiver w/four-lug rotating bolt and loaded chamber indicator. Checkered European walnut buttstock and forearm w/satin finish. Imported 1997-98.
Standard SV Model.................................. $435
Engraved SVL Model 475

Franchi Model 2003 Trap Over/Under $1075
Boxlock. Auto ejectors. Selective single trigger. 12 ga. Bbls.: 30-, 32-inch IM/F, F/F, high-vent rib. Weight (w/30-inch bbl.): 8.25 lbs. Checkered walnut beavertail forearm and stock w/straight or Monte Carlo comb, recoil pad. Luggage-type carrying case. Introduced 1976. Discontinued.

SHOTGUNS

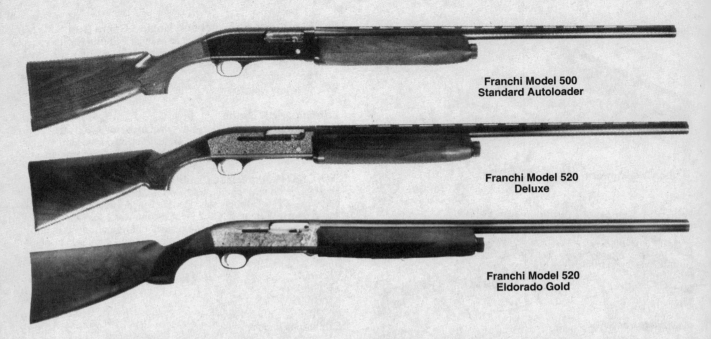

Franchi Model 500
Standard Autoloader

Franchi Model 520
Deluxe

Franchi Model 520
Eldorado Gold

Franchi Model 2004 Trap Single Barrel **$1095**
Same as Model 2003, except single bbl., 32- or 34-inch. Full choke.
Weight (w/32-inch bbl.): 8.25 lbs. Introduced 1976. Discontinued.

Franchi Model 2005 Combination Trap **$1595**
Model 2004/2005 type gun w/two sets of bbls., single and over/under. Introduced 1976. Discontinued.

Franchi Model 2005/3 Combination Trap **$1995**
Model 2004/2005 type gun w/three sets of bbls., any combination of single and over/under. Introduced 1976. Discontinued.

Franchi Model 3000/2 Combination Trap **$2495**
Boxlock. Automatic ejectors. Selective single trigger. 12 ga. only.
Bbls.: 32-inch over/under choked F/IM, 34-inch underbarrel M choke; high vent rib. Weight (w/32-inch bbls.): 8 lbs. 6 oz. Choice of six different castoff buttstocks. Introduced 1979. Discontinued.

Franchi Airone Hammerless Double **$1100**
Boxlock. Anson & Deeley system action. Auto ejectors. Double triggers. 12 ga. Various bbl. lengths, chokes, weights. Checkered straight-grip stock and forearm. Made 1940-50.

Franchi Alcione Over/Under Shotgun
Hammerless, takedown shotgun w/engraved receiver. Selective single trigger and ejectors. 12 ga. w/3-inch chambers. Bbls.: 26-inch (IC/M, 28-inch (M/F). Weight: 6.75 lbs. Checkered French walnut buttstock and forend. Imported from Italy 1982-89.
Standard Model. **$525**
SL Model (Disc. 1986) **895**

Franchi Alcione Field (97-12 IBS) O/U
Similar to the Standard Alcione Model, except w/nickel finished receiver. 26- or 28-inch bbls. w/Fankchoke tubes. Imported 1998 to date.
Standard Field Model **$675**
SL Field Model (w/sideplates discontinued.) **825**

Franchi Alcione Sport (SL IBS) O/U **$795**
Similar to the Alcione Field Model, except chambered 2.75 or 3-inch.
Ported 29-inch bbls. w/target vent rib and Fankchoke tubes.

Franchi Alcione 2000 SX O/U Shotgun **$1195**
Similar to the Standard Alcione Model, except w/ silver finished receiver and gold inlays. 28-inch bbls. w/Fankchoke tubes. Weight: 7.25 lbs. Imported 1996-97.

Franchi Aristocrat Field Model Over/Under **$465**
Boxlock. Selective auto ejectors. Selective single trigger. 12 ga.
Bbls.: 26-inch IC/M; 28- and 30-inch M/F choke, vent rib. Weight (w/26-inch bbls.): 7 lbs. Checkered pistol-grip stock and forearm. Made 1960-69.

Franchi Aristocrat Deluxe and Supreme Grades
Available in Field, Skeet and Trap Models w/the same general specifications as standard guns of these types. Deluxe and Supreme Grades are of higher quality w/stock and forearm of select walnut, elaborate relief engraving on receiver, trigger guard, tang and top lever. Supreme game birds inlaid in gold. Made 1960-66.
Deluxe Grade . . . . **$ 875**
Supreme Grade **1195**

Franchi Model 2004
Trap Single Barrel

**Franchi Model 2004
Trap Single Barrel**

Franchi Aristocrat Imperial and Monte Carlo Grades

Custom guns made in Field, Skeet and Trap Models w/the same general specifications as standard for these types. Imperial and Monte Carlo Grades are of highest quality w/stock and forearm of select walnut, fine engraving — elaborate on the latter grade. Made 1967-69.

Imperial Grade **$2195**
Monte Carlo Grade 3095

Franchi Aristocrat Magnum Model $475

Same as Field Model, except chambered for 3-inch shells, has 32-inch bbls. choked F/F; stock has recoil pad. Weight: about 8 lbs. Made 1962-65.

Franchi Aristocrat Silver King $575

Available in Field, Magnum, Meet and Trap models w/the same general specifications as standard guns of these types. Silver King has stock and forearm of select walnut more elaborately engraved silver-finished receiver. Made 1962-69.

Franchi Aristocrat Skeet Model $545

Same general specifications as Field Model, except made only w/26-inch vent-rib bbls. w/SK chokes #1 and #2, skeet-style stock and forearm. Weight: about 7.5 lbs. Later production had wider (10mm) rib. Made 1960-69.

Franchi Aristocrat Trap Model $550

Same general specifications as Field Model, except made only w/30-inch vent-rib bbls., M/F choke, trap-style stock w/recoil pad, beavertail forearm. Later production had Monte Carlo comb, 10mm rib. Made 1960-69.

Franchi Astore Hammerless Double $795

Boxlock. Anson & Deeley system action. Plain extractors. Double triggers. 12 ga. Various bbl. lengths, chokes, weights. Checkered straight-grip stock and forearm. Made 1937-60.

Franchi Astore II

Franchi Astore II $995

Similar to Astore S, but not as high grade. Furnished w/either plain extractors or auto ejectors, double triggers, pistol-grip stock. Bbls.: 27-inch IC/IM; 28-inch M/F chokes. Currently manufactured for Franchi in Spain.

Franchi Astore 5

Franchi Astore 5 . . . . $1795

Same as Astore, except has higher grade wood, fine engraving. Automatic ejectors, single trigger, 28-inch bbl. (M/F or IM/F choke) are standard on current production. Discontinued.

Franchi Crown, Diamond and Imperial Grade Autoloaders

Same general specifications as Standard Model, except these are custom guns of the highest quality. Crown Grade has hunting scene engraving, Diamond Grade has silver-inlaid scroll engraving; Imperial Grade has elaborately engraved hunting scenes w/figures inlaid in gold. Stock and forearm of fancy walnut. Made 1954-75.

Crown Grade **$1295**
Diamond Grade 1695
Imperial Grade 2150

Franchi Dynamic-12

Same general specifications and appearance as Standard Model, except 12 ga. only, has heavier steel receiver. Weight: about 7.25 lbs. Made 1965-72.

W/plain bbl. **$295**
W/ventilated rib 325

Franchi Dynamic-12 Slug Gun $335

Same as standard Slug Gun, except 12 ga. only, has heavier steel receiver. Made 1965-72.

Franchi Dynamic-12 Skeet Gun $395

Same general specifications and appearance as Standard Model, except has heavier steel receiver, made only in 12 ga. w/26-inch vent-rib bbl., SK choke, stock and forearm of extra fancy walnut. Made 1965-72.

Franchi Eldorado Model $435

Same general specifications as Standard Model except highest grade w/gold-filled engraving, stock and forearm of select walnut, furnished w/vent-rib bbl. only. Made 1954-75.

Franchi Falconet International Skeet Model $925

Similar to Standard Skeet Model, but higher grade. Made 1970-74.

Franchi Falconet International Trap Model $945

Similar to Standard Trap Model, but higher grade; w/straight or Monte Carlo comb stock. Made 1970-74.

SHOTGUNS

Franchi Crown Grade

Franchi Diamond Grade

Franchi Eldorado

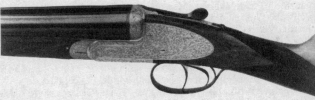

Imperiale Montecarlo Grade

Franchi Falconet Over/Under Field Models

Boxlock. Auto ejectors. Selective single trigger. Gauges: 12, 16, 20, 28, .410. Bbls.: 24-, 26-, 28-, 30-inch; vent rib. Chokes: C/IC, IC/M, M/F. Weight: from about 6 lbs. Engraved lightweight alloy receiver, light-colored in Buckskin Model, blued in Ebony Model, pickled silver in Silver Model. Checkered walnut stock and forearm. Made 1968-75.

Buckskin or Ebony Model **$495**
Silver Model **550**

Franchi Falconet Standard Skeet Model **$845**

Same general specifications as Field Models, except made only w/26-inch bbls. w/SK chokes #1 and #2, wide vent rib, color-casehardened receiver skeet-style stock and forearm. Weight: 12 ga., about 7.75 lbs. Made 1970-74.

Franchi Falconet Standard Trap Model **$835**

Same general specifications as Field Models, except made only in 12 ga. w/30-inch bbls., choked M/F, wide vent rib, color-casehardened receiver, Monte Carlo trap style stock and forearm, recoil pad. Weight: about 8 lbs. Made 1970-74.

Franchi Hammerless Sidelock Doubles

Hand-detachable locks. Self-opening action. Auto ejectors. Double triggers or single trigger. Gauges: 12,16, 20. Bbl. lengths, chokes, weights according to customer's specifications. Checkered stock and forend, straight or pistol grip. Made in six grades — Condor, Imperiale, Imperiale S, Imperiale Montecarlo No. 5, Imperiale Montecarlo No.11, Imperiale Montecarlo Extra — which differ chiefly in overall quality, engraving, grade of wood, checkering, etc.; general specifications are the same. Only the Imperiale Montecarlo Extra Grade is currently manufactured.

Condor Grade . . . . **$ 6,250**
Imperiale, Imperiale S Grades **8,925**
Imperiale Monte Carlo Grades No. 5, 11 . . . . **12,595**
Imperiale Monte Carlo Extra Grade. **14,650**

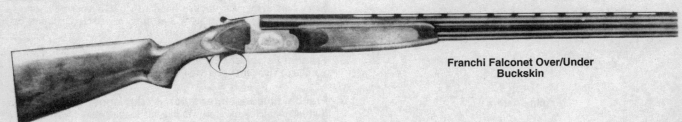

**Franchi Falconet Over/Under
Buckskin**

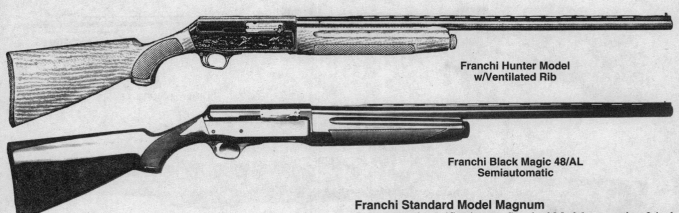

**Franchi Hunter Model
w/Ventilated Rib**

**Franchi Black Magic 48/AL
Semiautomatic**

Franchi Hunter Model
Same general specifications as Standard Model except higher grade w/engraved receiver; furnished w/ribbed bbl. only. Made 1950 to 90.
W/solid rib $325
W/ventilated rib 350

Franchi Hunter Model Magnum $395
Same as Standard Model Magnum, except higher grade w/engraved receiver, vent-rib bbl. only. Formerly designated "Wildfowler Model." Made 1954-73.

Franchi Peregrine Model 400 . . . . $595
Same general specifications as Model 451, except has steel receiver. Weight (w/26.5-inch bbl.): 6 lbs. 15 oz. Made 1975-78.

Franchi Peregrine Model 451 Over-and-Under $535
Boxlock. Lightweight alloy receiver. Automatic ejectors. Selective single trigger. 12 ga. Bbls.: 26.5-, 28-inch; choked C/IC, IC/M, M/F; vent rib. Weight (w/26.5-inch bbls.): 6 lbs. 1 oz. Checkered pistol-grip stock and forearm. Made 1975-78.

Franchi PG-80 Gas-Operated Semiautomatic Shotgun
Gas-operated, takedown, hammerless shotgun w/tubular magazine. 12 ga. w/2.75-inch chamber. 5-shot magazine. Bbls.: 24 to 30 inches w/vent rib. Weight: 7.5 lbs. Gold-plated trigger. Checkered pistol-grip stock and forend of European walnut. Imported from Italy 1985-90.
Prestige Model $425
Elite Model 475

Franchi Skeet Gun $365
Same general specifications and appearance as Standard Model, except made only w/26-inch vent-rib bbl., SK choke. Stock and forearm of extra fancy walnut. Made 1972-74.

Franchi Slug Gun $325
Same as Standard Model, except has 22-inch plain bbl., Cyl. bore, folding leaf open rear sight, gold bead front sight. Made 1960 to date.

Franchi Standard Model Autoloader
Recoil operated. Light alloy receiver. Gauges: 12, 20. 4-shot magazine. Bbls.: 26-, 28-, 30-inch; plain, solid or vent rib. IC/ M, F chokes. Weight: 12 ga., about 6.25 lbs. 20 ga., 5.13 lbs. Checkered pistol-grip stock and forearm. Made 1950 to date.
W/plain bbl. $315
W/solid rib 335
W/ventilated rib 360

Franchi Standard Model Magnum
Same general specifications as Standard Model, except has 3-inch chamber, 32-inch (12 ga.) or 28-inch (20 ga.) F choke bbl., recoil pad. Weight: 12 ga., 8.25 lbs.; 20 ga., 6 lbs. Formerly designated "Superange Model." Made 1954-88.
W/plain bbl. $355
W/vent rib 375

Franchi Turkey Gun $375
Same as Standard Model Magnum, except higher grade w/turkey scene engraved receiver, 12 ga. only, 36-inch matted-rib bbl., Extra Full choke. Made 1963-65.

Franchi Black Magic 48/AL Semiautomatic
Similar to the Franchi Model 48/AL, except w/Franchoke screw-in tubes and matte black receiver w/Black Magic logo. Gauge: 12 or 20, 2.75-inch chamber. Bbls.: 24-, 26-, 28-inch w/vent rib; 24-inch rifled slug w/sights. Weight: 5.2 lbs. (20 ga.). Checkered walnut buttstock and forend. Blued finish.
Standard Model $395
Slug bbl. Model 425

Franchi Falconet 2000 Over/Under $925
Boxlock. Single selective trigger. Selective automatic ejectors. Gauge: 12, 2.75-inch chambers. Bbls.: 26-inch w/Franchoke tubes; IC/M/F. Weight: 6 lbs. Checkered walnut stock and forearm. Engraved silver receiver w/gold-plated game scene. Imported from 1992-93.

Franchi LAW-12 Shotgun $475
Similar to the SPAS-12 Model, except gas-operated semiautomatic action only, ambidextrous safety, decocking lever and adj. sights. Made 1983-94.

Franchi SPAS-12 Shotgun
Selective operating system functions as a gas-operated semi-automatic or pump action. Gauge: 12, 2.75-inch chamber. 7-shot magazine. Bbl.: 21.5 inches w/cylinder bore and muzzle protector or optional screw-in choke tubes, matte finish. 41 inches overall w/fixed stock. Weight: 8.75 lbs. Blade front sight, aperture rear sight. Folding or black nylon buttstock w/pistol grip and forend, non-reflective anodized finish. Made 1983-94.
Fixed Stock Model $695
Folding Stock Model 795
Optional Choke Tubes, **add** 125

Franchi Sporting 2000 Over/Under $950
Similar to the Franchi Falconet 2000. Boxlock. Single selective trigger. Selective automatic ejectors. Gauge: 12; 2.75-inch chambers. Ported (1992-93) or unported 28-inch bbls., w/vent rib. Weight: 7.75 lbs. Blued receiver. Bead front sight. Checkered walnut stock and forearm; plastic composition buttplate. Imported from 1992-93 and 1997 to date.

Franchi LAW-12

Franchi SPAS-12

Franchi Sporting 2000

**Francotte
Model 10/18E628 and Model 9261**

AUGUSTE FRANCOTTE & CIE., S.A.
Liège, Belgium

Francotte shotguns for many years were distributed in the U.S. by Abercrombie & Fitch of New York City. This firm has used a series of model designations for Francotte guns which do not correspond to those of the manufacturer. Because so many Francotte owners refer to their guns by the A & F model names and numbers, the A & F series is included in a listing separate from that of the standard Francotte numbers.

**Francotte
Model 6886 and Model 8446**

Francotte Boxlock Hammerless Doubles

Anson & Deeley system. Side clips. Greener crossbolt on Models 6886, 8446, 4996 and 9261; square crossbolt on Model 6930, Greener-Scott crossbolt on Model 8457, Purdey bolt on Models 11/18E and 10/18E/628. Auto ejectors. Double triggers. Made in all standard gauges, barrel lengths, chokes, weights. Checkered stock and forend straight or pistol grip. The eight models listed vary chiefly in fastening as described above, finish and engraving, etc.; general specifications are the same. All except Model 10/18E/628 are disc..

Model 6886	**$2650**
Model 8446 ("Francotte Special"), 6930, 4996	**2995**
Model 8457, 9261 ("Francotte Original"), 11/18E	**3850**
Model 10/18E/628	**4895**

**Galef Silver Snipe
Over/Under Shotgun**

Francotte Boxlock Hammerless Doubles — A & F Series

Boxlock, Anson & Deeley type. Crossbolt. Sideplate on all except Knockabout Model. Side clips. Auto ejectors. Double triggers. Gauges: 12, 16, 20, 28, .410. Bbls.: 26- to 32-inch in 12 ga., 26- and 28-inch in other ga.; any boring. Weight: 4.75 to 8 lbs. depending on ga. and bbl. length. Checkered stock and forend; straight, half or full pistol grip. The seven grades — No. 45 Eagle Grade, No. 30, No. 25, No. 20, No. 14, Jubilee Model, Knockabout Model — differ chiefly in overall quality, engraving, grade of wood, checkering, etc.; general specifications are the same. Discontinued.

Jubilee Model No. 14	$2050
Jubilee Model No. 18	2395
Jubilee Model No. 20	2750
Jubilee Model No. 25	3295
Jubilee Model No. 30	4750
No. 45 Eagle Grade	5995
Knockabout Model	1750

Francotte Boxlock Hammerless Doubles (Sideplates)

Anson & Deeley system. Reinforced frame w/side clips. Purdey-type bolt except on Model 8455 which has Greener crossbolt. Auto ejectors. Double triggers. Made in all standard gauges, bbl. lengths, chokes, weights. Checkered stock and forend, straight or pistol grip. Models 10594, 8455 and 6982 are of equal quality, differing chiefly in style of engraving; Model 9/40E/38321 is a higher grade gun in all details and has fine English-style engraving. Currently manufactured.

Models 10594, 8455, 6982	$3895
Model 9/40E/38321	4750

Francotte Fine Over/Under Shotgun $9000

Model 9/40.SE. Boxlock, Anson & Deeley system. Auto ejectors. Double triggers. Made in all standard gauges; bbl. length, boring to order. Weight: about 6.75 lbs. 12 ga. Checkered stock and forend, straight or pistol grip. Currently manufactured.

Francotte Fine Sidelock Hammerless Double $19,250

Model 120.HE/328. Automatic ejectors. Double triggers. Made in all standard ga.; bbl. length, boring, weight to order. Checkered stock and forend, straight or pistol grip. Currently manufactured.

Francotte Half-Fine Over/Under Shotgun $7925

Model SOB.E/11082. Boxlock, Anson & Deeley system. Auto ejectors. Double triggers. Made in all standard gauges; barrel length, boring to order. Checkered stock and forend, straight or pistol grip. *Note:* This model is similar to No. 9/40.SE, except gen. quality lower. Discontinued.

GALEF SHOTGUNS
Manufactured for J. L. Galef & Son, Inc., New York, New York, by M. A. V. I., Gardone F. T., Italy, by Zabala Hermanos, Eiquetta, Spain, and by Antonio Zoli, Gardone V. T., Italy

**Galef Companion
Folding Single-Barrel Shotgun**

Galef Companion Folding Single-Barrel Shotgun

Hammerless. Underlever. Gauges: 12 Mag., 16, 20 Mag., 28, .410. Bbls.: 26-inch (.410 only), 28-inch (12, 16, 20, 28), 30-inch (12-ga. only); F choke; plain or vent rib. Weight: 4.5 lbs. for .410 to 5 lbs. 9 oz. for 12 ga. Checkered pistol-grip stock and forearm. Made by M. A. V. I. from 1968 to date.

W/plain bbl.	$ 95
W/ventilated rib	125

Galef Golden Snipe $445

Same as Silver Snipe, except has selective automatic ejectors. Made by Antonio Zoli 1968 to date.

Galef Monte Carlo Trap Single-Barrel Shotgun $195

Hammerless. Underlever. Plain extractor. 12 ga. 32-inch bbl., F choke, vent rib. Weight: about 8.25 lbs. Checkered pistol-grip stock w/Monte Carlo comb and recoil pad, beavertail forearm. Introduced by M. A. V. I. in 1968. Discontinued.

Galef Silver Hawk Hammerless Double.......... $375

Boxlock. Plain extractors. Double triggers. Gauges: 12, 20; 3-inch chambers. Bbls.: 26-, 28-, 30-inch (latter in 12 ga. only); IC/M, M/F chokes. Weight: 12 ga. w/26-inch bbls., 6 lbs. 6 oz. Checkered walnut pistol-grip stock and beavertail forearm. Made by Antonio Zoli 1968-72.

SHOTGUNS

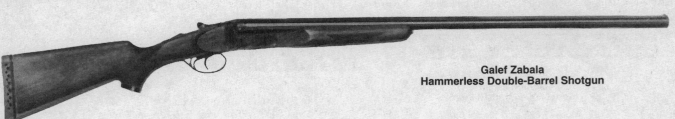

**Galef Zabala
Hammerless Double-Barrel Shotgun**

Garbi Model 200

Galef Silver Snipe Over/Under Shotgun **$450**
Boxlock. Plain extractors. Single trigger. Gauges: 12, 20; 3-inch chambers. Bbls: 26-, 28-, 30-inch (latter in 12 ga. only); IC/M, M/F chokes; vent rib. Weight: 12 ga. w/28-inch bbls., 6.5 lbs. Checkered walnut pistol-grip stock and forearm. Introduced by Antonio Zoli in 1968. Disc. *See* Illustration previous page.

Galef Zabala Hammerless Double-Barrel Shotgun
Boxlock. Plain extractors. Double triggers. Gauges: 10 Mag., 12 Mag., 16, 20 Mag., 28, .410. Bbls.: 22-, 26-, 28-, 30-, 32-inch; IC/IC, IC/M, M/F chokes. Weight: 12 ga. w/28-inch bbls., 7.75 lbs. Checkered walnut pistol-grip stock and beavertail forearm, recoil pad. Made by Zabala from 1972 to date. *See* Illustration previous page.
10 gauge **$225**
Other gauges **165**

GAMBA S. p. A.
Gardone V. T. (Brescia), Italy

Gamba Daytona Competition Over/Under
Boxlock w/Boss-style locking system. Anatomical single trigger; optional adj., single-selective release trigger. Selective automatic ejectors. Gauge: 12 or 20; 2.75- or 3-inch chambers. Bbls.: 26.75-, 28-, 30- or 32-inch choked SK/SK, IM/F or M/F. Weight: 7.5 to 8.5 lbs. Black or chrome receiver w/blued bbls. Checkered select walnut stock and forearm w/oil finish. Imported by Heckler & Koch until 1992.
American Trap Model . . . . **$ 4,250**
Pigeon, Skeet, Trap Models **4,895**
Sporting Model **4,050**
Sideplate Model **8,325**
Engraved Models . . . . **7,500 to 10,000**
Sidelock Model **21,500**

GARBI SHOTGUNS
Eibar, Spain

Garbi Model 100 Sidelock Shotgun **$2595**
Gauges: 12, 16, 20 and 28. Bbls.: 25-, 28-, 30-inch. Action: Holland & Holland pattern sidelock; automatic ejectors and double trigger. Weight: 5 lbs. 6 oz. to 7 lbs. 7 oz. English-style straight grip stock w/fine-line hand-checkered butt; classic forend. Made 1985 to date.

Garbi Model 101 Sidelock Shotgun **$3375**
Same general specifications as Model 100, except the sidelocks are handcrafted w/hand-engraved receiver; select walnut straight-grip stock.

Garbi Model 102 Sidelock Shotgun **$3595**
Similar to the Model 101, except w/large scroll engraving. Made 1985-93.

Garbi Model 103 Hammerless Double
Similar to Model 100, except w/Purdey-type, higher grade engraving.
Model 103A **$3750**
Model 103B **5795**

Garbi Model 200 Hammerless Double **$5495**
Similar to Model 100, except w/double heavy-duty locks. Continental-style floral and scroll engraving. Checkered deluxe walnut stock and forearm.

GARCIA CORPORATION
Teaneck, New Jersey

Garcia Bronco 22/.410 O/U Combination **$95**
Swing-out action. Takedown. 18.5-inch bbls.; 22 LR over, .410 ga. under. Weight: 4.5 lbs. One-piece stock and receiver, crackle finish. Intro. 1976. Discontinued.

Garcia Bronco .410 Single Shot **$75**
Swing-out action. Takedown. .410 ga. 18.5-inch bbl. Weight: 3.5 lbs. One-piece stock and receiver, crackle finish. Intro. In 1967. Discontinued.

GOLDEN EAGLE FIREARMS INC.
Houston, Texas
Mfd. By Nikko Firearms Ltd., Tochigi, Japan

Golden Eagle Model 5000 Grade I Field O/U **$725**
Receiver engraved and inlaid w/gold eagle head. Boxlock. Auto ejectors. Selective single trigger. Gauges: 12, 20; 2.75- or 3-inch chambers, 12 ga., 3-inch, 20 ga. Bbls.: 26-, 28-, 30-inch (latter only in 12-ga. 3-inch Mag.); IC/M, M/F chokes; vent rib. Weight: 6.25 lbs., 20 ga.; 7.25 lbs., 12 ga.; 8 lbs., 12-ga. Mag. Checkered pistol-grip stock and semibeavertail forearm. Imported 1975-1982. *Note:* Guns marketed 1975-76 under the Nikko brand name have white receivers; since 1976 are blued.

Golden Eagle
Model 5000 Grade II Field

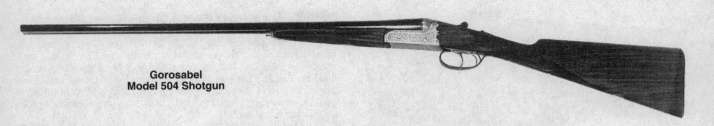

**Gorosabel
Model 504 Shotgun**

Golden Eagle Model 5000 Grade I Skeet **$750**
Same as Field Model, except has 26- or 28-inch bbls. w/wide (11 mm) vent rib, SK choked. Imported 1975-82.

Golden Eagle Model 5000 Grade I Trap **$755**
Same as Field Model, except has 30-, or 32-inch bbls. w/wide (11 mm) vent rib (M/F, IM/F, F/F chokes), trap-style stock w/recoil pad. Imported 1975-82.

Golden Eagle Model 5000 Grade II Field **$825**
Same as Grade I Field Model, except higher grade w/fancier wood, more elaborate engraving and "screaming eagle" inlaid in gold. Imported 1975-82.

Golden Eagle Model 5000 Grade II Skeet **$850**
Same as Grade I Skeet Model, except higher grade w/fancier wood, more elaborate engraving and "screaming eagle" inlaid in gold; inertia trigger, vent side ribs. Imported 1975-82.

Golden Eagle Model 5000 Grade II Trap **$865**
Same as Grade I Trap Model, except higher grade w/fancier wood, more elaborate engraving and "screaming eagle" inlaid in gold; inertia trigger, vent side ribs. Imported 1975-82.

Golden Eagle Model 5000 Grade III Grandee **$2150**
Best grade, available in Field, Skeet and Trap Models w/same general specifications as lower grades. Has sideplates w/game scene engraving, scroll on frame and bbls., fancy wood (Monte Carlo comb, full pistol grip and recoil pad on Trap Model). Made 1976-82.

GOROSABEL SHOTGUNS
Spain

Gorosabel Model 503 Shotgun **$695**
Gauges: 12, 16, 20 and .410. Action: Anson & Deely-style boxlock. Bbls.: 26-, 27-, and 28-inch. Select European walnut, English or pistol grip, sliver or beavertail forend, hand-checkering. Scalloped frame and scroll engraving. Intro. 1985; disc..

Gorosabel Model 504 Shotgun **$750**
Gauge: 12 or 20. Action: Holland & Holland-style sidelock. Bbl.: 26-, 27-, or 28-inch. Select European walnut, English or pistol grip, sliver or beavertail forend, hand-checkering. Holland-style large scroll engraving. Inro. 1985; disc..

Gorosabel Model 505 Shotgun **$995**
Gauge: 12 or 20. Action: Holland & Holland-style sidelock. Bbls.: 26-, 27-, or 28-inch. Select European walnut, English or pistol grip, sliver or beavertail forend, hand-checkering. Purdey-style fine scroll and rose engraving. Intro. 1985; disc..

STEPHEN GRANT
London, England

**Grant Best Quality Self-Opener Double-Barrel
Shotgun** **$10,225**
Sidelock, self-opener. Gauges: 12, 16 and 20. Bbls.: 25 to 30 inches standard. Highest grade English or European walnut straight-grip buttstock and forearm w/Greener type lever. Imported by Stoeger in the 1950s.

**Grant Best Quality Side-Lever Double-Barrel
Shotgun** **$9995**
Sidelock, self-lever. Gauges: 12, 16 and 20. Bbls.: 25 to 30 inches standard. Highest grade English or European walnut straight-grip buttstock and forearm w/Greener type lever. Imported by Stoeger in the 1950s.

W. W. GREENER, LTD.
Birmingham, England

**Greener
Empire Model Hammerless**

Greener Empire Model Hammerless Doubles
Boxlock. Non-ejector or w/automatic ejectors. Double triggers. 12 ga. only (2.75-inch or 3-inch chamber). Bbls.: 28- to 32-inch; any choke combination. Weight: from 7.25 to 7.75 lbs. depending on bbl. length. Checkered stock and forend, straight- or half-pistol grip. Also furnished in "Empire Deluxe Grade," this model has same general specs, but deluxe finish.
Empire Model, non-ejector. **$1550**
Empire Model, ejector **1695**
Empire Deluxe Model, non-ejector **1750**
Empire Deluxe Model, ejector **1995**

**Greener Far-Killer Model Grade FH35 Hammerless
Double-Barrel Shotgun**
Boxlock. Non-ejector or w/automatic ejectors. Double triggers. Gauges: 12 (2.75-inch or 3-inch), 10, 8. Bbls.: 28-, 30- or 32-inch. Weight: 7.5 to 9 lbs. in 12 ga. Checkered stock, forend; straight or half-pistol grip. *See* Illustration next page.
Non-ejector, 12 ga. **$2195**
Ejector, 12 ga. **2950**
Non-ejector, 10 or 8 ga. **2495**
Ejector, 10 or 8 ga. **3295**

SHOTGUNS

Greener Far-Killer

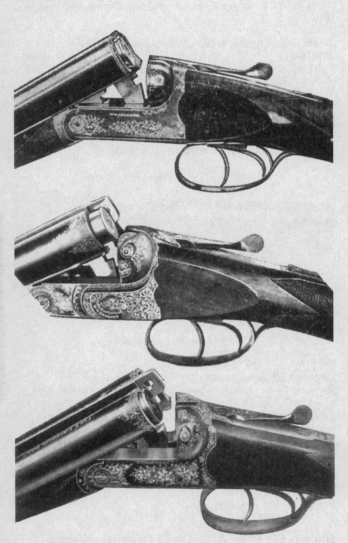

Greener Hammerless Ejector Double-Barrel Shotguns Jubilee, Royal and Sovereign

Greener G. P. (General Purpose) Single Barrel $350
Greener Improved Martini Lever Action. Takedown. Ejector. 12 ga. only. Bbl. lengths: 26-, 30-, 32-inch. M or F choke. Weight: 6.25 to 6.75 lbs. depending on bbl. length. Checkered straight-grip stock and forearm.

Greener Hammerless Ejector Double-Barrel Shotguns
Boxlock. Auto ejectors. Double triggers, non-selective or selective single trigger. Gauges: 12, 16, 20, 28, .410 (two latter gauges not supplied in Grades DH40 and DH35). Bbls.: 26-, 28-, 30-inch; any choke combination. Weight: from 4.75 to 8 lbs. Depending on ga. and bbl. length. Checkered stock and forend, straight- or half-pistol grip. The Royal, Crown, Sovereign and Jubilee Models differ in quality, engraving, grade of wood, checkering, etc. General specifications are the same.

Royal Model Grade DH75 . . . . **$3995**
Crown Model Grade DH55 . . . . **2950**
Sovereign Model Grade DH40. . . . . **2595**
Jubilee Model Grade DH35 . . . . **2150**
Selective single trigger, extra **330**
Non-selective single trigger, extra **250**

GREIFELT & COMPANY
Suhl, Germany

**Greifelt Grade No. 1
Over-and-Under Shotgun**

Greifelt Grade No. 1 Over-and-Under Shotgun
Anson & Deeley boxlock, Kersten fastening. Auto ejectors. Double triggers or single trigger. Elaborately engraved. Gauges: 12, 16, 20, 28, .410. Bbls.: 26- to 32-inch, any combination of chokes, vent or solid matted rib. Weight: 4.25 to 8.25 lbs. depending on ga. and bbl. length. Straight- or pistol-grip stock, Purdey-type forend, both checkered. Manufactured prior to World War II.

W/solid matted-rib bbl., except .410 & 28 ga. **$2995**
W/solid matted-rib bbl., .410 & 28 ga. **3750**
Extra for ventilated rib **395**
Extra for single trigger **425**

Greifelt Grade No. 3 Over-and-Under Shotgun
Same general specifications as Grade No. 1 except not as fancy engraving. Manufactured prior to World War II.

W/solid matted-rib bbl., except .410 & 28 ga. **$2395**
W/solid matted-rib bbl., .410 & 28 ga. **2850**
Extra for ventilated rib **395**
Extra for single trigger **425**

Greifelt Model 22 Hammerless Double **$1650**
Anson & Deeley boxlock. Plain extractors. Double triggers. Gauges: 12 and 16. Bbls.: 28- or 30-inch, M/F choke. Checkered stock and forend, pistol grip and cheekpiece standard, English-style stock also supplied. Manufactured since World War II.

Greifelt Model 22E Hammerless Double **$2195**
Same as Model 22, except has automatic ejectors.

Greifelt Model 103 Hammlerless Double **$1595**
Anson & Deeley boxlock. Plain extractors. Double triggers. Gauges: 12 and 16. Bbls.: 28- or 30-inch, M and F choke. Checkered stock and forend, pistol grip and cheekpiece standard, English-style stock also supplied. Manufactured since World War II.

Greifelt Model 103E Hammerless Double **$1695**
Same as Model 103, except has automatic ejectors.

Greifelt Model 143E Over-and-Under Shotgun
General specifications same as prewar Grade No. 1 Over-and-Under except this model is not supplied in 28 and .410 ga. or w/32-inch bbls. Model 143E is not as high quality as the Grade No. 1 gun. Mfd. Since World War II.
W/raised matted rib, double triggers **$1950**
W/ventilated rib, selective single trigger 2395

Greifelt Hammerless Drilling (Three Barrel Combination Gun) **$2950**
Boxlock. Plain extractors. Double triggers, front single set for rifle bbl. Gauges: 12, 16, 20; rifle bbl. in any caliber adapted to this type of gun. 26-inch bbls. Weight: about 7.5 lbs. Auto rear sight operated by rifle bbl. selector. Checkered stock and forearm, pistol grip and cheekpiece standard. Manufactured prior to WW II. *Note:* Value shown is for guns chambered for cartridges readily obtainable; if rifle bbl. is an odd foreign caliber, value will be considerably less.

Greifelt Over-and-Under Combination Gun
Similar in design to this maker's over-and-under shotguns. Gauges: 12, 16, 20, 28, .410; rifle bbl. in any caliber adapted to this type of gun. Bbls.: 24- or 26-inch, solid matted rib. Weight: from 4.75 to 7.25 lbs. Folding rear sight. Manufactured prior to WWII. *Note:* Values shown are for gauges other than .410 w/rifle bbl. Chambered for a cartridge readily obtainable; if in an odd foreign caliber, value will be considerably less. .410 ga. increases in value by about 50%.
W/nonautomatic ejector **$4595**
W/automatic ejector 5250

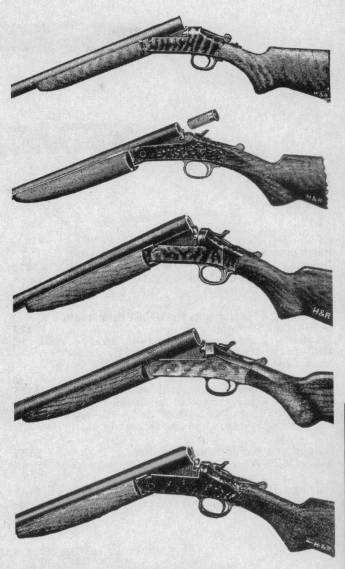

Harrington & Richardson
No. 3, 5, 6, 7 and 8 Shotguns

HARRINGTON & RICHARDSON ARMS COMPANY
Gardner, Massachusetts
Now H&R 1871, Inc.

In 1986, all H&R operations were discontinued. In 1992, the firm was purchased by New England Firearms of Gardner, Mass., and the H&R line was divided. Models are now manufactured under that banner as well as H&R 1871, Inc.

Harrington & Richardson No. 3 Hammerless Single-Shot Shotgun **$90**
Takedown. Automatic ejector. Gauges: 12, 16, 20, .410. Bbls.: plain, 26- to 32-inch, F choke. Weight: 6.5 to 7.25 lbs. depending on ga. and bbl. length. Plain pistol-grip stock and forend. Disc. 1942.

Harrington & Richardson No. 5 Standard Lightweight Hammer Single **$100**
Takedown. Auto ejector. Gauges: 24, 28, .410, 14mm. Bbls.: 26- or 28-inch, F choke. Weight: about 4 to 4.75 lbs. Plain pistol-grip stock/forend. Disc. 1942.

Harrington & Richardson No. 6 Heavy Breech Single-Shot Hammer Shotgun **$105**
Takedown. Automatic ejector. Gauges: 10, 12, 16, 20. Bbls.: plain, 28- to 36-inch, F choke. Weight: about 7 to 7.25 lbs. Plain stock and forend. Disc. 1942.

Harrington & Richardson No. 7 or 9 Bay State Single-Shot Hammer Shotgun **$110**
Takedown. Automatic ejector. Gauges: 12, 16, 20, .410. Bbls.: plain 26- to 32-inch, F choke. Weight: 5.5 to 6.5 lbs. depending on ga. and bbl. length. Plain pistol-grip stock and forend. Discontinued 1942.

Harrington & Richardson No. 8 Standard Single-Shot Hammer Shotgun **$135**
Takedown. Automatic ejector. Gauges: 12, 16, 20, 24, 28, .410. Bbl.: plain, 26- to 32-inch, F choke. Weight: 5.5 to 6.5 lbs. depending on ga. and bbl. length. Plain pistol-grip stock and forend. Made 1908-42.

Harrington & Richardson Model 348 Gamester Bolt-Action Shotgun **$95**
Takedown. 12 and 16 ga. 2-shot tubular magazine, 28-inch bbl, F choke. Plain pistol-grip stock. Weight: about 7.5 lbs. Made 1949-54.

SHOTGUNS

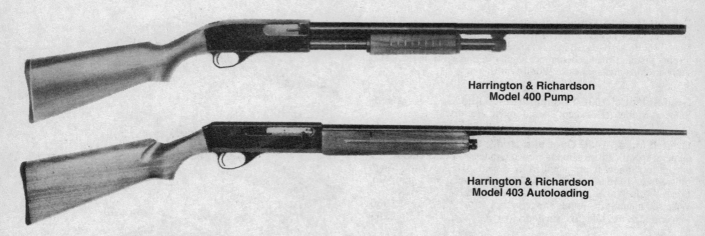

**Harrington & Richardson
Model 400 Pump**

**Harrington & Richardson
Model 403 Autoloading**

**Harrington & Richardson Model 349 Gamester
Deluxe** **$115**
Same as Model 348, except has 26-inch bbl. W/adj. choke device, recoil pad. Made 1953-55.

**Harrington & Richardson Model 351 Huntsman
Bolt-Action Shotgun** **$125**
Takedown. 12 and 16 ga. 2-shot tubular magazine. Pushbutton safety. 26-inch bbl. w/H&R variable choke. Weight: about 6.75 lbs. Monte Carlo stock w/recoil pad. Made 1956-58.

Harrington & Richardson Model 400 Pump **$175**
Hammerless. Gauges: 12, 16, 20. Tubular magazine holds 4 shells. 28-inch bbl., F choke. Weight: about 7.25 lbs. Plain pistol-grip stock (recoil pad in 12 and 16 ga.), grooved slide handle. Made 1955-67.

Harrington & Richardson Model 401 . . . . **$175**
Same as Model 400, except has H&R variable choke. Made 1956-63.

Harrington & Richardson Model 402 . . . . **$185**
Similar to Model 400, except .410 ga., weighs about 5.5 lbs. Made 1959-67.

**Harrington & Richardson Model 403
Autoloading Shotgun** **$200**
Takedown. .410 ga. Tubular magazine holds four shells. 26-inch bbl., F choke. Weight: about 5.75 lbs. Plain pistol-grip stock and forearm. Made in 1964.

Harrington & Richardson Model 404/404C **$225**
Boxlock. Plain extractors. Double triggers. Gauges: 12, 20, .410. Bbls.: 28-inch in 12 ga. (M/F choke), 26-inch in 20 ga. (IC/M and .410 (F/F). Weight: 5.5 to 7.25 lbs. Plain walnut-finished hardwood stock and forend on Model 404; 404C checkered. Made in Brazil by Amadeo Rossi 1969-1972.

Harrington & Richardson Model 440 Pump **$145**
Hammerless. Gauges: 12, 16, 20. 2.75-inch chamber in 16 ga., 3-inch in 12 and 20 ga. 3-shot magazine. Bbls.: 26-, 28-, 30-inch; IC, M, F choke. Weight: 6.25 lbs. Plain pistol-grip stock and slide handle, recoil pad. Made 1968-73. .

Harrington & Richardson Model 442 . . . . **$185**
Same as Model 440, except has vent-rib bbl., checkered stock and forearm, weighs 6.75 lbs. Made 1969-73.

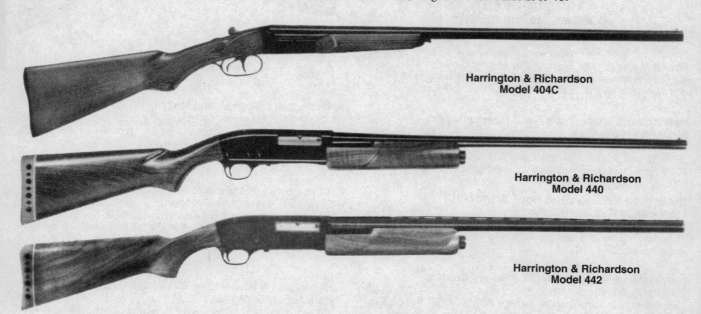

**Harrington & Richardson
Model 404C**

**Harrington & Richardson
Model 440**

**Harrington & Richardson
Model 442**

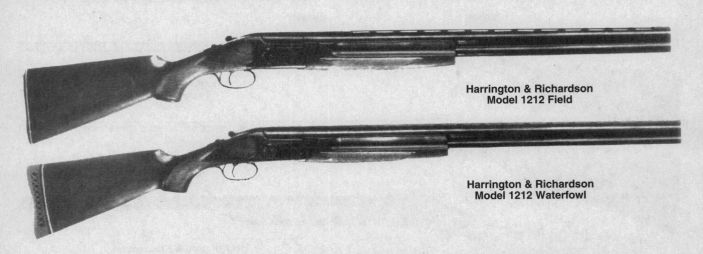

**Harrington & Richardson
Model 1212 Field**

**Harrington & Richardson
Model 1212 Waterfowl**

Harrington & Richardson Ultra Slug Series **$155**
Singel shot 12 or 20 ga w/3-inch chamber w/heavy-wall 24-inch
fully rifled bbl. w/scope . Weight: 9 lbs. Walnut-stained Monte
Carlo stock, sling swivels, black nylon sling. Made 1995 to date.

Harrington & Richardson Model 1212 Field **$295**
Boxlock. Plain extractors. Selective single trigger. 12 ga., 2.75-inch
chambers. 28-inch bbls., IC/IM, vent rib. Weight: 7 lbs. Checkered walnut pistol-gip stock and fluted forearm. Made 1976-80 by Lanber Arms
S. A., Zaldibar (Vizcaya), Spain.

**Harrington & Richardson Model 1212
Waterfowl Gun** **$325**
Same as Field Gun, except chambered for 12-ga, 3-inch Mag. Shells,
has 30-inch bbls., M/F chokes, stock and recoil pad, weighs, 7.5 lbs.
Made 1976-1980.

**Harrington & Richardson Model 1908
Single-Shot Shotgun** **$115**
Takedown. Automatic ejector. Gauges: 12, 16, 24 and 28. Bbls.: 26-
to 32-inch, F choke. Weight: 5.25 to 6.5 lbs., depending on ga. and
bbl. length. Casehardened receiver. Plain pistol-grip stock. Bead
front sight. Made 1908-1934.

**Harrington & Richardson Model 1908
.410 (12mm) Single-Shot Shotgun** **$125**
Same general specifications as standard Model 1908, except chambered for .410 or 12mm shot cartridge w/bbl. milled down at receiver
to give a more pleasing contour.

**Harrington & Richardson Model 1915
Single-Shot Shotgun**
Takedown. Both nonauto and auto ejectors available. Gauges: 24,
28, .410, 14mm and 12mm. Bbls.: 26- or 28-inch, F choke. Weight: 4
to 4.75 lbs., depending on ga. and bbl. length. Plain black walnut
stock w/semipistol grip.
24 ga. **$165**
28 , .410 ga. **135**

Harrington & Richardson Folding Gun **$135**
Single bbl. hammer shotgun hinged at the front of the frame, the bbl.
folds down against the stock. *Light Frame Model*: gauges — 28,
14mm, .410; 22-inch bbl.; weighs about 4.5 lbs. *Heavy Frame
Model*: gauges — 12, 16, 20, 28, .410; 26-inch bbl.; weighs from
5.75 to 6.5 lbs. Plain pistol-grip stock and forend. Discontinued
1942.

**Harrington & Richardson Golden Squire Model 159
Single-Barrel Hammer Shotgun** **$100**
Hammerless. Side lever. Automatic ejection. Gauges: 12, 20. Bbls.:
30-inch in 12 ga., 28-inch in 20 ga., both F choke. Weight: about 6.5
lbs. Straight-grip stock w/recoil pad, forearm w/schnabel. Made
1964-66.

**Harrington & Richardson Golden Squire Jr.
Model 459** **$115**
Same as Model 159, except gauges 20 and .410, 26-inch bbl., youth-
size stock. Made in 1964.

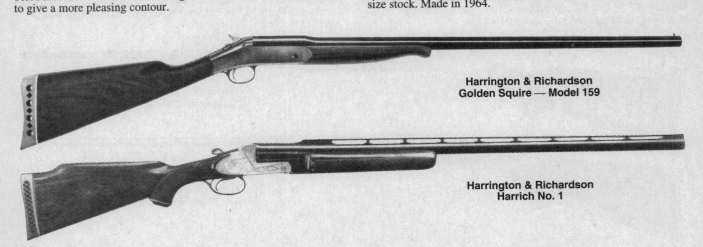

**Harrington & Richardson
Golden Squire — Model 159**

**Harrington & Richardson
Harrich No. 1**

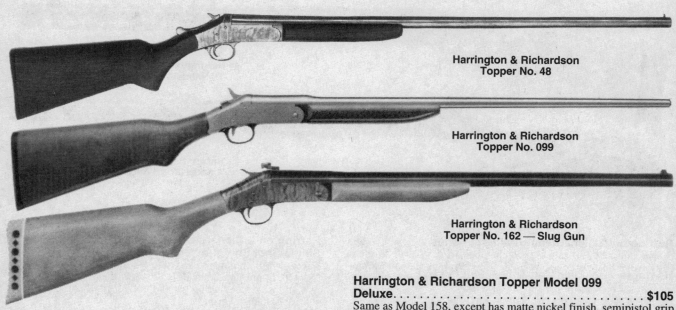

Harrington & Richardson
Topper No. 48

Harrington & Richardson
Topper No. 099

Harrington & Richardson
Topper No. 162 — Slug Gun

Harrington & Richardson Harrich No. 1 Single-Barrel Trap Gun $1595
Anson & Deeley-type locking system w/Kersten top locks and double underlocking lugs. Sideplates engraved w/hunting scenes. 12 ga. Bbls.: 32-, 34-inch; F choke; high vent rib. Weight: 8.5 lbs. Checkered monte Carlo stock w/pistol grip and recoil pad, beavertail forearm, of select walnut. Made in Ferlach, Austria, 1971-75. *See* Illustration previous page.

Harrington & Richardson "Top Rib" Single-Barrel Shotgun $165
Takedown. Auto ejector. Gauges: 12, 16 and 20. Bbls.: 28- to 30-inch, F choke w/full-length matted top rib. Weight: 6.5 to 7 lbs. depending on ga. and bbl. length. Black walnut pistol-grip stock (capped) and forend; both checkered. Flexible rubber buttplate. Made during 1930s.

Harrington & Richardson Topper No. 48 Single-Barrel Hammer Shotgun $125
Similar to old Model 8 Standard. Takedown. Top lever. Auto ejector. Gauges: 12, 16, 20, .410. Bbls.: plain; 26- to 30-inch; M or F choke. Weight: 5.5 to 6.5 lbs. depending on ga. and bbl. length. Plain pistol-grip stock and forend. Made 1946-57.

Harrington & Richardson Topper Model 099 Deluxe $105
Same as Model 158, except has matte nickel finish, semipistol grip walnut-finished American hardwood stock; semibeavertail forearm; 12, 16, 20, and .410 ga. Made 1982-86.

Harrington & Richardson Topper Model 148 Single-Shot Hammer Shotgun $110
Takedown. Side lever. Auto ejection. Gauges: 12, 16, 20, .410. Bbls.: 12 ga., 30-, 32- and 36-inch; 16 ga., 28- and 30-inch; 20 and .410 ga., 28-inch; F choke. Weight: 5 to 6.5 lbs. Plain pistol-grip stock and forend, recoil pad. Made 1958-61.

Harrington & Richardson Topper Model 158 (058) Single-Shot Hammer Shotgun $115
Takedown. Side lever. Automatic ejection. Gauges: 12, 20, .410 (2.75-inch and 3-inch shells); 16 (2.75-inch). bbl. length and choke combinations: 12 ga., 36-inch/F, 32-inch/F, 30-inch/F, 28-inch/F or M; .410, 28-inch/F. Weight: about 5.5 lbs. Plain pistol-grip stock and forend, recoil pad. Made 1962-81. *Note:* Designation changed to 058 in 1974.

Harrington & Richardson Topper Model 162 Slug Gun $135
Same as Topper Model 158, except has 24-inch bbl., Cyl. bore, w/rifle sights. Made 1968-86.

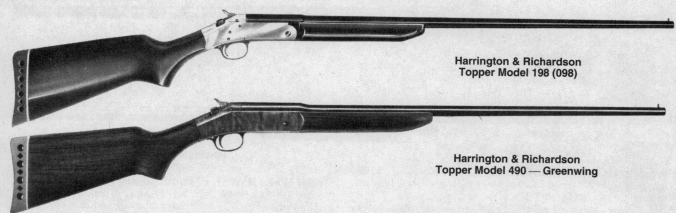

Harrington & Richardson
Topper Model 198 (098)

Harrington & Richardson
Topper Model 490 — Greenwing

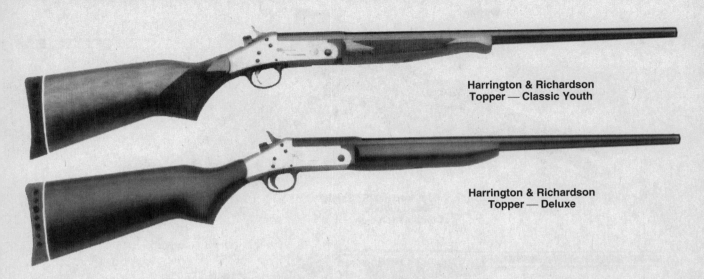

Harrington & Richardson
Topper — Classic Youth

Harrington & Richardson
Topper — Deluxe

**Harrington & Richardson Topper Model 176
10 Gauge Magnum** **$145**
Similar to Model 158, but has 36-inch heavy bbl. chambered for 3.5-inch 10- ga. Mag. shells, weighs 10 lbs.; stock w/Monte Carlo comb and recoil pad, longer and fuller forearm. Made 1977-86.

**Harrington & Richardson Topper Model 188
Deluxe.** **$125**
Same as standard Topper Model 148, except has chromed frame, stock and forend in black, red, yellow, blue, green, pink, or purple colored finish. .410 ga. only. Made 1958-61.

**Harrington & Richardson Topper Model
198 (098) Deluxe** **$130**
Same as Model 158, except has chrome-plated frame, black finished stock and forend; 12, 20 and .410 ga. Made 1962-81. *Note:* Designation changed to 098 in 1974.

Harrington & Richardson Topper Jr. Model 480 . . . . . **$115**
Similar to No. 48 Topper, except has youth-size stock, 26-inch bbl, .410 ga. only. Made 1958-61.

Harrington & Richardson Topper No. 488 Deluxe. **$110**
Same as standard No. 48 Topper, except chrome-plated frame, black lacquered stock and forend, recoil pad. Discontinued 1957.

Harrington & Richardson Topper Model 490 . . . . . . **$115**
Same as Model 158, except has youth-size stock (3 inches shorter), 26-inch bbl.; 20 and 28 gauge (M choke), .410 (F). Made 1962-86.

**Harrington & Richardson Topper Model 490
Greenwing** **$130**
Same as the Model 490, except has a special high-polished finish. Made 1981-86.

Harrington & Richardson Topper Jr. Model 580 . . . . . **$100**
Same as Model 480, except has colored stocks as on Model 188. Made 1958-61.

Harrington & Richardson Topper Model 590 . . . . . . **$105**
Same as Model 490, except has chrome-plated frame, black finished stock and forend. Made 1962-63.
The following models are manufactured and distributed by the re-organized company of H&R 1871, Inc.

H&R Model 098 Topper Classic Youth **$95**
Same as Topper Junior, except also available in 28 ga. and has checkered American black walnut stock/forend w/satin finish and recoil pad. Made 1991 to date.

H&R Model 098 Topper Deluxe **$95**
Same as Model 098 Single Shot Hammer, except in 12 ga., 3-inch chamber only. 28-inch bbl.; Mod. choke tube. Made 1992 to date.

**H&R Model 098 Topper Deluxe
Rifled Slug Gun** **$105**
Same as Topper Deluxe Shotgun, except has compensated 24-inch rifled slug bbl. Nickel plated receiver and blued bbl. Black finished hardwood stock. Made 1996 to date.

**H&R Model 098 Topper Hammer Single-Shot
Shotgun** **$80**
Side lever. Automatic ejector. Gauges: 12, 20 and .410; 3-inch chamber. Bbls.: 28-inch, (12 ga./M); 26-inch, (20 ga./M); 26-inch (.410/F). Weight: 5 to 6 lbs. Satin nickel receiver, blued bbl. Plain pistol-grip stock and semibeavertail forend w/black finish. ReIntro. 1992.

H&R Model 098 Topper Junior **$85**
Same as Model 098, except has youth-size stock and 22-inch bbl. 20 or .410 ga. only. Made 1991 to date.

H&R Model .410 Tamer Shotgun **$95**
Takedown. Topper-style single-shot, side lever action w/auto ejector. Gauge: .410; 3-inch chamber. 19.5-inch bbl. 33 inches overall. Weight: 5.75 lbs. Black polymer thumbhole stock designed to hold 4 extra shotshells. Matte electroless nickel finish. Made 1994 to date.

H&R Model N. W. T. F. Turkey Mag
Same as Model 098 Single-Shot Hammer, except has 24-inch bbl. chambered 10 or 12 ga. w/3.5-inch chamber w/screw-in choke tube. Weight: 6 lbs. American hardwood stock, Mossy Oak camo finish. Made 1991-96.
NWTF 10 ga. Turkey Mag (Made 1996) **$125**
NWTF 12 ga. Turkey Mag (Made 1991-95) **110**

H&R Model N. W. T. F. Youth Turkey Gun. **$100**
Same as Model N.W.T.F. Turkey Mag, except has 22-inch bbl. chambered 20 ga. w/3-inch chamber and fixed full choke. Realtree camo finish. Made 1994-95.

Right: Heym Model 22S "Safety"
Shotgun/Rifle
Combination Gun

Right: Heym Model 55 BF
Shotgun/Rifle

H&R Model SB1-920 Ultra Slug Hunter **$135**
Special 12 ga. action w/12 ga. bbl. blank underbored to 20 ga. to form a fully rifled slug bbl. Gauge: 20 w/3 inch chamber. 24-inch bbl. Weight: 8.5 lbs. Satin nickel receiver, blued bbl. Walnut finished hardwood Monte Carlo stock. Made 1996-98.

H&R Model Ultra Slug Hunter **$135**
Gauges: 12 or 20 ga. w/3 inch chamber. 22- or 24-inch rifled bbl. Weight: 9 lbs. Matte black receiver and bbl. Walnut finished hardwood Monte Carlo stock. Made 1997 to date.

H&R Model Ultra Slug Hunter Deluxe **$155**
Similar to Ultra Slug Hunter Model, except with compensated bbl. Made 1997 to date.

HERCULES SHOTGUNS

See Listings under "W" for Montgomery Ward.

HEYM SHOTGUNS
Münnerstadt, Germany

**Heym Model 22S "Safety" Shotgun/Rifle
Combination** **$2695**
Gauges: 16 and 20. Calibers: 22 Mag., 22 Hornet, 222 Rem., 222 Rem. Mag., 5.6×50R Mag., 6.5×57R, 7×57R, 243 Win. 24-inch bbls. 40 inches overall. Weight: about 5.5 lbs. Single-set trigger. Left-side bbl. selector. Integral dovetail base for scope mounting. Arabesque engraving. Walnut stock. Discontinued 1993.

Heym Model 55 BF Shotgun/Rifle Combo **$5250**
Gauges: 12, 16 and 20. Calibers: 5.6×50R Mag., 6.5×57R, 7×57R, 7×65R, 243 Win., 308 Win., 30-06. 25-inch bbls. 42 inches overall. Weight: about 6.75 lbs. Black satin-finished, corrosion-resistant bbls. of Krupps special steel. Hand-checkered walnut stock w/long pistol grip. Hand-engraved leaf scroll. German cheekpiece. Discontinued 1988.

J. C. HIGGINS SHOTGUNS

See Sears, Roebuck & Company.

HUGLU HUNTING FIREARMS
Huglu, Turkey
Imported by Turkish Firearms Corp.

HHF Model 101 B 12 AT-DT Combo O/U Trap **$1595**
Over/Under boxlock. 12 ga. w/3-inch chambers. Combination 30- or 32-inch top single & O/U bbls. w/fixed chokes or choke tubes. Weight: 8 lbs. Automatic ejectors or extractors. Single selective trigger. Manual safety. Circassian walnut Monte Carlo trap stock w/ palm swell grip and recoil pad. Silvered frame w/engraving. Imported 1993-97.

HHF Model 101 B 12 ST O/U Trap **$995**
Same as Model 101 AT-DT except in 32-inch O/U configuration only. Imported 1994-96.

HHF Model 103 B 12 ST O/U
Boxlock. Gauges: 12, 16, 20, 28 or .410. 28-inch bbls. w/fixed chokes. Engraved action w/inlaid game scene and dummy sideplates. Double triggers, extractors and manual safety. Weight: 7.5 lbs. Circassian walnut stock. Imported 1995-96.
Model 103B w/extractors **$750**
28 and 410, **add** **100**

HHF Model 103 C 12 ST O/U
Same general specs as Model 103 B 12 ST, except w/extractors or ejectors. 12 or 20 ga. w/3-inch chambers. Black receiver w/50% engraving coverage. Imported 1995-97.
Model 103C w/extractors **$795**
Model 103C w/ejectors **995**

HHF Model 103 D 12 ST O/U
Same general specs as Model 103 B 12 ST, except standard boxlock. Extractors or ejectors. 12 or 20 ga. w/3-inch chambers. 80% engraving coverage. Imported 1995-97.
Model 103D w/extractors **$775**
Model 103D w/ejectors **975**

**High Standard
Flite-King Brush**

HHF Model 103 F 12 ST O/U

Same as Model 103 B, except extractors or ejectors. 12 or 20 ga. only. 100% engraving coverage. Imported 1996-97.

Model 103F w/extractors **$850**
Model 103F w/ejectors **1050**

HHF Model 104 A 12 ST O/U

Boxlock. Gauges: 12, 20, 28 or .410. 28-inch bbls. w/fixed chokes or choke tubes. Silvered, engraved receiver w/15% engraving coverage. Double triggers, manual safety and extractors or ejectors. Weight: 7.5 lbs. Circassian walnut stock w/field dimensions. Imported 1995-97.

Model 104A w/extractors **$750**
Model 104A w/ejectors **950**
28 and 410, **add** **100**
W/Choke Tubes, **add** **50**

HHF Model 200 Series Double

Boxlock. Gauges: 12, 20, 28, or .410 w/3-inch chambers. 28-inch bbls. w/fixed chokes. Silvered, engraved receiver. Extractors, manual safety, single selective trigger or double triggers. Weight: 7.5 lbs. Circassion walnut stock. Imported 1995-97.

Model 200 (w/15% engraving coverage, SST) **$775**
Model 201 (w/30% engraving coverage, SST) **995**
Model 202 (w/Greener cross bolt, DT) **825**
28 and 410, **add** **100**

HIGH STANDARD SPORTING ARMS
East Hartford, Connecticut
Formerly High Standard Mfg. Corp. of
Hamden, Conn.

In 1966, High Standard introduced new series of Flite-King Pumps and Supermatic autoloaders, both readily identifiable by the damascened bolt and restyled checkering. To avoid confusion, these models are designated "Series II" in this text. This is not an official factory designation. Operation of this firm was discontinued in 1984.

High Standard Flite-King Brush—12 Ga. $200

Same as Flite-King Field 12, except has 18- or 20-inch bbl. (cylinder bore) w/rifle sights. Made 1962-64.

High Standard Flite-King Brush Deluxe $235

Same as Flite-King Brush, except has adj. peep rear sight, checkered pistol grip, recoil pad, fluted slide handle, swivels and sling. Not available w/18-inch bbl. Made 1964-66.

High Standard Flite-King Brush (Series II) $215

Same as Flite-King Deluxe 12 (II), except has 20-inch bbl., cylinder bore, w/rifle sights. Weight: 7 lbs. Made 1966-75.

High Standard Flite-King Brush Deluxe (II) $250

Same as Flite-King Brush (II), except has adj. peep rear sight, swivels and sling. Made 1966-75.

High Standard Flite-King Deluxe—12 Ga. (Series II)

Hammerless. 5-shot magazine. 27-inch plain bbls. w/adj. choke. 26-inch IC, 28-inch M or F. 30-inch F choke. Weight: about 7.25 lbs. Checkered pistol-grip stock and forearm, recoil pad. Made 1966-75.

W/adj. choke **$265**
W/O adj. choke **240**

High Standard Flite-King Deluxe—
20, 28, .410 ga. (Series II) $225

Same as Flite-King Deluxe 12 (II), except chambered for 20 and .410 ga. 3-inch shell, 28 ga. 2.75-inch shell w/20- or 28-inch plain bbl. Weight: about 6 lbs. Made 1966-75.

High Standard Flite-King Deluxe Rib—12 Ga. $275

Same as Flite-King Field 12, except vent-rib bbl. (28-inch M or F. 30-inch F). Checkered stock and forearm. Made 1961-66.

High Standard Flite-King Deluxe Rib—12 Gauge (II)

Same as Flite-King Deluxe 12 (II), except has vent-rib bbl., available in 27-inch w/adj. choke, 28-inch M or F, 30-inch F choke. Made 1966-75.

W/adj. choke **$295**
W/O adj. choke **275**

SHOTGUNS

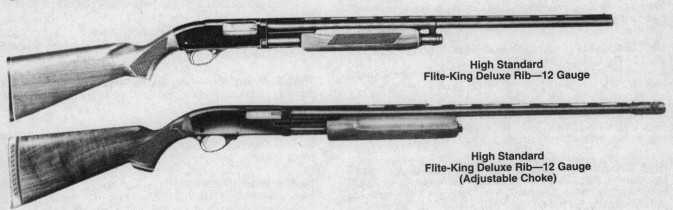

**High Standard
Flite-King Deluxe Rib—12 Gauge**

**High Standard
Flite-King Deluxe Rib—12 Gauge
(Adjustable Choke)**

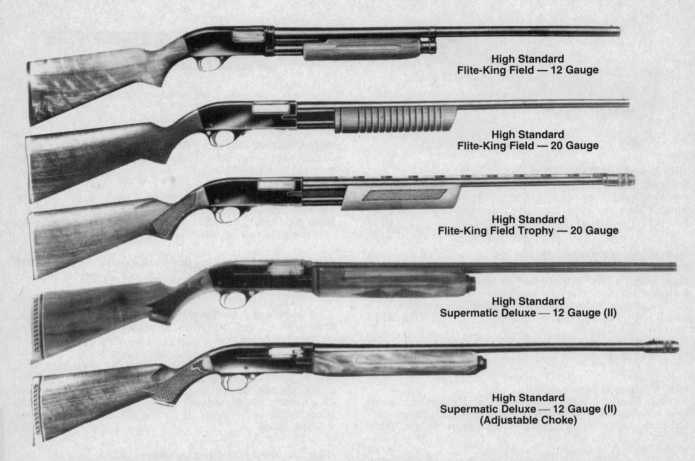

High Standard
Flite-King Field — 12 Gauge

High Standard
Flite-King Field — 20 Gauge

High Standard
Flite-King Field Trophy — 20 Gauge

High Standard
Supermatic Deluxe — 12 Gauge (II)

High Standard
Supermatic Deluxe — 12 Gauge (II)
(Adjustable Choke)

High Standard Flite-King Deluxe Rib—20 Ga. $265
Same as Flite-King Field 20, except vent-rib bbl. (28 inch M or F), checkered stock and slide handle. Made 1962-66.

**High Standard Flite-King Deluxe Rib—
20, 28, .410 Ga. (Series II)**
Same as Flite-King Deluxe 20, 28, .410 (II), except 20 ga. available w/27-inch adj. choke, 28-inch M or F choke. Weight: about 6.25 lbs. Made 1966-75.
W/adj. Choke **$295**
W/O adj. Choke **275**

**High Standard Flite-King Deluxe Skeet Gun—
12 Gauge (Series II)** **$255**
Same as Flite-King Deluxe Rib 12 (II), except available only w/26-inch vent-rib bbl., SK choke, recoil pad optional. Made 1966-75.

**High Standard Flite-King Deluxe Skeet Gun—
20, 28, 410 Gauge (Series II)** **$295**
Same as Flite-King Deluxe Rib 20, 28, .410 (II) except available only w/26-inch vent-rib bbl., SK choke. Made 1966-75.

High Standard Flite-King Deluxe Trap Gun (II) **$255**
Same as Flite-King Deluxe Rib 12 (II), except available only w/30-inch vent-rib bbl., F choke; trap-style stock. Made 1966-75.

High Standard Flite-King Field Pump—12 Ga. **$215**
Hammerless. Magazine holds 5 shells. Bbls.: 26-inch IC, 28-inch M or F, 30-inch F choke. Weight: 7.25 lbs. Plain pistol-grip stock and slide handle. Made 1960-66.

High Standard Flite-King Field Pump—20 Ga. **$195**
Hammerless. Chambered for 3-inch Magnum shells, also handles 2.75-inch. Magazine holds four shells. Bbls.: 26-inch IC, 28-inch M or F choke. Weight: about 6 lbs. Plain pistol-grip stock and slide handle. Made 1961-66.

High Standard Flite-King Pump Shotguns—16 Gauge
Same general specifications as Flite-King 12, except not available in Brush, Skeet and Trap Models, or 30-inch bbl. Values same as for 12-ga. guns. Made 1961-65.

High Standard Flite-King Pump Shotguns—.410 Ga.
Same general specifications as Flite-King 20, except not available in Special and Trophy Models, or w/other than 26-inch choke bbl. Values same as for 20-ga. guns. Made 1962-66.

High Standard Flite-King Skeet—12 Ga. **$295**
Same as Flite-King Deluxe Rib, except 26-inch vent-rib bbl., w/SK choke. Made 1962-66.

High Standard Flite-King Special—12 Ga. **$215**
Same as Flite-King Field 12, except has 27-inch bbl. w/adj. choke. Made 1960-66.

High Standard Flite-King Special—20 Ga. **$220**
Same as Flite-King Field 20, except has 27-inch bbl. w/adj. choke. Made 1961-66.

High Standard Flite-King Trap—12 Ga. **$295**
Same as Flite-King Deluxe Rib 12, except 30-inch vent-rib bbl., F choke, special trap stock w/recoil pad. Made 1962-66.

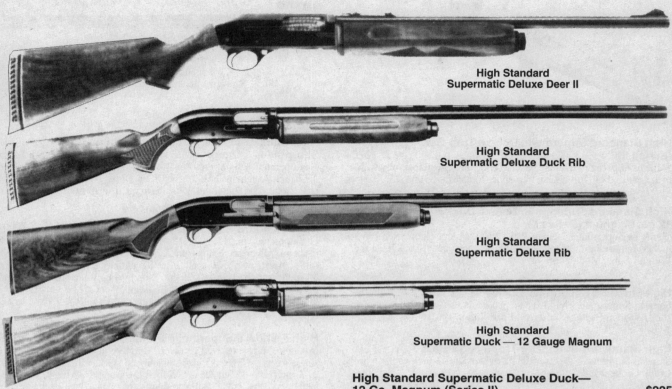

**High Standard
Supermatic Deluxe Deer II**

**High Standard
Supermatic Deluxe Duck Rib**

**High Standard
Supermatic Deluxe Rib**

**High Standard
Supermatic Duck — 12 Gauge Magnum**

High Standard Flite-King Trophy—12 Ga. **$275**
Same as Flite-King Deluxe Rib 12, except has 27-inch vent-rib bbl. w/adj. choke. Made 1960-66.

High Standard Flite-King Trophy—20 Ga. **$295**
Same as Flite-King Deluxe Rib 20, except has 27-inch vent-rib bbl. w/adj. choke. Made 1962-66.

High Standard Supermatic Deer Gun **$275**
Same as Supermatic Field 12, except has 22-inch bbl. (cylinder bore) w/rifle sights, checkered stock and forearm, recoil pad. Weight: 7.75 lbs. Made in 1965.

High Standard Supermatic Deluxe—12 Ga. (Series II)
Gas-operated autoloader. 4-shot magazine. Bbls.: plain; 27-inch w/adj. choke (disc. about 1970); 26-inch IC, 28-inch M or F. 30-inch F choke. Weight: about 7.5 lbs. Checkered pistol-grip stock and forearm, recoil pad. Made 1966-75.
W/adj. choke **$295**
W/O adj. choke 245

High Standard Supermatic Deluxe — 20 Ga. (Series II)
Same as Supermatic Deluxe 12 (II), except chambered for 20 ga. 3-inch shell; bbls. available in 27-inch w/adj. choke (disc. about 1970), 26-inch IC, 28-inch M or F choke. Weight: about 7 lbs. Made 1966-75.
W/adj. choke **$215**
W/O adj. choke 255

High Standard Supermatic Deluxe Deer Gun (II) **$325**
Same as Supermatic Deluxe 12 (II), except has 22-inch bbl., cylinder bore, w/rifle sights. Weight: 7.75 lbs. Made 1966-74.

**High Standard Supermatic Deluxe Duck—
12 Ga. Magnum (Series II)** **$225**
Same as Supermatic Deluxe 12 (II), except chambered for 3-inch Magnum shell, 3-shot magazine, 30-inch plain bbl., F choke. Weight: 8 lbs. Made 1966-74.

High Standard Supermatic Deluxe Rib — 12 Ga. . . **$225**
Same as Supermatic Field 12, except vent-rib bbl. (28-inch M or F, 30-inch F), checkered stock and forearm. Made 1961-66.

High Standard Supermatic Deluxe Rib — 12 Ga. (II)
Same as Supermatic Deluxe 12 (II), except has vent-rib bbl.; available in 27-inch w/adj. choke, 28-inch M or F, 30-inch F choke. Made 1966-75.
W/adj. choke **$265**
W/O adj. choke 240

High Standard Supermatic Deluxe Rib — 20 Ga. **$265**
Same as Supermatic Field 20, except vent-rib bbl. (28-inch M or F), checkered stock and forearm. Made 1963-66.

High Standard Supermatic Deluxe Rib — 20 Ga. (II)
Same as Supermatic Deluxe 20 (II), except has vent-rib bbl. Made 1966-75.
W/adj. choke **$255**
W/O adj. choke 235

**High Standard Supermatic Deluxe Skeet Gun —
12 Ga. (Series II)** **$255**
Same as Supermatic Deluxe Rib 12 (II), except available only w/26-inch vent-rib bbl., SK choke. Made 1966-75.

**High Standard Supermatic Deluxe Skeet Gun —
20 Ga. (Series II)** **$265**
Same as Supermatic Deluxe Rib 20 (II), except available only w/26-inch vent-rib bbl., SK choke. Made 1966-75.

**High Standard
Supermatic Duck Rib — 12 Gauge**

**High Standard Supermatic Deluxe Trap Gun
(Series II)** **$275**
Same as Supermatic Deluxe Rib 12 (II), except available only w/30-inch vent-rib bbl., full choke; trap-style stock. Made 1966-75.

**High Standard Supermatic Deluxe Duck Rib —
12 ga. Magnum (Series II)** **$255**
Same as Supermatic Deluxe Rib 12 (II), except chambered for 3-inch Magnum shell, 3-shot magazine; 30-inch vent-rib bbl., F choke. Weight: 8 lbs. Made 1966-75.

High Standard Supermatic Duck—12 Ga. Mag. **$255**
Same as Supermatic Field 12, except chambered for 3-inch Magnum shell, 30-inch F choke bbl., recoil pad. Made 1961-66.

High Standard Supermatic Trophy—12 Ga. **$215**
Same as Supermatic Deluxe Rib 12, except has 27-inch vent-rib bbl. w/adj. choke. Made 1961-66.

**High Standard Supermatic Duck Rib—
12 Ga. Magnum** **$275**
Same as Supermatic Duck 12 Magnum, except has vent-rib bbl., checkered stock and forearm. Made 1961-66.

**High Standard Supermatic Field Autoloading
Shotgun—12 Ga.** **$195**
Gas-operated. Magazine holds four shells. Bbls.: 26-inch IC, 28-inch M or F choke, 30-inch F choke. Weight: about 7.5 lbs. Plain pistol-grip stock and forearm. Made 1960-66.

**High Standard Supermatic Field Autoloading
Shotgun—20 Gauge.** **$215**
Gas-operated. Chambered for 3-inch Magnum shells, also handles 2.75-inch. Magazine holds three shells. Bbls.: 26-inch IC, 28-inch M or F choke. Weight: about 7 lbs. Plain pistol-grip stock and forearm. Made 1963-66.

High Standard Supermatic Shadow Automatic **$315**
Gas-operated. ga.: 12, 20, 2.75- or 3-inch chamber in 12 ga., 3-inch in 20 ga. Magazine holds four 2.75-inch shells, three 3-inch. Bbls.: full-size airflow rib; 26-inch (IC or SK choke), 28-inch (M, IM or F), 30-inch (trap or F choke), 12-ga. 3-inch Magnum available only in 30-inch F choke; 20 ga. not available in 30-inch. Weight: 12 ga., 7 lbs. Checkered walnut stock and forearm. Made 1974-75 by Caspoll Int'l., Inc., Tokyo.

High Standard Supermatic Shadow Indy O/U **$795**
Boxlock. Fully engraved receiver. Selective auto ejectors. Selective single trigger. 12 ga. 2.75-inch chambers. Bbls.: full-size airflow rib; 27.5 inch both SK choke, 29.75-inch IM/F or F/F. Weight: w/29.75-inch bbls., 8 lbs. 2 oz. Pistol-grip stock w/recoil pad, ventilated forearm, skip checkering. Made 1974-75 by Caspoll Int'l., Inc., Tokyo.

High Standard Supermatic Shadow Seven **$675**
Same general specifications as Shadow Indy, except has conventional vent rib, less elaborate engraving, standard checkering forearm is not vented, no recoil pad. 27.5-inch bbls.; also available in IC/M, M/F choke. Made 1974-75.

High Standard Supermatic Skeet — 12 Ga. **$215**
Same as Supermatic Deluxe Rib 12, except 26-inch vent-rib bbl. w/SK choke. Made 1962-66.

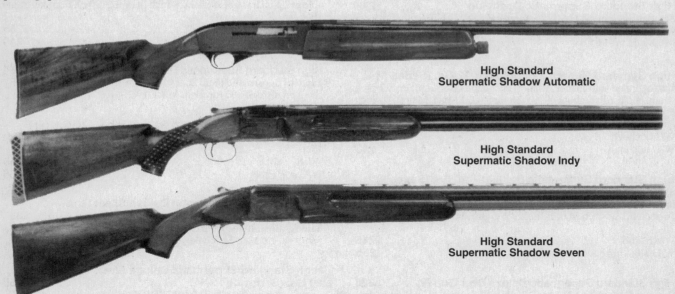

**High Standard
Supermatic Shadow Automatic**

**High Standard
Supermatic Shadow Indy**

**High Standard
Supermatic Shadow Seven**

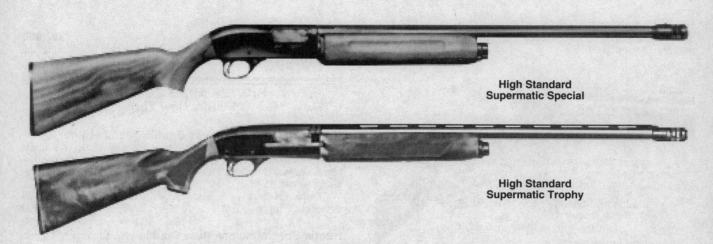

**High Standard
Supermatic Special**

**High Standard
Supermatic Trophy**

High Standard Supermatic Skeet — 20 Ga. **$230**
Same as Supermatic Deluxe Rib 20, except 26-inch vent-rib bbl.
w/SK choke. Made 1964-66.

High Standard Supermatic Special — 12 Ga. **$190**
Same as Supermatic Field 12, except has 27-inch bbl. w/adj. choke.
Made 1960-66.

High Standard Supermatic Special — 20 Ga. **$210**
Same as Supermatic Field 20 except has 27-inch bbl. w/adj. choke.
Made 1963-66.

High Standard Supermatic Trap — 12 Ga. **$215**
Same as Supermatic Deluxe Rib 12, except 30-inch vent-rib bbl., F
choke, special trap stock w/recoil pad. Made 1962-66.

High Standard Supermatic Trophy—20 Ga. **$225**
Same as Supermatic Deluxe Rib 20, except has 27-inch vent-rib bbl.
w/adj. choke. Made 1963-66.

HOLLAND & HOLLAND, LTD.
London, England

**Holland & Holland Badminton Model Hammerless
Double-Barrel Shotgun. Originally No. 2 Grade**
General specifications same as Royal Model except without self-
opening action. Made as a Game Gun or Pigeon and Wildfowl
Gun. Made from 1902 to date.
W/double triggers **$8350**
W/single trigger **9550**

**Holland & Holland Centenary Model Hammerless
Double-Barrel. Shotgun**
Lightweight (5.5 lbs.). 12 ga. game gun designed for 2-inch shell.
ade in four grades—Model Deluxe, Royal, Badminton, Dominion
— values same as shown for standard guns in those grades. Discontinued 1962.

**Holland & Holland Dominion Model Hammerless
Double-bbl. Shotgun** **$4695**
Game Gun. Sidelock. Auto ejectors. Double triggers. Gauges: 12,
16, 20. bbls. 25- to 30-inch, any standard boring. Checkered stock
and forend, straight grip standard. Discontinued 1967.

**Holland & Holland Model Deluxe
Hammerless Double**
Same as Royal Model, except has special engraving and exhibition
grade stock and forearm. Currently Manufactured.
W/Double Triggers **$27,500**
W/Single Trigger **29,750**

**Holland & Holland Northwood Model Hammerless
Double-Barrel Shotgun** **$5295**
Anson & Deeley system boxlock. Auto ejectors. Double triggers.
Gauges: 12, 16, 20, 28 in Game Model; 28 ga. not offered in Pigeon Model; Wildfowl Model in 12 ga. only (3-inch chambers
available). Bbls.: 28-inch standard in Game and Pigeon Models,
30-inch in Wildfowl Model; other lengths, any standard choke
combination available. Weight: from 5 to 7.75 lbs. depending on
ga. and bbls. Checkered straight-grip or pistol-grip stock and
forearm. Discontinued.

SHOTGUNS

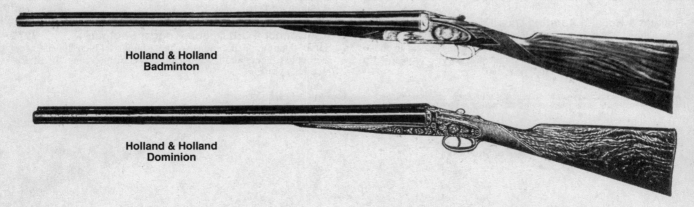

**Holland & Holland
Badminton**

**Holland & Holland
Dominion**

Holland & Holland Riviera Model Pigeon Gun **$10,500**
Same as Badminton Model but supplied w/two sets of bbls., double triggers. Discontinued 1967.

**Holland & Holland
Royal Model**

Holland & Holland Royal Model Hammerless Double
Self-opening. Sidelocks hand-detachable. Auto ejectors. Double triggers or single trigger. Gauges: 12, 16, 20, 28 .410. Built to customer's specifications as to bbl. length, chokes, etc. Made as a Game Gun or Pigeon and Wildfowl Gun, the latter having treble-grip action and side clips. Checkered stock and forend, straight grip standard. Made from 1885 to date.
W/Double Triggers **$25,250**
W/Single Trigger 28,750

Holland & Holland Royal Model Under-and-Over
Sidelocks, hand-detachable. Auto ejectors. Double triggers or single trigger. 12 ga. Built to customer's specifications as to bbl. length, chokes, etc. Made as a Game Gun or Pigeon and Wildfowl Gun. Checkered stock and forend, straight grip standard. *Note:* In 1951 Holland & Holland introduced its New Model Under/Over w/an improved, narrower action body. Discont. 1960.
New Model (Double Triggers) **$29,250**
New Model (Single Trigger) 30,500
Old Model (Double Triggers) 24,950
Old Model (Single Trigger) 26,500

Holland & Holland Single-Shot Super Trap Gun
Anson & Deeley system boxlock. Auto ejector. No safety. 12 ga. Bbls.: wide vent rib, 30- or 32-inch, w/Extra Full choke. Weight: about 8.75 lbs. Monte Carlo stock w/pistol grip and recoil pad, full beavertail forearm. Models differ in grade of engraving and wood. Discontinued.
Standard Grade **$4250**
Deluxe Grade 6750
Exhibition Grade 8250

Holland & Holland Sporting Over/Under **$21,500**
Blitz action. Auto ejectors; single selective trigger. Gauges: 12 or 20 w/2.75-inch chambers. Barrels: 28- to 32-inch w/screw-in choke tubes. Hand-checkered European walnut straight-grip or pistol grip stock, forearm. Made 1993 to date.

**Holland & Holland Sporting
Over/Under Deluxe** **$27,950**
Same general specs as Sporting Over/Under except better engraving and select wood. Made 1993 to date.

HUNTER ARMS COMPANY
Fulton, New York

Hunter Fulton Hammerless Double-bbl. Shotgun
Boxlock. Plain extractors. Double triggers or non-selective single trigger. Gauges: 12 16, 20. Bbls.: 26- to 32-inch various choke combinations. Weight: about 7 lbs. Checkered pistol-grip stock and forearm. Discont. 1948.
W/Double Triggers **$325**
W/Single Trigger 550

Hunter Special Hammerless Double-bbl. Shotgun
Boxlock. Plain extractors. Double triggers or non-selective single trigger. Gauges: 12,16, 20. Bbls.: 26- to 30-inch various choke combinations. Weight: 6.5 to 7.25 lbs. depending on bbl. length and ga. Checkered full pistol-grip stock and forearm. Discont. 1948.
W/Double Triggers **$525**
W/Single Trigger 650

IGA SHOTGUNS
Veranopolis, Brazil
Imported by Stoeger Industries, Inc. Wayne, NJ

IGA Coach Gun
Side-by-side Double. Gauges: 12, 20 and .410. 20-inch bbls. w/3-inch chambers. Fixed chokes (standard model) or screw-in tubes (deluxe model). Weight: 6.5 lbs. Double triggers. Ejector and automatic safety. Blued or nickel finish. Hand-rubbed oil-finished pistol grip stock and forend w/hand checkering (hardwood on standard model or Brazilian walnut (deluxe). Imported 1983 to date.
Standard Coach Gun (blued finish) **$265**
Standard Coach Gun (nickel finish) 350
Standard Coach Gun (engraved stock) 360
Deluxe Coach Gun (Intro. 1997) 375
Choke Tubes, **add** 20

IGA Condor I O/U Single-Trigger Shotgun
Gauges: 12 or 20. 26- or 28-inch bbls. of chrome-molybdenum steel. Chokes: Fixed — M/F or IC/M; screw-in choke tubes (12 and 20 ga.). 3-inch chambers. Weight: 6.75 to 7 lbs. Sighting rib w/anti-glare surface. Hand checkered hardwood pistol-grip stock and forend. Imported 1983 to date.
W/fixed chokes **$295**
W/Screw-in Tubes 350

IGA Condor II O/U Double-Trigger Shotgun **$275**
Same general specifications as the Condor I Over/Under except w/double triggers and fixed chokes only; 26-inch bbls., IC/M; 28-inch bbls., M/F.

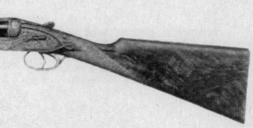

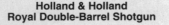

**Holland & Holland
Royal Double-Barrel Shotgun**

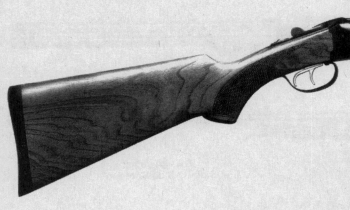

IGA Coach Gun

IGA Condor Supreme $465

Same general specifications as Condor I except upgraded w/fine-checkered Brazilian walnut buttstock and forend, a matte-laquered finish, and a massive monoblock that joins the bbls. in a solid one-piece assembly at the breech end. Bbls. w/recessed interchangeable choke tubes formulated for use w/steel shot. Automatic ejectors. Imported 1996 to date.

IGA Condor Turkey Model O/U Shotgun $495

12 gauge only. 26-inch vent-rib bbls. w/3-inch chambers fitted w/recessed interchangeable choke tubes . Weight: 8 lbs. Mechanical single trigger. Ejectors and automatic safety. Advantage™ Camouflage on stock and bbls. Made 1997 to date.

IGA Condor Waterfowl Model. $495

Similar to Condor Turkey-Advantage™ Camo Model, except w/30-inch bbls. Made 1998 to date.

IGA Deluxe Hunter Clay Shotgun

Same general specifications and values as IGA Condor Supreme. Imported 1997 to date.

IGA Era 2000 O/U Shotgun $395

Gauge: 12 w/3-inch chambers. 26- or 28-inch bbls. of chrome-molybdenum steel w/screw-in choke tubes. Extractors. Manual safety. (Mechanical triggers.) Weight: 7 lbs. Checkered Brazilian hardwood stock w/oil finish. Imported 1992-95.

IGA Reuna Single-Shot Shotgun

Visible hammer. Under-lever release. Gauges: 12, 20 and .410; 3-inch chambers. 26- or 28-inch bbls. w/fixed chokes or screw-in choke tubes (12 ga. only). Extractors. Weight: 5.25 to 6.5 lbs. Plain Brazilian hardwood stock and semibeavertail forend. Imported 1992 to date.

W/Fixed Choke. . . . . $ 85
W/Choke Tubes 145

IGA Uplander Side-by-Side Shotgun

Gauges: 12, 20, 28 and .410. 26- or 28-inch bbls. of chrome-molybdenum steel. Various fixed-choke combinations; screw-in choke tubes (12 and 20 ga.). 3-inch chambers (2.75-inch in 28 ga.). Weight: 6.25 to 7 lbs. Double triggers. Automatic safety. Matte-finished solid sighting rib. Hand checkered pistol-grip or straight stock and forend w/hand-rubbed, oil-finish. Imported 1983 to date.

W/fixed chokes. $275
W/Screw-in Tubes 305
English Model (straight grip) 310
Ladies Model 315
Youth Model. 285

SHOTGUNS

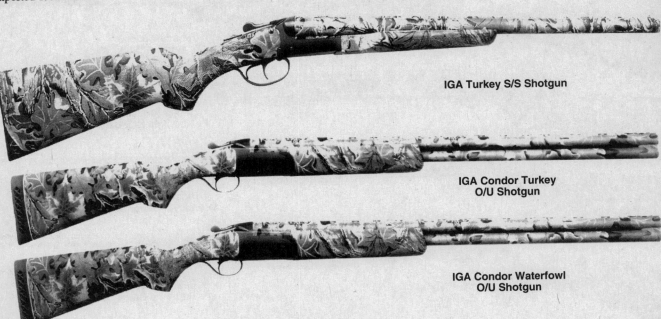

IGA Turkey S/S Shotgun

**IGA Condor Turkey
O/U Shotgun**

**IGA Condor Waterfowl
O/U Shotgun**

**IGA Deluxe
Hunter Clay Shotgun**

**IGA Uplander
Side-by-Side Shotgun**

IGA Uplander Turkey Model S/S Double **$350**
12 gauge only. 24-inch solid rib bbls. w/3-inch chambers choked F&F. Weight: 6.75 lbs. Double triggers. Automatic safety. Advantage™ Camouflage on stock and bbls. Made 1997 to date.

ITHACA GUN COMPANY
King Ferry (formerly Ithaca), New York
Now Ithaca Acquisition Corp./Ithaca Gun Co.

Ithaca Model 37 Bicentennial Commemorative **$425**
Limited to issue of 1976. Similar to Model 37 Supreme, except has special Bicentennial design etched on receiver, full-fancy walnut stock and slide handle. Serial numbers U.S.A. 0001 to U.S.A. 1976. Originally issued w/presentation case w/cast pewter belt buckle. Made in 1976. Value is for gun in new, unfired condition.

Ithaca Model 37 Deerslayer Deluxe
Formerly "Model 87 Deerslayer Deluxe" reintroduced under the original Model 37 designation w/the same specifications. Available w/smooth bore or rifled bbl. Reintroduced 1996.
Deluxe Model (smooth bore). **$335**
Deluxe Model (rifled bbl.) **375**

Ithaca Model 37 Deerslayer II **$420**
Gauges: 12 or 20 ga. 5-shot capacity. 20- or 25-inch rifled bbl. Weight: 7 lbs. Monte Carlo checkered walnut stock and forearm. Receiver drilled and tapped for scope mount. Made 1996 to date.

Ithaca Model 37 Deerslayer Standard **$285**
Same as Model 37 Standard, except has 20- or 26-inch bbl. bored for rifled slugs, rifle-type open rear sight and ramp front sight. Weight: 5.75 to 6.5 lbs. depending on ga. and bbl. length. Made 1959-86.

Ithaca Model 37 Deerslayer Super Deluxe **$325**
Formerly "Deluxe Deerslayer." Same as Model 37 Standard Deerslayer, except has stock and slide handle of fancy walnut. Made from 1962-86.

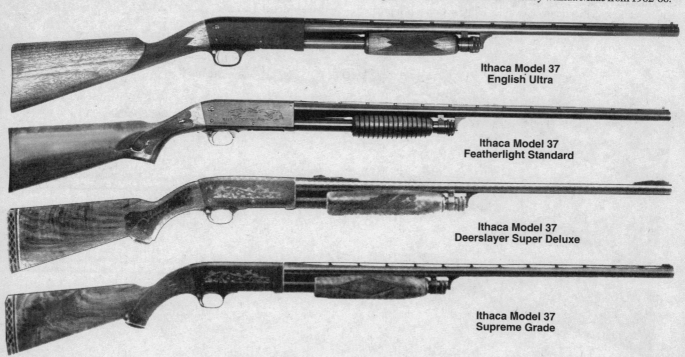

**Ithaca Model 37
English Ultra**

**Ithaca Model 37
Featherlight Standard**

**Ithaca Model 37
Deerslayer Super Deluxe**

**Ithaca Model 37
Supreme Grade**

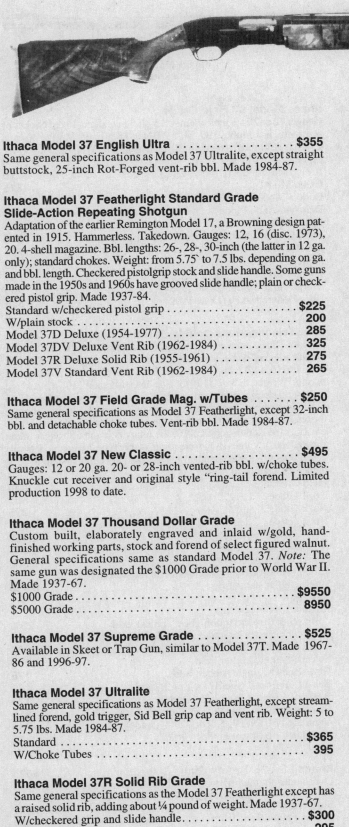

**Ithaca Model 51
Deluxe Trap**

Ithaca Model 37 English Ultra $355
Same general specifications as Model 37 Ultralite, except straight buttstock, 25-inch Rot-Forged vent-rib bbl. Made 1984-87.

Ithaca Model 37 Featherlight Standard Grade Slide-Action Repeating Shotgun
Adaptation of the earlier Remington Model 17, a Browning design patented in 1915. Hammerless. Takedown. Gauges: 12, 16 (disc. 1973), 20. 4-shell magazine. Bbl. lengths: 26-, 28-, 30-inch (the latter in 12 ga. only); standard chokes. Weight: from 5.75` to 7.5 lbs. depending on ga. and bbl. length. Checkered pistolgrip stock and slide handle. Some guns made in the 1950s and 1960s have grooved slide handle; plain or checkered pistol grip. Made 1937-84.

Standard w/checkered pistol grip	**$225**
W/plain stock	**200**
Model 37D Deluxe (1954-1977)	**285**
Model 37DV Deluxe Vent Rib (1962-1984)	**325**
Model 37R Deluxe Solid Rib (1955-1961)	**275**
Model 37V Standard Vent Rib (1962-1984)	**265**

Ithaca Model 37 Field Grade Mag. w/Tubes $250
Same general specifications as Model 37 Featherlight, except 32-inch bbl. and detachable choke tubes. Vent-rib bbl. Made 1984-87.

Ithaca Model 37 New Classic $495
Gauges: 12 or 20 ga. 20- or 28-inch vented-rib bbl. w/choke tubes. Knuckle cut receiver and original style "ring-tail forend. Limited production 1998 to date.

Ithaca Model 37 Thousand Dollar Grade
Custom built, elaborately engraved and inlaid w/gold, hand-finished working parts, stock and forend of select figured walnut. General specifications same as standard Model 37. *Note:* The same gun was designated the $1000 Grade prior to World War II. Made 1937-67.

$1000 Grade	**$9550**
$5000 Grade	**8950**

Ithaca Model 37 Supreme Grade $525
Available in Skeet or Trap Gun, similar to Model 37T. Made 1967-86 and 1996-97.

Ithaca Model 37 Ultralite
Same general specifications as Model 37 Featherlight, except streamlined forend, gold trigger, Sid Bell grip cap and vent rib. Weight: 5 to 5.75 lbs. Made 1984-87.

Standard	**$365**
W/Choke Tubes	**395**

Ithaca Model 37R Solid Rib Grade
Same general specifications as the Model 37 Featherlight except has a raised solid rib, adding about ¼ pound of weight. Made 1937-67.

W/checkered grip and slide handle.	**$300**
W/plain stock	**295**

Ithaca Model 37S Skeet Grade $425
Same general specifications as the Model 37 Featherlight, except has vent rib and large extension-type forend; weighs about ½ lb. more. Made 1937-55.

Ithaca Model 37T Target Grade $415
Same general specifications as Model 37 Featherlight, except has vent-rib bbl., checkered stock and slide handle of fancy walnut (choice of skeet- or trap-style stock). *Note:* This model replaced Model 37S Skeet and Model 37T Trap. Made 1955-61.

Ithaca Model 37T Trap Grade $425
Same general specifications as Model 37S, except has straighter trap-style stock of select walnut, recoil pad; weighs about ½ lb. more. Made 1937-55.

Ithaca Model 37 Turkeyslayer
Gauges: 12 ga. (Standard) or 20 ga. (youth). Slide action. 22-inch bbl. Extended choke tube. Weight: 7 lbs. Advantage™ Camoflauge or Realtree™ pattern. Made 1996 to date.

Standard Model	**$350**
Youth Model (intro. 1998)	**375**

Ithaca Model 37 Waterfowler $365
12 ga. only w/28-inch bbl. Wetlands™ Camoflauge. Made 1998 to date.

Ithaca Model 51 Deerslayer $295
Same as Model 51 Standard, except has 24-inch plain bbl. w/slug boring, rifle sights, recoil pad. Weight: about 7.25 lbs. Made 1972-84.

Ithaca Model 51 Deluxe Skeet Grade $395
Same as Model 51 Standard, except 26-inch vent-rib bbl. only, SK choke, skeet-style stock, semi-fancy wood. Weight: about 8 lbs. Made 1970-87.

Ithaca Model 51 Deluxe Trap Grade
Same as Model 51 Standard, except 12 ga. only, 30-inch bbl. w/broad floating rib, F choke, trap-style stock w/straight or Monte Carlo comb, semifancy wood, recoil pad. Weight: about 8 lbs. Made 1970-87.

W/straight stock	**$325**
W/Monte Carlo stock	**345**

Ithaca Model 51 Standard Automatic Shotgun
Gas-operated. Gauges: 12, 20. 3-shot. Bbls.: plain or vent rib, 30-inch F choke (12 ga. only), 28-inch F or M, 26-inch IC. Weight: 7.25-7.75 lbs. depending on ga. and bbl. Checkered pistol-grip stock, forearm. Made 1970-80. Still avail. in 12 and 20 ga., 28-inch M choke only.

W/plain barrel.	**$230**
W/ventilated rib	**275**

Ithaca Model 51 Standard Magnum
Same as Model 51 Standard, except has 3-inch chamber, handles Magnum shells only; 30-inch bbl. in 12 ga., 28-inch in 20 ga., F or M choke, stock w/recoil pad. Weight: 7.75-8 lbs. Made 1972 to date.

W/plain bbl. (disc. 1976)	**$235**
W/ventilated rib	**275**

**Ithaca Model 66RS
Buckbuster**

Ithaca Model 51A Turkey Gun $295
Same general specifications as standard Model 51 Magnum, except 26-inch bbl. and matte finish. Disc. 1986.

Ithaca Model 66 Long Tom $100
Same as Model 66 Standard, except has 36-inch F choke bbl., 12 ga. only, checkered stock and recoil pad standard. Made 1969-74.

Ithaca Model 66 Standard Supersingle Lever
Single shot. Hand-cocked hammer. Gauges: 12 (discont. 1974), 20, .410, 3-inch chambers. Bbls.: 12 ga., 30-inch F choke, 28-inch F or M; 20 ga., 28-inch F or M; .410, 26-inch F. Weight: about 7 lbs. Plain or checkered straight-grip stock, plain forend. Made 1963-78.
Standard Model $125
Vent Rib Model (20 ga., checkered stock, recoil
    pad, 1969-74) 145
Youth Model (20 & .410 ga., 26-inch bbl., shorter
    stock, recoil pad, 1965-78) 125

Ithaca Model 66RS Buckbuster $150
Same as Model 66 Standard, except has 22-inch bbl. cylinder bore w/rifle sights, later version has recoil pad. Originally offered in 12 and 20 ga.; the former was disc. in 1970. Made 1967-78.

NOTE

Previously issued as the Ithaca Model 37, the Model 87 guns listed below were made available through the Ithaca Acquisition Corp. From 1986-95. Production of the Model 37 resumed under the original logo in 1996.

Ithaca Model 87 Deerslayer Shotgun
Gauges: 12 or 20, 3-inch chamber. Bbls.: 18.5-, 20- or 25-inch (w/special or rifled bore). Weight: 6 to 6.75 lbs. Ramp blade front sight, adj. rear. Receiver grooved for scope. Checkered American walnut pistol-grip stock and forearm. Made 1988-96.
Basic Model $275
Basic Field Combo (w/extra 28-inch bbl.) 315
Deluxe Model 295
Deluxe Combo (w/extra 28-inch bbl.) 375
DSPS Model (8-shot) 280
Field Model 240
Monte Carlo Model 255
Ultra Model (disc. 1991) 350

Ithaca Model 87 Deerslayer II Rifled Shotgun $325
Similar to the Standard Deerslayer Model, except w/solid frame construction and 25-inch rifled bbl. Monte Carlo stock. Made 1988-96.

Ithaca Model 87 Ultralite Field Pump Shotgun $295
Gauges: 12 and 20; 2.75-inch chambers. 25-inch bbl. w/choke tube. Weight: 5 to 6 lbs. Made 1988-90.

Ithaca Model 87 Field Grade
Gauge: 12 or 20.; 3-inch chamber. 5-shot magazine. Fixed chokes or screw-in choke tubes (IC, M, F). Bbls.: 18.5-inch (M&P); 20- and 25-inch (Combo); 26-, 28-, 30-inch vent rib. Weight: 5 to 7 lbs. Made from 1988-96.
Basic Field Model (Disc. 1993) $290
Camo Model 315
Deluxe Model 340
Deluxe Combo Model 345
English Model 295
Hand Grip Model (w/polymer pistol-grip) 325
M&P Model (Disc. 1995) 280
Supreme Model 450
Turkey Model 295
Ultra Deluxe Model (Disc. 1992) 365
Ultra Deluxe Model (Disc. 1992) 335

Ithaca Hammerless Double-Barrel Shotguns
Boxlock. Plain extractors, auto ejectors standard on the "E" grades. Double triggers, non-selective or selective single trigger extra. Gauges: Magnum 10, 12; 12, 16, 20, 28, .410. Bbls.: 26- to 32-inch, any standard boring. Weight: 5.75 (.410) to 10.5 lbs. (Magnum 10). Checkered pistol-grip stock and forearm standard. Higher grades differ from Field Grade in quality of workmanship, grade of wood, checkering, engraving, etc.; general specifications are the same. Ithaca doubles made before 1925 (serial number 425,000) the rotary bolt and a stronger frame were adopted. Values shown are for this latter type; earlier models valued about 50% lower. Smaller gauge guns may command up to 75% higher. Discontinued 1948.
Field Grade . . . . $ 650
No. 1 Grade 695
No. 2 Grade 1,095
No. 3 Grade 1,295
No. 4E Grade (ejector) 2,495
No. 5E Grade (ejector) 3,350
No. 7E Grade (ejector) 12,500
$2000 (prewar $1000) Grade (ejector and selective
single trigger standard) 8,950
Extras:
Magnum 10 or 12 gauge
    (in other than the four highest grades), **add** 20%
Automatic ejectors (Grades No. 1, 2, 3, w/ejectors
    are designated No. 1E, 2E, 3E), **add** $200
Selective single trigger, **add** 250
Non-selective single trigger, **add** 175
Beavertail forend (Field No. 1 or 2), **add** 150
Beavertail forend (No. 3 or 4), **add** 175
Beavertail forend (No. 5, 7 or $2000 Grade), **add** 250
Ventilated rib (No. 4, 5, 7 or $2000 Grade), **add** 250
Ventilated rib (lower grades), **add** 175

Ithaca LSA-55 Turkey Gun $525
Over/under shotgun/rifle combination. Boxlock. Exposed hammer. Plain extractor. Single trigger. 12 ga./222 Rem. 24.5-inch ribbed bbls. (rifle bbl. has muzzle brake). Weight: about 7 lbs. Folding leaf rear sight, bead front sight. Checkered Monte Carlo stock and forearm. Made 1970-77 by Oy Tikkakoski AB, Finland.

**Ithaca Hammerless
Field Grade**

**Ithaca Hammerless
No. 2**

**Ithaca Hammerless
No. 4**

Ithaca Mag-10 Automatic Shotgun

Gas-operated. 10 ga. 3.5-inch Magnum. 3-shot. 32-inch plain (Standard Grade only) or vent-rib bbl. F choke. Weight: 11 lbs., plain bbl.; 11.5 lbs., vent rib. Standard Grade has plain stock and forearm. Deluxe and Supreme Grades have checkering, semi-fancy and fancy wood respectively, and stud swivel. All have recoil pad. Deluxe and Supreme Grades made 1974-1982. Standard Grade intro. in 1977. All grades disc. 1986.

Ithaca Mag-10 Automatic Shotgun *(Con't)*

Camo Model	**$595**
Deluxe Grade	625
Roadblocker	645
Standard Grade, plain barrel	530
Standard Grade, ventilated rib	625
Standard Grade, w/tubes	650
Supreme Grade	745

Ithaca Single-Shot Trap, Flues and Knick Models

Boxlock. Hammerless. Ejector. 12 ga. only. Bbl. lengths: 30-, 32-, 34-inch (32-inch only in Victory Grade). Vent rib. Weight: about 8 lbs. Checkered pistol-grip stock and forend. Grades differ only in quality of workmanship, engraving, checkering, wood, etc. Flues Model, serial numbers under 400,000, made 1908-1921. Triple-bolted Knick Model, serial numbers above 400,000, made since 1921. Victory Model disc. in 1938, No. 7-E in 1964, No. 4-E in 1976, No. 5-E in 1986, Dollar Grade in 1991. Values shown are for Knick Model; Flues Models about 50% lower.

Victory Grade	**$ 895**
No. 4-E	2,150
No. 5-E	2,495
No. 6-E	+12,000
No. 7-E	4,795
$5000 Grade (prewar $1000 Grade)	8,850
Sousa Grade	+10,000

> **NOTE**
>
> The following Ithaca-Perazzi shotguns were manufactured by Manifattura Armi Perazzi, Brescia, Italy. *See* also separate Perazzi listings.

Ithaca-Perazzi Competition I Skeet $2855

Boxlock. Auto ejectors. Single trigger. 12 ga. 26.75-inch vent-rib bbls. SK choke w/integral muzzle brake. Weight: about 7.75 lbs. Checkered skeet-style pistol-grip buttstock and forearm; recoil pad. Made 1969-74. *See* Illustration next page.

SHOTGUNS

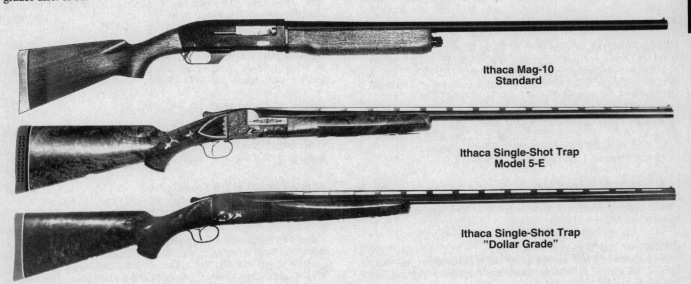

**Ithaca Mag-10
Standard**

**Ithaca Single-Shot Trap
Model 5-E**

**Ithaca Single-Shot Trap
"Dollar Grade"**

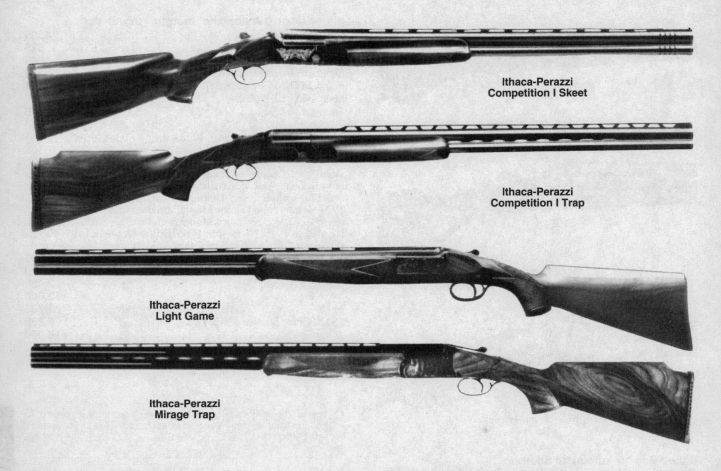

Ithaca-Perazzi
Competition I Skeet

Ithaca-Perazzi
Competition I Trap

Ithaca-Perazzi
Light Game

Ithaca-Perazzi
Mirage Trap

Ithaca-Perazzi Competition Trap I O/U **$2650**
Boxlock. Auto ejectors. Single trigger. 12 ga. 30- or 32-inch vent-rib bbls. IM/F choke. Weight: about 8.5 lbs. Checkered pistol-grip stock, forearm; recoil pad. Made 1969-74.

Ithaca-Perazzi Competition I Trap Single Barrel **$1950**
Boxlock. Auto ejection. 12 ga. 32- or 34-inch bbl., vent rib, F choke. Weight: 8.5 lbs. Checkered Monte Carlo stock and beavertail forearm, recoil pad. Made 1973-78.

Ithaca-Perazzi Competition IV Trap Gun **$2395**
Boxlock. Auto ejection. 12 ga. 32- or 34-inch bbl. With high, wide vent rib, four interchangeable choke tubes (Extra Full, F, IM, M). Weight: about 8.75 lbs. Checkered Monte Carlo stock and beavertail forearm, recoil pad. Fitted case. Made 1977-78.

Ithaca-Perazzi Light Game O/U Field **$2895**
Boxlock. Auto ejectaors. Single trigger. 12 ga. 27.5-inch vent-rib bbls., M/F or IC/M choke. Weight: 6.75 lbs. Checkered field-style stock and forearm. Made 1972-74.

Ithaca Perazzi Mirage Live Bird **$3250**
Same as Mirage Trap, except has 28-inch bbls., M and Extra Full choke, special stock and forearm for live bird shooting. Weight: about 8 lbs. Made 1973-78.

Ithaca-Perazzi Mirage Skeet **$2750**
Same as Mirage Trap, except has 28-inch bbls. w/integral muzzle brakes, SK choke, skeet-stype stock and forearm. Weight: about 8 lbs. Made 1973-78.

Ithaca-Perazzi Mirage Trap **$2850**
Same general specifications as MX-8 Trap, except has tapered rib. Made 1973-78.

Ithaca-Perazzi MT-6 Skeet **$2795**
Same as MT-6 Trap, except has 28-inch bbls. w/two skeet choke tubes instead of Extra Full and F, skeet-style stock and forearm. Weight: about 8 lbs. Made 1976-78.

Ithaca-Perazzi MT-6 Trap Combo **$3595**
MT-6 w/extra single under bbl. w/high-rise aluminum vent rib, 32- or 34-inch; seven interchanageable choke tubes (IC through Extra Full). Fitted case. Made 1977-78.

Ithaca-Perazzi MT-6 Trap Over/Under **$2495**
Boxlock. Auto selective ejectors. Non-selective single trigger. 12 ga. Barrels separated, wide vent rib, 30-or 32-inch, five interchangeable choke tubes (Extra full, F, IM, M, IC). Weight: about 8.5 lbs. Checkered pistol-grip stock/forearm, recoil pad. Fitted case. Made 1976-78.

Ithaca-Perazzi MX-8 Trap Combo **$3950**
MX-8 w/extra single bbl., vent rib, 32- or 34-inch, F choke, forearm; two trigger groups included. Made 1973-78.

Ithaca-Perazzi MX-8 Trap Over/Under **$2895**
Boxlock. Auto selective ejectors. Non-selective single trigger. 12 ga. Bbls.: high vent rib; 30- or 32-inch, IM/F choke. Weight: 8.25 to 8.5 lbs. Checkered Monte Carlo stock and forearm, recoil pad. Made 1969-78.

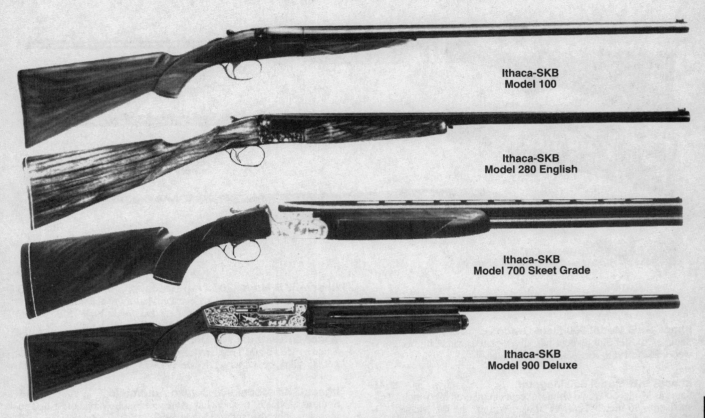

Ithaca-SKB
Model 100

Ithaca-SKB
Model 280 English

Ithaca-SKB
Model 700 Skeet Grade

Ithaca-SKB
Model 900 Deluxe

SHOTGUNS

Ithaca-Perazzi Single-Barrel Trap Gun **$1925**
Boxlock. Auto ejection. 12 ga. 34-inch vent-rib bbl., F choke. Weight: abaout 8.5 lbs. Checkered pistol-grip stock, forearm; recoil pad. Made 1971-72.

> **NOTE**
>
> The following Ithaca- SKB shotguns, manufactured by SKB Arms Company, Tokyo, Japan, were distributed in the U.S. by Ithaca Gun Company 1966-1976. *See* also listings under SKB.

Ithaca-SKB Model 100 Side-by-side **$425**
Boxlock. Plain extractors. Selective single trigger. Auto safety. Gauges: 12 and 20; 2.75-inch and 3-inch chambers respectively. Bbls.: 30-inch, F/F (12 ga. only); 28-inch, F/M; 26-inch, IC/M (12 ga. only); 25-inch, IC/M (20 ga. only). Weight: 12 ga., about 7 lbs.; 20 ga., about 6 lbs. Checkered stock and forend. Made 1966-76.

Ithaca-SKB Model 150 Field Grade **$450**
Same as Model 100, except has fancier scroll engraving, beavertail forearm. Made 1972-74.

Ithaca-SKB Model 200E Field Grade S/S **$550**
Same as Model 100, except auto selective ejectors, engraved and silver-plated frame, gold-plated nameplate and trigger, beavertail forearm. Made 1966-76.

Ithaca-SKB Model 200E Skeet Gun **$595**
Same as Model 200E Field Grade, except 26-inch (12 ga.) and 25-inch (20 ga./2.75-inch chambers) bbls., SK choke; nonautomatic safety and recoil pad. Made 1966-76.

Ithaca-SKB Model 280 English **$695**
Same as Model 200E, except has scrolled game scene engraving on frame, English-style straight-grip stock; 30-inch bbls. not available; special quail gun in 20 ga. has 25-inch bbls., both bored IC. Made 1971-76.

Ithaca-SKB Model 300 Standard Automatic Shotgun
Recoil-operated. Gauges: 12, 20 (3-inch). 5-shot. Bbls.: plain or vent rib; 30-inch F choke (12 ga. only), 28-inch F or M, 26-inch IC. Weight: about 7 lbs. Checkered pistol-grip stock and forearm. Made 1968-72.
W/plain barrel **$235**
W/ventilated rib 265

Ithaca-SKB Model 500 Field Grade O/U **$425**
Boxlock. Auto selective ejectors. Selective single trigger. Nonautomatic safety. Gauges: 12 and 20; 2.75-inch and 3-inch chambers respectively. Vent-rib bbls.: 30-inch M/F (12 ga. only); 28-inch M/F; 26-inch IC/M. Weight: 12 ga., about 7.5 lbs; 20 ga., about 6.5 lbs. Checkered stock and forearm. Made 1966-76.

Ithaca-SKB Model 500 Magnum **$475**
Same as Model 500 Field Grade, except chambered for 3-inch 12 ga. shells, has 30-inch bbls., IM/F choke. Weight: about 8 lbs. Made 1973-76.

Ithaca-SKB Model 600 Doubles Gun **$585**
Same as Model 600 Trap Grade, except specially choked for 21-yard first target, 30-yard second. Made 1973-75.

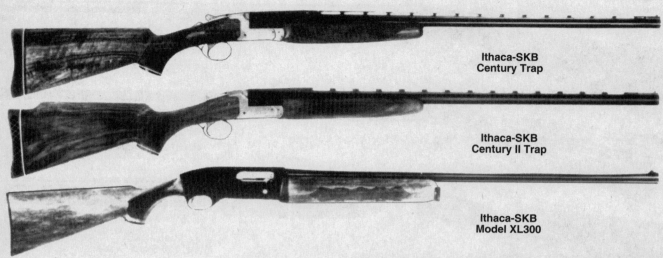

**Ithaca-SKB
Century Trap**

**Ithaca-SKB
Century II Trap**

**Ithaca-SKB
Model XL300**

Ithaca-SKB Model 600 Field Grade **$525**
Same as Model 500, except has silver-plated frame, higher grade wood. Made 1969-76.

Ithaca-SKB Model 600 Magnum **$545**
Same as Model 600 Field Grade, except chambered for 3-inch 12 ga. shells; has 30-inch bbls., IM/F choke. Weight: 8.5 lbs. Made 1969-72.

Ithaca-SKB Model 600 Skeet Grade
Same as Model 500, except also available in 28 and .410 ga., has silver-plated frame, higher grade wood, recoil pad, 26- or 28-inch bbls. (28-inch only in 28 and .410), SK choke. Weight: 7 to 7.75 lbs. depending on ga. and bbl. length. Made 1966-76.
12 or 20 gauge **$550**
28 or .410 gauge 625

Ithaca-SKB Model 600 Skeet Combo Set **$1595**
Model 600 Skeet Grade w/matched set of 20, 28 and .410 ga. bbls., 28-inch, fitted case. Made 1970-76.

Ithaca-SKB Model 600 Trap Grade O/U **$565**
Same as Model 500, except 12 ga. only, has silver-plated frame, 30- or 32-inch bbls. choked F/F or F/IM, choice of Monte Carlo or straight stock of higher grade wood, recoil pad. Weight: about 8 lbs. Made 1966-76.

Ithaca-SKB Model 680 English **$595**
Same as Model 600 Field Grade, except has intricate scroll engraving, English-style straight-grip stock and forearm of extra-fine walnut; 30-inch bbls. not available. Made 1973-76.

Ithaca-SKB Model 700 Skeet Combo Set **$1950**
Model 700 Skeet Grade w/matched set of 20, 28 and .410 ga. bbls., 28-inch fitted case. Made 1970-71.

Ithaca-SKB Model 700 Skeet Grade **$795**
Same as Model 600 Skeet Grade, except not available in 28 and .410 ga., has more elaborate scroll engraving, extra-wide rib, higher grade wood. Made 1969-75. *See* Illustration Previous Page.

Ithaca-SKB Model 700 Trap Grade **$785**
Same as Model 600 Trap Grade, except has more elaborate scroll engraving, extra-wide rib, higher grade wood. Made 1969-75.

Ithaca-SKB Model 700 Doubles Gun **$775**
Same as Model 700 Trap Grade, except choked for 21-yard first target, 30-yard second target. Made 1973-75.

Ithaca-SKB Model 900 Deluxe Automatic **$315**
Same as Model 300, except has game scene etched and gold-filled on receiver, vent rib standard. Made 1968-72.

Ithaca-SKB Model 900 Slug Gun **$285**
Same as Model 900 Deluxe, except has 24-inch plain bbl. w/slug boring, rifle sights. Weight: about 6.5 lbs. Made 1970-72.

Ithaca-SKB Century Single-Shot Trap Gun **$525**
Boxlock. Auto ejector. 12 ga. Bbls.: 32- or 34-inch, vent rib, F choke. Weight: about 8 lbs. Checkered walnut stock w/pistol grip, straight or Monte Carlo comb, recoil pad, beavertail forearm. Made 1973-74.

Ithaca-SKB Century II **$575**
Boxlock. Auto ejector. 12 ga. Bbls: 32- or 34-inch, vent rib, F choke. Weight: 8.25 lbs. Improved version of Century. Same general specifications, except has higher comb on checkered stock stock, reverse-taper beavertail forearm w/redesigned locking iron. Made 1975-76.

Ithaca-SKB Model XL300 Standard Automatic
Gas-operated. Gauges: 12, 20 (3-inch). 5-shot. Bbls.: plain or vent rib; 30-inch F choke (12 ga. only), 28-inch F or M, 26-inch IC. Weight: 6 to 7.5 lbs. depending on ga. and bbl. Checkered pistol-grip stock, forearm. Made 1972-76.
W/plain barrel **$215**
W/ventilated rib 235

Ithaca-SKB Model XL900 Deluxe Automatic **$275**
Same as Model XL300, except has game scene finished in silver on receiver, vent rib standard. Made 1972-76. *See* Illustration Previous Page.

Ithaca-SKB Model XL900 Skeet Grade **$335**
Gas-operated. Gauges: 12, 20 (3-inch). 5-shot tubular magazine. Same as Model XL900 Deluxe, except has scrolled receiver finished in black chrome, 26-inch bbl. only, SK choke, skeet-style stock. Weight: 7 or 7.5 lbs. depending on ga. Made 1972-76.

Ithaca-SKB Model XL900 Slug Gun **$295**
Same as Model XL900 Deluxe, except has 24-inch plain bbl. w/slug boring, rifle sights. Weight: 6.5 or 7 lbs. depending on ga. Made 1972-76.

Ithaca-SKB Model XL900 Trap Grade **$345**
Same as Model XL900 Deluxe, except 12 ga. only, has scrolled receiver finished in black chrome, 30-inch bbl. only, IM or F choke, trap style w/straight or Monte Carlo comb, recoil pad. Weight: about 7.75 lbs. Made 1972-76.

IVER JOHNSON'S ARMS & CYCLE WORKS
Fitchburg, Massachusetts
Currently a division of the American Military Arms Corp.
Jacksonville, Arkansas

**Iver Johnson
Champion Standard Garde**

Iver Johnson Champion Grade Single-Shot Hammer Shotgun
Auto ejector. Gauges: 12,16, 20, 28 and .410. Bbls.: 26- to 36-inch, F choke. Weight: 5.75 to 7.5 lbs. depending on ga. and bbl.length. Plain pistol-grip stock and forend. Extras include checkered stock and forend, pistol-grip cap and knob forend. Known as Model 36. Also made in a Semi-Octagon Breech, Top Matted and Jacketed Breech (extra heavy) model. Made in Champion. Lightweight as Model 39 in gauges 24, 28, 32 and .410, 44 and 45caliber, 12 and 14mm w/same extras — $200, add $100 in the smaller and obsolete gauges. Made 1909-73.
Standard Model. **$150**
Semi-Octagon Breech. 245
Top Matted Rib (Disc. 1948) 230

Iver Johnson Hercules Grade Hammerless Double
Boxlock. (Some made w/false sideplates.) Plain extractors and auto ejectors. Double or Miller Single triggers (both selective or non-selective). Gauges: 12, 16, 20 and .410. Bbl. lengths: 26- to 32-inch, all chokes. Weight: 5.75 to 7.75 lbs. depending on ga.e and bbl. length. Checkered stock and forend. Straight grip in .410 ga. w/both 2.5- and 3-inch chambers. Extras include Miller Single Trigger, Jostam Anti-Flinch Recoil Pad and Lyman Ivory Sights at extra cost. Discontinued 1946.
W/double triggers, extractors **$495**
W/double triggers, automatic ejectors 595
Extra for non-selective single trigger 100
Extra for selective single trigger. 135
Extra for .410 gauge 185

Iver Johnson Matted Rib Single-Shot Hammer Shotgun in Smaller Gauges **$275**
Same general specifications as Champion Grade except has solid matted top rib, checkered stock and forend. Weight: 6 to 6.75 lbs. Discontinued 1948.

Iver Johnson Silver Shadow Over/Under Shotgun
Boxlock. Plain extractors. Double triggers or non-selective single trigger.12 ga., 3-inch chambers. Bbls.: 26-inch IC/M; 28-inch IC/M, 28-inch M/F; 30-inch both F choke; vent rib. Weight: w/28-inch bbls., 7.5 lbs. Checkered pistol-grip stock/forearm. Made by F. Marocchi, Brescia Italy 1973-78.
Model 412 w/Double Triggers **$395**
Model 422 w/Single Trigger 495

Iver Johnson Skeeter Model Hammerless Double
Boxlock. Plain extractors or selective auto ejectors. Double triggers or Miller Single Trigger (selective or non-selective). Gauges: 12, 16, 20, 28 and .410. 26- or 28-inch bbls., skeet boring standard. Weight: about 7.5 lbs.; less in smaller gauges. Pistol- or straight-grip stock and beavertail forend, both checkered, of select fancy-figured black walnut. Extras include Miller Single Trigger, selective or non-selective, Jostam Anti-Flinch Recoil Pad and Lyman Ivory Rear Sight at additional cost. Discont. 1942.
W/Double Triggers, plain extractors . . . . **$ 895**
W/Double Triggers, automatic ejectors 1095
Extra for non-selective single trigger 100
Extra for selective single trigger. 135
Extra for .410 gauge 200
Extra for 28 gauge 300

Iver Johnaon Special Trap Single-Shot Hammer Shotgun **$295**
Auto ejector. 12 ga. only. 32-inch bbl. w/vent rib, F choke. Checkered pistol-grip stock and forend. Weight: about 7.5 lbs. Discontinued 1942.

Iver Johnson Super Trap Hammerless Double
Boxlock. Plain extractors. Double trigger or Miller single trigger (slective or non-selective), 12 ga. only, F choke 32-inch bbl., vent rib. Weight: 8.5 lbs. Checkered pistol-grip stock and bevertail forend, recoil pad, Discontinued 1942.
W/Double Triggers. **$850**
Extra for Non-selective Single Trigger. 250
Extra for Selective Single Trigger 300

KBI INC. SHOTGUNS

See listings under Armscor, Baikal, Charles Daly, Fias, & Omega

KESSLER ARMS CORP.
Silver Creek, New York

Kessler Lever-Matic Repeating Shotgun **$135**
Lever action. Takedown. Gauges: 12, 16, 20; three-shot magazine. Bbls.: 26-, 28-, 30-inch; F choke. Plain pistol-grip stock, recoil pad. Weight: 7 to 7.75 lbs. Discontinued 1953.

Kessler Three Shot Bolt-Action Repeater **$75**
Takedown. Gauges: 12, 16, 20. Two-shell detachable box magazine. Bbls.: 28-inch in 12 and 16 ga.; 26-inch in 20 ga.; F choke. Weight: 6.25 to 7.25 lbs. depending on ga. and bbl. length. Plain one-piece pistol-grip stock recoil pad. Made 1951-53.

H. KRIEGHOFF JAGD UND SPORTWAFFENFABRIK
Ulm (Donau), West Germany

SHOTGUNS

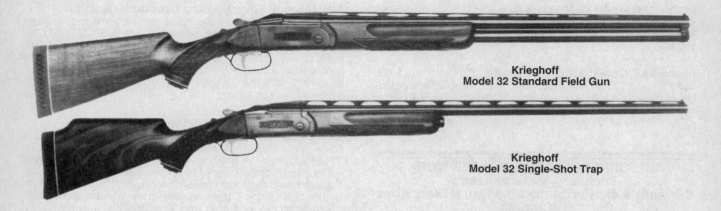

**Krieghoff
Model 32 Standard Field Gun**

**Krieghoff
Model 32 Single-Shot Trap**

Krieghoff Model 32 Four-Barrel Skeet Set
Over/under w/four sets of matched bbls.: 12, 20, 28 and .410 ga., in fitted case. Available in six grades that differ in quality of engraving and wood. Discontinued 1979.

Standard Grade	$ 8,950
München Grade	9,950
San Remo Grade	12,250
Monte Carlo Grade	15,950
Crown Grade	19,750
Super Crown Grade	21,500
Exhibition Grade	29,950

Krieghoff Model 32 Standard Grade Over/Under
Similar to prewar Remington Model 23. Boxlock. Auto ejector. Single trigger. Gauges: 12, 20, 28, .410. Bbls.: vent rib, 26.5- to 32-inch, any chokes. Weight: 12 ga. Field gun w/28-inch bbls., about 7.5 lbs. Checkered pistol-grip stock and forearm of select walnut; available in field, skeet and trap styles. Made 1958-1981.

W/one set of bbls.	$2950
Low-rib Two-Bbl. Trap Combo	3795
Vandalia (high-rib) Two-Bbl. Trap Combo	3995

Krieghoff Model 32 Standard Grade Single-Shot Trap Gun $1495
Same action as over/under. 12 ga. 32- or 34-inch bbl. w/high vent rib on bbl.; M, IM, or F choke. Checkered Monte Carlo buttstock w/thick cushioned recoil pad, beavertail forearm. Disc. 1979.

Krieghoff Model K-80
Refined and enhanced version of the Model 32. Single selective mechanical trigger, adj. for position; release trigger optional. Fixed chokes or screw-in choke tubes. Interchangeable front bbl. Hangers to adjust point of impact. Quick-removable stock. Color casehardened or satin grey finished receiver; aluminum alloy receiver on lightweight models. Available in standard plus five engraved grades. Made 1980 to date. Standard grade shown except where noted.

SKEET MODELS

Skeet International	$ 4,950
Skeet Special	4,595
Skeet Standard Model	3,950
Skeet w/Tubla Chokes	4,750

SKEET SETS

Standard Grade 2-Bbl. Set	$ 6,550
Standard Grade 4-Bbl. Set	8,595
Bavaria Grade 4-Bbl. Set	12,250
Danube Grade 4-Bbl. Set	15,950
Gold Target Grade 4-Bbl. Set	18,750

SPORTING MODELS

Pigeon	$ 4,695
Sporting Clays	4,950

TRAP MODELS

Trap Combo	$ 6,995
Trap Single	4,950
Trap Standard	4,595
Trap Unsingle	5,145
RT Models (Removable Trigger) **Add**	1,385

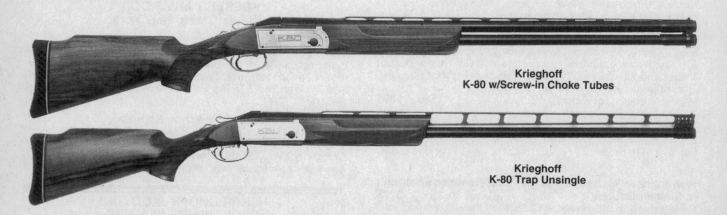

**Krieghoff
K-80 w/Screw-in Choke Tubes**

**Krieghoff
K-80 Trap Unsingle**

Krieghoff Model KS-5 Single-Barrel Trap
Boxlock w/no sliding top-latch. Adjustable or optional release trigger. Gauge: 12; 2.75-inch chamber. Bbl.: 32-, 34-inch w/fixed choke or screw-in tubes. Weight: 8.5 lbs. Adjustable or Monte Carlo European walnut stock. Blued or nickel receiver. Made from 1980 to date. Redesigned and streamlined in 1993.

Standard Model w/Fixed Chokes	**$2495**
Standard Model w/Tubes	**2950**
Special Model w/Adj. Rib & Stock	**3195**
Special Model w/Adj. Rib & Stock, Tubes	**3650**

**Krieghoff
ULM Over/Under**

**Krieghoff
Neptun Drilling**

Krieghoff Neptun Drilling **$9550**
Same general specifications as Trumpf model, except has sidelocks w/hunting scene engraving. Currently manufactured.

Krieghoff Neptun-Primus Drilling **$10,950**
Deluxe version of Neptun model; has detachable sidelocks, higher grade engraving and fancier wood. Currently manufactured.

Krieghoff Teck Over/Under Rifle-Shotgun **$5350**
Boxlock. Kersten double crossbolt system. Steel or dural receiver. Split extractor or ejector for shotgun bbl. Single or double triggers. Gauges: 12, 16, 20; latter w/either 2.75- or 3-inch chamber. Calibers: 22 Hornet, 222 Rem., 222 Rem. Mag., 7×57r5, 7×64, 7×65r5, 30-30, 300 Win. Mag., 30-6, 308, 9.3×74R. 25-inch bbls. With solid rib, folding leaf rear sight, post or bead front sight; over bbl. is shotgun, under bbl. rifle (later fixed or interchangeable; extra rifle bbl., $175). Weight: 7.9 to 9.5 lbs. depending on type of receiver and caliber. Checkered pistol-grip stock w/cheekpiece and semibeavertail forearm of figured walnut, sling swivels. Made 1967 to date. *Note:* This combination gun is similar in appearance to the same model shotgun.

Krieghoff Teck Over/Under Shotgun **$4550**
Boxlock. Kersten double crossbolt system. Auto ejector. Single or double triggers. Gauges: 12, 16, 20; latter w/either 2.75- or 3-inch chambers. 28-inch vent-rib bbl., M/F choke. Weight: about 7 lbs. Checkered walnut pistol-grip stock and forearm. Made 1967-89.

Krieghoff Trumpf Drilling **$5595**
Boxlock. Steel or dural receiver. Split extractor or ejector for shotgun bbls. Double triggers. Gauges: 12, 16, 20; latter w/either 2.75- or 3-inch chambers. Calibers: 243, 6.5×57r5, 7×57r5, 7×65r5, 30-06; other calibers available. 25-inch bbls. w/solid rib, folding leaf rear sight, post or bead front sight; rifle bbl. soldered or free floating. Weight: 6.6 to 7.5 lbs. depending on type of receiver, ga. And caliber. Checkered pistol-grip stock w/cheekpiece and forearm of figured walnut, sling swivels. Made 1953 to date.

Krieghoff Ulm Over/Under Rifle-Shotgun **$8750**
Same general specifications as Teck model, except has sidelocks w/leaf arabesque engraving. Made from 1963 to date. *Note:* This combination gun is similar in appearance to the same model shotgun.

Krieghoff Ulm Over/Under Shotgun **$7895**
Same general specifications as Teck model, except has sidelocks w/leaf arabesque engraving. Made 1958 to date.

Krieghoff ULM-P Live Pigeon Gun
Sidelock. Gauge: 12. 28- and 30-inch bbls. Chokes: F/IM. Weight: 8 lbs. Oil-finished, fancy English walnut stock w/semibeavertail forearm. Light scrollwork engraving. Tapered, vent rib. Made from 1983 to date.

Bavaria	**$12,500**
Standard	**9,350**

Krieghoff Ulm-Primus Over/Under **$9025**
Deluxe version of Ulm model; detachable sidelocks, higher grade engraving and fancier wood. Made 1958 to date.

Krieghoff Ulm-Primus O/U Rifle-Shotgun **$9255**
Deluxe version of Ulm model; has detachable sidelocks, higher grade engraving and fancier wood. Made 1963 to date. *Note:* This combination gun is similar in appearance to the same model shotgun.

Krieghoff ULM-S Skeet Gun
Sidelock. Gauge: 12. Bbl.: 28-inch. Chokes: skeet/skeet. Other specifications similar to the Model ULM-P. Made 1983-86.

Bavaria	**$7595**
Standard	**5995**

Krieghoff ULM-P O/U Live Trap Gun
Over/under sidelock. Gauge: 12. 30-inch bbl. Tapered vent rib. Chokes: IM/F; optional screw-in choke. Custom grade versions command a higher price. Discontinued 1986.

Bavaria	**$12,250**
Standard	**9,350**

Krieghoff Ultra O/U Rifle-Shotgun
Deluxe Over/Under combination w/25-inch vent-rib bbls. Chambered 12 ga. only and various rifle calibers for lower bbl. Kickspanner design permits cocking w/thumb safety. Satin receiver. Weight: 6 lbs. Made from 1985 to date.

Ultra O/U Combination	**$2595**
Ultra B w/Selective Front Trigger	**2795**

SHOTGUNS

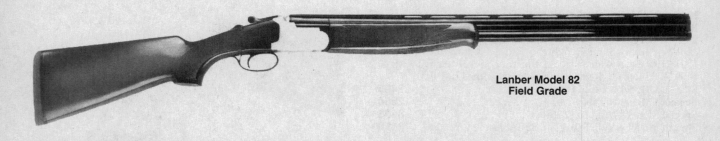

**Lanber Model 82
Field Grade**

LANBER SHOTGUNS
Spain

Lanber Model 82 O/U Shotgun **$425**
Boxlock. Gauge: 12 or 20; 3-inch chambers. 26- or 28-inch vent-rib bbls. w/ejectors and fixed chokes. Weight: 7 lbs. 2 oz. Double or single-selective trigger. Engraved silvered receiver. Checkered European walnut stock and forearm. Imported 1994.

Lanber Model 87 Deluxe **$625**
Over/Under; boxlock. Single selective trigger. 12 or 20 gauge w/3-inch chambers. Barrels: 26- or 28-inch w/choke tubes. Silvered engraved receiver. Imported 1994 only.

Lanber Model 97 Sporting Clays **$650**
Over/Under; boxlock. Single selective trigger. 12 ga. w/2.75-inch chambers. Bbls: 28-inch w/choke tubes. European walnut stock, forend. Engraved receiver. Imported 1994 only.

Lanber Model 844 MST Magnum O/U **$335**
Field grade. Gauge: 12. 3-inch Mag. chambers. 30-inch flat vent-rib bbls. Chokes: M/F. Weight: 7 lbs. 7 oz. Single selective trigger. Blued bbls. and engraved receiver. European walnut stock w/hand-checkered pistol grip and forend. Imported 1984-86.

Lanber Model 2004 LCH O/U **$475**
Field grade. Gauge: 12. 2.75-inch chambers. 28-inch flat vent-rib bbls. 5 interchangeable choke tubes: Cyl, IC, M, IM, F. Weight: about 7 lbs. Single selective trigger. Engraved silver receiver w/fine-line scroll. Walnut stock w/checkered pistol grip and forend. Rubber recoil pad. Imported 1984-86.

Lanber Model 2004 LCH O/U Skeet **$595**
Same as Model 2004 LCH except 28-inch bbls. w/5 interchangeable choke tubes. Imported 1984-86.

Lanber Model 2004 LCH O/U Trap **$585**
Gauge: 12. 30-inch vent-rib bbls. 3 interchangeable choke tubes: M, IM, F. Manual safety. Other specifications same as Model 2004 LCH O/U. Imported 1984-86.

CHARLES LANCASTER
London, England

**Lancaster "Twelve-Twenty" Double-Barrel
Shotgun** **$12,250**
Sidelock, self-opener. Gauge: 12. Bbls.: 24 to 30 inches standard. Weight: about 5.75 lbs. Elaborate metal engraving. Highest quality English or French walnut buttstock and forearm. Imported by Stoeger in the 1950s.

JOSEPH LANG & SONS
London, England

JLS Highest Quality Over/Under Shotgun **$21,600**
Sidelock. Gauges: 12, 16, 20, 28 and .410. Bbls.: 25 to 30 inches standard. Highest grade English or French walnut buttstock and forearm. Selective single trigger. Imported by Stoeger in 1950s.

LAURONA SHOTGUNS
Eibar, Spain

Laurona Grand Trap Combo
Same general specifications as Model 300, except supplied w/29-inch over/under bbls., screw-in choke tubes and 34-inch single barrel. Discontinued 1992.
Model GTO (top single) **$1695**
Model GTU (bottom single) 1795
Extra Field O/U bbls. (12 or 20 ga.), **add** 625

Laurona Silhouette 300 Over/Under
Boxlock. Single selective trigger. Selective automatic ejectors. Gauge: 12; 2.75-, 3- or 3.5-inch chambers. 28- or 29-inch vent-rib bbls. w/flush or knurled choke tubes. Weight: 7.75 to 8 lbs. Checkered pistol-grip European walnut stock and beavertail forend. Engraved receiver w/silvered finish and black chrome bbls. Made 1988-92.
Model 300 Sporting Clays . . . . $ 925
Model 300 Trap 950
Model 300 Trap, Single 975
Model 300 Ultra-Magnum 900

**Laurona
Grand Trap — GTO**

Laurona Super Model Over/Under Shotguns

Boxlock. Single selective or twin single triggers. Selective automatic ejectors. Gauges: 12 or 20; 2.75- or 3-inch chambers.26-, 28- or 29-inch vent-rib bbls. w/fixed chokes or screw-in choke tubes. Weight: 7 to 7.25 lbs. Checkered pistol-grip European walnut stock. Engraved receiver w/silvered finish and black chrome bbls. Made from 1985 to date.

Model 82 Super Game (disc.).	$ 695
Model 83 MG Super Game	725
Model 84 S Super Trap	950
Model 85 MG Super Game	735
Model 85 MG 2-Bbl. Set	1325
Model 85 MS Special Sporting (disc.)	925
Model 85 MS Super Trap	975
Model 85 MS Pigeon	995
Model 85 S Super Skeet	995

LEBEAU-COURALLY SHOTGUNS
Belgium

Lebeau-Courally Boxlock Side-by-Side

Shotguns $13,950
Gauges: 12, 16, 20 and 28. 26- to 30-inch bbls. Weight: 6.5 lbs. average. Checkered, hand-rubbed, oil-finished, straight-grip stock of French walnut. Classic forend. Made from 1986 to date.

LEFEVER ARMS COMPANY
Syracuse and Ithaca, N.Y.

Lefever sidelock hammerless double-barrel shotguns were made by Lefever Arms Company of Syracuse, New York, from about 1885-1915 (serial numbers 1 to 70,000) when the firm was sold to Ithaca Gun Company of Ithaca, New York. Production of these models was continued at the Ithaca plant until 1919 (serial numbers 70,001 to 72,000). Grades listed are those that appear in the last catalog of the Lefever Gun Company, Syracuse. In 1921, Ithaca introduced the boxlock Lefever Nitro Special double, followed in 1934 by the Lefever Grade A; there also were two single barrel Lefevers made from 1927-42. Manufacture of Lefever brand shotguns was disc. in 1948. Note: "New Lefever" boxlock shotguns made circa 1904-06 by D. M. Lefever Company, Bowling Green, Ohio, are included in a separate listing.

Lefever A Grade
Hammerless Double-Barrel Shotgun

Lefever A Grade Hammerless Double-Barrel Shotgun

Boxlock. Plain extractors or auto ejector. Single or double triggers. Gauges: 12, 16, 20, .410. Bbls.: 26-32 inches, standard chokes. Weight: about 7 lbs. in 12 ga. Checkered pistol-grip stock and forearm. Made 1934-42.

W/Plain Extractors, Double Triggers	$775
Extra for Automatic Ejector	250
Extra for Single Trigger	100
Extra for Beavertail Forearm	75

Lefever A Grade Skeet Model. $1095

Same as A Grade, except standard features include auto ejector, single trigger, beavertail forearm; 26-inch bbls., skeet boring. Discontinued 1942.

Lefever Hammerless Single-Shot Trap Gun $475

Boxlock. Ejector. 12 ga. only. 30- or 32-inch bbl.; vent rib. Weight: about 8 lbs. Checkered pistol-grip stock and forend, recoil pad. Made 1927-42.

Lefever Long Range Hammerless
Single-Barrel Field Gun $275

Boxlock. Plain extractor. Gauges: 12, 16, 20, .410. Bbl. lengths: 26-32 inches. Weight: 5.5 to 7 lbs. depending on ga. and bbl. length. Checkered pistol-grip stock and forend. Made 1927-42.

Lefever Nitro Special Hammerless Double

Boxlock. Plain extractors. Single or double triggers. Gauges: 12, 16, 20, .410. Bbls.: 26- to 32-inch, standard chokes. Weight: about 7 lbs. in 12 ga. Checkered pistol-grip stock and forend. Made 1921-48.

W/Double Triggers	$395
W/Single Trigger	495

Lefever Sidelock Hammerless Doubles

Plain extractors or auto ejectors. Double triggers or selective single trigger. Gauges: 10, 12, 16, 20. Bbls.: 26-32 inches; standard choke combinations. Weight: 5.75 to 10.5 lbs. depending on ga. and bbl. length. Checkered walnut straight-grip or pistol-grip stock and forearm. Grades differ chiefly in quality of workmanship, engraving, wood, checkering, etc.; general specifications are the same. DS and DSE Grade guns lack the cocking indicators found on all other models. Suffix "E" means model has auto ejector; also standard on A, AA, Optimus, and Thousand Dollar Grade guns. *See* Illustrations Next Page.

H Grade	$ 1,450
HE Grade	1,925
G Grade	1,695
GE Grade	2,350
F Grade	1,795
FE Grade	2,575
E Grade	2,450
EE Grade	3,395
D Grade	2,725
DE Grade	3,995
DS Grade	1,125
DSE Grade	1,395
C Grade	4,150
CE Grade	6,995
B Grade	4,995
BE Grade	8,550
A Grade	16,250
AA Grade	22,000
Optimus Grade	28,500
Thousand Dollar Grade	42,500
Extra for single trigger	650

D. M. LEFEVER COMPANY
Bowling Green, Ohio

In 1901, D. M. "Uncle Dan" Lefever, founder of the Lefever Arms Company, withdrew from that firm to organize D. M. Lefever, Sons & Company (later D. M. Lefever Company) to manufacture the "New Lefever" boxlock double- and single-barrel shotguns. These were produced at Bowling Green, Ohio, from about 1904-1906, when Dan Lefever died and the factory closed permanently. Grades listed are those that appear in the last catalog of D. M. Lefever Co.

SHOTGUNS

Lefever Sidelock
AA Grade

Lefever Sidelock
DE Grade

Lefever Sidelock
Thousand Dollar Grade

Lefever Sidelock
Sideplate BE Grade

Lefever Sidelock
Sideplate CE Grade

Lefever Sidelock
Optimus Grade

D. M. Lefever Hammerless Double-Barrel Shotguns

"New Lefever." Boxlock. Auto ejector standard on all grades except O Excelsior, which was regularly supplied w/plain extractors (auto ejector offered as an extra). Double triggers or selective single trigger (latter standard on Uncle Dan Grade, extra on all others). Gauges: 12, 16, 20. Bbls.: any length and choke combination. Weight: 5.5 to 8 lbs. depending on ga. and bbl. length. Checkered walnut straight-grip or pistol-grip stock and forearm. Grades differ chiefly in quality of workmanship, engraving, wood, checkering, etc.; general specifications are the same.

O Excelsior Grade w/plain extractors . . . . $ 2,550
O Excelsior Grade w/automatic ejectors.   2,895
No. 9, F Grade.   3,595
No. 8, E Grade.   4,375
No. 6, C Grade   5,195
No. 5, B Grade   6,895
No. 4, AA Grade 10,000
Uncle Dan Grade 15,000
Extra for single trigger   400

D. M. Lefever Single-Barrel Trap Gun $4995

Boxlock. Auto ejector. 12 ga. only. Bbls.: 26- to 32 inches, F choke. Weight: 6.5 to 8 lbs., depending on bbl. length. Checkered walnut pistol-grip stock and forearm.

MAGTECH SHOTGUNS
San Antonio, Texas
Mfd. By CBC in Brazil

Magtech Model 586.2 Slide-Action Shotgun

Gauge: 12; 3-inch chamber. 19-, 26- or 28-inch bbl.; fixed chokes or integral tubes. 46.5 inches overall. Weight: 8.5 lbs. Double-action slide bars. Brazilian hardwood stock. Polished blued finish. Imported 1992 to date.

Model 586.2 F (28-inch Bbl., Fixed Choke) **$175**
Model 586.2 P (19-inch Plain Bbl., Cyl. Bore) **185**
Model 586.2 S (24-inch Bbbl., Rifle Sights, Cyl. Bore **180**
Model 586.2 VR (Vent Rib w/Tubes) **195**

MARLIN FIREARMS CO.
North Haven (formerly New Haven), Conn.

Marlin Model 16 Visible Hammer Slide-Action Repeater

Takedown. 16 ga. 5-shell tubular magazine. Bbls.: 26- or 28-inch, standard chokes. Weight: about 6.25 lbs. Pistol-grip stock, grooved slide handle; checkering on higher grades. Difference among grades is in quality of wood, engraving on Grades C and D. Made 1904-10.

Grade A . . . . **$ 325**
Grade B **450**
Grade C **550**
Grade D **1125**

**Magtech
Model 586.2 Slide-Action**

Marlin Model 17 Brush Gun **$295**
Same as Model 17 Standard, except has 26-inch bbl., cylinder bore. Weight: about 7 lbs. Made 1906-08.

Marlin Model 17 Riot Gun **$315**
Same as Model 17 Standard, except has 20-inch bbl., cylinder bore. Weight: about 6.88 lbs. Made 1906-08.

Marlin Model 17 Standard Visible Hammer Slide-Action Repeater **$295**
Solid frame.12 ga. 5-shot tubular magazine. Bbls.: 30- or 32-inch, F choke. Weight: about 7.5 lbs. Straight-grip stock, grooved slide handle. Made 1906-08.

Marlin Model 19 Visible Hammer Slide-Action Repeater
Similar to Model 1898, but improved, lighter weight, w/two extractors, matted sighting groove on receiver top. Weight: about 7 lbs. Made 1906-07.
Grade A . . . . **$ 285**
Grade B **395**
Grade C **535**
Grade D **1095**

Marlin Model 21 Trap Visible Hammer Slide-Action Repeater
Similar to Model 19 w/same general specifications, except has straight-grip stock. Made 1907-09.
Grade A . . . . **$ 295**
Grade B **435**
Grade C **550**
Grade D **1050**

Marlin Model 24 Visible Hammer Slide-Action Repeater
Similar to Model 19, but has improved takedown system and auto recoil safety lock, solid matted rib on frame. Weight: about 7.5 lbs. Made 1908-15.
Grade A . . . . **$ 275**
Grade B **455**
Grade C **575**
Grade D **1195**

Marlin Model 26 Brush Gun **$250**
Same as Model 26 Standard, except has 26-inch bbl., cylinder bore. Weight: about 7 lbs. Made 1909-15.

SHOTGUNS

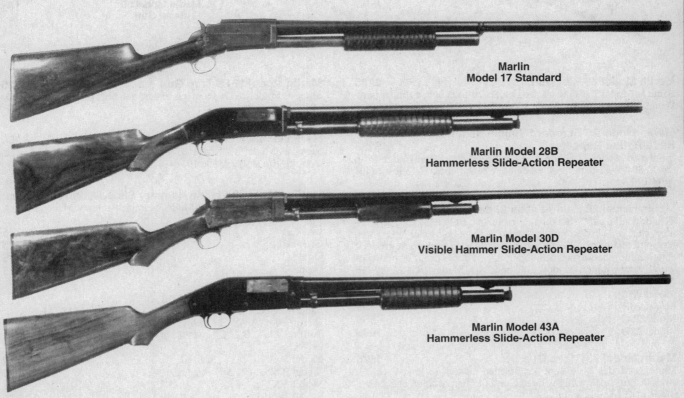

**Marlin
Model 17 Standard**

**Marlin Model 28B
Hammerless Slide-Action Repeater**

**Marlin Model 30D
Visible Hammer Slide-Action Repeater**

**Marlin Model 43A
Hammerless Slide-Action Repeater**

Marlin Model 43T
Hammerless Slide-Action Repeater

Marlin Model 53
Hammerless Slide-Action

Marlin Model 55
Goose Gun

Marlin Model 55
Hunter Bolt-Action Repeater

Marlin Model 55
Swap Gun

Marlin Model 26 Riot Gun **$225**
Same as Model 26 Standard, except has 20-inch bbl., cylinder bore.
Weight: about 6.88 lbs. Made 1909-15.

Marlin Model 26 Standard Visible Hammer
Slide-Action Repeater **$235**
Similar to Model 24 Grade A, except solid frame and straight-grip
stock. 30- or 32-inch full choke bbl. Weight: about 7.13 lbs. Made
from 1909-15.

Marlin Model 28 Hammerless Slide-Action Repeater
Takedown. 12 ga. 5-shot tubular magazine. Bbls.: 26-, 28-, 30-, 32-
inch, standard chokes; matted-top bbl. except on Model 28D which
has solid matted rib. Weight: about 8 lbs. Pistol-grip stock, grooved
slide handle; checkering on higher grades. Grades differ in quality of
wood, engraving on Models 28C and 28D. Made 1913-22; all but
Model 28A discont. in 1915. *See* Illustration Previous Page.
Model 28A . . . . **$ 295**
Model 28B **425**
Model 28C **550**
Model 28D **1095**

Marlin Model 28T Trap Gun **$525**
Same as Model 28, except has 30-inch matted-rib bbl., F choke,
straight-grip stock w/high-fluted comb of fancy walnut, checkered.
Made in 1915.

Marlin Model 28TS Trap Gun **$325**
Same as Model 28T, except has matted-top bbl., plainer stock. Made
in 1915.

Marlin Model 30 Field Gun **$250**
Same as Model 30 Grade B, except has 25-inch bbl., M choke,
straight-grip stock. Made 1913-14.

Marlin Model 30 Visible Hammer Slide-Action Repeater
Similar to Model 16, but w/Model 24 improvements. Made 1910-14.
See Illustration Previous Page.
Grade A . . . . **$ 275**
Grade B **425**
Grade C **545**
Grade D **1225**

Marlin Models 30A, 30B, 30C, 30D
Same as Model 30; designations were changed in 1915. Also avail-
able in 20 ga. w/25- or 28-inch bbl., matted-top bbl. on all grades.
Suffixes "A," "B," "C" and "D" correspond to former grades. Made
in 1915.
Model 30A . . . . **$ 320**
Model 30B **400**
Model 30C **595**
Model 30D **1075**

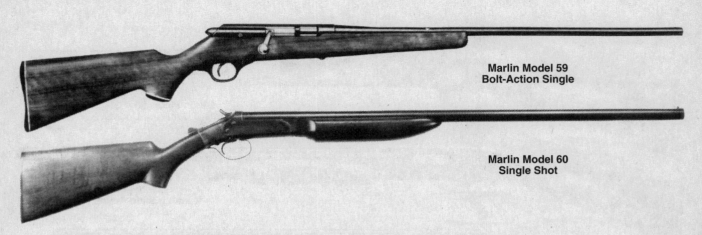

**Marlin Model 59
Bolt-Action Single**

**Marlin Model 60
Single Shot**

Marlin Model 31 Hammerless Slide-Action Repeater
Similar to Model 28, except scaled down for 16 and 20 ga. Bbls.: 25-inch (20 ga. only), 26-inch (16 ga. only), 28-inch, all w/matted top, standard chokes. Weight: 16 ga., about 6.75 lbs.; 20 ga., about 6 lbs. Pistol-grip stock, grooved slide handle; checkering on higher grades; straight-grip stock optional on Model 31D. Made 1915-17; Model 31A until 1922.

Model 31A	$ 315
Model 31B	425
Model 31C	550
Model 31D	1200

Marlin Model 31F Field Gun $375
Same as Model 31B, except has 25-inch bbl., M choke, straight- or pistol-grip stock. Made 1915-17.

Marlin Model 42A Visible Hammer Slide-Action Repeater $245
Similar to pre-World War I Model 24 Grade A w/same general specifications, but not of as high quality. Made from 1922-34.

Marlin Model 43 Hammerless Slide-Action Repeater
Similar to pre-World War I Models 28A, 28T and 28TS, w/same general specifications, but not of as high quality. Made 1923-30.

Model 43A	$250
Model 43T	495
Model 43TS	525

Marlin Model 44 Hammerless Slide-Action Repeater
Similar to pre-World War I Model 31A, w/same general specifications, but not of as high quality. 20 ga. only. Model 44A is a standard grade field gun. Model 44S Special Grade has checkered stock and slide handle of fancy walnut. Made 1923-35.

Model 44A	$325
Model 44S	395

Marlin Model 49 Visible Hammer Slide-Action Repeating Shotgun $375
Economy version of Model 42A, offered as a bonus on the purchase of four shares of Marlin stock. About 3000 were made 1925-28.

Marlin Model 50DL Bolt Action Shotgun $245
Gauge: 12 w/3-inch chamber. 2-shot magazine. 28-inch bbl. w/modified choke. 48.75 inches overall. Weight: 7.5 lbs. Checkered black synthetic stocks w/ventilated rubber recoil pad. Made 1997 to date.

Marlin Model 53 Hammerless Slide-Actlon Repeater $295
Similar to Model 43A, w/same general specifications. Made 1929-30.

Marlin Model 55 Goose Gun
Same as Model 55 Hunter, except chambered for 12-ga. 3-inch Magnum shell, has 36-inch bbl., F choke, swivels and sling. Weight: about 8 lbs. Walnut stock (standard model) or checkered black synthetic stock w/ventilated rubber recoil pad (GDL model). Made 1962 to date.

Model 55 Goose Gun	$175
Model 55GDL Goose Gun (Intro. 1997)	285

Marlin Model 55 Hunter Bolt-Action Repeater
Takedown. Gauges: 12, 16, 20. 2-shot clip magazine. 28-inch bbl. (26-inch in 20 ga.), F or adj. choke. Plain pistol-grip stock; 12 ga. has recoil pad. Weight: about 7.25 lbs.; 20 ga., 6.5 lbs. Made 1954-65.

W/plain bbl.	$ 65
W/adj. choke	80

Marlin Model 55 Swamp Gun $85
Same as Model 55 Hunter except chambered for 12-ga. 3-inch Magnum shell, has shorter 20.5-inch bbl. w/adj. choke, sling swivels and slightly better-quality stock. Weight: about 6.5 lbs. Made 1963-65.

Marlin Model 55S Slug Gun $115
Same as Model 55 Goose Gun, except has 24-inch bbl., cylinder bore, rifle sights. Weight: about 7.5 lbs. Made 1974-79.

Marlin Model 59 Auto-Safe Bolt-Action Single $85
Takedown. Auto thumb safety, .410 ga. 24-inch bbl., F choke. Weight: about 5 lbs. Plain pistol-grip stock. Made 1959-61.

Marlin Model 60 Single-Shot Shotgun $190
Visible hammer. Takedown. Boxlock. Automatic ejector. 12 ga. 30- or 32-inch bbl., F choke. Weight: about 6.5 lbs. Pistol-grip stock, beavertail forearm. *Note:* Only about 600 were produced in 1923.

Marlin Model 63 Hammerless Slide-Action Repeater
Similar to Models 43A and 43T w/same general specifications. Model 63TS Trap Special is same as Model 63T Trap Gun except stock style and dimensions to order. Made 1931-35.

Model 63A	$275
Model 63T or 63TS	350

SHOTGUNS

Marlin Model 90
Standard Over-and-Under

Marlin Model 120
Magnum Slide-Action Repeater

Marlin Model 410
Lever-Action Repeater

Marlin Model 512
Slugmaster

Marlin Model 55-10
Super Goose 10

Marlin Premier Mark I
Slide-Action Repeater

Marlin Premier Mark IV

Marlin-Glenfield
Model 50 Bolt-Action Repeater

**Marocchi Conquista
Sporting Clays**

Marlin Model 90 Standard Over-and-Under Shotgun

Hammerless. Boxlock. Double triggers; non-selective single trigger was available as an extra on prewar guns except .410. Gauges: 12, 16, 20, .410. Bbls.: plain; 26-, 28- or 30-inch; chokes IC/M or M/F; bbl. design changed in 1949, eliminating full-length rib between bbls. Weight: 12 ga., about 7.5 lbs.; 16 and 20 ga., about 6.25 lbs. Checkered pistol-grip stock and forearm, recoil pad standard on prewar guns. Postwar production: Model 90-DT (double trigger), Model 90-ST (single trigger). Made 1937-58.

W/double triggers **$395**
W/single trigger 495

Marlin Model 120 Magnum Slide-Action
Repeater **$235**

Hammerless. Takedown. 12 ga. (3-inch). 4-shot tubular magazine. Bbls.: 26-inch vent rib, IC; 28-inch vent rib M choke; 30-inch vent rib, F choke; 38-inch plain, F choke; 40-inch plain, F choke; 26-inch slug bbl. w/rifle sights, IC. Weight: about 7.75 lbs. Checkered pistol-grip stock and forearm, recoil pad. Made 1971-85.

Marlin Model 120 Slug Gun **$225**

Same general specifications as Model 120 Magnum, except w/20-inch bbl. and about .5 lbs. lighter in weight. No vent rib. Adj. rear rifle sights; hooded front sight. Discontinued 1990.

Marlin Model 410 Lever-Action Repeater **$895**

Action similar to that of Marlin Model 93 rifle. Visible hammer. Solid frame. .410 ga. (2.5-inch shell). 5-shot tubular magazine. 22-or 26-inch bbl., F choke. Weight: about 6 lbs. Plain pistol-grip stock and grooved beavertail forearm. Made 1929-32.

Marlin Model 512 Slugmaster Shotgun

Bolt-action repeater. Gauge: 12; 3-inch chamber, 2-shot magazine. 21-inch rifled bbl. w/adj. open sight. Weight: 8 lbs. Walnut-finished birch stock (standard model) or checkered black synthetic stock w/ventilated rubber recoil pad (GDL model). Made 1994 to date.

Model 512 Slugmaster **$275**
Model 512DL Slugmaster (Intro. 1998) 285

Marlin Model 1898 Visible Hammer Slide-Action
Repeater

Takedown. 12 ga. 5-shell tubular magazine. Bbls.: 26-, 28-, 30-, 32-inch; standard chokes. Weight: about 7.25 lbs. Pistol-grip stock, grooved slide handle; checkering on higher grades. Difference among grades is in quality of wood, engraving on Grades C and D. Made 1898-05. *Note:* This was the first Marlin shotgun.

Grade A (Field) .... **$ 325**
Grade B 525
Grade C 695
Grade D 1650

Marlin Model 55-10 Super Goose 10 .... **$165**

Similar to Model 55 Goose Gun, except chambered for 10 ga. 3.5-inch Magnum shell, has 34-inch heavy bbl., F choke. Weight: about 10.5 lbs. Made 1976-85.

Marlin Premier Mark I Slide-Action Repeater **$125**

Hammerless. Takedown. 12 ga. Magazine holds 3 shells. Bbls.: 30-inch F choke, 28-inch M, 26-inch IC or SK choke. Weight: about 6 lbs. Plain pistol-grip stock and forearm. Made in France 1960-63.

Marlin Premier Mark II and IV

Same action and mechanism as Premier Mark I, except engraved receiver (Mark IV is more elaborate), checkered stock and forearm, fancier wood, vent rib and similar refinements. Made 1960-63.

Premier Mark II **$225**
Premier Mark IV (plain barrel) 250
Premier Mark IV (vent-rib barrel) 295

Marlin-Glenfield Model 50 Bolt-Action Repeater **$65**

Similar to Model 55 Hunter, except chambered for 12-or 20-ga., 3-inch Magnum shell; has 28-inch bbl. in 12 ga., 26-inch in 20 ga., F choke. Made 1966-74.

Marlin-Glenfield 778 Slide-Action Repeater **$185**

Hammerless. 12 ga. 2.75-inch or 3-inch. 4-shot tubular magazine. Bbls.: 26-inch IC, 28-inch M, 30-inch F, 38-inch MXR, 20-inch slug bbl. Weight: 7.75 lbs. Checkered pistol grip. Made from 1979-84.

MAROCCHI SHOTGUNS
Brescia, Italy
Imported by Precision Sales International of
Westfield, MA

Marocchi Conquista Model O/U Shotgun

Boxlock. Gauge: 12; 2.75-inch chambers. 28-, 30- or 32-inch vent-rib bbl. Fixed choke or internal tubes. 44.38 to 48 inches overall. Weight: 7.5 to 8.25 lbs. Adj. single-selective trigger. Checkered American walnut stock w/recoil pad. Imported since 1994.

Lady Sport Garde I **$1295**
Lady Sport Grade II 1550
Lady Sport Grade III 2795
Skeet Model Grade I 1275
Skeet Model Grade II 1595
Skeet Model Grade III 2450
Sporting Clays Grade I 1275
Sporting Clays Grade II 1495
Sporting Clays Grade III 2295
Trap Model Grade I 1325
Trap Model Grade II 1575
Trap Model Grade III 2450
Left-Handed Model, **add** 10%

MAVERICK ARMS, INC.
Eagle Pass, Texas

Maverick Model 60 Autoloading Shotgun

Gauge: 12, 2.75- or 3-inch chamber. 5-round capacity. Bbls.: Magnum or non-Magnum; 24- or 28-inch w/fixed choke or screw-in tubes, plain or vent rib; blued. Weight: 7.25 lbs. Black synthetic buttstock and forend. Announced in 1993, but not produced.

Standard Model .... SRP **$279**
Combo Model w/extra 18.5" bbl. .... SRP 312
Turkey/Deer Model (w/Ghost Ring sights) .... SRP 324

SHOTGUNS

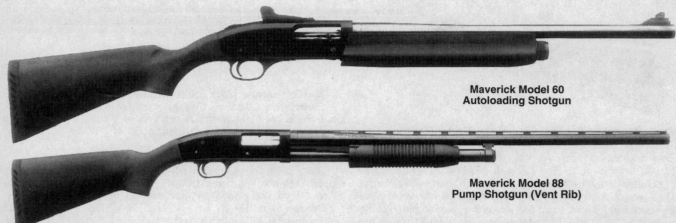

Maverick Model 60
Autoloading Shotgun

Maverick Model 88
Pump Shotgun (Vent Rib)

Maverick Model 88 Bullpup $225
Gauge: 12; 3-inch chamber. Bbl.: 18.5-inch w/fixed choke, blued. Weight: 9.5 lbs. Dual safeties: grip style and crossbolt. Fixed sights in carrying handle. High-impact black synthetic stock; trigger-forward bullpup configuration w/twin pistol-grip design. Made 1991-94.

Maverick Model 88 Deer Gun $145
Crossbolt safety and dual slide bars. Cylinder bore choke. Gauge: 12 only w/3-inch chamber. Bbl.: 24-inch. Weight: 7 lbs. Synthetic stock and forearm.

Maverick Model 88 Pump Shotgun
Gauge: 12; 2.75- or 3-inch chamber. Bbl.: 28 inches/M or 30 inches/F w/fixed choke or screw-in integral tubes; plain or vent rib, blued. Weight: 7.25 lbs. Bead front sight. Black synthetic or wood buttstock and forend; forend grooved. Made from 1989 to date.
Synthetic stock w/plain bbl. $150
Synthetic stock w/vent-rib bbl. 160
Synthetic Combo w/18.5" bbl. 180
Wood stock w/vent-rib bbl./tubes 175
Wood Combo w/vent-rib bbl./tubes 215

Maverick Model 88 Security $135
Crossbolt safety and dual slide bars. Optional heat shield. Cylinder bore choke. Gauge: 12 only w/3-inch chamber. Bbl.: 18.5-inches. Weight: 6 lbs., 8 ozs. Synthetic stock and forearm. Made 1993 to date.

Maverick Model 91 Pump Shotgun
Same as Model 88, except w/2.75-, 3- or 3.5-inch chamber, 28-inch bbl. W/ACCU-F choke, crossbolt safety and synthetic stock only. Made 1991-95.
Synthetic stock w/plain bbl. $185
Synthetic stock w/vent-rib bbl. 210

Maverick Model 95 Bolt Action $125
Modified, fixed choke. Built-in 2-round magazine. Gauge: 12 only. Bbl.: 25-inch. Weight: 6.75 lbs. Bead sight. Synthetic stock and rubber recoil pad. Made 1995 to date.

GEBRÜDER MERKEL
Suhl, Germany
Mfd. by Suhler Jagd-und Sportwaffen GmbH
Imported by GSI, Inc., Trussville, AL
(Previously by Armes de Chasse)

Merkel Model 8 Hammerless Double $895
Anson & Deeley boxlock action w/Greener double-bbl. hook lock. Double triggers. Extractors. Automatic safety. Gauges: 12, 16, 20; 2.75- or 3-inch chambers. 26-or 28-inch bbls. w/fixed standard chokes. Checkered European walnut stock, pistol-grip or English-style w/or w/o cheekpiece. Scroll engraved receiver w/tinted marble finish.

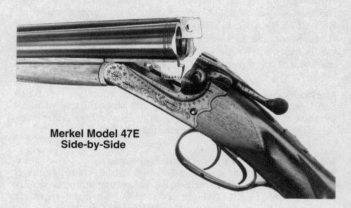

Merkel Model 47E
Side-by-Side

Merkel Side-by-Side Model 47E $1595
Boxlock; hammerless. Automatic ejectors; cocking indicators. Double hook bolting. Double triggers or single selective trigger. 12 or 20 ga. w/2.75-inch chambers. Standard bbl lengths, choke combos. Hand-checkered European walnut stock, forearm; pistol grip and cheekpiece or straight English style; sling swivels. Still in production.

Merkel Model 47LSC Sporting Clays S/S $2195
Anson & Deeley boxlock action w/single-selective adj. trigger, cocking indicators and manual safety. Gauge: 12; 3-inch chambers. 28-inch bbls.w/Briley choke tubes and H&H style ejectors. Weight: 7.25 lbs. Color cashardened receiver w/Arabesque engraving. Checkered select-grade walnut stock, beavertail forearm. Imported 1993-94.

Merkel Models 47S, 147S, 247S, 347S, 447S
Hammerless Sidelocks
Same general specifications as Model 147E except has sidelocks engraved w/arabesques, borders, scrolls or game scenes in varying degrees of elaborateness.
Model 47S $3595
Model 147S 4525
Model 247S 3995
Model 347S 4250
Model 447S 4595

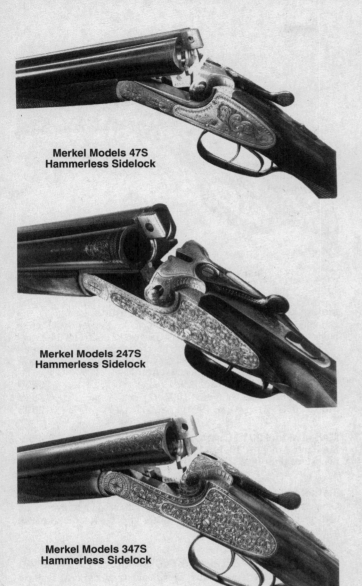

**Merkel Models 47S
Hammerless Sidelock**

**Merkel Models 247S
Hammerless Sidelock**

**Merkel Models 347S
Hammerless Sidelock**

Merkel Model 100 Over/Under Shotgun

Hammerless. Boxlock. Greener crossbolt. Plain extractor. Double triggers. Gauges: 12, 16, 20. Made w/plain or ribbed bbls. in various lengths and chokes. Plain finish no engraving. Checkered forend and stock w/pistol grip and cheekpiece or English style. Made prior to WWII.

W/plain bbl. $1295
W/ribbed bbl. 1350

Merkel Models 101 and 101E Over/Unders

Same as Model 100, except ribbed bbl. standard, has separate extractors (ejectors on Model 101E), English engraving. Made prior to World War II.

Model 101 .... $1395
Model 101E 1450

**Merkel Model 122
Hammerless Double**

Merkel Model 122 Hammerless Double $3525

Similar to the Model 147S except w/nonremovable sidelocks, in gauges 12, 16 or 20. Imported since 1993.

Merkel Model 122E Hammerless Sidelock $2595

Similar to the Model 122 except w/removable sidelocks and cocking indicators. Importation. Discontinued 1992.

Merkel Model 127 Hammerless Sidelock $22,950

Holland & Holland system, hand-detachable locks. Auto ejectors. Double triggers. Made in all standard gauges, bbl. lengths and chokes. Checkered forend and pistol-grip stock; available w/cheekpiece or English style buttstock. Elaborate engraving. Made prior to WW II.

Merkel Model 130 Hammerless Boxlock
Double $9750

Anson & Deeley system. Sideplates. Auto ejectors. Double triggers. Elaborate hunting scene or arabesque engraving. Made in all standard gauges, various bbl. lengths and chokes. Checkered forend and stock w/pistol grip and cheekpiece or English style. Made prior to WW II.

Merkel Model 147E Hammerless Boxlock
Double-Barrel Shotgun $2250

Anson & Deeley system. Auto ejectors. Double triggers. Gauges: 12, 16, 20 (3-inch chambers available in 12 and 20 ga.). Bbls.: 26-inch standard, other lengths available; any standard choke combination. Weight: about 6.5 lbs. Checkered straight-grip stock and forearm. Discontinued 1989. *See* Illustration Next Page.

SHOTGUNS

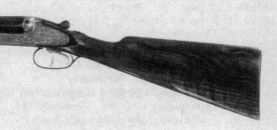

**Merkel Models 47LSC
Sporting Clay**

**Merkel Model 147E Hammerless Boxlock
Double-Barrel Shotgun**

**Merkel Model 200E
Over and Under Shotgun**

**Merkel Model 203E
Sidelock Over and Under Shotgun**

Merkel Models 200, 200E, 201, 201E, 202 and 202E Over/Under Shotguns

Hammerless. Boxlock. Kersten double crossbolt. Scalloped frame. Sideplates on Models 202 and 202E. Arabesque or hunting engraving supplied on all except Models 200 and 200E."E" models have ejectors, others have separate extractors. Signal pins. Double triggers. Gauges: 12, 16, 20, 24, 28, 32 (last three not available in post-war guns). Ribbed bbls. in various lengths and chokes. Weight: 5.75 to 7.5 lbs. depending on bbl. length and gauge. Checkered forend and stock w/pistol grip and cheekpiece or English style. The 200, 201, and 202 differ in overall quality, engraving, wood, checkering, etc.; aside from the faux sideplates on Models 202 and 202E, general specifications are the same. Models 200, 201, 202, and 202E, all made before WW II, are disc. Models 201E &202E in production w/revised 2000 series nomenclature.

Model 200	$1295
Model 200E	2495
Model 200 ES Skeet	3695
Model 200ET Trap	3550
Model 200 SC Sporting Clays	3955
Model 201 (Discontinued)	1625
Model 201E (Pre-WW II)	1895
Model 201E (Post-WW II)	3695
Model 201 ES Skeet	5595
Model 201 ET Trap	5550
Model 202 (Discontinued)	2195
Model 202E (Pre-WW II)	2650
Model 202 E (Post-WW II & 2002EL)	5495

Merkel Model 203E Sidelock O/U Shotguns

Hammerless action w/hand-detachable sidelocks. Kersten double cross bolt, auto ejectors and double triggers.Gauges: 12 or 20 (16, 24.28 and 32 disc.). 26.75- or 28- inch vent-rib bbls. Arabesque engraving standard or hunting engraving optional on coin finished receiver. Checkered English or pistol grip stock and forend.

Model 203E Sidelock (Disc. 1998)	$5250
Model 203ES Skeet (Imported 1993-97)	7550
Model 203ET Trap (Disc. 1997)	7595

Merkel Model 204E Over/Under Shotgun $5250
Similar to Model 203E; has Merkel sidelocks, fine English engraving. Made prior to World War II.

Merkel Model 210E Sidelock O/U Shotgun $4645
Gauges: 12 or 20; Kersten double cross-bolt, scroll engraved, case hardened receiver, double-triggers; pistol grip stock w/cheekpiece. Made 1998 to date.

Merkel Model 211E Sidelock O/U Shotgun $5620
Same specifications as Model 210E, except w/engraved hunting scenes on silver-gray receiver.

Merkel Models 300, 300E, 301, 301E and 302 O/U

Merkel-Anson system boxlock. Kersten double crossbolt, two underlugs, scalloped frame, sideplates on Model 302. Arabesque or hunting engraving. "E" models and Model 302 have auto ejectors, others have separate extractors. Signal pins. Double triggers. Gauges: 12, 16, 20, 24, 28, 32. Ribbed bbls. in various lengths and chokes. Checkered forend and stock w/pistol grip and cheekpiece or English style. Grades 300, 301 and 302 differ in overall quality, engraving, wood, checkering, etc.; aside from the dummy sideplates on Model 302, general specifications are the same. Manufactured prior to World War II.

Model 300	$1995
Model 300E	2350
Model 301	4395
Model 301E	5450
Model 302	8495

Merkel Model 303E Over/Under Shotgun $12,250
Similar to Model 203E. Has Kersten crossbolt, double underlugs, Holland & Holland-type hand-detachable sidelocks, auto ejectors. This is a finer gun than Model 203E. Currently manufactured.

**Merkel Model 303E
Over and Under Shotgun**

Merkel Model 304E Over/Under Shotgun **$14,750**
Special version of the Model 303E-type, but higher quality throughout. This is the top grade Merkel over/under. Currently manufactured.

Merkel Models 400, 400E, 401, 401E Over/Unders
Similar to Model 101 except have Kersten douhle crossbolt, arabesque engraving on Models 400 and 400E, hunting engraving on Models 401 and 401E, finer general quality. "E" models have Merkel ejectors, others have separate extractors. Made prior to World War II.
Model 400 .... **$1450**
Model 400E **1575**
Model 401 .... **1450**
Model 401E **1725**

Merkel Over/Under Combination Guns ("Bock-Büchsflinten")
Shotgun bbl. over, rifle bbl. under. Gauges: 12,16, 20; calibers: 5.6×35 Vierling, 7×57r5, 8×57JR, 8X60R Magnum, 9.3×53r5, 9.3×72r5, 9.3×74R and others. Various bbl. lengths, chokes and weights. Other specifications and values correspond to those of Merkel over/under shotguns listed below. Currently manufactured. Model 210 & 211 series disc. 1992.
Models 410, 410E, 411E .... *see* **shotgun Models 400, 400E, 401, 401E respectively**
Models 210, 210E, 211, 211E, 212, 212E .... *see* **shotgun Models 200, 200E, 201, 201E, 202, 202E**

NOTE

Merkel over/under guns were often supplied with accessory barrels, interchangeable to convert the gun into an arm of another type; for example, a set might consist of one pair each of shotgun, rifle and combination gun barrels. Each pair of interchangeable barrels has a value of approximately one-third that of the gun with which they are supplied.

Merkel Model 2000EL Over/Under Shotguns
Kersten double cross-bolt. Gauges: 12 and 20. 26.75- or 28-inch bbls. Weight: 6.4 to 7.28 lbs. Scroll engraved silver-gray receiver. Automatic ejectors and single selective or double triggers. Checkered forend and stock w/pistol grip and cheekpiece or English-style stock w/luxury grade wood. Imported 1998 to date.
Model 2000EL Standard **$3550**
Model 2000EL Sporter **3695**

Merkel Model 2001EL Over/Under Shotguns
Gauges: 12, 16, 20 and 28; Kersten double cross-bolt lock receiver. 26.75- or 28-inch IC/Mod, Mod/Full bbls. Weight: 6.4 to 7.28 lbs. Three piece forearm, automatic ejectors and single selective or double triggers. Imported 1993 to date.
Model 2001EL 12 ga. **$4420**
Model 2001EL 16 ga. (Disc. 1997) **4420**
Model 2001EL 20 ga. **4420**
Model 2001EL 28 ga. (Made 1995) **4870**

Merkel Model 2002EL **$7500**
Same specifications as Model 2000EL, except hunting scenes w/arabesque engraving.

Merkel Anson Drillings
Three-bbl. combination guns; usually made w/double shotgun bbls., over rifle bbl., although "Doppelbüchsdrillingen" were made w/two rifle bbls. over and shotgun bbl. under. Hammerless. Boxlock. Anson & Deeley system. Side clips. Plain extractors. Double triggers. Gauges: 12, 16, 20; rifle calibers: 7×57r5, 8×57JR and 9.3×74R are most common, but other calibers from 5.6mm to 10.75mm available. Bbls.: standard drilling 25.6 inches; short drilling, 21.6 inches. Checkered pistol-grip stock and forend. The three models listed differ chiefly in overall quality, grade of wood, etc.; general specifications are the same. Made prior to WW II.
Model 142 Engraved **$3950**
Model 142 Standard **3495**
Model 145 Field **2750**

**MIIDA SHOTGUNS
Manufactured for Marubeni America Corp., New York, N.Y., by Olin-Kodensha Co., Tochigi, Japan**

Miida Model 612 Field Grade Over-and-Under **$675**
Boxlock. Auto ejectors. Selective single trigger. 12 ga. Bbls.: vent rib; 26-inch, IC/M; 28-inch, M/F choke. Weight: w/26-inch bbl., 6 lbs.11 oz. Checkered pistol-grip stock and forearm. Made 1972-74. *See* Illustration Next Page.

Miida Model 2100 Skeet Gun **$725**
Similar to Model 612, except has more elaborate engraving on frame (50 percent coverage), skeet-style stock and forearm of select grade wood; 27-inch vent-rib bbls., SK choke. Weight: 7 lbs. 11 oz. Made 1972-74.

**Miida Model 2200T Trap Gun, Model 2200S
Skeet Gun** **$795**
Similar to Model 612, except more elaborate engraving on frame (60 percent coverage), trap- or skeet-style stock and semibeavertail forearm of fancy walnut, recoil pad on trap stock. Bbls.: wide vent rib; 29.75-inch, IM/F choke on Trap Gun; 27-inch, SK choke on Skeet Gun. Weight: Trap, 7 lbs. 14 oz.; Skeet, 7 lbs. 11 oz. Made 1972-74.

**Miida Model 2300T Trap Gun,
Model 2300S Skeet Gun** **$850**
Same as Models 2200T and 2200S, except more elaborate engraving on frame (70% coverage). Made 1972-74.

SHOTGUNS

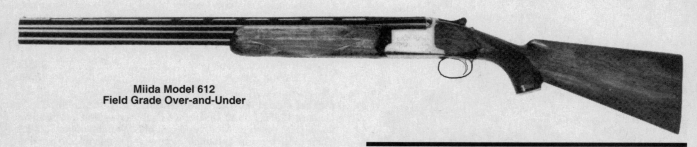

Miida Model 612
Field Grade Over-and-Under

Miida Grandee Model GRT/IRS Trap/Skeet Gun **$2095**
Boxlock w/sideplates. Frame, breech ends of bbls., trigger guard and locking lever fully engraved and gold inlaid. Auto ejectors. Selective single trigger. 12 ga. Bbls.: wide vent rib; 29-inch, F choke on Trap Gun; 27-inch, SK choke on Skeet Gun. Weight: Trap, 7 lbs. 14 oz.; Skeet, 7 lbs. 11 oz. Trap- or skeet-style stock and semibeavertail forearm of extra fancy wood, recoil pad on trap stock. Made 1972-74.

MITCHELL ARMS
Santa Ana, California

Mitchell Model 9104/9105 Pump Shotguns
Slide action in Field/Riot configuration. Gauge: 12; 5-shot tubular magazine. 20-inch bbl.; fixed choke or screw-in tubes. Weight: 6.5 lbs. Plain walnut stock. Made 1994 to date.
Model 9104 (w/Plain Bbl.) **$185**
Model 9105 (w/Rifle Sight) **195**
W/Choke Tubes, **add** **20**

Mitchell Model 9108/9109 Pump Shotgun
Slide action in Military/Police/Riot configuration. Gauge: 12, 7-shot tubular magazine. 20-inch bbl.; fixed choke or screw-in tubes. Weight: 6.5 lbs. Plain walnut stock and grooved slide handle w/brown, green or black finish. Blued metal. Made 1994 to date.
Model 9108 (w/Plain Bbl.) **$190**
Model 9109 (w/Rifle Sights) **205**
W/Choke tubes, **add** **20**

Mitchell Model 9111/9113 Pump Shotgun
Slide action in Military/Police/Riot configuration. Gauge: 12; 6-shot tubular magazine. 18.5-inch bbl.; fixed choke or screw-in tubes. Weight: 6.5 lbs. Synthetic or plain walnut stock and grooved slide handle w/brown, green or black finish. Blued metal. Made 1994 to date.
Model 9111 (w/Plain Bbl.) **$185**
Model 9113 (w/Rifle Sights) **205**
W/Choke Tubes, **add** **20**

Mitchell Model 9114/9114FS
Slide action in Military/Police/Riot configuration. Gauge: 12; 7-shot tubular magazine. 20-inch bbl.; fixed choke or screw-in tubes. Weight: 6.5-7 lbs. Synthetic pistol-grip or folding stock. Blued metal. Made 1994 to date.
Model 9114 . . . . **$225**
Model 9114FS **250**

Mitchell Model 9115/911FS Pump Shotgun **$250**
Slide action in Military/Police/Riot configuration. Gauge: 12; 6-shot tubular magazine. 18.5-inch bbl. w/heat-shield hand guard; Weight: 7 lbs. Gray synthetic stock and slide handle. Parkerized metal. Made 1994 to date.

MONTGOMERY WARD
See shotgun listings under "W"

MORRONE SHOTGUN
Manufactured by Rhode Island Arms Company
Hope Valley, RI

Morrone Standard Model 46 Over-and-Under **$725**
Boxlock. Plain extractors. Non-selective single trigger. Gauges: 12, 20. Bbls.: plain, vent rib; 26-inch IC/M; 28-inch M/F choke. Weight: about 7 lbs., 12 ga.; 6 lbs., 20 ga. Checkered straight- or pistol-grip stock and forearm. Made 1949-53. *Note:* Less than 500 of these guns were produced, about 50 in 20 ga., a few had vent-rib bbls. Value shown is for 12 ga. w/plain bbls.; the rare 20 ga. and vent-rib types should bring considerably more.

O. F. MOSSBERG & SONS, INC.
North Haven, Connecticut
Formerly of New Haven, Conn.

Mossberg Model 83D or 183D **$100**
3-shot. Takedown. .410-bore only. 2-shell fixed top-loading magazine. 23-inch bbl. w/two interchangeable choke tubes (M/F). Later production had 24-inch bbl. Plain one-piece pistol-grip stock. Weight: about 5.5 lbs. Originally designated Model 83D, changed in 1947 to Model 183D. Made 1940-71.

Mossberg Model 85D or 185D Bolt-Action
Repeating Shotgun **$85**
Takedown. 3-shot. 20 ga. only. 2-shell detachable box magazine. 25-inch bbl., three interchangeable choke tubes (F, M, IC). Later production had 26-inch bbl. w/F/IC choke tubes. Weight: about 6.25 lbs. Plain one-piece, pistol-grip stock. Originally designated Model 85D changed in 1947 to Model 185D. Made 1940-71.

Mossberg Model 183K **$95**
Same as Model 183D, except has 25-inch bbl. w/variable C-Lect-Choke instead of interchangeable choke tubes. Made 1953-86.

Mossberg Model 185K **$100**
Same as Model 185D, except has variable C-Lect-Choke instead of interchangeable choke tubes. Made 1950-63.

Mossberg Model 190D **$95**
Same as Model 185D, except in 16 ga. Weight: about 6 lbs. Made 1955-71.

Mossberg Model 190K **$105**
Takedown. 3-shot; 2-shell magazine and one shell in chamber. Same as Model 185K, except in 16 ga. Weight: about 6.75 lbs. Made 1956-63.

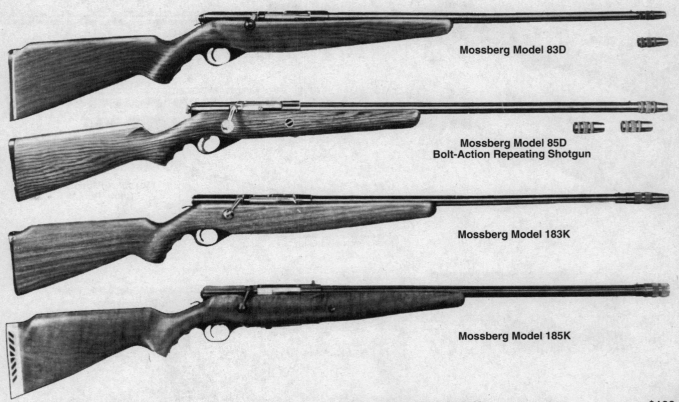

Mossberg Model 83D

Mossberg Model 85D
Bolt-Action Repeating Shotgun

Mossberg Model 183K

Mossberg Model 185K

Mossberg Model 195D **$110**
Takedown. 3-shot; 2-shell magazine and one shell in chamber. Same as Model 185D, except in 12 ga. Weight: about 6.75 lbs. Made 1955-71.

Mossberg Model 195K **$105**
Takedown. 3-shot; 2-shell magazine and one shell in chamber. Same as Model 185K, except in 12 ga. Weight: about 7.5 lbs. Made 1956-63.

Mossberg Model 200D **$115**
Same as Model 200K, except w/two interchangeable choke tubes instead of C-Lect-Choke. Made 1955-59.

Mossberg Model 200K Slide-Action Repeater **$115**
12 ga. 3-shot detachable box magazine. 28-inch bbl. C-Lect-Choke. Plain pistol-grip stock. Black nylon slide handle. Weight: about 7.5 lbs. Made 1955-59.

Mossberg Model 385K **$100**
Same as Model 395K, except 20 ga. (3-inch), 26-inch bbl. w/C-Lect-Choke. Weight: about 6.25 lbs.

Mossberg Model 390K **$115**
Same as Model 395K, except 16 ga. (2.75-inch). Made 1963-74.

Mossberg Model 395K Bolt-Action Repeater **$100**
Takedown. 3-shot (detachable-clip magazine holds two shells).12 ga. (3-inch chamber). 28-inch bbl. w/C-Lect-Choke. Weight: about 7.5 lbs. Monte Carlo stock w/recoil pad. Made 1963-83.

Mossberg Model 395S Slugster **$130**
Same as Model 395K, except has 24-inch bbl., cylinder bore, rifle sights, swivels and web sling. Weight: about 7 lbs. Made 1968-81.

SHOTGUNS

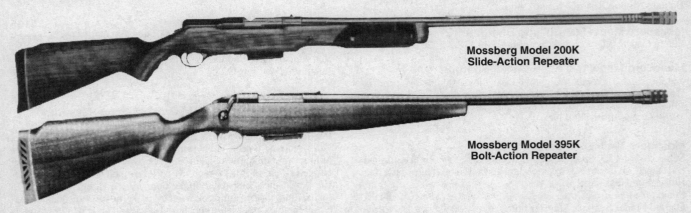

Mossberg Model 200K
Slide-Action Repeater

Mossberg Model 395K
Bolt-Action Repeater

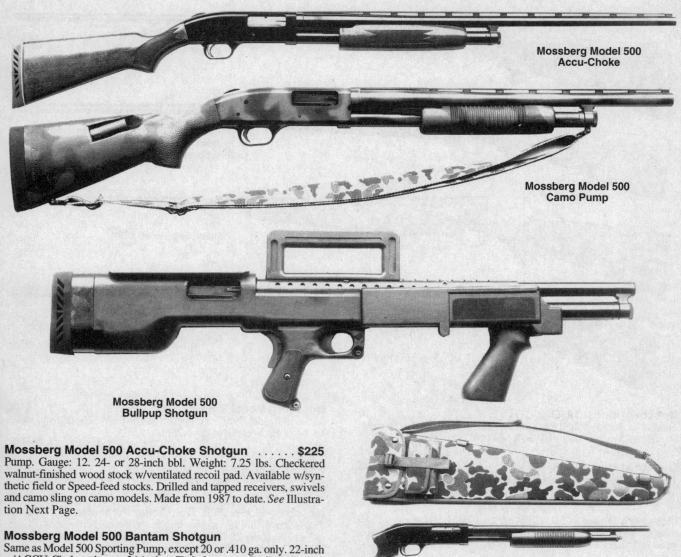

Mossberg Model 500
Accu-Choke

Mossberg Model 500
Camo Pump

Mossberg Model 500
Bullpup Shotgun

Mossberg Model 500
Camper

Mossberg Model 500 Accu-Choke Shotgun $225
Pump. Gauge: 12. 24- or 28-inch bbl. Weight: 7.25 lbs. Checkered walnut-finished wood stock w/ventilated recoil pad. Available w/synthetic field or Speed-feed stocks. Drilled and tapped receivers, swivels and camo sling on camo models. Made from 1987 to date. *See* Illustration Next Page.

Mossberg Model 500 Bantam Shotgun
Same as Model 500 Sporting Pump, except 20 or .410 ga. only. 22-inch w/ACCU-Choke tubes or 24-inch w/F choke; vent rib. Scaled-down checkered hardwood or synthetic stock w/standard or Realtree Camo finish. Made from 1990-96 and 1998-99.
Bantam Model (hardwood stock) **$175**
Bantam Model (synthetic stock) **165**
Bantam Model (Realtree Camo), **add** **50**

Mossberg Model 500 Bullpup Shotgun $395
Pump. Gauge: 12. 6- or 8-shot capacity. Bbl.: 18.5 to 20 inches. 26.5

Mossberg Model 500 Accu-Choke Shotgun *(Con't)*
and 28.5 inches overall. Weight: about 9.5 lbs. Multiple independent safety systems. Dual pistol grips, rubber recoil pad. Fully enclosed rifle-type sights. Synthetic stock. Ventilated bbl. heat shield. Made 1987-90. *See* Illustration Next Page.

Mossberg Model 500 Camo Pump
Same as Model 500 Sporting Pump, except 12 ga. only. Receiver drilled and tapped. QD swivels and camo sling. Special camouflage finish. *See* Illustration Next Page.
Standard Model **$195**
Combo Model (w/extra Slugster bbl.) **250**

Mossberg Model 500 Camper................... $195
Same general specifications as Model 500 Field Grade, except .410 bore, 6-shot magazine, 18.5-inch plain Cyl. bore bbl. Synthetic pistol grip and camo carrying case. Made 1986-90. *See* Illustration Next Page.

Mossberg Model 500 Field Grade Hammerless Slide Action Repeater
Pre-1977 type. Takedown. Gauges: 12, 16, 20, .410. 3-inch chamber (2.75-inch in 16 ga.). Tubular magazine holds five 2.75-inch shells or four three-inch. Bbls.: plain- 30-inch regular or heavy Magnum, F choke (12 ga. only); 28-inch, M or F; 26-inch, IC or adj. C-Lect-Choke; 24-inch Slugster, cylinder bore, w/rifle sights. Weight: 5.75 to lbs. Plain pistol-grip stock w/recoil pad grooved slide handle. After 1973, these guns have checkered stock and slide handle, Models

**Mossberg Model 500
Persuader Law Enforcement**

Mossberg Model 500 Field Grade *(Con't)*

500AM and 500AS have receivers etched w/game scenes. The latter has swivels and sling. Made 1962-76.

Model 500A, 12 gauge	**$185**
Model 500AM, 12 gauge, heavy Magnum bbl.	**200**
Model 500AK, 12 gauge, C-Lect-Choke	**225**
Model 500AS, 12 gauge, Slugster	**220**
Model 500B 16 gauge	**185**
Model 500BK, 16 gauge, C-Lect-Choke	**235**
Model 500BS, 16 gauge, Slugster	**225**
Model 500C 20 gauge	**195**
Model 500CK, 20 gauge, C-Lect-Choke	**240**
Model 500CS, 20 gauge, Slugster	**225**
Model 500E, .410 gauge	**245**
Model 500EK, .410 gauge, C-Lect-Choke	**275**

Mossberg Model 500 "L" Series

"L" in model designation. Same as pre-1977 Model 500 Field Grade, except not available in 16 ga., has receiver etched w/different game scenes; Accu-Choke w/three interchangeable tubes (IC, M, F) standard, restyled stock and slide handle. Bbls.: plain or vent rib; 30- or 32-inch, heavy, F choke (12 ga. Magnum and vent rib only); 28-inch, Accu-Choke (12 and 20 ga.); 26-inch F choke (.410 bore only); 18.5-inch (12 ga. only), 24-inch (12 and 20 ga.) Slugster w/rifle sights, cylinder bore. Weight: 6 to 8.5 lbs. Intro. 1977.

Model 500ALD, 12 gauge, plain bbl. (Disc. 1980)	**$210**
Model 500ALDR, 12 gauge, vent rib	**250**
Model 500ALMR, 12 ga., Heavy Duck Gun (Disc. 1980)	**250**
Model 500ALS, 12 gauge, Slugster (Disc. 1981)	**200**
Model 500CLD, 20 gauge, plain bbl. (Disc. 1980)	**195**
Model 500CLDR, 20 gauge, vent rib	**225**
Model 500CLS, 20 gauge, Slugster (Disc. 1980)	**250**
Model 500EL, .410 gauge, plain bbl. (Disc. 1980)	**215**
Model 500ELR, .410 gauge, vent rib	**250**

**Mossberg Model 500
Mariner**

Mossberg Model 500 Mariner Shotgun $265

Slide action. Gauge: 12. 18.5 or 20-inch bbl. 6-shot and 8-shot respectively. Weight: 7.25 lbs. High-strength synthetic buttstock and forend. Available in extra round-carrying speedfeed synthetic buttstock. All metal treated for protection against saltwater corrosion. Intro. 1987.

Mossberg Model 500 Muzzleloader Combo $275

Same as Model 500 Sporting Pump, except w/extra 24-inch rifled 50-caliber muzzleloading bbl. w/ram rod. Made from 1992 to date.

Mossberg Model 500 Persuader Law Enforcement

Similar to pre-1977 Model 500 Field Grade, except 12 ga. only, 6- or 8-shot, has 18.5- or 20-inch plain bbl., cylinder bore, either shotgun or rifle sights, plain pistol-grip stock and grooved slide handle, sling swivels. Special Model 500ATP8-SP has bayonet lug, Parkerized finish. Currently manufactured.

Model 500ATP6, 6-shot, 18.5-inch bbl., shotgun sights	**$185**
Model 500ATP6CN, 6-shot, nickel finish, "Cruiser" pistol grip	**195**
Model 500ATP6N, 6-shot, nickel finish, 2.75- or 3-inch Mag. shells	**190**
Model 500ATP6S, 6-shot, 18.5" bbl., rifle sights	**185**
Model 500ATP8, 8-shot, 20-inch bbl., shotgun sights	**205**
Model 500ATP8S, 8-shot, 20-inch bbl., rifle sights	**215**
Model 500ATP8-SP Special Enforcement	**250**
Model 500 Bullpup	**375**
Model 500 Security Combo Pack	**150**
Model 500 Cruiser w/pistol grip	**155**

Mossberg Model 500 Pigeon Grade

Same as Model 500 Super Grade, except higher quality w/fancy wood, floating vent rib; field gun hunting dog etching, trap and skeet guns have scroll etching. Bbls.: 30-inch, F choke (12 ga. only); 28-inch, M choke; 26-inch, SK choke or C-Lect-Choke. Made 1971-75.

Model 500APR, 12 ga., Field, Trap or Skeet	**$350**
Model 500APKR, 12 gauge, Field Gun, C-Lect-Choke	**360**
Model 500 APTR, 12 gauge, Trap Gun, Monte Carlo stock	**425**
Model 500CPR, 20 gauge, Field or Skeet Gun	**365**
Model 500EPR, .410 gauge, Field or Skeet Gun	**375**

Mossberg Model 500 Pump Combo Shotgun $245

Gauges: 12 and 20. 24- and 28-inch bbl. w/adj. rifle sights. Weight: 7 to 7.25 lbs. Available w/blued or camo finish. Drilled and tapped receiver w/sling swivels and a camo web sling. Made from 1987 to date.

Mossberg Model 500 Pump Slugster Shotgun

Gauges: 12 or 20 w/3-inch chamber. 24-inch smoothbore or rifled bbl. w/adj. rifle sights or intregral scope mount and optional muzzle break (1997 porting became standard). Weight: 7 to 7.25 lbs. Wood or synthetic stock w/standard or Woodland Camo finish. Blued, matte black or Marinecote metal finish. Drilled and tapped receiver w/camo sling and swivels. Made 1987 to date.

Slugster (w/cyl. bore, rifle sights)	**$175**
Slugster (w/rifled bore, ported)	**220**
Slugster (w/rifled bore, unported)	**195**
Slugster (w/rifled bore, ported, integral scope mount)	**245**
Slugster (w/Marinecote and synthetic stock), **add**	**275**
Slugster (w/Truglo fiber optics), **add**	**30**

Mossberg Model 500 Regal Slide-Action Repeater

Similar to regular Model 500 except higher quality workmanship throughout. Gauges: 12 and 20. Bbls.: 26- and 28-inch w/various chokes, or Accu-Choke. Weight: 6.75 to 7.5 lbs. Checkered walnut stock and forearm. Made 1985 to date.

Model 500 w/Accu-Choke	**$185**
Model 500 w/fixed choke	**170**

SHOTGUNS

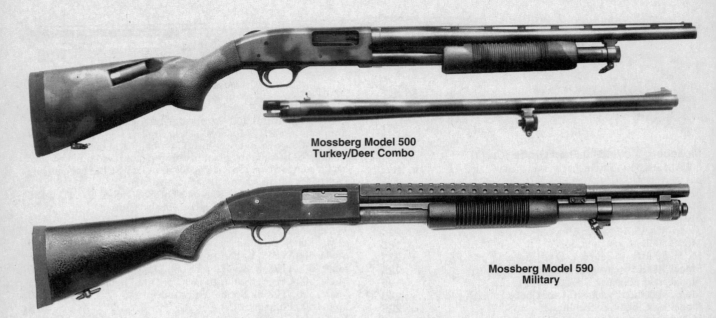

**Mossberg Model 500
Turkey/Deer Combo**

**Mossberg Model 590
Military**

Mossberg Model 500 Sporting Pump
Gauges: 12, 20 or .410, 2.75- or 3-inch chamber. Bbls.: 22 to 28 inches w/fixed choke or screw-in tubes; plain or vent rib. Weight: 6.25 to 7.25 lbs. White bead front sight, brass mid-bead. Checkered hardwood buttstock and forend w/walnut finish.
Standard Model **$195**
Field Combo (w/extra Slugster bbl.) **245**

Mossberg Model 500 Super Grade
Same as pre-1977 Model 500 Field Grade, except not made in 16 ga., has vent-rib bbl., checkered pistol grip and slide handle. Made 1965-76.
Model 500AR, 12 gauge **$195**
Model 500AMR, 12 gauge, heavy Magnum bbl. **220**
Model 500AKR, 12 gauge, C-Lect-Choke **225**
Model 500CR 20 gauge **210**
Model 500CKk, 20 gauge, C-Lect-Choke **295**
Model 500ER, .410 gauge **190**
Model 500EKR, .410 gauge, C-Lect-Choke **240**

Mossberg Model 500 Turkey/Deer Combo **$255**
Pump (slide action). Gauge: 12. 20- and 24-inch bbls. Weight: 7.25 lbs. Drilled and tapped receiver, camo sling and swivels. Adj. rifle sights and camo finish. Vent rib. Made from 1987 to date. *See* Illustration Previous Page.

Mossberg Model 500 Turkey Gun **$260**
Same as Model 500 Camo Pump, except w/24-inch ACCU-Choke bbl. w/extra full choke tube and Ghost Ring sights. Made from 1992 to date.

Mossberg Model 500 Viking Pump Shotgun
Gauges: 12 or 20 w/3-inch chamber. 24-, 26- or 28-inch bbls. available in smoothbore w/Accu-Choke and vent rib or rifled bore w/iron sights and optional muzzle break (1997 porting became standard). Optional optics: Slug Shooting System (SSS). Weight: 6.9 to 7.2 lbs. Moss-Green synthetic stock. Matte black metal finish. Made 1996-98.
Model 500 Viking (w/VR and choke tubes, unported). **$165**
Model 500 Viking (w/rifled bore, ported) **205**
Model 500 Viking (w/rifled bore, SSS and ported). **275**
Model 500 Viking (w/rifled bore, unported) **195**
Model 500 Viking Turkey Model (w/VR, tubes, ported). . . . **175**

Mossberg Model 500 Waterfowl/Deer Combo **$265**
Same general specifications as the Turkey/Deer Combo except w/either 28- or 30-inch bbl. along w/the 24-inch bbl. Made from 1987 to date.

Mossberg Model 500ATR Super Grade Trap **$290**
Same as pre-1977 Model 500 Field Grade, except 12 ga. only w/vent-rib bbl.; 30-inch F choke, checkered Monte Carlo stock w/recoil pad, beavertail forearm (slide handle). Made 1968-71.

**Mossberg Model 595
Bolt-Action Repeater**

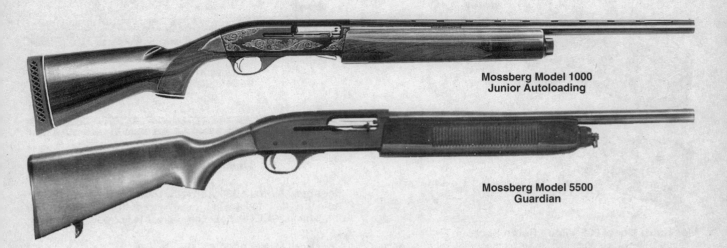

Mossberg Model 1000 Junior Autoloading

Mossberg Model 5500 Guardian

Mossberg Model 500DSPR Duck Stamp Commemorative $595
Limited edition of 1000 to commemorate the Migratory Bird Hunting Stamp Program. Same as Model 500DSPR Pigeon Grade 12-Gauge Magnum Heavy Duck Gun w/heavy 30-inch vent-rib bbl., F choke; receiver has special Wood Duck etching. Gun accompanied by a special wall plaque. Made in 1975. *Note:* Value is for gun in new, unfired condition.

Mossberg Model 590 Mariner Pump
Same general specifications as the Model 590 Military Security, except has Marinecote metal finish and field configuration synthetic stock w/pistiol-grip conversion included. Made from 1989 to date.
Model 590 Mariner (w/18.5-inch bbl.) $215
Model 590 Mariner (w/20-inch bbl.) 225
Model 590 Mariner (w/grip conversion), **add** 20
Model 590 Mariner (w/Ghost Ring sight), **add** 50

Mossberg Model 590 Military Security $275
Same general specifications as the Model 590 Military except there is no heat shield and gun has short pistol-grip style instead of buttstock. Weight: about 6.75 lbs. Made from 1987 to date.

Mossberg Model 590 Military Shotgun
Slide-action. Gauge: 12. 9-shot capacity. 20-inch bbl. Weight: about 7 lbs. Synthetic or hardwood buttstock and forend. Ventilated bbl. heat shields. Equipped w/bayonet lug. Blued or Parkerized finish. Made from 1987 to date. *See* Illustration Previous Page.
Synthetic Model, blued $265
Synthetic Model Parkerized 285
Speedfeed Model, blued 280
Speedfeed Model, Parkerized 285
Intimidator Model w/Laser Sight, blued 365
Intimidator Model w/Laser Sight, Parkerized 375
For Ghost Ring Sight, **add** 50

Mossberg Model 595 Bolt-Action Repeater $135
12 ga. only. 4-shot detachable magazine. 18.5-inch bbl. Weight: about 7 lbs. Walnut finished stock w/recoil pad and sling swivels. Made 1985-86.

Mossberg Model 695 Bolt Action Slugster
Gauge: 12 w/3-inch chamber. 2-round detachable magazine. 22-inch fully rifled and ported bbl. w/blade front and folding leaf rear sights. Receiver drilled and tapped for Weaver style scope bases. Also available w/1.5X-4.5X scope or fiber optics installed. Weight:

Mossberg Model 695 Bolt Action Slugster *(Con't)*
7.5 lbs. Black synthetic stock w/swivel studs and recoil pad. Made 1996 to date.
Model 695 (w/open sights) $195
Model 695 (w/1.5X-4.5X Bushnell scope) 275
Model 695 (w/Truglo fiber optics) 215
Model 695 OFM Camo 185

Mossberg Model 695 Bolt Action Turkey Gun $195
Similar to 695 Slugster Model, except has smoothbore 22-inch bbl. w/extra-full turkey Accu-Choke tube. Bead front and U-notch rear sights. Full OFM Camo finish. Made 1996 to date.

Mossberg Model 712 Autoloading Shotgun $250
Gas-operated, takedown, hammerless shotgun w/5-shot (4-shot w/3-inch chamber) tubular magazine. 12 ga. Bbls.: 28-inch vent rib or 24-inch plain bbl. Slugster w/rifle sights; ACCU-Choke tube system. Weight: 7.5 lbs. Plain alloy receiver, top-mounted ambidextrous safety. Checkered stained hardwood stock w/recoil pad. Imported from Japan 1986-90.

Mossberg Model 835 Field Pump Shotgun
Similar to the Model 9600 Regal, except has walnut-stained hardwood stock and one ACCU-Choke tube only.
Standard Model $215
Turkey Model 225
Combo Model (24- & 28-inch bbls.) 245

Mossberg Model 835 "NWTF" Ulti-Mag™ Shotgun
National Wild Turkey Federation pump-action. Gauge: 12, 3.5-inch chamber. 24-inch vent-rib bbl. w/four ACCU-MAG chokes. Realtree® Camo finish. QD swivel and post. Made 1989-93.
Limited Edition Model $350
Special Edition Model 295

Mossberg Model 835 Regal Ulti-Mag Pump
Gauge: 12, 3.5-inch chamber. Bbls.: 24- or 28-inch vent-rib w/ACCU-Choke screw-in tubes. Weight: 7.75 lbs. White bead front, brass mid-bead. Checkered hardwood or synthetic stock w/camo finish. Made 1991-96.
Special Model $225
Standard Model 280
Camo Synthetic Model 300
Combo Model 325

SHOTGUNS

Navy Arms Model 83
Bird Hunter Over and Under

Navy Arms Model 100
Field Hunter Double-Barrel Shotgun

Mossberg Model 835 Viking Pump Shotgun
Gauge: 12 w/3-inch chamber. 28-inch smoothbore bbl. w/Accu-Choke, vent rib and optional muzzle break (1997 porting became standard). Weight: 7.7 lbs. Green synthetic stock. Matte black metal finish. Made 1996-98.

Model 835 Viking (w/VR and choke tubes, ported)	**$205**
Model 835 Viking (w/VR and choke tubes, unported)	**$195**

Mossberg Model 1000 Autoloading Shotgun
Gas-operated, takedown, hammerless shotgun w/tubular magazine. Gauges: 12, 20; 2.75- or 3-inch chamber. Bbls.: 22- to 30-inch vent rib w/fixed choke or ACCU-Choke tubes; or 22-inch plain bbl, Slugster w/rifle sights. Weight: 6.5 to 7.5 lbs. Scroll-engraved alloy receiver, crossbolt-type safety. Checkered walnut buttstock and forend. Imported from Japan 1986-87.

Junior Model, 20 ga., 22-inch bbl.	**$345**
Standard Model w/Fixed Choke	355
Standard Model w/Choke Tubes	375

Mossberg Model 1000 Super Autoloading Shotgun
Similar to Model 1000, but in 12 ga. only w/3-inch chamber and new gas metering system. Bbls.: 26-, 28- or 30-inch vent rib w/ACCU-Choke tubes.

Standard Model w/Choke Tubes	**$395**
Waterfowler Model (Parkerized)	425

Mossberg Model 1000S Super Skeet $495
Similar to Model 1000 in 12 or 20 ga., except w/all-steel receiver and vented jug-type choke for reduced muzzle jump. Bright-point front sight and brass mid-bead. 1 and 2 oz. forend cap weights.

Mossberg Model 5500 Autoloading Shotgun
Gas-operated. Takedown. 12 ga. only. 4-shot magazine (3-shot w/3-inch shells). Bbls.: 18.5- to 30-inch; various chokes. Checkered walnut finished hardwood. Made 1985-86.

Model 5500 w/ACCU-Choke	**$250**
Model 5500 Modified Junior	255
Model 5500 Slugster	265
Model 5500 12 gauge	225
Model 5500 Guardian	215

Mossberg Model 5500 MKII Autoloading Shotgun
Same as Model 5500, except equipped w/two Accu-Choke bbls.: 26-inch ported for non-Magnum 2.75-inch shells; 28-inch for Magnum loads. Made 1988-93.

Standard Model	**$225**
Camo Model	250
NWTF Mossy Oak Model	275
USST Model (Made 1991-92)	235

Mossberg Model 6000 Auto Shotgun $235
Similar to the Model 9200 Regal, except has 28-inch vent-rib bbl. w/mod. ACCU-Choke tube only. Made from 1993 to date.

Mossberg Model 9200 A1 Jungle gun $355
Similar to the Model 9200 Persuader, except has mil-spec heavy wall 18.5-inch plain bbl. w/cyl. bore designed for 00 Buck shot. 12 ga. w/2.75-inch chamber. 5-shot magazine. 38.5 inches overall. Weight: 7 lbs. Black synthetic stock. Parkerized finish. Made 1998 to date.

Mossberg Model 9200 Camo Shotgun
Similar to the Model 9200 Regal, except has synthetic stock and forend and is completely finished in camouflage pattern (incl. bbl.). Made from 1993 to date.

Standard Model (OFM Camo)	**$285**
Turkey Model (Mossy Oak® camo)	315
Turkey Model (Shadow Branch® camo)	375
Combo Model (24- & 28-inch bbls. w/OFM Camo)	350

Mossberg Model 9200 Crown (Regal) Autoloader
Gauge: 12; 3-inch chamber. Bbls.: 18.5- to 28-inch w/ACCU-Choke tubes; plain or vent rib. Weight: 7.25 to 7.5 lbs. Checkered hardwood buttstock and forend w/walnut finish. Made from 1992 to date.

Model 9200 Bantam (w/1-inch shorter stock)	**$275**
Model 9200 w/ACCU-Choke	280
Model 9200 w/rifled bbl.	295
Model 9200 Combo (w/extra Slugster bbl.)	330
Model 9200 SP (w/matte blue finish, 18.5-inch bbl.)	225

Mossberg Model 9200 Persuader $265
Similar to the Model 9200 Regal, except has 18.5-inch plain bbl. w/fixed Mod. choke. Parkerized finish. Black synthetic stock w/sling swivels. Made 1996 to date.

Mossberg Model 9200 Special Hunter............ $285
Similar to the Model 9200 Regal, except has 28-inch vent-rib bbl. w/ACCU-Choke tubes. Parkerized finish. Black synthetic stock. Made from 1998 to date.

Mossberg Model 9200 Trophy
Similar to the Model 9200 Regal, except w/24-inch rifled bbl. or 24- or 28-inch vent-rib bbl. w/ACCU-Choke tubes. Checkered walnut stock w/sling swivels. Made from 1992-98.

Trophy (w/vent rib bbl.)	**$275**
Trophy (w/rifled bbl. & cantilever scope mount)	295
Trophy (w/rifled bbl. & rifle sights)	265

Mossberg Model 9200 USST Autoloader $275
Similar to the Model 9200 Regal, except has 26-inch vent-rib bbl. w/ACCU-Choke tubes. "United States Shooting Team" engraved on receiver. Made from 1993 to date.

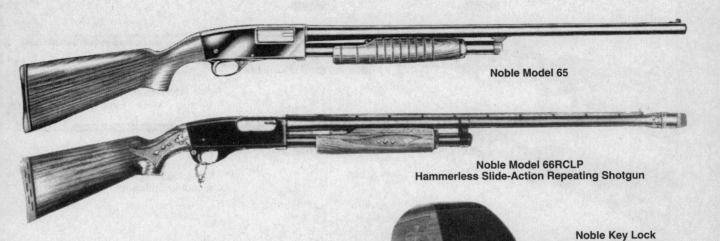

Noble Model 65

Noble Model 66RCLP
Hammerless Slide-Action Repeating Shotgun

Noble Key Lock
Fire Controll Mechanism
Supplied woth Models 66,
166L, and 602

Mossberg Model 9200 Viking Autoloader $275
Gauge: 12 w/3-inch chamber. 28-inch smoothbore bbl. W/Accu-Choke and vent rib. Weight: 7.7 lbs. Green synthetic stock. Matte black metal finish. Made 1996-98.

Mossberg Model HS410 Home Security Pump Shotgun
Gauge: .410; 3-inch chamber. Bbl.: 18.5-inch w/muzzle brake; blued. Weight: 6.25 lbs. Synthetic stock and pistol-grip slide. Optional laser sight. Made from 1990 to date. A similar version of this gun is marketed by Maverick Arms under the same model designation.
Standard Model $185
Laser Model 325

Mossberg Line Launhcer $675
Gauge: 12 w/blank cartridge. Projectile travel from 250 to 275 feet. 360 lbs. test floating line. 700 feet of coiled spool.

Mossberg "New Haven Brand" Shotguns
Promotional models, similar to their standard guns but plainer in finish, are marketed by Mossberg under the "New Haven" brand name. Values generally are about 20 percent lower than for corresponding standard models.

NAVY ARMS SHOTGUNS
Ridgefield, New Jersey

Navy Arms Model 83/93 Bird Hunter Over/Under
Hammerless. Boxlock, engraved receiver. Gauges: 12 and 20; 3--inch chambers. Bbls.: 28-inch chrome lined w/double vent-rib construction. Checkered European walnut stock and forearm. Gold plated triggers. Imported 1984-90.
Model 83 w/Extractors $225
Model 93 w/Ejectors 265

Navy Arms Model 95/96 Over/Under Shotgun
Same as the Model 83/93, except w/five interchangeable choke tubes. Imported 1984-90.
Model 95 w/Extractors $295
Model 96 w/Ejectors 415

Navy Arms Model 100/150 Field Hunter Double-Barrel Shotgun
Boxlock. Gauges: 12 and 20. Bbls.: 28-inch chrome lined. Checkered European walnut stock and forearm. Imported 1984-90.
Model 100 .... $295
Model 150 (auto ejectors) 395

Navy Arms Model 100 Over/Under Shotgun $195
Hammerless, takedown shotgun w/engraved chrome receiver. Single trigger. 12, 20, 28, or .410 ga. w/3-inch chambers. Bbls.: 26-inch (F/F or SK/SK); vent rib. Weight: 6.25 lbs. Checkered European walnut buttstock and forend. Imported 1986-90.

NEW ENGLAND FIREARMS
Gardner, Massachusetts

New England Firearms NWTF Turkey Special $150
Similar to Turkey and Goose Model except, has 24-inch plain bbl. w/screw-in Turkey full choke tube. Mossy Oak Camo finish on entire gun. Made from 1992-96.

New England Firearms Pardner Shotgun
Takedown. Side lever. Single bbl. Gauges: 12, 20 and .410 w/3-inch chamber; 16 and 28 w/2.75-inch chamber. Bbl.: 26- or 28-inch, plain; fixed choke. Weight: 5-6 lbs. Bead front sight. Pistol grip-style hardwood stock w/walnut finish. Made from 1988 to date.
Standard Model $75
Youth Model 80

New England Firearms Special Purpose Waterfowl Single-Shot $135
Similar to Turkey and Goose Model, except w/32-inch bbl. Mossy oak camo stock w/swivel and sling. Made 1988 to date.

New England Firearms Survivor Series
Break open w/side-lever release, automatic ejectior and patented transfer bar safety. Gauges: 12, 20, and .410/45 w/3-inch chamber. 22-inch bbl. w/modified choke and bead sight. Weight: 6 lbs. Polymer stock and forend w/hollow cavity for storage. Made 1992-93 and 1995 to date.
Blued finish.. $80
Nickel finish 95

New England Firearms Tracker Slug Gun
Similar to Pardner Model, except in 12 or 20 ga. only. Bbl.: 24-inch w/cylinder choke or rifled slug (Tracker II). Weight: 6 lbs. American

SHOTGUNS

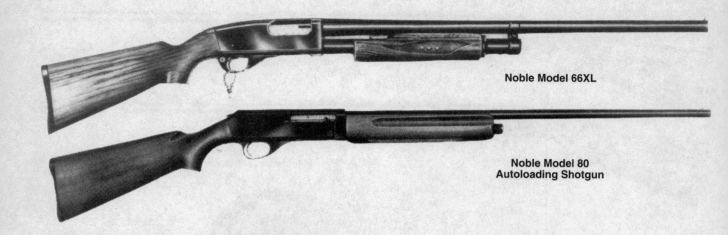

Noble Model 66XL

Noble Model 80
Autoloading Shotgun

New England Firearms Tracker Slug Gun *(Con't)*
hardwood stock w/walnut or camo finish, schnabel forend, sling swivel studs. Made from 1992 to date.

Tracker Slug . . . . $ 95
Tracker II (rifled bore) 100

New England Firearms Turkey and Goose Gun
Similar to Pardner Model, except chambered in 10 ga. w/3.5-inch chamber. 28-inch plain bbl. w/F choke. Weight: 9.5 lbs. American hardwood stock w/walnut or camo finish. Made from 1992 to date.

Standard Model $110
Camo Model 120

NIKKO FIREARMS LTD.
Tochigi, Japan

See listings under Golden Eagle Firearms, Inc.

NOBLE MANUFACTURING COMPANY
Haydenville, Massachusetts

Noble Series 602 and 70 are similar in appearance to the corresponding Model 66 guns.

Noble Model 40 Hammerless Slide-Action
Repeating Shotgun $115
Solid frame. 12 ga. only. 5-shell tubular magazine. 28-inch bbl. w/ventilated Multi-Choke. Weight: about 7.5 lbs. Plain pistol-grip stock, grooved slide handle. Made 1950-55.

Noble Model 50 Slide-Action. $105
Same as Model 40, except w/o Multi-Choke. M or F choke bbl. Made 1953-55.

Noble Model 60 Hammerless Slide-Action
Repeating Shotgun $155
Solid frame. 12 and 16 ga. 5-shot tubular magazine. 28-inch bbl. w/adj. choke. Plain pistol-grip stock w/recoil pad, grooved slide handle. Weight: about 7.5 lbs. Made 1955-66.

Noble Model 65 . . . . $135
Same as Model 60, except without adj. choke and recoil pad. M or F choke bbl. Made 1955-66.

Noble Model 66CLP $125
Same as Model 66RCLP, except has plain bbl. Introduced in 1967. Discontinued.

Noble Model 166L
Deergun

Noble Model 420
Hammerless Double

Noble Model 66RCLP Hammerless Slide-Action Repeating Shotgun **$150**
Solid frame. Key lock fire control mechanism. Gauges: 12, 16. 3-inch chamber in 12 ga. 5-shot tubular magazine. 28-inch bbl., vent rib, adj. choke. Weight: about 7.5 lbs. Checkered pistol-grip stock and slide handle, recoil pad. Made 1967-70.

Noble Model 66RLP **$145**
Same as Model 66RCLP, except w/F or M choke. Made 1967-1970.

Noble Model 66XL **$125**
Same as Model 66RCLP, except has plain bbl., F or M choke slide handle only checkered, no recoil pad. Made 1967-70.

Noble Model 70CLP Hammerless Slide-Action Repeating Shotgun **$145**
Solid frame. .410 gauge. Magazine holds 5 shells. 26-inch bbl. w/adj. choke. Weight: about 6 lbs. Checkered buttstock and forearm, recoil pad. Made 1958-70.

Noble Model 70RCLP **$150**
Same as Model 70CLP, except has vent rib. Made 1967-70.

Noble Model 70RLP **$140**
Same as Model 70CLP, except has vent rib and no adj. choke. Made 1967-70.

Noble Model 70XL **$105**
Same as Model 70CLP, except without adj. choke and checkering on buttstock. Made 1958-70.

Noble Model 80 Autoloading Shotgun **$175**
Recoil-operated. .410 ga. Magazine holds three 3-inch shells, four 2.5-inch shells. 26-inch bbl., full choke. Weight: about 6 lbs. Plain pistol-grip stock and fluted forearm. Made 1964-66.

Noble Model 166L Deergun **$195**
Solid frame. Key lock fire control mechanism. 12 ga. 2.75-inch chamber. 5-shot tubular magazine. 24-inch plain bbl., specially bored for rifled slug. Lyman peep rear sight, post ramp front sight. Receiver dovetailed for scope mounting. Weight: about 7.25 lbs. Checkered pistol-grip stock and slide handle, swivels and carrying strap. Made 1967-70.

Noble Model 420 Hammerless Double **$295**
Boxlock. Plain extractors. Double triggers. Gauges: 12 ga. 3-inch mag.; 16 ga.; 20 ga. 3-inch mag.; .410 ga. Bbls.: 28-inch, except .410 in 26-inch, M/F choke. Weight: about 6.75 lbs. Engraved frame. Checkered walnut stock and forearm. Made 1958-70.

Noble Model 450E Hammerless Double **$350**
Boxlock. Engraved frame. Selective auto ejectors. Double triggers. Gauges: 12, 16, 20. 3-inch chambers in 12 and 20 ga. 28-inch bbls., M/F choke. Weight: about 6 lbs. 14 oz., 12 ga. Checkered pistol-grip stock and beavertail forearm, recoil pad. Made 1967-70.

**Noble Model 450E
Hammerless Double**

Noble Model 602CLP **$155**
Same as Model 602RCLP, except has plain bbl. Made 1958-70.

Noble Model 602RCLP Hammerless Slide-Action Repeating Shotgun **$165**
Solid frame. Key lock fire control mechanism. 20 ga. 3-inch chamber. 5-shot tubular magazine. 28-inch bbl., vent rib, adj. choke. Weight: about 6.5 lbs. Checkered pistol-grip stock/slide handle, recoil pad. Made 1967-70.

Noble Model 602RLP **$150**
Same as Model 602RCLP, except without adj. choke, bored F or M choke. Made 1967-70.

Noble Model 602XL **$125**
Same as Model 602RCLP, except has plain bbl., F or M choke, slide handle only checkered, no recoil pad. Made 1958-70.

Noble Model 662 . . . . **$155**
Same as Model 602CLP, except has aluminum receiver and bbl. Weight: about 4.5 lbs. Made 1966-70.

OMEGA SHOTGUNS
Brescia, Italy, and Korea

Omega Folding Over/Under Shotgun, Standard
Hammerless Boxlock. Gauges: 12, 20, 28 w/2.75-inch chambers or .410 w/3-inch chambers. Bbls.: 26- or 28-inch vent-rib w/fixed chokes (IC/M, M/F or F/F (.410). Automatic safety. Single trigger. 40.5 inches overall (42.5 inches, 20 ga., 28-inch bbl.). Weight: 6 to 7.5 lbs. Checkered European walnut stock and forearm. Imported 1984-94.
Standard Model (12 ga.) **$300**
Standard Model (20 ga.) **325**
Standard Model (28 ga. & .410) **395**

**Omega
Side-by-Side Shotgun, Deluxe**

SHOTGUNS

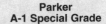

Parker
A-1 Special Grade

Parker
A.H. Grade

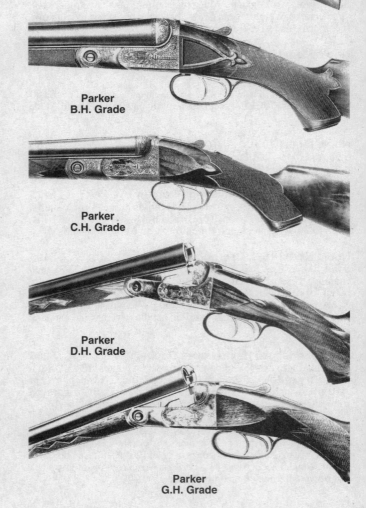

Parker
B.H. Grade

Parker
C.H. Grade

Parker
D.H. Grade

Parker
G.H. Grade

Omega Over/Under Shotgun, Deluxe $350
Gauges: 20, 28 and .410. 26- or 28-inch vent-rib bbls. 40.5 inches overall (42.5 inches, 20 ga., 28-inch bbl.). Chokes: IC/M, M/F or F/F (.410). Weight: about 5.5 -6 lbs. Single trigger. Automatic safety. European walnut stock w/checkered pistol grip and tulip forend. Imported from Italy 1984-90.

Omega Side-by-Side Shotgun, Deluxe $185
Same general specifications as the Standard Side-by-Side except has checkered European walnut stock and low bbl. rib. Made in Italy from 1984-89.

Omega Side-by-Side Shotgun, Standard $165
Gauge: .410. 26-inch bbl. 40.5 inches overall. Choked F/F. Weight: 5.5 lbs. Double trigger. Manual safety. Checkered beechwood stock and semi-pistol grip. Imported from Italy 1984-89.

Omega Single-Shot Shotgun, Deluxe $75
Same general specifications as the Standard Single Bbl. except has checkered walnut stock, top lever break, fully blued receiver, vent rib. Imported from Korea 1984-87.

Omega Single-Shot Shotgun, Standard
Gauges: 12, 16, 20, 28 and .410. Bbl. lengths: 26-,28- or 30-inch. Weight: 5 lbs. 4 oz.-5 lbs. 11 oz. Indonesian walnut stock. Matte-chromed receiver and top lever break. Imported from Korea 1984-87.
Standard Fixed . . . . $ 65
Standard Folding 125
Deluxe Folding 155

PARKER BROTHERS
Meriden, Connecticut

This firm was taken over by Remington Arms Company in 1934 and its production facilities moved to Remington's Ilion, New York, plant.

Parker
S.C. Grade

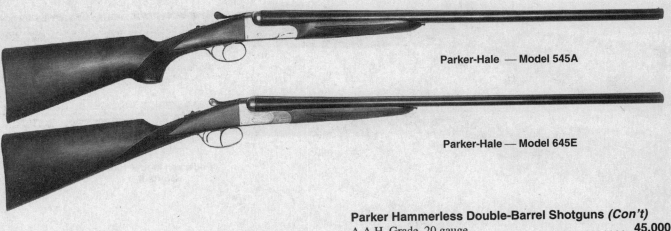

Parker-Hale — Model 545A

Parker-Hale — Model 645E

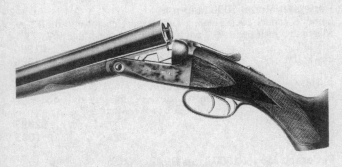

**Parker Trojan
Hammerless Double-Barrel Shotgun**

Parker Hammerless Double-Barrel Shotguns

Grades V.H. through A-1 Special. Boxlock. Auto ejectors. Double triggers or selective single trigger. Gauges: 10, 12, 16, 20, 28, .410. Bbls.: 26- to 32-inch, any standard boring. Weight: 6.88-8.5 lbs., 12 ga. Stock and forearm of select walnut, checkered; straight, half-or full-pistol grip. Grades differ only in quality of workmanship, grade of wood, engraving, checkering, etc.; general specifications are the same for all. Discontinued about 1940. *See* Illustrations Next page.

V.H. Grade, 12 or 16 gauge	$ 2,495
V.H. Grade, 20 gauge	3,650
V.H. Grade, 28 gauge	5,950
V.H. Grade, .410 gauge	14,950
G.H. Grade, 12 or 16 gauge	3,295
G.H. Grade, 20 gauge	3,850
G.H. Grade, 28 gauge	6,250
G.H. Grade, .410 gauge	15,950
D.H. Grade, 12 or 16 gauge	4,550
D.H. Grade, 20 gauge	5,850
D.H. Grade, 28 gauge	8,950
D.H. Grade, .410 gauge	29,500
C.H. Grade, 12 or 16 gauge	5,050
C.H. Grade, 20 gauge	6,550
C.H. Grade, 28 gauge	14,950
B.H. Grade, 12 or 16 gauge	7,550
B.H. Grade, 20 gauge	13,500
B.H. Grade, 28 gauge	24,750
A.H. Grade, 12 or 16 gauge	13,950
A.H. Grade, 20 gauge	17,750
A.H. Grade, 28 gauge	32,500
A.A.H. Grade, 12 or 16 gauge	23,950

Parker Hammerless Double-Barrel Shotguns *(Con't)*

A.A.H. Grade, 20 gauge	45,000
A-1 Special Grade, 12 or 16 gauge	68,000
A-1 Special Grade, 20 gauge	90,000
A-1 Special Grade, 28 gauge	130,000
W/Selective-Single Trigger, **add**	20%
W/Ventilated Rib, **add**	30%
W/Auto Ejectors, **add**	50%

For *non-ejector guns*, **deduct** 30 percent from values shown. Vent-rib barrels, **add** 20 percent to values shown.

Parker Single-Shot Trap Guns

Hammerless. Boxlock. Ejector. 12 ga. only. Bbl. lengths: 30-, 32-, 34-inch, any boring, vent rib. Weight: 7.5-8.5 lbs. Stock and forearm of select walnut, checkered; straight, half-or full-pistol grip. The five grades differ only in quality of workmanship, grade of wood, checkering, engraving, etc.; general specifications same for all. Discontinued about 1940.

S.C. Grade	$ 3,250
S.B. Grade	3,975
S.A. Grade	4,750
S.A.A. Grade	5,895
S.A.1 Special	19,000

Parker Skeet Gun

Same as other Parker doubles from Grade V.H.E. up, except selective single trigger and beavertail forearm are standard on this model, as are 26-inch bbls., SK choke. Discontinued about 1940. Values are 35 percent higher.

Parker Trojan Hammerless Double-Barrel Shotgun

Boxlock. Plain extractors. Double trigger or single trigger. Gauges: 12, 16, 20. Bbls.: 30-inch both F choke (12 ga. only), 26- or 28-inch M and F choke. Weight: 6.25-7.75 lbs. Checkered pistol-grip stock and forearm. Discont. 1939.

12 or 16 gauge	$1650
20 gauge	2850

PARKER REPRODUCTIONS
Middlesex, New Jersey

Parker Hammerless Double-Barrel Shotguns

Reproduction of the original Parker boxlock. Single selective trigger or double triggers. Selective automatic ejectors. Automatic safety. Gauges: 12, 16, 20 or 28; 2.75- or 3-inch chambers. Bbls.: 26- or 28-inch choked SK/SK, IC/M, M/F. Weight: 5.5-7 lbs. Checkered English-style or pistol-grip American walnut stock w/beavertail or splinter forend and checkered skeleton buttplate. Color casehardened receiver with game scenes and scroll engraving. Imported from Japan 1984 to date.

SHOTGUNS

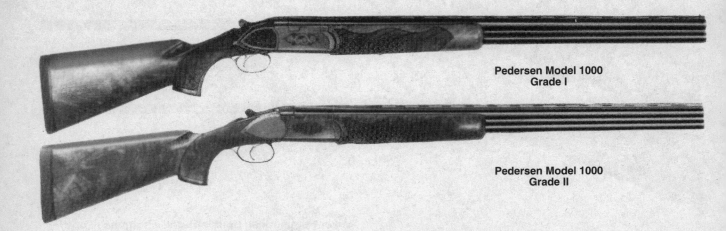

**Pedersen Model 1000
Grade I**

**Pedersen Model 1000
Grade II**

Parker Hammerless Double-Barrel Shotguns *(Con't)*

D Grade	$ 2,595
D Grade 2-barrel set	3,950
B Grade Bank Note Limited Edition	3,595
B Grade 2-barrel set	4,750
B Grade 3-barrel set	5,595
A-1 Special Grade	7,595
A-1 Special Grade 2-barrel set	8,550
A-1 Special Grade Custom Engraved	10,500
A-1 Special Grade 3-barrel set	12,750

PARKER-HALE SHOTGUNS
Mfd. by Ignacio Ugartechea, Spain

**Parker-Hale Model 645A (American)
Side-by-Side Shotgun** **$750**
Gauges: 12, 16 and 20. Boxlock action. 26- and 28-inch bbls.
Chokes: IC/M, M/F. Weight: 6 lbs. average. Single non-selective
trigger. Automatic safety. Hand-checkered pistol grip walnut stock
w/beavertail forend. Raised matted rib. English scroll-design engraved receiver. Discontinued 1990.

Parker-Hale Model 645E (English) Side-by-Side Shotgun
Same general specifications as the Model 645A, except double trig-
gers, straight grip, splinter forend, checkered butt and concave rib.
Discontinued 1990.
MODEL 645E

12, 16, 20 ga. with 26- or 28-inch bbl.	$795
28, .410 ga. with 27-inch bbl.	950

MODEL 645E-XXV

12, 16, 20 ga. with 25-inch bbl.	750
28, .410 ga. with 25-inch bbl.	895

PEDERSEN CUSTOM GUNS
North Haven, Connecticut
Division of O. F. Mossberg & Sons, Inc.

Pedersen Model 1000 Over/Under Hunting Shotgun
Boxlock. Auto ejectors. Selective single trigger. Gauges: 12, 20.
2.75-inch chambers in 12 ga., 3-inch in 20 ga. Bbls.: vent rib; 30-inch
M/F (12 ga. only); 28-inch IC/M (12 ga. only), M/F; 26-inch IC/M.
Checkered pistol-grip stock and forearm. Grade I is the higher quality gun with custom stock dimensions, fancier wood, more elaborate
engraving, silver inlays. Made 1973-75.

Grade I	$1895
Grade II	1525

Pedersen Model 1000 Magnum
Same as Model 1000 Hunting Gun, except chambered for 12-ga.
Magnum 3-inch shells, 30-inch bbls., IM/F choke. Made 1973-75.

Grade I	$1950
Grade II	1595

Pedersen Model 1000 Skeet Gun
Same as Model 1000 Hunting Gun, except has skeet-style stock; 26-
and 28-inch bbls. (12 ga. only), SK choke. Made 1973-75.

Grade I	$2095
Grade II	1695

Pedersen Model 1000 Trap Gun
Same as Model 1000 Hunting Gun, except 12 ga. only, has Monte
Carlo trap-style stock, 30- or 32-inch bbls., M/F or IM/F choke.
Made 1973-75.

Grade I	$1850
Grade II	1495

Pedersen Model 1500 O/U Hunting Shotgun **$550**
Boxlock. Auto ejectors. Selective single trigger. 12 ga. 2.75- or 3-inch
chambers. Bbls.: vent rib; 26-inch IC/M; 28- and 30-inch M/F; Magnum has 30-inch, IM/F choke. Weight. 7-7.5 lbs., depending on bbl.
length. Checkered pistol-grip stock and forearm. Made 1973-75.

Pedersen Model 1500 Skeet Gun **$595**
Same as Model 1500 Hunting Gun, except has skeet-style stock, 27-
inch bbls., SK choke. Made 1973-75.

Pedersen Model 1500 Trap Gun **$575**
Same as Model 1500 Hunting Gun, except has Monte Carlo trap-style
stock, 30- or 32-inch bbls., M/F or IM/F chokes. Made 1973-75.

Pedersen Model 2000 Hammerless Double
Boxlock. Auto ejectors. Selective single trigger. Gauges: 12, 20.
2.75-inch chambers in 12 ga., 3-inch in 20 ga. Bbls.: vent rib; 30-
inch M/F (12 ga. only); 28-inch M/F, 26-inch IC/M choke. Checkered pistol-grip stock and forearm. Grade I is the higher quality
gun w/custom dimensions, fancier wood, more elaborate engraving, silver inlays. Made 1973-74.

Grade I	$1950
Grade II	1795

Pedersen Model 2500 Hammerless Double **$385**
Boxlock. Auto ejectors. Selective single trigger. Gauges: 12, 20.
2.75-inch chambers in 12 ga., 3-inch in 20 ga. Bbls.: vent rib; 28-inch
M/F; 26-inch IC/M choke. Checkered pistol-grip stock and forearm.
Made 1973-74.

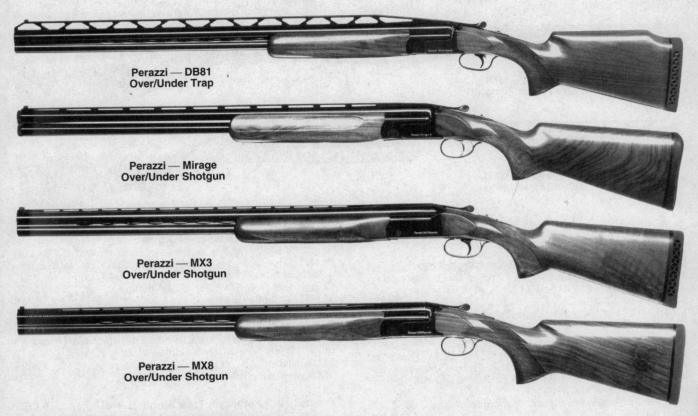

Perazzi — DB81
Over/Under Trap

Perazzi — Mirage
Over/Under Shotgun

Perazzi — MX3
Over/Under Shotgun

Perazzi — MX8
Over/Under Shotgun

Pedersen Model 4000 Hammerless Slide-Action Repeating Shotgun **$395**
Custom version of Mossberg Model 500. Full-coverage floral engraving on receiver. Gauges: 12, 20, .410. 3-inch chamber. Bbls.: vent rib; 26-inch IC or SK choke; 28-inch F or M; 30-inch F. Weight: 6-8 lbs. depending on ga. and bbl. Checkered stock and slide handle of select wood. Made in 1975.

Pedersen Model 4000 Trap Gun **$365**
Same as standard Model 4000, except 12 ga. only, has 30-inch F choke bbl., Monte Carlo trap-style stock w/recoil pad. Made in 1975.

Pedersen Model 4500 . . . . **$325**
Same as Model 4000, except has simpler scroll engraving. Made in 1975.

Pedersen Model 4500 Trap Gun **$350**
Same as Model 4000 Trap Gun, except has simpler scroll engraving. Made in 1975.

J. C. PENNEY CO., INC.
Dallas, Texas

J. C. Penney Model 4011 Autoloading Shotgun **$175**
Hammerless. 5-shot magazine. Bbls.: 26-inch IC; 28-inch M or F; 30-inch F choke. Weight: 7.25 lbs. Plain pistol-grip stock and slide handle.

J. C. Penney Model 6610 Single-Shot Shotgun **$80**
Hammerless. Takedown. Auto ejector. Gauges: 12, 16, 20 and .410. Bbl. length: 28-36 inches. Weight: about 6 lbs. Plain pistol-grip stock and forearm.

J. C. Penney Model 6630 Bolt-Action Shotgun **$110**
Takedown. Gauges: 12, 16, 20. 2-shot clip magazine. 26- and 28-inch bbl. lengths; with or without adj. choke. Plain pistol-grip stock. Weight: about 7.25 lbs.

J. C. Penney Model 6670 Slide-Action Shotgun **$130**
Hammerless. Gauges: 12, 16, 20, and .410. 3-shot tubular magazine. Bbls.: 26- to 30-inch; various chokes. Weight: 6.25-7.25 lbs. Walnut finished hardwood stock.

J. C. Penney Model 6870 Slide-Action Shotgun . . . **$195**
Hammerless. Gauges: 12, 16, 20, .410. 4-shot magazine. Bbls.: vent rib; 26- to 30-inch, various chokes. Weight: average 6.5 lbs. Plain pistol-grip stock.

PERAZZI SHOTGUNS
Manufactured by Manifattura Armi Perazzi, Brescia, Italy

See also listings under **Ithaca-Perazzi.**

Perazzi DB81 Over/Under Trap **$4750**
Gauge: 12; 2.75-inch chambers. 29.5- or 31.5-inch bbls. w/wide vent rib; M/F chokes. Weight: 8 lbs. 6 oz. Detachable and interchangeable trigger with flat V-springs. Bead front sight. Interchangeable and custom-made checkered stock; beavertail forend. Imported 1988 to date.

Perazzi DB81 Single-Shot Trap **$4250**
Same general specifications as the DB81 Over/Under, except in single bbl. version w/32- or 34-inch wide vent-rib bbl., F choke. Imported 1988 to date.

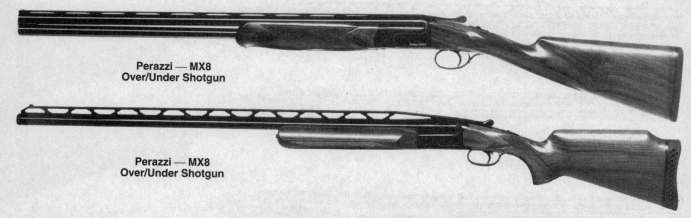

Perazzi — MX8
Over/Under Shotgun

Perazzi — MX8
Over/Under Shotgun

Perazzi Grand American 88 Special Single Trap
Same general specifications as MX8 Special Single Trap, except w/high ramped rib. Fixed choke or screw-in choke tubes.
Model 88 Standard **$4255**
Model 88 w/Interchangeable Choke Tubes **4595**

Perazzi Mirage Over/Under Shotgun
Gauge: 12; 2.75-inch chambers. Bbls.: 27.63-, 29.5- or 31.5-inch vent-rib w/fixed chokes or screw-in choke tubes. Single selective trigger. Weight: 7 to 7.75 lbs. Interchangeable and custom-made checkered buttstock and forend.
Competition Trap, Skeet, Pigeon, Sporting .... **$ 4,650**
Skeet 4-Barrel Sets **12,400**
Competition Special (w/adj. 4-position trigger) **add** **400**

Perazzi MX1 Over/Under Shotgun
Similar to Model MX8, except w/ramp-style, tapered rib and modified stock configuration.
Competition Trap, Skeet, Pigeon & Sporting **$3600**
MX1C (w/Choke Tubes) **3750**
MX1B (w/Flat Low-Rib) **3425**

Perazzi MX2 Over/Under Shotgun
Similar to Model MX8, except w/broad high-ramped competition rib.
Competition-Trap, Skeet, Pigeon & Sporting **$3750**
MX2C (w/choke tubes) **3950**

Perazzi MX3 Over/Under Shotgun
Similar to Model MX8, except w/ramp-style, tapered rib and modified stock configuration.
Competition Trap, Skeet, Pigeon & Sporting **$3595**
Competition Special (w/adj. 4-position trigger) **add** **300**
Game Models **3195**
Combo O/U plus SB **3950**
SB Trap 32- or 34-inch **2695**
Skeet 4-bbl. sets **8750**
Skeet Special 4-bbl. sets **8925**

Perazzi MX3 Special Pigeon Shotgun **$4250**
Gauge: 12; 2.75-inch chambers. 29.5- or 31.5-inch vent-rib bbl.; IC/M and extra full chokes. Weight: 8 lbs. 6 oz. Detachable and interchangeable trigger group w/flat V-springs. Bead front sight. Interchangeable and custom-made checkered stock for live pigeon shoots; splinter forend. Imported 1991 to date.

Perazzi MX4 Over/Under Shotgun
Similar to Model MX3 in appearance and shares the MX8 locking system. Detachable, adj. 4-position trigger standard. Interchangeable choke tubes optional.
Competition Trap, Skeet, Pigeon & Sporting **$3650**
MX4C (w/Choke Tubes) **3895**

Perazzi MX5 Over/Under Game Gun
Similar to Model MX8, except in hunting configuration chambered in 12 or 20 ga. Non-detachable single selective trigger.
MX5 Standard **$2495**
MX5C (w/Choke Tubes) **2750**

Perazzi MX6 American Trap Single Barrel **$3675**
Single shot. Removable trigger group. 12 gauge. Barrels: 32- or 34-inch with fixed or choke tubes. Raised vent rib. Checkered European walnut Monte Carlo stock, beavertail forend. Imported 1995 to date.

Perazzi MX6 Skeet O/U **$4350**
Same general specs as MX6 American Trap Single Barrel except over/under; boxlock. Barrels: 26.75- or 27.50-inch. Imported 1995 to date.

Perazzi MX6 Sporting O/U **$3650**
Same specs as MX6 American Trap Single Barrel except over/under; boxlock. Single selective trigger; external selector. Barrels: 28.38-, 29.50-, or 31.50-inch. Imported 1995 to date.

Perazzi MX6 Trap O/U **$4275**
Same general specs as MX6 Americna Trap Single Barrel except over/under; boxlock. Barrels: 29.50-, 30.75-, or 31.50-inch. Imported 1995 to date.

Perazzi MX7 Over/Under Shotgun **$4250**
Similar to Model MX12, except w/MX3-style receiver and top-mounted trigger selector. Bbls.: 28.73-, 2.5-, 31.5-inch w/vent rib; screw-in choke tubes. Imported 1992 to date.

Perazzi MX8 Over/Under Shotgun
Gauge: 12, 2.75-inch chambers. Bbls.: 27.63-, 29.5- or 31.5-inch vent-rib w/fixed chokes or screw-in choke tubes. Weight: 7 to 8.5 lbs. Interchangeable and custom-made checkered stock; beavertail forend. Special models have detachable and interchangeable 4-position trigger group w/flat V-springs. Imported 1988 to date.
MX8 Standard **$3695**
MX8 Special (adj. 4-pos. trigger) **3795**
MX8 Special Single (32- or 34-inch bbl.) **3595**
MX8 Special Combo **7390**

Piotti — Piuma
Boxlock Side-by-Side Shotgun

Perazzi MX8/20 Over/Under Snotgun **$3750**
Similar to the Model MX8, except w/smaller frame and custom stock. Available in sporting or game configurations with fixed chokes or screw-in tubes. Imported 1993 to date.

Perazzi MX9 Over/Under Shotgun **$5995**
Gauge: 12; 2.75-inch chambers. Bbls.: 29.5- or 30.5-inch w/choke tubes and vent side rib. Selective trigger. Checkered walnut stock w/adj. cheekpiece. Available in single bbl., combo, O/U trap, skeet, pigeon and sporting models. Imported 1993 to date.

Perazzi MX10 Over/Under Shotgun **$6595**
Similar to the Model MX9, except w/fixed chokes and different rib configuration. Imported 1993 to date.

Perazzi MX10 Pigeon-Electrocibles O/U **$6850**
Over/Under; boxlock. Removable trigger group; external selector. 12 gauge. Barrels: 27.50- or 29.50-inch. Checkered European walnut adjustable stock, beavertail forend. Imported 1995 to date.

Perazzi MX11 American Trap Combo **$6950**
Over/Under; boxlock. External selector. Removable trigger group; single selective trigger. 12 ga. Bbls: 29-1/2- to 34-inch with fixed or choke tubes; vent rib. European walnut Monte Carlo adjustable stock, beavertail forend. Imported 1995 to date.

Parazzi MX11 American Trap Single Barrel **$3695**
Same general specs as MX11 American Trap Combo except 32- or 34-inch single bbl. Imported 1995 to date.

Parazzi MX11 Pigeon-Electrocibles O/U **$3850**
Same specs as MX11 American Trap Combo except 27.50 O/U bbls. Checkered European walnut pistol grip adjustable stock, beavertail forend. Imported 1995 to date.

Perazzi MX11 Skeet O/U **$3895**
Same general specs as MX11 American Trap Combo except 26.75 or 27.50-inch O/U bbls. Checkered European walnut pistol-grip adjustable stock, beavertail forend. Imported 1995 to date.

Perazzi MX11 Sporting O/U **$4250**
Same general specs as MX11 American Trap Combo except 28.38, 29.50-, or 31.50-inch O/U bbls. Checkered European walnut pistol-grip adjustable stock, beavertail forend. Imported 1995 to date.

Perazzi MX11 Trap O/U **$3850**
Same general specs as MX11 American Trap Combo except 29.50,- 30.75, or 31.50-inch O/U bbls. Checkered European walnut pistol-grip adjustable stock, beavertail forend. Imported 1995 to date.

Perazzi MX12 Over/Under Game Gun
Gauge: 12, 2.75-inch chambers. Bbls.: 26-, 27.63-, 28.38- or 29.5-inch, vent-rib, fixed chokes or screw-in choke tubes. Non-detachable single selective trigger group w/coil springs. Weight: 7.25 lbs. Interchangeable and custom-made checkered stock; schnabel forend.
MX12 Standard **$3850**
MX12C (w/Choke Tubes) 4250

Perazzi MX14 American Trap Single-Barrel **$3550**
Single shot. Removable trigger group; unsingle configuration. 12 ga. Bbl: 34-inch with fixed or choke tubes; vent rib. Checkered European walnut Monte Carlo adjustable stock, beavertail forend. Imported 1995 to date.

Perazzi MX15 American Trap
Single-Barrel **$3795**
Full choke. Detachable trigger group. Gauge: 12 only with 2.75-inch chamber. Bbls: 32 and 34-inch. Weight: 8 lbs., 6 oz. Beavertail forend.

Perazzi MX20 O/U Game Gun
Gauges: 20, 28 and .410; 2.75- or 3-inch chambers. 26-inch vent-rib bbls., M/F chokes or screw-in chokes. Auto selective ejectors. Selective single trigger. Weight: 6 lbs. 6 oz. Non-detachable coil-spring trigger. Bead front sight. Interchangeable and custom-made checkered stock w/schnabel forend. Imported from 1988 to date.
Standard Grade . . . . $ 3,995
Standard Grade w/Gold Outline 6,950
MX20C w/Choke Tubes 4,295
SC3 Grade 7,500
SCO Grade 10,350

Perazzi MX28 Over/Under Game Gun **$10,950**
Similar to the Model MX12, except chambered in 28 ga. w/26-inch bbls. fitted to smaller frame. Imported 1993 to date.

Perazzi MX410 Over/Under Game Gun **$10,950**
Similar to the Model MX12, except in .410 bore w/3-inch chambers, 26-inch bbls. fitted to smaller frame. Imported 1993 to date.

Perazzi TM1 Special Single-Shot Trap **$3455**
Gauge: 12- 2.75-inch chambers. 32- or 34-inch bbl. w/wide vent rib; full choke. Weight: 8 lbs. 6 oz. Detachable and interchangeable trigger group with coil springs. Bead front sight. Interchangeable and custom-made stock w/checkered pistol grip and beavertail forend. Imported 1988 to date.

Perazzi TMX Special Single-Shot Trap **$3590**
Same general specifications as Model TM1 Special, except w/ultra-high rib. Interchangeable choke tubes optional.

PIOTTI SHOTGUNS
Italy

Piotti Boss Over/Under **$21,500**
Over/Under; sidelock. Gauges: 12 or 20. Barrels: 26- to 32-inch. Standard chokes. Best quality walnut. Custom-made to customer's specifications. Imported 1993 to date.

Piotti King No. 1 Sidelock **$13,750**
Gauges: 10, 12, 16, 20, 28 and .410. 25- to 30-inch bbls. (12 ga.), 25- to 28-inch (other ga.). Weight: about 5 lbs. (.410) to 8 lbs. (12 ga.) Holland & Holland pattern sidelock. Double triggers standard. Coin finish or color casehardened. Level file-cut rib. Full-coverage scroll engraving, gold inlays. Hand-rubbed, oil-finished, straight-grip stock with checkered butt, splinter forend.

SHOTGUNS

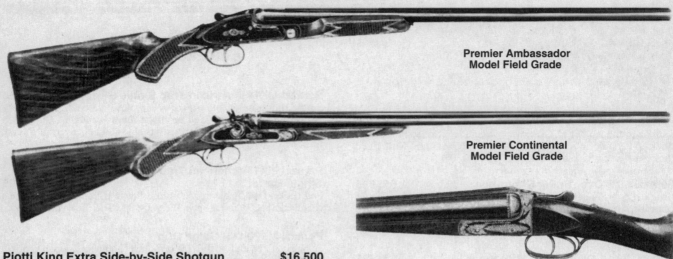

**Premier Ambassador
Model Field Grade**

**Premier Continental
Model Field Grade**

**Powell No. 7
Aristocrat Grade Double**

Piotti King Extra Side-by-Side Shotgun **$16,500**
Same general specifications as the Piotti King No. 1, except has choice of engraving, gold inlays plus stock is of exhibition grade wood.

Piotti Lunik Sidelock Shotgun **$13,950**
Same general specifications as the Monte Carlo model except has level, file-cut rib. Renaissance-style, large scroll engraving in relief, gold crown in top lever, gold name, and gold crest in forearm, finely figured wood.

Piotti Monte Carlo Sidelock Shotgun **$8750**
Gauges: 10, 12, 16, 20, 28 or .410. Bbls.: 25- to 30-inch. Holland & Holland pattern sidelock. Weight: 5-8 lbs. Automatic ejectors. Double triggers. Hand-rubbed oil-finished straight-grip stock with checkered butt. Choice of Purdey-style scroll and rosette or Holland & Holland-style large scroll engraving.

Piotti Piuma Boxlock Side-by-Side Shotgun **$8395**
Same general specifications as the Monte Carlo model, except has Anson & Deeley boxlock action w/demi-bloc bbls., scalloped frame. Standard scroll and rosette engraving. Hand-rubbed, oil-finished straight-grip stock.

WILLIAM POWELL & SON LTD.
Birmingham, England

**Powell No. 1 Best Grade Double-Barrel
Shotgun** **$25,000**
Sidelock. Gauges: Made to order, with 12, 16 and 20 the most common. Bbls.: Made to order in any length, but 28 inches was recommended. Highest grade French walnut buttstock and forearm with fine checkering. Metal elaborately engraved. Imported by Stoeger about 1938-51.

Powell No. 2 Best Grade Double **$20,000**
Same general specifications as the Powell No. 1, except plain finish without engraving. Imported by Stoeger about 1938-51.

Powell No. 6 Crown Grade Double **$8000**
Boxlock. Gauges: Made to order, with 12, 16 and 20 the most common. Bbls.: Made to order, but 28 inches was recommended. Highest grade French walnut buttstock and forearm w/fine checkering. Metal elaborately engraved. Uses Anson & Deeley locks. Imported by Stoeger about 1938-51.

Powell No. 7 Aristocrat Grade Double **$2500**
Same general specifications as the Powell No. 6, except w/lower quality wood and metal engraving.

PRECISION SPORTS SHOTGUNS
Cortland, New York
Manufactured by Ignacio Ugartechea, Spain

**Precision Sports 600 Series American Hammerless
Doubles**
Boxlock. Single selective trigger. Selective automatic ejectors. Automatic safety. Gauges: 12, 16, 20, 28, .410; 2.75- or 3-inch chambers. Bbls.: 26-,27- or 28-inch w/raised matte rib; choked IC/M or M/F. Weight: 5.75-7 lbs. Checkered pistol-grip walnut buttstock with beavertail forend. Engraved silvered receiver with blued bbls. Imported from Spain 1986-94.
640A (12, 16, 20 ga. w/extractors) **$625**
640A (28, .410 ga. w/extractors) 750
640 Slug Gun (12 ga. w/extractors) 735
645A (12, 16, 20 ga. w/ejectors) 695
645A (28, .410 ga. w/ejectors) 850
645A (20/28 ga. two-bbl. set) 995
650A (12 ga. w/extractors, choke tubes) 655
655A (12 ga. w/ejectors, choke tubes) 725

**Precision Sports 600 Series English Hammerless
Doubles**
Boxlock. Same general specifications as American 600 Series, except w/double triggers and concave rib. Checkered English-style walnut stock w/splinter forend, straight grip and oil finish.
640E (12, 16, 20 ga. w/extractors) **$555**
640E (28, .410 ga. w/extractors) 625
640 Slug Gun (12 ga. w/extractors) 735
645E (12, 16, 20 ga. w/ejectors) 750
645E (28, .410 ga. w/ejectors) 715
645E (20/28 ga. two-bbl. set) 945
650E (12 ga. w/extractors, choke tubes) 650
655E (12 ga. w/ejectors, choke tubes) 695

Precision Sports Model 640M Magnum 10 Hammerless Double

Similar to Model 640E, except in 10 ga. w/3.5-inch Mag. chambers. Bbls.: 26-, 30-, 32-inch choked F/F.
Model 640M Big Ten, Turkey **$655**
Model 640M Goose Gun **675**

Precision Sports Model 645E-XXV Hammerless Double

Similar to Model 645E, except w/25-inch bbl. and Churchill-style rib.
645E-XXV (12, 16, 20 ga. w/ejectors) **$725**
645E-XXV (28, .410 ga. w/ejectors) **795**

PREMIER SHOTGUNS

Premier shotguns have been produced by various gunmakers in Europe.

Premier Ambassador Model Field Grade Hammerless Double-Barrel Shotgun **$350**

Sidelock. Plain extractors. Double triggers. Gauges: 12, 16, 20, .410. 3-inch chambers in 20 and .410 ga., 2.75- inch in 12 and 16 ga. Bbls.: 26-inch in .410 ga., 28 inch in other ga.; choked M/F. Weight: 6 lbs.3 oz.-7 lbs.3 oz. depending on gauge. Checkered pistol-grip stock and beavertail forearm. Intro. in 1957; discontinued.

Premier Brush King **$250**

Same as standard Regent Model, except chambered for 12 (2.75-inch) and 20 ga. (3-inch) only; has 22-inch bbls., IC/M choke, straight-grip stock. Weight: 6 lbs. 3 oz. in 12 ga.; 5 lbs. 12 oz. in 20 ga. Introduced in 1959; discontinued.

Premier Continental Model Field Grade Hammer Double-Barrel Shotgun **$350**

Sidelock. Exposed hammers. Plain extractors. Double triggers. Gauges: 12, 16, 20, .410. 3-inch chambers in 20 and .410 ga., 2.75-inch in 12 and 16 ga. Bbls.: 26-inch in .410 ga.; 28-inch in other ga.; choked M/F. Weight: 6 lbs. 3 oz.-7 lbs. 3 oz. depending on gauge. Checkered pistol-grip stock and English-style forearm. Introduced in 1957; discontinued.

**Purdey Hammerless
Double-Barrel Shotgun**

**Purdey Hammerless
Double-Barrel Shotgun
w/Single Trigger**

Premier Monarch Supreme Grade Hammerless Double-Barrel Shotgun **$375**

Boxlock. Auto ejectors. Double triggers. Gauges: 12, 20. 2.75-inch chambers in 12 ga., 3-inch in 20 ga. Bbls.: 28-inch M/F; 26-inch IC/M choke. Weight: 6 lbs. 6 oz.-7 lbs. 2 oz. depending on gauge and bbl. Checkered pistol-grip stock and beavertail forearm of fancy walnut. Introduced in 1959; discontinued.

Premier Presentation Custom Grade **$950**

Similar to Monarch model, but made to order, of higher quality with hunting scene engraving gold and silver inlaid, fancier wood. Introduced in 1959; discontinued.

Premier Regent 10-Gauge Magnum Express **$295**

Same as standard Regent Model, except chambered for 10-ga. Magnum 3.5-inch shells, has heavier construction, 32-inch bbls. choked F/F, stock with recoil pad. Weight: 11.25 lbs. Introduced in 1957; discontinued.

Premier Regent 12-Gauge Magnum Express **$265**

Same as standard Regent Model, except chambered for 12-ga. Magnum 3-inch shells, has 30-inch bbls. choked F and F, stock with recoil pad. Weight: 7.25 lbs. Introduced in 1957; discontinued.

Premier Regent Model Field Grade Hammerless Double-Barrel Shotgun **$225**

Boxlock. Plain extractors. Double triggers. Gauges: 12,16, 20, 28, .410. 3-inch chambers in 20 and .410 ga., 2.75-inch in other gauges. Bbls.: 26-inch IC/M, M/F (28 and .410 ga. only); 28-inch M/F; 30-inch M/F (12 ga. only). Weight: 6 lbs. 2 oz.-7 lbs. 4 oz. depending on gauge and bbl. Checkered pistol-grip stock and beavertail forearm. Introduced in 1955; discontinued.

JAMES PURDEY & SONS, LTD.
London, England

Purdey Hammerless Double-Barrel Shotgun

Sidelock. Auto ejectors. Single or double triggers. Gauges: 12, 16, 20. Bbls.: 26-, 27-, 28-, 30-inch (latter in 12 ga. only);any boring, any shape or style of rib. Weight: 5.25 -5.5 lbs. depending on model, gauge and bbl length. Checkered stock and forearm, straight grip standard, pistol-grip also available. Purdey guns of this type have been made from about 1880 to date. Models include: Game Gun Featherweight Game Gun, Two-Inch Gun (chambered for 12 ga. 2-inch shells), Pigeon Gun (w/3rd fastening and side clips), values of all models are the same.
With double triggers **$31,500**
With single trigger **32,750**

Purdey Over/Under Shotgun

Sidelock. Auto ejectors. Single or double triggers. Gauges: 12 16, 20. Bbls.: 26-, 27-, 28-, 30-inch (latter in 12 ga. only); any boring, any style rib. Weight: 6-7.5 pounds depending on gauge and bbl. length. Checkered stock and forend, straight or pistol grip. Prior to WW II, the Purdey Over-and-Under Gun was made with a Purdey action; since the war James Purdey & Sons have acquired the business of James Woodward & Sons and all Purdey over/under guns are now built on the Woodward principle. General specifications of both types are the same.
With Purdey action, double triggers **$37,500**
With Woodward action, double triggers **42,000**
Single trigger, **extra** **1,500**

Purdey Single-Barrel Trap Gun **$9,300**

Sidelock. Mechanical features similar to those of the over/under model with Purdey action. 12 ga. only. Built to customer's specifications. Made prior to World War II.

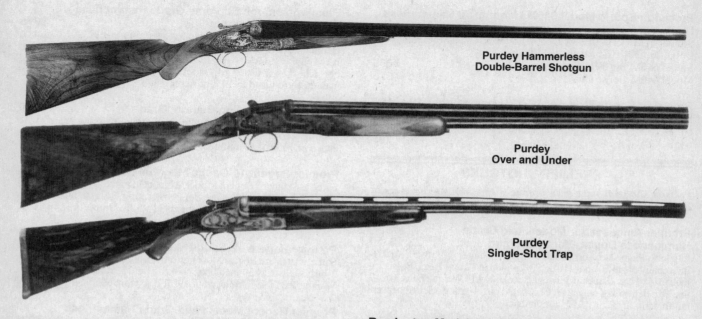

**Purdey Hammerless
Double-Barrel Shotgun**

**Purdey
Over and Under**

**Purdey
Single-Shot Trap**

REMINGTON ARMS CO.
Ilion, New York

Eliphalet Remington Jr. began making long arms with his father in 1816. In 1828 they moved their facility to Ilion, N.Y., where it remained a family-run business for decades. As the family died, other people bought controlling interests and today, still a successful gunmaking company, it is a subsidiary of the du Pont Corporation.

Remington Model 10A Standard Grade
Slide-Action Repeating Shotgun $275
Hammerless. Takedown. 6-shot. 12 ga. only. 5-shell tubular magazine. Bbls.: plain; 26- to 32-inch; choked F, M or Cyl. Weight: about 7.5 lbs. Plain pistol-grip stock, grooved slide handle. Made 1907-29.

Remington Model 11 Special, Tournament, Expert and Premier Grade Guns
These higher grade models differ from the Model 11A in general quality, grade of wood, checkering, engraving, etc. General specifications are the same.

Model 11B Special Grade	$ 425
Model 11D Tournament Grade	850
Model 11E Expert Grade	1195
Model 11F Premier Grade	1925

Remington Model 11A Standard Grade Autoloader
Hammerless Browning type. 5-shot. Takedown. Gauges: 12, 16, 20. Tubular magazine holds four shells. Bbls.: plain, solid or vent rib, lengths from 26-32 inches, F, M, IC, Cyl., SK chokes. Weight: about 8 lbs., 12 ga.; 7.5 lbs., 16 ga.; 7.25 lbs., 20 ga. Checkered pistol grip and forend. Made 1905-49.

With plain barrel	$245
With solid-rib barrel	325
With ventilated-rib barrel	365

Remington Model 11R Riot Gun $275
Same as Model 11A Standard Grade, except has 20-inch plain barrel, 12 ga. only. Remington Model 11-48. *See* Remington Sportsman-48 Series.

Remington Model 11-87 Premier Autoloader
Gas-operated. Hammerless. Gauge: 12; 3-inch chamber. Bbl.: 26-, 28- or 30-inch with REM choke. Weight: 8.13- 8.38 lbs., depending on bbl. length. Checkered walnut stock and forend in satin finish. Made 1987 to date.

Premier Deer Gun	$385
Premier Deer Gun w/Cantilever scope mount	435
Premier Skeet	495
Premier Sporting Clays	520
Premier Sporting Clays SCNP (nickel plated)	550
Premier Standard Autoloader	395
Premier Trap	465
For Left-Hand Models, add	50

**Remington Model 11-87
Premier Autoloader**

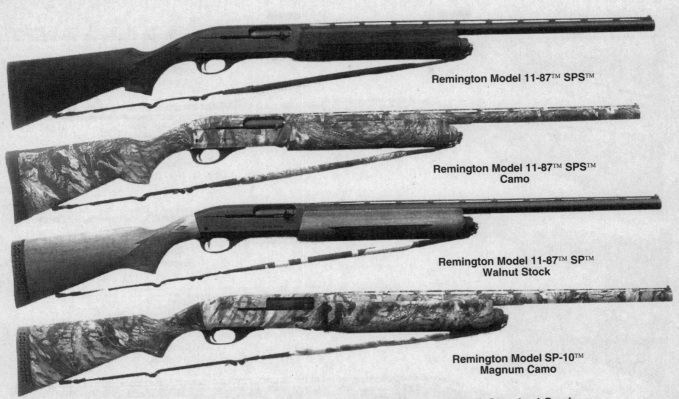

Remington Model 11-87™ SPS™

Remington Model 11-87™ SPS™ Camo

Remington Model 11-87™ SP™ Walnut Stock

Remington Model SP-10™ Magnum Camo

Remington Model 11-87 Special Purpose Magnum

Same general specifications Model 11-87 Premier, except with non-reflective wood finish and Parkerized metal. 21-, 26- or 28-inch vent-rib bbl. with REM Choke tubes. Made 1987-93.

Model 11-87 SP Field Magnum **$425**
Model 11-87 SP Deer Gun (w/21-inch bbl.) **385**
Model 11-87 SP Deer Gun w/cantilever scope mount **435**

Remington Model 11-87 SPS Magnum

Same general specifications Model 11-87 Special Purpose Magnum, except with synthetic buttstock and forend. 21-, 26- or 28-inch vent-rib bbl. with REM Choke tubes. Matte black or Mossy Oak Camo finish (except NWTF Turkey Gun). Made 1990 to date.

Model 11-87 SPS Magnum (Matte black) **$415**
Model 11-87 SPS Camo (Mossy Oak Camo) **435**
Model 11-87 SPS Deer Gun (w/21-inch bbl.) **375**
Model 11-87 SPS Deer Gun w/cantilever scope mount **425**
Model 11-87 NWTF Turkey Gun (Brown Trebark)
   Discontinued 1993 . . . . **495**
Model 11-87 NWTF Turkey Gun (Greenleaf)
   Discontinued 1996 . . . . **475**
Model 11-87 NWTF Turkey Gun (Mossy Oak)
   Discontinued 1996 . . . . **485**
Model 11-87 NWTF Turkey Gun (Mossy Oak breakup)
   Introduced 1999 . . . . **495**
Model 11-87 NWTF 20 ga. Turkey Gun (Mossy Oak Breakup)
   Produced 1998 only **465**
Model 11-87 SPST Turkey Gun (Matte black) **425**

Remington Model 11-96 EURO Lightweight Autoloading Shotgun **$565**

Lightweight version of Model 11-87 w/reprofiled receiver. 12 ga. only w/3-inch chamber. 26- or 28-inch bbl. w/6mm vent rib and REM Choke tubes. Semi-fancy Carlo walnut buttstock and forearm. Weight: 6.8 lbs. w/26-inch bbl. Made 1996 to date.

Remington Model 17A Standard Grade Slide-Action Repeating Shotgun

Hammerless. Takedown. 5-shot. 20 ga. only. 4-shell tubular magazine. Bbls.: plain; 26- to 32-inch; choked F, M or Cyl. Weight: about 5.75 lbs. Plain pistol-grip stock, grooved slide handle. Made 1921-33. *Note:* The present Ithaca Model 37 is an adaptation of this Browning design.

Plain Barrel **$295**
Solid Rib **395**

Remington Model 29A Standard Grade Slide-Action Repeating Shotgun **$350**

Hammerless. Takedown. 6-shot. 12 ga. only. 5-shell tubular magazine. Bbls.: plain- 26- to 32-inch, choked F, M or Cyl. Weight: about 7.5 lbs. Checkered pistol-grip stock and slide handle. Made 1929-33.

Remington Model 29T Target Grade **$365**

Same general specifications as Model 29A, except has trap-style stock with straight grip, extension slide handle, vent-rib bbl. Discontinued 1933.

Remington Model 31 and 31L Skeet Grade

Same general specifications as Model 31A, except has 26-inch bbl. with raised solid or vent rib, SK choke, checkered pistol-grip stock and beavertail forend. Weight: about 8 lbs., 12 ga. Made 1932-1939.

Model 31 Standard w/raised solid rib **$495**
Model 31 Standard w/ventilated rib **595**
Model 31L Lightweight w/raised solid rib **450**
Model 31L Lightweight w/ventilated rib **545**

Remington Model 31 Special, Tournament, Expert and Premier Grade Guns

These higher grade models differ from the Model 31A in general quality, grade of wood, checkering, engraving, etc. General specifications are the same.

Model 31B Special Grade . . . . **$ 475**

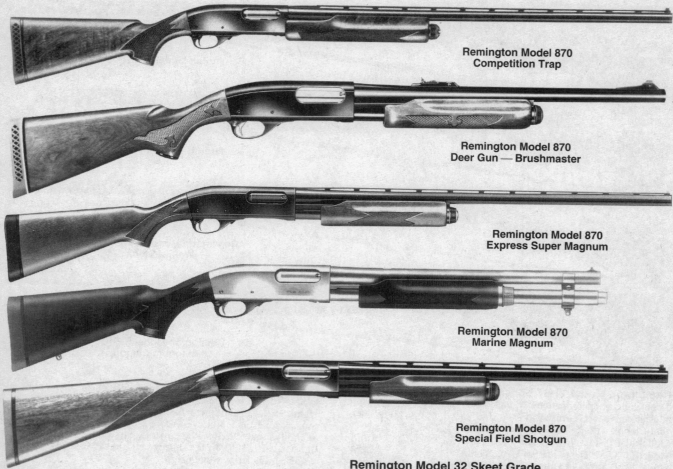

Remington Model 870
Competition Trap

Remington Model 870
Deer Gun — Brushmaster

Remington Model 870
Express Super Magnum

Remington Model 870
Marine Magnum

Remington Model 870
Special Field Shotgun

Remington Model 31 Special (Con't)
Model 31D Tournament Grade 795
Model 31E Expert Grade 1095
Model 31F Premier Grade 1695

Remington Model 31A Slide-Action Repeater
Hammerless. Takedown. 3- or 5-shot. Gauges: 12, 16, 20. Tubular magazine. Bbls.: plain, solid or vent rib; lengths from 26 -32 inches; F, M, IC, C or SK choke. Weight: about 7.5 lbs., 12 ga.; 6.75 lbs., 16 ga.; 6.5 lbs., 20 ga. Earlier models have checkered pistol-grip stock and slide handle; later models have plain stock and grooved slide handle. Made 1931-49.
Model 31A with plain barrel $325
Model 31A with solid-rib barrel 395
Model 31A with vent-rib barrel 425
Model 31H Hunters' Special w/sporting-style stock 395
Model 31R Riot Gun w/20-inch plain bbl., 12 ga. 250

Remington Model 31S Trap Special/31TC Trap Grade
Same general specifications as Model 31A, except 12 ga. only, has 30- or 32-inch vent-rib bbl., F choke, checkered trap stock with full pistol grip and recoil pad, checkered extension beavertail forend. Weight: about 8 lbs. (Trap Special has solid-rib bbl., half pistol-grip stock with standard walnut forend).
Model 31S Trap Special $425
Model 31TC Trap Grade 595

Remington Model 32 Skeet Grade
Same general specifications as Model 32A, except 26- or 28-inch bbl., SK choke, beavertail forend, selective single trigger only. Weight: about 7.5 lbs. Made 1932-42.

Remington Model 32 Tournament, Expert and Premier Grade Guns
These higher grade models differ from the Model 32A in general quality, grade of wood, checkering, engraving, etc. General specifications are the same. Made 1932-42.
Model 32D Tournament Grade $2995
Model 32E Expert Grade 3950
Model 32F Premier Grade 8500

Remington Model 32A Standard Grade Over/Under
Hammerless. Takedown. Auto ejectors. Early model had double triggers, later built with selective single trigger only. 12 ga. only. Bbls.: plain, raised matted solid or vent rib; 26-, 28-, 30-, 32-inch; F/M choke standard, option of any combination of F, M, IC, C, SK choke. Weight: about 7.75 lbs. Checkered pistol-grip stock and forend. Made 1932-42.
With double triggers $1550
With selective single trigger 1750
Extra for raised solid rib 200
Extra for ventilated rib 300

Remington Model 32TC Target (Trap) Grade $2695
Same general specifications as Model 32A, except 30- or 32-inch vent-rib bbl., F choke, trap-style stock with chleckered pistol-grip and beavertail forend. Weight: about 8 lbs. Made 1932-42.

Remington Model 89 (1989) **$950**
Hammers. Circular action. Gauges: 10, 12, 16, 28- to 32-inch bls.; steel or damascus twist. Weight 7-10 lbs. Made 1889-1908.

Remington Model 90-T Single-Shot Trap **$1995**
Gauge: 12; 2.75-inch chambers. 30-, 32- or 34-inch vent-rib bbl. with fixed chokes or screw-in REM Chokes; ported or non-ported. Weight: 8.25 lbs. Checkered American walnut standard or Monte Carlo stock with low-luster finish. Engraved sideplates and drop-out trigger group optional. Made 1990-97.

Remington Model 396 Over/Under
Boxlock. 12 ga. only w/2.75-inch chamber. 28- and 30-inch blued bbls. w/Rem chokes. Weight: 7.50 lbs. Nitride-grayed, engraved receiver, trigger guard, tang, hinge pins and forend metal. Engraved sideplates. Checkered satin-finished American walnut stock w/target style forend. Made 1996 to date.
396 Sporting Clays **$1350**
396 Skeet 1395

Remington Model 870
"All American" Trap Gun

Remington Model 870 "All American" Trap Gun **$725**
Same as Model 870TB, except custom grade with engraved receiver, trigger guard and bbl.; Monte Carlo or straight-comb stock and forend of fancy walnut; available only with 30-inch F choke bbl. Made 1972-77.

Remington Model 870 Competition Trap **$450**
Based on standard Model 870 receiver, except is single-shot with gas-assisted recoil-reducing system, new choke design, a high step-up vent rib, and redesigned stock, forend with cut checkering and a satin finish. Weight: 8.5 lbs. Made 1981-87.

Remington Model 870 Deer Gun Brushmaster Deluxe
Same as Model 870 Standard Deer Gun, except available in 20 ga. as well as 12, has cut-checkered, satin-finished American walnut stock and forend, recoil pad.
Right-Hand Model **$295**
Left-Hand Model 345

Remington Model 870 Deer Gun Standard **$295**
Same as Model 870 Wingmaster Riot Gun, except has rifle-type sights.

Remington Model 870 Express
Same general specifications Model 870 Wingmaster, except has low-luster walnut-finished hardwood stock with pressed checkering and black recoil pad. Gauges: 12, 20 or .410- 3-inch chambers. Bbls.: 26- or 28-inch vent-rib with REM Choke; 25-inch vent-rib with fixed choke (.410 only). Black oxide metal finish. Made 1987 to date.
Model 870 Express (12 or 20 ga., REM Choke) **$200**
Model 870 Express (.410 w/fixed choke) 215
Express Combo (w/extra 20-inch Deer Bbl.) 270

Remington Model 870 Express Deer Gun
Same general specifications as Model 870 Express, except in 12 ga. only, 20-inch bbl. with fixed IC choke, adj. rifle sights and Monte Carlo stock. Made 1991 to date.
Express Deer Gun w/standard barrel **$205**
Express Deer Gun w/fully rifled barrel 230

Remington Model 870 Express Super Magnum
Similar to Model 870 Express, except chambered 12 ga. mag. w/3.5-inch chamber. Bbls.: 23-, 26- or 28-inch vent rib w/REM Choke. Checkered low-luster walnut-finished hardwood, black synthetic or camo buttstock and forearm. Matte black oxide metal finish or full camo finish. Made 1998 to date.
Model 870 ESM (w/hardwood stock) **$185**
Model 870 ESM (w/black synthetic stock) 195
Model 870 ESM (w/camo synthetic stock) 265
Model 870 ESM Synthetic Turkey (w/synthetic stock) 195
Model 870 ESM Como Turkey (w/full camo) 250
Model 870 ESM Combo (w/full camo, extra bbl.) 295

**Remington Model 870 Express Synthetic Home
Defense** **$185**
Slide action, hammerless, takedown. 12 ga. only. 18-inch bbl. w/cylinder choke and bead front sight. Positive checkered synthetic stock and forend with non-reflective black finish. Made 1995 to date.

Remington Model 870 Express Turkey Gun **$195**
Same general specifications as Model 870 Express, except has 21-inch vent-rib bbl. and Turkey Extra-Full REM choke. Made 1991 to date.

Remington Model 870 Express Youth Gun **$185**
Same general specifications as Model 870 Express, except has scaled-down stock with 12.5-inch pull and 21-inch vent-rib bbl. with REM choke. Made 1991 to date.

Remington Model 870 Lightweight
Same as standard Model 870, but with scaled-down receiver and lightweight mahogany stock; 20 ga. only. 2.75-inch chamber. Bbls.: plain or vent rib; 26-inch, IC; 28-inch, M or F choke. REM choke available from 1987. Weight 5.75 lbs. w/26-inch plain bbl. American walnut stock and forend with satin or Hi-gloss finish. Made 1972-94.
With plain barrel **$235**
With ventilated-rib barrel 260
With REM choke barrel 295

Remington Model 870 Lightweight Magnum
Same as Model 870 Lightweight, but chambered for 20 ga. Magnum 3-inch shell; 28-inch bbl., plain or vent rib, F choke. Weight: 6 lbs. with plain bbl. Made 1972-94.
With plain barrel **$285**
With ventilated-rib barrel 325

Remington Model 870 Magnum Duck Gun
Same as Model 870 Field Gun, except has 3-inch chamber 12 and 20 gauge Magnum only. 28- or 30-inch bbl., plain or vent rib, M or F choke, recoil pad. Weight: about 7 or 6.75 lbs. Made 1964 to date.
With plain barrel **$295**
With ventilated-rib barrel 325

Remington Model 870 Marine Magnum **$295**
Same general specifications as Model 870 Wingmaster except with 7-shot magazine, 18-inch plain bbl. with fixed IC choke, bead front sight and electroless nickel finish. Made 1992 to date.

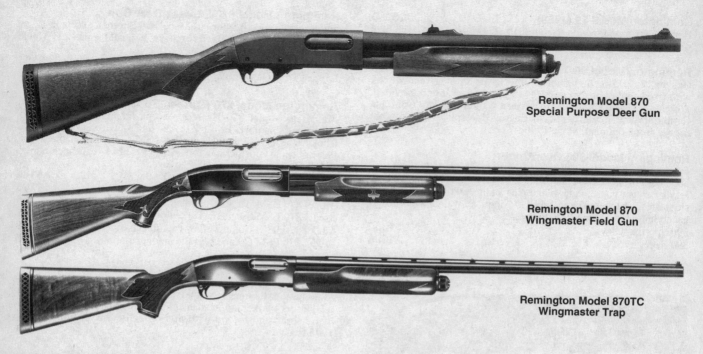

Remington Model 870
Special Purpose Deer Gun

Remington Model 870
Wingmaster Field Gun

Remington Model 870TC
Wingmaster Trap

Remington Model 870 SA Skeet Gun, Small Bore $325
Similar to Wingmaster Model 870SA, exeept chambered for 28 and .410 ga. (2.5-inch chamber for latter); 25- inch vent-rib bbl., SK choke. Weight: 6 lbs., 28 ga.; 6.5 lbs., .410. Made 1969-82.

Remington Model 870 Mississippi Magnum Duck Gun $350
Same as Remington Model 870 Magnum Duck Gun except has 32-inch bbl. Engraved receiver, "Ducks Unlimited." Made in 1983.

Remington Model 870 Special Field Shotgun $295
Pump action. Hammerless. Gauge: 12 or 20. 21-inch vent-rib bbl. with REM choke. 41.5 inches overall. Weight: 6-7 lbs. Straight-grip checkered walnut stock and forend. Made 1987-95.

Remington Model 870 Special Purpose Deer Gun
Similar to Special Purpose Magnum, except with 20-inch IC choke, rifle sights. Matte black oxide and Parkerized finish. Oil-finished, checkered buttstock and forend with recoil pad. Made 1986 to date.
Model 870 SP Deer Gun $275
Model 870 SP Deer Gun, cantilever scope mount 325

Remington Model 870 Special Purpose Magnum $295
Similar to the 870 Magnum Duck Gun, except with 26-, 28- or 30-inch vent-rib REM Choke bbl.12 ga. only; 3-inch chamber. Oil-finished field-grade stock with recoil pad, QD swivels and Cordura sling. Made 1985 to date.

Remington Model 870SPS Magnum
Same general specifications Model 870 Special Purpose Magnum, except with synthetic stock and forend. 26- or 28-inch vent-rib bbl. with REM Choke tubes. Matte black or Mossy Oak Camo finish. Made 1991 to date.
Model 870 SPS Magnum (black syn. stock) $270
Model 870 SPS-T Camo (Mossy Oak Camo) 280

Remington Model 870 Wingmaster Field Gun
Same general specifications as Model 870AP, except checkered stock and forend. Later models have REM choke systems in 12 ga. Made 1964 to date.
With plain barrel $250
With ventilated-rib barrel 295

Remington Model 870 Wingmaster Field Gun, Small Bore
Same as standard Model 870, except w/scaled down lightweight receivers. Gauges: 28 and .410. Plain or vent rib 25-inch bbl. choked IC, M or F. Weight: 5.5-6.25 lbs. depending on gauge and bbl. Made 1969-94.
With plain barrel $350
With ventilated-rib barrel 395

Remington Model 870 Wingmaster Magnum Deluxe Grade $325
Same as Model 870 Magnum Standard Grade, except has checkered stock and extension beavertail forearm, bbl. with matted top surface. Discontinued in 1963.

Remington Model 870 Wingmaster Magnum Standard Grade $285
Same as Model 870AP, except chambered for 12 ga. 3-inch Magnum, 30-inch F choke bbl., recoil pad. Weight: about 8.25 lbs. Made 1955-63.

Remington Model 870 Wingmaster REM Choke Series
Slide action, hammerless, takedown with blued all-steel receiver. Gauges: 12, 20; 3-inch chamber. Tubular magazine. Bbls.: 21-, 26-, 28-inch vent-rib with REM Choke. Weight: 7.5 lbs. (12 ga.). Satin-finished, checkered walnut buttstock and forend with recoil pad. Right- or left-hand models. Made 1986 to date.
Standard Model, 12 ga. $295
Standard Model, 20 ga. 315
Youth Model, 21-inch barrel 310

Remington Model 870ADL Wingmaster Deluxe Grade

Same general specifications as Wingmaster Model 870AP except has pistol-grip stock and extension beavertail forend, both finely checkered; matted top surface or vent-rib bbl. Made 1950-63.

With matted top-surface barrel **$235**
With ventilated-rib barrel **265**

Remington Model 870AP Wingmaster Standard Grade

Hammerless. Takedown. Gauges: 12, 16, 20. Tubular magazine holds four shells. Bbls.: plain, matted top surface or vent rib; 26-inch IC, 28-inch M or F choke, 30-inch F choke (12 ga. only). Weight: about 7 lbs., 12 ga.; 6.75 lbs., 16 ga.; 6.5 lbs., 20 ga. Plain pistol-grip stock, grooved forend. Made 1950-63.

With plain barrel **$200**
With matted top-surface barrel **210**
With ventilated-rib barrel **230**
Left-hand model **240**

Remington Model 870BDL Wingmaster Deluxe Special

Same as Model 870ADL, except select American walnut stock and forend. Made 1950-63.

With matted top-surface barrel **$275**
With ventilated-rib barrel **305**

Remington Model 870D, 870F Wingmaster Tournament and Premier Grade Guns

These higher grade models differ from the Model 870AP in general quality, grade of wood, checkering, engraving, etc. General operating specifications are essentially the same. Made 1950 to date.

Model 870D Tournament Grade **$1795**
Model 870F Premier Grade **3950**
Model 870F Premier Grade with gold inlay **5995**

Remington Model 870R Wingmaster Riot Gun $280

Same as Model 870AP, except 20-inch bbl., IC choke, 12 ga. only.

Remington Model 870SA Wingmaster Skeet Gun

Same general specifications as Model 870AP, except has 26-inch vent-rib bbl., SK choke, ivory bead front sight, metal bead rear sight, pistol-grip stock and extension beavertail forend. Weight: 6.75 to 7.5 lbs. depending on gauge. Made 1950-82.

Model 870SA Skeet Grade (Disc. 1982) **$275**
Model 870SC Skeet Target Grade (Disc 1980) **395**

Remington Model 870TB Wingmaster Trap Special $350

Same general specifications as Model 870AP Wingmaster except has 28- or 30-inch vent-rib bbl., F choke, metal bead front sight, no rear sight. "Special" grade trap-style stock and forend, both checkered, recoil pad. Weight: about 8 lbs. Made 1950-81.

Remington Model 870TC Trap Grade $495

Same as Model 870 Wingmaster TC, except has tourmament grade walnut in stock and forend w/satin finish. Over-bored 30-inch vent rib bbl. w/ 2.75-inch chamber and Rem-Choke tubes. Reissued in 1996. *See separate listing for earlier model.*

Remington Model 870TC Wingmaster Trap Grade

Same as Model 870TB, except higher grade walnut in stock and forend, has both front and rear sights. Made 1950-79. Model 870 TC reissued in 1996. *See separate listing for later model.*

Model 870 TC Trap (Standard) **$475**
Model 870 TC Trap (Monte Carlo) **495**

Remington Model 878A Automaster $225

Gas-operated Autoloader. 12 ga., 3-shot magazine. Bbls.: 26-inch IC, 28-inch M choke, 30-inch F choke. Weight: about 7 lbs. Plain pistol-grip stock and forearm. Made 1959-62.

> ### NOTE
>
> New stock checkering patterns and receiver scroll markings were incorporated on all standard Model 1100 field, magnum, skeet and trap models in 1979.

Remington Model 1100 Automatic Field Gun

Gas-operated. Hammerless. Takedown. Gauges: 12, 16, 20. Bbls.: plain or vent. rib; 30-inch F, 28-inch M or F, 26-inch IC; or REM choke tubes. Weight: average 7.25-7.5 lbs. depending on ga. and bbl. length. Checkered walnut pistol-grip stock and forearm in high-gloss finish. Made 1963 to date. 16 ga. discontinued.

With plain barrel **$265**
With ventilated-rib barrel **305**
REM choke model **335**
REM chokes, Left-hand action **365**

SHOTGUNS

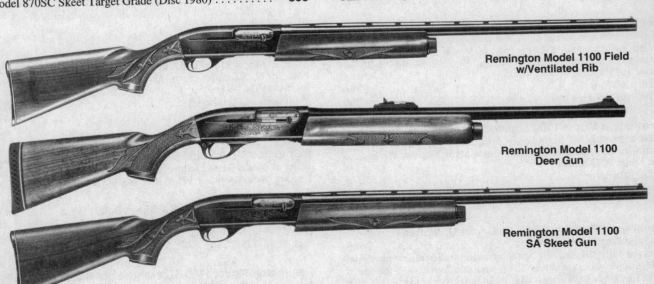

Remington Model 1100 Field
w/Ventilated Rib

Remington Model 1100
Deer Gun

Remington Model 1100
SA Skeet Gun

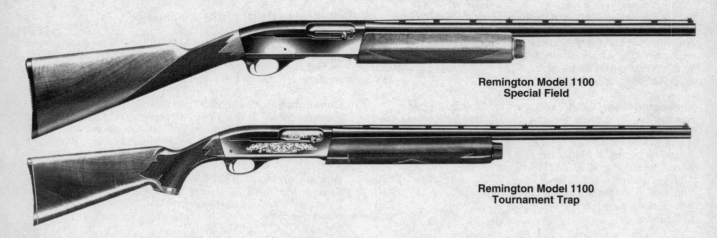

**Remington Model 1100
Special Field**

**Remington Model 1100
Tournament Trap**

Remington Model 1100 Deer Gun . . . . $3?5
Same as Model 1100 Field Gun, except has 22-inch barrel, IC, with rifle-type sights; 12 and 20 ga. only; recoil pad. Weight: about 7.25 lbs. Made 1963-96.

**Remington Model 1100 Ducks Unlimited Atlantic
Commemorative** **$695**
Limited production for one year. Similar specifications to Model 1100 Field, except with 32-inch F choke, vent-rib bbl. 12-ga. Magnum only. Made in 1982.

**Remington Model 1100 Ducks Unlimited "The
Chesapeake" Commemorative** **$525**
Limited edition 1 to 2400. Same general specifications as Model 1100 Field, except sequentially serial numbered with markings "The Chesapeake." 12-ga. Magnum with 30-inch F choke, vent-rib bbl. Made in 1981.

Remington Model 1100 Field Grade, Small Bore
Same as standard Model 1100, but scaled down. Gauges: 28, .410.25-inch bbl., plain or vent rib; IC, M or F choke. Weight: 6.25-7 lbs. depending on gauge and bbl. Made 1969-94.
With plain barrel **$385**
With ventilated rib **450**

Remington Model 1100 Lightweight
Same as standard Model 1100, but scaled-down receiver and lightweight mahogany stock; 20 ga. only, 2.75-inch chamber. Bbls.: plain or vent rib; 26-inch IC; 28-inch M and F choke. Weight: 6.25 lbs. Made 1971 to date.
With plain barrel **$350**
With ventilated rib **395**

Remington Model 1100 Lightweight Magnum
Same as Model 1100 Lightweight, but chambered for 20 gauge Magnum 3-inch shell; 28-inch bbl., plain or vent rib, F choke. Weight: 6.5 lbs. Made 1971 to date.
With plain barrel **$375**
With ventilated rib **420**
With choke tubes **450**

**Remington Model 1100 LT-20 Ducks Unlimited
Special Commemorative** **$495**
Limited edition 1 to 2400. Same general specifications as Model 1100 Field, except sequentially serial numbered with markings "The Chesapeake." 20 ga. only. 26-inch IC, vent-rib bbl. Made in 1981.

Remington Model 1100 LT-20 Series
Same as Model 1100 Field Gun, except in 20 ga. with shorter 23-inch vent-rib bbl., straight-grip stock. REM choke series has 21-inch vent-rib bbl., choke tubes. Weight: 6.25 lbs. Checkered grip and forearm. Made 1983 to date.
Model 1100 LT-20 Special **$385**
Model 1100 LT-20 Deer Gun **365**
Model 1100 LT-20 Youth **375**

Remington Model 1100 Magnum **$350**
Limited production. Similar to the Model 1100 Field, except with 26-inch F choke, vent-rib bbl. and 3-inch chamber. Made in 1981.

Remington Model 1100 Magnum Duck Gun
Same as Model 1100 Field Gun, except has 3-inch chamber, 12 and 20 ga. Mag. only. 30-inch plain or vent-rib bbl. in 12 ga., 28-inch in 20 ga.; M or F choke. Recoil pad. Weight: about 7.75 lbs. Made 1963-88.
With plain barrel **$295**
With ventilated-rib barrel **335**

Remington Model 1100 One of 3000 Field **$995**
Limited edition, numbered 1 to 3000. Similar to Model 1100 Field, except with fancy wood and gold-trimmed etched hunting scenes on receiver. 12 gauge with 28-inch Mod., vent-rib bbl. Made in 1980.

Remington Model 1100 SA Skeet Gun
Same as Model 1100 Field Gun, 12 and 20 ga., except has 26-inch vent-rib bbl., SK choke or with Cutts Compensator. Weight: 7.25-7.5 lbs. Made 1963-94.
With skeet-choked barrel **$390**
With Cutts Comp **410**
Left-hand action **425**

Remington Model 1100 SA Lightweight Skeet **$375**
Same as Model 1100 Lightweight, except has skeet-style stock and forearm, 26-inch vent-rib bbl., SK choke. Made 1971-97.

Remington Model 1100 SA Skeet Small Bore **$425**
Similar to standard Model 1100SA, except chambered for 28 and .410 ga. (2.5-inch chamber for latter); 25-inch vent-rib bbl., SK choke. Weight: 6.75 lbs., 28 ga.; 7.25 lbs., .410. Made 1969 to date.

Remington Model 1100 SB Lightweight Skeet **$395**
Same as Model 1100SA Lightweight, except has select wood. Introduced in 1977.

Remington Model 3200
"One of 1000" Skeet

Remington Model 1100 SB Skeet Gun $375
Same specifications as Model 1100SA, except has select wood.
Made 1963-97.

Remington Model 1100 Special Field Shotgun $395
Gas-operated. 5-shot. Hammerless. Gauges: 12 and 20. 21-inch
vent-rib bbl. with REM choke. Weight: 6.5-7.25 lbs. Straight-grip
checkered walnut stock and forend. Made 1987 to date.

Remington Model 1100 SP Magnum
Same as Model 1100 Field, except 12 ga. only with 3-inch chambers.
Bbls.: 26- or 30-inch F choke; or 26-inch with REM Choke tubes; vent
rib. Non-reflective matte black, Parkerized bbl. and receiver. Satin-
finished stock and forend. Made 1981 to date.
With Fixed Choke $325
With REM Choke 365

Remington Model 1100 Tournament and Premier
These higher grade guns differ from standard models in overall qual-
ity, grade of wood, checkering, engraving, gold inlays, etc. General
specifications are the same. Made 1963 to date.
Model 1100D Tournament $1695
Model 1100F Premier 3595
Model 1100F Premier with gold inlay 4595

Remington Model 1100 Tournament Skeet $475
Similar to Model 1100 Field, except with 26-inch bbl. SK choke.
Gauges: 12, LT-20, 28, and .410. Features select walnut stocks and
new cut-checkering patterns. Made 1979-89.

Remington Model 1100TA Trap Gun $325
Similar to Model 1100TB Trap Gun, except with regular-grade stocks.
Available in both left- and right-hand versions. Made 1979-86.

Remington Model 1100TB Trap Gun
Same as Model 1100 Field Gun, except has special trap stock,
straight or Monte Carlo comb, recoil pad; 30-inch vent-rib bbl., F
or M trap choke; 12 ga. only. Weight: 8.25 lbs. Made 1963-79.
With straight stock $395
With Monte Carlo stock 415

Remington Model 1900 Hammerless Double $850
Improved version of Model 1894. Boxlock. Auto ejector. Double triggers. Gauges: 10, 12, 16. Bbls.: 28 to 32 inches. Value shown is for
standard grade with ordnance steel bbls. Made 1900-10.

Remington Model 3200 Competition
Skeet Gun $1450
Same as Model 3200 Skeet Gun, except has gilded scrollwork on
frame, engraved forend latch plate and trigger guard, select fancy
wood. Made 1973-84.

Remington Model 3200 Competition Skeet Set $5395
Similar specifications to Model 3200 Field. 12-ga. O/U with additional,
interchangeable bbls. in 20, 28, and .410 ga. Cased. Made 1980-84.

Remington Model 3200 Competition Trap Gun $1395
Same as Model 3200 Trap Gun, except has gilded scrollwork on
frame, engraved forend latch plate and trigger guard, select fancy
wood. Made 1973-84.

Remington Model 3200 Field Grade Magnum $1195
Same as Model 3200 Field, except chambered for 12 ga. mag. 3-inch
shell- 30-inch bbls., M and F or both F choke. Made 1975-84.

Remington Model 3200 Field Grade O/U $995
Boxlock. Auto ejectors. Selective single trigger. 12 ga. 2.75-inch
chambers. Bbls.: vent rib, 26- and 28-inch M/F; 30-inch IC/M.
Weight: about 7.75 lbs. with 26-inch bbls. Checkered pistol-grip
stock/forearm. Made 1973-78.

Remington Model 3200 "One ol 1000" Skeet $1795
Same as Model 3200 "One of 1000" Trap, except has 26- or 28-inch
bbls., SK choke, skeet-style stock and forearm. Made in 1974.

Remington Model 3200 "One of 1000" Trap $1825
Limited edition numbered 1 to 1000. Same general specifications as
Model 3200 Trap Gun, but has frame, trigger guard and forend latch
elaborately engraved (designation "One of 1,000" on frame side),
stock and forearm of high grade walnut. Supplied in carrying case.
Made in 1973.

Remington Model 3200 Skeet Gun $1025
Same as Model 3200 Field Grade, except skeet-style stock and full
beavertail forearm, 26- or 28-inch bbls., SK choke. Made 1973-80.

Remington Model 3200 Special Trap Gun $1250
Same as Model 3200 Trap Gun, except has select fancy-grade wood
and other minor refinements. Made 1973-84.

Remington Model 3200 Trap Gun $1150
Same as Model 3200 Field Grade, except trap-style stock w/Monte
Carlo or straight comb, select wood, beavertail forearm, 30- or 32-
inch bbls. w/ventilated rib, IM/F or F/F chokes. Made 1973-77.

Remington Rider No. 9 Single-Shot Shotgun $325
Improved version of No. 3 Single Barrel Shotgun made in the late
1800s. Semihammerless. Gauges 10, 12, 16, 20, 24, 28. 30- to 32-
inch plain bbl. Weight: about 6 lbs. Plain pistol-grip stock and forearm. Auto ejector. Made 1902-10.

Remington SP-10 Magnum Autoloader $750
Takedown. Gas-operated with stainless steel piston. 10 ga., 3.5-
inch chamber. Bbls.: 26- or 30-inch vent-rib with REM Choke
screw-in tubes. Weight: 11 to 11.25 lbs. Metal bead front. Checkered walnut stock with satin finish. Made 1989 to date.

Remington SP-10 Magnum Turkey Combo $795
Same general specifications as Model SP-10 Magnum, except has
extra 22-inch REM Choke bbl. with M, F and Turkey Extra-Full
tubes. Rifle sights. QD swivels and camo sling. Made 1991 to date.

SHOTGUNS

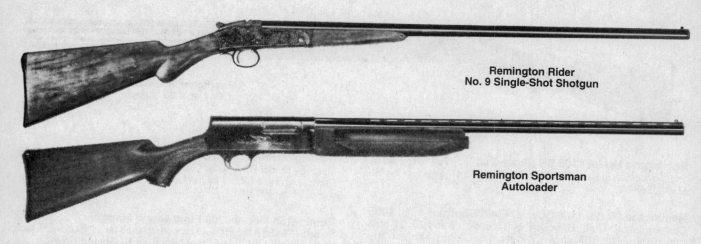

**Remington Rider
No. 9 Single-Shot Shotgun**

**Remington Sportsman
Autoloader**

Remington Peerless Over/Under $750
Boxlock action and removable, engraved sideplates. Gauge:12 only with 3-inch chambers. Barrels: 26-, 28-, or 30-inch with vent rib and REM choke system. Automatic safety and single selective trigger. Weight: 7.25 lbs. to 7.5 lbs. Blued receiver and bbls. Checkered American walnut stock. Made 1993 to date.

Remington Sportsman A Standard Grade Autoloader
Same general specifications as Model 11A, except magazine holds two shells. Also available in "B" Special Grade, "D" Tournament Grade, "E" Expert Grade, "F" Premier Grade. Made 1931-48. Same values as for Model 11A.

Remington Sportsman Skeet Gun
Same general specifications as the Sportsman A, except has 26-inch bbl. (plain, solid or vent rib), SK choke, beavertail forend. Discont. 1949.
With plain barrel $325
With solid-rib barrel 395
With ventilated-rib barrel 435

Remington Sportsman-48A Standard Grade 3-Shot Autoloader
Streamlined receiver. Hammerless. Takedown. Gauges: 12, 16, 20. Tubular magazine holds two shells. Bbls.: plain, matted top surface or vent rib; 26-inch IC, 28-inch M or F choke, 30-inch F choke (12 ga. only). Weight: about 7.5 lbs., 12 ga.; 6.25 lbs., 16 ga.; 6.5 lbs., 20 ga. Pistol-grip stock, grooved forend, both checkered. Made 1949-59.
With plain bbl. $295
With matted top-surface bbl. 325
With ventilated-rib bbl. 345

Remington Sportsman-48 B, D, F Special, Tournament and Premier Grade Guns
These higher grade models differ from the Sportsman-48A in general quality, grade of wood, checkering, engraving, etc. General specifications are the same. Made 1949-59.
Sportsman-48B Special Grade $355

Remington Sportsman-48 B, D, F Special *(Con't)*
Sportsman-48D Tournament Grade 775
Sportsman-48F Premier Grade 1695

Remington Sportsman-48SA Skeet Gun
Same general specifications as Sportsman-48A, except has 26-inch bbl. with matted top surface or vent rib, SK choke, ivory bead front sight, metal bead rear sight. Made 1949-60.
With matted top-surface barrel $275
With ventilated-rib barrel 325
Sportsman-48SC Skeet Target Grade 400
Sportsman-48SD Skeet Tournament Grade 725
Sportsman-48SF Skeet Premier Grade 1695

Remington Model 11-48A Riot Gun $255
Same as Model 11-48A, except 20-inch plain barrel and 12 ga. only. Discontinued in 1969.

Remington Model 11-48A Standard Grade 4-Shot Autoloader .410 & 28 Gauge
Same general specifications as Sportsman-48A, except gauge, 3-shell magazine, 25-inch bbl. Weight: about 6.25 lbs. 28 ga. introduced 1952, .410 in 1954. Discontinued in 1969. Values same as shown for Sportsman-48A.

Remington Model 11-48A Standard Grade 5-Shot Autoloader
Same general specifications as Sportsman-48A, except magazine holds four shells, forend not grooved. Also available in Special Grade (11-48B), Tournament Grade (11-48D) and Premier Grade (11-48F). Made 1949-69. Values same as shown for Sportsman-48A.

Remington Model 11-48SA .410E28 Ga. Skeet $325
Same general specifications as Model 11-48A 28 Gauge except has 25-inch vent-rib bbl., SK choke. 28 ga. introduced 1952, .410 in 1954.

Remington Sportsman — 48A

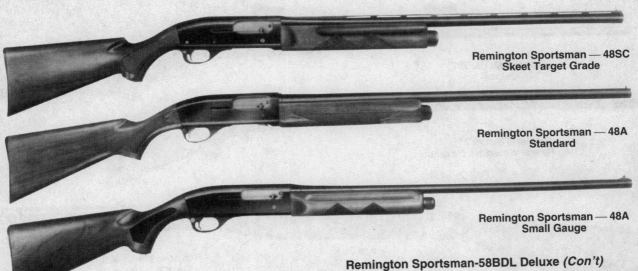

Remington Sportsman — 48SC
Skeet Target Grade

Remington Sportsman — 48A
Standard

Remington Sportsman — 48A
Small Gauge

Remington Sportsman-58 Skeet Target, Tournament and Premier Grades

These higher grade models differ from the Sportsman-58SA in general quality, grade of wood, checkering, engraving, and other refinements. General operating and physical specifications are the same.

Sportsman-58C Skeet Target Grade	**$450**
Sportsman-58D Skeet Tournament Grade	695
Sportsman-58SF Skeet Premier Grade	1295

Remington Sportsman-58 Tournament and Premier

These higher grade models differ from the Sportsman-58ADL with vent-rib bbl. in general quality, grade of wood, checkering, engraving, etc. General specifications are the same.

Sportsman-58D Tournament Grade	**$750**
Sportsman-58F Premier Grade	1395

Remington Sportsman-58ADL Autoloader

Deluxe Grade. Gas-operated. 12 ga. 3-shot magazine. Bbls.: plain or vent rib, 26-, 28- or 30-inch; IC, M or F choke, or Remington Special Skeet choke. Weight: about 7 lbs. Checkered pistol-grip stock and forearm. Made 1956-64.

With plain barrel	**$295**
With ventilated-rib barrel	325

Remington Sportsman-58BDL Deluxe Special Grade

Same as Model 58ADL, except select grade wood.

Remington Sportsman-58BDL Deluxe *(Con't)*

With plain barrel	**$295**
With ventilated-rib barrel	345

Remington Sportsman-58SA Skeet Grade $335

Same general specifications as Model 58ADL with vent-rib bbl., except special skeet stock and forearm.

REVELATION SHOTGUNS
See Western Auto listings.

RICHLAND ARMS COMPANY
Blissfield, Michigan
Manufactured in Italy and Spain

Richland Model 200 Field Grade Double **$275**
Hammerless, boxlock, Anson & Deeley-type. Plain extractors. Double triggers. Gauges: 12, 16, 20, 28, .410 (3-inch chambers in 20 and .410; others have 2.75-inch). Bbls.: 28-inch M/F choke, 26-inch IC/M; .410 with 26-inch M/F only; 22-inch IC/M in 20 ga. Weight: 6 lbs. 2 oz. to 7 lbs. 4 oz. Checkered walnut stock with cheekpiece, pistol grip, recoil pad; beavertail forend. Made in Spain 1963 to date.

Richland Model 202 All-Purpose Field Gun **$295**
Hammerless, boxlock, Anson & Deeley-type. Same as Model 200, except has two sets of barrels same gauge. 12 ga.: 30-inch bbls. F/F, 3-inch chambers; 26-inch bbls. IC/M, 2.75-inch chambers. 20 gauge: 28-inch bbls. M/F; 22-inch bbls. IC/M,3-inch chambers. Made 1963 to date.

Richland Model 200

Richland Model 202

SHOTGUNS

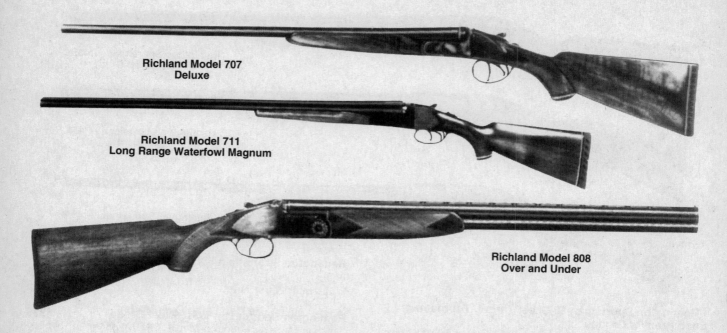

Richland Model 707
Deluxe

Richland Model 711
Long Range Waterfowl Magnum

Richland Model 808
Over and Under

Richland Model 707 Deluxe Field Gun $315
Hammerless, boxlock, triple bolting system. Plain extractors. Double triggers. Gauges: 12, 2.75-inch chambers; 20, 3-inch chambers. Bbls.: 12 ga., 28-inch M/F, 26-inch IC/M; 20 ga., 30-inch F/F, 28-inch M/F, 26-inch IC/M. Weight: 6 lbs. 4 oz. to 6 lbs. 15 oz. Checkered walnut stock and forend, recoil pad. Made 1963-72.

Richland Model 711 Long Range Waterfowl Magnum Double-Barrel Shotgun
Hammerless, boxlock, Anson & Deeley-type, Purdey triple lock. Plain extractors. Double triggers. Auto safety. Gauges: 10, 3.5-inch chambers; 12, 3-inch chambers. Bbls.: 10 ga., 32-inch; 12 ga., 30-inch; F/F. Weight: 10 ga., 11 pounds; 12 ga., 7.75 lbs. Checkered walnut stock and beavertail forend; recoil pad. Made in Spain 1963 to date.
10 Gauge Magnum $325
12 Gauge Magnum 275

Richland Model 808 Over-and-Under Gun $345
Boxlock. Plain extractors. Non-selective single trigger. 12 ga. only. Bbls. (Vickers steel): 30-inch F/F; 28-inch M/F; 26-inch IC/M. Weight: 6 lbs. 12 oz. to 7 lbs. 3 oz. Checkered walnut stock/forend. Made in Italy 1963-68.

JOHN RIGBY & COMPANY
London, England

Rigby
Regal Side Lock

Rigby Hammerless Box Lock Double-Barrel Shotguns
Auto ejectors. Double triggers. Made in all gauges, barrel lengths and chokes. Checkered stock and forend, straight grip standard. Made in two grades: Sackville and Chatsworth. These guns differ in general quality, engraving, etc.; specifications are the same.
Sackville Grade $4995
Chatsworth Grade 3695

Rigby Hammerless Side Lock Double-Barrel Shotguns
Auto ejectors. Double triggers. Made in all gauges, barrel lengths and chokes. Checkered stock and forend, straight grip standard. Made in two grades: Regal (best quality) and Sandringham; these guns differ in general quality, engraving, etc., specifications are the same.
Regal Grade $9250
Sandringham Grade 6595

AMADEO ROSSI, S.A.
Sao Leopoldo, Brazil

Rossi Hammerless Double-Barrel Shotgun $265
Boxlock. Plain extractors. Double triggers. 12 ga. 3-inch chambers. Bbls.: 26-inch IC/M; 28-inch M/F choke. Weight: 7 to 7.5 lbs. Pistol-grip stock and beavertail forearm, uncheckered. Made 1974 to date. *Note:* H&R Model 404 (1969-72) is same gun.

Rossi Overland Hammer Double $225
Sidelock. Plain extractors. Double triggers. Gauges: 12, .410; 3-inch chambers. Bbls.: 20-inch, IC/M in 12 g.; 26-inch, F/F choke in .410. Weight: 7 lbs., 12 ga.; 6 lbs. .410. Pistol-grip stock and beavertail forearm, uncheckered. *Note:* Because of its resemblance to the short-barreled doubles carried by guards riding shotgun on 19th-century stagecoaches, the 12 ga. version originally was called the "Coach Gun." Made 1968-89.

ROTTWEIL SHOTGUNS
West Germany

**Rossi Hammerless
Double-Barrel Shotgun**

**Rossi Overland
Hammer Double**

Rottweil Model 72 Over/Under Shotgun **$1595**
Hammerless, takedown with engraved receiver. 12 ga.; 2.75-inch chambers. 26.75-inch bbls. with SK/SK chokes. Weight: 7.5 lbs. Interchangeable trigger groups and buttstocks. Checkered French walnut buttstock and forend. Imported from West Germany.

Rottweil Model 650 Field O/U Shotgun **$625**
Breech action. Gauge: 12. 28-inch bbls. Six screw-in choke tubes. Automatic ejectors. Engraved receiver. Checkered pistol grip stock. Made 1984-86.

Rottweil American Skeet **$1550**
Boxlock action. Gauge: 12. 27-inch vent-rib bbls. 44.5 inches overall. SK chokes. Weight: 7.5 lbs. Designed for tube sets. Hand-checkered European walnut stock with modified forend. Made 1984-87.

Rottweil International Trap Shotgun **$1595**
Boxlock action. Gauge: 12. 30-inch bbls. 48.5 inches overall. Weight: 8 lbs. Choked IM/F. Selective single trigger. Metal bead front sight. Checkered European walnut stock w/pistol grip. Engraved action. Made 1984-87.

RUGER SHOTGUN
Southport, Connecticut
Manufactured by Sturm, Ruger & Company

Ruger Plain Grade Red Label Over/Under **$750**
Boxlock. Auto ejectors. Selective single trigger. 20 ga. 3-inch chambers. 26-inch vent-rib bbl., IC/M or SK choke. Weight: about 7 lbs. Checkered pistol-grip stock and forearm. Introduced in 1977, 12 ga. version 1982. Chambers: 2.75- and 3-inch. Bbls.: 26-, 28- and 30-inch. Weight: about 7.5 lbs.

Ruger Red Label Over/Under Stainless
Gauges: 12 and 20; 3-inch chambers. Bbls.: 26- or 28-inch. Various chokes, fixed or screw-in tubes. Weight: 7 to 7.5 lbs. Single selective trigger. Selective automatic ejectors. Automatic top safety. Standard gold bead front sight. Pistol-grip or English-style American walnut stock w/hand-cut checkering. Made 1985 to date.
W/Fixed Chokes **$795**
W/Screw-in Tubes 850

SHOTGUNS

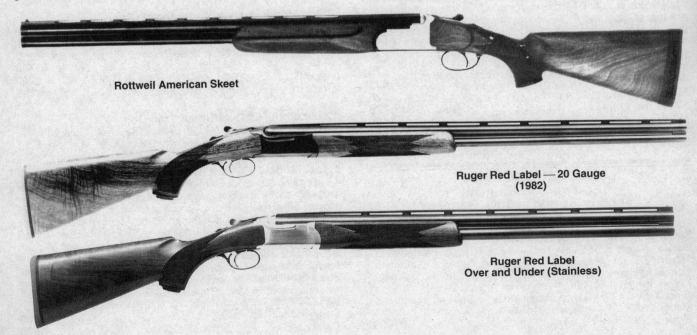

Rottweil American Skeet

**Ruger Red Label — 20 Gauge
(1982)**

**Ruger Red Label
Over and Under (Stainless)**

Ruger Red Label "Woodside" Over/Under **$995**
Similar to the Red Label O/U Stainless except in 12 ga. only with
wooden sideplaate extensions. Made 1995 to date.

Ruger Sporting Clays Over/Under Stainless **$825**
Similar to the Red Label O/U Stainless, except in 12 ga. only w/30-
inch vent-rib bbls., no side ribs; back-bored w/screw-in choke tubes
(not interchangeable w/other Red Label O/U models). Brass front
and mid-rib beads. Made 1992 to date.

VICTOR SARASQUETA, S. A.
Eibar, Spain

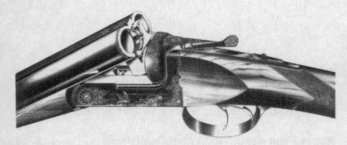

**Sarasqueta Model 3
Hammerless Boxlock**

Sarasqueta Model 3 Hammerless Boxlock Double-Barrel Shotgun
Plain extractors or auto ejectors. Double triggers. Gauges: 12, 16,
20. Made in various bbl. lengths, chokes and weights. Checkered
stock and forend, straight grip standard. Currently manufactured.
Model 3, plain extractors **$395**
Model 3E, automatic ejectors **495**

Sarasqueta Hammerless Sidelock Doubles
Automatic ejectors (except on Models 4 and 203 which have plain
extractors). Double triggers. Gauges: 12, 16, 20. Barrel lengths,
chokes and weights made to order. Checkered stock and forend,
straight grip standard. Models differ chiefly in overall quality, en-
graving, grade of wood, checkering, etc.; general specifications are
the same. Currently manufactured.
Model 4 . . . . **$ 575**
Model 4E **625**
Model 203 . . . . **575**
Model 203E **625**
Model 6E **725**
Model 7E **775**
Model 10E **1595**
Model 11E **1650**
Model 12E **1895**

J. P. SAUER & SOHN
Eckernförde, Germany
Formerly located in Suhl, Germany

Sauer Model 66 Over/Under Field Gun
Purdey-system action with Holland & Holland-type sidelocks.
Selective single trigger. Selective auto ejectors. Automatic
safety. Available in three grades of engraving. 12 ga. only.
Krupp-Special steel bbls. w/vent rib 28-inch, M/F choke. Weight:
about 7.25 lbs. Checkered walnut stock and forend; recoil pad.
Made 1966-75.
Grade I **$1695**

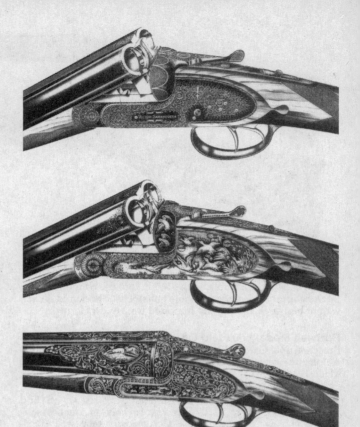

Sarasqueta Models 6E, 11E and 12E

Sauer Model 66 Over/Under Field Gun *(Con't)*
Grade II **2150**
Grade III **2995**

Sauer Model 66 Over/Under Skeet Gun
Same as Model 66 Field Gun, except 26-inch bbls. with wide vent
rib, SK choked- skeet-style stock and ventilated beavertail forearm;
nonautomatic safety. Made 1966-75.
Grade I **$1725**
Grade II **2250**
Grade III **3195**

Sauer Model 66 Over/Under Trap Gun
Same as Model 66 Skeet Gun, except has 30-inch bbls. choked F/F or
M/F; trap-style stock. Values same as for Skeet Model. Made 1966-75.

Sauer Model 3000E Drilling
Combination rifle and double barrel shotgun. Blitz action with Greener
crossbolt, double underlugs, separate rifle cartridge extractor, front set
trigger, firing pin indicators, Greener side safety, sear slide selector
locks right shotgun bbl. for firing rifle bbl. Gauge/calibers: 12 ga.
(2.75-inch chambers); 222, 243, 30-06, 7×65R. 25-inch Krupp-Special
steel bbls.; M/F choke automatic folding leaf rear rifle sight. Weight:
6.5 to 7.25 lbs. depending on rifle caliber. Checkered walnut stock and
forend; pistol grip, M Monte Carlo comb and cheekpiece, sling swivels.
Standard Model with arabesque engraving; Deluxe Model with hunting
scenes engraved on action. Currently manufactured. *Note: Also see listing under Colt.*
Standard Model **$3250**
Deluxe Model **4595**

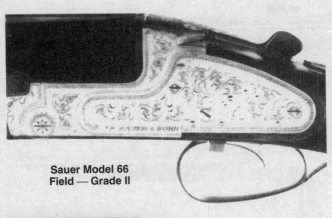

Sauer Model 66
Field — Grade II

Sauer Model 66
Field — Grade III

Sauer Model 3000E — Drilling

Sauer Artemis Double-Barrel Shotgun

Holland & Holland-type sidelock with Greener crossbolt double underlugs, double sear safeties, selective single trigger, selective auto ejectors. Grade I with fine-line engraving, Grade II with full English arabesque engraving. 12 ga. (2.75-inch chambers). Krupp-Special steel bbls., 28-inch, M/F choke. Weight: about 6.5 lbs. Checkered walnut pistol-grip stock and beavertail forend; recoil pad. Made 1966-77.

Grade I $3995
Grade II 4950

Sauer BBF 54
Combination Rifle/Shotgun

Sauer BBF 54 Over/Under Combination Rifle/Shotgun

Blitz action with Kersten lock, front set trigger fires rifle bbl., slide-operated sear safety. Gauge/calibers: 16 ga.; 30-30, 30-06, 7×65R. 25-inch Krupp-Special steel bbls.; shotgun bbl. F choke, folding-leaf rear sight. Weight: about 6 lbs. Checkered walnut stock and forend; pistol grip, mod. Monte Carlo comb and cheekpiece, sling swivels. Standard Model with arabesque engraving; Deluxe Model with hunting scenes engraved on action. Currently manufactured.

Standard Model $1895
Deluxe Model 2195

Sauer Royal
Double-Barrel Shotgun

Sauer Royal Double-Barrel Shotguns

Anson & Deeley action (boxlock) with Greener crossbolt, double underlugs, signal pins, selective single trigger, selective auto ejectors, auto safety. Scalloped frame with arabesque engraving. Krupp-Special steel bbls. Gauges: 12, 2.75-inch chambers, 20, 3-inch chambers. Bbls.: 30-inch (12 ga. only) and 28-inch, M/F- 26-inch (20 ga. only), IC/M. Weight: 12 ga., about 6.5 lbs.; 20 ga., 6 lbs. Checkered walnut pistol-grip stock and beavertail forend; recoil pad. Made 1955-77.

Standard Model $1295
20 Gauge 1795

SAVAGE ARMS
Westfield, Massachusetts
Formerly located in Utica, New York

SHOTGUNS

**Savage 22-.410
Over and Under Combination**

**Savage Model 24-VS
Camper/Survival/Centerfire Rifle/Shotgun**

Savage Model 24 22-.410 O/U Combination **$130**
Same as Stevens No. 22-.410, has walnut stock and forearm. Made 1950-65.

Savage Model 24C Camper's Companion **$145**
Same as Model 24FG, except made in 22 Magnum/20 ga. only; has 20-inch bbls., shotgun tube Cyl. bore. Weight: 5.75 lbs. Trap in butt provides ammunition storage; comes with carrying case. Made 1972-89.

Savage Model 24D................................ **$195**
Same as Models 24DL and 24MDL, except frame has black or case-hardened finish. Game scene decoration of frame eliminated in 1974; forearm uncheckered after 1976. Made 1970-88.

Savage Model 24DL............................... **$145**
Same general specifications as Model 24S, except top lever opening; satin-chrome-finished frame decorated with game scenes, checkered Monte Carlo stock and forearm. Made 1965-69.

Savage Model 24F-12T Turkey Gun **$295**
12- or 20-ga. shotgun bbl./22 Hornet, 223 or 30-30 caliber rifle. 24-inch blued bbls., 3-inch chambers, extra removable F choke tube. Hammer block safety. Color casehardened frame. du Pont Rynite® camo stock. Swivel studs. Made 1989 to date.

Savage Model 24FG Field Grade **$125**
Same general specifications as Model 24S, except top lever opening. Made 1972 to date.

Savage Model 24MDL........................... **$140**
Same as Model 24DL, except rifle bbl. chambered for 22 WMR. Made 1965-69.

Savage Model 24MS **$125**
Same as Model 24S, except rifle bbl. chambered for 22 WMR. Made 1965-71.

Savage Model 24S Over/Under Combination **$145**
Boxlock. Visible hammer. Side lever opening. Plain extractors. Single trigger. 20 ga. or .410 bore shotgun bbl. under 22 LR bbl., 24-inch. Open rear sight, ramp front, dovetail for scope mounting. Weight: about 6.75 lbs. Plain pistol-grip stock and forearm. Made 1965-71.

Savage Model 24V **$245**
Similar to Model 24D, except 20 ga. under 222 Rem., 22 Rem., 357 Mag., 22 Hornet or 30-30 rifle bbl. Made 1971-89.

Savage Model 24-VS Camper/Survival/Centerfire Rifle/Shotgun
.... **$195**
Similar to Model 24V except 357 Rem. Mag. over 20 ga. Nickel finish full-length stock and accessory pistol-grip stock. Overall length: 36 inches with full stock; 26 inches w/pistol grip. Weight: about 6.5 lbs. Made 1983-88.

Savage Model 28A Standard Grade Slide-Action Repeating Shotgun **$250**
Hammerless. Takedown. 12 ga. 5-shell tubular magazine. Plain bbl., lengths: 26-,28-, 30-, 32-inches, choked C/M/F. Weight: about 7.5 lbs. with 30-inch bbl. Plain pistol-grip stock, grooved slide handle. Made 1928-31.

Savage Model 28B **$265**
Raised matted rib; otherwise the same as Model 28A.

Savage Model 28D Trap Grade **$285**
Same general specifications as Model 28A, except has 30-inch F choke bbl. w/matted rib, trap-style stock w/checkered pistol grip, checkered slide handle of select walnut.

Savage Model 30 Solid Frame Hammerless Slide-Action Shotgun **$180**
Gauges: 12, 16, 20, .410. 2.75-inch chamber in 16 ga., 3- inch in other ga. Magazine holds four 2.75-inch shells or three 3-inch shells. Bbls.: vent rib; 26-, 28-, 30-inch; IC, M, F choke. Weight: average 6.25 to 6.75 lbs. depending on ga. Plain pistol-grip stock (checkered on later production), grooved slide handle. Made 1958-70.

Savage Model 30 Takedown Slug Gun **$175**
Same as Model 30FG, except 21-inch Cyl. bore bbl. with rifle sights. Made 1971-79.

Savage Model 30AC Solid Frame **$195**
Same as Model 30 Solid Frame, except has 26-inch bbl. with adj. choke; 12 ga. only. Made 1959-70.

Savage Model 30AC Takedown **$175**
Same as Model 30FG, except has 26-inch bbl. with adj. choke; 12 and 20 ga. only. Made 1971-72.

Savage Model 30ACL Solid Frame **$195**
Same as Model 30AC Solid Frame, except left-hand model with ejection port and safety on left side; 12 ga. only. Made 1960-64.

Savage Model 30D Takedown **$175**
Deluxe Grade. Same as Model 30FG, except has receiver engraved with game scene, vent-rib bbl., recoil pad. Made 1971 to date.

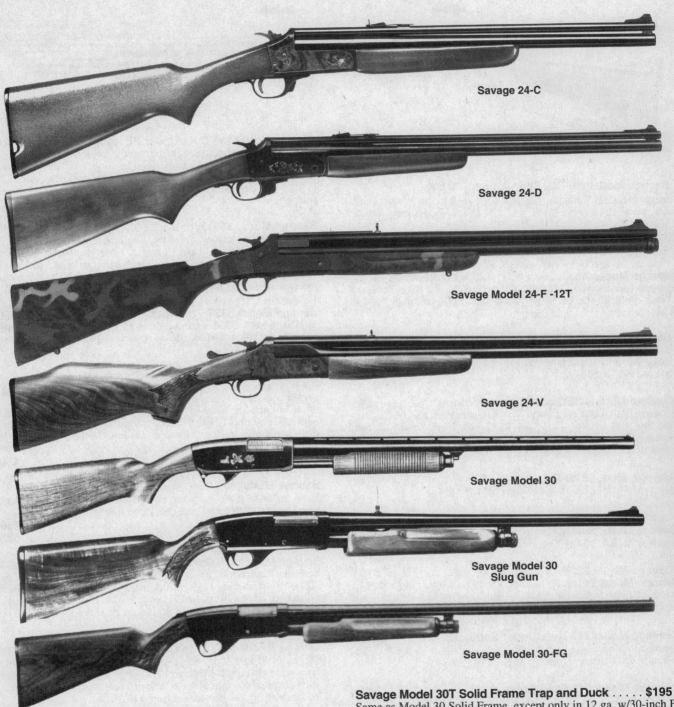

Savage 24-C

Savage 24-D

Savage Model 24-F -12T

Savage 24-V

Savage Model 30

Savage Model 30
Slug Gun

Savage Model 30-FG

Savage Model 30FG Takedown Hammerless Slide-Action Shotgun $145
Field Grade. Gauges: 12, 20, .410. 3-inch chamber. Magazine holds four 2.75-inch shells or three 3-inch shells. Bbls.: plain; 26-inch F choke (.410 ga. only); 28-inch M/F choke; 30-inch F choke (12 ga. only). Weight: average 7 to 7.75 lbs. depending on gauge. Checkered pistol-grip stock, fluted slide handle. Made 1970-79.

Savage Model 30L Solid Frame $170
Same as Model 30 Solid Frame, except left-handed model with ejection port and safety on left side; 12 ga. only. Made 1959-70.

Savage Model 30T Solid Frame Trap and Duck $195
Same as Model 30 Solid Frame, except only in 12 ga. w/30-inch F choke bbl.; has Monte Carlo stock with recoil pad, weighs about 8 lbs. Made 1963-70.

Savage Model 30T Takedown Trap Gun $175
Same as Model 30D, except only in 12 ga. w/30-inch F choke bbl. Monte Carlo stock with recoil pad. Made 1970-73.

Savage Model 69-RXL Slide-Action Shotgun $165
Similar to Model 67 (law enforcement configuration). Hammerless, side ejection top tang safe for left- or right-hand use. 12 ga. chambered for 2.75- and 3-inch magnum shells. 18.25-inch bbl. Tubular

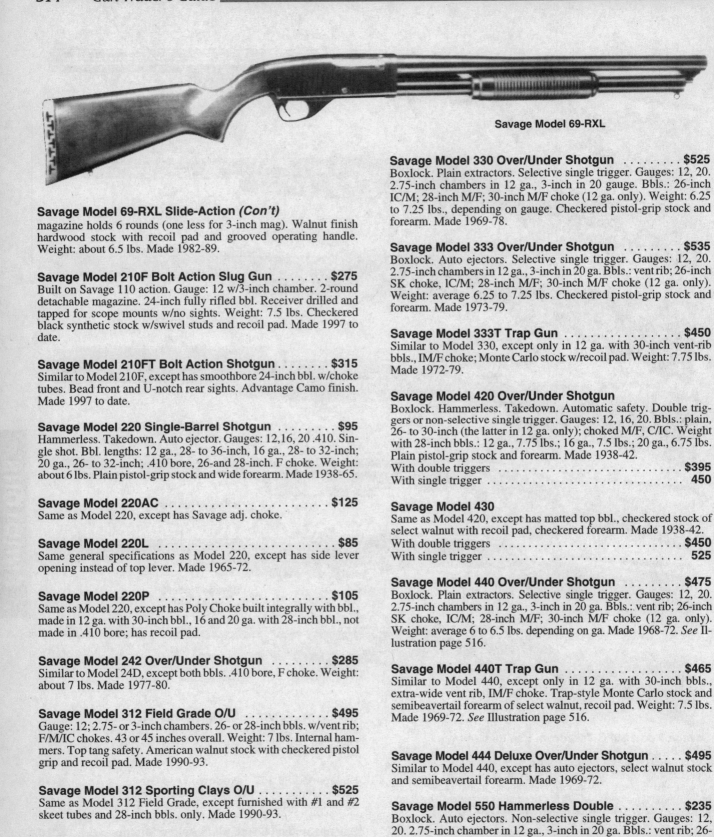

Savage Model 69-RXL

Savage Model 69-RXL Slide-Action *(Con't)*
magazine holds 6 rounds (one less for 3-inch mag). Walnut finish hardwood stock with recoil pad and grooved operating handle. Weight: about 6.5 lbs. Made 1982-89.

Savage Model 210F Bolt Action Slug Gun $275
Built on Savage 110 action. Gauge: 12 w/3-inch chamber. 2-round detachable magazine. 24-inch fully rifled bbl. Receiver drilled and tapped for scope mounts w/no sights. Weight: 7.5 lbs. Checkered black synthetic stock w/swivel studs and recoil pad. Made 1997 to date.

Savage Model 210FT Bolt Action Shotgun $315
Similar to Model 210F, except has smoothbore 24-inch bbl. w/choke tubes. Bead front and U-notch rear sights. Advantage Camo finish. Made 1997 to date.

Savage Model 220 Single-Barrel Shotgun $95
Hammerless. Takedown. Auto ejector. Gauges: 12,16, 20 .410. Single shot. Bbl. lengths: 12 ga., 28- to 36-inch, 16 ga., 28- to 32-inch; 20 ga., 26- to 32-inch; .410 bore, 26-and 28-inch. F choke. Weight: about 6 lbs. Plain pistol-grip stock and wide forearm. Made 1938-65.

Savage Model 220AC $125
Same as Model 220, except has Savage adj. choke.

Savage Model 220L $85
Same general specifications as Model 220, except has side lever opening instead of top lever. Made 1965-72.

Savage Model 220P $105
Same as Model 220, except has Poly Choke built integrally with bbl., made in 12 ga. with 30-inch bbl., 16 and 20 ga. with 28-inch bbl., not made in .410 bore; has recoil pad.

Savage Model 242 Over/Under Shotgun $285
Similar to Model 24D, except both bbls. .410 bore, F choke. Weight: about 7 lbs. Made 1977-80.

Savage Model 312 Field Grade O/U $495
Gauge: 12; 2.75- or 3-inch chambers. 26- or 28-inch bbls. w/vent rib; F/M/IC chokes. 43 or 45 inches overall. Weight: 7 lbs. Internal hammers. Top tang safety. American walnut stock with checkered pistol grip and recoil pad. Made 1990-93.

Savage Model 312 Sporting Clays O/U $525
Same as Model 312 Field Grade, except furnished with #1 and #2 skeet tubes and 28-inch bbls. only. Made 1990-93.

Savage Model 312 Trap Over/Under $530
Same as Model 312 Field Grade, except with 30-inch bbls. only, Monte Carlo buttstock and weight of 7.5 lbs. Made 1990-93.

Savage Model 330 Over/Under Shotgun $525
Boxlock. Plain extractors. Selective single trigger. Gauges: 12, 20. 2.75-inch chambers in 12 ga., 3-inch in 20 gauge. Bbls.: 26-inch IC/M; 28-inch M/F; 30-inch M/F choke (12 ga. only). Weight: 6.25 to 7.25 lbs., depending on gauge. Checkered pistol-grip stock and forearm. Made 1969-78.

Savage Model 333 Over/Under Shotgun $535
Boxlock. Auto ejectors. Selective single trigger. Gauges: 12, 20. 2.75-inch chambers in 12 ga., 3-inch in 20 ga. Bbls.: vent rib; 26-inch SK choke, lC/M; 28-inch M/F; 30-inch M/F choke (12 ga. only). Weight: average 6.25 to 7.25 lbs. Checkered pistol-grip stock and forearm. Made 1973-79.

Savage Model 333T Trap Gun $450
Similar to Model 330, except only in 12 ga. with 30-inch vent-rib bbls., IM/F choke; Monte Carlo stock w/recoil pad. Weight: 7.75 lbs. Made 1972-79.

Savage Model 420 Over/Under Shotgun
Boxlock. Hammerless. Takedown. Automatic safety. Double triggers or non-selective single trigger. Gauges: 12, 16, 20. Bbls.: plain, 26- to 30-inch (the latter in 12 ga. only); choked M/F, C/IC. Weight with 28-inch bbls.: 12 ga., 7.75 lbs.; 16 ga., 7.5 lbs.; 20 ga., 6.75 lbs. Plain pistol-grip stock and forearm. Made 1938-42.
With double triggers $395
With single trigger 450

Savage Model 430
Same as Model 420, except has matted top bbl., checkered stock of select walnut with recoil pad, checkered forearm. Made 1938-42.
With double triggers $450
With single trigger 525

Savage Model 440 Over/Under Shotgun $475
Boxlock. Plain extractors. Selective single trigger. Gauges: 12, 20. 2.75-inch chambers in 12 ga., 3-inch in 20 ga. Bbls.: vent rib; 26-inch SK choke, IC/M; 28-inch M/F; 30-inch M/F choke (12 ga. only). Weight: average 6 to 6.5 lbs. depending on ga. Made 1968-72. *See Illustration page 516.*

Savage Model 440T Trap Gun $465
Similar to Model 440, except only in 12 ga. with 30-inch bbls., extra-wide vent rib, IM/F choke. Trap-style Monte Carlo stock and semibeavertail forearm of select walnut, recoil pad. Weight: 7.5 lbs. Made 1969-72. *See Illustration page 516.*

Savage Model 444 Deluxe Over/Under Shotgun $495
Similar to Model 440, except has auto ejectors, select walnut stock and semibeavertail forearm. Made 1969-72.

Savage Model 550 Hammerless Double $235
Boxlock. Auto ejectors. Non-selective single trigger. Gauges: 12, 20. 2.75-inch chamber in 12 ga., 3-inch in 20 ga. Bbls.: vent rib; 26-inch IC/M; 28-inch M/F; 30-inch M/F choke (12 ga. only). Weight: 7 to 8 lbs. Checkered pistol-grip stock and semibeavertail forearm. Made 1971-73. *See Illustration page 516.*

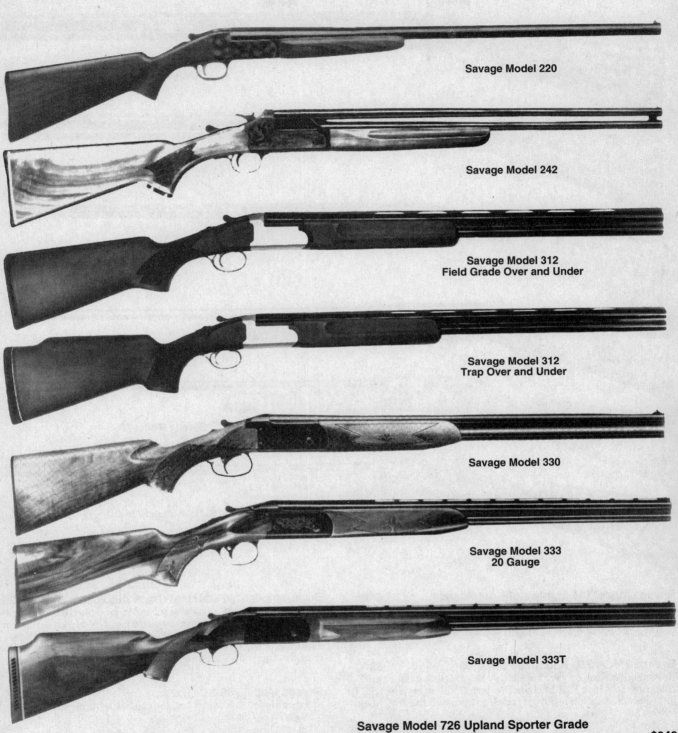

Savage Model 220

Savage Model 242

Savage Model 312
Field Grade Over and Under

Savage Model 312
Trap Over and Under

Savage Model 330

Savage Model 333
20 Gauge

Savage Model 333T

**Savage Model 726 Upland Sporter Grade
3-Shot Autoloading Shotgun** **$240**
Same as Model 720, except has 2-shell magazine capacity. Made
1931-49.

Savage Model 740C Skeet Gun **$295**
Same as Model 726, except has special skeet stock and full beavertail forearm, equipped with Cutts Compensator bbl. length overall
with spreader tube is about 24.5 inches. Made 1936-49.

**Savage Model 720 Standard Grade 5-Shot Autoloading
Shotgun** **$225**
Browning type. Takedown. 12 and 16 ga. 4-shell tubular magazine.
Bbl.: plain; 26- to 32-inch (the latter in 12 ga. only); choked C, M, F.
Weight: about 8.25 lbs., 12 ga. with 30-inch bbl.; 16 ga., about .5 lb.
lighter. Checkered pistol-grip stock and forearm. Made 1930-49.

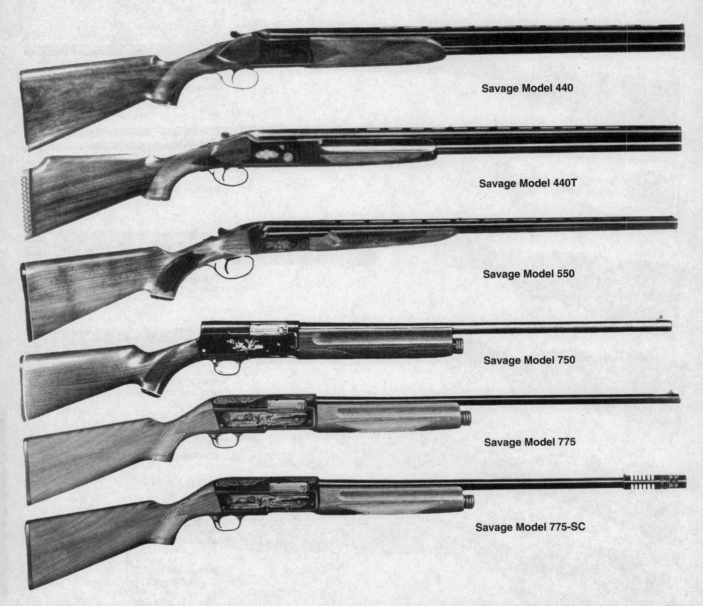

Savage Model 440

Savage Model 440T

Savage Model 550

Savage Model 750

Savage Model 775

Savage Model 775-SC

Savage Model 745 Lightweight Autoloader $225
Three- or five-shot model. Same general specifications as Model 720, except has lightweight alloy receiver, 12 ga.only, 28-inch plain bbl. Weight: about 6.75 lbs. Made 1940-49.

Savage Model 750 Automatic Shotgun $220
Browning-type autoloader. Takedown. 12 ga. 4-shot tubular magazine. Bbls.: 28-inch F or M choke; 26-inch IC. Weight: about 7.25 lbs. Checkered walnut pistol-grip stock and grooved forearm. Made 1960-67.

Savage Model 750-AC $275
Same as Model 750, except has 26-inch bbl. with adj. choke. Made 1964-67.

Savage Model 750-SC $250
Same as Model 750, except has 26-inch bbl. with Savage Super Choke. Made 1962-63.

Savage Model 755 Standard Grade Autoloader $225
Streamlined receiver. Takedown. 12 and 16 ga. 4-shell tubular magazine (a three-shot model with magazine capacity of two shells was also produced until 1951). Bbl.: plain, 30-inch F choke (12 ga. only), 28-inch F or M, 26-inch IC. Weight: about 8.25 lbs., 12 ga. Checkered pistol-grip stock and forearm. Made 1949-58.

Savage Model 755-SC $200
Same as Model 755, except has 26-inch bbl. w/recoil-reducing, adj. Savage Super Choke.

Savage Model 775 Lightweight $215
Same general specifications as Model 755, except has lightweight alloy receiver and weighs about 6.73 lbs. Made 1950-65.

Savage Model 775-SC $225
Same as Model 775, except has 26-inch bbl. with Savage Super Choke.

**Savage Model 2400
Over and Under Combination Gun**

Savage Model 2400 Over/Under Combination **$595**
Boxlock action similar to that of Model 330. Plain extractors. Selective single trigger. 12-ga. (2.75-inch chamber) shotgun bbl., F choke, over 308 Win. or 222 Rem. rifle bbl.; 23.5-inch; solid matted rib with blade front sight and folding leaf rear, dovetail for scope mounting. Weight: about 7.5 lbs. Monte Carlo stock w/pistol grip and recoil pad, semibea-vertail forearm, checkered. Made 1975-79 by Valmet.

SEARS, ROEBUCK & COMPANY
Chicago, Illinois
J. C. Higgins and Ted Williams Models

Although they do not correspond to specific models below the names Ted Williams and J. C. Higgins have been used to designate various Sears shotguns at various times.

Sears Model 18 Bolt-Action Repeater **$80**
Takedown. 3-shot top-loading magazine. Gauge: .410 only. Bbl.: 25-inch w/variable choke. Weight: about 5.75 lbs.

Sears Model 20 Slide-Action Repeater **$160**
Hammerless. 5-shot magazine. Bbls.: 26- to 30-inch w/various chokes. Weight: 7.25 lbs. Plain pistol-grip stock and slide handle.

Sears Model 21 Slide-Action Repeater **$185**
Same general specifications as the Model 20 except vent rib and adj. choke.

Sears Model 30 Slide-Action Repeater **$175**
Hammerless. Gauges: 12, 16, 20 and .410. 4-shot magazine. Bbls.: 26- to 30-inch, various chokes. Weight: 6.5 lbs. Plain pistol-grip stock, grooved slide handle.

Sears Model 97 Single-Shot Shotgun **$65**
Takedown. Visible hammer. Automatic ejector. Gauges: 12, 16, 20 and .410. Bbls.: 26- to 36-inch, F choke. Weight: average 6 lbs. Plain pistol-grip stock and forearm.

Sears Model 97-AC Single-Shot Shotgun **$80**
Same general specifications as Model 97 except fancier stock and forearm.

Sears Model 101.7 Double-Barrel Shotgun **$160**
Boxlock. Double triggers. Gauges: 12, 16, 20, .410. Bbls.: 26- to 32-inch, choked M and F. Weight: from 6 to 7.5 lbs. Plain stock and forend.

Sears Model 101.7C Double-Barrel Shotgun **$175**
Same general specifications as Model 101.7, except checkered stock and forearm.

Sears Model 101.25 Bolt-Action Shotgun **$80**
Takedown. .410 gauge. 5-shell tubular magazine. 24-inch bbl., F choke. Weight: about 6 lbs. Plain, one-piece pistol-grip stock.

Sears Model 101.40 Single-Shot Shotgun **$65**
Takedown. Visible hammer. Automatic ejector. Gauges: 12, 16, 20 and .410. Bbls.: 26- to 36-inch, F choke. Weight: average 6 lbs. Plain pistol-grip stock and forearm.

Sears Model 101.1120 Bolt-Action Repeater **$80**
Takedown. .410 ga. 24-inch bbl., F choke. Weight: about 5 lbs. Plain one-piece pistol-grip stock.

Sears Model 101.1380 Bolt-Action Repeater **$90**
Takedown. Gauges: 12, 16, 20. 2-shell detachable box magazine. 26-inch bbl., F choke. Weight: about 7 lbs. Plain one-piece pistol-grip stock.

Sears Model 101.1610 Double-Barrel Shotgun **$235**
Boxlock. Double triggers. Plain extractors. Gauges: 12, 16, 20 and .410. Bbls.: 24- to 30-inch. Various chokes, but mostly M and F. Weight: about 7.5 lbs, 12 ga. Checkered pistol-grip stock and forearm.

Sears Model 101.1701 Double-Barrel Shotgun **$245**
Same general specifications as Model 101.1610 except satin chrome frame and select walnut stock and forearm.

Sears Model 101.5350-D Bolt-Action Repeater **$80**
Takedown. Gauges: 12, 16, 20. 2-shell detachable box magazine. 26-inch bbl., F choke. Weight: about 7.25 lbs. Plain one-piece pistol-grip stock.

Sears Model 101.5410 Bolt-Action Repeater **$80**
Same general specifications as Model 101.5350-D.

SKB ARMS COMPANY
Tokyo, Japan
Imported by G.U. Inc., Omaha, Nebraska

SKB Model 385 Side-by-Side **$1325**
Boxlock action w/double locking lugs. Gauges: 12, 20 and 28 w/2.75- and 3-inch chambers. 26- or 28-inch bbls. w/inter-choke tube system. Single selective trigger. Selective automatic ejectors and automatic safety. Weight: 6 lbs. 10 oz. Silver nitride receiver w/engraved scroll and game scene. Solid rib w/flat matte finish and metal front bead sight. Checkered American walnut English or pistol-grip stock. Imported 1992 to date.

SKB Models 300 and 400 Side-by-Side Doubles
Similar to Model 200E, except higher grade. Models 300 and 400 differ in that the latter has more elaborate engraving and fancier wood.
Model 300 . . . . **$625**
Model 400 . . . . **795**

SHOTGUNS

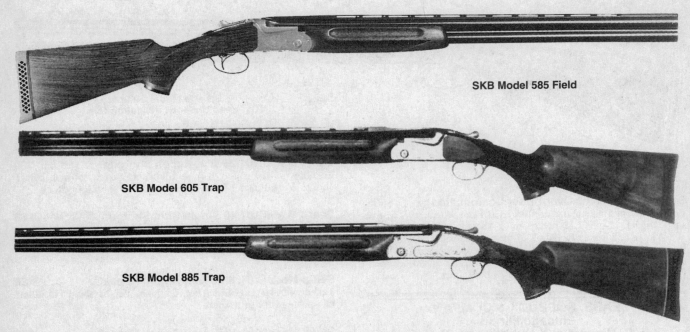

SKB Model 585 Field

SKB Model 605 Trap

SKB Model 885 Trap

SKB Model 400 Skeet $825
Similar to Model 200E Skeet, except higher grade with more elaborate engraving and full fancy wood.

SKB Model 480 English $895
Similar to Model 280 English, except higher grade with more elaborate engraving and full fancy wood.

SKB Model 500 Small Gauge O/U Shotgun $550
Similar to Model 500, except gauges 28 and .410; has 28-inch vent-rib bbls., M/F chokes. Weight: about 6.5 lbs.

SKB Model 505 O/U Shotgun
Blued boxlock action. Gauge: 12, 20, 28 and .410. Bbls.: 26-, 28, 30-inch; IC/M, M/F or inner choke tubes. 45.19 inches overall. Weight: 6.6 to 7.4 lbs. Hand checkered walnut stock. Metal bead front sight, ejectors, single selective trigger. Introduced 1988.
Standard Model . . . . $ 795
Two-bbl. Field Set 1250

SKB Model 585 Deluxe Over/Under Shotgun
Boxlock. Gauges: 12, 20, 28 and .410; 2.75-or 3-inch chambers. Bbls.: 26-, 28-, 30-, 32- or 34-inch with vent rib; fixed chokes or Inter-choke tubes. Weight: 6.5 to 8.5 lbs. Single selective trigger. Selective automatic ejectors. Manual safety. Checkered walnut stock in standard or Monte Carlo style. Silver nitride finish with engraved game scenes. Made 1987 to date.
Field, Skeet, Trap Grade . . . . $ 790
Field Grade, two-bbl. set 1395
Skeet Set (20, 28, .410 ga.) 1895
Sporting Clays 825
Trap Combo (two-bbl.) 1395

SKB Model 600 Small Gauge $675
Same as Model 500 Small Gauge, except higher grade with more elaborate engraving and fancier wood.

SKB Model 605 Trap O/U Shotgun. $895
Similar to the Model 505 except has silvered, engraved receiver. Introduced 1988.

SKB Model 685 Deluxe Over/Under
Similar to the 585 Deluxe, except with semi-fancy American walnut stock. Gold trigger and jeweled barrel block. Silvered receiver with fine engraving.
Field, Skeet, Trap Grade . . . . $ 895
Field Grade, two-bbl. set 1395
Skeet Set 1995
Sporting Clays 925
Trap Combo. two bbl. 1595

SKB Model 800 Skeet/Trap Over/Under
Similar to Model 700 Skeet and Trap, except higher grade with more elaborate engraving and fancier wood.
Model 800 Skeet $850
Model 800 Trap 925

SKB Model 880 Skeet/Trap
Similar to Model 800 Skeet, except has sideplates.
Model 880 Skeet $1195
Model 880 Trap 1225

SKB Model 885 Deluxe Over/Under
Similar to the 685 Deluxe, except with engraved sideplates.
Field, Skeet, Trap Grade $1025
Field Grade, two-bbl. Set 1395
Skeet Set 2595
Sporting Clays 1045
Trap Combo 1695

The following SKB shotguns were distributed by Ithaca Gun Co. 1966-76. For specific data, please see corresponding listings under Ithaca.

SKB Century Single-Barrel Trap Gun
The SKB catalog does not differentiate between Century and Century II; however, specifications of current Century are those of Ithaca-SKB Century II.
Century (505) $625
Century II (605) 725

Sile Field Master II

SKB Gas-operated Automatic Shotguns

Model XL300 with plain barrel	$265
Model XL300 with vent rib	295
Model XL900	275
Model XL900 Trap	345
Model XL900 Skeet	335
Model XL900 Slug	295
Model 1300 Upland, Slug	375
Model 1900 Field, Trap, Slug	445

SKB Over/Under Shotguns

Model 500 Field	$ 425
Model 500 Magnum	475
Model 600 Field	525
Model 600 Magnum	545
Model 600 Trap	565
Model 600 Doubles	585
Model 600 Skeet—12 or 20 gauge	550
Model 600 Skeet—28 or .410	625
Model 600 Skeet Combo	1595
Model 600 English	595
Model 700 Trap	785
Model 700 Doubles	775
Model 700 Skeet	795
Model 700 Skeet Combo	1950

SKB Recoil-Operated Automatic Shotguns

Model 300—with plain barrel	$235
Model 300—with vent rib	265
Model 900	315
Model 900 Slug	285

SKB Side-by-Side Double-Barrel Shotguns

Model 100	$425
Model 150	450
Model 200E	550
Model 200E Skeet	595
Model 280 English	695

SIG SAUER
(SIG) Schweizerische Industrie-Gesellschaft
Neuhausen, Switzerland

Sigarms Model SA3 Over/Under Shotgun
Monobloc boxlock action. Single selective trigger. Automatic ejectors. Gauges: 12 or 20 w/3- inch chambers. 26-, 28- or 30-inch vent rib bbls. w/choke tubes. Weight: 6.8 to 7.1 lbs. Checkered select walnut stock and forearm. Satin nickel finished receiver w/game scene and blued bbls. Imported 1997-98.

Field Model	$795
Sporting Clays Model	875

Sigarms Model SA5 Over/Under Shotgun
Similar to SA3 Model, except w/detachable sideplates. Gauges: 12 or 20 w/3- inch chambers. 26.5-, 28- or 30-inch vent rib bbls. w/choke tubes. Imported 1997 to date.

Field Model	$1495
Sporting Clays Model	1650

SILE SHOTGUNS
Sile Distributors
New York, NY

Sile Field Master II O/U Shotgun	$450

Gauge: 12, 3-inch chambers. 28-inch bbl., IC, M, IM, F choke tubes. 45.25 inches overall. Weight: 7.25 lbs. Satin-finished walnut, checkered stock and forend. Introduced 1989.

L. C. SMITH SHOTGUNS
Made 1890-1945 by Hunter Arms Company, Fulton, N.Y.; 1946-51 and 1968-73 by Marlin Firearms Company, New Haven, Conn.

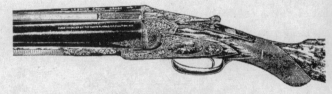

L.C. Smith Crown Grade

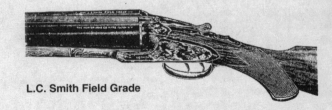

L.C. Smith Field Grade

L. C. Smith Double-Barrel Shotguns
Values shown are for L. C. Smith doubles made by Hunter. Those of 1946-51 Marlin manufacture generally bring prices about $\frac{1}{3}$ lower. Smaller gauge models, especially in the higher grades, command premium prices: up to 50 percent more for 20 gauge, up to 200 percent for .410 gauge.

Crown Grade, double triggers, automatic ejectors	$ 4,395
Crown Grade, selective single trigger, automatic ejectors	4,595
Deluxe Grade, selective single trigger, automatic ejectors	+16,500
Field Grade, double triggers, plain extractors	895
Field Grade, double triggers, auto ejectors	1,195
Field Grade, non-selective single trigger, plain extractors	1,125

SHOTGUNS

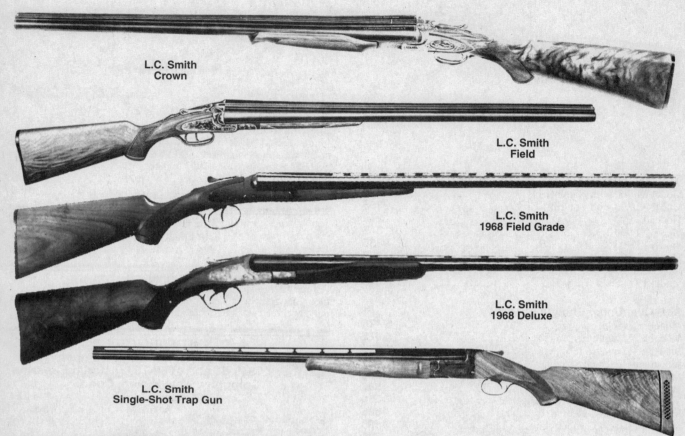

L.C. Smith
Crown

L.C. Smith
Field

L.C. Smith
1968 Field Grade

L.C. Smith
1968 Deluxe

L.C. Smith
Single-Shot Trap Gun

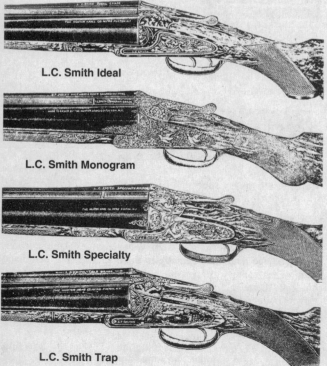

L.C. Smith Ideal

L.C. Smith Monogram

L.C. Smith Specialty

L.C. Smith Trap

L. C. Smith Double-Barrel Shotguns *(Con't)*

Field Grade, selective single trigger, automatic ejectors	**1,195**
Ideal Grade, double triggers, plain extractors	**1,250**
Ideal Grade, double triggers, automatic ejectors	**1,695**
Ideal Grade, selective single trigger, automatic ejectors	**1,950**
Monogram Grade, selective single trigger, automatic ejectors	**9,595**
Olympic Grade, selective single trigger, automatic ejectors	**1,790**
Premier Grade, selective single trigger automatic ejectors	**+12,500**
Skeet Special, non-selective single trigger, automatic ejectors	**1,695**
Skeet Special, sel. single trigger, auto ejectors	**2,250**
.410 Ga.	**10,500**
Specialty Grade, double triggers, auto ejectors	**2,895**
Specialty Grade, selective single trigger, automatic ejectors	**3,095**
Trap Grade, sel. single trigger, auto ejectors	**1,295**

L C. Smith Hammerless Double-Barrel Shotguns

Sidelock. Auto ejectors standard on higher grades, extra on Field and Ideal Grades. Double triggers or Hunter single trigger (non-selective or selective). Gauges: 12, 16, 20, .410. Bbls.: 26- to 32-inch, any standard boring. Weight: 6.5 to 8.25 lbs., 12 ga. Checkered stock and forend; choice of straight, half or full pistol grip, beavertail or standard-type forend. Grades differ only in quality of workmanship, wood, checkering, engraving, etc. Same general specifications apply to all. Manufacture of these L. C. Smith guns was discontinued in 1951. Production of Field Grade 12 ga. was resumed 1968-73. *Note:*

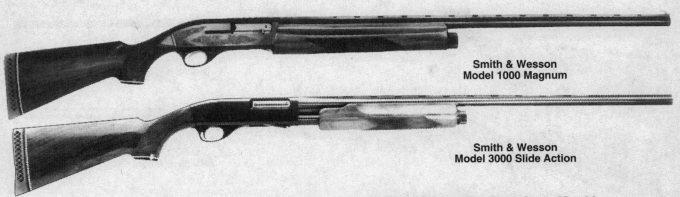

**Smith & Wesson
Model 1000 Magnum**

**Smith & Wesson
Model 3000 Slide Action**

L. C. Smith Hammerless Double-Barrel *(Con't)*

L. C. Smith Shotguns manufactured by the Hunter Arms Co. 1890-13 were designated by numerals to indicate grade, with the exception of Pigeon and Monogram.

00 Grade	$ 1,025
0 Grade	1,350
1 Grade	1,550
2 Grade	1,850
3 Grade	2,500
Pigeon	3,000
4 Grade	5,500
5 Grade	5,750
Monogram	8,500
A1	3,500
A2	8,500
A3	+15,000

L. C. Smith Hammerless Double Model 1968
Field Grade **$595**
"Re-creation" of the original L. C. Smith double. Sidelock. Plain extractors. Double triggers. 12 ga. 28-inch vent-rib bbls., M/F choke. Weight: about 6.75 lbs. Checkered pistol-grip stock and forearm. Made 1968-73.

L. C. Smith Hammerless Double
Model 1968 Deluxe **$750**
Same as 1968 Field Grade, except has Simmons floating vent rib, beavertail forearm. Made 1971-73.

L. C. Smith Single-Shot Trap Guns

Boxlock. Hammerless. Auto ejector. 12 gauge only. Bbl. lengths: 32- or 34-inch. Vent rib. Weight: 8 to 8.25 lbs. Checkered pistol-grip stock and forend, recoil pad. Grades vary in quality of workmanship, wood, engraving, etc.; general specifications are the same. Discont. 1951. *Note:* Values shown are for L. C. Smith single-barrel trap guns made by Hunter. Those of Marlin manufacture generally bring prices about one-third lower.

Olympic Grade	$ 1,375
Specialty Grade	1,695

L. C. Smith Single-Shot Trap Guns *(Con't)*

Crown Grade	2,950
Monogram Grade	4,395
Premier Grade	6,995
Deluxe Grade	10,750

SMITH & WESSON SHOTGUNS
Springfield, Massachusetts
Mfd. by Howa Machinery, Ltd., Nagoya, Japan

In 1985 Smith and Wesson sold its shotgun operation to O. F. Mossberg & Sons, Inc.

Smith & Wesson Model 916 Slide-Action Repeater
Hammerless. Solid frame. Gauges: 12, 16, 20. 3-inch chamber in 12 and 20 ga. 5-shot tubular magazine. Bbls.: plain or vent rib; 20-inch C (12 ga., plain only); 26-inch IC- 28-inch M or F; 30-inch F choke (12 ga. only). Weight: with 28-inch plain bbl., 7.25 lbs. Plain pistol-grip stock, fluted slide handle. Made 1972-81.

With plain bbl.	$145
With ventilated-rib bbl.	175

Smith & Wesson Model 916T
Same as Model 916, except takedown, 12 ga. only. Not available with 20-inch bbl. Made 1976-81.

With plain bbl.	$165
With ventilated-rib bbl.	195

Smith & Wesson Model 1000 Autoloader $325
Gas-operated. Takedown. Gauges: 12, 20. 2.75-inch chamber in 12 ga., 3-inch in 20 ga. 4-shot magazine. Bbls.: vent rib, 26-inch SK choke, IC; 28-inch M or F; 30-inch F choke (12 ga. only). Weight: with 28-inch bbl., 6.5 lbs. in 20 ga.,7.5 lbs. in 12 ga. Checkered pistol-grip stock and forearm. Made 1972 to date.

Smith & Wesson Model 1000 Magnum $425
Same as standard Model 1000, except chambered for 12 ga. magnum, 3-inch shells; 30-inch bbl. only, M or F choke; stock with recoil pad. Weight: about 8 lbs. Introduced in 1977.

**Squires Bingham
Model 30 Pump Shotgun**

SHOTGUNS

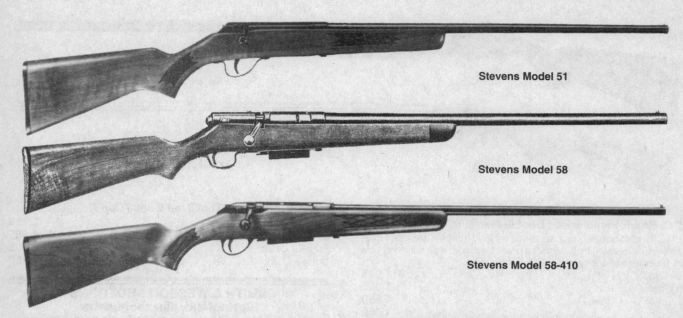

Stevens Model 51

Stevens Model 58

Stevens Model 58-410

Smith & Wesson Model 1000P **$325**
Same general specifications as Model C 3000 Slide Action, but an
earlier version.

Smith & Wesson Model 3000 Slide Action **$295**
Hammerless. 20-ga. Bbls.: 26-inch IC; 28-inch M or F. Chambered for
3-inch magnum and 2.75-inch loads. American walnut stock and forearm. Checkered pistol grip and forearm. Introduced 1982.

SPRINGFIELD ARMS
Built by Savage Arms Company, Utica, New York

Springfield Double-Barrel Hammer Shotgun **$325**
Gauges: 12 and 16. Bbls.: 28 to 32 inches. In 12 ga., 32-inch model,
both bbls. have F choke. All other gauges and barrel lengths are left
barrel Full, right barrel Mod. Weight: 7.25 to 8.25 lbs., depending on
gauge and barrel length. Black walnut checkered buttstock and
forend. Discontinued 1934.

SQUIRES BINGHAM CO., INC.
Makati, Rizal, Philippines

Squires Bingham Model 30 Pump Shotgun **$160**

Stevens Model 67 Pump Shotgun *(Con't)*
Hammerless. 12 ga. 5-shot magazine. Bbl.: 20-inch Cyl.; 28-inch M;
30-inch F choke. Weight: about 7 lbs. Pulong Dalaga stock and slide
handle. Currently manufactured.

J. STEVENS ARMS COMPANY
Chicopee Falls, Massachusetts
Division of Savage Arms Corporation

Stevens No. 20 "Favorite" Shotgun **$350**
Calibers: 22 and 32 Shot. Smoothbore bbl. Blade front sight; no rear.
Made 1893-1939.

Stevens No. 39 New Model Pocket Shotgun **$595**
Gauge: .410. Calibers: 38-40 Shot, 44-40 Shot. Bbls.: 10, 12, 15 or 18
inches, half-octagonal smoothbore. Shotgun sights. Made 1895-1906.

Stevens No. 22-.410 Over/Under Combination Gun
22 caliber rifle barrel over .410 bore shotgun barrel. Visible hammer.
Takedown. Single trigger. 24-inch bbls., shotgun bbl. F choke.
Weight: about 6 lbs. Open rear sight and ramp front sight of sporting
rifle type. Plain pistol-grip stock and forearm; originally supplied
with walnut stock and forearm. "Tenite" (plastic) was used in later

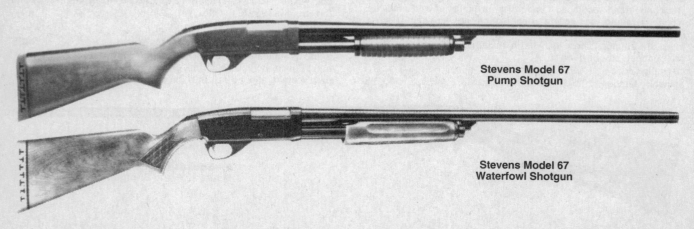

Stevens Model 67
Pump Shotgun

Stevens Model 67
Waterfowl Shotgun

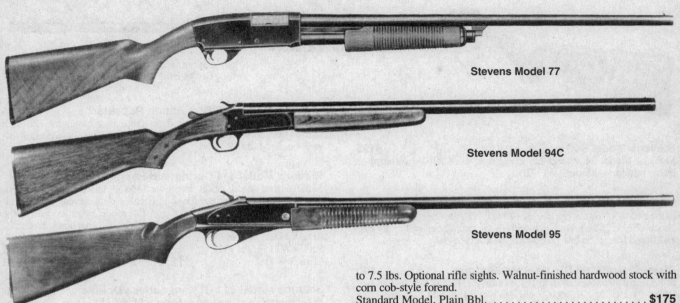

Stevens Model 77

Stevens Model 94C

Stevens Model 95

production. Made 1938-50. *Note:* This gun is now manufactured as the Savage Model 24.

With wood stock and forearm **$195**

With Tenite stock and forearm **145**

Stevens Model 51 Bolt-Action Shotgun **$85**
Single shot. Takedown. .410 ga. 24-inch bbl., F choke. Weight: about 4.75 lbs. Plain one-piece pistol-grip stock. checkered on later models. Made 1962-71.

Stevens Model 58 Bolt-Action Repeater **$105**
Takedown. Gauges: 12, 16, 20. 2-shell detachable box magazine. 26-inch bbl., F choke. Weight: about 7.25 lbs. Plain one-piece pistol-grip stock on early models w/takedown screw on bottom of forend. Made 1933-81. *Note:* Later production models have 3-inch chamber in 20 ga., checkered stock with recoil pad.

Stevens Model 58-.410 Bolt-Action Repeater **$95**
Takedown. .410 ga. 3-shell detachable box magazine. 24-inch bbl., F choke. Weight: about 5.5 lbs. Plain one-piece pistol-grip stock, checkered on later production. Made 1937-81.

Stevens Model 59 Bolt-Action Repeater **$135**
Takedown. .410 ga. 5-shell tubular magazine. 24-inch bbl., F choke. Weight: about 6 lbs. Plain, one-piece pistol-grip stock, checkered on later production. Made 1934-73.

Stevens Model 67 Pump Shotgun
Hammerless, side-ejection solid-steel receiver. Gauges: 12, 20 and .410, 2.75- or 3-inch shells. Bbls.: 21-, 26-, 28- 30-inch with fixed chokes or interchangeable choke tubes, plain or vent rib. Weight: 6.25

to 7.5 lbs. Optional rifle sights. Walnut-finished hardwood stock with corn cob-style forend.

Standard Model, Plain Bbl. **$175**

Standard Model, Vent Rib **185**

Standard Model, w/Choke Tubes **195**

Slug Model w/Rifle Sights **165**

Lobo Model, Matte Finish **215**

Youth Model, 20 ga. **185**

Camo Model. w/Choke Tubes **225**

Stevens Model 67 Waterfowl Shotgun **$195**
Hammerless. Gauge: 12. 3-shot tubular magazine. Walnut finished hardwood stock. Weight: about 7.5 lbs. Made 1972-89. *See* Illustration Previous Page.

Stevens Model 77 Slide-Action Repeater **$175**
Solid frame. Gauges: 12, 16, 20. 5-shot tubular magazine. Bbls.: 26-inch IC, 28-inch M or F choke. Weight: about 7.5 lbs. Plain pistol-grip stock with recoil pad, grooved slide handle. Made 1954-71.

Stevens Model 77-AC **$195**
Same as Model 77, except has Savage Super Choke.

Stevens Model 79-VR Super Value **$185**
Hammerless, side ejection. Bbl.: chambered for 2.75-inch and 3-inch mag. shells. 12, 20, and .410 ga. vent rib. Walnut finished hardwood stock with checkering on grip. Weight: 6.75-7 lbs. Made 1979-90.

Stevens Model 94 Single-Shot Shotgun **$95**
Takedown. Visible hammer. Auto ejector. Gauges: 12, 16, 20, 28, .410. Bbls.: 26-, 28-, 30-, 32-, 36-inch, F choke. Weight: about 6 lbs. depending on gauge and barrel. Plain pistol-grip stock and forearm. Made 1939-61.

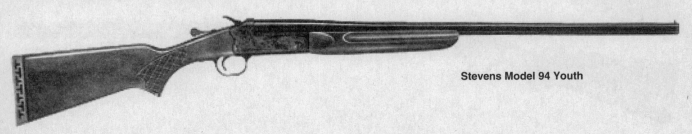

Stevens Model 94 Youth

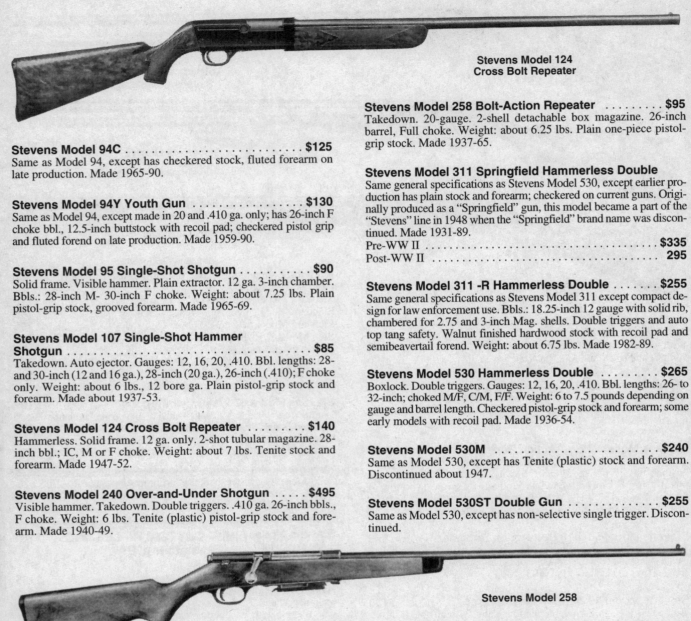

**Stevens Model 124
Cross Bolt Repeater**

Stevens Model 94C **$125**
Same as Model 94, except has checkered stock, fluted forearm on late production. Made 1965-90.

Stevens Model 94Y Youth Gun **$130**
Same as Model 94, except made in 20 and .410 ga. only; has 26-inch F choke bbl., 12.5-inch buttstock with recoil pad; checkered pistol grip and fluted forend on late production. Made 1959-90.

Stevens Model 95 Single-Shot Shotgun **$90**
Solid frame. Visible hammer. Plain extractor. 12 ga. 3-inch chamber. Bbls.: 28-inch M- 30-inch F choke. Weight: about 7.25 lbs. Plain pistol-grip stock, grooved forearm. Made 1965-69.

Stevens Model 107 Single-Shot Hammer Shotgun **$85**
Takedown. Auto ejector. Gauges: 12, 16, 20, .410. Bbl. lengths: 28- and 30-inch (12 and 16 ga.), 28-inch (20 ga.), 26-inch (.410); F choke only. Weight: about 6 lbs., 12 bore ga. Plain pistol-grip stock and forearm. Made about 1937-53.

Stevens Model 124 Cross Bolt Repeater **$140**
Hammerless. Solid frame. 12 ga. only. 2-shot tubular magazine. 28-inch bbl.; IC, M or F choke. Weight: about 7 lbs. Tenite stock and forearm. Made 1947-52.

Stevens Model 240 Over-and-Under Shotgun **$495**
Visible hammer. Takedown. Double triggers. .410 ga. 26-inch bbls., F choke. Weight: 6 lbs. Tenite (plastic) pistol-grip stock and forearm. Made 1940-49.

Stevens Model 258 Bolt-Action Repeater **$95**
Takedown. 20-gauge. 2-shell detachable box magazine. 26-inch barrel, Full choke. Weight: about 6.25 lbs. Plain one-piece pistol-grip stock. Made 1937-65.

Stevens Model 311 Springfield Hammerless Double
Same general specifications as Stevens Model 530, except earlier production has plain stock and forearm; checkered on current guns. Originally produced as a "Springfield" gun, this model became a part of the "Stevens" line in 1948 when the "Springfield" brand name was discontinued. Made 1931-89.
Pre-WW II **$335**
Post-WW II **295**

Stevens Model 311 -R Hammerless Double **$255**
Same general specifications as Stevens Model 311 except compact design for law enforcement use. Bbls.: 18.25-inch 12 gauge with solid rib, chambered for 2.75 and 3-inch Mag. shells. Double triggers and auto top tang safety. Walnut finished hardwood stock with recoil pad and semibeavertail forend. Weight: about 6.75 lbs. Made 1982-89.

Stevens Model 530 Hammerless Double **$265**
Boxlock. Double triggers. Gauges: 12, 16, 20, .410. Bbl. lengths: 26- to 32-inch; choked M/F, C/M, F/F. Weight: 6 to 7.5 pounds depending on gauge and barrel length. Checkered pistol-grip stock and forearm; some early models with recoil pad. Made 1936-54.

Stevens Model 530M **$240**
Same as Model 530, except has Tenite (plastic) stock and forearm. Discontinued about 1947.

Stevens Model 530ST Double Gun **$255**
Same as Model 530, except has non-selective single trigger. Discontinued.

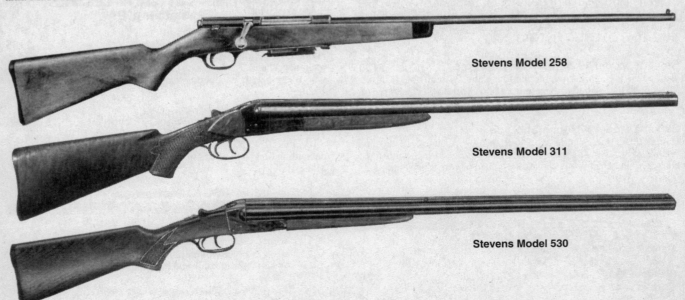

Stevens Model 258

Stevens Model 311

Stevens Model 530

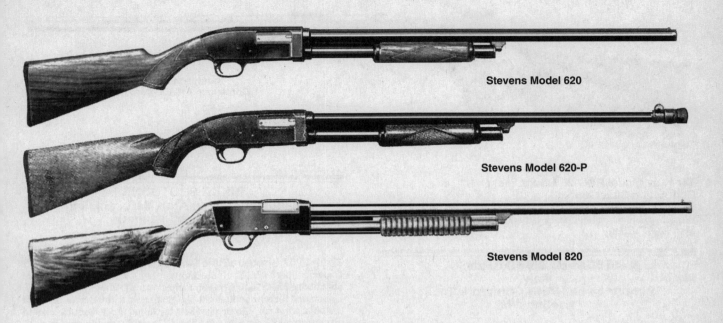

Stevens Model 620

Stevens Model 620-P

Stevens Model 820

Stevens Model 620 Hammerless Slide Action Repeating Shotgun **$225**
Takedown. Gauges: 12, 16, 20. 5-shell tubular magazine. Bbl. lengths: 26-, 28-, 30-, 32-inch; choked F, M IC, C. Weight: about 7.75 lbs., 12 ga.; 7.25 lbs., 16 ga.- 6 lbs., 20 ga. Checkered pistol-grip stock and slide handle. Made 1927-53.

Stevens Model 620-C **$225**
Same specifications as Model 620 except equipped with Cutts Compensator and two choke tubes.

Stevens Model 620-P **$235**
Same specifications as Model 620 equipped with Aero-Dyne Poly Choke and 27-inch bbl.

Stevens Model 620-PV **$235**
Same specifications as Model 620 except equipped with ventilated Poly Choke and 27-inch bbl.

Stevens Model 621 . . . . **$275**
Same as Model 620, except has raised solid matted-rib barrel. Discontinued.

Stevens Model 820 Hammerless Slide-Action Repeating Shotgun **$195**
Solid frame. 12 gauge only. 5-shell tubular magazine. 28-inch barrel; IC, M or F choke. Weight: about 7.5 lbs. Plain pistol-grip stock, grooved slide handle. Early models furnished w/Tenite buttstock and forend. Made 1949-54.

Stevens Model 820-SC **$215**
Same as Model 820, except has Savage Super Choke.

Stevens Model 940 Single-Shot Shotgun **$95**
Same general specifications as Model 94, except has side lever opening instead of top lever. Made 1961-70.

Stevens Model 940Y Youth Gun **$110**
Same general specifications as Model 94Y, except has side lever opening instead of top lever. Made 1961-70.

Stevens Model 9478 . . . . **$95**
Takedown. Visible hammer. Automatic ejector. Gauges: 12, 20, .410. Bbls.: 26-, 28-, 30-, 36-inch; Full choke. Weight: average 6 pounds depending on gauge and barrel. Plain pistol-grip stock and forearm. Made 1978-85.

Stevens Model 5151 Springfield **$325**
Same specifications as the Stevens Model 311 except with checkered grip and forend; equipped with recoil pad and two Ivory sights.

STOEGER SHOTGUNS

See **IGA and Tikka Shotguns**

TAR-HUNT CUSTOM RIFLES, INC.
Bloomsburg, PA

Tar-Hunt Model RSG-12 Machless
Bolt Action Slug Gun **$1295**
Similar to Professional Model, except has McMillan Fibergrain stock and deluxe blue finish. Made 1995 to date.

Tar-Hunt Model RSG-12 Peerless
Bolt Action Slug Gun **$1350**
Similar to Professional Model, except has McMillan Fibergrain stock and deluxe NP-3 (Nickel/Teflon) metal finish. Made 1995 to date.

Tar-Hunt Model RSG-12 Professional
Bolt Action Slug Gun
Bolt action 12 ga. w/2.75-inch chamber. 2-round detachable magazine. 21.5-inch fully rifled bbl. w/ or w/o muzzle break. Receiver drilled and tapped for scope mounts w/no sights. Weight: 7.75 lbs. 41.5 inches overall. Checkered black McMillan fiberglass stock w/swivel studs and Pachmayr Deacelerator pad. Made 1991 to date.
RSG-12 Model w/o muzzle break (Disc. 1993) **$995**
RSG-12 Model w/muzzle break **1095**

SHOTGUNS

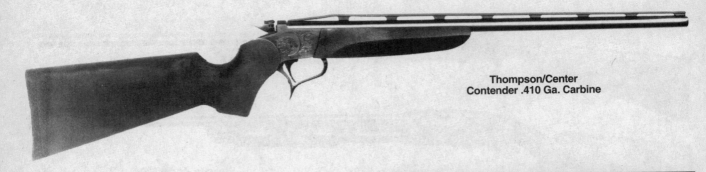

**Thompson/Center
Contender .410 Ga. Carbine**

**Tar-Hunt Model RSG-20 Mountaineer
Bolt Action Slug Gun** **$895**
Similar to Professional Model, except 20 ga. w/2.75 inch chamber. Black McMillan synthetic stock w/blind magazine. Weight: 6.5 lbs. Made 1997 to date.

TECNI-MEC SHOTGUNS
Italy
Imported by RAHN Gun Work, Inc., Hastings, MI

Tecni-Mec Model SPL 640 Folding Shotgun
Gauges: 12, 16, 20, 24, 28, 32 and .410 bore. 26-inch bbl. Chokes: IC/IM. Weight: 6.5 lbs. Checkered walnut pistol-grip stock and forend. Engraved receiver. Available with single or double triggers. Imported 1988-94.
640 w/Single Trigger **$350**
640 w/Double Trigger. **395**

THOMPSON/CENTER ARMS
Rochester, New Hampshire

Thompson/Center Contender .410 Ga. Carbine **$295**
Gauge: .410 smoothbore. 21-inch vent-rib bbl. 34.75 inches overall. Weight: about 5.25 lbs. Bead front sight. Rynite® stock and forend. Made 1991 to date.

Thompson/Center Model '87 Hunter Shotgun **$365**
Single shot. Gauge: 10 or 12; 3.5-inch chamber. 25-inch field bbl. with F choke. Weight 8 lbs. Bead front sight. American black walnut stock with recoil pad. Drop at heel .9 inch. Made 1987 to 1992. *See* Illustration next page.

Thompson/Center Model '87 Hunter Slug **$375**
Gauge: 10 (3.5-inch chamber) or 12 (3-inch chamber). Same general specifications as Model '87 Hunter Shotgun, except with 22-inch slug (rifled) bbl. and rifle sights. Made 1987 to 1992.

TIKKA SHOTGUNS
Manufactured by: Armi Marocchi, Italy
(formerly by Valmet)
Riihimaki, Finland

Tikka M 07 Shotgun/Rifle Combination **$795**
Gauge/caliber: 12/222 Rem. Shotgun bbl.: about 25 inches; rifle bbl.: about 22.75 inches. 40.66 inches overall. Weight: about 7 lbs. Dovetailed for telescopic sight. Single trigger with selector between the bbls. Vent rib. Monte Carlo-style walnut stock with checkered pistol grip and forend. Made 1965-87. *See* Illustration next page.

Tikka M 77 Over/Under Shotgun **$995**
Gauge: 12. 27-inch vent-rib bbls. Approx. 44 inches overall. Weight: about 7.25 lbs. Bbl. selector. Ejectors. Monte Carlo-style walnut stock with checkered pistol grip and forend; rollover cheekpiece. Made 1977-87.

Tikka M 77K Shotgun/Rifle Combination **$1050**
Gauge: 12/70. Calibers: 222 Rem., 5.6×52r5, 6.5×55, 7×57r5, 7×65r5, 308 Win. Vent-rib bbls.: about 25 inches (shotgun; almost 23 inches (rifle). 42.3 inches overall. Weight: about 7.5 lbs. Double triggers. Monte Carlo-style walnut stock with checkered pistol grip and forend; rollover cheekpiece. Made 1977-86.

Tikka Model 412S Over/Under
Gauge: 12; 3-inch chambers. 24-, 26-, 28- or 30-inch blued chrome-lined bbls. with five integral stainless steel choke tubes. Weight: 7.25 to 7.5 lbs. Matte nickel receiver. Select American walnut stock with checkered pistol grip and forend. (Same as the former Valmet Model 412.) Manufactured in Italy by arrangement with Marocchi from 1990 to 1993.
Field Model **$825**
Sporting Clays Model **895**

TRISTAR SPORTING ARMS
North Kansas City, MO

**Tecni-Mec Model SPL 640
Folding Shotgun**

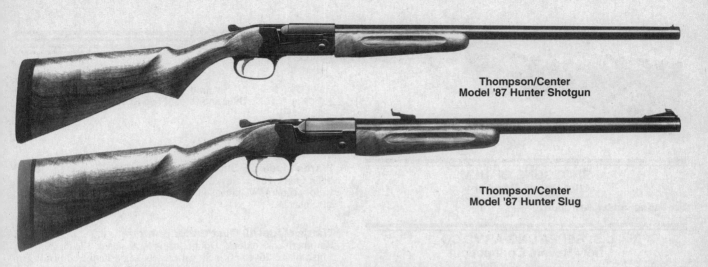

**Thompson/Center
Model '87 Hunter Shotgun**

**Thompson/Center
Model '87 Hunter Slug**

Tristar Model 1887 Lever-Action Repeater **$450**
Copy of John Browning's Winchester Model 1887 lever-action shotgun. 12 ga. only. 30-inch bbl. which may be cut down to any desired length of 18 inches or more. Version shown has 20-inch bbl. Imported 1997-98.

Tristar Model 300 O/U Shotgun **$325**
Similar to the Model 333, except 12 ga. only w/3-inch chambers. 26- or 28-inch vent rib bbls. w/extractors and fixed chokes. Etched receiver w/double triggers and standard walnut stock. Imported 1994-98.

Tristar Model 311 Side-by-side Shotgun
Boxlock action w/underlug and Greener cross bolt. 12 or 20 ga. w/3-inch chambers. 20, 28- or 30-inch bbls. w/choke tubes or fixed chokes (311R). Double triggers. Extractors. Black chrome finish. Checkered Turkish walnut buttstock and forend. Weight: 6.9 to 7.2 lbs. Imported 1994-97.
311 Model (w/extractors and choke tubes). **$450**
311R Model (w/20-inch bbls. and fixed chokes) **325**

Tristar Model 330 O/U Shotgun
Similar to the Model 333, except 12 ga. only w/3-inch chambers. 26-, 28- or 30-inch vent rib bbls w/extractors or ejectors and fixed chokes or choke tubes. Etched receiver and standard walnut stock. Imported 1994 to date.
330 Model (w/extractors and fixed chokes) **$395**
330 D Model (w/ejectors and choke tubes) **525**

Tristar Model 333 O/U Shotgun
Boxlock action. 12 or 20 ga. w/3-inch chambers. 26-, 28- or 30-inch vent rib bbls. w/choke tubes. Single selective trigger. Selective automatic ejectors. Engraved receiver w/satin nickel finish. Checkered Turkish fancy walnut buttstock and forend. Weight: 7.5 to 7.75 lbs. Imported 1994 to date..
333 Field Model **$575**
333 Sporting Clays Model (1994-97) **650**
333 TRL Ladies Field Model **585**
333 SCL Ladies Sporting Clays Model (1994-97) **665**

SHOTGUNS

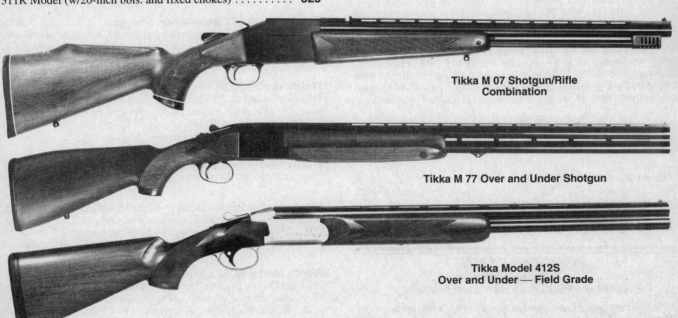

**Tikka M 07 Shotgun/Rifle
Combination**

Tikka M 77 Over and Under Shotgun

**Tikka Model 412S
Over and Under — Field Grade**

**Valmet Model 412 K
Over and Under Field Shotgun**

SHOTGUNS OF ULM
Ulm, West Germany

See listings under Krieghoff.

U.S. REPEATING ARMS CO.
New Haven, Connecticut

See Winchester Shotgun listings.

VALMET OY
Jyväskylä, Finland

See also Savage Models 330, 333T, 333 and 2400, which are Valmet guns.

Valmet Lion Over-and-Under Shotgun **$365**
Boxlock. Selective single trigger. Plain extractors.12 ga. only. Bbls.: 26-inch IC/M; 28-inch M/F, 30-inch M/F, F/F. Weight: about 7 lbs. Checkered pistol-grip stock and forearm. Imported 1947-68.

Valmet Model 412 K Over/Under Field Shotgun **$645**
Hammerless. 12-ga., 3-inch chamber. 36-inch bbl., F/F chokes. American walnut Monte Carlo stock. Made 1982-87.

Valmet Model 412 K Shotgun/Rifle Combination **$775**
Similar to Model 412 K, except bottom bbl. chambered for either 222 Rem., 223 Rem., 243 Win., 308 Win. or 30-06. 12-ga. shotgun bbl. with IM choke. Monte Carlo American walnut stock, recoil pad.

Valmet Model 412 KE Over/Under Field Shotgun **$625**
12-ga. chambered for 2.75-inch shells 26-inch bbl., IC/M chokes; 28-inch bbl., M/F chokes; 12-ga. chambered for 3-inch shells, 30-inch bbl., M/F chokes. 20-ga. (3-inch shells); 26-inch bbl. IC/M chokes; 28-inch bbl., M/ F chokes. American walnut Monte Carlo stock.

Valmet Model 412 KE Skeet **$665**
Similar to Model 412 K, except Skeet stock and chokes. 12 and 20 ga. Discontinued 1989.

Valmet Model 412 KE Trap **$695**
Similar to Model 412 K Field, except trap stock, recoil pad. 30-inch bbls., IM/F chokes. Discontinued 1989.

Valmet Model 412 3-Barrel Set **$1995**

MONTGOMERY WARD
Chicago, Illinois
Western Field and Hercules Models

Although they do not correspond to specific models below, the names Western Field and Hercules have been used to designate various Montgomery Ward shotguns at various times.

Wards Model 25 Slide-Action Repeater **$160**
Solid frame. 12 ga. only. 2- or 5-shot tubular magazine. 28-inch bbl., various chokes. Weight: about 7.5 lbs. Plain pistol-grip stock, grooved slide handle.

Wards Model 40 Over/Under Shotgun **$525**
Hammerless. Boxlock. Double triggers. Gauges: 12, 16, 20, .410. Bbls.: plain; 26- to 30-inch, various chokes. Checkered pistol-grip stock and forearm.

Wards Model 40N Slide-Action Repeater **$165**
Same general specifications as Model 25.

Wards (Western Field) Model 50 Pumpgun **$170**
Solid frame. Gauges: 12 and 16. 2- and 5-shot magazine. 26-, 28- or 30-inch bbl., 48 inches overall w/28-inch bbl. Weight: 7.25 - 7.75 lbs. Metal bead front sight. Walnut stock and grooved forend. *See* Illustration next page.

**Wards (Western Field) Model 52
Double-Barrel Shotgun** **$185**
Hammerless coil-spring action. Gauges: 12, 16, 20 and .410. 26-, 28-, or 30-inch bbls., 42 to 46 inches overall, depending upon bbl. length. Weight: 6 (.410 ga. w/26-inch bbl.) to 7.25 lbs. (12 ga. w/30-inch bbls.), depending upon gauge and bbl. length. Case-hardened receiver; blued bbls. Plain buttstock and forend. Made circa 1954. *See* Illustration next page.

Wards Model 172 Bolt-Action Shotgun **$85**
Takedown. 2-shot detachable clip magazine.12 ga. 28-inch bbl. with variable choke. Weight: about 7.5 lbs. Monte Carlo stock with recoil pad.

Wards Model 550A Slide-Action Repeater **$195**
Takedown. Gauges: 12, 16, 20, .410. 5-shot tubular magazine. Bbls.: plain, 26- to 30-inch, various chokes. Weight: 6 (.410 ga. w/26-inch bbl.) to 8 lbs. (12 ga. w/30-inch bbls.). Plain pistol-grip stock and grooved slide handle.

Wards Model SB300 Double-Barrel Shotgun **$215**
Same general specifications as Model SD52A.

Wards Model SB312 Double-Barrel Shotgun **$225**
Boxlock. Double triggers. Plain extractors. Gauges: 12, 16, 20, .410. Bbls.: 24- to 30-inch. Various chokes. Weight, about 7.5 lbs. in 12 ga.; 6.5 lbs in .410 ga. Checkered pistol-grip stock and forearm.

Wards Model SD52A Double-Barrel Shotgun **$175**
Boxlock. Double triggers. Plain extractors. Gauges: 12, 16, 20, .410. Bbls.: 26- to 32-inch, various chokes. Plain forend and pistol-grip buttstock. Weight: 6 (.410 ga., 26-inch bbls.) to 7.5 lbs. (12 ga., 32-inch bbls.).

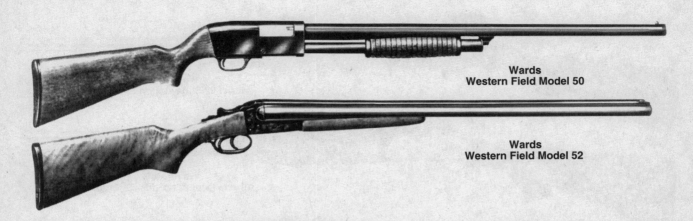

**Wards
Western Field Model 50**

**Wards
Western Field Model 52**

WEATHERBY, INC.
South Gate, California

Weatherby Model 82 Autoloading Shotgun
Hammerless, gas-operated. 12 ga. only. Bbls.: 22- to 30-inch, various integral chokes. Weight: 7.5 lbs. Checkered walnut stock and forearm. Imported 1982-89.
Standard Autoloading Shotgun $345
BuckMaster Auto Slug w/rifle sights (1986-90) 350

Weatherby Orion Over/Under Shotgun *(Con't)*

Weatherby Model 92 Slide-Action Shotgun
Hammerless, short-stroke action. 12 ga.; 3-inch chamber. Tubular magazine. Bbls.: 22-, 26-, 28-, 30-inch with fixed choke or IMC choke tubes; plain or vent rib with rifle sights. Weight: 7.5 lbs. Engraved, matte black receiver and blued barrel. Checkered high-gloss buttstock and forend. Imported from Japan since 1982. *See* Illustration Next Page.
Standard Model 92 . . . . $275
BuckMaster Pump Slug w/rifle sights (intro. 1986) 295

Weatherby Athena Over/Under Shotgun
Engraved boxlock action with Greener crossbolt and sideplates. Gauges: 12, 20, 28 and .410; 2.75- or 3.5-inch chambers. Bbls.: 26-, 28-, 30- or 32-inch with fixed or IMC Multi-choke tubes. Weight: 6.75 to 7.38 lbs. Single selective trigger. Selective auto ejectors. Top tang safety. Checkered Claro walnut stock and forearm with high-luster finish. Imported 1982 to date. See Illustration Next Page.
Field Model w/IMC Multi-choke (12 or 20 ga.) $1195
Field Model w/Fixed Chokes (28 or .410 ga.) 1295
Skeet Model w/Fixed Chokes (12 or 20 ga.) 1275
Skeet Model w/Fixed Chokes (28 or .410 ga.) 1425
Master Skeet Tube Set 2350
Trap Model w/IMC Tubes 1195
Grade V (1993 to date) 1595

Weatherby Centurion Automatic Shotgun
Gas-operated. Takedown. 12 ga. 2.75-inch chamber. 3-shot magazine. Bbls.: vent ribs; 26-inch SK, IC or M 28-inch M or F; 30-inch Full choke. Weight: with 28-inch bbl., 7 lbs. 10.5 oz. Checkered pistol-grip stock and forearm, recoil pad. Made in Japan 1972-81.
Centurion Field Grade $295
Centurion Trap Gun (30-inch Full choke bbl.) 325
Centurion Deluxe (etched receiver, fancy wood) 350

Weatherby Olympian Over/Under Shotgun
Gauges: 12 and 20. 2.75- (12 ga.) and 3-inch (20 ga.) chambers. Bbls.: 26-, 28-, 30, and 32-inch. Weight: 6.75 - 8.75 lbs. America walnut stock and forend.
Field Model. $745
Skeet Model 765
Trap Model 750

Weatherby Orion Over/Under Shotgun
Boxlock with Greener crossbolt. Gauges: 12, 20, 28 and .410; 2.75- or 3-inch chambers. Bbls.: 26-, 28, 30-, 32- or 34-inch with fixed or IMC Multi-choke tubes. Weight: 6.5 to 9 lbs. Single selective trigger. Selective auto ejectors. Top tang safety. Checkered, high-gloss pistol-grip Claro walnut stock and forearm. Finish: Grade I, plain blued receive; Grade II, engraved blued receiver; Grade III, silver gray receiver. Imported 1982 to date.
Orion I Field w/IMC (12 or 20 ga.) $770
Orion II Field w/IMC (12 or 20 ga.) 845
Orion III Field w/IMC (12 or 20 ga.) 870
Skeet II w/Fixed Chokes 895
Sporting Clays II 945
Trap II 920

Weatherby Patrician Slide-Action Shotgun
Hammerless. Takedown. 12 ga. 2.75-inch chamber. 4 shot tubular magazine. Bbls.: vent rib; 26-inch, SK, IC M; 28-inch, M F; 30-inch, F choke. Weight: with 28-inch bbl., 7 lbs. 7 oz. Checkered pistol-grip stock and slide handle, recoil pad. Made in Japan 1972-82.

SHOTGUNS

**Weatherby
Model 82 Autoloading Shotgun**

Weatherby
Model 92 Slide-Action Shotgun

Weatherby
Athena Over and Under Shotgun

Weatherby
Centurion Automatic Shotgun

Weatherby Olympian Trap
Over and Under Shotgun

Weatherby Orion
Over and Under Shotgun

Weatherby Patrician
Deluxe Grade

Weatherby
Regency Trap Gun

Weatherby Patrician Slide-Action Shotgun (Con't)

Patrician Field Grade $235
Patrician Deluxe (etched receiver, fancy grade wood) 275
Patrician Trap Gun (30-inch F choke bbl.) 250

Weatherby Regency Field Grade O/U Shotgun $775

Boxlock with sideplates, elaborately engraved. Auto ejectors. Selective single trigger. Gauges: 12, 20. 2.75-inch chamber in 12 ga., 3-
inch in 20 ga. Bbls.: vent rib; 26-inch SK, IC/M, M/F (20 ga. only);
28-inch SK, IC/M, M/F; 30-inch M/F (12 ga only). Weight with 28-
inch Bbls.: 7 lbs. 6 oz., 12 ga.; 6 lbs. 14 oz., 20 ga. Checkered pistol-
grip stock and forearm of fancy walnut. Made in Italy 1965-82.

Weatherby Regency Trap Gun $795

Similar to Regency Field Grade, except has trap-style stock with
straight or Monte Carlo comb. Bbls. have vent side ribs and high,
wide vent top rib; 30- or 32-inch, M/F, IM/F or F/F chokes.
Weight: with 32-inch bbls., 8 lbs. Made in Italy 1965-1982.

WESTERN ARMS CORP.
Ithaca, New York
Division of Ithaca Gun Company

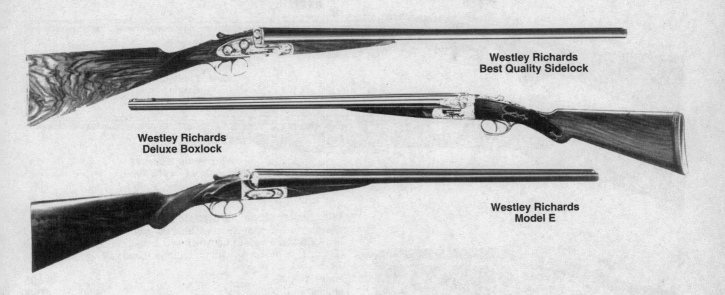

Westley Richards Best Quality Sidelock

Westley Richards Deluxe Boxlock

Westley Richards Model E

Western Long Range Hammerless Double

Boxlock. Plain extractors. Single or double triggers. Gauges: 12, 16, 20, .410. Bbls.: 26- to 32-inch, M/F choke standard. Weight: 7.5 lbs., 12 ga. Plain pistol-grip stock and forend. Made 1929-46.

With double triggers **$275**

With single trigger **350**

WESTERN AUTO SHOTGUNS
Kansas City, Missouri

Revelation Model 300H Slide-Action Repeater **$195**
Gauges: 12,16, 20, .410. 4-shot tubular magazine. Bbls.: 26- to 30-inch, various chokes. Weight: about 7 lbs. Plain pistol-grip stock, grooved slide handle.

Revelation Model 310A Slide-Action Repeater **$185**
Takedown. 12 ga. 5-shot tubular magazine. Bbls.: 28- and 30-inch. Weight: about 7.5 lbs. Plain pistol-grip stock.

Revelation Model 310B Slide-Action Repeater **$175**
Same general specifications as Model 310A except chambered for 16 ga.

Revelation Model 310C Slide-Action Repeater **$195**
Same general specifications as Model 310A except chambered for 20 ga.

Revelation Model 310E Slide-Action Repeater **$225**
Same general specifications as Model 310A except chambered for .410 bore.

Revelation Model 325BK Bolt-Action Repeater **$85**
Takedown. 2-shot detachable clip magazine. 20 ga. 26-inch bbl. with variable choke. Weight: 6.25 lbs.

WESTERN FIELD SHOTGUNS

See "W" for listings under Montgomery Ward.

WESTLEY RICHARDS & CO., LTD.
Birmingham, England

The Pigeon and Wildfowl Gun, available in all of the Westley Richards models except the Ovundo, has the same general specifications as the corresponding standard field gun, except has magnum action of extra strength and treble bolting, chambered for 12 gauge only (2.75- or 3-inch); 30-inch Full choke barrels standard. Weight: about 8 lbs. The manufacturer warns that 12-gauge magnum shells should not be used in their standard weight double-barrel shotguns.

Westley Richards Best Quality Boxlock Hammerless Double-Barrel Shotgun
Boxlock. Hand-detachable locks and hinged cover plate. Selective ejectors. Double triggers or selective single trigger. Gauges: 12, 16, 20. Barrel lengths and boring to order. Weight: 5.5 to 6.25 lbs. depending on ga. and bbl. length. Checkered stock and forend, straight or half-pistol grip. Also supplied in Pigeon and Wildfowl Model with same values. Made from 1899 to date.

With double triggers **$15,500**

With selective single trigger **17,500**

Westley Richards Best Quality Sidelock Hammerless Double-Barrel Shotgun
Hand-detachable sidelocks. Selective ejectors. Double triggers or selective single trigger. Gauges: 12, 16, 20, 28, .410. Bbl. lengths and boring to order. Weight: 4.75 to 6.75 lbs., depending on ga. and bbl. length. Checkered stock and forend, straight or half-pistol grip. Also supplied in Pigeon and Wildfowl Model with same values. Currently manufactured.

With double triggers **$19,225**

With selective single trigger **21,000**

Westley Richards Model Deluxe Boxlock Hammerless Double-Barrel Shotgun
Same general specifications as standard Best Quality gun except higher quality throughout. Has Westley Richards top-projection and treble-bite lever-work, hand-detachable locks. Also supplied in Pigeon and Wildfowl Model with same values. Currently manufactured.

With double triggers . . . . **$ 8,950**

With selective single trigger **10,000**

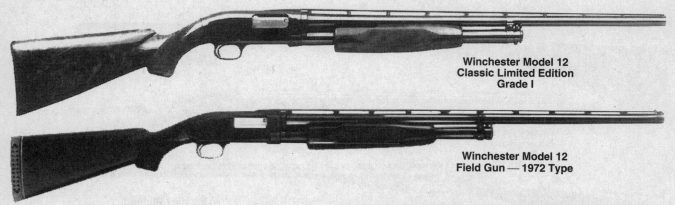

Winchester Model 12
Classic Limited Edition
Grade I

Winchester Model 12
Field Gun — 1972 Type

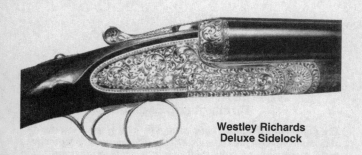

Westley Richards
Deluxe Sidelock

Westley Richards Model Deluxe Sidelock
Same as Best Quality Sidelock, except higher grade engraving and wood. Currently manufactured.
With double triggers **$20,000**
With single trigger **24,000**

Westley Richards Model E Hammerless Double
Anson & Deeley-type boxlock action. Selective ejector or non-ejector. Double triggers. Gauges: 12, 16, 20. Barrel lengths and boring to order. Weight: 5.5 to 7.25 lbs. depending on type, ga. and bbl. length. Checkered stock and forend, straight or half-pistol grip. Also supplied in Pigeon and Wildfowl Model with same values. Currently manufactured.
Ejector model **$3795**
Non-ejector model **3300**

Westley Richards Ovundo (Over/Under) **$14,995**
Hammerless. Boxlock. Hand-detachable locks. Dummy sideplates. Selective ejectors. Selective single trigger. 12 ga. Barrel lengths and boring to order. Checkered stock/ forend, straight or half-pistol grip. Mfd. before WW II.

TED WILLIAMS SHOTGUNS

See Sears shotguns.

WINCHESTER SHOTGUNS
New Haven, Connecticut

Formerly Winchester Repeating Arms Co. Now mfd. by Winchester-Western Div., Olin Corp., and by U.S. Repeating Arms company In 1999, production rights were acquired Browning Arms.

Winchester Model 12 Classic Limited Edition
Gauge: 20; 2.75-inch chamber. Bbl.: 26-inch vent rib; IC. Weight: 7 lbs. Checkered walnut buttstock and forend. Polished blue finish (Grade I) or engraved with gold inlays (Grade IV). Made 1993 to 1995.
Grade I (4000) . . . . **$ 695**
Grade IV (1000) **1095**

Winchester Model 12 Featherweight **$475**
Same as Model 12 Standard w/plain barrel, except has alloy trigger guard. Modified takedown w/redesigned magazize tube, cap and slide handle. 12 ga. only. Bbls.: 26-inch IC; 28-inch M or F; 30-inch F choke. Serial numbers with "F" prefix. Weight: about 6.75 lbs. Made 1959-62.

Winchester Model 12 Field Gun, 1972 Type **$550**
Same general specifications as Standard Model 12. 12 ga. only. 26-, 28- or 30-inch vent-rib bbl., standard chokes. Engine-turned bolt and carrier. Hand-checkered stock/slide handle of semifancy walnut. Made 1972-75.

Winchester Model 12 Heavy Duck Gun
Same general specifications as Standard Grade, except 12 ga. only, chambered for 3-inch shells. 30- or 32-inch plain, solid or vent rib bbl. w/full choke only. 3-shot magazine. Checkered slide handle and pistol-grip walnut buttstock w/recoil pad. Weight: 8.5 to 8.75 lbs. Made 1935-63.
Heavy Duck Gun, Plain bbl. **$725**
Heavy Duck Gun, Solid rib (Disc. 1959) **995**
Heavy Duck Gun, Vent rib (Special order only) **1295**

Winchester Model 12 Pigeon Grade
Deluxe versions of the regular Model 12 Standard or Field Gun, Duck Gun, Skeet Gun and Trap Gun made on special order. This grade has finer finish throughout, hand-smoothed action, engine-turned breech bolt and carrier, stock and extension slide handle of high grade walnut, fancy checkering, stock dimensions to individual specifications. Engraving and carving available at extra cost ranging from about $135 to over $1000. Discont. 1965.
Field Gun, plain bbl. **$1250**
Field Gun, vent rib **1495**
Skeet Gun, matted rib **1595**
Skeet Gun, vent rib **1750**
Skeet Gun, Cutts Compensator **1025**
Trap Gun, matted rib **1495**
Trap Gun, vent rib **1650**

Winchester Model 12 Riot Gun **$950**
Same general specifications as Plain Barrel Model 12 Standard, except has 20-inch cylinder bore bbl.,12 gauge only. Made 1918-1963.

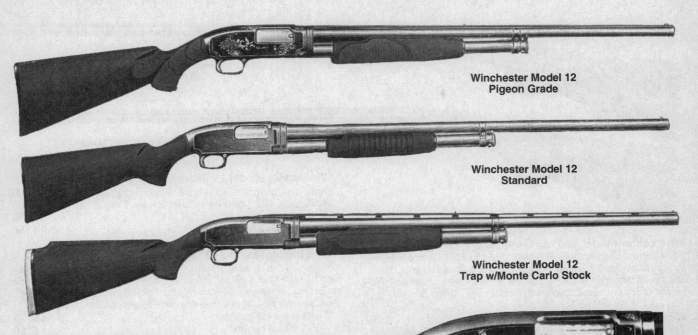

**Winchester Model 12
Pigeon Grade**

**Winchester Model 12
Standard**

**Winchester Model 12
Trap w/Monte Carlo Stock**

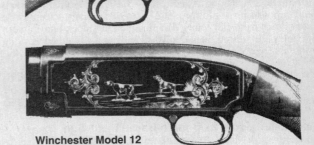

**Winchester Model 12
12-4 Engraving**

Winchester Model 12 Skeet Gun $850
Gauges: 12, 16, 20, 28. 5-shot tubular magazine. 26-inch matted rib bbl., SK choke. Weight: about 7.75 lbs., 12 ga.; 6.75 lbs., other gauges. Bradley red or ivory bead front sight. Winchester 94B middle sight. Checkered pistol-grip stock and extension slide handle. Discont. after WWII.

Winchester Model 12 Skeet Gun, Cutts Compensator $750
Same general specifications as standard Model 12 Skeet Gun, except has plain bbl. fitted with Cutts Compensator, 26 inches overall. Discontinued 1954.

Winchester Model 12 Skeet Gun, Plain Barrel $695
Same general specifications as standard Model 12 Skeet except w/no rib.

Winchester Model 12 Skeet Gun, Vent Rib $1150
Same general specifications as standard Model 12 Skeet Gun, except has 26-inch bbl. with vent rib, 12 and 20 ga. Discontinued in 1965.

Winchester Model 12 Skeet Gun, 1972 Type $725
Same gen. specifications as Standard Model 12. 12 ga. only. 26-inch vent-rib bbl., SK choke. Engine-turned bolt and carrier. Hand-checkered skeet-style stock and slide handle of choice walnut, recoil pad. Made 1972-75.

Winchester Model 12 Standard Gr., Matted Rib $750
Same general specifications as Plain Bbl. Model 12 Standard, except has solid raised matted rib. Discontinued after World War II.

Winchester Model 12 Standard Gr., Vent Rib $825
Same general specifications as Plain Barrel Model 12 Standard, except has vent rib. 26.75- or 30-inch bbl.,12 ga. only. Discont. after World War II.

Winchester Model 12 Standard Slide-Action Repeater
Hammerless. Takedown. Gauges: 12, 16, 20, 28. 6-shell tubular magazine. Plain bbl. Lengths: 26- to 32-inches; choked F to Cyl. Weight: about 7.5 lbs., 12 ga. 30-inch, 6.5 lbs. in other ga. with 28-inch bbl. Plain pistol-grip stock, grooved slide handle. Made 1912-64.

Winchester Model 12 Standard Slide *(Con't)*
12-Gauge, 28-inch bbl. (Full)	$595
16-Gauge.	525
20-Gauge.	695
28-Gauge.	2995

Winchester Model 12 Super Pigeon Grade $1995
Custom version of Model 12 with same general specifications as standard models. 12 ga. only. 26-, 28- or 30-inch vent-rib bbl., any standard choke. Engraved receiver. Hand-smoothed and fitted action. Full fancy walnut stock and forearm made to individual order. Made 1965-72.

Winchester Model 12 Trap Gun
Same general specifications as standard Model 12, except has straighter stock, checkered pistol grip and extension slide handle, recoil pad, 30-inch matted-rib bbl., F choke, 12 ga. only. Discont. after World War II; vent-rib model discontinued 1965.
Matted-rib bbl.	$850
With straight stock, vent rib	950
With Monte Carlo stock, vent rib	995

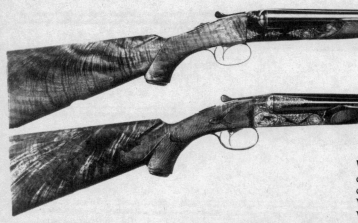

**Winchester Model 21
Custom Grade**

**Winchester Model 21
Pigeon Grade**

Winchester Model 12 Trap Gun, 1972 Type $695
Same general specifications as Standard Model 12. 12 gauge only. 30-inch vent-rib bbl., F choke. Engine-turned bolt and carrier. Hand-checkered trap-style stock (straight or Monte Carlo comb) and slide handle of select walnut, recoil pad. Intro. in 1972. Discontinued.

Winchester Model 20 Single-Shot Hammer Gun $550
Takedown. .410 bore. 2.5-inch chamber. 26-inch bbl., F choke. Checkered pistol-grip stock and forearm. Weight: about 6 lbs. Made 1919-24.

Winchester Model 21 Custom, Pigeon, Grand American
Since 1959, the Model 21 has been offered only in deluxe models: Custom, Pigeon, Grand American — on special order. General specifications same as for Model 21 standard models, except these custom guns have full fancy American walnut stock and forearm with fancy checkering, finely polished and hand-smoothed working parts, etc.; engraving inlays, carved stocks and other extras are available at additional cost. Made 1959-88.
Custom Grade .... $ 5,695
Pigeon Grade 8,550
Grand American 12,950

Winchester Model 21 Double-Barrel Field Gun
Hammerless. Boxlock. Automatic safety. Double triggers or selective single trigger, selective or non-selective ejection (all postwar Model 21 shotguns have selective single trigger and selective ejection). Gauges: 12,16,20. Bbls.: raised matted rib or vent rib; 26-, 28-, 30-, 32-inch, the latter in 12 ga. only; F, IM, M, IC, SK chokes. Weight: 7.5 lbs., 12 ga. w/30-inch bbl.; about 6.5 lbs. 16 or 20 ga. w/28-inch bbl. Checkered pistol- or straight-grip stock, regular or beavertail forend. Made 1930-58.
With double trigger, non-selective ejection $2795
With double trigger, selective ejection 2995
With selective single trigger, non-selective ejection 3250
With selective single trigger, selective ejection 3400
Extra for vent rib 500

Winchester Model 21 Duck Gun
Same general specifications as Model 21 Field Gun, except chambered for 12 ga. 3-inch shells, 30- or 32-inch bbls. only, F choke, selective single trigger, selective ejection, pistol-grip stock with recoil pad, beavertail forearm, both checkered. Discont. 1958.
With matted-rib barrels $2995
With vent-rib barrels 3250

Winchester Model 21 Skeet Gun
Same general specifications as Model 21 Standard, except has 26- or 28-inch bbls. only, SK chokes No. 1 and 2 Bradley red bead front sight, selective single trigger, selective ejection, nonauto safety, check-

Winchester Model 21 Skeet Gun *(Con't)*
ered pistol- or straight-grip stock without buttplate or pad (wood butt checkered), checkered beavertail forearm. Discont. 1958.
With matted-rib barrels $3550
With vent-rib barrels 4350

Winchester Model 21 Trap Gun
Same general specifications as Model 21 Standard, except has 30- or 32-inch bbls. only, F choke, selective single trigger, selective ejection, nonauto safety, checkered pistol-or straight-grip stock with recoil pad, checkered beavertail forearm. Discont. 1958.
With matted-rib barrels $3575
With vent-rib barrels 4350

Winchester Model 23 Side-by-Side Shotgun
Boxlock. Single trigger. Automatic safety. Gauges: 12, 20, 28, .410. Bbls.: 25.5-, 26-, 28-inch with fixed chokes or Winchoke tubes. Weight: 5.88 to 7 lbs. Checkered American walnut buttstock and forend. Made in 1979 for Olin at its Olin-Kodensha facility, Japan.
Classic 23 — Gold inlay, Engraved $1595
Custom 23 — Plain receiver, Winchoke system 895
Heavy Duck 23 — Standard 1225
Lightweight 23 — Classic Style 1215
Light Duck 23 — Standard 1150
Light Duck 23 — 12 ga. Golden Quail 1350
Light Duck 23 — .410 Golden Quail 2350
Custom Set 23 — 20 & 28 gauge 3695

Winchester Model 24 Hammerless Double $495
Boxlock. Double triggers. Plain extractors. Auto safety. Gauges: 12, 16, 20. Bbls.: 26-inch IC/M; 28-inch M/F (also IC/M in 12 ga. only); 30-inch M and F in 12 ga. only. Weight: about 7.5 lbs., 12 ga. Metal bead front sight. Plain pistol-grip stock, semibeavertail forearm. Made 1939-57.

Winchester Model 25 Riot Gun $375
Same as Model 25 Standard, except has 20-inch cylinder bore bbl., 12 ga. only. Made 1949-55.

Winchester Model 25 Slide-Action Repeater $395
Hammerless. Solid frame. 12 ga. only. 4-shell tubular magazine. 28-in. Plain bbl.; IC, M or F choke. Weight: about 7.5 lbs. Metal bead front sight. Plain pistol-grip stock, grooved slide handle. Made 1949-55.

Winchester Model 36 Single-Shot Bolt Action $450
Takedown. Uses 9mm Short or Long shot or ball cartridges interchangeably. 18-inch bbl. Plain stock. Weight: about 3 lbs. Made 1920-27.

Winchester Model 37 Single-Shot Shotgun
Semi-hammerless. Auto ejection. Takedown. Gauges: 12, 16, 20, 28, .410. Bbl. lengths: 28-, 30-, 32-inch in all gauges except .410; 26- or 28-inch in .410; all barrels plain with F choke. Weight: about 6.5 pounds, 12 ga. Made 1937-63.

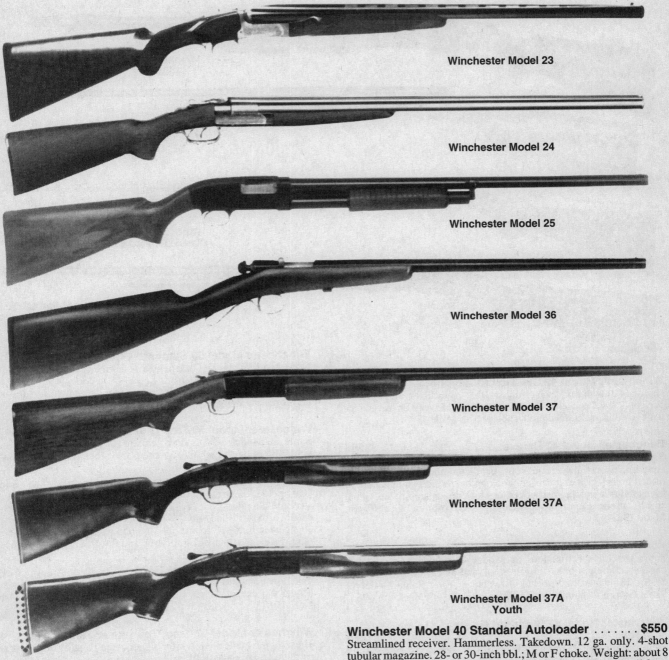

Winchester Model 23

Winchester Model 24

Winchester Model 25

Winchester Model 36

Winchester Model 37

Winchester Model 37A

Winchester Model 37A Youth

Winchester Model 37 Single-Shot Shotgun *(Con't)*

12 Gauge **$225**
28 Gauge 850
Other Gauges 175

Winchester Model 37A Single Shot Shotgun **$125**
Similar to Model 370, except has engraved receiver and gold trigger, checkered pistol-grip stock, fluted forearm; 16 ga. available with 30-inch bbl. only. Made 1973-80.

Winchester Model 37A Youth **$135**
Similar to Model 370 Youth, except has engraved receiver and gold trigger, checkered pistol-grip stock, fluted forearm. Made 1973-80.

Winchester Model 40 Standard Autoloader **$550**
Streamlined receiver. Hammerless. Takedown. 12 ga. only. 4-shot tubular magazine. 28- or 30-inch bbl.; M or F choke. Weight: about 8 lbs. Bead sight on ramp. Plain pistol-grip stock, semibeavertail forearm. Made 1940-41.

Winchester Model 40 Skeet Gun **$795**
Same general specifications as Model 40 Standard, except has 24-inch plain bbl. w/Cutts Compensator and screw-in choke tube, checkered forearm and pistol grip, grip cap. Made 1940-1941.

Winchester Model 41 Single-Shot Bolt Action
Takedown. .410 bore. 2.5-inch chamber (chambered for 3-inch shells after 1932). 24-inch bbl., F choke. Plain straight stock standard. Also made in deluxe version. Made 1920-34.
Standard Model. **$525**
Deluxe Model 600

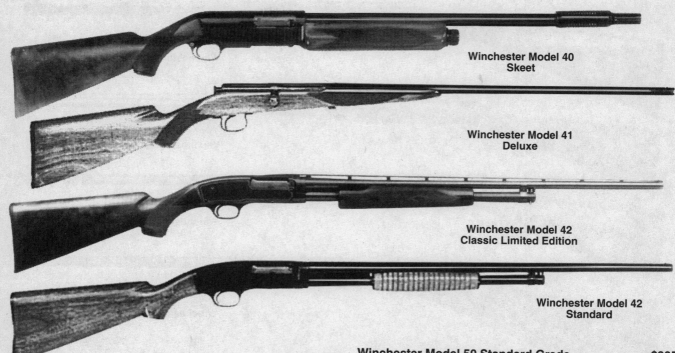

**Winchester Model 40
Skeet**

**Winchester Model 41
Deluxe**

**Winchester Model 42
Classic Limited Edition**

**Winchester Model 42
Standard**

Winchester Model 42 Classic Ltd. Edition **$1250**
Gauge: .410 with 2.75-inch chamber. Bbl.: 26-inch vent rib; F choke. Weight: 7 lbs. Checkered walnut buttstock and forend. Engraved blue with gold inlays. Limited production of 850. Made 1993.

Winchester Model 42 Deluxe **$2295**
Same general specifications as the Model 42 Standard, except has vent rib, finer finish throughout, hand-smoothed action, engine-turned breech bolt and carrier, stock and extension slide handle of high grade walnut, fancy checkering, stock dimensions to individual specifications. Engraving and carving were offered at extra cost. Made 1933-63.

Winchester Model 42 Skeet Gun **$1850**
Same general specifications as Model 42 Standard, except has checkered straight or pistol-grip stock and extension slide handle, 26- or 28-inch matted-rib bbl., SK choke. *Note:* Some Model 42 Skeet Guns are chambered for 2.5-inch shells only. Discont. 1963.

**Winchester Model 42 Standard Grade,
Matt Rib** **$1450**
Same general specifications as Plain Bbl. Model 42, except has solid raised matted rib. Discont. 1963.

Winchester Model 42 Standard **$995**
Hammerless. Takedown. .410 bore (3- or 2.5-inch shell). Tubular magazine holds five 3-inch or six 2.5-inch shells. 26- or 28-inch plain bbl.; cylinder bore, M or F choke. Weight: about 6 lbs. Plain pistol-grip stock; grooved slide handle. Made 1933-63.

Winchester Model 50 Field Gun, Vent Rib **$395**
Same as Model 50 Standard, except has vent rib.

Winchester Model 50 Skeet Gun **$475**
Same as Model 50 Standard, except has 26-inch vent-rib bbl. with SK choke, skeet-style stock of select walnut.

Winchester Model 50 Standard Grade **$335**
Non-recoiling bbl. and independent chamber. Gauges: 12 and 20. 2-shot tubular magazine. Bbl.: 12 ga. — 26-, 28-, 30-inch; 20 ga. — 26-, 28-inch; IC, SK, M, F choke. Checkered pistol-grip stock and forearm. Weight: about 7.75 lbs. Made 1954-61.

Winchester Model 50 Trap Gun **$475**
Same as Model 50 Standard, except 12 ga. only, has 30-inch vent-rib bbl. with F choke, Monte Carlo stock of select walnut.

Winchester Model 59 Autoloading Shotgun **$450**
Gauge: 12. Magazine holds two shells. Alloy receiver. Win-Lite steel and fiberglass bbl.: 26-inch IC, 28-inch M or F choke, 30-inch F choke; also furnished with 26-inch bbl. with Versalite choke (interchangeable F, M, IC tubes; one supplied with gun). Weight: about 6.5 lbs. Checkered pistol-grip stock and forearm. Made 1959-65.

Winchester Model 97 Riot Gun **$895**
Takedown or solid frame. Same general specifications as standard Model 97, except 12 ga. only, 20-inch cylinder bore bbl. Made 1897-1957.

Winchester Model 97 Trap, Tournament and Pigeon
These higher grade models offer higher overall quality than the standard grade. Discont. 1939.
Trap Gun **$695**
Tournament Grade 825
Pigeon Grade 995

Winchester Model 97 Trench Gun **$1595**
Solid frame. Same as Model 97 Riot Gun, except has handguard and is equipped with a bayonet. World War I government issue, 1917-18.

Winchester Model 97 Slide-Action Repeater **$475**
Standard Grade. Takedown or solid frame. Gauges: 12 and 16. 5-shell tubular magazine. Bbl.: plain; 26 to 32 inches, the latter in 12 ga. only; choked F to Cyl. Weight: about 7.75 lbs. (12 ga. w/28-inch barrel). Plain pistol-grip stock, grooved slide handle. Made 1897-1957.

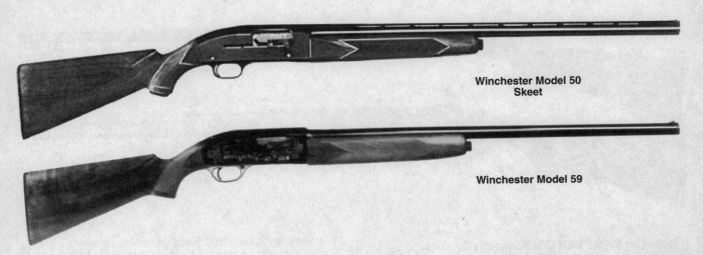

Winchester Model 50
Skeet

Winchester Model 59

Winchester Model 101 Diamond Grade Target $1325
Similar to Model 101 Standard except silvered frame and Winchoke interchangeable choke tubes. Made 1981-90.

Winchester Model 101 Field Gun Over/Under
Boxlock. Engraved receiver. Auto ejectors. Single selective trigger. Combination bbl. selector and safety. Gauges: 12 and 28, 2.75-inch chambers; 20 and .410, 3-inch chambers. Vent-rib bbls.: 30- (12 ga. only) and 26.5-inch, IC/M. Weight: 6.25 to 7.75 lbs. depending on gauge and bbl. length. Hand-checkered French walnut and forearm. Made 1963-81. Gauges other than 12 intoduced 1966.
12 and 20 Gauge $750
28 and .410 Gauge 925

Winchester Model 101 Quail Special O/U $1595
Same specifications as small-frame Model 101, except in 28 and .410 ga. with 3-inch chambers. 25.5-inch bbls. with choke tubes (28 ga.) or M/F chokes (.410). Imported from Japan in 1987.

Winchester Model 101 Shotgun/Rifle Combination Gun $1695
12-ga. Winchoke bbl. on top and rifle bbl. chambered for 30-06 on bottom (over/under). 25-inch bbls. Engraved receiver. Hand checkered walnut stock and forend. Weight: 8.5 lbs. Mfd. for Olin Corp. in Japan.

Winchester Model 370 Single-Shot Shotgun $95
Visible hammer. Auto ejector. Takedown. Gauges: 12, 16, 20, 28, .410. 2.75-inch chambers in 16 and 28 ga., 3-inch in other ga. Bbls.: 12 ga., 30-, 32- or 36-inch,16 ga; 30- or 32-inch; 20 and 28 ga., 28-inch; .410 bore, 26-inch, all F choke. Weight: 5.5-6.25 lbs. Plain pistol-grip stock and forearm. Made 1968-73.

Winchester Model 370 Youth $105
Same as standard Model 370, except has 26-inch bbl. and 12.5-inch stock with recoil pad; 20 gauge with IM choke, .410 bore with F choke. Made 1968-73.

Winchester Model 1001 O/U Shotgun
Boxlock. 12 ga., 2.75- or 3-inch chambers. Bbls.: 28- or 30-inch vent rib; WinPlus choke tubes. Weight: 7-7.75 lbs. Checkered walnut

SHOTGUNS

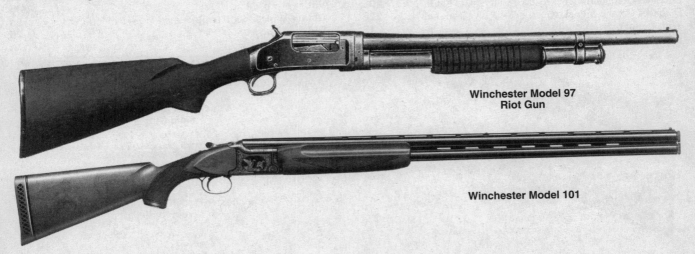

Winchester Model 97
Riot Gun

Winchester Model 101

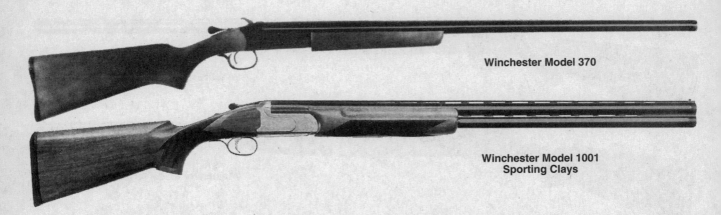

Winchester Model 370

Winchester Model 1001
Sporting Clays

Winchester Model 1001 O/U Shotgun *(Con't)*

buttstock and forend. Blued finish with scroll engraved receiver. Made 1993 to date.

Field Model (28" bbl., 3") $795
Sporting Clays 850
Sporting Clays Lite 825

Winchester Model 1200 Deer Gun $195

Same as standard Model 1200, except has special 22-inch bbl. with rifle-type sights, for rifled slug or buckshot; 12 ga. only. Weight: 6.5 lbs. Made 1965-74.

Winchester Model 1200 Defender Series
Slide-Action Security Shotguns

Hammerless. 12 ga. w/3-inch chamber. 18-inch bbl. w/cylinder bore and metal front bead or rifle sights. 4- or 7-shot magazine. Weight: 5.5 to 6.75 lbs. 25.6 inches (PG Model) or 38.6 inches overall. Matte blue finish. Synthetic pistol grip or walnut finished hardwood buttstock w/grooved synthetic or hardwood slide handle. **NOTE:** *Even though the 1200 series was introduced in 1964 and was surplanted by the Model 1300 in 1978, the Security Series (including the Defender Model) was marketed under 1200 series alpha-numeric product codes (G1200DM2R) until 1989. In 1990, the same Defender model was marketed under a 4-digit code (7715) and was then advertised in the 1300 series.*

Defender w/hardwood stock, bead sight. $165
Defender w/hardwood stock, rifle sights 175
Defender Model w/pistol grip stock 180
Defender Combo Model (w/extra 28-inch plain bbl.) 205
Defender Combo Model (w/extra 28-inch vent rib bbl.) 220
With 7-shot magazine, **add** 15

Winchester Model 1200 Field Gun — Magnum

Same as standard Model 1200, except chambered for 3-inch 12 and 20 ga. magnum shells; plain or vent-rib bbl., 28- or 30-inch, F choke. Weight: 7.38 to 7.88 lbs. Made 1964-83.

With plain bbl. $200
With vent-rib bbl. 225
Add for Winchester Recoil Reduction System 50

Winchester Model 1200 Ranger
Slide-Action Shotgun $180

Hammerless. 12 and 20 ga.; 3-inch chambers. Walnut finished hardwood stock, ribbed forearm. 28-inch vent-rib bbl.; Winchoke system. Weight: 7.25 lbs. Made 1982 to 1990 by U. S. Repeating Arms.

Winchester Model 1300 Featherweight *(Con't)*

Winchester Model 1200 Ranger
Youth Slide-Action Shotgun. $175

Same general specifications as standard Ranger Slide-Action except chambered for 20 ga. only, has 4-shot magazine, recoil pad on buttstock and weighs 6.5 lbs. Mfd. by U. S. Repeating Arms.

Winchester Model 1200 Slide-Action Field Gun

Front-locking rotary bolt. Takedown. 4-shot magazine. Gauges: 12, 16, 20 (2.75-inch chamber). Bbl.: plain or vent rib; 26-, 28-, 30-inch; IC, M, F choke or with Win-choke (interchangeable tubes IC-M-F). Weight: 6.5 to 7.25 lbs. Checkered pistol-grip stock and forearm (slide handle), recoil pad; also avail. 1966-70 w/Winchester Recoil Reduction System (Cycolac stock). Made 1964-83.

With plain bbl. $185
With vent-rib bbl. 205
Add for Winchester Recoil Reduction System 50
Add for Winchoke 25

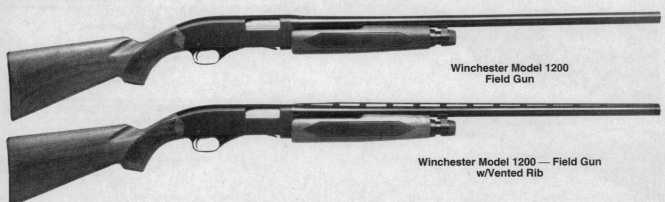

Winchester Model 1200
Field Gun

Winchester Model 1200 — Field Gun
w/Vented Rib

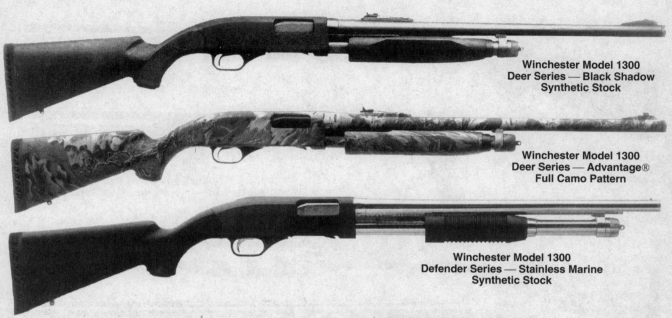

**Winchester Model 1300
Deer Series — Black Shadow
Synthetic Stock**

**Winchester Model 1300
Deer Series — Advantage®
Full Camo Pattern**

**Winchester Model 1300
Defender Series — Stainless Marine
Synthetic Stock**

**Winchester Model 1200
Stainless**

Winchester Model 1200 Stainless Marine Series Slide-Action Security Shotgun

Similar to Model 1200 Defender, except w/6-shot magazine. 18-inch bbl. of ordnance stainless steel w/cylinder bore and rifle sights. Weight: 7 lbs. Bright chrome finish. Synthetic pistol grip or walnut finished hardwood buttstock w/grooved synthetic or hardwood slide handle.Made 1984-90.

Marine Model w/hardwood stock. **$225**
Marine Model w/pistol grip stock **275**

Winchester Model 1200 Stainless Police Series Slide-Action Security Shotgun

Similar to Model 1200 Defender, except w/6-shot magazine. 18-inch bbl. of ordnance stainless steel w/cylinder bore and rifle sights. Weight: 7 lbs. Matte chrome finish. Synthetic pistol grip or walnut finished hardwood buttstock w/grooved synthetic or hardwood slide handle.Made 1984-90.

Police Model w/hardwood stock **$225**
Police Model w/pistol grip stock **275**

Winchester Model 1200 Trap Gun

Same as standard Model 1200, except 12 gauge only. Has 2-shot magazine, 30-inch vent-rib bbl., Full choke or 28-inch with Winchoke. Semi-fancy walnut stock, straight Made 1965-73. Also available 1966-70 with Winchester Recoil Reduction System.

With straight-trap stock **$265**
With Monte Carlo stock **330**
Add for Winchester Recoil Reduction System **75**
Add for Winchoke **25**

Winchester Model 1200 Ranger Combination Shotgun **$225**

Same as Ranger Deer Combination, except has one 28-inch vent-rib bbl. with M choke and one 18-inch Police Cyl. bore. Made 1987-90.

Winchester Model 1200 Skeet Gun **$255**

Same as standard Model 1200, except 12 and 20 ga. only; has 2-shot magazine, specially tuned trigger, 26-inch vent-rib bbl. SK choke, semi-fancy walnut stock and forearm. Weight: 7.25 to 7.5 lbs. Made 1965-73. Also avail. 1966-70 with Winchester Recoil Reduction System (**add** $50 to value).

Winchester Model 1300 CamoPack **$315**

Gauge: 12.3-inch Magnum. 4-shot magazine. Bbls.: 30-and 22-inch with Winchoke system. Weight: 7 lbs. Laminated stock with Win-Cam camouflage green, cut checkering, recoil pad, swivels and sling. Made 1987-88.

Winchester Model 1300 Deer Series

Similar to standard Model 1300, except 12 or 20 ga. only w/special 22-inch cyl. bore or rifled bbl. and rifle-type sights. Weight: 6.5 lbs. Checkered walnut or synthetic stock w/satin walnut, black or Advantage® Full Camo Pattern finish. Matte blue or full-camo metal finish. Made 1994 to date.

Deer Model w/walnut stock (Intro. 1994) **$265**
Black Shadow Deer w/synthetic stock (Intro. 1994) **165**
Advantage® Full Camo Pattern (1995-98) **255**
Deer Combo w/22- and 28-inch bbls. (1994-98) **225**
W/rifled bbl. (Intro. 1996), **add** **20**

Winchester Model 1300 Defender Series

Gauges: 12 or 20 ga. 18- 24- 28-inch VR bbl. w/3-inch chamber. 4- 7- or 8- shot magazine. Weight: 5.6 to 7.4 lbs. Blued, chrome or matte stainless finish. Wood or synthetic stock. Made 1985 to date.

Combo Model. **$295**
Hardwood Stock Model **185**
Synthetic Pistol Grip Model **205**
Synthetic Stock **195**
Lady Defender Synthetic Stock (Made 1996) **200**
Lady Defender Synthetic Pistol Grip (Made 1996) **195**
Stainless Marine Synthetic Stock **325**

SHOTGUNS

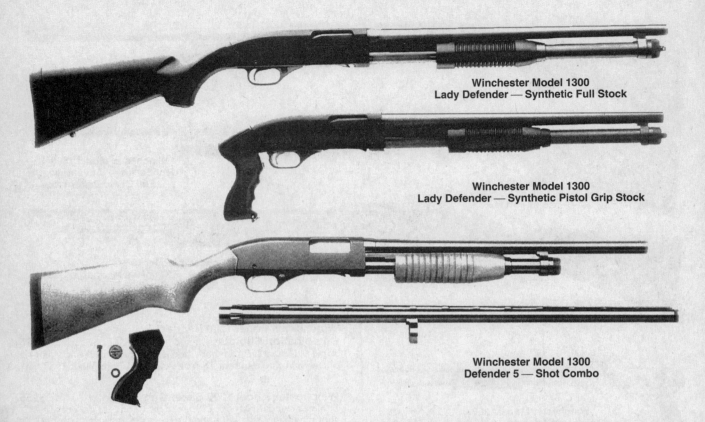

Winchester Model 1300
Lady Defender — Synthetic Full Stock

Winchester Model 1300
Lady Defender — Synthetic Pistol Grip Stock

Winchester Model 1300
Defender 5 — Shot Combo

Winchester Model 1300 Deluxe Slide-Action
Gauges: 12 and 20 w/3-inch chamber. 4-shot magazine. Bbl.:22, 26 or 28 inch vent rib bbl. w/Winchoke tubes. Weight: 6.5 lbs. Checkered walnut buttstock and forend w/high luster finish. Polished blue metal finish with roll-engraved receiver. Made 1984 to date.
Model 1300 Deluxe w/high gloss finish **$280**
Model 1300 Ladies/Youth w/22-inch bbl.(Disc. 1992) **235**

Winchester Model 1300 Featherweight
Slide-Action Shotgun
Hammerless. Takedown. 4-shot magazine. Gauges: 12 and 20 (3-inch chambers). Bbls.: 22, 26 or 28 inches w/plain or vent rib w/Winchoke tubes. Weight: 6.38 to 7 lbs. Checkered walnut buttstock, grooved slide handle. Made 1978-94.
Model 1300 FW (plain bbl.) **$200**
Model 1300 FW (vent rib) **235**
Model 1300 FW XTR **245**

Winchester Model 1300 Ranger Series
Gauges: 12 or 20 ga. w/3-inch chamber. 5-shot magazine. 22- (Ri-fled), 26- or 28-inch vent-rib bbl. w/Winchoke tubes. Weight: 7.25 lbs. Blued finish. Walnut-finished hardwood buttstock and forend. Made 1984 to date.
Standard Model **$225**
Combo Model **275**
Ranger Deer Combo (D&T w/rings & bases) **290**
Ranger Ladies/Youth **230**

Winchester Model 1300 Slide-Action Field Gun
Takedown w/front-locking rotary bolt. Gauges: 12, 20 w/3-inch chamber. 4-shot magazine. Bbl.: vent rib; 26-, 28-, 30-inch w/Win-choke tubes IC-M-F). Weight: 7.25 lbs. Checkered walnut or synthetic stock w/standard, black or Advantage® Full Camo

Winchester Model 1300 Slide-Action Field Gun *(Con't)*
Pattern finish. Matte blue or full-camo metal finish. Made 1994 to date.
Model 1300 Standard Field (w/walnut stock) **$185**
Model 1300 Black Shadow (w/black synthetic stock) **165**
Model 1300 Advantage Camo® **240**

Winchester Model 1300 Slug Hunter Series
Similar to standard Model 1300, except chambered 12 ga. only w/special 22-inch smoothbore w/Sabot-rifled choke tube or fully ri-fled bbl. w/rifle-type sights. Weight: 6.5 lbs. Checkered walnut, hardwood or laminated stock w/satin walnut or WinTuff finish. Matte blue metal finish. Made 1988-94.
Slug Hunter w/hardwood stock **$225**
Slug Hunter w/laminated stock. **265**
Slug Hunter w/walnut stock **245**
Whitetails Unlimited w/beavertail forend **250**
W/Sabot-rifled choke tubes, **add** **15**

Winchester Model 1300 Turkey Series
Gauges: 12 or 20 ga. 22-inch VR bbl. w/3-inch chamber. 4-shot magazine. 43 inches overall. Weight: 6.4 to 6.75 lbs. Buttstock and magazine cap sling studs w/Cordura sling. Drilled and tapped to ex-cept scope base. Checkered walnut, synthetic or laminated wood stock w/low luster finish. Matte blue or full camo finish. Made 1985 to date.
Synthetic Configurations
Turkey Advantage® Full Camo **$275**
Turkey Realtree® All-Purpose Full Camo **285**
Turkey Realtree® Gray All-Purpose Full Camo. **295**
Turkey Realtree® All-Purpose Camo (matte metal). **225**
Turkey Black Shadow (black synthetic stock) **155**
Wood and Laminated Configurations (Disc.)

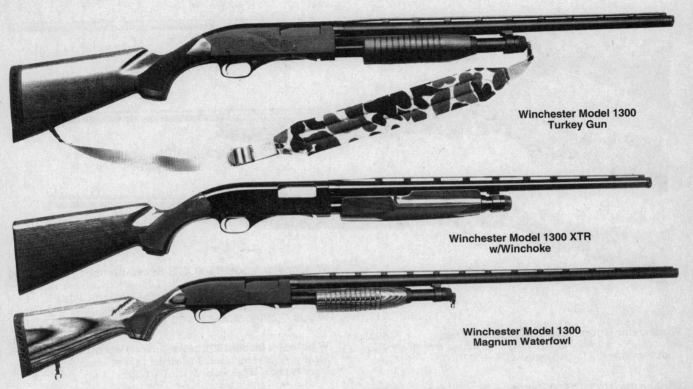

Winchester Model 1300 Turkey Gun

Winchester Model 1300 XTR w/Winchoke

Winchester Model 1300 Magnum Waterfowl

Winchester Model 1300 Turkey Series *(Con't)*

Turkey Win-Cam® (green laminated wood)	265
Turkey Win-Cam® Combo (22- and 30-inch bbls.)	295
Turkey Win-Cam® NWTF (22- and 30-inch bbls.)	250
Turkey Win-Cam® Youth/Ladies Model (20 ga.)	295
Turkey Win-Tuff® (brown laminated wood)	245

Winchester Model 1300 Waterfowl Slide-Action Shotgun

Similar to 1300 Standard Model, except has 28- or 30-inch vent rib bbl. w/Winchoke tubes. Weight: 7 lbs. Matte blue metal finish. Checkered walnut finished hardwood or brown laminated Win-Tuffwood stock w/camo sling, swivels and recoil pad. Made 1984-92.

Model 1300 Waterfowl w/hardwood stock	$225
Model 1300 Waterfowl w/laminated stock	245

Winchester Model 1300 XTR Slide-Action $295

Hammerless. Takedown. 4-shot magazine. Gauges: 12 and 20 (3-inch chambers). Bbl.: plain or vent rib; 28-inch bbls.; Winchoke (interchangeable tubes IC-M-F). Weight: about 7 lbs.

Winchester Model 1400 Automatic Field Gun

Gas-operated. Front-locking rotary bolt. Takedown. 2-shot magazine. Gauges: 12, 16, 20 (2.75-inch chamber). Bbl.: plain or vent rib; 26-,28-,30-inch; IC, M, F choke, or with Winchoke (interchangeable tubes IC-M-F). Weight: 6.5 to 7.25 lbs. Checkered pistol-grip stock and forearm, recoil pad, also available with Winchester Recoil Reduction System (Cycolac stock). Made 1964-68. *See Illustration Next page.*

With plain bbl.	$250
With vent-rib bbl.	275
Add for Winchester Recoil Reduction System	75
Add for Winchoke	25

Winchester Model 1400 Deer Gun $225

Same as standard Model 1400, except has special 22-inch bbl. with rifle-type sights, for rifle slug or buckshot; 12 ga. only. Weight: 6.25 lbs. Made 1965-68.

Winchester Model 1400 Mark II Deer Gun $275

Same general specifications as Model 1400 Deer Gun. Made 1968-73.

Winchester Model 1400 Mark II Field Gun

Same general specifications as Model 1400 Field Gun, except not chambered for 16 gauge; Winchester Recoil Reduction System not available after 1970- only 28-inch barrels w/Winchoke offered after 1973. Made 1968-78.

With plain bbl.	$240
With plain bbl. and Winchoke	260
With vent-rib bbl.	275
With vent-rib bbl. and Winchoke	295
Add for Winchester Recoil Reduction System	75

Winchester Model 1400 Mark II Skeet Gun $325

Same general specifications as Model 1400 Skeet Gun. Made 1968-73.

Winchester Model 1400 Mark II Trap Gun

Same general specifications as Model 1400 Trap Gun, except also furnished with 28-inch bbl. and Winchoke. Winchester Recoil Reduction System not available after 1970. Made 1968-73.

With straight stock	$320
With Monte Carlo stock	350
With Winchester Recoil Reduction System, **add**	75
With Winchoke, **add**	25

Winchester Model 1400 Mark II Utility Skeet $245

Same general specifications as Model 1400 Mark II Skeet Gun, except has stock and forearm of field grade walnut. Made 1970-73.

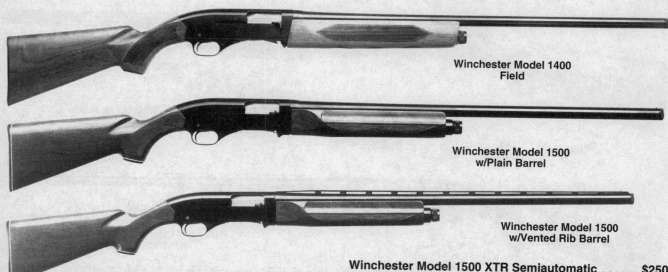

**Winchester Model 1400
Field**

**Winchester Model 1500
w/Plain Barrel**

**Winchester Model 1500
w/Vented Rib Barrel**

Winchester Model 1400 Mark II Utility Trap $265
Same as Model 1400 Mark II Trap Gun, except has Monte Carlo stock/forearm of field grade walnut. Made 1970-73.

Winchester Model 1400 Ranger Semiautomatic Shotgun $195
Gauges: 12, 20. 2-shot magazine. 28-inch vent-rib bbl. with F choke. Overall length: 48.63 inches. Weight: 7 to 7.25 lbs. Walnut finish, hardwood stock and forearm with cut checkering. Made 1984 to 1990 by U. S. Repeating Arms.

Winchester Model 1400 Ranger Semiautomatic Deer Shotgun $215
Same general specifications as Ranger Semiautomatic except 24.13-inch plain bbl. with rifle sights. Mfd. by U.S. Repeating Arms.

Winchester Model 1400 Skeet Gun $275
Same as standard Model 1400, except 12 and 20 ga. only, 26-inch vent-rib bbl., SK choke, semi-fancy walnut stock and forearm. Weight: 7.25 to 7.5 lbs. Made 1965-68. Also available with Winchester Recoil Reduction System (**add** $50 to value).

Winchester Model 1400 Trap Gun
Same as standard Model 1400, except 12 ga. only with 30-inch vent-rib bbl., F choke. Semi-fancy walnut stock, straight or Monte Carlo trap style. Also available with Winchester Recoil Reduction System. Weight: about 8.25 lbs. Made 1965-68.
With straight stock $325
With Monte Carlo stock 350
Add for Winchester Recoil Reduction System 75

Winchester Model 1500 XTR Semiautomatic $250
Gas-operated. Gauges: 12 and 20 (2.75-inch chambers). Bbl.: plain or vent rib; 28-inch; Winchoke (interchangeable tubes IC-M-F). American walnut stock and forend; checkered grip and forend. Weight: 7.25 lbs. Made 1978-82.

Winchester Model 1901 Lever-Action Repeater $1295
Same general specifications as Model 1887 of which this is a redesigned version. 10 ga. only. Made 1901-20.

Winchester Model 1911 Autoloading Shotgun $425
Hammerless. Takedown. 12 gauge only. 4-shell tubular magazine. Bbl.: plain, 26- to 32-inch, standard borings. Weight: about 8.5 lbs. Plain or checkered pistol-grip stock and forearm. Made 1911-25.

Winchester Ranger Deer Combination $240
Gauge: 12, 3-inch Magnum. 3-shot magazine. Bbl.: 24-inch Cyl. bore deer bbl. and 28-inch vent-rib bbl. with Winchoke system. Weight: 7.25 lbs. Made 1987-90.

Winchester Model Single-Shot 1885 Shotgun $2995
Falling-block action, same as Model 1885 Rifle. Highwall receiver. Solid frame or takedown. 20 ga. 3-inch chamber. 26-inch bbl.; plain, matted or matted rib; Cyl. bore, M or F choke. Weight: about 5.5 lbs. Straught-grip stock and forearm. Made 1914-16.

Winchester Super-X Model I Auto Field Gun $375
Gas-operated. Takedown. 12 ga. 2.75-inch chamber 4-shot magazine. Bbl.: vent-rib 26-inch IC; 28-inch M or F; 30-inch F choke. Weight about 7 lbs. Checkered pistol-grip stock. Made 1974-84.

Winchester Super-X Model I Skeet Gun $595
Same as Super-X Field Gun, except has 26-inch bbl., SK choke, skeet-style stock and forearm of select walnut. Made 1974-84.

**Winchester Model 1901
Lever-Action Repeater**

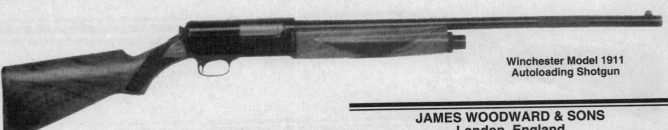

**Winchester Model 1911
Autoloading Shotgun**

Winchester Super-X Model I Trap Gun
Same as Super-X Field Gun, except has 30-inch bbl., IM or F choke, trap-style stock (straight or Monte Carlo comb) and forearm of select walnut, recoil pad. Made 1974-84.

With straight stock **$425**
With Monte Carlo stock **495**

Winchester Xpert Model 96 O/U Field Gun **$675**
Boxlock action similar to Model 101. Plain receiver. Auto ejectors. Selective single trigger. Gauges: 12, 20. 3-inch chambers. Bbl.: vent rib; 26-inch IC/M; 28-inch M/F, 30-inch F/F choke (12 ga. only). Weight: 6.25 to 8.25 lbs. depending on ga. and bbls. Checkered pistol-grip stock and forearm. Made 1976-81 for Olin Corp. at its Olin-Kodensha facility in Japan.

Winchester Xpert Model 96 Skeet Gun **$750**
Same as Xpert Field Gun, except has 2.75-inch chambers, 27-inch bbls., SK choke, skeet-style stock and forearm. Made 1976-1981.

Winchester Xpert Model 96 Trap Gun
Same as Xpert Field Gun, except 12 ga. only, 2.75-inch chambers, has 30-inch bbls., IM/F or F/F choke, trap-style stock (straight or Monte Carlo comb) with recoil pad. Made 1976-81.

With straight stock **$725**
With Monte Carlo stock **745**

JAMES WOODWARD & SONS
London, England

The business of James Woodward & Sons was acquired by James Purdey & Sons after World War II.

Woodward Best Quality Hammerless Double
Sidelock. Automatic ejectors. Double triggers or single trigger. Built to order in all standard gauges, bbl. lengths, boring and other specifications; made as a field gun, pigeon and wildfowl gun, skeet gun or trap gun. Manufactured prior to World War II.

With double triggers **$20,950**
With single trigger **21,995**

Woodward Best Quality Single-Shot Trap **$10,925**
Sidelock. Mechanical features of the Under and Over Gun. vent-rib bbl. 12 ga. only. Built to customers' specifications and measurements, including type and amount of checkering, carving, and engraving. Made prior to World War II.

Woodward Best Quality Over/Under Shotgun
Sidelock. Automatic ejectors. Double triggers or single trigger. Built to order in all standard gauges, bbl. lengths, boring and other specifications, including Special Trap Grade with vent rib. Woodward introduced this type of gun in 1908. Made until World War II. *See* listing of Purdey Over/Under Gun.

With Double Triggers **$24,500**
With Single Trigger. **25,750**

SHOTGUNS

Winchester — Ranger Deer

**Winchester — Ranger
Slide-Action**

**Winchester — Model 1885
Single-Shot**

**Winchester — Model 1 Field
Super-X**

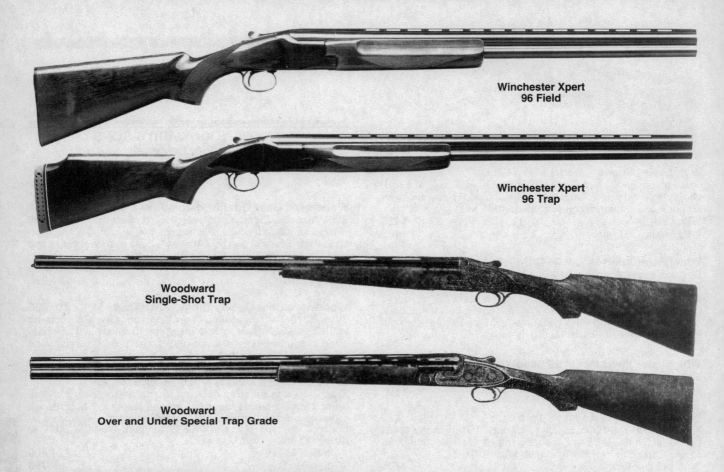

Winchester Xpert 96 Field

Winchester Xpert 96 Trap

Woodward Single-Shot Trap

Woodward Over and Under Special Trap Grade

ZEPHYR SHOTGUNS
Manufactured by Victor Sarasqueta Company, Eibar, Spain

Zephyr Model 1 Over/Under Shotgun **$925**
Same general specifications as Field Model O/U, except with more elaborate engraving, finer wood and checkering. Imported by Stoeger 1930s-51.

Zephyr Model 2 Over/Under Shotgun **$1155**
Sidelock. Auto ejectors. Gauges: 12, 16, 20, 28 and .410. Bbls.: 25 to 30 inches most common. Modest scroll engraving on receiver and sideplates. Checkered, straight-grain select walnut buttstock and forend. Imported by Stoeger 1930s-51.

Zephyr Model 3 Over/Under Shotgun **$1595**
Same general specifications as Zephyr Model 2 O/U, except with more elaborate engraving, finer wood and checkering. Imported by Stoeger 1930s-51.

Zephyr Model 400E Field Grade Double-Barrel Shotgun
Anson & Deeley boxlock system. Gauges: 12 16, 20, 28 and .410. Bbls.:25 to 30 inches. Weight: 4.5 lbs. (.410) to 6.25 lbs. (12 ga.). Checkered French walnut buttstock and forearm. Modest scroll engraving on bbls., receiver and trigger guard. Imported by Stoeger 1930s-50s.

12, 16 or 20 gauge	**$1150**
28 or .410 gauge	1295
Add for selective single trigger	200

Zephyr Model 401 E Skeet Grade Double-Barrel Shotgun
Same general specifications as Field Grade (above), except with beavertail forearm. Bbls.:25 to 28 inches. Imported by Stoeger 1930s-50s.

12, 16 or 20 gauge	**$1450**
28 or .410 gauge	**1595**
Extra for selective single trigger	**350**
Extra for nonselective single trigger	**250**

Zephyr Model 402E Deluxe Double-Barrel Shotgun **$1750**
Same general specifications as Model 400E Field Grade except for custom refinements. The action was carefully hand-honed for smoother operation; finer, elaborate engraving throughout plus higher quality wood in stock and forearm. Imported by Stoeger 1930s-50s.

Zephyr Crown Grade **$1250**
Boxlock. Gauges: 12, 16, 20, 28 and .410. Bbls.: 25 to 30 inches standard, but any lengths could be ordered. Weight: 6 lbs. 4 oz. (.410) to 7 lbs. 4 oz. (12 ga.). Checkered Spanish walnut stock and beavertail forearm. Receiver engraved with scroll patterns. Imported by Stoeger 1938-51.

Zephyr Field Model Over/Under Shotgun **$645**
Anson & Deeley boxlock. Auto ejectors. Gauges: 12, 16 and 20. Bbls.: 25 to 30 inches standard- full-length matt rib. Double triggers. Checkered buttstock and forend. Light scroll engraving on receiver. Imported by Stoeger 1930s-51.

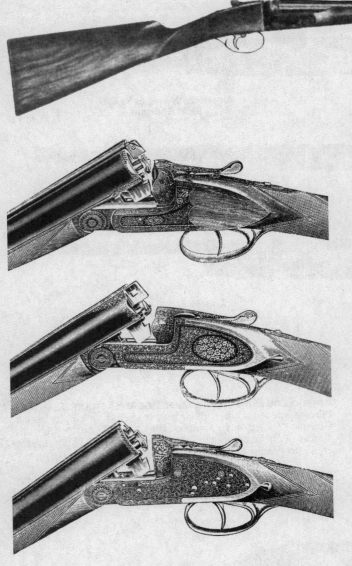

**Angelo Zoli — Patricia
Side-by Side**

**Zephyr
Crown, Premier and Royal Grades**

Zephyr Honker Single-Shot Shotgun **$495**
Sidelock. Gauge: 10; 3.5-inch magnum. 36-inch vent-rib barrel w/F choke. Weight: 10.5 lbs. Checkered select Spanish walnut buttstock and beavertail forend; recoil pad. Imported by Stoeger 1950s-72.

Zephyr Pinehurst Double-Barrel Shotgun **$895**
Boxlock. Gauges: 12, 16, 20, 28 and .410. Bbls.: 25 to 28 inches most common. Checkered, select walnut buttstock and forend. Selective single trigger and auto ejectors. Imported by Stoeger 1950s-72.

Zephyr Premier Grade Double-Barrel Shotgun **$2050**
Sidelock. Gauges: 12, 16, 20, 28 and .410. Bbls.: any length, but 25 to 30 inches most popular. Weight: 4.5 lbs. (.410) to 7 lbs. (12 ga.). Checkered high-grade French walnut buttstock and forend. Imported by Stoeger 1930s-51.

Zephyr Royal Grade Double-Barrel Shotgun **$2995**
Same general specifications as the Premier Grade, except with more elaborate engraving, finer checkering and wood. Imported by Stoeger 1930s-51.

Zephyr Sterlingworth II Double-Barrel Shotgun **$795**
Genuine sidelocks with color-casehardened sideplates. Gauges: 12, 16, 20 and .410. Bbls.: 25 to 30 inches. Weight: 6 lbs. 4 oz. (.410) to 7 lbs. 4 oz. (12 ga.). Select Spanish walnut buttstock and beavertail forearm. Light scroll engraving on receiver and sideplates. Automatic, sliding-tang safety. Imported by Stoeger 1950s-72.

Zephyr Thunderbird Double-Barrel Shotgun **$895**
Sidelock. Gauges: 12 and 10 Magnum. Bbls.: 32-inch, both F choke. Weight: 8 lbs. 8 oz. (12 ga.), 12 lbs. (10 ga.). Receiver elaborately engraved with waterfowl scenes. Checkered select Spanish walnut buttstock and beavertail forend. Plain extractors, double triggers. Imported by Stoeger 1950-72.

Zephyr Upland King Double-Barrel Shotgun **$925**
Sidelock. Gauges: 12, 16, 20, 28 and .410. Bbls.: 25 to 28 inches most popular. Checkered buttstock and forend of select walnut. Selective single trigger and auto ejectors. Imported by Stoeger 1950-72.

Zephyr Uplander 4E Double-Barrel Shotgun **$750**
Same general specifications as the Zephyr Sterlingworth II, except with selective auto ejectors and highly polished sideplates. Imported by Stoeger 1951-72.

Zephyr Woodlander II Double-Barrel Shotgun **$450**
Boxlock. Gauges: 12, 20 and .410. Bbls.:25 to 30 inches. Weight: 6 lbs. 4 oz. (.410) to 7 lbs. 4 oz. (12 ga.). Checkered Spanish walnut stock and beavertail forearm. Engraved receiver. Imported by Stoeger 1950-72.

ANGELO ZOLI
Mississauga, Ontario, Canada

Angelo Zoli Alley Cleaner S/S Shotgun **$550**
Gauges: 12 and 20. 20-inch bbl. 26.5 inches overall. Weight: 7 lbs. average. Chokes: F/M, F, IM, M, IC, SK. Chrome-lined barrels. Walnut stock and engraved action. Made 1986-87.

Angelo Zoli Apache O/U Field Shotgun **$425**
Gauge: 12. Chokes: F/M, F, IM, M, IC, SK. 20-inch bbl. Short vent rib. Weight: 7.5 lbs. Checkered walnut stock and forearm. Lever action. Pistol grip. Made 1986-88.

Angelo Zoli Daino I Folding Single Shot **$125**
Gauges: 12, 20 and .410. 28- or 30-inch bbl. Weight: 6 lbs. average. Choke: Full. Chrome-lined barrel, vent rib. Pistol-grip walnut stock. Engraved action. Made 1986-88.

Angelo Zoli HK 2000 Semiautomatic Shotgun **$525**
Gauge: 12. 5-shot capacity. Bbls.: 24-, 26-, 28- and 30-inch; vent rib. Weight: 6.5 to 7.5 lbs. Checkered walnut stock and forearm. Glossy finish. Pistol grip. Engraved receiver. Made 1988.

SHOTGUNS

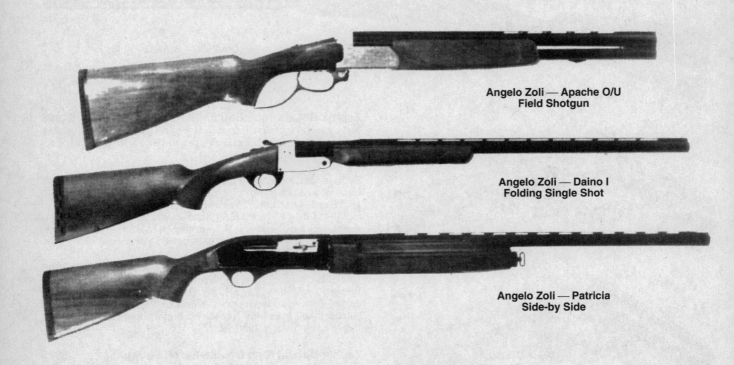

Angelo Zoli — Apache O/U Field Shotgun

Angelo Zoli — Daino I Folding Single Shot

Angelo Zoli — Patricia Side-by Side

Angelo Zoli Patricia Side-by-Side Shotguns **$1095**
Gauge: .410 only. 28-inch bbl. Choke: F/M. Weight: 5.5 lbs. Automatic ejectors; Zoli single selective trigger, boxlock. Hand-checkered walnut stock with English straight grip, splinter forearm. Made 1986-88.

Angelo Zoli Saint George O/U Competition **$1695**
Gauge: 12. 30- and 32-inch vent rib barrels. Weight: 8 lbs. Single selective trigger. Oil-finished, pistol-grip walnut stock. Made 1986-88.

Angelo Zoli Silver Snipe Over/Under Shotgun **$625**
Purdey-type boxlock with crossbolt. Selective single trigger. Gauges: 12, 20; 3-inch chambers. Bbls.: 26-, 28- or 30-inch with a variety of choke combinations. Weight: 5.75 to 6.75 lbs. Checkered European walnut buttstock and forend. Made in Italy.

Angelo Zoli Z43 Standard O/U Shotgun **$425**
Gauge: 12 or 20. 26-, 28- or 30-inch bbls. Chokes: F/M, M/IC Weight: 6.75 to 8 lbs. vent lateral ribs, standard extractors, single

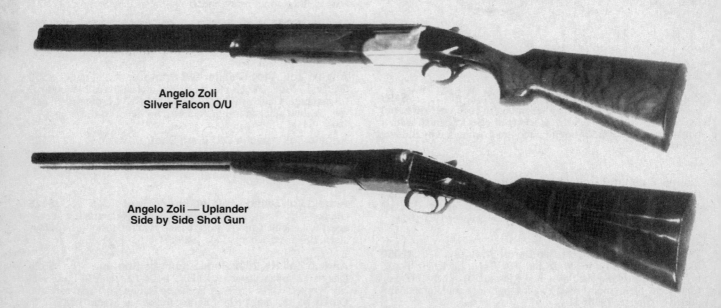

Angelo Zoli Silver Falcon O/U

Angelo Zoli — Uplander Side by Side Shot Gun

Angelo Zoli Z43 Standard O/U Shotgun *(Con't)*
non-selective trigger. Glossy-finished walnut stock and forearm. Automatic safety. Made 1986-88.

ANTONIO ZOLI, U.S.A., INC.
Fort Wayne, Indiana
Manufactured in Italy

Antonio Zoli Silver Falcon O/U **$795**
Gauges: 12 and 20; 3-inch chambers. 26- or 28-inch blued barrels. Weight: 6.25 to 7.25 lbs. Antiqued silver finish on receiver. Pistol-grip stock of Turkish Circassian walnut with polyurethane-type finish. Imported 1989-1990.

Antonio Zoli Uplander Side/Side Shotgun **$650**
Gauges: 12 and 20. Casehardened receiver. Checkered oil-finished, hand-rubbed stock of Turkish Circassian walnut; splinter forend. Imported 1989-90.

SHOTGUNS

INDEX

B

BAER
HANDGUNS
1911 Concept Series , *25*
1911 Custom Carry Series , *25*
1911 Premier Series , *25*
1911 Ultimate Master Combat Series , *25*
S.R.P. Automatic Pistol, *25*

BAIKAL
SHOTGUNS
Model IJ-18M Single Shot, *403*
Model IJ-27 Field Over/Under, *403*
Model IJ-27 Over/Under, *403*
Model IJ-43 Field Side-by-side, *403*
IZH-43 Series Side-by-side, *403*

BAKER
SHOTGUNS
Batavia Ejector, *403*
Batavia Leader Hammerless Double, *403*
Batavia Special, *403*
Black Beauty, *403*
Grade R, *404*
Grade S, *404*
Paragon, Expert and Deluxe Grades, *404*

BARRETT FIREARMS
RIFLES
Model 82 A-1 Semiawtomatic Rifle, *193*
Model 90 Bolt-Action Rifle, *193*
Model 95 Bolt Action, *193*

BAUER FIREARMS CORPORATION
25 Automatic Pistol, *25*

BAYARD PISTOLS
Model 1908 Pocket , *25*
Model 1923 Pocket , *25*
Model 1923 Pocket 25 , *25*
Model 1930 Pocket 25 , *25*

BEEMAN PRECISION ARMS
HANDGUNS
Mini P08 Automatic Pistol, *25*
P08 Automatic Pistol, *25*
SP Metallic Silhouette Pistols, *26*
RIFLES
/Feinwerkbau Model 2600 Series, *194*
/Weihrauch HW Model 60M, *194*
/Weihrauch HW Model 660 Match, *194*
/Weihrauch HW Models 60J and 60J-ST
Bolt-Action Rifles, *194*

BEHOLLA
PISTOL
Pocket Automatic Pistol, *26*

BELGIAN
MILITARY RIFLES
Model 1889 Mauser Military Rifle , *194*

Model 1916 Mauser Carbine , *194*
Model 1935 Mauser Military Rifle , *194*
Model 1936 Mauser Military Rifle , *194*

BELKNAP
SHOTGUNS
Model B-63 Single-Shot Shotgun, *404*
Model B-63E Single-Shot Shotgun, *404*
Model B-64 Slide-Action Shotgun, *404*
Model B-65C Autoloading Shotgun, *404*
Model B-68 Single-Shot Shotgun, *404*

BENELLI
PISTOLS
MP90S World Cup Target Pistol , *26*
MP95E Sport Target Pistol , *26*
SHOTGUNS
Black Eagle Autoloading Shotgun, *404*
Black Eagle Executive, *404*
Lagecy Autoloading Shotgun, *405*
M1 Super 90 Autoloading Shotgun, *405*
M1 Super 90 Defense Autoloader, *405*
M1 Super 90 Entry Autoloader, *405*
M1 Super 90 Field, *405*
M1 Super 90 Slug Autoloader, *405*
M1 Super 90 Sporting Special
Autoloader, *405*
M1 Super 90 Tactical Autoloader, *405*
M3 Super 90 Pump/Autoloader, *405*
Model 121 M1 Military/Police, *404*
Montefeltro Super 90 Semiautomatic, *405*
SL 121V Semiauto Shotgun, *405*
SL 121V Slug Shotgun, *405*
SL 123V Semiauto Shotgun, *406*
SL 201 Semiautomatic Shotgun, *406*
Sport Autoloading Shotgun, *406*
Super Black Eagle Autoloading, *406*

BENTON & BROWN FIREARMS
Model 93 Bolt-Action Rifle , *194*

BERETTA USA
HANDGUNS
Model 20 Double-Action, *26*
Model 21 Double-Action, *26*
Model 70 Automatic, *26*
Model 70S, *27*
Model 70T, *27*
Model 71, *27*
Model 72, *27*
Model 76 Auto Target Pistol, *27*
Model 81 Double-Action, *27*
Model 84 Double-Action, *27*
Model 84(BB) Double-Action, *28*
Model 84B DA Auto Pistol, *27*
Model 85 Double-Action, *28*
Model 85B DA, *28*
Model 85BB Double-Action, *28*
Model 85F Double-Action, *28*

Model 86 Double-Action, *28*
Model 87, *28*
Model 89 Target, *28*
Model 90 DA, *28*
Model 92 DA Auto Pistol (1st series), *29*
Model 92 SB-F DA Auto Pistol, *29*
Model 92D, *29*
Model 92F Compact , *29*
Model 92F DA, *29*
Model 92S DA (2nd series), *29*
Model 92SB DA (3rd series), *29*
Model 96 Double-Action, *29*
Model 101, *30*
Model 318 (1934), *30*
Model 1915, *31*
Model 1923, *31*
Model 1934, *31*
Model 1935, *31*
Model 3032 DA Tomcat, *31*
Model 8000/8040 Cougar , *31*
Model 8000/8040 Mini-Cougar, *31*
Model 949 Olimpionico, *30*
Model 950 BS Single Action, *30*
Model 950B, *30*
Model 950CC, *30*
Model 950CC Special, *30*
Model 951 (1951) Military, *30*
RIFLES
455 SxS Express Double Rifle, *194*
500 Bolt-Action Sporter, *194*
501 Bolt-Action Sporter, *195*
502 Bolt-Action Sporter, *195*
AR-70 Semiautomatic Rifle, *195*
Express SSO O/U Express Double, *195*
Mato, *195*
Small Bore Sporting Carbine, *195*
SHOTGUNS
Model 57E Over-and-Under, *406*
Model 409PB Hammerless Double, *406*
Model 410, 10-Gauge Magnum, *406*
Model 410E, *406*
Model 411E, *406*
Model 424 Hammerless Double, *406*
Model 426E , *407*
Model 450 Series
Hammerless Doubles, *407*
Model 625 S/S Hammerless Double, *407*
Model 626 S/S Hammerless Double, *407*
Model 627 S/S Hammerless Double, *407*
Model 682 Over/Under Shotgun, *407*
Model 686 Over/Under Shotgun, *407*
Model 687 Over/Under Shotgun, *408*
Model 1200 Semiautoloading, *408*
Model A-301 Autoloading Shotgun, *408*
Model A-301 Magnum, *408*
Model A-301 Skeet Gun, *408*
Model A-301 Slug Gun, *408*

F

K

LUNA
PISTOLS
RIFLES

M.A.C. (Military Armament Corp.)

MAGNUM RESEARCH
HANDGUNS
RIFLES

MAGTECH
RIFLES
SHOTGUNS

MANNLICHER SPORTING RIFLES

MARLIN FIREARMS
RIFLES

MOSSBERG & SONS

PISTOLS
Brownie "Pepperbox" Pistol, *125*

RIFLES
Model 10 , *279*
Model 14 , *279*
Model 20 , *279*
Model 25/25A , *279*
Model 26B/26C , *279*
Model 30, *279*
Model 34, *279*
Model 35 Target Grade , *279*
Model 35A , *279*
Model 35A-LS , *279*
Model 35B, *279*
Model 40, *279*
Model 42, *279*
Model 42A/L42A , *281*
Model 42B/42C , *281*
Model 42M , *281*
Model 43/L43, *281*
Model 44, *281*
Model 44B, *281*
Model 45 , *281*
Model 45A, L45A, 45AC, *281*
Model 45B/45C , *281*
Model 46 , *281*
Model 46A, 46A-LS, L46A-LS , *281*
Model 46B , *281*
Model 46BT, *281*
Model 46M, *281*
Model 50, *281*
Model 51 , *281*
Model 51M , *281*
Model 140B , *281*
Model 140K, *282*
Model 142-A, *282*
Model 142K, *282*
Model 144 , *282*
Model 144LS, *282*
Model 146B , *282*
Model 151K, *283*
Model 151M , *283*
Model 152, *283*
Model 152K, *283*
Model 320B Boy Scout Target , *283*
Model 320K Hammerless , *283*
Model 321B, *283*
Model 321K , *283*
Model 333, *283*
Model 340B Target Sporter , *283*
Model 340K Hammerless, *283*
Model 340M, *283*
Model 341 , *283*
Model 342K Hammerless, *283*
Model 346B , *283*
Model 346K Hammerless , *283*
Model 350K, *283*
Model 351C , *283*

Model 351K , *283*
Model 352K, *283*
Model 353 , *283*
Model 377 Plinkster , *283*
Model 380, *285*
Model 400 Palomino, *285*
Model 402 Palomino , *285*
Model 430 , *285*
Model 432 Western-Style , *285*
Model 43B, *281*
Model 472 , *285*
Model 472 Brush Gun , *285*
Model 472 One in Five Thousand , *286*
Model 472 Rifle , *286*
Model 479 , *286*
Model 620K Hammerless , *286*
Model 620K-A , *286*
Model 640K Chuckster Hammerless, *286*
Model 640KS, *286*
Model 640M , *286*
Model 642K, *286*
Model 800, *286*
Model 800D Super Grade , *286*
Model 800M, *286*
Model 800VT Varmint/Target , *286*
Model 810 , *286*
Model 1500 Mountaineer Grade I Centerfire, *286*
Model 1500 Varmint , *286*
Model 1700LS Classic Hunter , *288*
Model B, *288*
Model K , *288*
Model L , *288*
Model M , *288*
Model R , *288*
Models L42A, L43, L45A, L46A-LS, *288*

SHOTGUNS
Model 83D or 183D, *480*
Model 85D or 185D Bolt-Action Repeating Shotgun, *480*
Model 183K , *480*
Model 185K , *480*
Model 190D , *480*
Model 190K , *480*
Model 195D, *481*
Model 195K , *481*
Model 200D, *481*
Model 200K Slide-Action Repeater , *481*
Model 385K , *481*
Model 390K , *481*
Model 395K Bolt-Action Repeater , *481*
Model 395S Slugster , *481*
Model 500 Accu-Choke Shotgun , *482*
Model 500 Bantam Shotgun , *482*
Model 500 Bullpup Shotgun , *482*
Model 500 Camo Pump, *482*
Model 500 Camper, *482*
Model 500 Field Grade Hammerless Slide Action Repeater, *482*

Model 500 "L" Series, *483*
Model 500 Mariner Shotgun , *483*
Model 500 Muzzleloader Combo, *483*
Model 500 Persuader Law Enforcement Shotgun , *483*
Model 500 Pigeon Grade, *483*
Model 500 Pump Combo Shotgun , *483*
Model 500 Pump Slugster Shotgun, *483*
Model 500 Regal Slide-Action Repeater, *483*
Model 500 Sporting Pump, *484*
Model 500 Super Grade, *484*
Model 500 Turkey Gun , *484*
Model 500 Turkey/Deer Combo, *484*
Model 500 Viking Pump Shotgun, *484*
Model 500 Waterfowl/Deer Combo , *484*
Model 500ATR Super Grade Trap , *484*
Model 500DSPR Duck Stamp Commemorative , *485*
Model 590 Mariner Pump, *485*
Model 590 Military Security , *485*
Model 590 Military Shotgun, *485*
Model 595 Bolt-Action Repeater , *485*
Model 6000 Auto Shotgun , *485*
Model 695 Bolt Action Slugster, *485*
Model 695 Bolt Action Turkey Gun, *485*
Model 712 Autoloading Shotgun, *485*
Model 835 "NWTF" Ulti-Mag™, *485*
Model 835 Field Pump Shotgun, *485*
Model 835 Regal Ulti-Mag Pump, *485*
Model 835 Viking Pump Shotgun, *486*
Model 1000 Autoloading Shotgun, *486*
Model 1000 Super Autoloading, *486*
Model 1000S Super Skeet , *486*
Model 5500 Autoloading Shotgun, *486*
Model 5500 MKII Autoloading, *486*
Model 9200 A1 Jungle gun , *486*
Model 9200 Camo Shotgun, *486*
Model 9200 Crown (Regal) Autoloader, *486*
Model 9200 Persuader, *486*
Model 9200 Special Hunter, *486*
Model 9200 Trophy, *486*
Model 9200 USST Autoloader , *486*
Model 9200 Viking Autoloader , *487*
Model HS410 Home Security Pump Shotgun, *487*
"New Haven Brand" Shotguns, *487*
Line Launhcer, *487*

MUSGRAVE RIFLES, *288*
Premier NR5 BA Hunting Rifle, *288*
RSA NR1 BA - SSTarget Rifle , *289*
Valiant NR6 Hunting Rifle , *289*

MUSKETEER RIFLES
Mauser Sporter , *289*

N

NAMBU PISTOLS

Model 1200 Super BA Sporting Rifle , *294*
Model 1200 Super Clip BA Rifle , *294*
Model 1200 Super Magnum , *294*
Model 1200P Presentation , *294*
Model 1200V Varmint , *294*
Model 1300C Scout , *294*
Model 2100 Midland Bolt Action Rifle, *294*
Model 2700 Lightweight , *294*
Model 2800 Midland, *295*
SHOTGUNS
Model 645A (American) Side-by-Side, *492*
Model 645E (English) Side-by-Side, *492*

PEDERSEN CUSTOM GUNS
RIFLES
Model 3000 Grade I Bolt-Action Rifle , *295*
Model 3000 Grade II , *295*
Model 3000 Grade III , *295*
Model 4700 Custom Deluxe LA Rifle , *295*
SHOTGUNS
Model 1000 Trap Gun, *492*
Model 1000 Magnum, *492*
Model 1000 Over/Under Hunting, *492*
Model 1000 Skeet Gun, *492*
Model 1500 O/U Hunting Shotgun, *492*
Model 1500 Skeet Gun , *492*
Model 1500 Trap Gun , *492*
Model 2000 Hammerless Double, *492*
Model 2500 Hammerless Double, *492*
Model 4000 Hammerless Slide-Action
Repeating Shotgun , *493*
Model 4000 Trap Gun , *493*
Model 4500 , *493*
Model 4500 Trap Gun , *493*

PENNEY, J.C. COMPANY
RIFLES
Model 2025 Bolt-Action Repeater, *295*
Model 2035 Bolt-Action Repeater , *295*
Model 2935 Lever-Action Rifle, *295*
Model 6400 BA Centerfire Rifle, *295*
Model 6660 Autoloading Rifle , *295*
SHOTGUNS
Model 4011 Autoloading Shotgun, *493*
Model 6610 Single-Shot Shotgun , *493*
Model 6630 Bolt-Action Shotgun , *493*
Model 6670 Slide-Action Shotgun, *493*
Model 6870 Slide-Action Shotgun, *493*

PHOENIX ARMS
Model "Raven" SA , *127*
Model HP Rangemaster
  Deluxe Target SA , *127*
Model HP Rangemaster Target SA , *127*
Model HP22/HP25 SA , *127*
**PERAZZI SHOTGUNS — *See also*
listings under Ithaca — Perazzi**
DB81 Over/Under Trap , *493*
DB81 Single-Shot Trap, *493*
Grand American 88 Special
  Single Trap, *494*

Mirage Over/Under Shotgun, *494*
MX1 Over/Under Shotgun, *494*
MX2 Over/Under Shotgun, *494*
MX3 Over/Under Shotgun, *494*
MX3 Special Pigeon Shotgun , *494*
MX4 Over/Under Shotgun, *494*
MX5 Over/Under Game Gun, *494*
MX6 American Trap Single Barrel, *494*
MX6 Skeet O/U, *494*
MX6 Sporting O/U, *494*
MX6 Trap O/U, *494*
MX7 Over/Under Shotgun , *494*
MX8 Over/Under Shotgun, *494*
MX8/20 Over/Under Snotgun , *495*
MX9 Over/Under Shotgun , *495*
MX10 Over/Under Shotgun , *495*
MX10 Pigeon-Electrocibles O/U, *495*
MX11 American Trap Combo, *495*
MX11 American Trap Single Barrel, *495*
MX11 Pigeon-Electrocibles O/U, *495*
MX11 Skeet O/U, *495*
MX11 Sporting O/U, *495*
MX11 Trap O/U, *495*
MX12 Over/Under Game Gun, *495*
MX14 American Trap Single-Barrel, *495*
MX15 American Trap Single-Barrel, *495*
MX20 O/U Game Gun, *495*
MX28 Over/Under Game Gun , *495*
MX410 Over/Under Game Gun , *495*
TM1 Special Single-Shot Trap , *495*
TMX Special Single-Shot Trap , *495*

PIOTTI SHOTGUNS
Boss Over/Under, *495*
King No. 1 Sidelock, *495*
King Extra Side-by-Side Shotgun, *496*
Lunik Sidelock Shotgun, *496*
Monte Carlo Sidelock Shotgun, *496*
Piuma Boxlock Side-by-Side, *496*

PLAINFIELD MACHINE COMPANY
HANDGUNS
Model 71 , *128*
Model 72 , *128*
RIFLES
M-1 Carbine, *295*
M-1 Carbine, Commando Model , *295*
M-1 Carbine, Military Sporter , *295*
M-1 Deluxe Sporter , *295*

POWELL, William & Sons, Ltd.
RIFLES
Bolt-Action Rifle, *296*
Double-Barrel Rifle, *296*
SHOTGUNS
No. 1 Best Grade Double-Barrel, *496*
No. 2 Best Grade Double, *496*
No. 6 Crown Grade Double, *496*
No. 7 Aristocrat Grade Double, *496*

PRECISION SPORTS SHOTGUNS

Sports 600 Series American
  Hammerless Doubles , *496*
Sports 600 Series English
  Hammerless Doubles, *496*
Sports Model 640M Magnum 10
  Hammerless Double, *497*
Sports Model 645E-XXV
  Hammerless Double, *497*

PREMIER SHOTGUNS
Ambassador Model Field Grade Ham-
  merless Double-Barrel, *497*
Brush King , *497*
Continental Model Field Grade Hammer
  Double-Barrel Shotgun , *497*
Monarch Supreme Grade Hammerless
  Double-Barrel Shotgun , *497*
Presentation Custom Grade , *497*
Regent 10-Gauge Magnum Express , *497*
Regent 12-Gauge Magnum Express, *497*
Regent Model Field Grade
  Hammerless, *497*

PROFESSIONAL ORDNANCE
Model Carbon-15 Type 20, *129*

PURDEY, James & Sons
RIFLES
Bolt-Action Rifle , *296*
Double Rifle, *296*
SHOTGUNS
Hammerless Double-Barrel Shotgun, *497*
Over/Under Shotgun, *497*
Single-Barrel Trap Gun , *497*

R

RADOM PISTOL
P-35 Automatic Pistol, *129*

RAPTOR ARMS COMPANY
Bolt-Action Rifle , *296*

RECORD-MATCH PISTOLS
Model 200 Free Pistol, *129*
Model 210 Free Pistol, *129*

REISING ARMS
Target Automatic Pistol, *129*

REMINGTON ARMS COMPANY
HANDGUNS
Model 51 Automatic Pistol, *129*
Model 95 Double Derringer, *129*
Model XP-100 Custom Pistol, *130*
Model XP-100 Silhouette, *130*
Model XP-100 Single-Shot Pistol, *130*
Model XP-100 Varmint Special, *130*
Model XP-100R Custom Repeater, *130*
New Model Single-Shot
  Target Plstol, *130*
XP22R Rimfire Repeater, *130*
RIFLES
Remington Single-Shot Rifles

REVELATION SHOTGUNS — *See*
Western Auto listings.
RG REVOLVERS

S

Model 500 Small Gauge O/U Shotgun , *517*
Model 505 O/U Shotgun, *517*
Model 585 Deluxe Over/Under Shotgun, *517*
Model 600 Small Gauge , *517*
Model 605 Trap O/U Shotgun, *517*
Model 685 Deluxe Over/Under, *517*
Model 800 Skeet/Trap Over/Under, *518*
Model 880 Skeet/Trap , *518*
Model 885 Deluxe Over/Under, *518*
Century Single-Barrel Trap Gun, *518*
Gas-operated Automatic Shotguns, *518*
Model 1300 Upland, Slug, *518*
Model 1900 Field, Trap, Slug, *518*
Model XL300, *518*
Model XL900, *518*
Model XL900 Skeet, *518*
Model XL900 Slug, *518*
Model XL900 Trap, *518*

Over/Under Shotguns
Model 500 Field, *518*
Model 500 Magnum, *518*
Model 600 Doubles, *518*
Model 600 English, *518*
Model 600 Field, *518*
Model 600 Magnum, *518*
Model 600 Skeet, *518*
Model 600 Skeet Combo, *518*
Model 600 Trap, *518*
Model 700 Doubles, *518*
Model 700 Skeet, *518*
Model 700 Skeet Combo, *518*
Model 700 Trap, *518*

Recoil-Operated Automatic Shotguns
Model 300, *518*
Model 900, *518*

Side-by-Side Double-Barrel Shotguns
Model 100, *518*
Model 150, *518*
Model 200E, *518*
Model 200E Skeet, *518*
Model 280 English, *518*

SECURITY IND. OF AMERICA

Model PM357 DA Revolver, *140*
Model PPM357 DA Revolver, *140*
Model PSS38 DA Revolver, *140*

SEDGLEY

HANDGUNS
Baby Hammerless Ejector Revolver, *140*
RIFLES
Springfield Sporter, *343*
Springfield Left-Hand Sporter , *343*
Springfield Mannlicher-Type Sporter , *343*

SEECAMP

Model LWS 25 DAO Pistol, *140*
Model LWS 32 DAO Pistol, *140*

SHERIDAN PRODUCTS

Knocabout Single-Shot Pistol, *140*

SHILEN RIFLES

DGA Benchrest Rifle, *343*
DGA Sporter , *343*
DGA Varminter , *343*

SHILOH RIFLE MFG

Sharps Model 1874 Business Rifle, *343*
Sharps Model 1874
  Long Range Express Rifle, *343*
Sharps Model 1874
  Montana Roughrider, *344*
Sharps Model 1874 Saddle Rifle , *343*
Sharps Model 1874
  Sporting Rifle No. 1 , *343*
Sharps Model 1874
  Sporting Rifle No. 3 , *344*

SHOTGUNS OF ULM — *See listings under* **Krieghoff**

SIG PISTOLS

Model P210-2, *141*
Model P210-5 Target Pistol, *141*
Model P210-6 Target Pistol, *141*
P210 22 Conversion Unit, *141*
P210-1 Automatic Pistol, *141*

SIG SAUER

HANDGUNS
Model P220 DA, *141*
Model P225 DA , *141*
Model P226 DA, *141*
Model P228 DA , *141*
Model P229 DA , *141*
Model P230 DA , *142*
Model P232 , *142*
Model P239, *142*
SHOTGUNS
Model SA3 Over/Under Shotgun , *519*
Model SA5 Over/Under Shotgun , *519*

SIG — SWISS AMT

AMT Semiautomatic Rifle , *344*
AMT Sporting Rifle, *344*
PE57 Semiautomatic Rifle , *344*

SILE SHOTGUNS

Field Master II O/U Shotgun, *519*

SMITH, L.C.

SHOTGUNS
Double-Barrel Shotguns
  Crown Grade, *519*
  Deluxe Grade, *519*
  Field Grade, *519*
  Ideal Grade, *520*
  Monogram Grade, *520*
  Olympic Grade, *520*
  Premier Grade, *520*
  Skeet Special, *520*
  Specialty Grade, *520*
  Trap Grade, *520*

Hammerless Double
    Model 1968 Deluxe, *520*
Hammerless Double
    Model 1968 Field Grade , *521*
Hammerless Double-Barrel Shotguns
  0 Grade, *521*
  00 Grade, *521*
  1 Grade, *521*
  2 Grade , *521*
  3 Grade, *521*
  4 Grade, *521*
  5 Grade, *521*
  A1, *521*
  A2, *521*
  A3, *521*
  Monogram, *521*
  Pigeon, *521*
Single-Shot Trap Guns, *520*

SMITH & WESSON

AUTOMATIC/SINGLE-SHOT PISTOLS
32 Automatic Pistol, *142*
35 (1913) Automatic Pistol , *142*
Model 1000 Series DA Auto, *145*
Model 1891 Target Pistol,
  First Model , *147*
Model 1891 Target Pistol,
  Second Model, *147*
Model 22 Sport Series, *142*
Model 2206 SA Automatic Pistol, *145*
Model 2213 Sportsman Auto, *145*
Model 2214 Sportsman Auto, *145*
Model 22A Semi-automatic
  Target Pistol, *142*
Model 39 9mm DA Auto Pistol, *142*
Model 3904/3906 DA Auto Pistol, *145*
Model 3913/3914 DA Automatic, *146*
Model 3953/3954 DA Auto Pistol, *146*
Model 4000 Series DA Auto, *146*
Model 4013/4014 DA Automatic, *146*
Model 4053/4054 DA Auto Pistol, *146*
Model 41 22 Automatic Pistol, *142*
Model 410 Auto Pistol, *143*
Model 411 Auto Pistol , *143*
Model 422 SA Auto Pistol, *143*
Model 439 9mm Automatic, *143*
Model 4500 Series DA Automatic, *146*
Model 457Compact Auto Pistol, *144*
Model 459 DA Automatic, *144*
Model 46 22 Auto Pistol , *142*
Model 469 9mm Automatic, *144*
Model 52 38 Master Auto, *143*
Model 59 9mm DA Auto , *143*
Model 5900 Series DA Automatic, *146*
Model 61 Escort Pocket, *143*
Model 622 SA Auto Pistol, *144*
Model 639 Automatic , *144*
Model 645 DA Automatic , *144*
Model 659 9mm Automatic, *144*
Model 669 Automatic, *144*

W

Z